D0203216

Hydrologic Analysis and Design

Second Edition

Richard H. McCuen
Department of Civil Engineering
University of Maryland

Prentice Hall
Upper Saddle River, New Jersey 07458

Library of Congress Cataloging-in-Publication Data

McCuen, Richard H.
Hydrologic analysis and design / Richard H. McCuen
p. cm.
Bibliography: p.
Includes index.
ISBN 0-13-134958-9
1. Hydraulic engineering. 2. Hydrology. I. Title.
TC145.M384 1998 88-22397
627—dc 19 CIP

Acquisitions editor: *Bill Stenquist*
Editor-in-Chief: *Marcia Horton*
Assistant Vice President of Production and Manufacturing: *David W. Riccardi*
Managing Editor: *Bayani Mendoza de Leon*
Manufacturing Buyer: *Donna Sullivan*
Manufacturing Manager: *Trudy Pisciotti*
Creative Director: *Paula Maylahn*
Art Director: *Jayne Conte*
Cover Designer: *Bruce Kenselaar*
Editorial Assistant: *Meg Weist*
Compositor/Production services: *Pine Tree Composition, Inc.*

Printed in the United States of America

10 9 8 7 6 5 4 3 2 1

ISBN: 0-13-134958-9

Prentice-Hall International (UK) Limited, *London*
Prentice-Hall of Australia Pty. Limited, *Sydney*
Prentice-Hall Canada Inc., *Toronto*
Prentice-Hall Hispanoamericana, S.A., *Mexico*
Prentice-Hall of India Private Limited, *New Delhi*
Prentice-Hall of Japan, Inc., *Tokyo*
Simon & Schuster Asia Pte. Ltd., *Singapore*
Editora Prentice-Hall do Brasil, Ltda., *Rio de Janeiro*

Contents

Preface

Hydrologic Analysis and Design is intended for a first course in hydrology. It introduces the student to the physical processes of the hydrologic cycle, the computational fundamentals of hydrologic analysis, and the elements of design hydrology. Although the sections of the book that introduce engineering design methods are intended for the engineering student, the concepts and methods will also be of interest to, and readily understood by, students in other disciplines including geology, geography, forestry, and planning. The material is appropriate for undergraduate and graduate students as well as for the practitioner interested in reviewing the fundamentals of hydrology. While computers are widely used in the practice of hydrology, knowledge of computer hardware, software, or programming languages is not necessary to use this book; however, computer literacy will certainly facilitate problem-solving.

A number of factors influenced the structure of the book. Chapter 3 introduces watershed characteristics, Chapter 4 addresses precipitation, Chapter 6 deals with topics in ground water hydrology, Chapter 14 provides a brief overview of evaporation, and Chapter 15 discusses erosion. Given the interdependence of these processes, other factors also determined the book's structure. Because of the diverse ways that hydrology can be taught, flexibility was a criterion used in structuring the book. The need for flexibility led to a book comprising fifteen chapters. To the extent possible, I tried to make the chapters self-contained so that the order in which the topics are presented is not constrained by the book's structure. Given the orientation of the book towards design, I carefully examined the type of design problems that were to be emphasized. Land development and urban hydrology are very important aspects of hydrology, and therefore many of the topics that are used in such design work were placed at the beginning of the book. Thus, topics such as flood frequency analysis (Chapter 5) and hydrograph separation and synthesis (Chapter 9) were placed before snowmelt runoff estimation (Chapter 12) and evaporation (Chapter 14). I hope that the criteria that influenced my decision on how to structure the book has resulted in a favorable balance for both teaching and learning hydrology.

Material for most topics in the fundamentals of hydrology can be divided into four areas: physical processes, measurement and collection of data, model conceptualization and data analysis, and design or solution synthesis. A basic knowledge of the physical processes is necessary to properly analyze data or perform design work. Thus, I have tried to provide a discussion of the physical processes associated with the components of the hydrologic cycle. The discussion is limited to the dominant factors, especially those measurable characteristics that are used in design. The measurement and collection of data is one area of hydrology where many hydrologists are employed. This fact is not emphasized here because the book is intended to be a student's first exposure to topics in hydrology. The title, *Hydrologic Analysis and Design,* indicates that the emphasis has been placed on the last two of the four areas identified, namely model conceptualization and data analysis, and design or solution synthesis. In most chapters a brief section that discusses analysis and synthesis is included. Emphasis has been placed on analysis and synthesis because I believe that knowledge of these subjects is most important for the beginning student. Also, it is these topics that students usually find most interesting; thus, by emphasizing these areas I hope that the book will motivate students toward hydrologic practice. I believe that detailed discussions of the physical processes and data collection should follow the introductory discussion of the methods of analysis and design. This belief shaped the emphasis that is placed on the topics covered.

Educational programs in engineering and the sciences are often criticized as being too theoretical. Practitioners often complain that recent graduates of engineering programs cannot perform basic design computations. *Hydrologic Analysis and Design* has been structured to overcome this problem. While the text is not intended to be a design manual, a number of design methods are included. Although all of the details of these design methods are not given, the material provided should demonstrate to the student that design computations are not difficult and that the design methods have a basis in the fundamental concepts of the hydrologic cycle.

Students of engineering hydrology and the hydrologic sciences often fail to recognize the interconnection between topics discussed in an introductory course in hydrologic analysis and design. Engineering design requires knowledge of watershed characteristics, rainfall, frequency concepts, and the processes that control runoff. In an attempt to alleviate this problem, I have developed a series of projects that are presented in a supplementary manual. I have used these projects over the last few years with much success. While a few students complain of the effort required, almost all of them agree that the project approach is both interesting and educational. The "real world" flavor of the project approach to learning hydrology definitely serves to inspire interest in hydrologic analysis and design. I highly recommend the project approach.

In addition to the project approach to learning, I have included two other items that are different from most books on hydrologic analysis and design. Each chapter includes a series of multiple-choice review questions. These are designed to encourage the student to understand that the qualitative concepts of hydrology are just as important as the quantitative design methods. If these review questions are assigned at the end of the discussion of a chapter, they can also serve to tie together the different concepts discussed in the chapter. I have included material on ethical issues in hydrology (the last few sections of Chapter 1). The purpose of the material is to introduce the student to another side of hydrologic practice, namely the societal or value element that accompanies every hydrologic analysis and design. While the book is devoted to the quantitative aspects of hydrology, the practitioner should always give consideration to the societal effects of a project. Those involved in the practice of hydrologic analysis and design do not work in a social vacuum: The presented material will make this point to the student. Students always enjoy the lecture that is associated with Chapter 1, and some former students have told me that it was the most important lecture. Additionally, each chapter includes a discussion question that poses a situation that requires knowledge of values and value decision making.

I want to acknowledge the help of the following people who reviewed drafts of selected chapters of the first edition: Henry Anderson, Mark Hawley, David Murray, Walter Rawls, and Peter Waldo. Their comments contributed to the success of the first edition. I also greatly appreciate the helpful suggestions of the following reviewers of a prospectus of this second edition: Robert M. Ayer; Thomas Ballestero, University of New Hampshire; Oktay Gúven, Auburn University; Mark H. Houck, George Mason University; and Jose D. Salas, Colorado State University. Of course, I remain responsible for all errors or omission of topics. I also want to acknowledge the many students who have read drafts of various sections of the manuscript. The final product is much improved because of their efforts. I am very indebted to Ms. Florence Kemerer for her efforts in typing many drafts of this second edition.

Richard H. McCuen

1

Introduction to Hydrology

CHAPTER OBJECTIVES

1. Define hydrology.
2. Introduce the hydrologic cycle.
3. Develop a conceptual understanding of society's effect on the hydrologic cycle.
4. Distinguish between hydrologic analysis and synthesis.
5. Introduce the linear storage equation.
6. Demonstrate that hydrologic practice involves more than just solving technical design problems; hydrologists do not work in a social vacuum.
7. Introduce the fundamental components of professionalism and place them in the context of hydrologic practice.
8. Identify human and social values that are important for the hydrologist to consider.
9. Provide a framework for understanding and solving value dilemmas in a professional setting.

1.0 NOTATION

b_o	= model coefficient (intercept)	O	= outflow
b_1	= model coefficient (slope)	S	= storage
E	= evaporation	t	= time
I	= inflow	T	= temperature

1.1 HYDROLOGY: THE STUDY OF WATER

The word *hydrology* combines the Greek and word *hudōr*, which means "water," and the term *-logy*, which designates "a study of." It also has origins in the New Latin word *hydrologia*. More specifically, the general word *hydrology* refers to the scientific study of water and its properties, distribution, and effects on Earth's surface, soil, and atmosphere. The study of water can mean different things to different professions. To a chemist, a water molecule is a stable chemical bond of two atoms of hydrogen and one atom of oxygen; the chemist will be interested in the properties of water and its role in chemical reactions. The climatologist will be interested in the effect of the water stored in the soil and lakes on climatic processes. To those involved in the design of hydraulic machinery, the study of the properties of water will concentrate on the forces exerted by water in a dynamic state. To the mechanical engineer, the properties of water in the form of steam can be important. The ground water hydrologist will be interested in the movement of water in transporting pollutants. Even geographers and historians may be interested in water, at least in terms of how its availability and accessibility has shaped development and culture. However, our interest herein is in the narrow field of hydrologic engineering analysis and design. Engineering hydrology encompasses those aspects of hydrology that relate to the design and operation of engineering projects for the control and use of water. Aspects that relate to our nation's infrastructure are of special interest because of the decay of our infrastructure. Whereas the primary interest will be to the engineer, others, such as planners, environmentalists, water managers, and meteorologists, may find the knowledge of hydrologic engineering to be of interest.

Earth's atmosphere, oceans, ice masses, lakes, rivers, streams, and soil contain over 50 billion cubic feet of water. In spite of this abundance, problems are created by either too much or too little water at a given location; that is, problems are caused by the spatial variation of water. For example, people living in southern California and other areas of the arid southwest show concern over the lack of an inexpensive source of water supply. Problems also result from variations in the time distribution of water. An overabundance of water at one time or an undersupply at other times can have serious consequences to both agriculture and manufacturing, as well as inconveniencing the public. Occasional flooding is a problem to homeowners and to entire cities. Crops do not grow at the optimum rate when the soil is either too wet or too dry. Manufacturing operations require a consistent water supply over time for a variety of purposes, such as to provide cooling water and to assimilate wastes. Thus although Earth's total volume of water may be adequate to meet all needs, problems are created by variations in both the spatial and temporal distributions of water availability. Extreme problems, including life-threatening situations, can result from extreme variations in either the spatial or temporal distribution of water, or both.

In an attempt to overcome the problems created by these variations in the temporal and spatial variations in water availability, engineers and hydrologists attempt to make predictions of water availability. These predictions are used in the evaluation of alternative means of preventing or solving problems. A number of factors contribute to the ineffectiveness of these engineering designs. First, the occurrence of rainfall cannot be predicted with certainty. That is, it is not possible to predict exactly how much rain will occur in one time period (for example, day, month, year). The uncertainty of extreme variation in rainfall amounts is even greater than the uncertainty in the rainfall volumes occurring in the more frequent storm events. It is difficult to design engineering works that will control the water under all conditions of variation in both the time and spatial distribution. Second, even if we had perfect information, the cost of all of the worthwhile projects needed to provide the optimum availability of water is still prohibitive. Therefore, only the most efficient and necessary projects can be constructed. Third, hydrologic processes such as rainfall and runoff are very complex and a complete, unified theory of hydrology does not exist. Therefore, measurements of observed occurrences are used to supplement the scant theoretical understanding of hydrologic processes that exists. However, given the limited records of data, the accuracy of many engineering designs is less than we would like. These three factors (hydrologic uncertainty, economic limitations, and lack of theory and observed data) are just some of the reasons that we cannot provide solutions to all problems created by undesirable variations in the spatial and temporal distributions of water.

In spite of the inherent uncertainty in precipitation, the economic constraints, and the bounds on our theoretical understanding of hydrometeorological processes, solutions to the problems that are created by the temporal and spatial variations in water availability must be provided. Estimates of hydrologic quantities such as streamflow are required as input to engineering designs, which represent the engineer's attempt to solve the problem. All engineering designs should, at the minimum, be rational. An understanding of the physical processes is a necessary prerequisite to the development of rational designs.

1.2 THE HYDROLOGIC CYCLE

The physical processes controlling the distribution and movement of water are best understood in terms of the hydrologic cycle. Although there is no real beginning or ending point of the hydrologic cycle, we can begin the discussion with precipitation. For the purposes of this discussion, we will assume that precipitation consists of rainfall and snowfall. A schematic of the hydrologic cycle for a natural environment is shown in Figure 1–1. Rain falling on Earth may enter a water body directly, travel over the land surface from the point of impact to a watercourse, or infiltrate into the ground. Some rain is intercepted by vegetation; the intercepted water is temporarily stored on the vegetation until it evaporates back to the atmosphere. Some rain is stored in surface depressions, with almost all of the depression storage infiltrating into the ground. Water stored in depressions, water intercepted by vegetation, and water that infiltrates into the soil during the early part of a storm represent the initial losses. The loss is water that does not appear as runoff during or immediately following a rainfall event. Water entering the upland streams travels to increasingly larger rivers and then to the seas and oceans. The water that infiltrates into the ground may percolate to the water table or

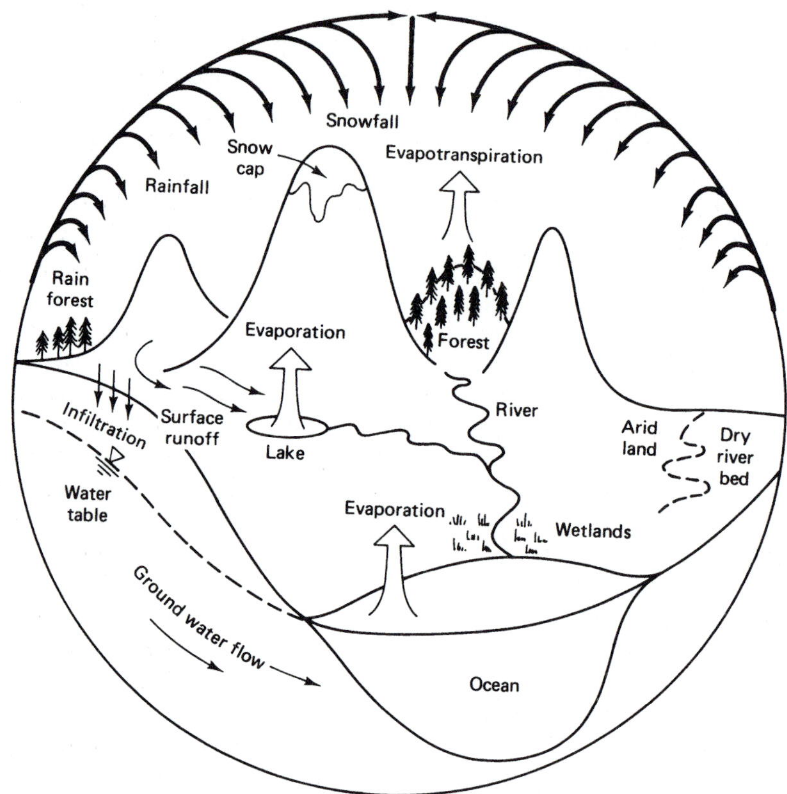

FIGURE 1–1 Hydrologic cycle of a natural environment.

travel in the unsaturated zone until it reappears as surface flow. The amount of water stored in the soil determines, in part, the amount of rain that will infiltrate during the next storm event. Water stored in lakes, seas, and oceans evaporates back to the atmosphere, where it completes the cycle and is available for rainfall. Water also evaporates from soil devoid of vegetation. Rain that falls on vegetated surfaces may be intercepted; however, after the storage that is available for interception is filled, the water will immediately fall from the plant surfaces to the ground and infiltrate into the soil in a similar manner as the water falling on bare ground infiltrates. Some of the water stored in the soil near plants is taken up by the roots of the vegetation, and subsequently passed back to the atmosphere from the leaves of the plants; this process is called *transpiration*.

Although it may appear that the hydrologic cycle of a natural environment is static, it is important to recognize that the landscape is constantly undergoing a transformation. High-intensity storms cause erosion of the land surface. Flood runoff from large-volume storms causes bankfull and high-velocity flows in streams with the potential for large amounts of channel erosion. During periods of extreme drought the perimeter of desert lands may increase. Forest fires caused either by natural means such as electrical storms or by the carelessness of human beings cause significant decreases in the available storage and decrease the surface roughness, both of which contribute to increases in surface runoff rates and volumes, as well as surface erosion. When mud flows are a potential problem, forest fires consume the vegetation, which contributes significantly to the increased production of debris. In many parts of the world, the largest floods result from the rapid melting of snow; such events can be as devastating as floods produced by large rainfall events. Flooding accompanying hurricanes and monsoons can also cause significant changes to the landscape, which by itself affects runoff rates and volumes from storms occurring long after the hurricane or monsoon. In summary, even in a natural environment, rainfall and runoff cause major changes in the watershed.

As the population of the world has increased, changes to the land have often been significant, with major changes to the runoff characteristics of a watershed as a result. Land clearing for agricultural development increases the amount of exposed soil, with obvious decreases in the protective covering of the natural vegetation. This loss of protective covering decreases the potential for infiltration, increases surface runoff, and can result in significant soil losses. Over the last two centuries, urbanization has caused significant changes to the landscape surrounding these urban centers. Urbanization has had significant effects on the processes of the hydrologic cycle for watersheds subject to the urban development. Clearing of the land has reduced the vegetation and therefore the availability of interception storage; grading of land surfaces reduces the available volume of depression storage. Impervious surfaces reduce the potential for infiltration and the resulting recharge of ground water storage. Impervious surfaces are also less rough than the natural surfaces and thus offer less resistance to the runoff; this change in roughness can increase runoff velocities and surface erosion. These changes to the processes of the hydrologic cycle, which are shown schematically in Figure 1–2, cause significant changes in runoff characteristics. The reduced storage results in increased volumes of surface runoff. The reduced surface roughness decreases the travel time of runoff. The reductions in both storage and travel time result in increased peak rates of runoff, which increase both flood damages due to overbank flows and channel erosion.

In an attempt to compensate for the lost natural storage, many localities require the replacement of the lost natural storage with human-made storage. While the storm water detention basin is the most frequently used method of storm water management, other methods are used, such as infiltration pits, rooftop and parking lot storage, and porous pavement. These engineering works do not always return the runoff characteristics to those that existed in the natural environment. In fact, poorly conceived methods of control have, in some cases, made flood runoff conditions worse.

1.3 HYDROLOGIC DESIGNS

The average American drives about twenty-five miles per day. While many of these miles involve travel on the same route, say to or from school or a job, the individual traveler will encounter a surprisingly large number of engineering designs where hydrologic analyses were required. In fact, hydrology was considered in the design of the very twenty-five miles of highway covered in an average day. Failure to consider drainage of the highway subbase can lead to premature failure of the highway. In addition to subbase design, hydrologic analyses

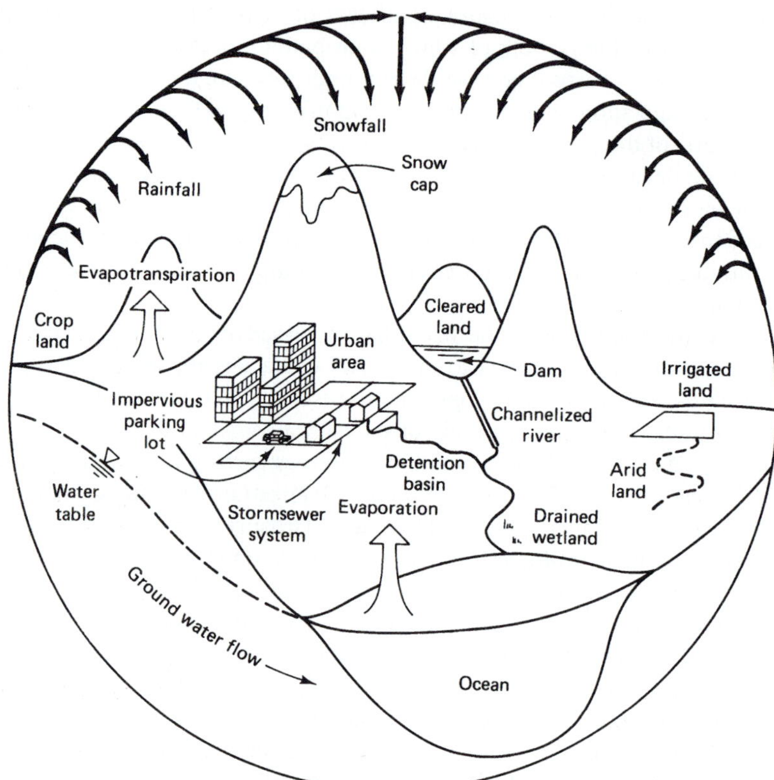

FIGURE 1–2 Hydrologic cycle of a developed environment.

are required for the design of culverts (for example, a pipe that crosses under a road or embankment), surface drainage inlets, and bridges that cross over rivers and streams. It should be evident that those involved in highway design must understand the basic concepts of hydrologic analysis since the design of every mile of highway requires consideration of the fundamental concepts of hydrology. Other elements of transportation systems require hydrologic analyses. Those involved in the design of parking lots, airport runways and aprons, rapid mass transit lines, and train rights-of-way must give just as much consideration to the proper drainage of storm runoff as those involved in the design of highways.

Those involved in the design of transportation facilities are not the only ones who must consider the natural passage of water resulting from storm events. Anyone involved in land development and the construction of homes, as well as commercial, industrial, and institutional buildings, must give consideration to storm runoff. Obviously, those involved in home design must provide gutters and down spouts. Buildings in commercial and industrial developments also require roof drainage.

There are many other hydrologic analyses required in building construction. When clearing land for development, it is important to provide sediment control facilities to ensure that eroded soil does not enter into waterways and wetlands. Sediment control depends on the area of the land being cleared, the amount of rainfall that can be expected during the period where the soil will be exposed to rainfall impact, and site characteristics such as the slope and soil type. In addition to hydrologic considerations during the land development stage, site development must consider drainage patterns after development.

Site development usually results in significant increases in impervious surfaces, which results in increased surface runoff rates and volumes. In many localities, storm water control facilities are required. In the upper reaches of a site, swales can be used to move water away from buildings and transportation facilities. Concentrated runoff from swales may enter gutters and drainage ditches along roadways where the runoff may drain into highway inlets or small streams. At many sites where land development has resulted in large amounts of imperviousness, on-site detention basins may be required to control the increased runoff. The design of storm water detention basins requires knowledge of routing of water through the hydraulic outlet structure, as well as knowledge about surface runoff into the detention basin. The design must consider meteorological factors, geomorphological factors, and the economic value of the land, as well as human value considerations such as aesthetic and public safety aspects of the design. The design of a storm water detention basin should also consider the possible effects of inadequate maintenance of the facility.

The hydrologic designs discussed in the preceding paragraphs are based primarily on rainfall and the resulting surface runoff. Dams and the water stored in the reservoirs behind the dams provide many benefits, such as power generation, recreation, flood control, irrigation, and the maintenance of low flows for water quality control. In addition to estimating the volume of inflow into the reservoir, dam design requires assessment of the evaporation losses from the reservoir. For reservoirs with large surface areas, evaporation losses can be significant. Failure to consider evaporation losses during the design could result in overestimating the water that would be available for the purposes stated above. Thus, failure to understand the processes of the hydrologic cycle may render the design inadequate.

1.4 ANALYSIS VERSUS SYNTHESIS

Like most of the basic sciences, hydrology requires both analysis and synthesis to use the fundamental concepts in the solution of engineering problems. The word *analysis* is derived from the Greek word *analusis*, which means "a releasing," and from *analuein*, which means "to undo." In practical terms, it means "to break apart" or "to separate into its fundamental constituents." Analysis should be compared with the word synthesis. The word *synthesis* comes from the Latin word *suntithenai*, which means "to put together." In practical terms, it means "to combine separate elements to form a whole."

Because of the complexity of most hydrologic engineering design problems, the fundamental elements of the hydrologic sciences cannot be used directly. Instead, it is necessary to take measurements of the response of a hydrologic process and analyze the measurements in an attempt to understand how the process functions. Quite frequently, a model is formulated on the basis of the physical concepts that underlie the process, and the fitting of the model with the measurements provides the basis for understanding how the physical process varies as the input to the process varies. After the measurements have been analyzed (taken apart) to fit the model, the model can be used to synthesize (put together) design rules. That is, the analysis leads to a set of systematic rules that explain how the underlying hydrologic process will function in the future. We recognize that the act of synthesizing is not a total reproduction of the original process. It is a simplification. As with any simplification, it will not provide a totally precise representation of the physical process under all conditions. But in general, it should provide reasonable solutions, especially when many designs based on the same design rules are considered.

It should be emphasized that almost every hydrologic design (or synthesis) was preceded by a hydrologic analysis. Most often, one hydrologic analysis is used as the basis for many, many hydrologic designs. But the important point is that the designer must understand the basis for the analysis that underlies any design method; otherwise, the designer may not apply the design procedure in a way that is compatible with the underlying analysis. This is not to say that a design method cannot be applied without knowing the underlying analysis, only that it is best when the design engineer fully understands the analysis that led to development of the design rules. Anyone can substitute the values of input variables into a design method. But when a design is used under circumstances for which it was not intended to be used, inaccurate designs can be the result. Without further consideration, the engineer uses these estimates as the basis of the design.

1.4.1 Water Budget Analysis and Synthesis

A simplified example may help illustrate the concepts of analysis and synthesis. Let's assume that an engineer has the task of designing a reservoir, and the design requires estimates of the water loss by evaporation from the reservoir. While the design period will be from June 1 to July 31, the engineer is assigned the project just before October 1, with a completion date of April 15 of the following year.

Recognizing his lack of experience in making hydrologic evaluations, the engineer correctly decides to hire a hydrologist to perform an analysis to provide a means of making estimates of evaporation. Recognizing that temperature is an important determinant in the evaporation process, the hydrologist decides to develop a design procedure based on a model relating evaporation to temperature. Since there are no meteorological data collection stations in the region of the design, the hydrologist decides to collect data of the monthly evaporation from a nearby lake and compute the mean monthly temperature. The following data are the result of the data collection effort:

Month	T (°F)	E (in.)
October	70	4.3
November	55	3.3
December	51	3.2
January	50	2.9
February	53	3.1
March	60	3.6

An examination of the literature suggests that evaporation can be related to temperature using a power model $E = b_0T^{b1}$ with the values of b_0 and b_1 computed using regression analysis. The analysis of the data above led to the model $E = 0.043687T^{1.0792}$ with a correlation coefficient of 0.98, which suggests that the model should be accurate. To summarize, the analysis problem involved (1) deciding on the variables to use, (2) deciding on the form of the model, (3) collecting the matrix of E and T data shown above, and (4) fitting the model to the data and evaluating the expected accuracy of the design model. The hydrologist provides the engineer with the foregoing model.

Given the results of the analysis above, the engineer must now "put together" a design. The engineer knows that the mean temperatures for June and July would be about 85°F and 90°F, respectively. Using these temperatures as input to the design model above, evaporation estimates of 5.3 and 5.6 in., respectively, are obtained. Without further consideration, the engineer uses these estimates as the basis of the design. The high correlation of 0.98 provides the engineer with a false sense of comfort about the accuracy of estimates.

The problem with this description is that the engineer used the results of the analysis (the model) without understanding the analysis. If the hydrologist had known the engineer's objective, the hydrologist would have told the engineer that the model was based on data for the period from October to March, or if the engineer had inquired about the basis for the analysis, the engineer could have made an independent assessment of the results. For example, for the design location the engineer could have compared the design evaporation estimates of 10.9 in. with the mean May–October lake evaporation of 50 in. reported in a climatic atlas. The estimate of 11 in. would then have been rejected as being unreasonably low.

Although this is a simplified example, it is not uncommon for a hydrologic model to be used without the user taking the time to determine the analysis that underlies the model. In cases where the user is fortunate enough to be applying the model within the proper bounds of the analysis, the accuracy of the design is probably within the limits established by the analysis; however, all too often inaccurate designs result because the assumptions used in the analysis are not valid for the particular design. The moral of the story is that those involved in the analysis phase should clearly define the limits of the model, and those involved in synthesis or design should make sure that the design does not require using the model outside the bounds established by the analysis.

1.4.2 A Conceptual Representation

Because of the importance of the concepts of analysis and synthesis, it will be worthwhile to place the problem in a conceptual framework. We will conceptualize the hydrologic system to consist of three parts: the input, the output, and the transfer function. This conceptual framework is shown schematically in Figure 1–3. In the analysis phase, the input and output are known and the analyst must find a rational model of the transfer function. When the analysis phase is completed, either the model of the transfer function or design tools developed from the model are ready to be used in the synthesis phase. In the synthesis or design phase, the design input and the model of the transfer function are known, and the predicted

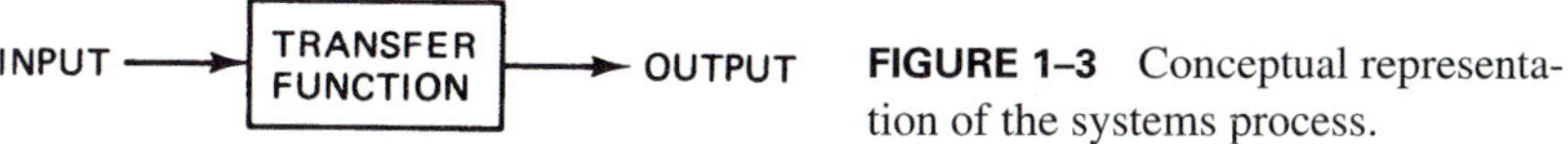

FIGURE 1–3 Conceptual representation of the systems process.

system output must be computed; the true system output is unknown. The designer predicts the response of the system using the model and bases the engineering design solution of the predicted or synthesized response.

The reservoir design problem can be used to illustrate this conceptual framework of hydrologic analysis and design. In this simplified problem, the evaporation is needed for design; thus the evaporation serves as the output variable. Temperature serves as the input. In the analysis phase, the model for predicting evaporation is evaluated; therefore, the model represents all of the hydrologic processes involved in converting temperature (for example, heat energy) to evaporation. Obviously, temperature is only a single measurement of the processes that effect evaporation, but it is a surrogate variable for these processes in the conceptual representation of the system. In the design phase, the designer uses other temperature conditions as input to the model to predict the response (for example, the output) of the system. It is this synthesis of the system output that is used for the design. In summary, in the analysis phase, the temperature (input) and evaporation (output) were known, while in the synthesis phase, the temperature (input) and the model (transfer function) were known. Examples of the analysis/synthesis representation that will be discussed in other chapters are given in Table 1–1.

1.5 THE HYDROLOGIC BUDGET

The systems concept of Figure 1–3 can be applied to the elements of the hydrologic cycle, which can be viewed as inputs, outputs, and storages. Rainfall can be viewed as an input to the surface of Earth. The surface can be viewed as a series of storage elements, such as storage on the surface of vegetation and depression storage. Runoff from the surface can be viewed as an output from surface storage elements. This would be a systems representation of the physical processes controlling surface runoff.

If river channel processes are the important elements of the hydrologic design, then the surface runoff can be viewed as the input, the channel itself as the storage element, and the runoff out of the channel (into another channel, a lake, or an ocean) as the output from the system. Hydrologic analysts and designers make use of this systems representation of the elements of the hydrologic cycle to develop design methods to solve engineering problems.

The conceptual representation of hydrologic systems can be stated in mathematical terms. Letting I, O, S, and t denote the input, output, storage, and time, respectively, the following equation is known as the linear storage equation:

$$I - O = \frac{dS}{dt} \tag{1–1}$$

The derivative on the right-hand side of Equation 1–1 can be approximated by the numerical equivalent $\Delta S/\Delta t$, when one wishes to examine the change in storage between two times, say t_2 and t_1. In this case, Equation 1–1 becomes

$$I - O = \frac{\Delta S}{\Delta t} = \frac{S_2 - S_1}{t_2 - t_1} \tag{1–2}$$

in which S_2 and S_1 are the storages at times t_2 and t_1, respectively.

TABLE 1–1 Examples of Hydrologic Analysis and Synthesis

Input	Model or Transfer Function	Output	Analysis		Synthesis	
			Known	Unknown	Known	Unknown
1. Runoff coefficient (C) Rainfall intensity (i) Drainage area (A)	Rational formula $Q_p = CiA$ (Eq. 7–20)	Peak discharge (Q_p)	i, A, Q_p	C	C, i, A	Q_p
2. Peak inflow rate (Q_i) Allowable outflow (Q_o) Runoff Volume (V_r)	Figure 8–17	Volume of storage (V_s)	Q_i, Q_o, V_r, V_s	Figure 8–17	Q_i, Q_o, V_r	V_s
3. Upstream hydrograph (I)	Muskingum equation $O_2 = C_0 I_2 + C_1 I_1 + C_2 O_1$ (Eq. 10–53)	Downstream hydrograph (O_2)	I, O_2	C_0, C_1, C_2	I, C_0, C_1, C_2	O_2
4. Erosivity index (R) Soil erodibility factor (K) Topographic factor (T) Cover factor (C) Support factor (P)	Universal soil loss equation $E = RKTCP$ (Eq. 15–14)	Soil loss (E)	E, R, K, T	C, P	R, K, T, C, P	E
5. Length (HL) Slope (Y) Storage (S)	SCS lag equation $L = \dfrac{HL^{0.8}(S+1)^{0.7}}{1900Y^{0.5}}$ (Eq. 3–56)	Watershed time lag (L)	L, HL, Y, S	Coefficients of Eq. 3–56	HL, Y, S	L
6. Storage (S) Rainfall (P)	$Q = \dfrac{(P - 0.2S)^2}{P + 0.8S}$ (Eq. 7–42)	Runoff Volume (Q)	Q, P	S	P, S	Q

Equation 1–1 can be illustrated using the flow in a river as the input and output. Assume that the inflow and outflow rates to a river reach of 1500 ft in length are 40 ft^3/sec and 30 ft^3/sec, respectively. Therefore, the rate of change in storage equals 10 ft^3/sec or 6000 ft^3 during a 10 min period. Since the inflow is greater than the outflow, the water surface elevation will rise. Assuming an average width of 16 ft, the increased storage would cause the water surface elevation to rise by 6000 ft^3/[16 ft (1500 ft)] = 0.25 ft over the 10 min period.

The storage equation of Equation 1–1 can be used for other types of hydrologic problems. Estimates of evaporation losses from a lake could be made by measuring: all inputs, such as rainfall (I_1), inflow from streams (I_2), and ground-water inflow (I_3); all outputs, such as streamflow out of the lake (O_1), ground-water flow out of the lake (O_2), and evaporation from the lake (O_3); and the change in storage between two time periods, which could be evaluated using the lake levels measured at the beginning and end of the time period. In this case, both the input and the output involve multiple measurements, that is, they are vector quantities with the elements of each vector I_i and O_j representing the different components of the inflow and outflow from the system. The unknown is one of the elements of the outflow vector. Mathematically, the water balance is

$$(I_1 + I_2 + I_3) - (O_1 + O_2 + O_3) = \frac{dS}{dt} \tag{1–3}$$

Estimating evaporation losses (O_3) would require measurements of each of the inflows, the other outflows, and lake levels; obviously, any errors in the measurements of the elements of Equation 1–3 would effect the accuracy of the estimated evaporation. Also, if there are any inflows other than I_1, I_2, and I_3 or any outflows other than O_1 and O_2, the estimated evaporation will be in error by such amounts.

In summary, the hydrologic budget is a convenient way of modeling the elements of the hydrologic cycle. It will be used frequently in describing the problems of analysis and design.

1.6 INTRODUCTION TO PROFESSIONAL ETHICS

The chapters of this book detail various methods of hydrologic analysis and design. This may imply that the heart of hydrologic work is the number crunching. While the hydrologist must certainly understand the physical processes and know the design procedures, he or she must also recognize that designs are necessary because society has needs and problems. Very often these needs or problems arise because of value issues. An adequate water supply is important for both the public health and the happiness of citizens. But a dam and reservoir may represent a hazard to public safety. The hydrologist will be involved in the technical aspects of solving the problem, but the hydrologist has social responsibilities related to the value issues. These value issues should be just as important to the hydrologist as the technical details are. Sound hydrologic practice is not limited to providing correct evaluations of the design elements. Hydrologists must be able to recognize value issues, be willing and able to consider the value issues along with the technical issues, and know how to incorporate these value issues into decision making. These value-based responsibilities are just as important as the design responsibilities.

1.6.1 Case Studies: Ethical Dilemmas in a Professional Setting

To create a foundation for a discussion of the social responsibilities of professionals who are involved in hydrologic analysis and design, it is worthwhile to sketch four hypothetical situations that involve value issues in a professional setting. The initial emphasis should be on identifying the value issues. The task of solving value dilemmas that arise in professional practice will be discussed later.

Case Study 1. The firm for which you work obtains a contract for designing a pipe to carry a liquid hazardous waste between the plant where it is produced as a by-product and a storage facility where it is held until it can be transported to a disposal site. The two sites are 1.2 miles apart. During the site investigation you find that the route proposed by the client passes through a very permeable geologic formation (see Chapter 6). You recognize that a leak or rupture of the pipe in the area of the permeable formation could cause significant problems with the quality of the water withdrawn from the aquifer.

You bring the problem to the attention of your supervisor and propose an alternative route for the pipe. Your proposal would increase both the length of the pipe and the cost of the project, as well as delay the completion date of the project. Your supervisor tells you that the project is to be completed as planned and that the potential hazard is not your concern. What are the values that are in conflict? What action, if any, should you take?

Case Study 2. The firm that you work for obtains a contract for constructing a water supply reservoir. You find out that the proposed site is above an area that is zoned for commercial and high-density residential land uses, although the areas are currently undeveloped. Based on this, you recognize that state policy would require the hydrologic design to be based on a probable maximum flood analysis (see Chapter 4). You indicate to your project manager that you have no experience with PMF analyses and that no one else in the firm has the experience. After you recommend that the firm hire an outside consultant who has the necessary experience, the project manager indicates that your proposal is not economically feasible and that you will be responsible for the design. What are the value issues involved in this case? Recognizing that the effort is outside your area of professional competency, what action, if any, should you take?

Case Study 3. A public agency issues a request for proposals on a design project that is in your area of professional expertise. Your supervisor indicates that you will be in charge of writing the proposal, which you believe will be a three- to four-week effort. Your supervisor tells you to charge the time to a project that is currently in progress but to which you have no involvement and is totally unrelated to the project for which you will be writing the proposal. You believe that your time should be reimbursed using company overhead funds, rather than funds from another client. Your supervisor rejects your concern about charging time to a project on which you were not involved and states that if you want to get paid, you must charge your time as instructed. What are the value issues that are in conflict? What action, if any, should you take?

Case Study 4. The firm that you work for discharges wastes into a local stream. It is your responsibility to take samples for specific water quality indicators, make the necessary laboratory tests, and report the results to the local public health agency. On a regular basis, you inform your supervisor of the test results prior to providing a summary of the results to the local public health agency. At one reporting date you find that the effluent from your firm's plant does not meet state water quality standards. When you present the results to your supervisor, he or she indicates that the test results are probably wrong and that you should send in a report to the public health agency stating that the effluent meets the state water quality standards for the reporting period. Failure to meet the standards can result in a fine or a plant process shutdown. What action should you take? Do you report the true test results to the public health agency or falsify the report as ordered by your supervisor?

1.6.2 Case Studies: A Question of Values

The four case studies presented in Section 1.6.1 should not necessarily be viewed as a problem of employee versus employer. The heart of each of the dilemmas is the employee's struggle with his or her own value system. Before the employee further confronts the supervisor or project manager, the employee must evaluate the value conflicts inherent in each situation. Such evaluation is a prerequisite to finding an optimal solution. The employee must also know what the profession and society expect of him or her.

To solve each of these dilemmas, the employee must address a number of concerns. These concerns can be summarized by the following four questions:

1. What are the value issues involved?
2. How important are the consequences associated with the issues?
3. How can the issues and consequences be weighted so that a decision can be reached?
4. What procedure should be followed to arrive at a solution?

These four questions form the theme of the following sections of this chapter.

1.7 THE DIMENSIONS OF PROFESSIONALISM

When we hear the word *professional*, we may think of a professional athlete, or maybe even a hired killer. After, all, the most general definition of the word *professional* provided by a dictionary would be something like the following: one who is engaged in a specific activity as a source of livelihood. Using this definition, almost all people who work could be classed as professionals. Plumbers, grave diggers, used-car salespersons, and chimney sweepers are engaged in specific activities as a source of livelihood. The intent here is not to say that the grave diggers and football players are not professionals; they certainly are needed citizens and have skills that are used in service to society. But certain specific activities are accorded the status of professions; for example, doctors, members of the clergy, and teachers perform activities that indicate that they belong to professions. These professions are usually thought to require a higher level of knowledge than others that just satisfy the dictionary definition of a professional because they serve society in special ways. It is in this sense that we want to classify hydrologists as professionals; while hydrologists engage in specific activities as a source of livelihood, their work carries special responsibilities to society that accords their work the status of a profession.

In his discussion of professional development, Dougherty (1961) defines a professional as

> one who uses specialized knowledge and skill in the solution of problems that cannot be standardized. He [or she] is actuated by a service motive; he [or she] works in a relation of confidence, and observes an acceptable code of ethical conduct.

But it must be emphasized that the term professional does not apply just to someone who has specialized knowledge and skill. Other dimensions of professionalism are important. In the definition above, Dougherty recognized the multifaceted nature of professionalism, and from this he made the following conclusion:

> Professionalism is a way of thinking and living rather than an accumulation of knowledge and power. Knowledge and power are essential and when actuated by the professional spirit they produce the leaders and torch-bearers.

What are the fundamental dimensions of professionalism, all of which are necessary for an individual to be considered a professional? NSPE (1976) identified five characteristics of a professional engineer; these also apply to those practicing hydrologic design work:

1. One who possesses a body of distinctive technical knowledge and art gained by education, research, and experience.
2. One who recognizes a service motive to society in vital and honorable activities.
3. One who believes in standards of conduct, such as represented by ethical rules.
4. One who supports the legal status of admission to the profession. The most common forms are registration to practice engineering and graduation from an accredited engineering curriculum.
5. One who has pride in the profession and a desire to promote technical knowledge and professional ideals.

To these five characteristics, Dougherty's definition identifies a sixth factor that should be emphasized as a fundamental characteristic, even though it may be inherent in the five characteristics listed.

6. One who works in a relation of confidence to his employer, the public, and all who use his or her works.

Considerable time, education, and professional practice are required to obtain the specialized knowledge and skill that characterize a professional. The professional, in contrast to the artisan or technician, is capable of synthesizing past experience and fundamental principles in solving new problems. Such problem solving requires independent thought and motivation that is not self-serving. The motivation must stem from a dedication to service of both society and the profession; otherwise, the knowledge and skill does not maximize human welfare. Because of the interaction of a professional with society and the advantage that the professional has through his or her specialized knowledge and skill, it is important for a professional to have sound moral judgment and be able and willing to translate this moral judgment into principled professional conduct. While the six dimensions of professionalism are independent in that any person may possess one or a few of these characteristics, a true professional must possess all six of the fundamental dimensions.

1.8 HUMAN VALUES: A DEFINITION

A *value* can be defined as a principle or quality considered worthwhile or desirable. As used here, a *human value* is a quality that relates directly to the potential for improving the conditions of humankind. The ability to make a value decision is one of the characteristics that distinguishes human beings from other living things. Along with this ability goes the responsibility to make value decisions in an unbiased and precise way. To make accurate value decisions, it is necessary for the decision maker to understand his or her value system and establish value priorities. He or she must also know and understand the values held important by the profession. This requires a certain commitment to examine critically one's own value priorities. Unfortunately, too few people take the time to make an assessment of their value system. Individuals are guided by standards established by their family and friends; peer pressure is all too often used as an excuse for improper conduct.

While "life, liberty, and the pursuit of happiness" is probably the most famous value statement and "I love you" is the most frequently used value statement, there are numerous other values. The following is a partial list of human values:

beauty/aesthetics	honor	mercy
cleanliness	honesty	pleasure/happiness
courage	kindness	public health/safety
diligence	knowledge	respect
efficiency	life	security
equality	love	truth
freedom/liberty	loyalty	wisdom

These values represent feelings or attitudes rather than physical entities. This does not mean that a scale could not be developed in an attempt to reflect variations in a feeling or attitude. Industrial engineers are constantly trying to find rules and yardsticks to measure efficiency. Environmental engineers develop standards that define when minimum levels of public health are not being met. Engineering planners attempt to define indices that measure the beauty of an engineering project, and wildlife biologists develop indices to measure changes in the quality of life for animals whose habitat is affected by engineering projects. While such attempts make it possible to incorporate value issues into the quantitative decision-making framework, all too often hydrologists, as well as those in other professions, fail to recognize that everyone affected by a project may not agree with the scale used to quantify a human value. Such disagreement is the basis for value conflicts in a professional setting. But at least the hydrologist can attempt to incorporate value issues into the decision-making process. The relative effort that a hydrologist expends in developing a scale to measure the various states of the value issue may be an indication of the sensitivity of the hydrologist to human values.

1.9 VALUES IN HYDROLOGY

While all hydrologists recognize the importance of the technical issues in the workplace, they should also recognize that hydrologists do not practice in a value vacuum. Decisions involving human values are important to the individual, the firm, the profession, and society. In making value-based decisions, the individual usually places far greater weight on the consequences to himself or herself, such as the potential loss of employment or stature within the professional community, than on the consequences to the firm, the profession, or society. Also, the person quickly becomes aware of the personal stress that accompanies value conflicts. The person has moral obligations to his or her family, and these personal responsibilities cannot be ignored when solving ethical dilemmas in the workplace. Although professional codes of ethics do not require the hydrologist to be self-sacrificing, the individual must be willing to consider values important to the firm, the profession, and society in making value decisions, not just the values that are important to him or herself. Proper value decision making requires the person to give weight to and properly balance value responsibilities to each of the parties.

To make value decisions that best meet the ethical rights of all parties involved, the values that are important to all parties need to be identified and respected in the decision. In the case of the hydrologist who was told to file false time reports, he/she placed emphasis on values such as honesty and security. He/she may not have recognized the values that were important to his or her supervisor. The supervisor placed emphasis on values such as loyalty, performance, and respect, but showed little regard to the values that were important to either the hydrologist or society. Neither the hydrologist nor the supervisor recognized that their actions impacted on values that are important either to the profession, such as honor, or to society, such as truth in matters that affect society. Such failure to recognize values that are important to all parties involved may lead to nonoptimal decisions. In such cases, incorrect value decisions can lead to expulsion from the profession.

Values that should be important to the firm include honesty, responsibility, loyalty, efficiency, and harmony. Failure of an individual hydrologist to consider the value rights of the firm in solving ethical conflicts reduces the effectiveness of the firm in meeting its societal responsibilities. When a person fails to recognize the values that are important to the firm, it is often viewed by the firm to reflect disloyalty.

While ethical problems such as those discussed in Section 1.6.1 may appear to have effects only to the person and the firm, they, in fact, can impact on the profession and society. Knowledge, honor, and integrity are just three human values that are important to the profession, and incorrect decisions to an ethical conflict can adversely affect these values. For example, the use of computer software for design by a hydrologist who is not familiar with the underlying theory might imply to society that knowledge, as well as the societal value of public safety, is not important to the profession. In this case, the design hydrologist and the firm appear to be insensitive to values that are considered to be important to both the profession and society. Failed designs can lead to unnecessary regulation of the profession by societal institutions. Society places high value on public safety and efficiency, with the value of efficiency being interpreted as human welfare and the optimum use of public resources. If a number of hydrologic designs were to fail because hydrologists were using software for designs outside their area of technical competency, society's image of both the firms and the profession would be tarnished.

The practice of hydrology is not just technical decision making. Nonquantifiable value goals are just as important as the benefit-cost ratios that reflect quantifiable monetary goals. Therefore, value decision making is important both within the firm and in the actions of hydrologists in their relations with the public.

From this discussion it should be evident just why value decision making in one's professional life is more complex than many of the value decisions made in one's personal life. When making a personal value decision, just one or two parties are usually involved, for example, oneself and a friend. When making value decisions in professional life, numerous parties are involved. In addition to ethical responsibilities to oneself and family, the professional has legitimate and important ethical responsibilities to the employer, the client, the profession, and society. Failure to recognize and give proper consideration to each of these responsibilities can lead to nonoptimum decisions and even the appearance of professional misconduct.

1.10 SOLVING VALUE DILEMMAS

What should a young hydrologist do when instructed to falsify time reports or reports on water quality test results? What should the hydrologist who is responsible for designing the pipe to carry a hazardous waste do when he or she is told not to worry about failure of the pipe? It should be evident that the mishandling of an ethical dilemma can have adverse effects on the individual, both professionally and personally; furthermore, there may be negative effects on the firm, the profession, and society. On the other hand, following the proper procedure may actually result in many human value benefits.

The most important element in solving ethical conflicts within the workplace is to distinguish between internal and external appeal. *Internal appeal* refers to actions taken within the firm to solve an ethical dilemma. *External appeal* refers to actions with respect to parties outside the firm. Except under unusual circumstances, all internal options should be considered in attempting to resolve an ethical conflict before seeking an external solution. The internal options are separated into three levels (see Table 1–2): individual preparation, communication with supervisor, and initiation of the formal internal appeals process. It is important to follow the appeals process in the sequence shown. As a first step, it is to the advantage of the employee to develop accurate records of the facts and details of the conflict, familiarize himself or herself with the appeals process of solving ethical conflicts within the company, and identify alternative courses of action. At this point, it is also extremely important for the individual to know exactly what he or she expects any appeal to accomplish. It is inadequate just to state that there appears to be a problem; ideally, the employee can propose a solution that can resolve the problem. As stated previously, the person must recognize the moral rights of the firm, the profession, and society, not just show concern for the effects on himself or herself.

TABLE 1–2 Procedure for Solving Ethical Conflicts

- I. Internal appeal options
  - A. Individual preparation
    1. Maintain a record of events and details
    2. Examine the firm's internal appeals process
    3. Know federal and state laws
    4. Identify alternative courses of action
    5. Specify the outcome that you expect the appeal to accomplish
  - B. Communicate with immediate supervisor
    1. Initiate informal discussion
    2. Make a formal appeal
    3. Indicate that you intend to begin the firm's internal process of appeal
  - C. Initiate appeal through the internal chain of command
    1. Maintain formal communication appeal
    2. Formally inform the company that you intend to pursue an external solution
- II. External appeal options
  - A. Personal options
    1. Engage legal counsel
    2. Contact a professional society
  - B. Communication with client
  - C. Contact the media

Once the person has studied and documented the facts and formulated a plan for the appeal, the matter should be informally discussed with his or her immediate supervisor. In most cases, problems can be resolved at this level, which is to the benefit of all parties involved. In some cases, a more formal appeal to the immediate supervisor is necessary; this is usually in the form of a memorandum that clearly states the facts, the individual's concerns, and the actions that would be necessary to resolve the concerns. It should be evident that good communication skills are important in solving ethical conflicts because a failure to clearly state the concerns, whether orally or in writing, may hinder the resolution of the problem. Appealing the problem to higher levels before completing the appeal to the immediate supervisor are viewed negatively by all involved and will decrease the likelihood of a favorable resolution. Finally, before continuing the appeal beyond the immediate supervisor, an individual should inform the supervisor in writing of the intention to continue the appeal to higher authorities.

The process of appealing an ethical conflict within the firm is usually quite similar to the interaction with the immediate supervisor. Formal steps should follow informal discussions, and steps within the process should not be bypassed. If the internal appeals process is completed without a resolution of the conflict, the person should formally notify the company that he or she intends to continue with an external review of the problem.

Before expressing any concern to the client or making concerns public, a lawyer and/or professional society should be consulted. Lawyers can identify courses of action and pitfalls in external appeal. Whereas lawyers understand legal constraints and consequences, they often lack the technical expertise that would be necessary to evaluate the technical adequacy of your arguments. Professional societies have such technical expertise available and may be willing to informally discuss the case and make recommendations; however, they may not have a complete understanding of the legal implications.

After engaging legal counsel and/or obtaining advice from a professional society, the individual may elect to approach the client. In many cases, the client lacks the expertise to judge the accuracy of the claims and will rely on the advice of the firm in evaluating the concerns; however, if the issues involved in the conflict can be accurately communicated, the client may pressure the firm to resolve the issue internally. Alternatively, the client may expend the resources that would be necessary to obtain an unbiased review of the matter. While it is extremely rare that it will be necessary to carry the appeals process to the client, the effects on one's professional career can be very detrimental if the results fail to find validity in the arguments.

The last resort in solving ethical conflicts is public disclosure. The benefits and costs in terms of both economic criteria and nonquantifiable value goals should be rigorously assessed before contacting the media. It is not even in the best interest of society to make public disclosure prior to seeking an internal resolution. The only case where public disclosure is acceptable prior to following the appeals process (Table 1–2) is when an immediate danger to the public exists.

Four situations in which ethical dilemmas existed were introduced in Section 1.6.1. While the four situations were quite different in many respects, the process of Table 1–2 can be used in resolving each of the conflicts. The problem of falsifying time reports is best handled with the hydrologist informally discussing the situation with the supervisor. In discussion, the hydrologist should clearly state the value conflict created by the supervisor's request to falsify the time sheets. He or she might try to persuade the supervisor to change the way in which project accounts are handled so that actions that violate an employee's value system are not necessary. Furthermore, the employee might point out that such a requirement reduces the employee's respect for the firm and may affect his or her sense of loyalty. By appealing to the supervisor in terms of the values that are of interest to the firm, the supervisor may be more responsive than if the hydrologist only showed concern with his or her own position.

1.11 RATIONALIZATION

In Case Study 4 of Section 1.6.1, how does the supervisor justify asking the employee to falsify the water quality report? In Case Study 2, how does the project manager justify asking someone to perform hydrologic engineering work that is clearly outside the individual's competency? If instead of these professional case studies, we were to ask how a student justifies cheating on a test, the response would probably involve "rationalization." In many cases, this is a primary factor in unethical professional conduct.

Rationalization is the mental process of completely falsifying the results of one's actions so that he or she is incapable of regretting the action (Ross, 1972). It is a form of self-deception, for the purpose of justifying one's actions. Rationalization serves two purposes (Ross, 1972). First, it serves as an excuse to avoid guilt feelings and the scorn of others. Second, it provides self-justification of an action, thus excusing the individual from making a change in his or her moral decision process.

What are possible rationalizations that professionals could employ to provide self-justification for their actions? In Case Study 4, the supervisor justifies the action of requesting the engineer to file a false report by arguing that the test results are probably incorrect and that, if additional tests were made, they would prove that the standards are being met. The supervisor believes that his or her professional experience is more accurate than some laboratory tests.

Other common rationalizations to justify unethical conduct are that, "everybody is doing it" or, "if I don't do it, someone will." If the individual can convince him or herself that everyone is doing it, then the individual will feel less guilt because he or she believes that they are on the same moral level as others in comparable positions. Using such logic, it follows that it is not necessary to rise to a level of ethical conduct that is higher than that of his or her peers. The rationalization makes it unnecessary to consider the value rights of others and enables the supervisor to repeat the conduct in the future.

What rationalization might the young engineer make in order to charge the time to the incorrect account? The following are a few responses:

1. It is standard business practice.
2. I am only doing what I am told to do.
3. It is only a small amount of money.
4. It is the manager's responsibility to decide money matters.
5. I want to be loyal to the company.

Any one of these might serve the immediate needs of the individual and the firm, but does it meet the needs of the profession, society, or the client who is paying for someone to work on another client's project? Society values truth. The profession values the integrity of its members, as well as its reputation. The client values trust. These values are being given zero weight in the decision. Instead, the young engineer is placing value on personal needs and the supervisor is valuing efficiency within the firm. Neither the engineer nor the supervisor is properly balancing the value issues. Their rationalizations are serving as the means to place selfish ends before the legitimate expectations of the profession and society.

1.12 CONCLUDING THOUGHTS

Hydrologists do not work in a value vacuum, and it is unreasonable to believe that you will never be confronted by a value conflict in the workplace. If the appeals are handled properly, the ethical conflicts will usually be resolved without incident. The solution of more complex ethical conflicts requires a good understanding of the systematic process of Table 1–2 which is designed to give the optimum weight to the moral rights of all the parties involved. Resolution of ethical conflicts should not be viewed solely from the viewpoint of the effects on the individual. Just as the firm, the profession, and society have moral responsibilities, they have value rights that must be respected. Proper respect for these values is in the best interest of society as well as the individual, and the responsibilities are best met when a procedure such as that outlined in Table 1–2 is followed.

Solutions to ethical dilemmas require knowledge of three aspects of ethical decision making. First, the person involved must have an appreciation for human value issues and be able to identify the values that are in conflict. Second, the person must recognize that the effects of value conflicts extend beyond himself or herself. Value conflicts within the workplace impact on the firm, the profession, and society, not just the individual. The individual, the firm, the profession, and society have ethical responsibilities as well as ethical rights. Moral rights and responsibilities to each of these parties must be considered in finding the optimal solution to an ethical dilemma. Third, the person must know the steps to follow that will provide the optimum solution to an ethical dilemma. These steps are summarized in Table 1–2.

The discussion in this chapter has tried to show that the hydrologist has both technical and value professional responsibilities, both of which require the hydrologist to show concern for the public. While I may have overtly ignored the employer, the hydrologist has a direct responsibility to the employer; this responsibility goes beyond just fulfilling the technical requirements of the job. Important employer-related responsibilities must be considered in decision making. Before going beyond the firm to protest corporate actions, the hydrologist should make every attempt to view the problem and the value issues from the perspective of the employer. The professional reputation of both the firm and its other employees should be important to the hydrologist even when he or she disagrees with the employer's decision. The hydrologist should also respect the collective knowledge and experience of the managers of the firm. The hydrologist should recognize that the employer has a responsibility to provide long-term employment for the employees of the firm. I personally know of many employers and managers in engineering firms who place high priority on the welfare of the employees and personally agonize over decisions to discharge employees because the funding is no longer there to support them. These are important value issues that increase the complexity of a decision to protest the action of an employer.

The main point of this discussion is that hydrologic design requires more than just technical knowledge of the design methods. Human and societal values must be considered in the practice of hydrology. This should be quite evident from the four case studies presented in Section 1.6.1. While the technical details are often very standardized, it is rarely possible to standardize the value issues. Value issues cannot be easily quantified and numerically combined with quantifiable goals such as maximizing benefit-cost ratios. Our inability to quantify these value issues does not mean that we should not incorporate them into our decisions. They must be considered; otherwise, the hydrologist will be viewed as lacking a social conscience. In the first case study of Section 1.6.1, how can we quantify the increased risk to the health of the local citizens? In deciding on a course of action in the fourth case study, can the damage to the environment and the potential reduction in the aesthetical qualities of the riverine environment be quantified with the cost of cleanup?

I have been asked on numerous occasions for advice on how to handle ethical dilemmas. In the initial stage of one such problem, I discussed the issue with both parties and thought that I could resolve the issue. Therefore, I ignored Step IA1 of Table 1–2. Instead of making notes after each of the discussions, I mistakenly assumed that the situation was being resolved. I made a written statement only after I was so instructed to do so by a lawyer who was approached by one of the parties. At that point I could not remember all of the details and left out what turned out to be some very important events. In retrospect, it may have hindered my ability to help the individual and resolve the issue. Thus, learn from my experience. When a problem arises, no matter how minor it seems at the time, maintain a record of it. The details could prove to be very valuable at a later time. A tape recorder is a very efficient means of recording the details of events. However, whatever medium is used to record the information, make sure that it is placed in a secure environment.

In summary, every student of hydrology must recognize that the study of hydrology is more than just developing skills related to the technical aspects of hydrologic design; the hydrologist must also consider the social impact of a design.

PROBLEMS

1–1. Identify the processes of the hydrologic cycle that affect flood runoff from a 3-acre forested lot. Discuss the relative importance of each of the processes. If the lot is cleared, what changes in the processes and their importance will occur? If a single-family residence is constructed on the cleared lot, what processes will control flood runoff from the lot?

1–2. If the drainage area of the site of Problem 1–1 is 50 acres and 150 residences are to be built, identify the processes of the hydrologic cycle that will affect flood runoff for the three watershed states: forested, cleared, and developed. In each case, discuss the relative importance of the processes.

1–3. If a 25-acre commercial shopping center was to be constructed at one end of a 50-acre watershed, identify the processes of the hydrologic cycle that will affect flood runoff after development. Discuss the relative importance of the processes.

1–4. Would you expect a 25-acre commercial shopping center to affect the flood runoff from a 500-acre watershed, assuming that the remainder of the watershed is forested? Discuss your responses in terms of the hydrologic cycle.

1–5. Identify the hydrologic processes that affect flood runoff from a 3000-ft section of interstate highway and the accompanying grass-covered right-of-way. Which processes are most important in determining the flood runoff from the site?

1–6. A wet basement (for example, moisture on the floor and water seeping in through cracks in the wall and floor) is not an uncommon problem in single-family residential homes. From the standpoint of the hydrologic cycle, briefly discuss the problem and possible solutions.

1–7. A single-family residential development lies down slope from a small suburban shopping center that has no facilities for controlling flood runoff. Several of the homes experience erosion of the lawns during moderate and large storm events. From the standpoint of the hydrologic cycle, briefly discuss the problem and possible solutions.

1–8. For a circular pipe flowing full, the head loss (h_L) can be computed using the Darcy-Weisbach equation:

$$h_L = f\left(\frac{L}{D}\right)\frac{V^2}{2g}$$

in which f is the dimensionless friction factor that is a function of the Reynolds number and pipe roughness, L and D are the length and diameter of the pipe, respectively, and V is the velocity of flow. A firm intends to develop a pipe made of a new ceramic material for which f values are not available. Discuss the task in terms of analysis and synthesis.

1–9. Design an experiment in which the efficiency of a pump at some specified Q could be determined. Discuss the problem in terms of the concepts of analysis and synthesis.

1–10. A storm with a uniform depth of 1 in. falls on a 25-acre watershed. Determine the total volume of rainfall in acre-feet and cubic feet. If all of the water were collected in a storage basin having vertical walls and a floor area of 25,000 ft^2, how deep of a basin would be needed?

1–11. A hurricane has an average rainfall depth of 9 in. over a 200-mi^2 watershed. What size reservoir in acre-ft would be required to contain completely 20% of the rain?

1–12. The following table gives the measured rainfall and surface runoff from a 65-acre watershed. Values given are averages over the time interval. Compute the accumulated storage (ft^3) of water within the watershed. Plot on the same graph the inflow, outflow, and cumulative storage as a function of time.

Time (min)	Rainfall (inches)	Runoff (ft^3/sec)
0	0	0
10	0.07	2.1
20	0.15	9.5
30	0.21	18.9
40	0.16	37.3
50	0.09	48.8
60	0.05	52.6

1–13. For the six dimensions of professionalism given in Section 1.7, discuss how each applies to the practicing hydrologist. Provide examples to illustrate each point.

1–14. Discuss the importance of maintaining professional competence to the professional hydrologist, including the importance to the individual, the firm, the profession, the client, and society. Also, discuss ways that the hydrologist can maintain professional competence.

1–15. Select five values from the list provided in Section 1.8 and discuss how each can be important to society. Provide illustrating examples from hydrologic practice to illustrate each point.

1–16. Obtain a copy of the code of ethics for a professional society to which hydrologists might belong. Evaluate the human values basis that underlies each section of the code.

For each of the examples given in Problems 1–17 to 1–23 provide an evaluation based on the four questions given at the end of Section 1.6.2. Use Table 1–2 to arrive at a recommended course of action.

1–17. The manager of a firm that performs hydrologic design work underbids on a contract in order to improve chances of being awarded the contract, knowing that he will later submit requests for contract "add-ons" with higher than reasonable budget amendments.

1–18. A hydrologist makes extensive personal use of the company's computer system without approval or any expectation of reimbursing the firm.

1–19. A hydrologist, who is also a registered professional engineer, signs design work completed by unregistered subordinates without checking the work for accuracy.

1–20. A research hydrologist writes a paper that details personal research and submits the paper to two different professional journals without informing them of the dual submission. The paper is accepted for publication and subsequently published in both journals.

1–21. A hydrologist, who serves as office manager for a large engineering firm, hires two new hydrologists, one man and one woman; both are recent college graduates with no experience. While both show equal ability and effort, the manager continually gives the better assignments to the man and limits the professional opportunities of the woman.

1–22. A college professor, who has a full-time position teaching hydrology at a state university, spends almost all of her time at the firm where she serves as a consulting hydrologist. While she makes most of her classes, she greatly exceeds the university norm of "one day per week of outside consulting."

1–23. A hydrologist whose only professional experience is in surface water hydrology purchases a piece of computer software that can be used for solving problems in ground water hydrology. The hydrologist then uses the software while selling his services as a consulting ground water hydrologist.

1–24. For each of the cases in Problems 1–17 through 1–23, give a rationalization that the individual might use to justify the improper conduct.

1–25. Give three (or more) rationalizations used by students who cheat on examinations. For each rationalization, show the irrationality of the reasoning.

1–26. Are students who cheat on tests and falsify lab results while in college more likely to participate in unprofessional conduct after they have graduated and are working in professional life? Support your response.

1–27. Critique the following statement of an engineering student: "I'm ethical, so I don't need to hear any lectures on professional conduct."

1–28. A woman actively participated in a research project while she was a student. Two years later, she realized that her advisor published the research results but did not include her as a co-author. What action, if any, should she take?

REVIEW QUESTIONS

1–1. Which one of the following tasks would <u>not</u> be an element of engineering hydrology? (a) Evaluate the effect of the size (capacity) of the outlet from a reservoir on recreation facilities at the reservoir site; (b) estimate the size of a pipe that would be required to carry flood water under a highway; (c) evaluate the effect of water supply on the distribution and movement of people in the United States; (d) estimate rates of storm runoff from rooftops.

1–2. Which one of the following is <u>not</u> considered as part of hydrologic design? (a) The spatial variation of storm rainfall; (b) the economic value of the water; (c) the quality of the water supply; (d) the temporal variation of rainfall; (e) all of the above are considered.

1–3. Which one of the following is <u>not</u> a major factor contributing to uncertainty in hydrologic designs? (a) The spatial and temporal variability of rainfall; (b) the effect of the soil and vegetation in converting rainfall to flood runoff; (c) the storage of flood runoff in river channels and floodplains; (d) our inability to accurately forecast future occurrences of rainfall; (e) all of the above are important factors.

1–4. Which one of the following elements of the hydrologic cycle is <u>not</u> affected by urbanization? (a) Rainfall; (b) depression storage; (c) interception storage; (d) channel storage; (e) all of the above are affected.

1–5. Consider a small forested area as a system. The rate of rainfall is measured with a rain gage in an open area about 100 ft from the forest. A second gage that measures the depth of rain that passes through the tree canopy is located under one of the trees near the center of the forest. In a problem of analysis, the objective would be to estimate (a) the rate of evaporation from the leaves of the tree; (b) the portion of the rain that is available for infiltration; (c) the water available for runoff; (d) the volume of interception storage.

1–6. If water travels at 6 ft/sec through a 24-in. diameter culvert that passes under a 30-ft roadway, the travel time in minutes of a particle of water as it passes through the culvert is (a) 0.1; (b) 1; (c) 6; (d) 120.

1–7. For the data of Review Question 1–6, what is the volume rate of flow (ft^3/sec) that passes through the culvert when it flows full? (a) Insufficient information is provided to compute it; (b) 18.3; (c) 75.4; (d) 27.4.

1–8. Floodwater with an average depth of 1 in. over a 96-acre parking lot drains into a small river during a storm of 10 min. The average rate of water discharging from the river during the storm is 360 ft^3/sec. The volume (in acre-ft) of water in the river will increase (+) or decrease (–) by (a) –7.5; (b) 3; (c) 7.5; (d) 96.

2

Statistical Methods in Hydrology

$$\bar{x} = \int x f(x)\, dx$$

CHAPTER OBJECTIVES

1. Provide statistical fundamentals that are important to a wide variety of analysis problems in hydrology.
2. Show that statistical methods are not a set of unrelated analysis tools but are meant to be applied sequentially in solving hydrologic problems.
3. Introduce concept of randomness and the use of probability to measure the likelihood of occurrence of random variables.
4. Demonstrate the use of the fundamentals of hypothesis testing in making decisions.
5. Introduce concepts related to the calibration of simple empirical models.

2.0 NOTATION

A = drainage area
b = regression coefficient for zero-intercept model
b_i = slope coefficient for variable i
b_j = j^{th} partial regression coefficient
b_o = regression coefficient (intercept)
b_1 = regression coefficient (slope)
e_i = error (or residual) for observation i
E = daily evaporation rate
E_i = expected frequency in cell i for chi-square test
F = value of least-squares objective function
F = value of random variable having an F distribution
F_z = value of least-squares objective function for the standardized model
g = sample skew
H_A = alternative hypothesis
H_o = null hypothesis
K = recession coefficient for snowmelt runoff
M_k = k^{th} moment about the origin
M_k = k^{th} moment about the mean
n = sample size
N = total number of observations
O_i = observed frequency in cell i for chi-square test
p = number of predictor variables in multiple regression equation
p = probability
P = precipitation
q = number of predictor variables in the multiple linear model
q = $1 - p$ = probability of an event not occurring
Q = mean annual flood
R = correlation coefficient
R = runoff
R_{jY} = correlation coefficient between predictor j and Y
R_q = correlation coefficient for model with q predictor variables
S = standard deviation
S_e = standard error of estimate
S_Q = standard deviation of the mean annual flood
S_y = standard deviation of random variable y
t = value of a random variable having a t distribution
t_i = standardized partial regression coefficient for variable i
T = temperature
w_i = transformed predictor variable
x = random variable
x_k = specific value of x
$\bar{x}$ = sample mean
y = random variable
$\bar{y}$ = mean value of y
$\hat{y}$ = predicted value of y
z = logarithmic transform of criterion variable y
z = standard normal deviate
α = level of significance
β = probability of a type II error
β_j = population value of the jth partial regression coefficient
γ = population skew
μ = population mean
μ_0 = specific value of μ for hypothesis testing
ν = degrees of freedom
σ = standard deviation of the population

2.1 INTRODUCTION

Although many design methods are based partly on theoretical principles, the development of almost all hydrologic design methods uses some form of data analysis. In some cases, a theoretical analysis is used to identify the input variables and the form of the underlying prediction equation, but the design method is made operational only after analyzing measured hydrologic data. Given the widespread use of data analysis techniques in hydrology, it is important to be familiar with the data analysis methods used most frequently. Knowledge of data analysis methods is necessary in order to appreciate the limitations and accuracy of existing hydrologic design methods that are the result of data analysis and to develop new hydrologic design methods that require calibration to measured hydrologic data.

For the objectives of this book it is only necessary for the reader to be knowledgeable about a few probability and statistical concepts. The methods discussed in this chapter are (1) probability concepts, (2) statistical moments, (3) hypothesis tests, and (4) regression analysis. A fifth statistical method, frequency analysis, is discussed in Chapter 5. Knowledge of these methods can provide a basic understanding of the potential limitations and accuracy of existing hydrologic methods, as well as the basis for developing new design methods.

This chapter represents a significant departure from the general thrust of this book. Most of the remainder of the book deals with elements of the hydrologic cycle; this chapter deals with statistical methods. Given the importance of these methods to the development and understanding of hydrologic design and given the fact that these statistical methods are used in the development of the methods discussed in most of the other chapters, a general discussion of the statistical methods is provided here. Thus, we can concentrate on the hydrologic concepts in the other chapters.

2.2 STATISTICAL TERMINOLOGY

Each year, considerable effort and resources are expended on the collection of hydrologic and meteorologic data. These data represent sample observations from a population. There are two key words here: sample and population. The set of past observations represents a sample from the population, not the population itself. It is important to understand that the sample is a random collection or subset of the population. As an example, we might set up an instrument to record the highest flood level during the year at a particular location along a stream reach. The largest observed flood for that year is an observation on a random variable because we can be assured that the annual peak flood will be different each year. If we obtain this measurement each year, then after n years we will have a sample size of n. The objective of the data analysis will be to use the sample to identify and draw inferences about the population. After the population has been estimated, the assumed population will be the basis for hydrologic designs. The population would consist of all values that could have been observed in the past and will occur in the future. Unfortunately, we cannot make measurements after the event has passed, nor can we measure events that have not yet occurred. We cannot know the population, and we have to use the sample to draw inferences about what will happen in the future.

Before providing a more formal discussion of these statistical terms, it is important to point out the basis for much of statistical analysis. Statistical inference is concerned with methods that permit an investigator to make generalizations about populations using information obtained from random samples. That is, samples are subsets of the population. If knowledge of the population represents complete information, statistical analyses of sample data represent decision making under conditions of incomplete information (for example, under conditions of uncertainty). Statistical methods are used for collecting, organizing, summarizing, and analyzing quantitative information.

2.2.1 Probability

Probability is a scale of measurement that is used to describe the likelihood of an event where an event is defined as the occurrence of a specified value of the random variable. The scale on which probability is measured extends from 0 to 1, inclusive, where a value of 1 indicates a certainty of occurrence of the event and a value of 0 indicates a certainty of failure to occur or nonoccurrence of the event. Quite often, probability is specified as a percentage; for example, when the weatherperson indicates that there is a 30% chance of rain, past experience indicates that under similar meteorological conditions it has rained 3 out of 10 times. In this example, the probability was estimated empirically using the concept of relative frequency:

$$p(x = x_0) = \frac{n}{N} \tag{2–1}$$

in which n is the number of observations on the random variable x that result in outcome x_0, and N is the total number of observations on x.

As an example, consider the case of the design of a coffer dam around the site where a bridge pier is being constructed. The height to which the coffer dam is constructed depends, in part, on the level of flooding that can be expected during the period of construction. For example, we will assume that construction of the bridge pier will take one year. Let's assume that we have a flood record at the site consisting of the largest flood event in each of the last 10 years. Let's also assume that a benefit-cost analysis suggests that we should set the height of the coffer dam so that there is only a 10% chance that the flood level will exceed the height of the coffer dam during the 1-yr period of construction. Using Equation 2–1 we know the probability, $p(x = x_0) = 0.10$, and $N = 10$; therefore, in the sample of 10 we would want to set the height of the coffer dam so that only 1 of 10 measured peak discharges will exceed the height of the coffer dam. Thus we would set the height of the dam at some point between the largest and second largest of the ten floods of record. In this simplified example, we have used sample information to draw inferences about the population. Since we are uncertain about what will occur in the future, we make decisions under conditions of uncertainty.

Hydrologists and engineers must make decisions under conditions of uncertainty. For example, engineers who have the responsibility of monitoring the quality of water in streams and bodies of water estimate pollution levels using samples collected from the water. The samples are then analyzed in a laboratory and the results are used to make a decision. Many water quality sampling programs involve ten or fewer measurements. Uncertainty arises because of the highly variable nature of pollution; that is, the concentration of a pollutant may vary with time, the degree of turbulence in the water, and the magnitude of and the frequency with which wastes are discharged into the water. These sources of variation must be accounted for when the engineer makes a decision about water quality.

2.2.2 Random Variables

Two types of random variables must be considered. First, a *discrete random variable* is one that may only take on distinct, usually integer, values; for example, the outcome of a roll of a die may only take on the integer values from 1 to 6 and is therefore a discrete random variable. If we define a random variable to be whether or not a river has overflowed its banks in any one year, then the variable is discrete because only discrete outcomes are possible; the out-of-bank flood occurred or it did not occur in any one year.

Second, the outcome of an event may take on any value within a continuum of values; such a random variable is called *continuous*. For example, the average of all scores on a test having a maximum possible score of 100 may take on any value, including nonintegers, between 0 and 100; thus the class average would be a continuous random variable. The volume of rainfall occurring during storm events is an example of a continuous random variable because it can be any value on a continuous scale from 0 inches to some very large value. A distinction is made between discrete and continuous random variables, because the computation of probabilities is different for the two types.

2.2.3 Probability for Discrete Random Variables

The probability of a discrete random variable is given by a mass function. A *mass function* specifies the probability that the discrete random variable x equals the value x_k and is denoted by

$$p(x_k) = p(x = x_k) \tag{2–2}$$

Probability has two important boundary conditions. First, the probability of an event x_k must be less than or equal to 1, and second, it must be greater than or equal to zero:

$$0 \leq p(x_k) \leq 1 \tag{2–3}$$

This property is valid for all possible values of k. Additionally, the sum of all possible probabilities must equal 1:

$$\sum_{k=1}^{N} p(x_k) = 1 \tag{2–4}$$

in which N is the total number of possible outcomes; for the case of the roll of a die, N equals 6.

It is often useful to present the likelihood of the outcome for a discrete random variable using the cumulative mass function, which is given by

$$p(x \leq x_k) = \sum_{j=1}^{k} p(x_j) \tag{2–5}$$

The cumulative mass function is used to indicate the probability that the random variable x is less than or equal to x_k. It is inherent in the definition (Equation 2–5) that the cumulative probability is defined as 0 for all values less than the smallest x_k and 1 for all values greater than the largest value.

To illustrate the concepts introduced for discrete random variables, assume that the random variable is the outcome of the roll of a die. The sample space consists of the collection of all possible values of the random variable; the probability of each element of the sample space is given by

$$p(x_k) = \frac{1}{6} \qquad \text{for } k = 1, 2, 3, 4, 5, 6 \tag{2–6}$$

The mass function is shown in Figure 2–1a. The cumulative function, which is shown in Figure 2–1b, is given by

$$p(x \leqslant k) = \frac{k}{6} \qquad \text{for } k = 1, 2, 3, 4, 5, 6 \tag{2–7}$$

The number of floods of a certain magnitude during the design life of an engineering structure would be an example of a discrete random variable. Using the method to be described in Section 2.4.1, the probability of exactly x 2-yr floods occurring in a period of 10 yr is shown by the mass function of Figure 2–1c. We would most likely expect five such floods

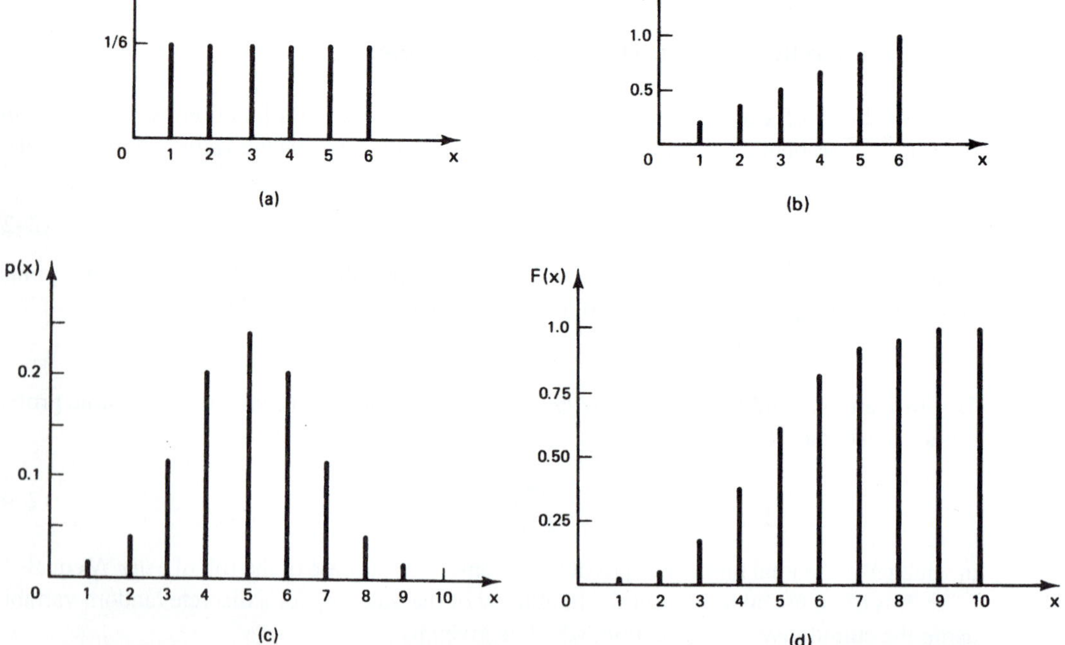

FIGURE 2–1 Probabilites for discrete random variables—the role of a die: (a) mass function and (b) the cumulative function. Two-year floods in a period of 10 years: (c) mass function of exactly x floods and (d) cumulative function.

in a 10-yr period, although the probability of either four or six events during a period of 10 yr is just slightly smaller than the occurrence of exactly five floods. The cumulative function is shown in Figure 2–1d.

2.2.4 Probability for Continuous Random Variables

A probability density function (PDF) is used to define the likelihood of occurrence of a continuous random variable. Specifically, the probability that the random variable x lies within the interval from x_1 to x_2 is given by the integral of the density function from x_1 to x_2 over all values within the interval:

$$p(x_1 \leqslant x \leqslant x_2) = \int_{x_1}^{x_2} f(x)dx \tag{2–8}$$

in which $f(x)$ is the density function. If the interval is made infinitesimally small, x_1 approaches x_2 and $P(x_1 \leq x \leq x_2)$ approaches 0. This illustrates a property that distinguishes discrete random variables from continuous random variables. Specifically, the probability that a continuous random variable takes on a specific value equals 0; that is, probabilities for continuous random variables must be defined over an interval. Discrete random variables can take on a specific value so it will have a non-zero probability.

It is also important to note that the integral of the PDF from $-\infty$ to ∞ equals 1:

$$\int_{-\infty}^{\infty} f(x)dx = 1 \tag{2–9}$$

Also, because of Equation 2–9, the following is valid:

$$p(x > x_k) = \int_{x_k}^{\infty} f(x)dx = 1 - p(x \leqslant x_k) \tag{2–10}$$

The cumulative distribution function (CDF) of a continuous random variable is defined by

$$p(x \leqslant x_k) = \int_{-\infty}^{x_k} f(\xi)d\xi \tag{2–11}$$

in which ξ is the dummy argument of integration. The CDF is a nondecreasing function in that $p(x \leq x_1) \leq p(x \leq x_2)$ when $x_1 \leq x_2$. The CDF equals 0 at $-\infty$ and 1 at ∞.

Example 2–1

Assume that the probability distribution of evaporation E on any day during the year is given by

$$f(E) = \begin{cases} 4 & 0 \leqslant E \leqslant 0.25 \text{ in./day} \\ 0 & \text{otherwise} \end{cases} \tag{2–12}$$

To show that this is a legitimate PDF, Equation 2–9 gives

$$\int_0^{0.25} 4dx = 4x \,\Big|_0^{0.25} = 1.0$$

The probability that E is between 0.1 and 0.2 in./day can be found by integration:

$$\int_{0.1}^{0.2} f(E)dE = \int_{0.1}^{0.2} 4dE = 4E\Big|_{0.1}^{0.2} = 0.4 \qquad (2\text{–}13a)$$

The cumulative function $F(E)$ defines the probability that E is less than or equal to some value X:

$$F(E) = \int_{0}^{x} f(E)dE = \int_{0}^{x} 4dE = 4X \qquad (2\text{–}13b)$$

Thus the probability that E is less than 0.15 in./day is $F(0.15) = 0.6$. The probability that E is between 0.1 and 0.2 in./day is $F(0.2) - F(0.1) = 0.8 - 0.4$, which is the same probability derived by integrating the density function $f(E)$. The probability that E is greater than X is given by $1 - F(E)$; for example, the probability that E is greater than 0.05 in./day is

$$1 - F(0.05) = 1 - 0.2 = 0.8.$$

2.3 MOMENTS OF A SAMPLE OR DISTRIBUTION FUNCTION

Whether summarizing a data set or attempting to find the population, one must characterize the sample. Moments are useful descriptors of data; for example, the mean, which is a moment, is an important characteristic of a set of observations on a random variable, such as rainfall volume or the concentration of a water pollutant. A moment can be referenced to any point on the measurement axis; however, the origin (for example, zero point) and the mean are the two most common reference points.

Although most data analyses use only two moments, in some statistical studies it is important to examine three moments:

1. The mean is the first moment of values measured about the origin.
2. The variance is the second moment of values measured about the mean.
3. The skew is the third moment of values measured about the mean.

The moments can be computed using the equations that follow. For a continuous random variable x, the kth moment about the origin is given by

$$M'_k = \int_{-\infty}^{\infty} x^k f(x)dx \qquad (2\text{–}14a)$$

in which x is the random variable and $f(x)$ is its density function. The corresponding equation for a discrete random variable is

$$M'_k = \sum_{i=1}^{n} x_i^k f(x_i) \qquad (2\text{–}14b)$$

in which n is the number of elements in the sample and $f(x)$ is the mass function.

For a continuous random variable the kth moment about the mean is given by

$$M_k = \int_{-\infty}^{\infty} (x - \mu)^k f(x)dx \tag{2–15a}$$

in which μ is the population mean of the random variable x. The superscript of M'_k in Equation 2–14a indicates that the moment is calculated about the origin. The lack of a superscript on M_k in Equation 2–15a indicates that the moment is calculated about the mean. The corresponding equation for a discrete random variable is

$$M_k = \sum_{i=1}^{n} (x_i - \mu)^k f(x_i) \tag{2–15b}$$

In Equations 2–15, the sample mean $\bar{x}$ can be substituted for the population mean μ when μ is not known.

2.3.1 Mean

The *mean* is the first moment measured about the origin; it is also the average of all observations on a random variable. It is important to note that the population mean is most often denoted as μ, while the sample mean is denoted by $\bar{x}$. For a continuous random variable, it is computed as

$$(\bar{x} \text{ or } \mu) = \int_{-\infty}^{\infty} xf(x)dx \tag{2–16}$$

For a discrete random variable, the mean is given by

$$(\bar{x} \text{ or } \mu) = \sum_{i=1}^{n} x_i f(x_i) \tag{2–17}$$

If each observation is given equal weight, then $f(x_i)$ equals $1/n$, and the mean of observations on a discrete random variable is given by

$$(\bar{x} \text{ or } \mu) = \frac{1}{n}\sum_{i=1}^{n} x_i \tag{2–18}$$

The mean value shows where the average value lies along the measurement axis; this is most easily represented by the two density functions, $f(x_i)$, which are shown in Figure 2–2. If the random variable x is the magnitude of a flood, $f(x_1)$ and $f(x_2)$ could represent the distributions of floods on two rivers. Obviously, the floods on the river characterized by

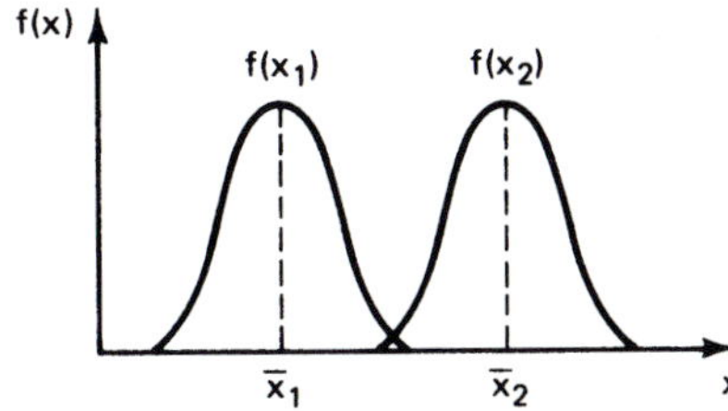

FIGURE 2–2 Comparison of distributions with different means.

$f(x_2)$ are usually larger than floods on the river characterized by $f(x_1)$; however, there is some chance that at any one point in time, the flow in river 2 will be smaller than the flow in river 1.

Although the mean conveys certain information about either a sample or population, it does not completely characterize a random variable. The distributions shown in Figure 2–3 are obviously not identical, even though they have the same mean.

The probability mass function of Figure 2–1c can be used to illustrate the use of Equation 2–17 to compute the mean for a discrete random variable such as the number of floods X in a 10-yr period:

$$\begin{aligned} u &= 0\,(0.0010) + 1\,(0.0097) + 2\,(0.0440) + 3\,(0.1172) \\ &\quad + 4\,(0.2051) + 5\,(0.2460) + 6\,(0.2051) + 7\,(0.1172) \\ &\quad + 8\,(0.0440) + 9\,(0.0097) + 10\,(0.0010) \\ &= 5 \end{aligned}$$

Thus the mean number of floods in a 10-yr period is 5; however, Figure 2–1c shows that in any 10-yr period, any number of floods from 0 to 10 may occur.

Equation 2–18 can be used to compute the mean areal rainfall when the catches at four rain gages in a region measured 1.7, 2.1, 1.8, and 2.4 in. If the catch at each gage is given equal weight, the mean depth of rainfall is

$$\begin{aligned} \bar{x} &= \frac{1}{4}(1.7 + 2.1 + 1.8 + 2.4) \\ &= 2.0 \text{ in.} \end{aligned}$$

The mean of a density function of a continuous random variable can be estimated with Equation 2–16. For example, the mean evaporation for the density function of Equation 2–12 is

$$\mu = \int_0^{0.25} Ef(E)dE = \int_0^{0.25} E(4.0)\,dE = 2E^2\big|_0^{0.25} = 0.125 \text{ in./day}$$

2.3.2 Variance

The variance is the second moment about the mean. The variances of the population and sample are denoted by σ^2 and S^2, respectively. The units of the variance are the square of the units of the random variable; for example, if the random variable is measured in feet, the

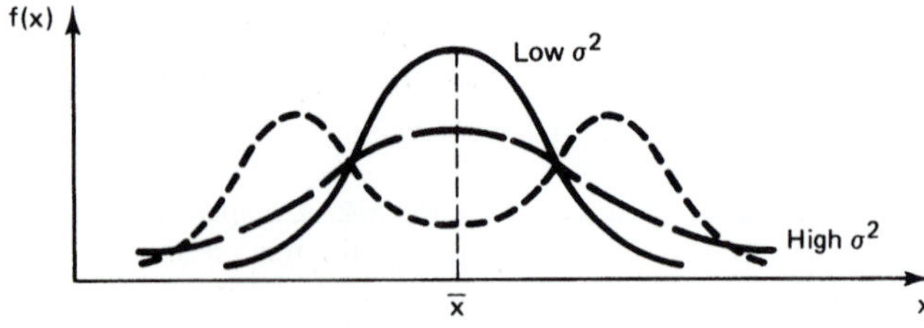

FIGURE 2–3 Comparison of distributions with the same mean but different variances.

variance will have units of ft^2. For a continuous random variable, the variance is computed by

$$(S^2 \text{ or } \sigma^2) = \int_{-\infty}^{\infty} (x - \mu)^2 f(x)dx \tag{2–19}$$

For a discrete variable, the variance is computed by

$$(S^2 \text{ or } \sigma^2) = \sum_{i=1}^{n} (x_i - \mu)^2 f(x_i) \tag{2–20}$$

If each of the n observations in a sample are given equal weight, $f(x) = 1/(n - 1)$, then the variance is given by

$$(S^2 \text{ or } \sigma^2) = \frac{1}{n-1} \sum_{i=1}^{n} (x_i - \mu)^2 \tag{2–21}$$

Equation 2–21 provides an unbiased estimate because the average value of many sample estimates of σ^2 would approach the true value σ^2. The average value of many sample estimates of σ^2 obtained if $1/n$ were used in place of $1/(n-1)$ would not approach the true value of σ^2. The variance of a sample can be computed using the following alternative equation:

$$S^2 = \frac{1}{n-1}\left[\sum_{i=1}^{n} x_i^2 - \frac{1}{n}\left(\sum_{i=1}^{n} x_i\right)^2\right] \tag{2–22}$$

Equation 2–22 will provide the same answer as Equation 2–21 when computations are made using an appropriate number of significant digits. In Equation 2–19 to 2–21, the sample mean ($\bar{x}$) may be substituted for μ when μ is not known.

Variance is the most important concept in statistics because almost all statistical methods require some measure of variance. Therefore, it is important to have a conceptual understanding of this moment. In general, it is an indicator of the closeness of the values of a sample or population to the mean. If all values in a sample equaled the mean, the sample variance would equal zero. Figure 2–3 illustrates density functions with different variances.

Although the variance is used in all aspects of statistical analysis, its use as a descriptor is limited because of its units; specifically, the units of the variance are not the same as those of either the random variable or the mean.

The variance of the mass function of Figure 2–1c can be computed by Equation 2–20. For a mean of x of 5, the variance is

$$\begin{aligned}\sigma^2 &= (0-5)^2(0.0010) + (1-5)^2(0.0097) + (2-5)^2(0.0440) \\ &\quad + (3-5)^2(0.1172) + (4-5)^2(0.2051) + (5-5)^2(0.2460) \\ &\quad + (6-5)^2(0.2051) + (7-5)^2(0.1172) + (8-5)^2(0.0440) \\ &\quad + (9-5)^2(0.0097) + (10-5)^2(0.0010) \\ &= 2.50\end{aligned} \tag{2–23}$$

The variance of the four rain gage catches of 1.7, 2.1, 1.8, and 2.4 in. can be computed with either Equation 2–21:

$$S^2 = \frac{1}{4-1}[(1.7-2)^2 + (2.1-2)^2 + (1.8-2)^2 + (2.4-2)^2]$$
$$= 0.10 \text{ in.}^2$$

or Equation 2–22:

$$S^2 = \frac{1}{4-1}\left[(1.7)^2 + (2.1)^2 + (1.8)^2 + (2.4)^2 - \frac{1}{4}(1.7 + 2.1 + 1.8 + 2.4)^2\right]$$
$$= \frac{1}{3}\left[16.3 - \frac{1}{4}(8)^2\right]$$
$$= 0.10 \text{ in.}^2$$

2.3.3 Standard Deviation

By definition, the standard deviation is the square root of the variance. It has the same units as both the random variable and the mean; therefore, it is a useful descriptor of the dispersion or spread of either a sample of data or a distribution function. The standard deviation of the population is denoted by σ, while the sample value is denoted by S.

Example 2–2

During a particularly heavy rainstorm, the depth of rain measured at five rain gages in the city was 2.7, 2.9, 3.4, 3.1, and 2.9 in. The computations for the mean, variance, and standard deviation of the five catches are as follows:

Gage	x	x^2	$(x - \bar{x})$	$(x - \bar{x})^2$
1	2.7	7.29	−0.3	0.09
2	2.9	8.41	−0.1	0.01
3	3.4	11.56	0.4	0.16
4	3.1	9.61	0.1	0.01
5	2.9	8.41	−0.1	0.01
	15.0	45.28	0.0	0.28

From Equation 2–18 the mean is

$$\bar{x} = \frac{1}{5}(15) = 3.0 \text{ in.}$$

From Equation 2–21 the sample variance is

$$S^2 = \frac{1}{5-1}(0.28) = 0.07 \text{ in.}^2$$

Therefore, the sample estimate of the standard deviation is 0.26 in. Equation 2–22 can also be used to compute the variance:

$$S^2 = \frac{1}{5-1}\left[45.28 - \frac{1}{5}(15)^2\right] = 0.07 \text{ in.}^2$$

2.3.4 Skew

The skew is the third moment measured about the mean. In this text, the sample skew will be denoted by g, while γ will be used to indicate the skew of the population. Mathematically, the skew is given as follows for a continuous random variable:

$$(g \text{ or } \gamma) = \int_{-\infty}^{\infty} (x - \mu)^3 f(x)dx \tag{2–24a}$$

For a discrete random variable, the skew can be computed by

$$(g \text{ or } \gamma) = \sum_{i=1}^{n} (x_i - \mu)^3 f(x_i) \tag{2–24b}$$

Skew has units of the cube of the random variable; thus if the random variable has units of pounds, the skew has units of lb^3. A standardized estimate of the skew is

$$g = \frac{n\Sigma(x - \bar{x})^3}{(n - 1)(n - 2)\, S^3} \tag{2–25}$$

Skew is a measure of symmetry. A symmetric distribution will have a skew of zero, while a nonsymmetric distribution will have a positive or negative skew depending on the location of the tail of the distribution. If the more extreme tail of the distribution is to the right, the skew is positive; the skew is negative when the more extreme tail is to the left of the mean. This is illustrated in Figure 2–4.

Example 2–3

An 18-yr record of total precipitation for the month of June on the Wildcat Creek watershed is given in Table 2–1. Values of $(x - \bar{x})$, $(x - \bar{x})^2$, and $(x - \bar{x})^3$ are given for each year of record; the sum of the annual values are given at the bottom of each column. The mean is computed with Equation 2–18:

$$\bar{x} = \frac{1}{n}\Sigma x_i = \frac{1}{18}(68.12) = 3.78 \text{ in.}$$

The sum of $(x - \bar{x})$ is 0.008. This sum will always equal 0, and the non-zero value only reflects roundoff error. The variance is computed using Equation 2–21 and the summation of $(x - \bar{x})^2$ from Table 2–1:

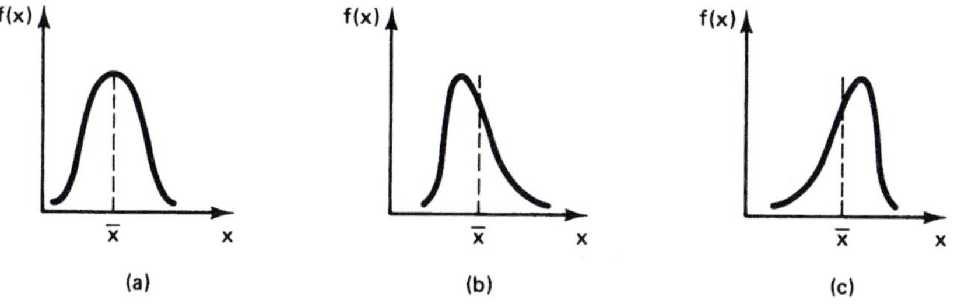

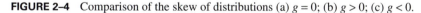

FIGURE 2–4 Comparison of the skew of distributions (a) $g = 0$; (b) $g > 0$; (c) $g < 0$.

TABLE 2–1 Computations of Moments of June Precipitation (in.) for Wildcat Creek near Lawrenceville, GA

Year	X	$X - \bar{x}$	$(X - \bar{X})^2$	$(X - \bar{X})^3$
1954	4.27	0.486	0.236	0.1145
1955	2.92	–0.864	0.747	–0.6460
1956	2.00	–1.784	3.184	–5.6821
1957	3.34	–0.444	0.198	–0.0878
1958	2.32	–1.464	2.145	–3.1406
1959	2.89	–0.894	0.800	–0.7156
1960	2.79	–0.994	0.989	–0.9834
1961	5.52	1.736	3.012	5.2278
1962	4.09	0.306	0.093	0.0285
1963	9.86	6.076	36.912	224.2632
1964	3.90	0.116	0.013	0.0015
1965	6.85	3.066	9.398	28.8090
1966	2.33	–1.454	2.115	–3.0767
1967	4.16	0.376	0.141	0.0530
1968	2.10	–1.684	2.837	–4.7794
1969	0.85	–2.934	8.611	–25.2684
1970	5.00	1.216	1.478	1.7961
1971	2.93	–0.854	0.730	–0.6238
Sum	68.12	0.008	73.639	215.2898

$$S^2 = \frac{1}{18 - 1}(73.639) = 4.332 \text{ in.}^2$$

which gives a standard deviation of 2.081 in. The skew is estimated using Equation 2–24b and the summation from Table 2–1, with $f(x_i)$ set equal to $1/n$:

$$g = \frac{1}{n}\Sigma\,(x - x_i)^3 = \frac{1}{18}(215.2898) = 11.96$$

The positive skew reflects primarily the effect of the one extreme event of the 1963 rainfall. While ten of the eighteen values are below the mean, the one large deviation from the mean offsets the greater number of negative values.

2.4 DISTRIBUTION FUNCTIONS

As indicated, random variables can be measured either at discrete values or over a continuous scale. Every random variable, whether discrete or continuous, has an underlying probability function. A number of different probability functions can be used to represent a random variable and to determine the probability of occurrence. Although there are many probability functions, for our purposes, only three will be discussed in this section: binomial, normal, and F. Several other probability functions will be introduced where they are needed.

2.4.1 Binomial Distribution

The binomial distribution is used to define probabilities of discrete events. It is applicable to random variables that satisfy the following four assumptions:

1. There are n occurrences, or trials, of the random variable.
2. The n trials are independent.
3. There are only two possible outcomes for each trial.
4. The probability of each outcome is constant from trial to trial.

The probabilities of occurrence of any random variable satisfying these four assumptions can be computed using the binomial distribution. For example, in flipping a coin n times, the n trials are independent, there are only two possible outcomes (one must assume that the coin will not stand on edge), and the probability of a head remains 0.5; thus the probability of getting three heads, or no heads, in five tosses could be computed using the binomial distribution. One could not use the binomial distribution to compute probabilities for the toss of a die because there are six possible outcomes; however, the four assumptions would apply if one defines the two outcomes as either an even or an odd number of dots.

We can denote the two outcomes of a binomial process as A and B, with the probability of A equal to p and the probability of B occurring equal to $(1 - p)$, which we can denote as q (that is, $q = 1 - p$). If we let x be the number of occurrences of A, then B occurs $(n - x)$ times in n trials. One possible sequence would be

$$\underbrace{AAAA\cdots A}_{x} \qquad \underbrace{BBBB\cdots B}_{n-x}$$

Since the trials are independent, the probability of this sequence is the product of the probabilities of the n outcomes:

$$\underbrace{ppp\cdots p}_{x} \qquad \underbrace{(1-p)\cdots(1-p)}_{n-x} \tag{2–26}$$

which is equal to

$$p^x(1-p)^{n-x} = p^x q^{n-x} \tag{2–27}$$

But many other possible sequences of the outcomes could occur; for example:

$$\underbrace{AAA\cdots A}_{x-1} \qquad BA \qquad \underbrace{BBB\cdots B}_{n-x-1}$$

It would be easy to show that the probability of this sequence occurring is also given by Equation 2–27. In fact, any sequence involving x occurrences of A and $(n - x)$ occurrences of B would have the probability given by Equation 2–27. Thus, it is only necessary to determine how many different sequences of x occurrences of A and $(n - x)$ occurrences of B are possible. It can be shown that the number of occurrences is

$$\frac{n!}{x!\,(n-x)!} \tag{2–28}$$

where $n!$ is read "n factorial" and equals

$$n! = n(n-1)\,(n-2)...(2)\,(1) \tag{2–29}$$

The factorial 0! is defined as 1. Computationally, the value of Equation 2–28 can be found from

$$\frac{n\,(n-1)...(n-x+1)}{x!} \tag{2–30}$$

The quantity given by Equation 2–28 is computed so frequently that it is often abbreviated as $\binom{n}{x}$ and called the binomial coefficient. It represents the number of ways that sequences involving events A and B can occur with x occurrences of A and $(n-x)$ occurrences of B. Combining Equations 2–27 and 2–28, we get the probability of getting exactly x occurrences of A in n trials, given that the probability of event A occurring on any trial is p:

$$\binom{n}{x}\,p^x(1-p)^{n-x} \text{ for } x = 0, 1, 2, ..., n \tag{2–31}$$

This is a binomial probability, and the probabilities defined by Equation 2–31 represent the distribution of binomial probabilities. It is often denoted as $b(x;\, n,\, p)$, which is read, "the probability of getting exactly x occurrences of a random variable in n trials when the probability of the event occurring on any one trial is p."

For example, if $n = 4$ and $x = 2$, Equation 2–28 would suggest six possible sequences:

$$\frac{4!}{2!\,(4-2)!} = \frac{4\,(3)\,(2)\,(1)}{2\,(1)\,(2)\,(1)} = 6 \tag{2–32}$$

The six possible sequences are ($AABB$), ($ABAB$), ($ABBA$), ($BAAB$), ($BABA$), and ($BBAA$). Thus if the probability of A occurring on any one trial is 0.3, then the probability of exactly two occurrences in four trials is

$$b\,(2;4,\,0.3) = \binom{4}{2}\,(0.3)^2\,(1-0.3)^{4-2} = 0.2646 \tag{2–33}$$

Similarly, if p equals 0.5, the probability of getting exactly two occurrences of event A would be

$$b\,(2;4,\,0.5) = \binom{4}{2}\,(0.5)^2\,(1-0.5)^{4-2} = 0.375 \tag{2–34}$$

It is easy to show that for four trials there is only one way of getting either zero or four occurrences of A, there are four ways of getting either one or three occurrences of A, and there are six ways of getting two occurrences of A. Thus with a total of sixteen possible outcomes, the value given by Equation 2–32 for the number of ways of getting two occurrences divided by the total of sixteen possible outcomes supports the computed probability of Equation 2–34.

Example 2–4

A coffer dam is to be built on a river bank so that a bridge pier can be built. The dam is designed to prevent flow from the river from interfering with the construction of the pier. The cost of the dam is related to the height of the dam; as the height increases, the cost increases. But as the height is increased, the potential for flood damage decreases. The level of flow in the stream varies weekly and can be considered as a random variable. However, the design engineer is interested only in two states, the overtopping of the dam during a one-workweek period or non-overtopping. If construction of the pier is to require two years for completion, the time period consists of 104 independent "trials." If the probability of the flood that would cause overtopping remains constant (p), the problem satisfies the four assumptions required to use the binomial distribution for computing probabilities.

If x is defined as an occurrence of overtopping and the height of the dam is such that the probability of overtopping during any 1-week period is 0.05, then for a 104-week period ($n = 104$) the probability that the dam will not be overtopped ($x = 0$) is computed using Equation 2–31:

$$\begin{aligned} p \text{ (no overtopping)} &= b(0;\ 104,\ 0.05) \\ &= \binom{104}{0} (0.05)^0 (0.95)^{104} = 0.0048 \end{aligned} \tag{2–35}$$

The probability of exactly one overtopping is

$$b\,(1;\ 104,\ 0.05) = \binom{104}{1} (0.05)^1 (0.95)^{103} = 0.0264 \tag{2–36}$$

Thus the probability of more than one overtopping is

$$1 - b\,(0;\ 104,\ 0.05) - b\,(1;\ 104,\ 0.05) = 0.9688 \tag{2–37}$$

The probability of the dam not being overtopped can be increased by increasing the height of the dam. If the height of the dam is increased so that the probability of overtopping in a 1-week period is decreased to 0.02, the probability of no overtoppings increases to

$$\begin{aligned} p \text{ (no overtoppings)} &= b(0;\ 104,\ 0.02) \\ &= \binom{104}{0} (0.02)^0 (0.98)^{104} = 0.1223 \end{aligned} \tag{2–38}$$

Thus the probability of no overtopping during the 104-week period increased 25 times when the probability of overtopping during one week was decreased from 0.05 to 0.02.

2.4.2 Normal Distribution

The most frequently used density function is the normal or Gaussian distribution, which is a symmetric distribution with a bell-shaped appearance. The density function is given by

$$f(x) = \frac{1}{\sigma\sqrt{2\pi}} \exp\left[-\frac{1}{2}\left(\frac{x - \mu}{\sigma}\right)^2\right] \quad \text{for} \quad -\infty \leqslant x \leqslant \infty \tag{2–39}$$

where μ and σ are called the location and scale parameters, respectively. Because there are an infinite number of possible values of μ and σ, there are an infinite number of possible configurations of the normal distribution. The cumulative distribution function for the normal distribution can be determined by integrating Equation 2–39.

Because of the extensive use of the normal distribution and the infinite number of possible values of μ and σ, probabilities for a normal curve are usually evaluated using the standard normal distribution. The standard normal distribution is derived by transforming the random variable x to the random variable z using the following equation:

$$z = \frac{x - \mu}{\sigma} \tag{2–40}$$

In practice, the sample moments $\bar{x}$ and S may be used in place of μ and σ in Equation 2–40. The density function for the standard normal curve is

$$f(z) = \frac{1}{\sqrt{2\pi}} e^{-0.5z^2} \tag{2–41}$$

The transformed variable z has a mean of 0 and a standard deviation of 1.

Probability estimates for the standard normal curve can be evaluated by integrating Equation 2–41 between specific limits. For example, the probability that z is between 0.5 and 2.25 could be determined using

$$p(0.5 \leq z \leq 2.25) = \int_{0.5}^{2.25} f(z)dz = \int_{0.5}^{2.25} \frac{1}{\sqrt{2\pi}} e^{-0.5z^2} dz \tag{2–42}$$

Because probabilities for the standard normal distribution are frequently required, integrals, such as that of Equation 2–42, have been computed and placed in tabular form. Table 2–2 gives the probability that z is between $-\infty$ and any value of z from −3.49 to 3.49, in increments of z of 0.01. Probabilities for cases in which the lower limit is different from $-\infty$ can be determined using the following identity:

$$p(z_L < z < z_0) = p(-\infty < z < z_0) - p(-\infty < z < z_L) \tag{2–43}$$

For example, the probability corresponding to the integral of Equation 2–42 is given by

$$p(0.55 < z < 2.25) = p(z < 2.25) - p(z < 0.55) \tag{2–44a}$$

$$= 0.9878 - 0.7088 = 0.2790 \tag{2–44b}$$

By transforming values using Equation 2–40 the standard normal curve can be used to determine probabilities for all values of x with any values of μ and σ.

Example 2–5

Assume that the distribution of soil erosion x from small urban construction sites is normally distributed with a mean of 100 tons/acre/yr and a standard deviation of 20 tons/acre/yr. The probability of soil erosion rates being greater than 150 tons/acre/yr can be computed either by integrating the density function of Equation 2–39 or using the transformation of Equation 2–40 with Equation 2–43. Specifically,

$$P(x > 150) = P\left(z > \frac{150 - 100}{20}\right) = P(z > 2.5)$$
$$= 1 - P(z < 2.5) = 1 - 0.9938 = 0.0062$$

2.4.3 *F* Distribution

The F distribution is used quite frequently in statistical analysis. It is a function of two shape parameters, ν_1 and ν_2. It is bounded on the left tail at zero and unbounded on the right tail, with a range of zero to infinity. Because of its dependency on two parameters, tables of F values (Table 2–3) are limited. The horizontal and vertical margins are for values of ν_1 and ν_2, respectively. It should be emphasized that the axes cannot be transposed; that is, $F(\nu_1, \nu_2)$

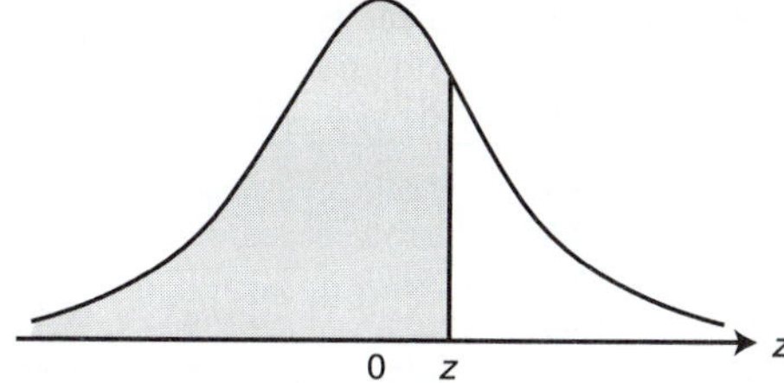

TABLE 2–2 Areas Under the Normal Curve

z	0.00	0.01	0.02	0.03	0.04	0.05	0.06	0.07	0.08	0.09
−3.4	.0003	.0003	.0003	.0003	.0003	.0003	.0003	.0003	.0003	.0002
−3.3	.0005	.0005	.0005	.0004	.0004	.0004	.0004	.0004	.0004	.0003
−3.2	.0007	.0007	.0006	.0006	.0006	.0006	.0006	.0005	.0005	.0005
−3.1	.0010	.0009	.0009	.0009	.0008	.0008	.0008	.0008	.0007	.0007
−3.0	.0013	.0013	.0013	.0012	.0012	.0011	.0011	.0011	.0010	.0010
−2.9	.0019	.0018	.0018	.0017	.0016	.0016	.0015	.0015	.0014	.0014
−2.8	.0026	.0025	.0024	.0023	.0023	.0022	.0021	.0021	.0020	.0019
−2.7	.0035	.0034	.0033	.0032	.0031	.0030	.0029	.0028	.0027	.0026
−2.6	.0047	.0045	.0044	.0043	.0041	.0040	.0039	.0038	.0037	.0036
−2.5	.0062	.0060	.0059	.0057	.0055	.0054	.0052	.0051	.0049	.0048
−2.4	.0082	.0080	.0078	.0075	.0073	.0071	.0069	.0068	.0066	.0064
−2.3	.0107	.0104	.0102	.0099	.0096	.0094	.0091	.0089	.0087	.0084
−2.2	.0139	.0136	.0132	.0129	.0125	.0122	.0119	.0116	.0113	.0110
−2.1	.0179	.0174	.0170	.0166	.0162	.0158	.0154	.0150	.0146	.0143
−2.0	.0228	.0222	.0217	.0212	.0207	.0202	.0197	.0192	.0188	.0183
−1.9	.0287	.0281	.0274	.0268	.0262	.0256	.0250	.0244	.0239	.0233
−1.8	.0359	.0351	.0344	.0336	.0329	.0322	.0314	.0307	.0301	.0294
−1.7	.0446	.0436	.0427	.0418	.0409	.0401	.0392	.0384	.0375	.0367
−1.6	.0548	.0537	.0526	.0516	.0505	.0495	.0485	.0475	.0465	.0455
−1.5	.0668	.0655	.0643	.0630	.0618	.0606	.0594	.0582	.0571	.0559
−1.4	.0808	.0793	.0778	.0764	.0749	.0735	.0721	.0708	.0694	.0681
−1.3	.0968	.0951	.0934	.0918	.0901	.0885	.0869	.0853	.0838	.0823
−1.2	.1151	.1131	.1112	.1093	.1075	.1056	.1038	.1020	.1003	.0985
−1.1	.1357	.1335	.1314	.1292	.1271	.1251	.1230	.1210	.1190	.1170
−1.0	.1587	.1562	.1539	.1515	.1492	.1469	.1446	.1423	.1401	.1379
−.9	.1841	.1814	.1788	.1762	.1736	.1711	.1685	.1660	.1635	.1611
−.8	.2119	.2090	.2061	.2033	.2005	.1977	.1949	.1922	.1894	.1867
−.7	.2420	.2389	.2358	.2327	.2296	.2266	.2236	.2206	.2177	.2148
−.6	.2743	.2709	.2676	.2643	.2611	.2578	.2546	.2514	.2483	.2451
−.5	.3085	.3050	.3015	.2981	.2946	.2912	.2877	.2843	.2810	.2776
−.4	.3446	.3409	.3372	.3336	.3300	.3264	.3228	.3192	.3156	.3121
−.3	.3821	.3783	.3745	.3707	.3669	.3632	.3594	.3557	.3520	.3483
−.2	.4207	.4168	.4129	.4090	.4052	.4013	.3974	.3936	.3897	.3859
−.1	.4602	.4562	.4522	.4483	.4443	.4404	.4364	.4325	.4286	.4247
.0	.5000	.4960	.4920	.4880	.4840	.4801	.4761	.4721	.4681	.4641

(continued)

TABLE 2–2 Areas Under the Normal Curve (*Continued*)

z	0.00	0.01	0.02	0.03	0.04	0.05	0.06	0.07	0.08	0.09
.0	.5000	.5040	.5080	.5120	.5160	.5199	.5239	.5279	.5319	.5359
.1	.5398	.5438	.5478	.5517	.5557	.5596	.5636	.5675	.5714	.5753
.2	.5793	.5832	.5871	.5910	.5948	.5987	.6026	.6064	.6103	.6141
.3	.6179	.6217	.6255	.6293	.6331	.6368	.6406	.6443	.6480	.6517
.4	.6554	.6591	.6628	.6664	.6700	.6736	.6772	.6808	.6844	.6879
.5	.6915	.6950	.6985	.7019	.7054	.7088	.7123	.7157	.7190	.7224
.6	.7257	.7291	.7324	.7357	.7389	.7422	.7454	.7486	.7517	.7549
.7	.7580	.7611	.7642	.7673	.7704	.7734	.7764	.7794	.7823	.7852
.8	.7881	.7910	.7939	.7967	.7995	.8023	.8051	.8078	.8106	.8133
.9	.8159	.8186	.8212	.8238	.8264	.8289	.8315	.8340	.8365	.8389
1.0	.8413	.8438	.8461	.8485	.8508	.8531	.8554	.8577	.8599	.8621
1.1	.8643	.8665	.8686	.8708	.8729	.8749	.8770	.8790	.8810	.8830
1.2	.8849	.8869	.8888	.8907	.8925	.8944	.8962	.8980	.8997	.9015
1.3	.9032	.9049	.9066	.9082	.9099	.9115	.9131	.9147	.9162	.9177
1.4	.9192	.9207	.9222	.9236	.9251	.9265	.9279	.9292	.9306	.9319
1.5	.9332	.9345	.9357	.9370	.9382	.9394	.9406	.9418	.9429	.9441
1.6	.9452	.9463	.9474	.9484	.9495	.9505	.9515	.9525	.9535	.9545
1.7	.9554	.9564	.9573	.9582	.9591	.9599	.9608	.9616	.9625	.9633
1.8	.9641	.9649	.9656	.9664	.9671	.9678	.9686	.9693	.9699	.9706
1.9	.9713	.9719	.9726	.9732	.9738	.9744	.9750	.9756	.9761	.9767
2.0	.9772	.9778	.9783	.9788	.9793	.9798	.9803	.9808	.9812	.9817
2.1	.9821	.9826	.9830	.9834	.9838	.9842	.9846	.9850	.9854	.9857
2.2	.9861	.9864	.9868	.9871	.9875	.9878	.9881	.9884	.9887	.9890
2.3	.9893	.9896	.9898	.9901	.9904	.9906	.9909	.9911	.9913	.9916
2.4	.9918	.9920	.9922	.9925	.9927	.9929	.9931	.9932	.9934	.9936
2.5	.9938	.9940	.9941	.9943	.9945	.9946	.9948	.9949	.9951	.9952
2.6	.9953	.9955	.9956	.9957	.9959	.9960	.9961	.9962	.9963	.9964
2.7	.9965	.9966	.9967	.9968	.9969	.9970	.9971	.9972	.9973	.9974
2.8	.9974	.9975	.9976	.9977	.9977	.9978	.9979	.9979	.9980	.9981
2.9	.9981	.9982	.9982	.9983	.9984	.9984	.9985	.9985	.9986	.9986
3.0	.9987	.9987	.9987	.9988	.9988	.9989	.9989	.9989	.9990	.9990
3.1	.9990	.9991	.9991	.9991	.9992	.9992	.9992	.9992	.9993	.9993
3.2	.9993	.9993	.9994	.9994	.9994	.9994	.9994	.9995	.9995	.9995
3.3	.9995	.9995	.9995	.9996	.9996	.9996	.9996	.9996	.9996	.9997
3.4	.9997	.9997	.9997	.9997	.9997	.9997	.9997	.9997	.9997	.9998

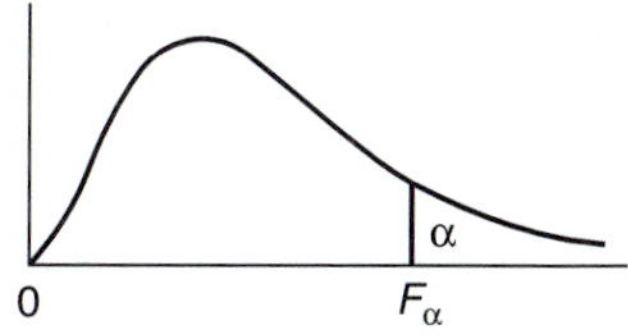

TABLE 2–3 Critical Values of the F Distribution (one-tailed test)

$F_{0.05}(\nu_1,\nu_2)$

ν_2	ν_1 = 1	2	3	4	5	6	7	8	9
1	161.4	199.5	215.7	224.6	230.2	234.0	236.8	238.9	240.5
2	18.51	19.00	19.16	19.25	19.30	19.33	19.35	19.37	19.38
3	10.13	9.55	9.28	9.12	9.01	8.94	8.89	8.85	8.81
4	7.71	6.94	6.59	6.39	6.26	6.16	6.09	6.04	6.00
5	6.61	5.79	5.41	5.19	5.05	4.95	4.88	4.82	4.77
6	5.99	5.14	4.76	4.53	4.39	4.28	4.21	4.15	4.10
7	5.59	1.74	4.35	4.12	3.97	3.87	3.79	3.73	3.68
8	5.32	4.46	4.07	3.84	3.69	3.58	3.50	3.44	3.39
9	5.12	4.26	3.86	3.63	3.48	3.37	3.29	3.23	3.18
10	4.96	4.10	3.71	3.48	3.33	3.22	3.14	3.07	3.02
11	4.84	3.98	3.59	3.36	3.20	3.09	3.01	2.95	2.90
12	4.75	3.89	3.49	3.26	3.11	3.00	2.91	2.85	2.80
13	4.67	3.81	3.41	3.18	3.03	2.92	2.83	2.77	2.71
14	4.60	3.74	3.34	3.11	2.96	2.85	2.76	2.70	2.65
15	4.54	3.68	3.29	3.06	2.90	2.79	2.71	2.64	2.59
16	4.49	3.63	3.24	3.01	2.85	2.74	2.66	2.59	2.54
17	4.45	3.59	3.20	2.96	2.81	2.70	2.61	2.55	2.49
18	4.41	3.55	3.16	2.93	2.77	2.66	2.58	2.51	2.46
19	4.38	3.52	3.13	2.90	2.74	2.63	2.54	2.48	2.42
20	4.35	3.49	3.10	2.87	2.71	2.60	2.51	2.45	2.39
21	4.32	3.47	3.07	2.84	2.68	2.57	2.49	2.42	2.37
22	4.30	3.44	3.05	2.82	2.66	2.55	2.46	2.40	2.34
23	4.28	3.42	3.03	2.80	2.64	2.53	2.44	2.37	2.32
24	4.26	3.40	3.01	2.78	2.62	2.51	2.42	2.36	2.30
25	4.24	3.39	2.99	2.76	2.60	2.49	2.40	2.34	2.28
26	4.23	3.37	2.98	2.74	2.59	2.47	2.39	2.32	2.27
27	4.21	3.35	2.96	2.73	2.57	2.46	2.37	2.31	2.25
28	4.20	3.34	2.95	2.71	2.56	2.45	2.36	2.29	2.24
29	4.18	3.33	2.93	2.70	2.55	2.43	2.35	2.28	2.22
30	4.17	3.32	2.92	2.69	2.53	2.42	2.33	2.27	2.21
40	4.08	3.23	2.84	2.61	2.45	2.34	2.25	2.18	2.12
60	4.00	3.15	2.76	2.53	2.37	2.25	2.17	2.10	2.04
120	3.92	3.07	2.68	2.45	2.29	2.17	2.09	2.02	1.96
∞	3.84	3.00	2.60	2.37	2.21	2.10	2.01	1.94	1.88

TABLE 2–3 Critical Values of the F Distribution (*Continued*)

$F_{0.05}(\nu_1,\nu_2)$

ν_2	ν_1 = 10	12	15	20	24	30	40	60	120	∞
1	241.9	243.9	245.9	248.0	249.1	250.1	251.1	252.2	253.3	254.3
2	19.40	19.41	19.43	19.45	19.45	19.46	19.47	19.48	19.49	19.50
3	8.79	8.74	8.70	8.66	8.64	8.62	8.59	8.57	8.55	8.53
4	5.96	5.91	5.86	5.80	5.77	5.75	5.72	5.69	5.66	5.63
5	4.74	4.68	4.62	4.56	4.53	4.50	4.46	4.43	4.40	4.36
6	4.06	4.00	3.94	3.87	3.84	3.81	3.77	3.74	3.70	3.67
7	3.64	3.57	3.51	3.44	3.41	3.38	3.34	3.30	3.27	3.23
8	3.35	3.28	3.22	3.15	3.12	3.08	3.04	3.01	2.97	2.93
9	3.14	3.07	3.01	2.94	2.90	2.86	2.83	2.79	2.75	2.71
10	2.98	2.91	2.85	2.77	2.74	2.70	2.66	2.62	2.58	2.54
11	2.85	2.79	2.72	2.65	2.61	2.57	2.53	2.49	2.45	2.40
12	2.75	2.69	2.62	2.54	2.51	2.47	2.43	2.38	2.34	2.30
13	2.67	2.60	2.53	2.46	2.42	2.38	2.34	2.30	2.25	2.21
14	2.60	2.53	2.46	2.39	2.35	2.31	2.27	2.22	2.18	2.13
15	2.54	2.48	2.40	2.33	2.29	2.25	2.20	2.16	2.11	2.07
16	2.49	2.42	2.35	2.28	2.24	2.19	2.15	2.11	2.06	2.01
17	2.45	2.38	2.31	2.23	2.19	2.15	2.10	2.06	2.01	1.96
18	2.41	2.34	2.27	2.19	2.15	2.11	2.06	2.02	1.97	1.92
19	2.38	2.31	2.23	2.16	2.11	2.07	2.03	1.98	1.93	1.88
20	2.35	2.28	2.20	2.12	2.08	2.04	1.99	1.95	1.90	1.84
21	2.32	2.25	2.18	2.10	2.05	2.01	1.96	1.92	1.87	1.81
22	2.30	2.23	2.15	2.07	2.03	1.98	1.94	1.89	1.84	1.78
23	2.27	2.20	2.13	2.05	2.01	1.96	1.91	1.86	1.81	1.76
24	2.25	2.18	2.11	2.03	1.98	1.94	1.89	1.84	1.79	1.73
25	2.24	2.16	2.09	2.01	1.96	1.92	1.87	1.82	1.77	1.71
26	2.22	2.15	2.07	1.99	1.95	1.90	1.85	1.80	1.75	1.69
27	2.20	2.13	2.06	1.97	1.93	1.88	1.84	1.79	1.73	1.67
28	2.19	2.12	2.04	1.96	1.91	1.87	1.82	1.77	1.71	1.65
29	2.18	2.10	2.03	1.94	1.90	1.85	1.81	1.75	1.70	1.64
30	2.16	2.09	2.01	1.93	1.89	1.84	1.79	1.74	1.68	1.62
40	2.08	2.00	1.92	1.84	1.79	1.74	1.69	1.64	1.58	1.51
60	1.99	1.92	1.84	1.75	1.70	1.65	1.59	1.53	1.47	1.39
120	1.91	1.83	1.75	1.66	1.61	1.55	1.50	1.43	1.35	1.25
∞	1.83	1.75	1.67	1.57	1.52	1.46	1.39	1.32	1.22	1.00

TABLE 2–3 Critical Values of the F Distribution (*Continued*)

$F_{0.01}(\nu_1,\nu_2)$

ν_2	ν_1 = 1	2	3	4	5	6	7	8	9
1	4052	4999.5	5403	5625	5764	5859	5928	5981	6022
2	98.50	99.00	99.17	99.25	99.30	99.33	99.36	99.37	99.39
3	34.12	30.82	29.46	28.71	28.24	27.91	27.67	27.49	27.35
4	21.20	18.00	16.69	15.98	15.52	15.21	14.98	14.80	14.66
5	16.26	13.27	12.06	11.39	10.97	10.67	10.46	10.29	10.16
6	13.75	10.92	9.78	9.15	8.75	8.47	8.26	8.10	7.98
7	12.25	9.55	8.45	7.85	7.46	7.19	6.99	6.84	6.72
8	11.26	8.65	7.59	7.01	6.63	6.37	6.18	6.03	5.91
9	10.56	8.02	6.99	6.42	6.06	5.80	5.61	5.47	5.35
10	10.04	7.56	6.55	5.99	5.64	5.39	5.20	5.06	4.94
11	9.65	7.21	6.22	5.67	5.32	5.07	4.89	4.74	4.63
12	9.33	6.93	5.95	5.41	5.06	4.82	4.64	4.50	4.39
13	9.07	6.70	5.74	5.21	4.86	4.62	4.44	4.30	4.19
14	8.86	6.51	5.56	5.04	4.69	4.46	4.28	4.14	4.03
15	8.68	6.36	5.42	4.89	4.56	4.32	4.14	4.00	3.89
16	8.53	6.23	5.29	4.77	4.44	4.20	4.03	3.89	3.78
17	8.40	6.11	5.18	4.67	4.34	4.10	3.93	3.79	3.68
18	8.29	6.01	5.09	4.58	4.25	4.01	3.84	3.71	3.60
19	8.18	5.93	5.01	4.50	4.17	3.94	3.77	3.63	3.52
20	8.10	5.85	4.94	4.43	4.10	3.87	3.70	3.56	3.46
21	8.02	5.78	4.87	4.37	4.04	3.81	3.64	3.51	3.40
22	7.95	5.72	4.82	4.31	3.99	3.76	3.59	3.45	3.35
23	7.88	5.66	4.76	4.26	3.94	3.71	3.54	3.41	3.30
24	7.82	5.61	4.72	4.22	3.90	3.67	3.50	3.36	3.26
25	7.77	5.57	4.68	4.18	3.85	3.63	3.46	3.32	3.22
26	7.72	5.53	4.64	4.14	3.82	3.59	3.42	3.29	3.18
27	7.68	5.49	4.60	4.11	3.78	3.56	3.39	3.26	3.15
28	7.64	5.45	4.57	4.07	3.75	3.53	3.36	3.23	3.12
29	7.60	5.42	4.54	4.04	3.73	3.50	3.33	3.20	3.09
30	7.56	5.39	4.51	4.02	3.70	3.47	3.30	3.17	3.07
40	7.31	5.18	4.31	3.83	3.51	3.29	3.12	2.99	2.89
60	7.08	4.98	4.13	3.65	3.34	3.12	2.95	2.82	2.72
120	6.85	4.79	3.95	3.48	3.17	2.96	2.79	2.66	2.56
∞	6.63	4.61	3.78	3.32	3.02	2.80	2.64	2.51	2.41

TABLE 2–3 Critical Values of the F Distribution (*Continued*)

$F_{0.01}(\nu_1,\nu_2)$

ν_2	ν_1 = 10	12	15	20	24	30	40	60	120	∞
1	6056	6106	6157	6209	6235	6261	6287	6313	6339	6366
2	99.40	99.42	99.43	99.45	99.46	99.47	99.47	99.48	99.49	99.50
3	27.23	27.05	26.87	26.69	26.60	26.50	26.41	26.32	26.22	26.13
4	14.55	14.37	14.20	14.02	13.93	13.84	13.75	13.65	13.56	13.46
5	10.05	9.89	9.72	9.55	9.47	9.38	9.29	9.20	9.11	9.02
6	7.87	7.72	7.56	7.40	7.31	7.23	7.14	7.06	6.97	6.88
7	6.62	6.47	6.31	6.16	6.07	5.99	5.91	5.82	5.74	5.65
8	5.81	5.67	5.52	5.36	5.28	5.20	5.12	5.03	4.95	4.86
9	5.26	5.11	4.96	4.81	4.73	4.65	4.57	4.48	4.40	4.31
10	4.85	4.71	4.56	4.41	4.33	4.25	4.17	4.08	4.00	3.91
11	4.54	4.40	4.25	4.10	4.02	3.94	3.86	3.78	3.69	3.60
12	4.30	4.16	4.01	3.86	3.78	3.70	3.62	3.54	3.45	3.36
13	4.10	3.96	3.82	3.66	3.59	3.51	3.43	3.34	3.25	3.17
14	3.94	3.80	3.66	3.51	3.43	3.35	3.27	3.18	3.09	3.00
15	3.80	3.67	3.52	3.37	3.29	3.21	3.13	3.05	2.96	2.87
16	3.69	3.55	3.41	3.26	3.18	3.10	3.02	2.93	2.84	2.75
17	3.59	3.46	3.31	3.16	3.08	3.00	2.92	2.83	2.75	2.65
18	3.51	3.37	3.23	3.08	3.00	2.92	2.84	2.75	2.66	2.57
19	3.43	3.30	3.15	3.00	2.92	2.84	2.76	2.67	2.58	2.49
20	3.37	3.23	3.09	2.94	2.86	2.78	2.69	2.61	2.52	2.42
21	3.31	3.17	3.03	2.88	2.80	2.72	2.64	2.55	2.46	2.36
22	3.26	3.12	2.98	2.83	2.75	2.67	2.58	2.50	2.40	2.31
23	3.21	3.07	2.93	2.78	2.70	2.62	2.54	2.45	2.35	2.26
24	3.17	3.03	2.89	2.74	2.66	2.58	2.49	2.40	2.31	2.21
25	3.13	2.99	2.85	2.70	2.62	2.54	2.45	2.36	2.27	2.17
26	3.09	2.96	2.81	2.66	2.58	2.50	2.42	2.33	2.23	2.13
27	3.06	2.93	2.78	2.63	2.55	2.47	2.38	2.29	2.20	2.10
28	3.03	2.90	2.75	2.60	2.52	2.44	2.35	2.26	2.17	2.06
29	3.00	2.87	2.73	2.57	2.49	2.41	2.33	2.23	2.14	2.03
30	2.98	2.84	2.70	2.55	2.47	2.39	2.30	2.21	2.11	2.01
40	2.80	2.66	2.52	2.37	2.29	2.20	2.11	2.02	1.92	1.80
60	2.63	2.50	2.35	2.20	2.12	2.03	1.94	1.84	1.73	1.60
120	2.47	2.34	2.19	2.03	1.95	1.86	1.76	1.66	1.53	1.38
∞	2.32	2.18	2.14	1.88	1.79	1.70	1.59	1.47	1.32	1.00

is not equal to $F(\nu_2, \nu_1)$. Tables are available for different probabilities, with 5% and 1% being the most frequent; tables for these two values are provided in this book as Table 2–3. As an example, for $\nu_1 = 20$ and $\nu_2 = 10$, the F statistic that cuts off 5% of the right tail is 2.77. For $\nu_1 = 10$ and $\nu_2 = 20$, the corresponding value is 2.35. For a probability of 1%, the F value for $\nu_1 = 20$ and $\nu_2 = 10$, $F(20, 10)$ equals 4.41.

2.5 HYPOTHESIS TESTING

In problems of hydrologic analysis, it is frequently desirable to make a statistical test of some hypothesis. Specifically, we may want to use sample data to draw inferences about the underlying population. Decisions should be made with the population parameters, not the sample statistics. Where the population parameter is computed from sample information, a statistical hypothesis test may be made to provide verification of some hypothesis. For example, water quality laws may require the mean concentration of a pollutant to be below some water quality standard. A sample of five may be taken from either a river or a well and the mean value computed. All five values will not be the same, with the variation in the five values reflecting the uncertainty associated with samples; this constitutes decision making under conditions of uncertainty. We cannot automatically conclude that the water does not meet the state water standard just because the mean of the sample of five measurements is greater than the standard. Our decision must take the sample variation into account. A statistical hypothesis test of the mean would be the appropriate test for such decisions.

Other types of hydrologic decisions require other statistical tests, including tests on a single mean, two means, groups of means, a single variance, two variances, and groups of variances. Some hydrologic decisions require statistical tests on the underlying probability distribution function. Because of the large number of hypothesis tests available, the discussion of hypothesis tests will be limited to the general procedure. Since all tests follow the same six-step procedure, knowledge of the process will facilitate the understanding and application of any such test.

2.5.1 Steps of a Hypothesis Test

The following six steps can be used to perform a statistical analysis of a hypothesis:

1. Formulate hypotheses expressed using population descriptors.
2. Select the appropriate statistical model (theorem) that identifies the test statistic.
3. Specify the level of significance, which is a measure of risk.
4. Collect a data sample and compute the test statistic.
5. Find the critical value of the test statistic and define the region of rejection.
6. Make a decision by selecting the appropriate hypothesis.

Each of these six steps is discussed in more detail in the following paragraphs and then illustrated with an example.

Step 1: Formulation of Hypotheses. Hypothesis testing represents a class of statistical techniques that are designed to extrapolate information from samples of data to make inferences about populations. The first step is to formulate two or more hypotheses for testing. The hypotheses will depend on the problem under investigation. Specifically, if the objective is to make inferences about a single population, the hypotheses will usually be statements indicating that a random variable has a specific probability distribution or that a population parameter has some specific value. If the objective is to compare two or more specific parameters, such as the means of two populations, the hypotheses will be statements formulated to indicate the absence or presence of differences. It is important to note that the hypotheses are composed of statements involving either population distributions or population parameters; hypotheses should not be expressed in terms of sample statistics.

The first hypothesis is called the null hypothesis, is denoted by H_0, and is always formulated to indicate that a difference does not exist; that is, H_0 is expressed as an equality. The second hypothesis, which is called the alternative hypothesis, is formulated to indicate that a difference does exist; thus, it indicates an inequality. The alternative hypothesis is denoted by either H_1 or H_A. The null and alternative hypotheses should be expressed both grammatically and in mathematical terms and should represent mutually exclusive conditions. Thus when a statistical analysis of sampled data suggests that the null hypothesis should be rejected, the alternative hypothesis must be accepted.

Step 2: The Test Statistic and its Sampling Distribution. The alternative hypothesis of Step 1 provides for a difference between specified populations or parameters. To test the hypotheses of Step 1, it is necessary to develop a test statistic that reflects the difference suggested by the alternative hypothesis. The value of a test statistic will vary from one sample to the next. Therefore, the test statistic will be a random variable and have a sampling distribution. A hypothesis test should be based on a theoretical model that defines the probability distribution function of the test statistic and the parameters of the distribution. Based on the distribution of the test statistic, probability statements about computed values may be made.

Theoretical models are available for all of the more frequently used hypothesis tests. In cases where theoretical models are not available, empirical approximations have usually been developed. In any case, a model or theorem that specifies the test statistic, its distribution, and its parameters must be identified. The test statistic reflects the hypotheses and the data that are usually available. Also, the test statistic is a random variable, and thus it has a distribution function that is defined by a functional form and one or more parameters. It is important to remember that more than one random variable is involved. For example, if we are dealing with a random variable x and are interested in testing a hypothesis about the mean of x, there are three random variables involved in the analysis: x, the mean of x, and the test statistic; each of these are random variables and each has a different probability distribution.

Step 3: The Level of Significance. A set of hypotheses were formulated in Step 1. In Step 2, a test statistic and its distribution were identified to reflect the problem for which the hypotheses were formulated. In Step 4, data will be collected to test the hypotheses. Therefore, before the data are collected, it is necessary to provide a probabilistic framework for accepting or rejecting the null hypothesis and, subsequently, making a decision; the framework will reflect the allowance to be made for the chance variation that can be expected in a sample of data.

Table 2–4 shows the available situations and the potential decisions involved in a hypothesis test. The decision table suggests two types of error:

Type I Error: Reject H_0 when, in fact, H_0 is true.

Type II Error: Accept H_0 when, in fact, H_0 is false.

These two types of incorrect decisions are not independent; however, the decision process is most often discussed with reference to only one of the decisions.

The level of significance, which is a primary element of the decision process in hypothesis testing, represents the probability of making a Type I error and is denoted by the Greek lowercase letter alpha, α. The probability of a Type II error is denoted by the Greek lowercase letter beta, β. The two possible incorrect decisions are not independent. The level of significance should not be made exceptionally small, because the probability of making a Type II error will then increase. Selection of the level of significance should, therefore, be based on a rational analysis of the effect on decisions and should be selected prior to the collection and analysis of the sample data (for example, before Step 4). Specifically, one would expect the level of significance to be different when considering a case involving the loss of human life than when the case involves minor property damage. However, the value chosen for α is often based on convention and the availability of statistical tables, with values for α of 0.05 and 0.01 being selected frequently. However, the implications of this traditional means of specifying α should be understood.

Because α and β are not independent of each other, it is necessary to consider the implications of both types of errors in selecting a level of significance. The concept of the power of a statistical test is important when discussing a Type II error. The power is defined as the probability of rejecting H_0 when, in fact, it is false:

$$\text{power} = 1 - \beta \tag{2–45}$$

The concept of the power of a test will not be discussed in more detail here; the reader should consult statistical texts for a more complete discussion of the relationship between α, β, and n.

TABLE 2–4 Decision Table for Hypothesis Testing

	Situation	
Decision	H_0 is true	H_0 is false
Accept H_0	Correct decision	Incorrect decision: type II error
Reject H_0	Incorrect decision: type II error	Correct decision

Step 4: Data Analysis. Given α and β, it is possible to determine the sample size required to meet any rejection criterion. After obtaining the necessary data, the sample is used to provide an estimate of the test statistic. In most cases, the data are also used to provide estimates of other parameters, such as the degrees of freedom, required to define the sampling distribution of the test statistic.

Step 5: The Region of Rejection. The region of rejection consists of those values of the test statistic that would be unlikely to occur in a sample when the null hypothesis is, in fact, true. Conversely, the region of acceptance consists of those values of the test statistic that can be expected when the null hypothesis is, in fact, true. Extreme values of the test statistic are least likely to occur when the null hypothesis is true. Thus the region of rejection is usually represented by one or both tails of the distribution of the test statistic.

The critical value of the test statistic is defined as the value that separates the region of rejection from the region of acceptance. The critical value of the test statistic depends on (1) the statement of the alternative hypothesis, (2) the distribution of the test statistic, (3) the level of significance, and (4) characteristics of the sample of data. These four components represent the first four steps of a hypothesis test.

Depending on the statement of the alternative hypothesis, the region of rejection may consist of values associated with either one or both tails of the distribution of the test statistic. This may best be illustrated with examples. Consider the case of a manufacturer of cold drinks. It must be recognized that the accuracy of the bottling process is not sufficient to assure that every bottle will contain exactly 12 oz. Some bottles will contain less, whereas others will contain more. If the label on the bottle indicates that the bottle contains 12 oz, the

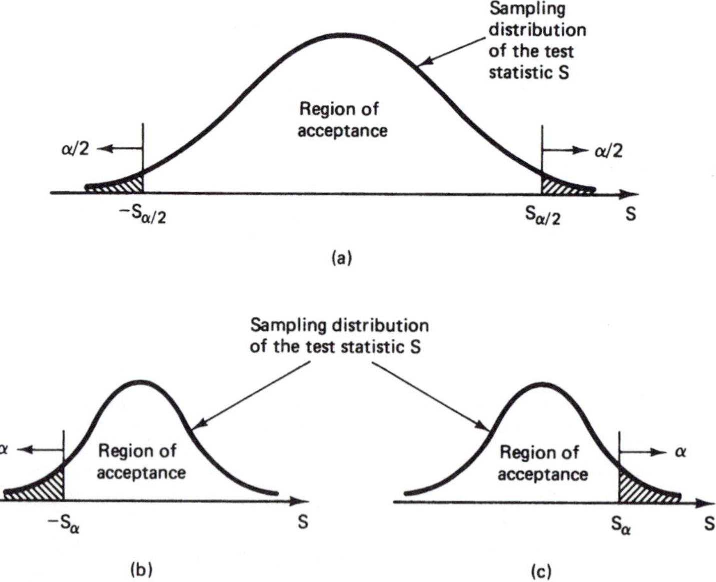

FIGURE 2–5 Representation of the region of rejection (crosshatched area), region of acceptance, and the critical value (S_α): (a) H_A: $\mu \neq \mu_0$; (b) H_A: $\mu < \mu_0$; (c) H_A: $\mu > \mu_0$.

manufacturer wants to be assured that, on the average, each bottle contains at least that amount; otherwise, the manufacturer may be subject to a lawsuit for false advertising. But if the bottles contain, on the average, more than 12 oz, then the manufacturer may be losing money. Thus the manufacturer is interested in both extremely large and extremely small deviations from 12 oz and would be interested in a two-tailed test:

$$H_0\colon \mu = \mu_0 \tag{2–46a}$$

$$H_A\colon \mu \neq \mu_0 \tag{2–46b}$$

where μ is the population mean and μ_0 is the standard of comparison (for example, 12 oz). In this case, the region of rejection will consist of values in both tails. This is illustrated in Figure 2–5a.

In other cases, the region of rejection may consist of values in only one tail of the distribution. For example, if the mean dissolved oxygen (*DO*) concentration from ten samples is used as an indicator of water quality, the regulatory agency would only be interested in whether or not the state water quality standard for dissolved oxygen is being met. Specifically, if a manufacturing plant is discharging waste that contains a pollutant, the state may fine the manufacturer or close down the plant if the effluent causes the *DO* level to go below the standard. Thus the regulatory agency is only interested in whether or not the quality is less than the standard limit. It is not concerned with the degree to which the standard is exceeded since levels of *DO* that are higher than the standard are not considered to be undesirable. This situation represents a one-tailed hypothesis test because the alternative hypothesis is directional. The following hypotheses reflect this problem:

$$H_0\colon \mu = \mu_0 \tag{2–47a}$$

$$H_A\colon \mu < \mu_0 \tag{2–47b}$$

In this case, the region of rejection is associated with values in only one tail of the distribution, as illustrated in Figure 2–5b.

Although the region of rejection should be defined in terms of values of the test statistic, it is often pictorially associated with an area of the sampling distribution that is equal to the level of significance. The region of rejection, region of acceptance, and the critical value are shown in Figure 2–5 for both two-tailed and one-tailed tests.

Step 6: Select the Appropriate Hypothesis. A decision on whether or not to accept the null hypothesis depends on a comparison of the computed value of the test statistic and the critical value. The null hypothesis is rejected when the computed value lies in the region of rejection. Rejection of the null hypothesis implies acceptance of the alternative hypothesis. When a computed value of the test statistic lies in the region of rejection, there are two possible explanations. First, the sampling procedure may have resulted in an extreme value purely by chance; although this a very unlikely event, this corresponds to the Type I error of Table 2–4. Because the probability of this event is relatively small, this explanation is most often rejected. Second, the extreme value of the test statistic may have occurred because the null hypothesis is false; this is the explanation most often accepted and forms the basis for statistical inference.

2.5.2 Summary of Common Hypothesis Tests

It would not be possible to summarize all of the hypothesis tests that are available; in fact, if an entire textbook were written that did nothing more than summarize the available hypothesis tests, the book would be very large. Instead, a few of the more frequently used tests will be listed here just to show that all tests follow the same six steps and that it is not difficult to apply these tests. Table 2–5 provides a summary of frequently used tests. The table identifies the statistical characteristic involved, the null hypothesis for each test, the alternative hypothesis or hypotheses if more than one alternative is possible, the test statistic, and the decision criterion. The tests that are summarized in Table 2–5 are as follows:

(1) The test of a single mean with σ known, which is sometimes referred to as the z test
(2) The test of a single mean with σ unknown, which is sometimes referred to as the t test
(3) The test of two means with the random variables having identical population variances which are not known
(4) The test of a single variance to some standard value σ_0^2
(5) The chi-square goodness-of-fit test for the probability distribution of a random variable x.

TABLE 2–5 Summary of Hypothesis Tests

Statistical Characteristic	H_0	Test Statistic	H_A	Reject H_0 if:
One Mean (σ known)	$\mu = \mu_0$	$z = \dfrac{\overline{X} - \mu_0}{\sigma/\sqrt{n}}$ where $z = N(0, 1)$	$\mu < \mu_0$ $\mu > \mu_0$ $\mu \neq \mu_0$	$z < -z_\alpha$ $z > z_\alpha$ $z < -z_{\alpha/2}$ or $z > z_{\alpha/2}$
One Mean (σ unknown)	$\mu = \mu_0$	$t = \dfrac{\overline{X} - \mu_0}{S/\sqrt{n}}$ with $\nu = n - 1$	$\mu < \mu_0$ $\mu > \mu_0$ $\mu \neq \mu_0$	$t < -t_\alpha$ $t > t_\alpha$ $t < -t_{\alpha/2}$ or $t > t_{\alpha/2}$
Two Means ($\sigma_1 = \sigma_2$, but unknown)	$\mu_1 = \mu_2$	$t = \dfrac{\overline{X_1} - \overline{X_2}}{S(1/n_1 + 1/n_2)^{0.5}}$ with $\nu = n_1 + n_2 - 2$ $S^2 = \dfrac{(n_1 - 1)\, S_1^2 + (n_2 - 1)\, S_2^2}{n_1 + n_2 - 2}$	$\mu_1 < \mu_2$ $\mu_1 > \mu_2$ $\mu_1 \neq \mu_2$	$t < -t_\alpha$ $t > t_\alpha$ $t < -t_{\alpha/2}$ or $t > t_{\alpha/2}$
One Variance	$\sigma^2 = \sigma_0^2$	$\chi^2 = \dfrac{(n - 1)\, s^2}{\sigma_0^2}$ with $\nu = n - 1$	$\sigma^2 < \sigma_0^2$ $\sigma^2 < \sigma_0^2$ $\sigma^2 \neq \sigma_0^2$	$\chi^2 < \chi^2_{1-\alpha}$ $\chi^2 > \chi^2_\alpha$ $\chi^2 < \chi^2_{1-\alpha/2}$ or $\chi^2 > \chi^2_{\alpha/2}$
Probability Distribution	$x \sim f(x, p)$ where x = random variable; p = parameter(s)	$\chi^2 = \displaystyle\sum_{i=1}^{k} \frac{(O_i - E_i)^2}{E_i}$ with $\nu = k - 1$	$x \neq f(x, p)$	$\chi^2 > \chi^2_\alpha$

In reviewing Table 2–5, several points need to be made. First, for sample sizes greater than 30, the z test can be used in place of the t test with little loss of accuracy. Second, the critical values for the tests using the t and chi-square statistics can be obtained from Tables 2–6 and 2–7, respectively. In these tables, the critical value is obtained by entering the table with the degrees of freedom (ν) on the left-hand margin and the probability argument (α, $1 - \alpha$, $\alpha/2$, or $1 - \alpha/2$) along the top of the table. Third, except for the chi-square goodness-of-fit test, all of the tests can be applied as either one-sided or two-sided tests. The chi-square goodness-of-fit test is always one-sided with the region of rejection in the upper tail of the χ^2 distribution. Fourth, to apply the chi-square test the sample values of the random variable x are separated into k cells and the observed number of values (O_i) in each cell i is determined. The expected number of values (E_i) is computed from the probability function identified in the null hypothesis. Each value of E_i should be at least 5; therefore, the test requires a moderate-size sample.

TABLE 2–6 Critical Values of the t Distribution

	α				
ν	0.10	0.05	0.025	0.01	0.005
1	3.078	6.314	12.706	31.821	63.657
2	1.886	2.920	4.303	6.965	9.925
3	1.638	2.353	3.182	4.541	5.841
4	1.533	2.132	2.776	3.747	4.604
5	1.476	2.015	2.571	3.365	4.032
6	1.440	1.943	2.447	3.143	3.707
7	1.415	1.895	2.365	2.998	3.499
8	1.397	1.860	2.306	2.896	3.335
9	1.383	1.833	2.262	2.821	3.250
10	1.372	1.812	2.228	2.764	3.169
11	1.363	1.796	2.201	2.718	3.106
12	1.356	1.782	2.179	2.681	3.055
13	1.350	1.771	2.160	2.650	3.012
14	1.345	1.761	2.145	2.624	2.977
15	1.341	1.753	2.131	2.602	2.947
16	1.337	1.746	2.120	2.583	2.921
17	1.333	1.740	2.110	2.567	2.898
18	1.330	1.734	2.101	2.552	2.878
19	1.328	1.729	2.093	2.539	2.861
20	1.325	1.725	2.086	2.528	2.845
21	1.323	1.721	2.080	2.518	2.831
22	1.321	1.717	2.074	2.508	2.819
23	1.319	1.714	2.069	2.500	2.807
24	1.318	1.711	2.064	2.492	2.797
25	1.316	1.708	2.060	2.485	2.787
26	1.315	1.706	2.056	2.479	2.779
27	1.314	1.703	2.052	2.473	2.771
28	1.313	1.701	2.048	2.467	2.763
29	1.311	1.699	2.045	2.462	2.756
∞	1.282	1.645	1.960	2.326	2.576

TABLE 2–7 Critical Values of the Chi-Square Distribution

	α							
ν	0.995	0.990	0.975	0.95	0.050	0.025	0.010	0.005
1	0.0^4393	0.0^3157	0.0^3982	0.0^3393	3.84	5.02	6.63	7.88
2	0.0100	0.0201	0.0506	0.103	5.99	7.38	9.21	10.60
3	0.0717	0.115	0.216	0.352	7.81	9.35	11.34	12.84
4	0.207	0.297	0.484	0.711	9.49	11.14	13.28	14.86
5	0.412	0.554	0.831	1.145	11.07	12.83	15.09	16.75
6	0.676	0.872	1.237	1.635	12.59	14.45	16.81	18.55
7	0.989	1.239	1.690	2.167	14.07	16.01	18.48	20.28
8	1.344	1.646	2.180	2.733	15.51	17.53	20.09	21.96
9	1.735	2.088	2.700	3.325	16.92	19.02	21.67	23.59
10	2.156	2.558	3.247	3.940	18.31	20.48	23.21	25.19
11	2.603	3.053	3.816	4.575	19.68	21.92	24.72	26.76
12	3.074	3.571	4.404	5.226	21.03	23.34	26.22	28.30
13	3.565	4.107	5.009	5.892	22.36	24.74	27.69	29.82
14	4.075	4.660	5.629	6.571	23.68	26.12	29.14	31.32
15	4.601	5.229	6.262	7.261	25.00	27.49	30.58	32.80
16	5.142	5.812	6.908	7.962	26.30	28.85	32.00	34.27
17	5.697	6.408	7.564	8.672	27.59	30.19	33.41	35.72
18	6.265	7.015	8.231	9.390	28.87	31.53	34.81	37.16
19	6.844	7.633	8.907	10.117	30.14	32.85	36.19	38.58
20	7.434	8.260	9.591	10.851	31.41	34.17	37.57	40.00
21	8.034	8.897	10.283	11.591	32.67	35.48	38.93	41.40
22	8.643	9.542	10.982	12.338	33.92	36.78	40.29	42.80
23	9.260	10.196	11.689	13.091	35.17	38.08	41.64	44.18
24	9.886	10.856	12.401	13.848	36.42	39.36	42.98	45.56
25	10.520	11.524	13.120	14.611	37.65	40.65	44.31	46.93
26	11.160	12.198	13.844	15.379	38.89	41.92	45.64	48.29
27	11.808	12.879	14.573	16.151	40.11	43.19	46.96	49.64
28	12.461	13.565	15.308	16.928	41.34	44.46	48.28	50.99
29	13.121	14.256	16.047	17.708	42.56	45.72	49.59	52.34
30	13.787	14.953	16.791	18.493	43.77	46.98	50.89	53.67

Example 2–6

Table 2–8 lists the monthly precipitation as a depth in inches for the Wildcat Creek near Lawrenceville, Georgia, for June, July, and August from 1954 to 1971. The monthly means and standard deviations are also shown in Table 2–8. From these data, can we conclude that the mean precipitation for July is 5 in.? Even though the sample mean of 4.51 in. is less than the hypothesized population mean of 5 in., the monthly precipitation exceeded 5 in. in 7 of the 18 years. Thus it is reasonable to test the one-sided alternative hypothesis:

$$H_0\colon\ \mu = 5 \text{ in.} \quad H_A\colon\ \mu < 5 \text{ in.}$$

Since the standard deviation of the population, s, is not known, the *t* test should be used. In this case, the sample value of the test statistic is

TABLE 2–8 Monthly Precipitation (in.) for Wildcat Creek near Lawrenceville, Georgia

Year	June	July	August
1954	4.27	2.63	1.62
1955	2.92	3.61	2.14
1956	2.00	4.83	3.19
1957	3.34	2.55	0.92
1958	2.32	5.93	2.23
1959	2.89	3.89	2.02
1960	2.79	1.02	4.40
1961	5.52	4.34	5.16
1962	4.09	7.13	3.46
1963	9.86	3.84	0.50
1964	3.90	6.95	3.44
1965	6.85	2.11	2.64
1966	2.33	2.05	3.02
1967	4.16	10.03	6.47
1968	2.10	5.14	1.20
1969	0.85	4.27	3.94
1970	5.00	5.19	0.50
1971	2.93	5.67	5.69
Mean	3.78	4.51	2.92
Std. dev.	2.08	2.18	1.74

$$t = \frac{4.51 - 5.00}{2.18/\sqrt{18}} = -0.954$$

Since the sample size equals 18, the t distribution has 17 degrees of freedom. Using Table 2–6 with a level of significance of 5% (chosen by convention) the critical t value is -1.74. Since the computed t is not less than the critical value, the null hypothesis cannot be rejected. In this case, the computed value lies in the region of acceptance rather than the region of rejection. This conclusion states only that there is not sufficient evidence to conclude that the population mean is significantly less than 5 in. While the null hypothesis states that the population mean equals 5 in., the statistical test certainly does not imply that the true mean is exactly 5 in. It only implies that we cannot conclude that the population mean is significantly less than 5 in.

Example 2–7

If prior to organizing the data of Table 2–8, one wished to test the hypothesis that the mean monthly precipitations for July and August were equal, the following hypotheses could be stated:

$$H_0\colon \mu_J = \mu_A$$

$$H_A\colon \mu_J \neq \mu_A$$

In this case, the two-sided alternative is selected because our interest is not directional. It is important to note that it would be inappropriate to test the one-sided alternative that $\mu_J < \mu_A$ after the data had been examined because then the sample data would be dictating the experimental test. To compare the two means, the appropriate test statistic of Table 2–5 is used. The best estimate to the sample variance is

$$S^2 = \frac{(18 - 1)(2.18)^2 + (18 - 1)(1.74)^2}{18 + 18 - 2} = 3.89 \text{ in.}$$

Therefore, the test statistic is

$$t = \frac{4.51 - 2.92}{1.97\left(\frac{1}{18} + \frac{1}{18}\right)^{0.5}} = 2.42$$

For $\upsilon = n_1 + n_2 - 2 = 34$ and a 5% level of significance, the critical t value is 2.032 (see Table 2–6). For the two-tailed test, the critical t value is found from Table 2–6 using $\alpha/2$. Therefore, for the two-tailed test, the region of rejection consists of all computed t values less than −2.032 or greater than 2.032. Since the computed value is outside the range from −2.032 to 2.032, the null hypothesis should be rejected. Based on these assumptions and data, we would conclude that the mean monthly precipitations are different.

It is important to note that the test was conducted at the 5% level of significance and that this level is selected somewhat arbitrarily. If a 1% had been selected, the critical value for 34 degrees of freedom is 2.729 (see Table 2–6). In this case, the region of rejection would be all values of t less than −2.729 or greater than 2.729. In this case, the null hypothesis could not be rejected because the computed value of 2.42 lies in the region of acceptance. This fact illustrates the problem with the arbitrary selection of the level of significance.

Example 2–8

Assuming that there is no reason to believe that the monthly precipitations for June, July, and August are not independent measurements of the random variable, the data of Table 2–8 can be used to test a hypothesis about the probability distribution of the random variable. For purposes of illustration, the null hypothesis that the monthly precipitation during June, July, and August is normally distributed with a mean of 4 in. and a standard deviation of 2 in. will be tested:

$$H_0: x \sim N(\mu = 4,\ \sigma = 2)$$

$$H_A: x \neq N(\mu = 4,\ \sigma = 2)$$

In the event that the null hypothesis is rejected, it could be because the assumed distribution is incorrect, the assumed population mean is incorrect, the assumed population variance is incorrect, or any combination of these.

To compute the test statistic given in Table 2–5, it is necessary to establish cells and form a histogram of both the observed and expected frequencies. Since there are 54 values in Table 2–8, 6 cells will be used; the x values separating the cells are chosen so that we can expect similar expected frequencies in each cell. For this purpose, the x values are 2, 3, 4, 5, and 6 in. The probabilities for a standard normal distribution were obtained from Table 2–2 after converting the x values to z values using Equation 2–40. The calculations are shown in Table 2–9. The expected frequencies are computed by multiplying the cell probabilities by the sample size of 54. Both the expected (E_i) and observed (O_i) cell frequencies must sum to the sample size. The computed value of the test statistic (see Table 2–9) equals 9.88. For a 5% level of significance and $\nu = k - 1 = 5$ degrees of freedom, the critical χ^2 value is 11.073 (see Table 2–7). Since the computed value of the test statistic is less than the critical value, the null hypothesis is accepted. The sample information is not sufficient enough to reject the assumption of a normal population with

TABLE 2–9 Chi-Square Goodness-of-Fit for Monthly Precipitation for Wildcat Creek

Precipitation (in.)	Normal z	Cumulative Probability	Cell Probability	E_i	O_i	$\frac{(O_i - E_i)^2}{E_i}$
			0.1587	8.57	7	0.29
2	−1.0	0.1587				
			0.1498	8.09	16	7.73
3	−0.5	0.3085				
			0.1915	10.34	10	0.01
4	0.0	0.5000				
			0.1915	10.34	7	1.08
5	0.5	0.6915				
			0.1498	8.09	8	0.00
6	1.0	0.8413				
			0.1587	8.57	6	0.77
			1.0000	54.00	54	9.88

$\mu = 4$ and $\sigma = 2$. Again, it should be obvious that we could accept other null hypotheses based on the sample of data, so we can only conclude that there is insufficient sample evidence to reject the stated null hypothesis.

2.6 REGRESSION ANALYSIS

Regression analysis is a procedure for fitting an equation to a set of data. Specifically, given a set of measurements on two random variables, y and x, regression provides a means for finding the values of the coefficients, b_0 and b_1, for the straight line ($y = b_0 + b_1 x$) that best fits the data. Statistical optimization has three elements: an objective function, a model, and a sample of y and x values. The function to be optimized is called the objective function, which is an explicit mathematical function that describes what is considered the optimal solution. The mathematical model relates a random variable, called the criterion or dependent variable, to the unknowns and the predictor variable x, which is sometimes called the independent variable. The predictor variable usually has a causal relationship with the criterion variable. For example, if one is interested in using measurements of air temperature to predict evaporation rates, the evaporation would be the criterion variable and the temperature would be the predictor variable; alternative predictor variables might be the relative humidity and the wind speed. It is important to note that the objective function and the mathematical model are two separate explicit functions. The third element of statistical optimization is a data set. The data set consists of measured values of the criterion variable and corresponding values of the predictor variable.

As an example, one may attempt to relate evaporation (E), the criterion variable, to temperature (T), which is the predictor variable, using the linear equation

$$\hat{E} = b_0 + b_1 T \tag{2–48}$$

in which b_0 and b_1 are the unknown coefficients and $\hat{E}$ is the predicted value of E. Because E must be positive and E should increase with increases in T, the solution is considered rational only if b_0 and b_1 are nonnegative. To evaluate the unknowns, a set of simultaneous measurements on E and T would be made. For example, if we were interested in daily evaporation rates, we may measure the total evaporation for each day in a year and the corresponding average temperature; this would give us 365 observations from which we could estimate the best values of b_0 and b_1. Equation 2–48 could then be used to predict E for any value of T.

The most frequently used linear model relates a criterion variable y to a single predictor variable x by the equation

$$\hat{y} = b_0 + b_1 x \tag{2–49}$$

in which b_0 is the intercept coefficient and b_1 is the slope coefficient; b_0 and b_1 are often called regression coefficients because they are obtained from a regression analysis. Because Equation 2–49 involves two variables, y and x, it is sometimes referred to as the bivariate model. The intercept coefficient represents the value of y when x equals zero. The slope coefficient represents the rate of change in y with respect to change in x. Whereas b_0 has the same units as y, the units of b_1 are the ratio of the units of y to x. For Equation 2–48, E might be expressed in inches/day and T in °F; therefore, b_0 has units of inches/day and b_1 has units of inches/(day-°F).

The linear multivariate model relates a criterion variable to two or more predictor variables:

$$\hat{y} = b_0 + b_1 x_1 + b_2 x_2 + \ldots + b_p x_p \tag{2–50}$$

in which p is the number of predictor variables, x_i the ith predictor variable, b_i the ith slope coefficient, and b_0 the intercept coefficient, where $i = 1, 2, \ldots, p$. The coefficients b_i are often called partial regression coefficients and have units equal to the ratio of the units of y to x_i. Equation 2–49 is a special case of Equation 2–50.

2.6.1 Principle of Least Squares

The values of the slope and intercept coefficients of Equations 2–49 and 2–50 can be computed using the principle of least squares. The principle of least squares, which is sometimes referred to as regression analysis, is a process of obtaining "best" estimates of the coefficients and is referred to as a regression method. The principle of least squares is used to regress y on either x of Equation 2–49 or the x_i values of Equation 2–50. To express the principle of least squares, it is necessary to define the error, e, or residual, as the difference between the predicted and measured values of the criterion variable:

$$e_i = \hat{y}_i - y_i \tag{2–51}$$

in which $\hat{y}_i$ is the ith predicted value of the criterion variable, y_i is the ith measured value of y, and e_i is the ith error. It is important to note that the error is defined as the measured value of y subtracted from the predicted value. Some computer programs use the measured value minus the predicted value. I avoid this definition because it then indicates that a positive residual implies underprediction. With Equation 2–51 a positive residual indicates overprediction, while a negative residual indicates underprediction.

The objective function for the principle of least squares is to minimize the sum of the squares of the errors:

$$F = \min \sum_{i=1}^{n} e_i^2 = \min \sum_{i=1}^{n} (\hat{y}_i - y_i)^2 \tag{2–52}$$

in which n is the number of observations on the criterion variable (for example, the sample size).

Solution procedure. The objective function of Equation 2–52 can be minimized by inserting the model (such as Equation 2–49) into Equation 2–52 for $\hat{y}_i$, taking the derivatives of F with respect to each unknown, setting the derivatives equal to zero, and then solving for the unknowns. The solution requires the model for predicting y_i to be substituted into the objective function. It is important to note that the derivatives are taken with respect to the unknowns b_i and not the predictor variables x_i.

To illustrate the solution procedure, the model of Equation 2–49 is substituted into the objective function of Equation 2–52, which yields

$$F = \sum_{i=1}^{n} (b_0 + b_1 x_i - y_i)^2 \tag{2–53}$$

The derivatives of Equation 2–53 with respect to the unknowns are

$$\frac{\partial F}{\partial b_0} = 2 \sum_{i=1}^{n} (b_0 + b_1 x_i - y_i) = 0 \tag{2–54a}$$

$$\frac{\partial F}{\partial b_1} = 2 \sum_{i=1}^{n} (b_0 + b_1 x_i - y_i) x_i = 0 \tag{2–54b}$$

Dividing the equations by 2, separating the terms in the summations, and rearranging yields the set of normal equations:

$$n b_0 + b_1 \Sigma x = \Sigma y \tag{2–55a}$$

$$b_0 \Sigma x + b_1 \Sigma x^2 = \Sigma xy \tag{2–55b}$$

All of the summations in Equations 2–55 are calculated over the n values of the sample; in future equations, the index values of the summations will be omitted when they refer to summation over all elements in a sample of size n. The subscripts for x and y have been omitted but are inferred. The two unknowns b_0 and b_1 can be evaluated by solving the two simultaneous equations (Equation 2–55):

$$b_1 = \frac{\Sigma xy - \Sigma x \Sigma y / n}{\Sigma x^2 - (\Sigma x)^2 / n} \tag{2–56a}$$

$$b_0 = \bar{y} - b_1 \bar{x} = \frac{1}{n} \Sigma y - \frac{b_1}{n} \Sigma x \tag{2–56b}$$

Example 2–9

The data of Table 2–10 can be used to illustrate the solution of the least-squares principle for a data set of five observations on two variables. Substituting the sums of Table 2–10 into the normal equations of Equations 2–55 gives

TABLE 2–10 Data Base and Computations

i	x_i	y_i	x_i^2	y_i^2	x_iy_i	e_i	e_i^2
1	12	4	144	16	48	−0.2	0.04
2	11	3	121	9	33	−0.1	0.01
3	10	2	100	4	20	0.0	0.00
4	9	0	81	0	0	1.1	1.21
5	8	1	64	1	8	−0.8	0.64
	50	10	510	30	109	0.0	1.90

$$5b_0 + 50b_1 = 10 \tag{2–57a}$$

$$50b_0 + 510b_1 = 109 \tag{2–57b}$$

Solving Equations 2–57 for the unknowns yields the model

$$\hat{y} = -7 + 0.9x \tag{2–58}$$

The errors, which are shown in Table 2–10, can be computed by substituting the measured values of x into Equation 2–58. In solving a linear model using the least squares principle, the sum of the errors always equals 0, which is shown in Table 2–10. The sum of the squares of the errors equals 1.9.

It is interesting to note that if a line is passed through the first three observations of Table 2–10, the error for each of these three observations is 0. Such a model would be

$$\hat{y} = -8 + 1.0x \tag{2–59}$$

While three of the residuals are 0, the sum of the squares of the errors for Equation 2–59 equals 2.0, which is greater than the corresponding sum for Equation 2–58. When using either of the models of Equations 2–49 and 2–50, the least squares solution will always provide the equation with the smallest sum of squares of errors.

2.6.2 Reliability of the Regression Equation

Having evaluated the coefficients of the regression equation, it is of interest to evaluate the reliability of the regression equation. The following criteria should be assessed in evaluating the model: (1) the correlation coefficient, (2) the standard error of estimate, and (3) the rationality of the coefficients and the relative importance of the predictor variables, both of which can be assessed using the standardized partial regression coefficients.

Graphical analysis. The first step in examining the relationship between variables is to perform a graphical analysis. Visual inspection of the data can provide the following information:

1. Identify the degree of common variation, which is an indication of the degree to which the two variables are related.
2. Identify the range and distribution of the sample data points.

3. Identify the presence of extreme events.
4. Identify the form of the relationship between the two variables.
5. Identify the type of the relationship.

Each of these factors is of importance in the statistical analysis of sample data and decision making.

When variables show a high degree of common variation, one assumes that there is a causal relationship. If there is a physical reason to suspect that a causal relationship exists, the association demonstrated by the sample data provides empirical support for the assumed relationship. Common variation implies that when the value of one of the random variables is changed, the value of the other variable will change in a systematic manner. For example, an increase in the value of one variable occurs when the value of the other variable increases. Using the evaporation versus temperature model of Equation 2–48, a high degree of common variation in the sample measurements would be expected because of the physical relationship between the two variables. If the change in the one variable is highly predictable from a given change in the other variable, then a high degree of common variation exists. Figure 2–6 shows graphs of different samples of data for two variables having different degrees of common variation. In parts a and e of Figure 2–6, the degree of common variation is very high; thus the variables are said to be correlated. In Figure 2–6c there is no correlation between the two variables because as the value of x is increased it is not certain whether or not y will increase or decrease. In Figure 2–6b and d the degree of correlation is moderate; in Figure 2–6b, it is evident that y will increase as x is increased, but the exact change in y for a

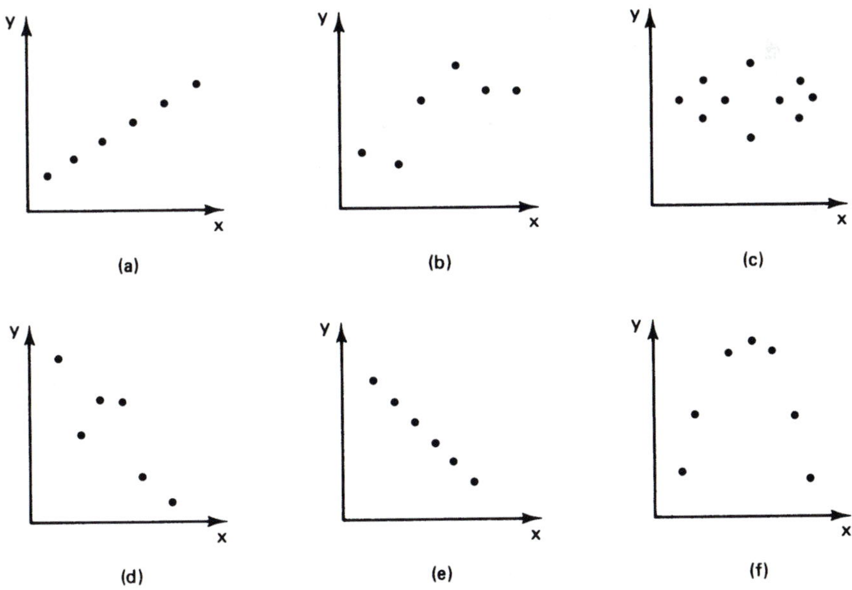

FIGURE 2–6 Different degrees of correlation between two random variables (x and y): (a) $R = 1.0$; (b) $R = 0.5$; (c) $R = 0.0$; (d) $R = -0.5$; (e) $R = -1.0$; (f) $R = 0.3$.

change in x is difficult to estimate. A more quantitative discussion of the concept of common variation will be presented.

It is important to use a graphical analysis to identify the range and distribution of the sample data points so that the stability of the relationship can be assessed and so that one can assess the ability of the data sample to represent the distribution of the population. If the range of the data is limited, a computed relationship may not be stable; that is, it may not apply to the distribution of the population. The data set of Figure 2–7a shows a case where the range of the sample is much smaller than the expected range of the population. If an attempt is made to use the sample to extrapolate the relationship between the two random variables, a small change in the slope of the relationship would cause a large change in the predicted estimate of y for values of x at the extremes of the range of the population. A graph of two random variables might alert the investigator that there may be stability problems in a derived relationship between two random variables, especially when the relationship is extrapolated beyond the range of the sample data.

It is important to identify extreme events in a sample of data for several reasons. First, extreme events can dominate a computed relationship between two variables. For example, in Figure 2–7b, the extreme point will suggest a high correlation between x and y; in this case, the cluster of points acts like a single observation. In Figure 2–7c, the extreme point causes a poor correlation between the two random variables; since the cluster of points has the same mean value of y as the value of y of the extreme point, the data of Figure 2–7c suggest that a change in y is not associated with a change in x. A correlation coefficient is more sensitive to an extreme point when the sample size is small. An extreme event may be due to (1) errors in recording or plotting the data or (2) a legitimate observation in the tail of the dis-

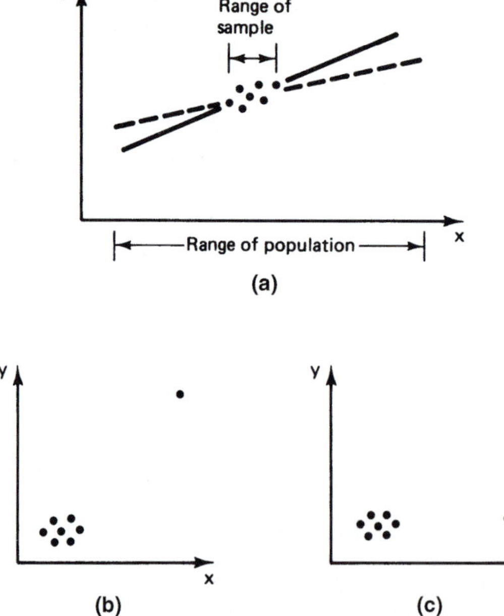

FIGURE 2–7 (a) Instability in the relationship between random variables. Effect of an extreme event in a data sample on the correlation: (b) high correlation; (c) low correlation.

tribution. Therefore, an extreme event must be identified and the reason for the event determined. Otherwise, it will not be possible to properly interpret the results of the correlation analysis.

Relationships can be linear or nonlinear. Since the statistical methods to be used for the two forms of a relationship differ, it is important to identify the form. Additionally, the most frequently used correlation coefficient depends on a linear relationship existing between the two random variables; thus low correlations may result from a nonlinear relationship even when the relationship is obvious. For example, the bivariate relationship of Figure 2–6f suggests a very predictable trend in the relationship between y and x; however, the correlation is poor because of the nonlinearity of the relationship and it is certainly not as good as that in Figure 2–6a.

Graphs relating pairs of variables can be used to identify the type of the relationship. Linear trends can be either direct or indirect, with an indirect relationship indicating a decrease in y as x increases. This information is useful for checking the rationality of the relationship, especially when dealing with data sets that include more than two variables. A variable that is not dominant in the physical processes that underlie the data may demonstrate a physically irrational relationship with another variable only because of the values of the other variables affecting the physical relationship.

Correlation analysis. Correlation is the degree of association between the elements of two samples of data, that is, between observations on two variables. Correlation coefficients provide a quantitative index of the degree of association. Examples of variables that are assumed to have a causal relationship and significant correlation are the cost of living and wages, examination grades and IQ, and the volumes of rainfall and streamflow. However, examples that have no cause-and-effect relationship but may have significant correlation are (1) the crime rate and the sale of chewing gum over the past two decades, and (2) annual population growth rates in nineteenth century France and annual cancer death rates in the United States in the twentieth century. Assessing causality is just as important as assessing the numerical value of a computed correlation coefficient.

A set of observations on a random variable y has a certain amount of variation, which may be characterized by the variance of the sample. The variance equals the sum of squares of the deviations of the observations from the mean of the observations divided by the degrees of freedom. Ignoring the degrees of freedom, this variation can be separated into two parts: that which is associated with a second variable x and that which is not. The total variation (TV), which equals the sum of the squares of the differences between the sample values and the mean of the data points, is separated into the variation that is explained by the second variable (EV) and the variation that is not explained, that is, the unexplained variation (UV). Thus we get

$$TV = EV + UV \tag{2–60}$$

Using the general form of the variation of a random variable, each of the three terms in this equation can be represented by a sum of squares:

$$\sum_{i=1}^{n} (y_i - \bar{y})^2 = \sum_{i=1}^{n} (\hat{y}_i - \bar{y})^2 + \sum_{i=1}^{n} (y_i - \hat{y}_i)^2 \tag{2–61}$$

where y is an observation on the random variable, $\hat{y}$ is the value of y estimated from the best linear relationship with the second variable x, and $\bar{y}$ is the mean of the observations on y. These variation components are shown in Figure 2–8; the dashed lines illustrate the deviations of Equation 2–61. Figure 2–8a shows the variation of the observations about the mean; this represents the variation that exists in the sample and that may potentially be explained by the variable x. It reflects the left side of Equation 2–61. The variation in Figure 2–8b represents the variation of the line about the mean and thus corresponds to the first term of the right side of Equation 2–61. If all of the sample points fall on the line, the explained variation will equal the total variation. The variation in Figure 2–8c represents the variation of the points about the line; therefore, it corresponds to the second term on the right side of Equation 2–61. This is the variation that is not explained by the relationship between y and x. If the unexplained variation equals the total variation, the line does not explain any variation and the relationship between x and y is not significant.

This separation of variation concept is useful for quantifying the correlation coefficient developed by Karl Pearson. Specifically, dividing both sides of Equation 2–60 by the total variation (TV) gives

$$1 = \frac{EV}{TV} + \frac{UV}{TV} \tag{2–62}$$

The ratio EV/TV represents the fraction of the total variation that is explained by the linear relationship between y and x; this is called the coefficient of determination:

$$R^2 = \frac{EV}{TV} = \frac{\sum_{i=1}^{n} (\hat{y}_i - \bar{y})^2}{\sum_{i=1}^{n} (y_i - \bar{y})^2} \tag{2–63}$$

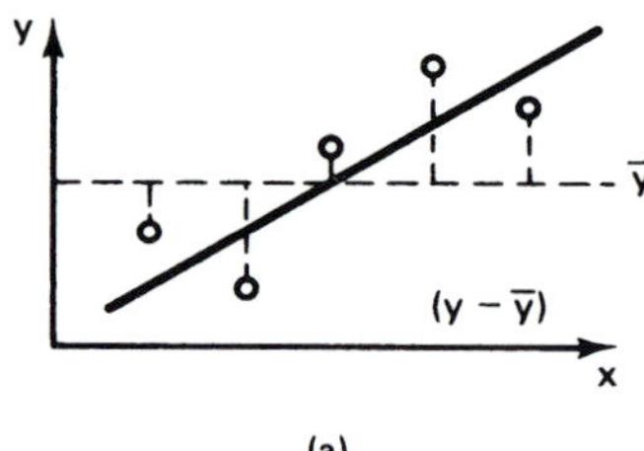

(a)

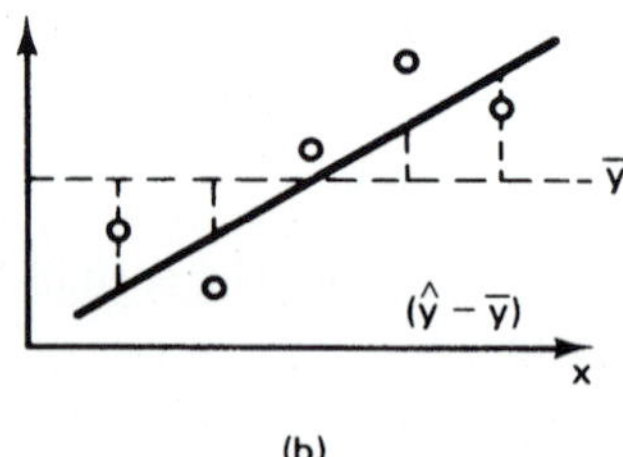

(b)

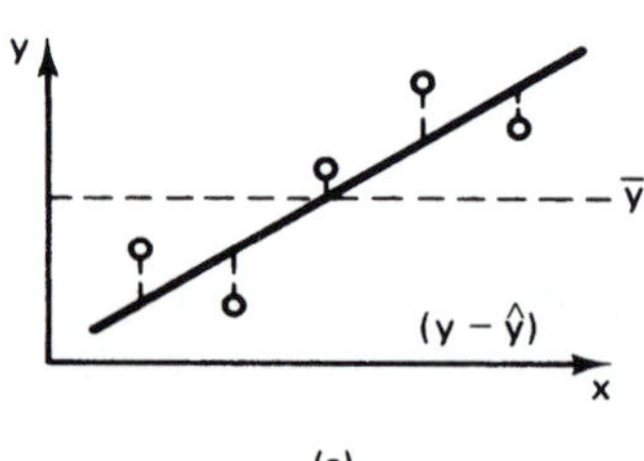

(c)

FIGURE 2–8 Separation of variation: (a) total variation; (b) explained variation; (c) unexplained variation.

The square root of the ratio is the correlation coefficient, R. If the explained variation equals the total variation, the correlation coefficient will equal 1. If the relationship is inverse and the explained variation equals the total variation in magnitude, R will equal −1. These represent the extremes but both values indicate a perfect association, with the sign only indicating the direction of the relationship. If the explained variation equals 0, R equals 0. Thus a correlation coefficient of 0, which is sometimes called the null correlation, indicates the lack of a linear association between the two variables x and y.

While Equation 2–63 provides the means to compute a value of the correlation coefficient, it is easily shown that Equation 2–63 can be arranged to

$$R = \frac{\sum_{i=1}^{n} x_i y_i - \left(\sum_{i=1}^{n} x_i\right)\left(\sum_{i=1}^{n} y_i\right)/n}{\left[\sum_{i=1}^{n} x_i^2 - \left(\sum_{i=1}^{n} x_i\right)^2/n\right]^{0.5}\left[\sum_{i=1}^{n} y_i^2 - \left(\sum_{i=1}^{n} y_i\right)^2/n\right]^{0.5}} \tag{2–64}$$

This equation is used most often because it does not require prior computation of the means and thus the computational algorithm is easily programmed for a computer.

The square of the correlation coefficient (R^2) equals the percentage of the variance in the criterion variable that is explained by variance of the predictor variable. Because of this physical interpretation, R^2 is a meaningful indicator of the accuracy of predictions.

The graphs of Figure 2–9 are useful for summarizing the major points about correlation coefficients. It should be evident from the following discussion that the value of a correlation coefficient, which is a single-valued index of common variation, can be misleading. Computing the correlation coefficient is not a substitute for plotting the data. It might be interesting to try to assess the correlation coefficient for each of the graphs of Figure 2–9 before reading the discussion in the following paragraphs.

The graph of Figure 2–9a suggests a potential for accurate predictions. Because the relationship is nonlinear, the correlation coefficient equals 0.7, or $R^2 = 0.49$. Figures 2–9b, 2–9c, 2–9d, and 2–9e show graphs that appear to have a decreasing amount of scatter, which would suggest an increase in R; however, they also show a decrease in the slope of the relationships and a corresponding decrease in the total variation. For each of the four graphs, the correlation coefficient equals 0.7, which indicates that while the scatter decreases, the proportional decrease in the total variation causes the four sets of points to have the same degree of correlation.

Figures 2–9f, 2–9g, and 2–9h are characterized by the presence of an extreme event. In f and g, the scatter of the remaining points is relatively small. In h, the scatter of the remaining points is relatively large. Yet, all three graphs have the same correlation coefficient, yes the magical 0.7. While the extreme point of Figure 2–9g appears to be farther from the mass of points than the extreme point of Figure 2–9f, the correlation is the same because the extreme point of Figure 2–9g is within the range of the other values of y and does not increase the total variation as much as the extreme point of Figure 2–9f. Thus, the unexplained variation in part f is larger than that in part g, but it is offset by the larger total variation; thus, the correlations are the same. In part h, the location of the extreme point increases the total variation and offsets the relatively large unexplained variation of the fourteen points near the origin of the graph. In this case, the extreme point significantly influences the location of the straight line.

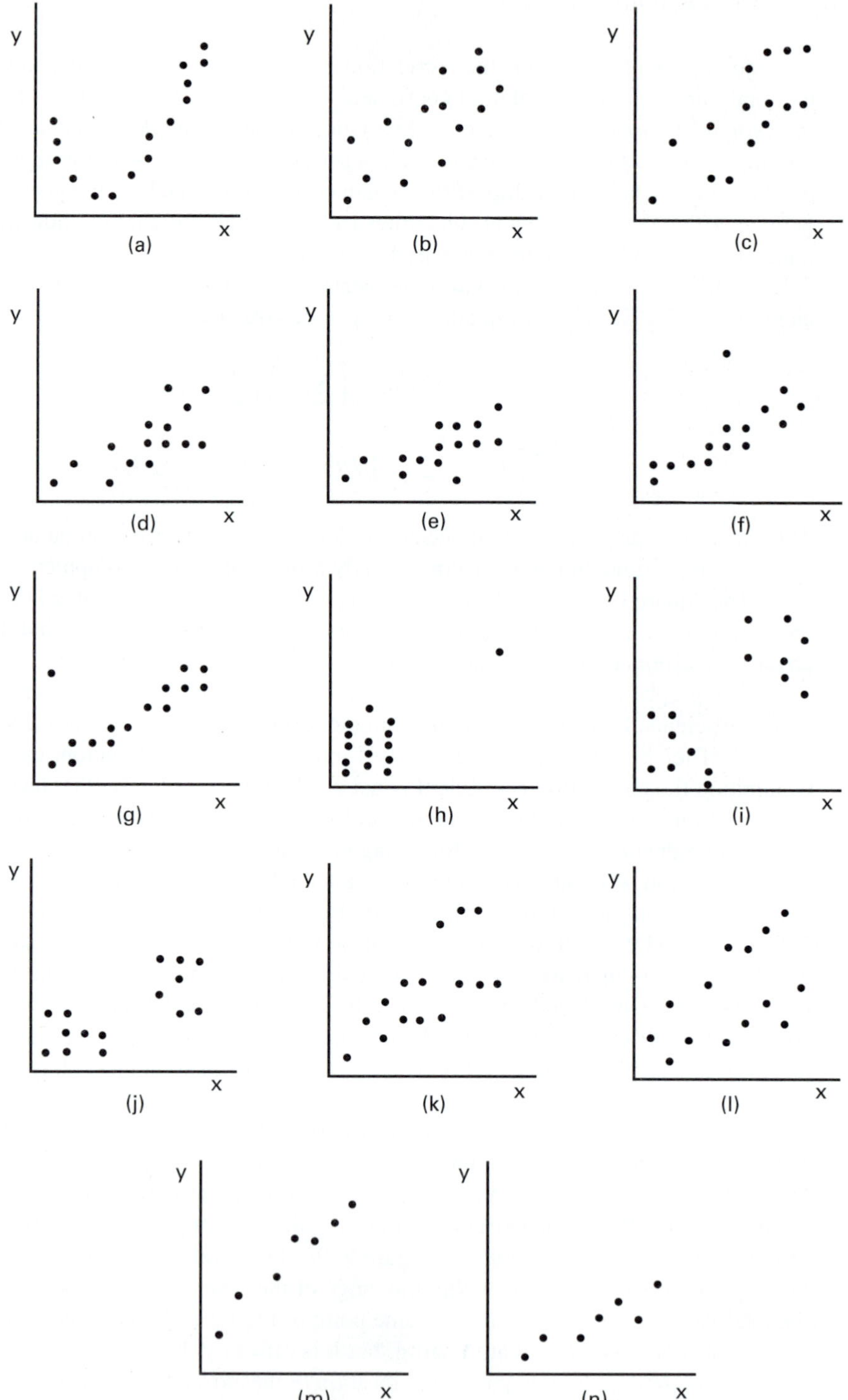

FIGURE 2–9 Graphical analysis and correlation.

Figures 2–9i and j show clusters of points. While part i shows greater scatter within the clusters than for part j, the lower slope in part j means less total variation. Thus, the ratios of the unexplained to total variations are the same, which yields equal correlation coefficients, in this case 0.7.

It is interesting to note that the ten graphs of Figures 2–9a to 2–9j have identical correlation coefficients in spite of the different graphical characteristics. Again, this illustrates the importance of constructing the graph rather than trying to interpret the quality of a relationship from the computed correlation coefficient.

Figures 2–9k and 2–9l are plots that suggest the effect of a third variable. In each case, if two lines are drawn with positive slopes and intercepts near the origin, with one line having a low slope and the other a higher slope, the points local to each line suggest a very high correlation for the local points. The two lines can reflect two levels of the third random variable. When the points are combined as they are in each plot, the scatter is large, which produces a lower correlation. The correlation for Figure 2–9k is 0.7, while for part l it is 0.5. The effect of the third variable is easily illustrated by separating the points of Figure 2–9l, as shown in parts m and n. Both of these plots suggest high correlation, which they are: 0.98 for Figure 2–9m and 0.90 for Figure 2–9n. These four graphs show that, if the effect of a third variable is ignored it can distort the value of the correlation between two variables.

Standard error of estimate. In the absence of additional information, the mean ($\overline{y}$) is the best estimate of the criterion variable; the standard deviation S_y of y is an indication of the prediction accuracy of the mean. If y is related to a predictor variable x and y is regressed on x using the linear model of Equation 2–49, the error of prediction is reduced from S_y to the standard error of estimate, S_e. Mathematically, the standard error of estimate equals the standard deviation of the errors and has the same units as y:

$$S_e = \left[\frac{1}{\upsilon}\sum_{i=1}^{n}(\hat{y}_i - y_i)^2\right]^{0.5} \tag{2–65}$$

in which ν is the degrees of freedom, which equals the sample size minus the number of unknowns. For the bivariate model of Equation 2–49, $p = 1$ and $\nu = n - 2$. For the general linear model with an intercept, Equation 2–50, there are $(p + 1)$ unknowns; thus $\nu = n - p - 1$. In terms of the separation of variation concept discussed previously, the standard error of estimate equals the square root of the ratio of the unexplained variation to the degrees of freedom. To assess the reliability of the regression equation, S_e should be compared with the bounds of 0 and S_y. If S_e is near S_y, the regression equation has not made a significant improvement in prediction accuracy. If S_e is much smaller than S_y and is near 0, the regression analysis has improved the reliability of prediction. It is important to note that S_e is based on $(n - p - 1)$ degrees of freedom, the error S_y is based on $(n - 1)$ degrees of freedom, and the correlation coefficient is based on n degrees of freedom. Thus, in some cases S_e may be greater than S_y, which indicates that the mean $\overline{y}$ is a better means of prediction than is the regression equation.

The standard error of estimate is sometimes computed by

$$S_e = S_y\sqrt{1 - R^2} \tag{2–66}$$

Equation 2–66 must be considered as only an approximation to Equation 2–65 because R is based on n degrees of freedom and S_y is based on $(n - 1)$ degrees of freedom. Using the separation of variation concept, the exact relationship between S_e, S_y, and R can be computed. Given

$$TV = EV + UV \tag{2–67}$$

the total variation TV is related to the variance of y by

$$S_y^2 = \frac{TV}{n - 1} \tag{2–68}$$

The square of the correlation coefficient is the ratio of the explained variation EV to the total variation TV:

$$R^2 = \frac{EV}{TV} \tag{2–69}$$

The standard error of estimate is related to the unexplained variation UV by

$$S_e^2 = \frac{UV}{n - p - 1} \tag{2–70}$$

Equation 2–68 can be solved for TV, which can then be substituted into Equations 2–67 and 2–69. Equation 2–69 can then be solved for EV, which can also be substituted into Equation 2–67. Equation 2–70 can be solved for UV, which is also substituted into Equation 2–67. Solving for S_e^2 yields

$$S_e^2 = \frac{n - 1}{n - p - 1} S_y^2 (1 - R^2) \tag{2–71}$$

Thus Equation 2–71 is a more exact relationship than Equation 2–66; however, for large sample sizes the difference between the two estimates will be small.

Although S_e may actually be greater than S_y, in general, S_e will be within the range from zero to S_y. When the equation fits the data points exactly, S_e equals 0; this corresponds to a correlation coefficient of 1. When the correlation coefficient equals 0, S_e equals S_y; as indicated previously, S_e may actually exceed S_y when the degrees of freedom have a significant effect. The standard error of estimate is often preferred to the correlation coefficient because S_e has the same units as the criterion variable and its magnitude is a physical indicator of the error.

2.6.3 Correlation versus Regression

Before discussion of other topics in correlation and regression it is important to emphasize the differences between correlation and regression. Regression is a means of fitting the coefficients of a prediction equation; correlation provides a measure of goodness of fit. Regression is a method for model calibration, while correlation would have usefulness in model formulation and model verification. It is improper to refer to an equation fitted with least squares as a correlation equation. When using regression, it is necessary to specify which variable is the criterion and which is the predictor. When using correlation it is not necessary to make such a distinction. The correlation coefficients for y on x and x on y are the same, but the regression relationships are not the same. The distinction is necessary with regression because a regression equation is not transformable, unless the correlation coefficient equals 1.0. That is, if y is the criterion variable when the equation is calibrated, the equation cannot be rearranged algebraically to get an equation for predicting x. Specifically, Equation 2–49 can be algebraically rearranged to predict x:

$$y = b_0 + b_1 x \tag{2–72a}$$

$$y - b_0 = b_1 x \tag{2–72b}$$

$$x = \frac{y - b_0}{b_1} \tag{2–72c}$$

$$x = -\frac{b_0}{b_1} + \frac{1}{b_1} y \tag{2–72d}$$

$$x = a_0 + a_1 y \tag{2–72e}$$

If values of a_0 and a_1 are obtained by regressing values of x on y, the resulting coefficients of Equation 2–72e will not be the same as the coefficients obtained by regressing y on x (Equation 2–72a) and then setting $a_0 = -b_0/b_1$ and $a_1 = 1/b_1$. The only exception to this is when the correlation coefficient equals 1.0. For a given data set, the correlation coefficient for the regression of y on x will be the same as the correlation for the regression of x on y. Computationally, the correlation coefficient is a function of the explained and total variation; the slope coefficient of a regression equation (Equation 2–49) is related to the correlation coefficient by

$$b_1 = \frac{RS_y}{S_x} \tag{2–73}$$

in which S_x and S_y are the standard deviations of x and y, respectively.

In summary, correlation and regression differ in their use, computation, and interpretation. The two statistical methods are quite different.

Example 2–10

The data of Table 2–11 are the drainage area A (mi^2) and mean annual flood Q (ft^3/sec) for six basins in the Sleepers River Experimental Watershed, Vermont. With the drainage area as the predictor variable, the regression coefficients can be computed from Equations 2–56:

$$b_1 = \frac{\Sigma AQ - \Sigma A \Sigma Q/n}{\Sigma A^2 - (\Sigma A)^2/n} = \frac{61,974.83 - 93.91(2472)/6}{2515.04 - (93.91)^2/6} = 22.277$$

$$b_0 = \frac{1}{n}\Sigma Q - b_1 \frac{\Sigma A}{n} = \frac{1}{6}(2472) - 22.277\left(\frac{93.91}{6}\right) = 63.3258$$

TABLE 2–11 Regression of Drainage Drea (A) on the Mean Annual Flood (Q) for Six Watersheds in the Sleepers River Experimental Watershed, Vermont

A (mi^2)	Q (ft^3/sec)	A^2	Q^2	AQ	$\hat{Q}$	e	e^2
16.58	455	274.90	207025	7543.90	432.7	−22.3	498.15
3.23	105	10.43	11025	339.15	135.3	30.3	916.91
16.80	465	282.24	216225	7812.00	437.6	−27.4	751.77
42.91	1000	1841.27	1000000	42910.00	1019.2	19.2	370.11
8.35	290	69.72	84100	2,421.50	249.3	−40.7	1653.27
6.04	157	36.48	24649	948.28	197.9	40.9	1671.12
93.91	2472	2515.04	1543024	61974.83		0.0	5861.33

This yields the model

$$\hat{Q} = 63.326 + 22.277A$$

The predicted values $\hat{Q}$ for each of the six watersheds are given in Table 2–11. The errors e_i are also shown; the sum of the errors equals 0, as it always will with a linear regression analysis. The sum of the squares of the errors can be used with Equation 2–65 to compute the standard error of estimate:

$$S_e = \left(\frac{1}{n-2}\Sigma e^2\right)^{0.5} = \left[\frac{1}{4}(5861.33)\right]^{0.5} = 38.28 \text{ ft}^3/\text{sec}$$

The correlation coefficient can be computed with Equation 2–64 and the sums given in Table 2–11:

$$R = \frac{61,974.83 - 93.91(2472)/6}{[2515.04 - (93.91)^2/6]^{0.5}[1,543,024 - (2472)^2/6]^{0.5}} = 0.994$$

The standard deviation of the mean annual flood can be computed using Equation 2–23:

$$S_Q = \left(\frac{1}{5}\left[(1,543,024) - \frac{1}{6}(2472)^2\right]\right)^{0.5} = 323.9 \text{ ft}^3/\text{sec}$$

Thus the ratio S_e/S_y equals 11.8%. For small samples the S_e/S_y ratio is a better indicator of reliability than the correlation coefficient. In either case, the prediction equation provides accurate estimates of the mean annual discharge for watersheds similar to the six watersheds used to compute the regression equation.

Example 2–11

Precipitation (P) is a primary determinant of runoff (R), and so a number of equations have been developed for predicting R from measurements of P. Table 2–12 gives the total monthly precipitation and runoff, both presented as depths in inches, for December during the ten years from 1960 to 1969 on Wildcat Creek near Lawrenceville, Georgia. Figure 2–10 shows the data.

The data were analyzed using regression analysis. Based on the summations given in Table 2–12, the slope and intercept coefficients are

$$b_1 = \frac{\Sigma PR - \Sigma P \Sigma R/n}{\Sigma P^2 - (\Sigma P)^2/n} = \frac{89.26 - 50.20(14.23)/10}{319.11 - (50.20)^2/10} = 0.2657$$

$$b_0 = \overline{R} - b_1\overline{P} = \frac{14.23}{10} - 0.2657\,\frac{50.20}{10} = 0.08918$$

TABLE 2–12 December Precipitation (in.) and Runoff (in.) for Wildcat Creek near Lawrenceville, Georgia, 1960–1969

Year	P	R	P^2	R^2	PR	$\hat{R}$	e
1960	1.95	0.46	3.80	0.21	0.90	0.61	0.147
1961	10.82	2.85	117.07	8.12	30.84	2.96	0.114
1962	3.22	0.99	10.37	0.98	3.19	0.94	–0.045
1963	4.51	1.40	20.34	1.96	6.31	1.29	–0.113
1964	6.71	1.98	45.02	3.92	13.29	1.87	–0.108
1965	1.18	0.45	1.39	0.20	0.53	0.40	–0.047
1966	4.82	1.31	23.23	1.72	6.31	1.37	0.060
1967	6.38	2.22	40.70	4.93	14.16	1.78	–0.436
1968	5.97	1.36	35.64	1.85	8.12	1.68	0.315
1969	4.64	1.21	21.53	1.46	5.61	1.32	0.112
	50.20	14.23	319.11	25.36	89.26		–0.001

The resulting regression line derived from these coefficients is also shown in Figure 2–10. The intercept is nearly 0, which is rational since runoff would not be expected when the total monthly precipitation is 0. A hypothesis test could be made to determine whether or not the intercept is statistically different from 0. The slightly positive value of b_0 might suggest that ground water flowing into the channel contributes to the total monthly runoff. The regression equation was used to predict monthly values of runoff, $\hat{R}$, and compute the residuals, e, which are also given on Table 2–12. Based on the computed residuals the standard error of estimate is

$$S_e = \left(\frac{1}{n-2}\Sigma e^2\right)^{0.5} = \left[\frac{1}{8}(0.3687)\right]^{0.5} = 0.2147 \text{ in.}$$

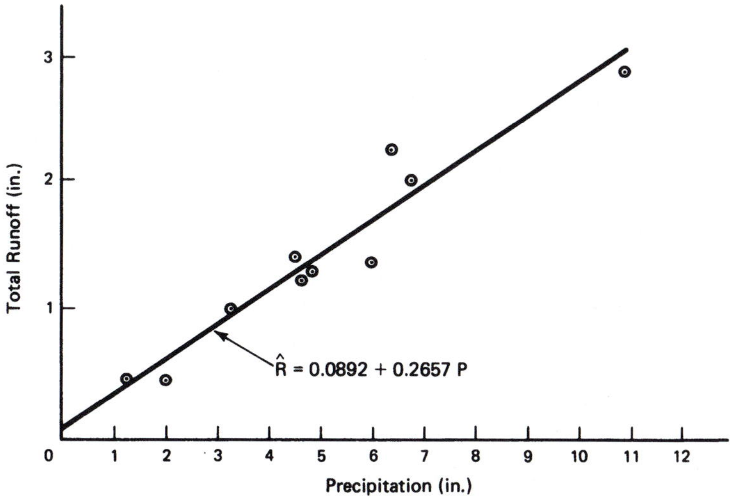

FIGURE 2–10 December precipitation and runoff for Wildcat Creek near Lawrenceville, GA, 1960–69.

The standard deviation of the measured runoff values can be computed using Equation 2–22:

$$S_R = \left(\frac{1}{n-1}\left[\Sigma R^2 - \frac{(\Sigma R)^2}{n}\right]\right)^{0.5} = \left(\frac{1}{9}\left[25.36 - \frac{(14.23)^2}{10}\right]\right)^{0.5} = 0.7536 \text{ in.}$$

Thus the standard error is about 28.5% of the initial error, S_R, and thus the regression equation should provide reasonably accurate estimates of the monthly runoff. The correlation coefficient can be computed using Equation 2–64:

$$R = \frac{\Sigma PR - \Sigma P \Sigma R/n}{[\Sigma P^2 - (\Sigma P)^2/n]^{0.5}\,[\Sigma R^2 - (\Sigma R)^2/n]^{0.5}}$$

$$= \frac{89.26 - 50.20(14.23)/10}{[319.11 - (50.20)^2/10]^{0.5}\,[25.36 - (14.23)^2/10]^{0.5}} = 0.963$$

Thus the correlation coefficient supports the conclusion drawn from the standard error of estimate that the model provides reasonably accurate estimates of runoff.

2.6.4 Calibration of the Multiple Linear Regression Model

The components of an unconstrained regression analysis include the model, the objective function, and the data set. The data set consists of a set of n observations on p predictor variables and one criterion variable, where n should be, if possible, at least four times greater than p. The data set can be viewed as a matrix having dimensions of n by $(p + 1)$. The principle of least squares is used as the objective function. The model, in raw-score form is

$$\hat{y} = b_0 + b_1x_1 + b_2x_2 + \ldots + b_px_p \tag{2–74}$$

in which x_j $(j = 1, 2, \ldots, p)$ are the predictor variables, b_j $(j = 1, 2, \ldots, p)$ are the partial regression coefficients, b_0 is the intercept coefficient, and y is the criterion variable. Using the least squares principle and the model of Equation 2–74, the objective function becomes

$$F = \min \sum_{i=1}^{n} e_i^2 = \min \sum_{i=1}^{n} \left(b_0 + \sum_{j=1}^{p} b_jx_{ij} - y_i\right)^2 \tag{2–75}$$

in which F is the value of the objective function. It should be noted that the predictor variables include two subscripts, with i indicating the observation and j the specific predictor variable.

The method of solution is to take the $(p + 1)$ derivatives of the objective function, Equation 2–75, with respect to the unknowns, b_j $(j = 0, 1, \ldots, p)$, setting the derivatives equal to 0, and solving for the unknowns. A set of $(p + 1)$ normal equations is an intermediate result of this process.

As an example, consider the case where $p = 2$; thus Equation 2–74 reduces to

$$\hat{y} = b_0 + b_1x_1 + b_2x_2 \tag{2–76}$$

Also, the objective function, Equation 2–75, is given by

$$F = \min \sum_{i=1}^{n} (b_0 + b_1x_{i1} + b_2x_{i2} - y_i)^2 \tag{2–77}$$

The resulting derivatives are

$$\frac{\partial F}{\partial b_0} = 2\sum_{i=1}^{n}(b_0 + b_1x_{i1} + b_2x_{i2} - y_i)(1) = 0 \tag{2–78a}$$

$$\frac{\partial F}{\partial b_1} = 2\sum_{i=1}^{n}(b_0 + b_1x_{i1} + b_2x_{i2} - y_i)(x_{i1}) = 0 \tag{2–78b}$$

$$\frac{\partial F}{\partial b_2} = 2\sum_{i=1}^{n}(b_0 + b_1x_{i1} + b_2x_{i2} - y_i)(x_{i2}) = 0 \tag{2–78c}$$

Rearranging Equations 2–78 yields a set of normal equations:

$$nb_0 + b_1\sum_{i=1}^{n} x_{i1} + b_2\sum_{i=1}^{n} x_{i2} = \sum_{i=1}^{n} y_i \tag{2–79a}$$

$$b_0\sum_{i=1}^{n} x_{i1} + b_1\sum_{i=1}^{n} x_{i1}^2 + b_2\sum_{i=1}^{n} x_{i1}x_{i2} = \sum_{i=1}^{n} x_{i1}y_i \tag{2–79b}$$

$$b_0\sum_{i=1}^{n} x_{i2} + b_1\sum_{i=1}^{n} x_{i1}x_{i2} + b_2\sum_{i=1}^{n} x_{i2}^2 = \sum_{i=1}^{n} x_{i2}y_i \tag{2–79c}$$

The solution of the three simultaneous equations would yield values of b_0, b_1, and b_2.

Standardized model. When the means and standard deviations of the predictor and criterion variables are significantly different, roundoff error, which results from the inability to maintain a sufficient number of significant digits in the computations, may cause the partial regression coefficients computed from Equations 2–79 to be erroneous. Thus most multiple regression analyses are computed using a standardized model:

$$z_y = t_1z_1 + t_zz_2 + \ldots + t_pz_p \tag{2–80}$$

in which t_j ($j = 1, 2, \ldots, p$) are called standardized partial regression coefficients, and z_y and the z_j ($j = 1, 2, \ldots, p$) are the criterion variable and the predictor variables, respectively, expressed in standardized form. For $i = 1, 2, \ldots, n$, they are computed by

$$z_{iy} = \frac{y_i - \bar{y}}{S_y} \tag{2–81}$$

and

$$z_{ij} = \frac{x_{ij} - \bar{x}_j}{S_j} \tag{2–82}$$

in which S_y is the standard deviation of the criterion variable and S_j ($j = 1, 2, \ldots, p$) are the standard deviations of the predictor variables. It can be shown that the standardized partial regression coefficients (for example, the t_j's) and the partial regression coefficients (for example, the b_j's) are related by

$$b_j = \frac{t_j S_y}{S_j} \tag{2–83}$$

The intercept coefficient can be computed by

$$b_0 = \bar{y} - \sum_{j=1}^{p} b_j \bar{x}_j \tag{2–84}$$

Thus the raw score model of Equation 2–74 can be computed directly from the standardized model (Equation 2–80) and Equations 2–83 and 2–84.

2.6.5 Criteria for Evaluating a Multiple Regression Model

After a multiple regression model has been calibrated, we may ask how well the linear model represents the observed data. The following criteria should be used in answering this question: (1) the rationality of the coefficients, (2) the coefficient of multiple determination (R^2), and (3) the standard error of estimate (S_e), which is usually compared with S_y. The relative importance of each of these criteria may vary with the problem as well as with the person.

Coefficient of multiple determination. The coefficient of multiple determination equals the fraction of the variation in the criterion variable that is explained by the regression equation. It is the square of the correlation coefficient. Mathematically, it is defined as

$$R^2 = \frac{\sum_{i=1}^{n} (\hat{y} - \bar{y})^2}{\sum_{i=1}^{n} (y - \bar{y})^2} \tag{2–85}$$

It may also be computed by

$$R^2 = \sum_{j=1}^{p} t_j R_{jy} \tag{2–86}$$

in which R_{jy} is the predictor-criterion correlation for predictor j. The value of R^2 is always in the range from 0.0 to 1.0, with a value of 0 indicating that y is not related to any of the predictor variables. The coefficient of multiple determination must be at least as large as the square of the largest predictor-criterion correlation; if the intercorrelations are high, there will be little difference between the two, which would indicate that including more than one predictor variable in the regression equation does little to improve the accuracy of prediction. If the intercorrelations are low, more than one predictor variable will probably be necessary.

Standard error of estimate. The standard error of estimate, which was defined previously as the standard deviation of the errors, is defined here as the square root of the sum of the squares of the errors divided by the degrees of freedom (ν):

$$S_e = \left[\frac{1}{\nu}\sum_{i=1}^{n} (\hat{y}_i - y_i)^2\right]^{0.5} \tag{2–87}$$

Previously, the degrees of freedom used to compute S_e was $(n - 2)$; however, in general, the degrees of freedom is defined as

$$\upsilon = n - q \tag{2–88}$$

in which q is the number of unknowns. For the case where there are p partial regression coefficients and one intercept coefficient, $q = p + 1$ and $\nu = n - p - 1$. The S_e has the same units as the criterion variable and should be compared with S_y in assessing the accuracy of prediction. In some cases, S_e is divided by S_y to provide a standardized measure of fit.

2.6.6 Two-Predictor Case

A multiple regression model that includes two predictor variables can be used to illustrate the computational procedure. The two-predictor model in raw-score form is given by Equation 2–76. For the raw-score model the least squares objective function is

$$F = \min \sum_{i=1}^{n} e_i^2 = \min \sum_{i=1}^{n} (\hat{y}_i - y_i)^2 \tag{2–89a}$$

$$= \min \sum_{i=1}^{n} (b_0 + b_1 x_{i1} + b_2 x_{i2} - y_i)^2 \tag{2–89b}$$

Equation 2–89b can be differentiated with respect to each of the unknowns. After setting the three derivatives equal to 0, the following normal equations result:

$$b_0 \Sigma x_1 + b_1 \Sigma x_1^2 \quad + b_2 \Sigma x_1 x_2 = \Sigma x_1 y \tag{2–90a}$$

$$b_0 \Sigma x_2 + b_1 \Sigma x_1 x_2 + b_2 \Sigma x_2^2 \quad = \Sigma x_2 y \tag{2–90b}$$

$$b_0 n + b_1 \Sigma x_1 \quad + b_2 \Sigma x_2 \quad = \Sigma y \tag{2–90c}$$

in which each Σ is a summation over the n observations. The normal equations of Equations 2–90 are solved as three simultaneous equations with the three unknowns, b_0, b_1, and b_2.

The standardized model is given by

$$z_{iy} = t_1 z_{i1} + t_2 z_{i2} \tag{2–91}$$

Expressing the objective function in terms of the standardized variables gives

$$F_z = \min \sum_{i=1}^{n} e_{zi}^2 = \min \sum_{i=1}^{n} (\hat{z}_{iy} - z_{iy})^2 \tag{2–92a}$$

$$= \min \sum_{i=1}^{n} (t_1 z_{i1} + t_2 z_{i2} - z_{iy})^2 \tag{2–92b}$$

Recognizing the relationship between the correlation coefficient and the standardized variables, Equation 2–92b can be differentiated with respect to the unknowns (for example, t_1 and t_2). Setting the derivatives equal to 0 results in the normal equations

$$t_1 + r_{12}t_2 = r_{1y} \tag{2–93a}$$

$$r_{12}t_1 + t_2 = r_{2y} \tag{2–93b}$$

Solving for the unknowns yields

$$t_1 = \frac{r_{1y} - r_{12}r_{2y}}{1 - r_{12}^2} \tag{2–94a}$$

and

$$t_2 = \frac{r_{2y} - r_{12}r_{1y}}{1 - r_{12}^2} \tag{2–94b}$$

It can be shown that the coefficient of multiple determination for a predictor model can be computed by

$$R^2 = \sum_{j=1}^{p} t_j r_{jy} \tag{2–95}$$

For the two-predictor model, R^2 is given by

$$R^2 = t_1 r_{iy} + t_2 r_{2y} \tag{2–96}$$

The standard error of estimate is given as

$$S_e = S_y \left[\frac{n-1}{n-p-1}(1 - R^2)\right]^{0.5} \tag{2–97a}$$

For the two-predictor model, this is

$$S_e = S_y \left[\frac{n-1}{n-3}(1 - R^2)\right]^{0.5} \tag{2–97b}$$

The standard error of estimate should be compared with S_y to decide whether or not the regression equation has resulted in a model that will provide sufficient accuracy for predicting future values of y.

Example 2–12

Table 2–13 lists the precipitation for the months of February and March and runoff for the month of March measured at a forest research watershed. We consider that some runoff measured in March is the result of rainfall in March, but some of this runoff also is the result of delayed response of the stream to rain that fell in February. This concept may be expressed as the simple linear model

$$RO_m = b_1(P_m - I_{am}) + b_2(P_f - I_{af}) \tag{2–98a}$$

in which RO_m is the March runoff; P_m and P_f are March and February precipitation, respectively; I_{am} represents the initial abstraction of March rainfall that does not become runoff; and I_{af} represents the similar abstraction of February rainfall. The coefficients b_1 and b_2 are the regression coefficients that scale the rainfall to the runoff. Equation 2–98a may be expanded to

TABLE 2–13 Selected Monthly Rainfall for February (P_f) and March (P_m) and Runoff (RO_m), White Hollow Watershed (in.)

Year	P_m	P_f	RO_m
1935	9.74	4.11	6.15
1936	6.01	3.33	4.93
1937	1.30	5.08	1.42
1938	4.80	2.41	3.60
1939	4.15	9.64	3.54
1940	5.94	4.04	2.26
1941	2.99	0.73	0.81
1942	5.11	3.41	2.68
1943	7.06	3.89	4.68
1944	6.38	8.68	5.18
1945	1.92	6.83	2.91
1946	2.82	5.21	2.84
1947	2.51	1.78	2.02
1948	5.07	8.39	3.27
1949	4.63	3.25	3.05
1950	4.24	5.62	2.59
1951	6.38	8.56	4.66
1952	7.01	1.96	5.40
1953	4.15	5.57	2.60
1954	4.91	2.48	2.52
1955	8.18	5.72	6.09
1956	5.85	10.19	4.58
1957	2.14	5.66	2.02
1958	3.06	3.04	2.59
Sum	116.35	119.58	82.39
$\Sigma X_1 X_j$	663.1355	589.8177	458.9312
$\Sigma X_2 X_j$		753.1048	435.9246
ΣY^2			331.4729

$$RO_m = b_1 P_m + b_2 P_f - b_1 I_{am} - b_2 I_{af}$$

If a is substituted for the term $-(b_1 I_{am} + b_2 I_{af})$, we get

$$RO_m = a + b_1 P_m + b_2 P_f \tag{2–98b}$$

The sums of variates and sums of squares and products necessary to apply the model, expressed as Equation 2–98b, to the data are given at the bottom of the respective columns in Table 2–13. Thus the simultaneous normal equations are

$$\begin{aligned} 24a \quad &+ 116.35b_1 \quad + 119.58b_2 \quad = 82.39 \\ 116.35a &+ 663.1355b_1 + 589.8177b_2 = 458.9312 \\ 119.58a &+ 589.8177b_1 + 753.1048b_2 = 435.9246 \end{aligned}$$

which upon solution yields the following model:

$$\hat{RO}_m = -0.0346 + 0.5880P_m + 0.1238P_f$$

This model has a correlation coefficient of 0.8852 and a standard error of estimate of 0.7081 in., which can be compared to a S_y of 1.4542 in. The runoff and precipitation for March are plotted in Figure 2–11. The points are labeled with truncated values of February precipitation. The quantified model plots as a family of parallel lines. The slope of these lines is the coefficient b_1, 0.5880. The coefficient b_2, 0.1238, sets the spacing of the lines.

The threshold values of the model, I_{am} and I_{af}, cannot be determined exactly. However, these values should be nearly the same for the months of February and March, which are part of the dormant season of the year. Assuming that I_{af} equals I_{am}, then

$$a = (0.5880I_{am} + 0.1238I_{am})$$

From the value for a of –0.0348, I_{am} is found to be 0.049 inches. The quantified physical interpretation of the wintertime monthly runoff process is, therefore, as follows. A potential for runoff is generated as soon as monthly rainfall exceeds about 0.05 in. About 59% of this potential becomes runoff in the month during which the rain fell. An additional 12% becomes runoff the following month. In Figure 2–11 it should be noted that the proportions 59% and 12% are set reasonably well by the data. The threshold value of 0.05 in. seems small. Figure 2–11 shows that this is literally an extrapolation of the lines back to zero runoff. Such an extrapolation may be invalid. However, this structure is incorporated in the conceptualization of the model, and the model can only do what it was designed to do. If the model is unsatisfactory, the hydrologist must design a new model structure that is more applicable to the problem.

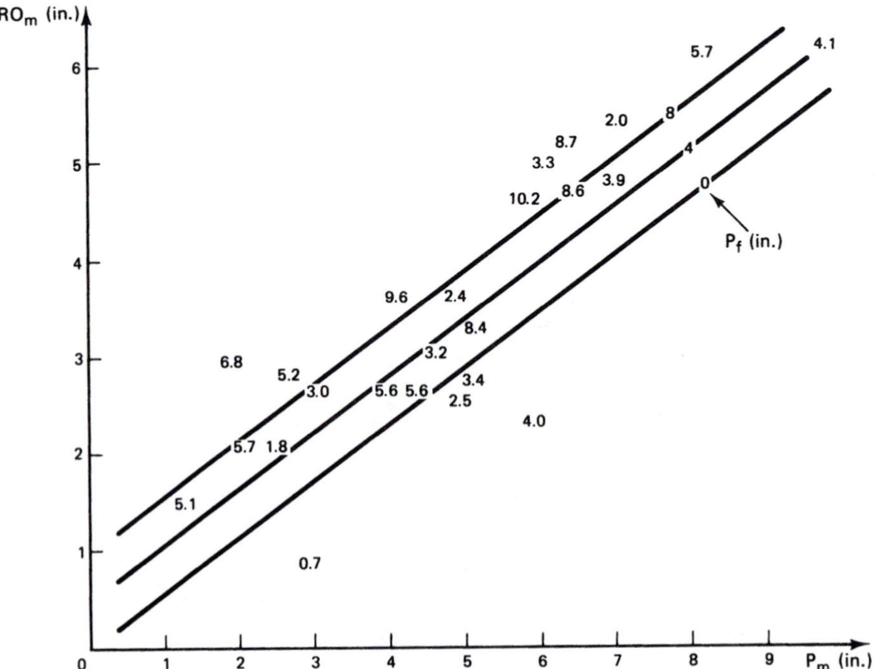

FIGURE 2–11 Rainfall versus runoff for March, White Hollow watershed.

2.6.7 Zero-Intercept Regression

Many problems arise where the regression line would be expected to pass through the origin of the graph of y versus x; that is, the intercept coefficient of Equation 2–49 is 0. In such case, the linear model is

$$\hat{y} = bx \tag{2–99}$$

The least squares estimator of b can be found by substituting Equation 2–99 into Equation 2–52 and finding the normal equation, which is

$$b = \frac{\Sigma\, x_i y_i}{\Sigma\, x_i^2} \tag{2–100}$$

Example 2–13

The zero-intercept regression analysis can be illustrated using snowmelt runoff data for the problem of estimating the recession coefficient K for the recession model

$$Q_{t+1} = KQ_t \tag{2–101}$$

in which Q_t and Q_{t+1} are the streamflow rates for days t and $t + 1$, respectively.

The calibration of Equation 2–101 using Equation 2–100 can be illustrated using data for the Conejos River. Table 2–14 gives the streamflow for the period from June 8 to June 18, 1976. Using the streamflow for June 8 to predict streamflow for June 9 requires evaluation of K. The summations of Equation 2–100 were evaluated using the data of Table 2–14 and provide the following estimate of K:

$$K = \frac{\Sigma\, Q_t Q_{t+1}}{\Sigma\, Q_t^2} = \frac{17,197,050}{18,823,550} = 0.9136 \tag{2–102}$$

which yields the following model:

$$\hat{Q}_{t+1} = 0.9136 Q_t \tag{2–103}$$

TABLE 2–14 Snowmelt Runoff Recession Data for the Conejos River near Magote, Colorado, for June 8–18, 1976

Day	Streamflow Q_t (ft^3/sec)	Q_{t+1}	$\hat{Q}_{t+1}$	e
8	1870	1840	1708	−131.6
9	1840	1790	1681	−109.0
10	1790	1570	1635	65.3
11	1570	1240	1434	194.4
12	1240	1110	1133	22.9
13	1110	1055	1014	−40.9
14	1055	940	964	23.8
15	940	885	859	−26.2
16	885	850	809	−41.5
17	850	685	777	91.6
18	685	—		
				48.8

The predicted values of Q_{t+1} and the residuals (e) are also given in Table 2–14. Since the sum of the residuals does not equal 0, the zero-intercept model provides a biased estimate. For this example, the mean error is 4.88 ft^3/sec, which suggests that estimates made with Equation 2–103 tend to overestimate the value of Q_{t+1} by about 5 ft^3/sec. Thus, the prediction equation of Equation 2–103 has a bias of 0.4% of the mean of the measured values of Q_t; this bias is insignificant.

2.7 STEPWISE REGRESSION

Stepwise regression is a term that is applied to several algorithms for simultaneously selecting variables and calibrating a regression equation. The algorithms differ in the sequence of selecting variables. Forward regression starts with the predictor variable having the highest correlation with the criterion variable and continues adding variables so that the explained variance (Equation 2–69) is maximized at each step; a test of hypothesis is performed at each step and computation ends when all statistically significant predictor variables have been included. Backward regression, a second form of stepwise regression, begins with an equation that includes all of the predictor variables in the analysis and sequentially deletes variables, with the predictor variable contributing the least explained variance being deleted first. The third and fourth forms of stepwise regression are variations of the first two forms. Forward stepwise regression with deletion, which is the method discussed in this chapter, adds variables to maximize the total variation at each step; however, at each step those predictor variables that are already included in the equation are checked to ensure that they are still statistically significant. A variable found to be no longer significant is deleted. The fourth form of stepwise regression is backward regression with addition. This method is similar to the second form except that at each step all of the predictor variables that have been deleted are checked again for statistical significance; if any variable that was previously deleted is found to be significant, it is added back into the equation. All stepwise regression algorithms are based on the same statistical concepts.

2.7.1 Total *F* Test

The objective of a total F test is to determine whether or not the criterion variable is significantly related to the predictor variables that have been included in the equation. It is a test of significance of the following null (H_0) and alternative hypotheses (H_A):

$$H_0\colon \beta_1 = \beta_2 = \dots = \beta_q = 0 \tag{2–104a}$$

$$H_A\colon \text{at least one regression coefficient is significantly different from zero} \tag{2–104b}$$

in which q is the number of predictor variables included in the equation, and β_i (i = 1, 2, ..., q) are the population regression coefficients. The null hypothesis is tested using the following test statistic F:

$$F = \frac{R_q^2/q}{(1 - R_q^2)/(n - q - 1)} \tag{2–105}$$

in which R_q is the multiple correlation coefficient for the equation containing q predictor variables, and n is the number of observations on the criterion variable (for example, the sample size). The null hypothesis is accepted if F is less than or equal to the critical F value, F_α, which is defined by the selected level of significance α and the degrees of freedom $(q, n - q - 1)$ for the numerator and denominator, respectively. If the null hypothesis is accepted, one must conclude that the criterion variable is not related to any of the predictor variables that are included in the equation. If the null hypothesis is rejected, one or more of the q predictor variables are statistically related to the criterion variable; this does not imply that all of the predictor variables that are in the equation are necessary, only that at least one predictor variable is significant.

2.7.2 Partial *F* Test

The partial F test is used to test the significance of one predictor variable. It can be used to test the significance of either the last variable added to the equation or for deleting any one of the variables that is already in the equation; the second case is required to check whether or not the reliability of prediction will be improved if a variable is deleted from the equation. The null and alternative hypotheses are

$$H_0\colon \beta_k = 0 \tag{2–106a}$$

$$H_A\colon \beta_k \neq 0 \tag{2–106b}$$

where β_k is the regression coefficient for the predictor variable under consideration. The hypothesis is tested using the following test statistic:

$$F = \frac{\text{fraction increase in explained variation due to subject variable}/\nu_1}{\text{fraction of unexplained variation of the prediction equation}/\nu_2} \tag{2–107}$$

in which ν_1 and ν_2 are the degrees of freedom associated with the quantities in the numerator and denominator, respectively. In general, $\nu_1 = 1$ and $\nu_2 = n - q - 1$. For example, when selecting the first predictor variable, the partial F test statistic is

$$F = \frac{(R_1^2 - R_0^2)/1}{(1 - R_1^2)/(n - 2)} = \frac{R_1^2}{(1 - R_1^2)/(n - 2)} \tag{2–108}$$

in which R_1 and R_0 are the correlation coefficients between the criterion variable and the first predictor variable and no predictor variables, respectively, and n is the sample size. The test statistic, which can be used for either the case when a second predictor variable is being added to the equation or the case when the two variables already in an equation are being tested for deletion is

$$F = \frac{(R_2^2 - R_1^2)/1}{(1 - R_2^2)/(n - 3)} \tag{2–109}$$

in which R_1 and R_2 are the correlation coefficients between the criterion variable and a linear regression equation having one and two predictor variables, respectively. The null hypothesis (Equation 2–106a) is accepted if the test statistic F is less than or equal to the critical

F value, F_α. If $F < F_\alpha$, the predictor variable (X_k) that corresponds to β_k is not significantly related to the criterion variable when the other $(k - 1)$ predictor variables are in the equation; in other words, adding X_k to the prediction equation does not result in a significant increase in the explained variation. If the null hypothesis is rejected (for example, $F > F_\alpha$), the variable X_k makes a significant contribution toward the prediction accuracy that is significant even beyond the contribution made by the other $(k - 1)$ predictor variables. When the partial F test is used to check for the addition of a new predictor variable, a significant F value (for example, $F > F_\alpha$) indicates that the variable should be added. When the partial F test is used to check for deletion of a predictor variable, a significant F value (for example, $F > F_\alpha$) indicates that the variable should not be dropped from the equation. Partial F statistics are computed for every variable at each step of a stepwise regression.

2.7.3 Procedure

The forward stepwise regression with deletion procedure to be illustrated herein calibrates an equation by step-forward insertion, with predictor variables deleted when appropriate. There are three basic elements: (1) partial F tests for insertion, (2) partial F tests for deletion, and (3) total F tests. The general procedure is outlined in Figure 2–12 and discussed in the following paragraphs.

Partial *F* test for insertion (Step 1). The partial F values for all predictor variables that are not included in the equation should be computed. The variable with the largest partial F should be selected to enter the equation. On the first step, the partial correlations equal the predictor-criterion correlations, so the predictor variable that has the largest predictor-criterion correlation is added first. The partial F test for insertion is made:

(a) If $F < F_\alpha$, the variable is not significant and the equation from the previous iteration is the final model (if this option occurs with the first predictor variable tested, then the criterion variable is not significantly related to any of the predictor variables).

(b) If $F > F_\alpha$, the variable is statistically significant and the variable should be included in the equation; the process should proceed to the total F test (Step 2).

Total *F* test (Step 2). An equation that includes all predictor variables that have been inserted and not subsequently deleted is calibrated. The total F value is calculated and compared with the critical value F_α.

(a) If $F < F_\alpha$, the equation calibrated for the previous iteration should be used as the final model.

(b) If $F > F_\alpha$, the entire model is significant, and all of the predictor variables that are in the equation should be tested for significance; control should then proceed to Step 3.

Partial *F* test for deletion (Step 3). Partial F values for all predictor variables that are included in the equation are computed. The variable that has the smallest F value is compared with the critical F value, F_α.

(a) If $F < F_\alpha$, the predictor variable is not significant and it is deleted from the equation; control then passes to the total F test (Step 2).

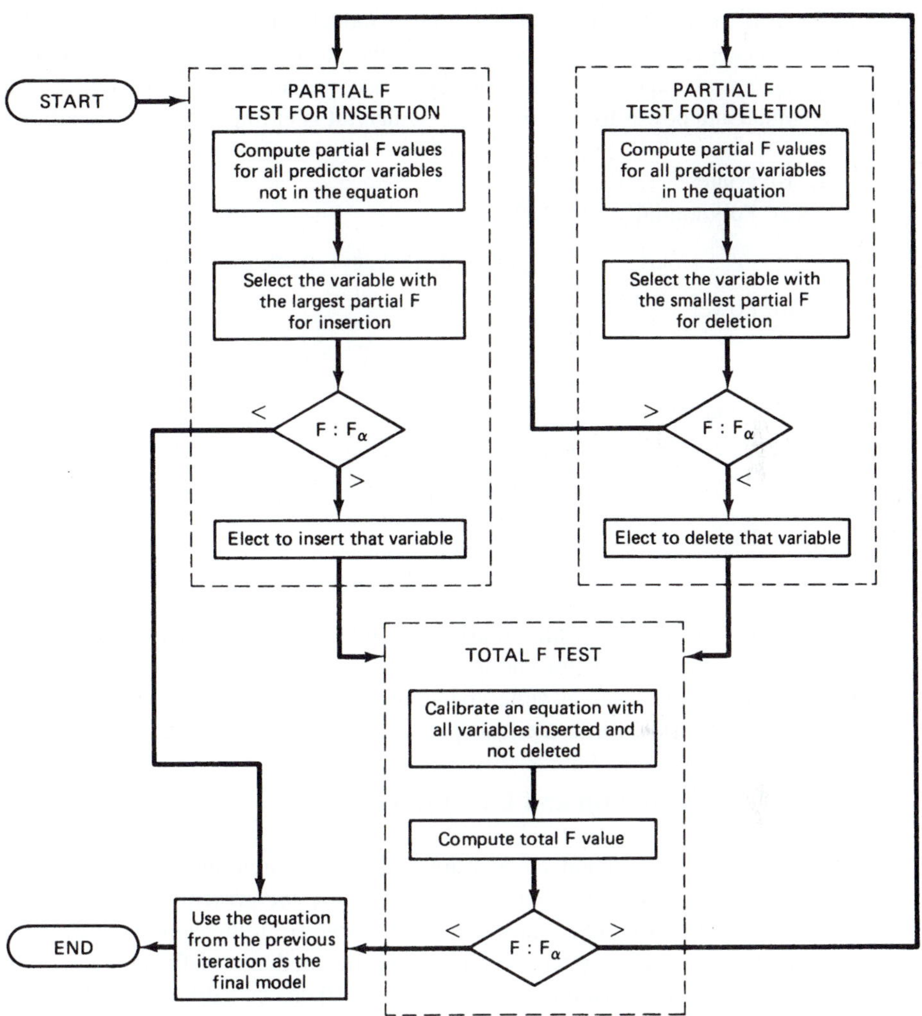

FIGURE 2–12 Flowchart for stepwise regression procedure.

(b) If $F > F_\alpha$, all predictor variables included in the equation are significant; control passes to the partial F tests for insertion (Step 1).

2.8 ANALYSIS OF NONLINEAR EQUATIONS

Linear models were separated into bivariate and multivariate; the same separation is applicable to nonlinear models. It is also necessary to separate nonlinear models on the basis of the functional form. Although polynomial and power models are the most frequently used nonlinear forms, it is important to recognize that other model structures are available and may actually be more accurate in structure. In addition to the power and polynomial structures, forms such as a square root, exponential, or logarithmic may provide the best fit to a set of data. Since the polynomial and power forms are used so widely, it is of importance to identify their structure:

1. Bivariate
(a) Polynomial

$$\hat{y} = b_0 + b_1 x + b_2 x^2 + \cdots + b_p x^p \tag{2–110}$$

(b) Power

$$\hat{y}_0 = b_0 x^{b_1} \tag{2–111}$$

2. Multivariate
(a) Polynomial

$$\hat{y} = b_0 + b_1 x_1 + b_2 x_2 + b_3 x_1^2 + b_4 x_2^2 + b_5 x_1 x_2 + \cdots \tag{2–112}$$

(b) Power

$$\hat{y} = b_0 x_1^{b_1} x_2^{b_2} \cdots x_p^{b_p} \tag{2–113}$$

In these relationships, y is the criterion variable, x is the predictor variable in the bivariate case and x_i is the ith predictor variable in the multivariate case, $b_j (j = 0, 1, ..., p)$ is the jth regression coefficient, and p is either the number of predictor variables or the order of the polynomial. The bivariate forms are just special cases of the multivariate form.

2.8.1 Transformation and Calibration

The power and polynomial models are widely used nonlinear forms because they can be transformed in a way that makes it possible to use the principle of least squares. Although the transformation to a linear structure is desirable from the standpoint of calibration, it has important consequences in terms of assessing the goodness-of-fit statistics.

The bivariate polynomial of Equation 2–110 can be calibrated by forming a new set of predictor variables:

$$w_i = x^i \quad (i = 1, 2, \cdots, p) \tag{2–114}$$

This results in the model

$$\hat{y} = b_0 + b_1 w_1 + b_2 w_2 + \cdots + b_p w_p \tag{2–115}$$

The coefficients b_j can be estimated using a standard multiple regression analysis.

The bivariate power model of Equation 2–111 can be calibrated by forming the following set of variables:

$$z = \log y \tag{2–116}$$

$$c = \log b_0 \tag{2–117}$$

$$w = \log x \tag{2–118}$$

These transformed variables form the following linear equation, which can be calibrated using bivariate linear regression analysis:

$$z = c + b_1 w \tag{2–119}$$

After values of c and b_1 are obtained, the coefficient b_0 can be determined using

$$b_0 = 10^c \tag{2–120}$$

Natural logarithms can also be used; in such a case, Equations 2–116 to 2–118 will use the base e logarithm and Equation 2–120 will use a base e rather than 10.

The multivariate polynomial of Equation 2–112 can also be solved using a multiple regression analysis for a set of transformed variables. For the model given by Equation 2–112, the predictor variables are transformed as follows:

$$w_1 = x_1 \tag{2–121}$$

$$w_2 = x_2 \tag{2–122}$$

$$w_3 = x_1^2 \tag{2–123}$$

$$w_4 = x_2^2 \tag{2–124}$$

$$w_5 = x_1 x_2 \tag{2–125}$$

The revised model has the form

$$\hat{y} = b_0 + \sum_{i=1}^{5} b_i w_i \tag{2–126}$$

It is important to note that the polynomial models do not require a transformation of the criterion variable.

The model given by Equation 2–112 has only two predictor variables (for example, x_1 and x_2) and is a second-order equation. In practice, a model may have more predictor variables and may be of higher order. In some cases the interaction terms (for example, $x_i x_j$) may be omitted to decrease the number of coefficients that must be calibrated. However, if the interaction terms are omitted when they are actually significant, the goodness-of-fit statistics may suffer unless the variation is explained by the other terms in the model; when this occurs, the coefficients may lack physical significance.

The multivariate power model of Equation 2–113 can be evaluated by making a logarithmic transformation of both the criterion and the predictor variables:

$$z = \log y \tag{2–127}$$

$$c = \log b_0 \tag{2–128}$$

$$w_i = \log x_i \tag{2–129}$$

The resulting model has the form

$$z = c + \sum_{i=1}^{p} b_i w_i \tag{2–130}$$

The coefficients of Equation 2–130 can be evaluated using a multiple regression analysis. The value of b_0 can be determined by making the transformation of Equation 2–120. Again, it is possible to use a base e transformation rather than a base 10 log transformation.

PROBLEMS

2–1. A farmer maintained records of the number (N) of times per year that flow in a specific irrigation canal exceeded its capacity. Based on the concept of relative frequency, what is the probability that in any one year the capacity will be exceeded (a) three times (b) at least five times?

1980	2	1984	4	1988	0	1992	5
1981	6	1985	9	1989	2	1993	7
1982	7	1986	3	1990	8	1994	4
1983	4	1987	7	1991	3	1995	1

2–2. The number (N) of floods expected to occur during the project life of 50 years is as follows: $p(0 \text{ floods}) = 0.2$, $P(1) = 0.25$, $P(2) = 0.2$, $P(3) = 0.15$, $P(4) = 0.1$, $P(5) = 0.05$, $P(6) = 0.03$, $P(7) = 0.02$. Graph the mass and cumulative mass curves. Based on the expected frequency concept, what are the following probabilities? (a) $P(N > 3)$; (b) $P(N < 3)$; (c) $P(2 \leq N < 5)$.

2–3. The probability that a roadway will be flooded i times in any one year is given by the following mass function:

$$P(i) = \begin{cases} k & \text{for } i = 0 \\ k/i & \text{for } i = \; = 1, 2, \ldots, 5 \\ 0 & \text{otherwise} \end{cases}$$

Find the value of k that makes this a legitimate probability mass function. Plot the mass and cumulative functions and find the following probabilities: (a) $P(2 \text{ floods})$; (b) $P(i \leq 3)$; (c) $P(i > 1)$; (d) $P(2 \leq i < 4)$.

2–4. The distribution of floods (Q) on two watersheds are given in the following figure:

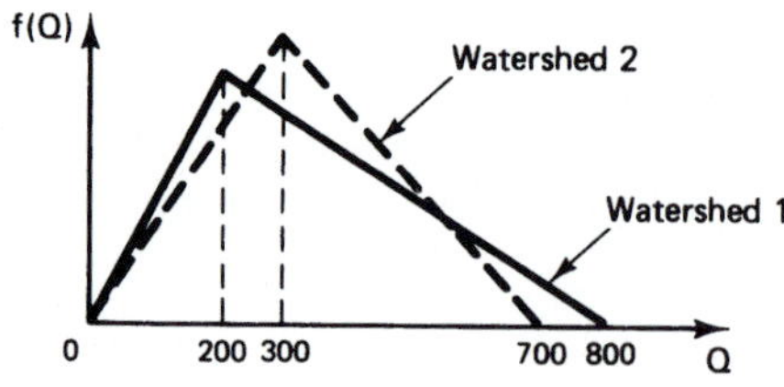

For watershed 1 find the following probabilities: (a) $P(Q > 300)$; (b) $P(Q < 400)$; (c) $P(Q = 200)$; (d) $P(100 < Q < 450)$.

2–5. For the distribution of floods for watershed 2 shown in Problem 2–4 find the following probabilities: (a) $P(Q < 200)$; (b) $P(Q > 600)$; (c) $P(150 < Q < 500)$.

2–6. Find the mean of the distribution of floods for watershed 1 of Problem 2–4.

2–7. Find the mean of the distribution of floods for watershed 2 of Problem 2–4.

2–8. Measurements of rainfall *P* (in.) on seven consecutive days are: 2.1, 0.0, 1.2, 0.3, 0.0, 0.5, 0.8. Find the sample mean, standard deviation, and standardized skew.

2–9. The bifurcation ratio of ten watersheds in a region are as follows: 2.3, 3.6, 2.8, 3.6, 3.4, 2.8, 2.6, 3.3, 3.2, 2.4. Find the mean and standard deviation.

2–10. Annual measurements of sediment yield in tons/acre for a given watershed are as follows: 12.8, 8.5, 16.5, 14.2, 10.7, 20.3, 9.6, 11.4. Find the mean, standard deviation, and standardized skew.

2–11. Compute the mean and standard deviation of the March runoff for the data of Table 2–14.

2–12. Compute the mean and standard deviation of the March precipitation for the data of Table 2–13.

2–13. If a flood of a specified magnitude has a probability of 0.02 of occurring in any one year, find the following probabilities: (a) The flood occurring in two consecutive years; (b) the flood not occurring in a 5-yr period; (c) the flood occurring two or more times in a 10-yr period.

2–14. If a levee is designed to control floods with an annual exceedence probability of 0.05, find the following probabilities: (a) The levee will not be overtopped in any one year; (b) the levee will not be overtopped during a 2-yr period; (c) the levee will be overtopped exactly once in a 6-yr period; (d) the levee will be overtopped three or more times in a 10-yr period.

2–15. If we define a random variable to have two possible outcomes, March runoff 4 in. or above and March runoff less than 4 in., use the data of Table 2–13 to estimate the probability of the event that March runoff is 4 in. or greater. Find the probabilities: (a) In the next 5 years, at least 3 of the years will have a March runoff that is 4 in. or greater; (b) exactly 2 of the next 10 years will have a March runoff of 4 in. or more; (c) in the next 6 years, no March runoff will be 4 in. or more; (d) in the next 8 years, the runoff in March for at least 6 of the years will be less than 4 in.

2–16. If we define a random variable to have two possible outcomes, the monthly rainfall in June, July, or August to be less than 1 in. and the monthly rainfall to be 1 in. or greater, use the data of Table 2–8 to estimate the probabilities of the two events. Find the following probabilities: (a) Exactly 2 months in the months of June, July, or August in the next 5 years will have less than 1 in. of rainfall; (b) in the next 2 years, there will be no monthly rainfall during June, July, or August that is less than 1 in.; (c) in any one year, all monthly rainfalls during June, July, and August are less than 1 in.; (d) in any 7-yr period, there will be no more than two months in June, July, and August where the rainfall is less than 1 in.

2–17. Assume that runoff into a storm drain inlet follows a normal distribution with a mean of 12 ft^3/sec and a standard deviation of 3 ft^3/sec. What is the probability that ponding will occur at the inlet if the inlet can accommodate flows of 20 ft^3/sec?

2–18. Assume that evaporation *E* rates from an open water surface during June follow a normal distribution with a mean of 0.20 in./day and a standard deviation of 0.04 in./day. Find the following probabilities: (a) $P(E > 0.25$ in./day); (b) $P(E < 0.10$ in./day): (c) $P(0.15 < E < 0.25$ in./day).

2–19. Based on the sediment yield data of Problem 2–10 and assuming that the sediment yield (*Y*) follows a normal distribution, find the following probabilities: (a) $P(Y > 22$ tons/acre/yr); (b) $P(Y < 8$ tons/acre/yr); (c) $P(10 < Y < 20$ tons/acre/yr).

2–20. Find the *F* value corresponding to a probability of 5% for (a) $\nu_1 = 6$, $\nu_2 = 10$; (b) $\nu_1 = 3$, $\nu_2 = 7$.

2–21. Find the *F* value corresponding to a probability of 1% for (a) $\nu_1 = 1$, $\nu_2 = 6$; (b) $\nu_1 = 4$, $\nu_2 = 12$; (c) $\nu_1 = \nu_2 = 5$.

2–22. A number of studies have suggested that the average bifurcation ratio r_b is 3.5. Would it be safe to conclude from the data of Problem 2–9 that the region in which the ten watersheds are located also has a mean r_b of 3.5 if a level of significance of 5% is used? Assume the standard deviation of the population is 0.5.

2–23. From the measurements of Problem 2–10, could one conclude that the mean annual sediment yield is greater than 10 tons/acre? Use a 1% level of significance and assume that the population has a standard deviation of 5 tons/acre.

2–24. For the eighteen values of June precipitation in Table 2–8, test the following hypotheses using a 1% level of significance: H_0: $\mu = 4$ in., H_A: $\mu < 4$ in.

2–25. For the eighteen values of August precipitation in Table 2–8, test the following hypotheses using a 5% level of significance: H_0: $\mu = 4$ in., H_A: $\mu \neq 4$ in.

2–26. For the eighteen values of July precipitation in Table 2–8 and the hypotheses H_0: $\mu = \mu_0$; H_A: $\mu > \mu_0$, find the value of μ_0 at which the null hypothesis must be rejected for levels of significance of 10, 5, and 1%.

2–27. For the runoff (RO_m) data in Table 2–13, make separate tests of the following hypotheses: (a) H_0: $\mu = 3.0$ in., H_A: $\mu \neq 3.0$ in.; (b) H_0: $\mu = 2.75$ in., H_A: $\mu \neq 2.75$ in. Compare the results and discuss the implications. Use a 5% level of significance.

2–28. For the June precipitation data of Table 2–8, make separate tests of the following hypotheses: (a) H_0: $\mu = 2.5$ in., H_A: $\mu \neq 2.5$ in.; (b) H_0: $\mu = 5.0$ in., H_A: $\mu \neq 5.0$ in. Compare the results and discuss the implications. Use a 1% level of significance.

2–29. For the data of Table 2–8 test the two-sided hypothesis that the means for June and August are different. Use a 10% level of significance.

2–30. For the data of Table 2–8, test the two-sided hypothesis that the means for June and July are different. Use a 10% level of significance.

2–31. Test the one-sided hypothesis that the variance of the March precipitation (Table 2–13) is less than 6 in.^{2} Use a 1% level of significance.

2–32. Test the one-sided hypothesis that the variance of the March runoff (Table 2–13) is greater than 1 in.^{2} Use a 1% level of significance.

2–33. Test the two-sided hypothesis that the variance of the June precipitation (Table 2–8) is not equal to 9 in.^{2} Use a 5% level of significance.

2–34. Using a 1% level of significance, test the one-sided hypothesis that the variance of the July precipitation (Table 2–8) is greater than 1 in.^{2}

2–35. Using the forty-eight values of precipitation for February and March (Table 2–13), test the hypothesis that the values are normally distributed with $\mu = 5$ in. and $\sigma = 2$ in. Use a 5% level of significance.

2–36. For the thirty-six values of precipitation for June and July (Table 2–8), test the hypothesis that the values are normally distributed with $\mu = 4$ in. and $\sigma = 1.75$ in. Use a 5% level of significance.

For Problems 2–37 to 2–39, evaluate the linear model $\hat{y} = b_0 + b_1 x$ *using the principle of least squares and compute the correlation coefficient (R), the proportion of explained variance* (R^2)*, the standard error of estimate* (S_e)*, and the ratio of* S_e *to the standard deviation of the variable* y*. Discuss the reliability of the model.*

2–37. Let y and x be the rainfall intensity (in./hr) and the storm duration (min), respectively.

y	9.5	8.8	8.7	8.2	6.2	5.2
x	6	10	12	15	23	30

2–38. Let y and x be the peak discharge (ft^3/sec) and the drainage area (acres), respectively:

y	21	24	52	41	43	71	63
x	6	10	13	17	23	26	28

2–39. Let y and x be the depression storage (in.) and imperviousness (%), respectively:

y	1.2	0.7	0.4	0.5	0.3	0.3	0.2
x	2	5	21	32	41	49	68

For Problems 2–40 to 2–42, evaluate the linear multiple regression model with two or three predictor variables. Compute the multiple correlation coefficient, the standard error of estimate, and the standard deviation of the variable y. Discuss the reliability of the model.

2–40. Let y, x_1, x_2, and x_3 be the time of concentration (t_c in hr), watershed length (L in ft), watershed slope (S in %), and runoff curve number (CN):

t_c	0.2	0.2	0.2	0.3	0.4
L	800	1200	2100	2000	1500
S	2	3	4	6	1
CN	75	84	88	70	85

2–41. Let y, x_1, and x_2 be the evaporation rate (E in in./day), the mean daily temperature (T in °F), and the relative humidity (H in %):

E	0.11	0.12	0.18	0.19	0.13	0.18
T	55	64	72	83	68	77
H	60	74	50	66	82	48

2–42. Let y, x_1, and x_2 be the peak discharge (q_p in ft^3/sec), drainage area (A in acres), and the rainfall intensity (i in in./hr), respectively:

q_p	23	45	44	64	68	62
i	3.2	4.6	5.1	3.8	6.1	7.4
A	12	21	18	32	24	16

2–43. Using the data of Problem 2–41, fit a multiple-variable power model for predicting E. Compute the correlation coefficient, standard error of estimate, and the standard deviation of E.

2–44. Show the calculations for deriving Equation 2–55 from Equation 2–54.

2–45. Show the development of Equation 2–71 from Equations 2–67 to 2–70.

REVIEW QUESTIONS

2–1. Which one of the following responses is not true? Probability is a scale of measurement (a) that can be expressed as a percentage; (b) that can be used only with discrete random variables; (c) that ranges from 0 to 1; (d) that provides a basis for making decisions under conditions of uncertainty; (e) that is used to describe the likelihood of an event.

2–2. A mass function is used with (a) a cumulative density function; (b) continuous random variables; (c) probability density functions; (d) discrete random variables.

2–3. Which one of the following responses is not true? For a probability mass function the probability of an event or probabilities of all possible outcomes must (a) be less than or equal to 1; (b) sum to 1; (c) equal an integer value; (d) be greater than or equal to 0.

2–4. Assume that the discrete random variable X can take on values of 2, 4, 6, and 8, with probabilities of 0.3, 0.2, 0.4, and 0.1, respectively. At $X = 4$, the value of the cumulative mass function is (a) 0.0; (b) 0.2; (c) 0.5; (d) 0.9.

2–5. Which one of the following is a discrete random variable? (a) The volume of infiltration; (b) the percentage of urban land use in a watershed; (c) the daily rate of evaporation; (d) the number of floods in a period of ten years.

2–6. From Figure 2–1c, the probability of three or fewer floods in a period of ten years is approximately (a) 0.04; (b) 0.12; (c) 0.17; (d) 0.38.

2–7. Which one of the following responses is not true? For a continuous random variable X (a) $P(-\infty \le X \le \infty) = 1$; (b) $P(X \le X_1) < P(X \le X_2)$ when $X_1 \le X_2$; (c) $P(X = X_0) = 0$; (d) $\int_{-\infty}^{\infty} f(x)\,dx = 1$.

2–8. If the density function $f(i)$ of the infiltration rate i is 2.5 for i from 0 to 0.4 in./hr and 0 otherwise, then the probability that i is between 0.15 and 0.3 in./hr is (a) 0.15; (b) 0.30; (c) 0.375; (d) 0.4; (e) none of the above.

2–9. The skew of a distribution of measured peak flood flow rates would reflect (a) the symmetry of large and small flood flows about the mean; (b) the closeness of the flood peaks to the mean; (c) the uniformity of the distribution of flood peaks; (d) the magnitude of the mean flow; (e) none of the above.

2–10. Which one of the following responses is not an assumption of the binomial probability function? (a) There are only two trials per experiment; (b) the trials are independent; (c) each trial has only two possible outcomes; (d) for each trial the probability of each outcome is constant.

2–11. How many different sequences involving outcomes X and Y are possible in five trials? (a) 8; (b) 20; (c) 32; (d) 64.

2–12. For a normal distribution, what percent of outcomes will be within one standard deviation of the mean? (a) 34; (b) 68; (c) 95; (d) 99.

2–13. The scale parameter for a normal distribution is (a) μ; (b) σ; (c) σ^2; (d) $\sigma/\sqrt{n}$.

2–14. Which of the following is not a decision parameter when performing a statistical test of a hypothesis? (a) The null hypothesis; (b) the sample size; (c) the level of significance; (d) the power of the test.

2–15. For the chi-square goodness-of-fit test, the degrees of freedom are (a) $n - 1$; (b) $k - 1$; (c) χ^2; (d) α; (e) none of the above.

2–16. For a multiple regression analysis, which one of the following is a measure of reliability? (a) n; (b) R^2; (c) S_e; (d) $\sum e^2$; (e) t_i; (f) all of the above.

2–17. The number of normal equations in a multiple regression analysis equals: (a) the number of unknowns; (b) the sample size, n; (c) $n + 1$; (d) the number of predictor variables, p; (e) $p + 1$.

2–18. Which one of the following responses is not a reason for graphical analysis? (a) Identify the presence of extreme events; (b) identify the form and type of the relationship; (c) identify the range of the sample data; (d) identify the existence of a causal relationship.

2–19. Both correlation and bivariate linear regression analysis (a) provide a means of predicting the value of a variable; (b) are independent of scale; (c) are transformable; (d) assume a linear separation of variation.

2–20. Which one of the following is not true? The standard error of estimate (a) may be greater than S_y; (b) is a function of the sample size; (c) cannot equal 0; (d) is a function of the error variation.

2–21. The total F test is a ratio of the explained variation to the (a) total variation; (b) unexplained variation; (c) variation of prediction; (d) none of the above.

2–22. The numerator of the partial F statistic is a function of (a) the change in the coefficient of multiple determination; (b) the change in the explained variation; (c) the change in the correlation coefficient; (d) all of the above.

2–23. The correlation coefficient is not applicable to regression analyses with the power model structure because (a) there may be non-zero correlation between the predictor variables; (b) the logarithms of the predictor variables are used; (c) there is more than one predictor variable in the equation; (d) a nonlinear transformation of the criterion variable is made in calibration of the coefficients; (e) none of the above.

2–24. Polynomial models often yield irrational coefficients because (a) terms such as x and x^2 are highly correlated; (b) the polynomial model is nonlinear; (c) the predictor-variable terms must be replaced by new variables such as $w_2 = x^2$ and $w_3 = x^3$; (d) none of the above.

DISCUSSION QUESTION

The technical content of this chapter is important to the professional hydrologist, but practice is not confined to making technical decisions. The intent of this discussion question is to show that hydrologists must often address situations where value issues intermingle with the technical aspects of a project. In discussing the stated problem, at a minimum include responses to the following questions:

1. What value issues are involved, and how are they in conflict?
2. Are technical issues involved with the value issues? If so, how are they in conflict with the value issues?
3. If the hydrologist attempted to rationalize the situation, what rationalizations might he or she use? Provide arguments to suggest why the excuses represent rationalization.
4. What are the hydrologist's alternative courses of action? Identify all alternatives, regardless of the ethical implications.
5. How should the conflict be resolved?

You may want to review Sections 1.6 to 1.12 in Chapter 1 in responding to the problem statement.

Case. A research hydrologist makes a statistical analysis of data collected as part of a research program. Part of the analysis involves a hypothesis test. The test leads to the decision to reject the research null hypothesis that the researcher wanted and expected to be accepted. In reviewing the statistical test, the researcher recognizes that the null hypothesis would have been accepted if a more stringent level of significance had been used. While the commonly used level of significance was initially employed, the researcher reports the results using the more stringent level of significance so that the finding supports the initial expectations.

3

Watershed Characteristics

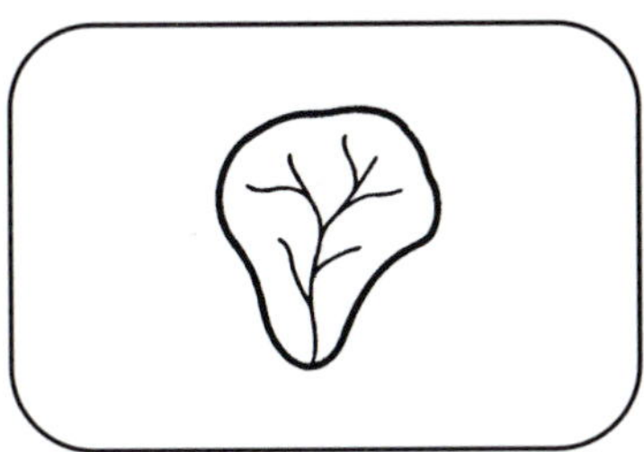

CHAPTER OBJECTIVES

1. Define the concept of a watershed.
2. Discuss quantification of watershed, soil, and channel characteristics.
3. Provide an understanding of the importance of watershed characteristics in controlling flood runoff.
4. Discuss simplifying assumptions made in characterizing flood runoff.
5. Characterize the timing of flood runoff.

3.0 NOTATION

a = area of watershed below elevation Δe
A = cross-sectional area of channel
A = drainage area
A_i = average drainage area of streams of order i
A_0 = area of a circle with a perimeter equal to the watershed perimeter
A_1 = average drainage area of first-order streams
C = Chezy coefficient
C = runoff coefficient
C_f = proportionality constant
C_1 = aquifer constant
C_2 = aquifer constant
CN = runoff curve number
CN_p = pervious area CN
CN_w = weighted curve number
d = depth of flow in channel
d = mean soil particle diameter
D = drainage density
D_m = maximum deviation of hypsometric curve from the 45° line
$E_{10\text{–}85}$ = difference in elevation between 10 and 85 points
f = fraction of impervious cover
F_c = circularity ratio
F_f = friction force
F_p = profile factor based on hypsometric curve
F_1, F_2 = hydrostatic pressure forces on end of control volume
H_a = hypsometric area ratio
i = rainfall intensity
I = percentage of imperviousness
k = coefficient of velocity-slope relationship
k = variable used in estimating channel slope to weight the slopes of subreaches
l_i = length of subreach i
l_m = total meander length
l_s = straight-line length of stream
L = watershed length
L_c = channel length
L_d = length of line for hypsometric index
L_f = length of main sewer
L_i = average length of streams of order i
L_i = length of subreach i
L_l = shape factor
L_m = length of longest channel section
L_m = maximum length of basin parallel to principal drainage line
L_m = total basin length
L_t = total length of drainageways and storm sewers
L_t = total length of streams
L_{ca} = length to the center of area of a watershed (measured along channel)
L_1 = average length of first-order streams
$L_{10\text{–}85}$ = length of channel within 10% and 85% points
n = Manning's roughness coefficient
n = number of segments into which a channel was segmented
n_1 = basic coefficient for roughness
n_2 = correction for channel irregularity
n_3 = correction for cross-section variation
n_4 = correction for obstructions
n_5 = correction for vegetation
n_6 = correction for meandering
N_i = number of streams of order i
P = watershed perimeter
P = wetted perimeter of channel cross-section
P_e = specific yield
P_v = porosity of soil mass
P_2 = 2-year, 24-hr rainfall (inches)
q_p = peak discharge
Q = channel discharge rate
r_a = stream area ratio

r_L = stream length ratio
r_s = stream slope ratio
R_b = bifurcation ratio
R_c = circularity ratio
R_e = elongation ratio
R_h = hydraulic radius
R_s = specific retention
S = average slope
S = friction slope
S = slope of overland flow length
S_c = channel slope
S_f = slope of main sewer
S_i = average slope of streams of order i
S_i = channel slope index
S_m = slope of longest channel segment
S_s = specific storage
S_0 = slope of channel bed
S_1 = average slope of first-order streams
S_{10-85} = channel slope between 10% and 85% points
t_c = time of concentration
T_t = travel time
V = channel velocity
V = volume of soil mass
V_g = volume of water drained by gravity
V_i = velocity of overland flow, section i
V_r = volume of water retained by surface tension
V_w = volume of water in saturated soil mass
w = bottom width of channel
W = weight of soil mass
W = weight of water in control volume
Δe_c = elevation difference between channel end points
Δe_i = difference in elevation of ends of subreach i
ΔE = difference in elevation of upper and lower ends of channel or watershed
Δx = length of incremental channel section
ΔX = length of channel section
γ = specific weight of water
ρ_b = bulk density
τ = shear stress of water on channel bed
θ = channel slope
ϕ = Espey-Winslow channelization factor

3.1 INTRODUCTION

Every hydrologic design is different because the factors that affect the design vary with location. It is necessary to make measurements at the design site of factors that affect the design. The factors to be measured characterize the processes of the hydrologic cycle that are important to the design problem. Factors such as the following may be important: the size, slope, and land use of the watershed, as well as the amount of storage and vegetation within the channel. Given the importance of such factors as input to a hydrologic design, the accuracy with which the measurements of these factors are made will partly determine the accuracy of a design.

The objective of this chapter is to define some of the more important watershed characteristics and discuss alternative methods for representing each characteristic. Watershed characteristics that are used as inputs to hydrologic design methods include the drainage area, linear measurements such as the watershed or channel length, the shape of the watershed, the slope of the watershed or channel, the drainage pattern, channel roughness and cross-sectional properties, time of flow parameters, and the land cover. The soil type is also important.

3.1.1 Analysis versus Synthesis

In hydrologic analysis, the watershed characteristics are used in defining the nature of the transfer function (Figure 1–3). In hydrologic synthesis or design, the characteristics are measured at the design site in order to define the transfer function that is necessary to compute the output function. When using a hydrologic model for design, the watershed characteristics that are required as input to the design method should be similar in magnitude to the values that were used in development of the design method (i.e., the analysis phase). When the values of the characteristics used for design are much different from those used in developing the design method, the design represents an extrapolation of the design method and may not be accurate because of the extrapolation. Therefore, it is important for the design engineer to be knowledgeable about the development of the design method and not just be able to make the measurements of the input characteristics of the design method.

As an example, consider the problem of estimating the maximum flood flow, which is the peak discharge q_p of Chapter 7. Many investigators have developed simple prediction equations that relate the peak discharge to the drainage area A. If, for example, the parameter C in the equation $q_p = CiA$ is based on watersheds for drainage areas from 1 to 200 acres, the equation may not provide accurate design estimates of q_p for drainage areas outside this range. In this case, the use of the equation for synthesis or design should be based on knowledge of the values of the watershed characteristics used in the analysis phase.

3.2 WATERSHED: DEFINITION AND DELINEATION

The concept of a watershed is basic to all hydrologic designs. Since large watersheds are made up of many smaller watersheds, it is necessary to define the watershed in terms of a point; this point is usually the location at which the design is being made and is referred to as the watershed "outlet." With respect to the outlet, the watershed consists of all land area that sheds water to the outlet during a rainstorm. Using the concept that "water runs downhill," a watershed is defined by all points enclosed within an area from which rain falling at these points will contribute water to the outlet. This is best shown pictorially, as in Figure 3–1. The shaded area of Figure 3–1 represents the watershed for the outlet at point A. Water is contributed to the outlet from many smaller areas, which are also watersheds. For example, if a design were being made at point B rather than point A, the watershed would be the small area enclosed within the dashed lines. The watershed for point B is made up of smaller watersheds, with the two stream tributaries reflecting the collecting areas for water resulting from rain on the watershed. The concept of smaller watersheds making bigger watersheds can be carried to the extreme. The watershed of the Mississippi River drains almost fourteen million square miles, and the continental divide, which runs from northern Montana to southern New Mexico, divides the continuous part of the United States into two watersheds, with one watershed directing runoff to the Pacific Ocean and the other watershed to the Atlantic Ocean.

Given this definition of a watershed and assuming that the delineation of a watershed is important to hydrologic design, it is necessary to show how the boundary of a watershed is delineated. Keeping in mind that the boundary of a watershed is defined by all points that will shed water to the outlet, it is only necessary to decide which points in a region will con-

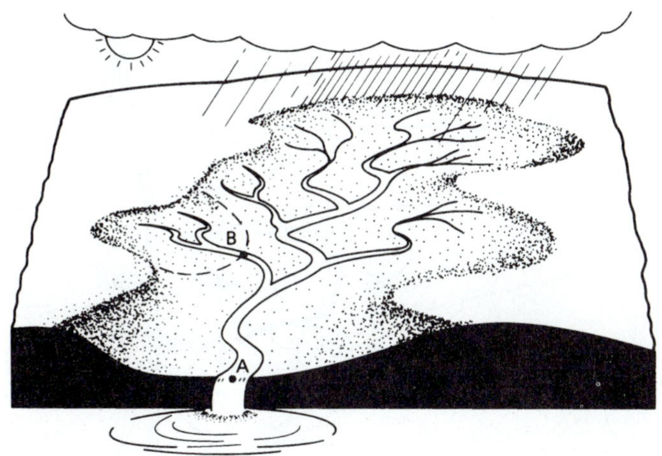

FIGURE 3–1 Delineation of watershed boundary.

tribute water to the outlet; the most extreme of these points represents the watershed boundary.

Consider the hypothetical schematic of a topographic map that is shown in Figure 3–2. If the point of design is at *A*, which is the outlet of the watershed and the downstream end of reach *AB*, then all points bounded by the elevation contours *CDEF* would contribute runoff to point *A*. Water will travel perpendicular to the elevation contours, which is the direction that maximizes the slope. Rain falling at points *C* and *F* would travel overland directly to point *A*. Rain falling at points *D* and *E* would travel toward the upper end of the channel (i.e., point B), and from there the water would travel in the channel to point *A*. Rain falling at points *G*, *H*, and *K* would not flow toward point *A* because the elevations of these points are less than that of the watershed divide, which has an elevation of 70; thus the rain falling at these three points would travel to the outlet of other watersheds.

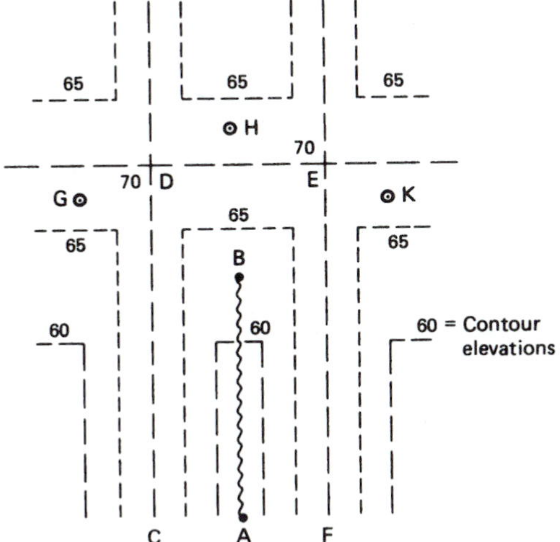

FIGURE 3–2 Delineation of hypothetical watershed.

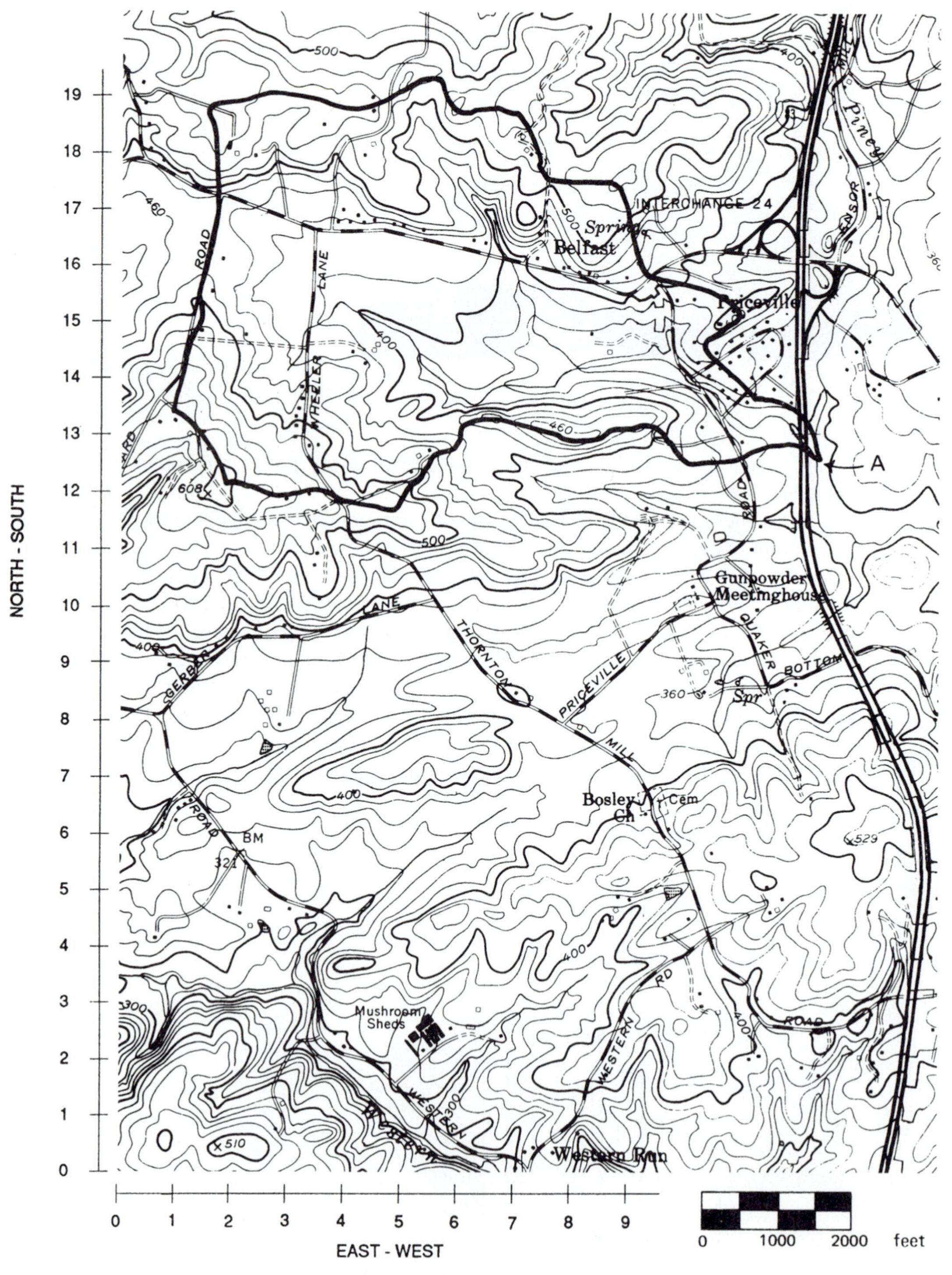

FIGURE 3–3(a) Topographic map.

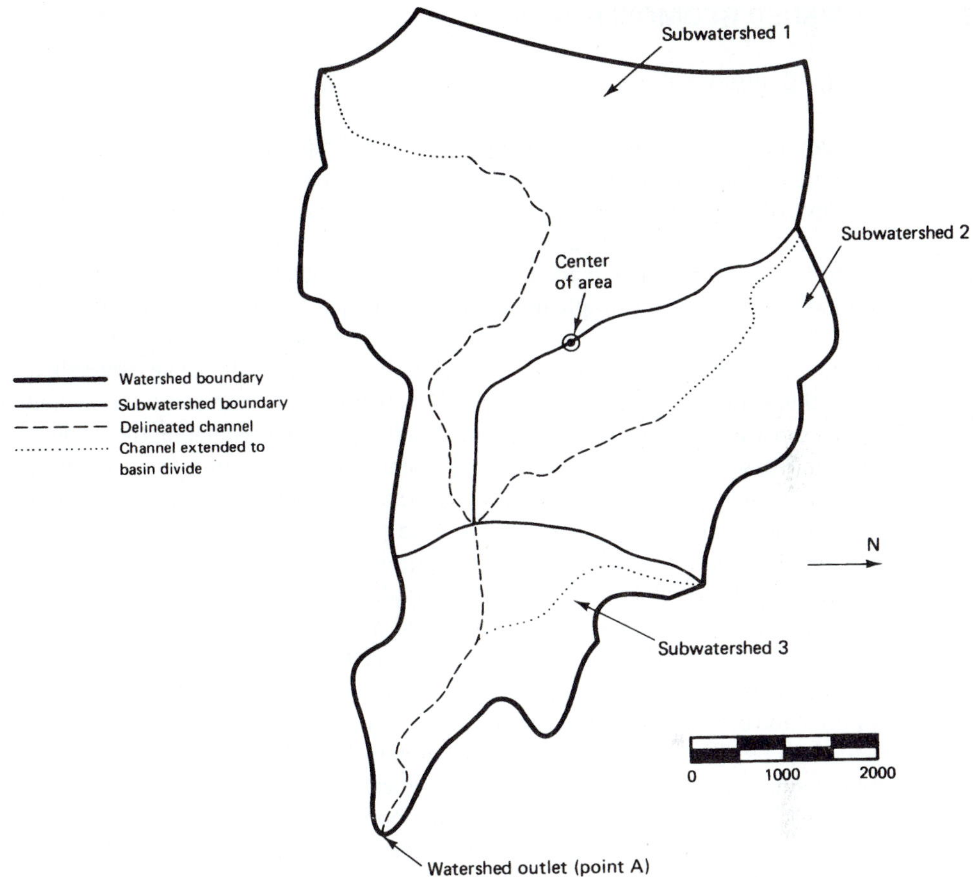

FIGURE 3–3(b) Delineation of watershed boundary.

As indicated previously, the watershed boundary is defined by identifying all points within an area from which rain will contribute water to the outlet. The hypothetical watershed of Figure 3–2 suggests that watersheds have regular geometric shapes such as rectangles. This is not generally the case. Figure 3–3a shows a topographic map of an area. If one were interested in making a design at point *A*, the watershed boundary would be delineated by drawing lines perpendicular to the elevation contour lines for land that drains to point *A*. The resulting watershed boundary is shown in Figure 3–3b. Approximately 3600 feet upstream of point *A* the channel divides, with one stream channel coming from the northwest and the other from a direction slightly south of west. Using the confluence of the two tributaries as a point, the subwatersheds that drain the tributaries can be delineated, as shown in Figure 3–3b. Thus the watershed can be viewed as consisting of three subwatersheds. Subdividing a watershed is often necessary for modeling, such as when only a portion of the watershed is being developed and stormwater management is necessary.

3.3 WATERSHED GEOMORPHOLOGY

3.3.1 Drainage Area

The drainage area (A) is probably the single most important watershed characteristic for hydrologic design. It reflects the volume of water that can be generated from rainfall. It is common in hydrologic design to assume a constant depth of rainfall occurring uniformly over the watershed. Under this assumption, the volume of water available for runoff would be the product of the rainfall depth and the drainage area. Thus the drainage area is required as input to models ranging from simple linear prediction equations to complex computer models.

The drainage area of a watershed requires the delineation of the watershed boundary. Geographic information systems are commonly used to delineate watershed boundaries. Once this is done, the area is automatically computed. The area can be computed manually with an instrument called a planimeter. Where this is not available, the "stone-age" method of counting squares can be used. A transparency showing a grid, usually square, is laid over the map showing the drainage boundaries, and the number of grid blocks within the boundary is counted. The drainage area equals the product of the number of grid blocks and the area of each grid block. The area of each grid block is computed using the scale of the topographic map from which the watershed boundary was delineated. The accuracy of the estimated value of the drainage area will depend on the care taken in counting the grid blocks, especially the partial grid blocks along the watershed boundary. Given the importance of the drainage area in hydrologic design, special care should be made to ensure that the estimated value of the drainage area is accurate.

As computers become more prevalent in hydrologic design, they are being used to compute drainage areas. The watershed boundary can be delineated by indicating the latitude and longitude of points along the boundary; the boundary is assumed to be linear between each pair of points and the drainage area is the area enclosed by the series of linear segments. The drainage area is computed internally by the computer using basic trigonometry. Some software packages do not even require the user to specify points; from the user-specified watershed outlet, the watershed boundary is constructed from an analysis of computed slopes, with the slopes evaluated numerically from the topographic map that is stored within the computer.

Example 3–1

To illustrate the calculation of the drainage area, Figure 3–4 shows the watershed of Figure 3–3 with a square-grid overlay. The scale of the topo map is 1 in. = 2000 ft. Since each side of the grid square is 0.1 in., then each square of the grid represents an area of 40,000 ft^2, or slightly less than 1 acre. The watershed boundary encloses 733 squares. Thus the total drainage area is 673 acres. The areas of the three subwatersheds can be determined at the same time. They have areas of 375, 179, and 119 acres.

3.3.2 Watershed Length

The length (L) of a watershed is the second watershed characteristic of interest. While the length increases as the drainage area increases, the length of a watershed is important in hydrologic computations; for example, it is used in time-of-concentration calculations. The wa-

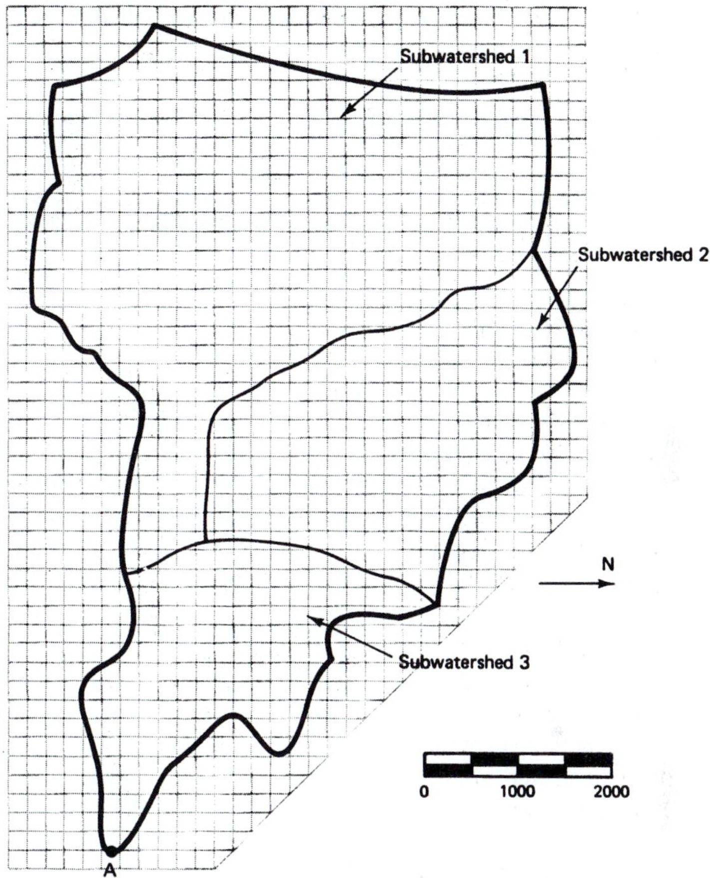

FIGURE 3–4 Delineation of watershed boundary.

tershed length is also highly correlated with channel length, which is discussed in Section 3.5.1.

Watershed length is usually defined as the distance measured along the main channel from the watershed outlet to the basin divide. Since the channel does not extend to the basin divide, it is necessary to extend a line from the end of the channel to the basin divide following a path where the greatest volume of water would travel. The straight-line distance from the outlet to the farthest point on the watershed divide is not usually used to compute L because the travel distance of flood waters is conceptually the length of interest. Thus, the length is measured along the principal flow path. Since it will be used for hydrologic calculations, this length is more appropriately labeled the hydrologic length.

While the drainage area and length are both measures of watershed size, they may reflect different aspects of size. The drainage area is used to indicate the potential for rainfall to provide a volume of water. The length is usually used in computing a time parameter, which is a measure of the travel time of water through the watershed. Specific time parameters will be discussed in Section 3.6.

Example 3–2

The lengths of the channel and total watershed for the basins of Figure 3–4 are delineated in Figure 3–3b. Based on a map scale of 1 in. equals 2000 ft, the lengths were computed using a map wheel (see Table 3–1). The channel lengths as a proportion of the total length of the subarea or watershed range from 50% for subarea 2, 73% for subareas 1 and 3, and 82% for the total watershed.

3.3.3 Watershed Slope

Flood magnitudes reflect the momentum of the runoff. Slope is an important factor in the momentum. Both watershed and channel slope may be of interest. Watershed slope reflects the rate of change of elevation with respect to distance along the principal flow path. Typically, the principal flow path is delineated, and the watershed slope (S) is computed as the difference in elevation (ΔE) between the end points of the principal flow path divided by the hydrologic length of the flow path (L):

$$S = \frac{\Delta E}{L} \tag{3–1}$$

The elevation difference ΔE may not necessarily be the maximum elevation difference within the watershed since the point of highest elevation may occur along a side boundary of the watershed rather than at the end of the principal flow path.

Where the design work requires the watershed to be subdivided, it will be necessary to compute the slopes of each subarea. It may also be necessary to compute the channel slopes for the individual sections of the streams that flow through the subareas. When computing the slope of a subarea, the principal flow path for that subarea must be delineated. It should reflect flow only for that subarea rather than flow that enters the subarea in a channel. The stream-reach slope may also be necessary for computing reach travel times.

Example 3–3

The upper and lower elevations for the watersheds of Figure 3–3b are given in Table 3–1. They are used with the watershed lengths to compute watershed slopes. The watershed slopes range from 3% for the entire watershed to 5% for subarea 3.

TABLE 3–1 Lengths and Slopes of Watersheds of Figure 3–4

	Length (ft)		Channel Elevation (ft)			Watershed Elevation (ft)			Slope (ft/ft)	
Area	Channel	Watershed	Upper	Lower	Difference	Upper	Lower	Difference	Channel	Watershed
Sub 1	4940	6810	450	340	110	608	340	268	0.022	0.039
Sub 2	2440	4875	400	340	60	545	340	205	0.025	0.042
Sub 3	3670	5000	340	295	45	545	295	250	0.012	0.050
Total	8610	10480	450	295	155	608	295	313	0.018	0.030

3.3.4 Hypsometric Curve

The hypsometric curve is a description of the cumulative relationship between elevation and the area within elevation intervals. The curve is plotted with the elevation plotted as the ordinate and the area within the watershed above the elevation plotted as the abscissa. The hypsometric curve can also be represented in standardized form, with the cumulative fractions plotted rather than the actual values. This is shown in Figure 3–5b for a hypothetical watershed (Figure 3–5a). The standardized form is useful for comparing the area-elevation characteristics of watersheds. Such comparison might be useful if a regional hypsometric curve were being developed from the analysis of hypsometric curves of watersheds in the region; after standardizing the hypsometric curves for each watershed, an "average" curve could be constructed for the region. Where the necessary data are not available at other watersheds in the region, the regional hypsometric curve could be used to represent the area-elevation characteristics of the watershed.

A number of indices have been developed to convert the hypsometric curve into a single-valued index of the area-elevation characteristics of a watershed. For example, the hypsometric area ratio (H_a) is the ratio of the area under the hypsometric curve to the area of the square formed by the points (0, 0), (0, 1), (1, 1), and (1, 0) on the hypsometric curve of Figure 3–5b. For the hypsometric curve of Figure 3–5b, H_a equals 0.43.

The profile factor (F_p) is a second example of a single-valued index that can be derived from a hypsometric curve. The profile factor is defined as the ratio of the maximum deviation (D_m) of the hypsometric curve from a line connecting the points (0, 1) and (1, 0) to the length of the line (L_d); this is shown in Figure 3–6. For the hypsometric curve of Figure 3–5b, the profile factor equals 0.1, with typical values ranging from 0.01 to 0.15.

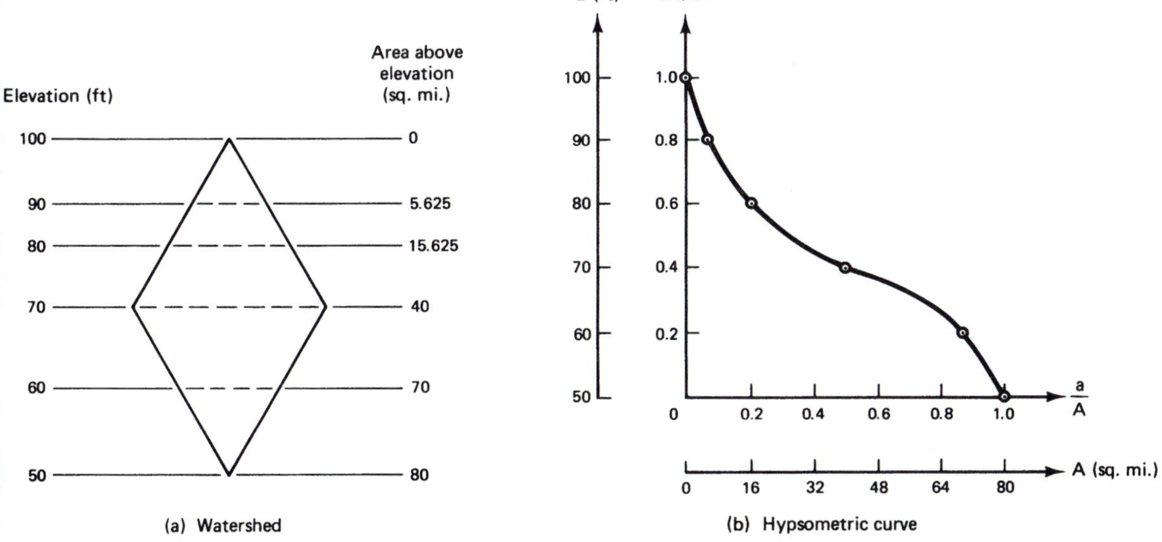

FIGURE 3–5 Construction of a hypsometric curve for a hypothetical watershed.

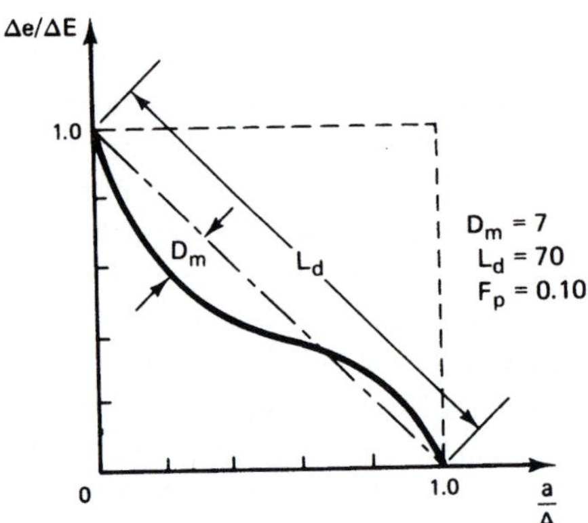

FIGURE 3–6 Estimation of the profile factor for the hyposometric curve of Figure 3–5.

Example 3–4

The hypsometric curve for the 673–acre watershed of Figure 3–3 was computed using a 50-ft elevation grid. The computed values of a/A and $\Delta e/\Delta E$ are given with the hypsometric curve in Figure 3–7. It is similar in shape to the curve of Figure 3–6, although it does not show as large of a maximum deviation as the curve of Figure 3–7. The hypsometric area ratio can be estimated using the trapezoidal rule:

$$
\begin{aligned}
H_a = \; & 0.5(0.002-0.0)(1+0.984) + 0.5(0.068-0.002)(0.984+0.824) \\
& + 0.5(0.249-0.068)(0.824+0.665) + 0.5(0.563-0.249)(0.665+0.505) \\
& + 0.5(0.766-0.563)(0.505+0.345) + 0.5(0.945-0.766)(0.345+0.185) \\
& + 0.5(0.998-0.945)(0.185+0.026) + 0.5(1.0-0.998)(0.026+0.0) \\
= \; & 0.519
\end{aligned}
$$

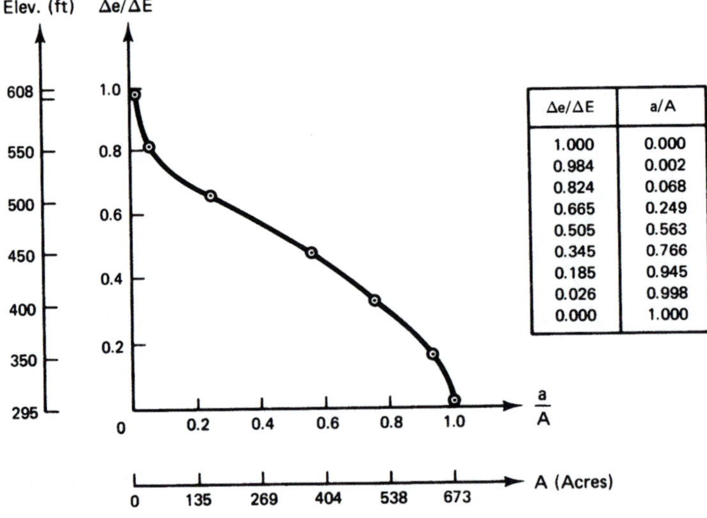

Δe/ΔE	a/A
1.000	0.000
0.984	0.002
0.824	0.068
0.665	0.249
0.505	0.563
0.345	0.766
0.185	0.945
0.026	0.998
0.000	1.000

FIGURE 3–7 Hyposometric curve for the watershed of Figure 3–3.

The profile factor is

$$F_p = \frac{0.4}{6.0} = 0.067$$

which is a typical value.

3.3.5 Watershed Shape

Basin shape is not usually used directly in hydrologic design methods; however, parameters that reflect basin shape are used occasionally and have a conceptual basis, so a few words about it are in order. Watersheds have an infinite variety of shapes, and the shape supposedly reflects the way that runoff will "bunch up" at the outlet. A circular watershed would result in runoff from various parts of the watershed reaching the outlet at the same time. An elliptical watershed having the outlet at one end of the major axis and having the same area as the circular watershed would cause the runoff to be spread out over time, thus producing a smaller flood peak than that of the circular watershed. The importance of watershed shape will be more apparent after the concept of a time-area diagram is introduced and discussed in Chapter 9.

A number of watershed parameters has been developed to reflect basin shape. The following are a few typical parameters:

1. Length to the center of area (L_{ca}): the distance in miles measured along the main channel from the basin outlet to the point on the main channel opposite the center of area. The center of area for the watershed of Figure 3–3b is shown on the map.
2. Shape factor (L_l):

$$L_l = (LL_{ca})^{0.3} \tag{3–2}$$

where L is the length of the watershed in miles.

3. Circularity ratio (F_c):

$$F_c = \frac{P}{(4\pi A)^{0.5}} \tag{3–3}$$

where P and A are the perimeter (ft) and area (ft^2) of the watershed, respectively.

4. Circularity ratio (R_c):

$$R_c = \frac{A}{A_o} \tag{3–4}$$

where A_o is the area of a circle having a perimeter equal to the perimeter of the basin.

5. Elongation ratio (R_e):

$$R_e = \frac{2}{L_m}\left(\frac{A}{\pi}\right)^{0.5} \tag{3–5}$$

where L_m is the maximum length (ft) of the basin parallel to the principal drainage lines.

The circularity ratios of Equations 3–3 and 3–4 are related by

$$R_c = \frac{1}{F_c^2} \tag{3–6}$$

The length to the center of area is used in deriving Snyder's synthetic unit hydrograph of the HEC-1 computer program (see Chapter 9). The distance L_{ca} may be estimated by the mean distance from a graph of the cumulative area versus distance. As an alternative, L_{ca} can be estimated by the distance measured along the principal channel to a point approximately opposite (i.e., measured perpendicular to) the center of area. The location of the center of area of the drainage area may be determined using another "stone-age" method of suspending a cardboard outline of the drainage area by means of a straight pin inserted into the edge, drawing a vertical line, and then rotating the cardboard board approximately 90° and drawing a second vertical line. The intersection of the two lines is the center of area of the watershed. In practice, three lines rotated approximately 60° may be used to provide a slightly greater level of accuracy.

Example 3–5

For the watershed of Figure 3–3, the location of the center of area is shown on Figure 3–3b. Measuring along the channel to a point opposite the center of area, a value of 6400 ft is measured for L_{ca} Thus the shape factor L_l is

$$L_l = \left[\left(\frac{10{,}350}{5280}\right)\left(\frac{6400}{5280}\right)\right]^{0.3} = 1.30 \tag{3–7}$$

Values for the circularity ratios of Equations 3–3 and 3–4 can also be computed. The perimeter was measured as 26,500 ft; thus, from Equation 3–3,

$$F_c = \frac{26{,}500}{[4\pi(2.932 \times 10^7)]^{0.5}} = 1.38 \tag{3–8}$$

The equivalent area, A_o (Equation 3–4), is

$$A_o = \pi\left(\frac{26{,}500}{2\pi}\right)^2 = 5.588 \times 10^7 \text{ ft}^2 \tag{3–9}$$

Thus the value of the circularity ratio R_c of Equation 3–4 is

$$R_c = \frac{A}{A_o} = \frac{2.932 \times 10^7}{5.588 \times 10^7} = 0.52 \tag{3–10}$$

Example 3–6

Shape factors were originally developed because it was believed that they were correlated with peak discharge. To examine the association, seven hypothetical watersheds were evaluated using a time-area analysis (discussed in Chapter 9) and the peak discharge obtained from the linear-reservoir routed, time-area unit hydrograph. The watersheds and assumed contours are given in Figure 3–8, with the values of the parameters given in Table 3–2. The peak discharges (q_p) were correlated with each of the five shape parameters (see Table 3–2). L_{ca} and L_l show negative correlations with q_p. Positive correlations would normally be expected because longer lengths are associated with larger areas. However, in this case, all of the watersheds have the same area so

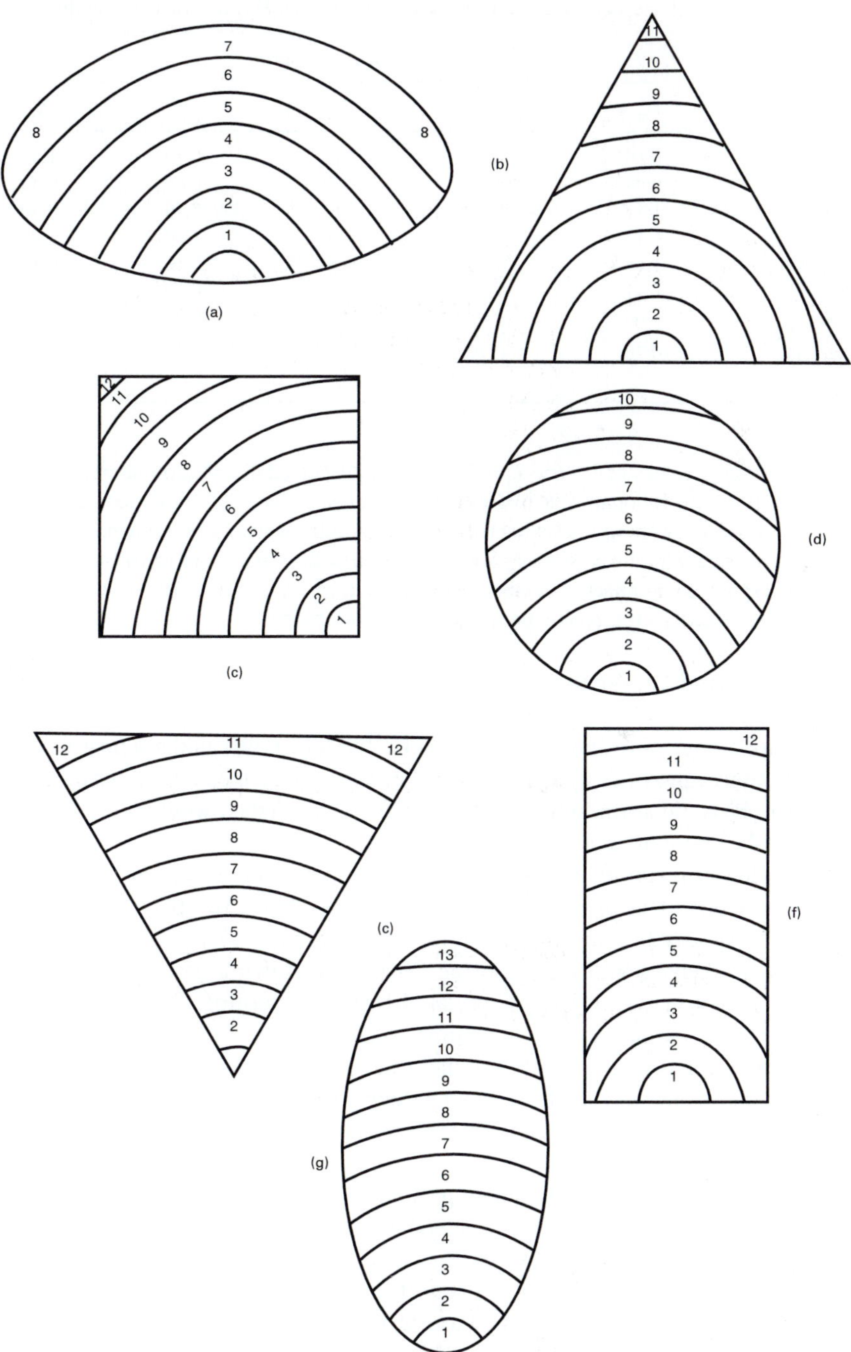

FIGURE 3–8 Hypothetical watersheds. (a) Ellipse: side; (b) triangle: center; (c) square: corner; (d) circle; (e) triangle: vertex; (f) rectangle; (g) ellipse: end.

TABLE 3–2 Values of Shape Parameters and Peak Discharges for Hypothetical Watersheds of Figure 3–8

Watershed Shape	Shape Parameters					Peak Discharge
	L_{ca}	L_l	F_c	R_c	R_e	
Ellipse: side	0.80	1.08	1.13	0.79	1.41	0.22
Triangle: center	0.88	1.29	1.29	0.60	0.86	0.22
Square: corner	1.41	1.52	1.13	0.79	0.80	0.18
Circle	1.13	1.32	1.00	1.00	1.00	0.16
Triangle: vertex	1.75	1.58	1.29	0.60	0.86	0.16
Rectangle	1.41	1.52	1.20	0.70	0.80	0.12
Ellipse: end	1.60	1.63	1.13	0.79	0.71	0.10
Correlation with peak discharge	–0.76	–0.79	0.19	–0.17	0.64	

longer values of L_{ca} and L_l reflect a more dispersed distribution of area and the negative correlations are rational. The two circularity ratios, F_c and R_c, show poor correlation with peak discharge, as should be expected because circular watersheds are average in peak discharge. The elongation ratio R_e shows good correlation with peak discharge (0.64). Large values of R_e correspond to compact watersheds which produce high discharges. These results suggest that LL_{ca} represents a reasonable predictor of peak discharge.

3.3.6 Land Cover and Use

During a brief rain shower, people will often take cover under a nearby tree; they recognize that the tree provides a temporary shelter since it intercepts rain during the initial part of the storm. It would seem to follow that a forested watershed would have less flood runoff than a watershed with no tree cover.

Rooftop runoff provides another example of the effect of land cover on runoff rates and volumes. During a rainstorm it is obvious that flow from the downspouts on houses starts very shortly after the start of rain. Rooftops are impervious, steeply sloped, planar surfaces, so there is little to retard the flow. Flow down a grassy hill of the same size as the rooftop will begin long after similar flow over a rooftop. The grassy hill sheds water at a slower rate and has a smaller volume because some of the water infiltrates into the topsoil and the grass is hydraulically rougher than the shingles on the roof. It would seem to follow from this example that flow from impervious surfaces would have greater volumes and smaller travel times than flow over pervious surfaces having similar size, shape, and slope characteristics.

These two conceptual examples should illustrate that the land cover significantly affects the runoff characteristics of a watershed. When watershed characteristics other than the land cover are held constant, the runoff characteristics of a watershed, which includes the volume of runoff, the timing of runoff, and maximum flood flow rates, can differ significantly. Therefore, land cover and use serve as inputs for problems in hydrologic analysis and design.

Many descriptors of land cover/use are used in hydrologic design. Most often, a qualitative description of land cover is transformed into a quantitative index of runoff potential.

For example, the Rational Method (see Chapter 7) uses a runoff coefficient C to reflect the runoff potential of a watershed. Larger values of C reflect increased runoff potential. The values for commercial properties ($C = 0.75$) have a greater runoff potential than residential areas ($C = 0.3$), which in turn has a greater runoff potential than forested areas ($C = 0.15$). From this it should be evident that the runoff potential increases as a watershed is transformed from a forested cover to an urban land cover.

The Soil Conservation Service (SCS) uses a different land cover/use index in their models than is used in the Rational Method. The runoff curve number (*CN*) is the SCS cover type index. It is an integration of the hydrologic effects of land use, soil type, and antecedent moisture. The runoff curve number is discussed in Section 3.7.

Urban land covers are especially important in hydrology. Many hydrologic design problems result from urban expansion. The percentage of imperviousness is a commonly used index of the level of urban development. High-density residential areas characteristically have percentages of imperviousness from 40% to 70%. Commercial and industrial areas are characterized by impervious cover often from 70% to 90%. Impervious covers in urban areas are not confined to the watershed surface. Channels are often lined with concrete to increase the flow capacity of the channel cross section and to quickly remove flood waters. Channel lining is often criticized because it can transfer the flooding problem from an upstream reach to a reach downstream.

3.3.7 Surface Roughness

Roughness implies unevenness of texture. Sandpaper is thought of as being rough. Yet there are degrees of roughness as evidenced by the different grades of sandpaper. Surface roughness is also important in hydrologic design. As indicated in the previous section, a grass surface is hydrologically rougher then a shingled roof. The grass retards the flow to a greater extent than the shingles.

Manning's roughness coefficient (n) is the most frequently used index of surface roughness. While the units of n are rarely shown, its units are dependent on the dimension system used. In general, its dimensions are $TL^{-1/3}$, with units of sec·ft$^{-1/3}$ in the English system. Values of n for hydrologic overland flow surfaces are given in Table 3–3. Quite frequently, tables of n values give a range of values for each surface. This reflects the uncertainty in the value. Unless there is a specific reason for selecting a value within the stated range, the mean value of the endpoints of the range should be used.

3.4 SOIL CHARACTERISTICS

3.4.1 The Soil Profile

It is widely recognized that soil properties vary spatially; however, the hydrologic characteristics of the soil also vary with depth. If one were to take a core of soil and characterize the properties over the depth of the soil profile, a number of horizontal layers would be evident. These layers would vary in composition, structure, texture, and color. Also, these character-

TABLE 3–3 Manning's Roughness Coefficient (n) for Overland Flow Surfaces

Surface	n
Plastic, glass	0.009
Fallow	0.010
Bare sand	0.010
Graveled surface	0.012
Smooth concrete	0.011
Asphalt	0.012
Bare clay	0.012
Ordinary concrete lining	0.013
Good wood	0.014
Brick with cement mortar	0.014
Unplanned timber	0.014
Vitrified clay	0.015
Cast iron	0.015
Smooth earth	0.018
Corrugated metal pipes	0.023
Cement rubble surface	0.024
Conventional tillage	
no residue	0.09
with residue	0.19
Grass	
Short	0.15
Dense	0.24
Bermudagrass	0.41
Woods	
No underbrush	0.20
Light underbrush	0.40
Dense underbrush	0.80
Rangeland	0.13

istics and the number of layers would vary from site to site. Soil horizons are referenced as follows:

O-Horizon: surface litter consisting primarily of organic matter

A-Horizon: topsoil consisting of humus and inorganic minerals

E-Horizon: the zone of leaching where percolating water dissolves water-soluble matter

B-Horizon: the subsoil below the A- or E-horizons that contains minerals and humic compounds

C-Horizon: a zone consisting primarily of undercomposed mineral particles and rock fragments

R-Horizon: bedrock, an impermeable layer

The E-horizon is sometimes referred to as the A_2-horizon.

3.4.2 Soil Texture

Soil texture is a physical characteristic of a soil; it refers to the size of the mineral particles and the fraction of the particles in different size classes. Particles of soil can be separated into three classes on the basis of the mean diameter (d) in millimeters:

Clay: $d < 0.002$ mm

Silt: $0.002 \le d \le 0.02$ mm

Sand: $0.02 \le d \le 2$ mm

Gravel is sometimes considered to be a fourth class, with gravel consisting of particles with diameters greater than 2 mm. Also, the sand class can be divided into coarse sand (0.2 to 2 mm) and fine sand (0.02 to 0.2 mm). The texture is further divided on the basis of the percentage of each of these soil classes. Mineral particles have diameters (d) that typically fall into size ranges. While a number of classification systems have been proposed, Figure 3–9 shows the U.S. System for Texture Designations. The texture class can be determined from the percentages of sand, silt, and clay. The texture triangle of Figure 3–9 shows eleven texture groups, which are a function of the mixture of the three soil classes.

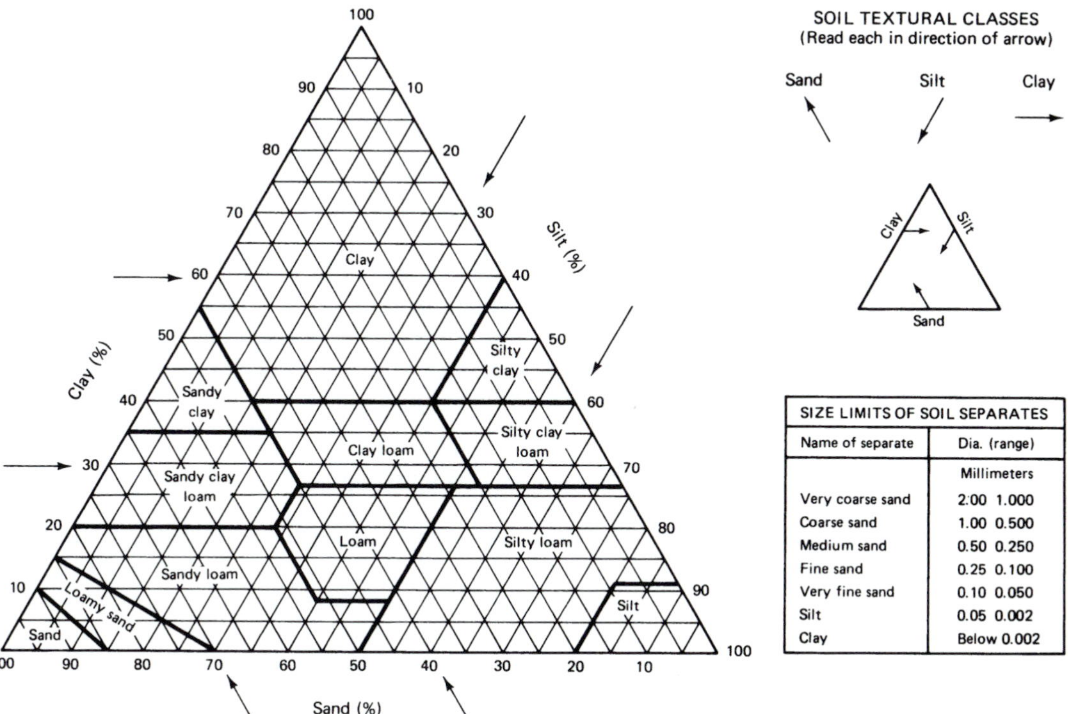

SIZE LIMITS OF SOIL SEPARATES	
Name of separate	Dia. (range)
	Millimeters
Very coarse sand	2.00 1.000
Coarse sand	1.00 0.500
Medium sand	0.50 0.250
Fine sand	0.25 0.100
Very fine sand	0.10 0.050
Silt	0.05 0.002
Clay	Below 0.002

FIGURE 3–9 Guide for textural classification by the U.S. System for Texture Designations.

The soil texture is an important factor in determining the water-holding characteristics of the soil and therefore the infiltration capacity of a soil layer. As the diameters of the soil particles increase, the pore spaces increase in size, which increases the capacity of the soil to pass and store infiltrating water through the soil profile. However, between storms, especially when the intervals between storms are long, the soils with high percentages of sand pass the water quickly and may not retain sufficient water to fulfill the needs of vegetation.

3.4.3 Soil Structure

Soil structure is another property of a soil that affects the hydrologic response of a watershed. Soil structure refers to the tendency of soil particles to aggregate into lumps and clods. The structure is a function of the soil texture, the texture and type of minerals present, and the amount of biological activity in the soil column. The structure of the soil influences the amount of biological activity in the soil column. The structure of the soil influences the amount of pore space in the soil column, which, in turn, affects infiltration, soil moisture retention, and other water movement through the column. The average soil column may have a pore space of about 45%, with a variation from 30% to 70%.

3.4.4 Volumetric Characteristics of Soils

A geologic formation consists of solid matter, water, and air, with the water and air located in the spaces between the particles of the solid matter. For a volume V of the geologic formation, there will be volumes of solid matter (V_s), water (V_W), and air (V_a), such that

$$V = V_s + V_W + V_a \tag{3–11}$$

The hydrologic response of a geologic formation will depend on the distribution of these volumes. Therefore, it is useful to define soil characteristics that depend on the volumetric distribution of Equation 3–11.

Pores or void space are part of geologic formations. The pores offer the opportunity for the passage of water. Therefore, the volumetric proportion of geologic formation that is void space is of importance in subsurface drainage. The porosity of a soil mass defines this volumetric proportion, and thus it is an index of the water-bearing capacity of an aquifer. Specifically, the porosity (P_v) is given by

$$P_v = \frac{V_W}{V} \tag{3–12}$$

in which V_W is the volume of water that would be required to saturate the pore space, and V is the total volume of the geologic formation. Figure 3–10 shows typical values of the porosity. Where materials are mixed, the porosity is usually lower because the smaller particles tend to fill up the void space created by the material having the larger pore space. For example, if a 6 in.^{3} sample of soil is taken, poured into a graduated cylinder, and displaces 3.75 in.^{3} of water, then the porosity is

$$P_v = \frac{V - V_s}{V} = \frac{6 - 3.75}{6} = 0.375 \text{ or } 37.5\% \tag{3–13}$$

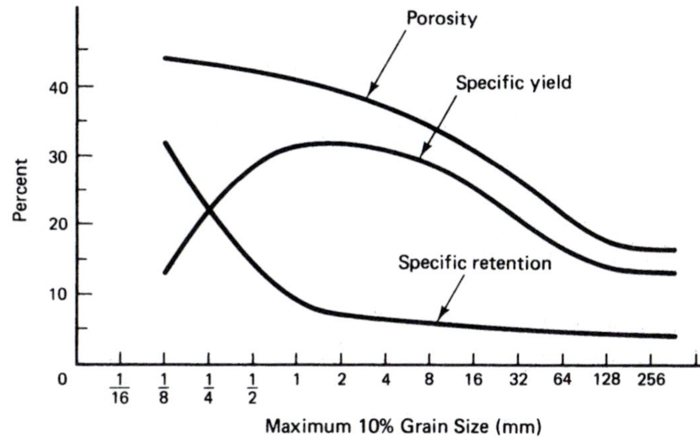

FIGURE 3–10 Variation of specific retention, specific yield, and porosity with the grain size for which the cumulative total (starting with the coarsest material) reaches 10% of the total.

Water in a geologic formation may freely drain from the formation under the force of gravity. The proportion of the water that can drain under the force of gravity relative to the total volume is called the effective porosity, or specific yield (P_e):

$$P_e = \frac{V_g}{V} \tag{3–14}$$

in which V_g is the volume of water drained by gravity.

Some water in a geologic formation will not freely drain under the force of gravity because of surface tension and molecular forces. The specific retention (R_s) is the ratio of the volume of water retained by these forces to the total volume:

$$R_s = \frac{V_r}{V} \tag{3–15}$$

in which V_r is the volume of water retained by surface tension and molecular forces.

The porosity, effective porosity, and specific retention are related by

$$P_v = P_e + R_s \tag{3–16}$$

Figure 3–10 shows the porosity, specific yield, and specific retention for a range of soil types.

The porosity does not fully characterize the ability of a geologic formation to pass water. The ability to pass water through a geologic formation also depends on the degree to which the pore spaces are connected. This ability is called the permeability of a soil. The term hydraulic conductivity is currently used to represent the permeability of a soil. A specific measure of the permeability of a soil will be mathematically defined later.

In addition to soil characteristics based on the volumetric proportions, the mass of a soil is an important characteristic. The bulk density is the mass of an oven dried soil per unit volume of the soil. After drying a soil mass, the bulk density ρ_b is computed as

$$\rho_b = \frac{W/V}{\gamma} = \frac{W/V}{62.4\ \text{lb/ft}^3} \tag{3–17}$$

in which W is the weight of the soil (lb), V is the *in situ* volume of soil (ft^3), and γ is the specific weight of water (lb/ft^3). It is evident that ρ_b is dimensionless and measures the density of the soil relative to the density of an equal volume of water.

While the term *specific yield* applies to an unconfined aquifer, the underlying concept can be applied to a confined aquifer; however, the term *storativity* or *storage coefficient* is used when referring to the concept for a confined aquifer. Specifically, the storage coefficient is the volume of water released from a confined aquifer when the pressure head decreases by a unit depth over a unit horizontal area. Where a confined aquifer is subject to recharge, the storage coefficient would be defined by the water taken into a unit area of the aquifer per unit increase in the pressure head.

The specific storage S_s of a confined aquifer equals the storage coefficient divided by the thickness (or depth) of the aquifer. DeWiest provided the following relationship for estimating the specific storage:

$$S_s = C_1(1 - P_v) + C_2P_v \tag{3–18}$$

in which C_1 and C_2 are constants for the aquifer.

3.4.5 Soil Moisture

While water in the zone of saturation is referred to as ground water, water in the zone of aeration is referred to as soil moisture. During a storm, the upper layers of the zone of aeration may actually become saturated. After the rainfall ceases, the water in the zone of aeration is acted upon by gravitational, surface tension, and molecular forces. The drainage of water by gravitational forces will cease when the surface tension forces equal the force of gravity. The soil moisture content at this point is called the field capacity of the soil. Unfortunately, a specific percentage cannot be specified to delimit field capacity. While most of the gravity water will drain from a soil mass in a few days, some water may continue to drain for months. Thus field capacity refers to the soil moisture content at which rapid drainage has ceased. Gravitational water is water that drains under the force of gravity prior to the time at which field capacity is reached.

After field capacity has been reached, the soil moisture may continue to decrease because of evaporation through the pores to the atmosphere or the removal of water by the roots of vegetation. As the soil moisture decreases, it becomes more difficult for plants to obtain the moisture that is necessary for sustaining growth. At this point, the plants start to wilt. This level of soil moisture where plants wilt and cannot sustain growth is called the permanent wilting point. Water below this level of soil moisture is referred to as unavailable water; the only moisture present in the soil is the thin film or layer of moisture attached to the soil particles, which is also referred to as hygroscopic moisture. Water that exists between field capacity and the permanent wilting point is also referred to as available water; it represents the available water supply, which is greatest for soil textures in the loam texture classes (see Figure 3–11).

While soil moisture is not a principal factor in engineering design, it does have important implications in agriculture and forestry. Plants have varying needs for water during their growth cycle, and plant growth can be severely affected when the soil moisture content is inadequate to meet the needs of the plants. Since the root zone extends down only a few feet

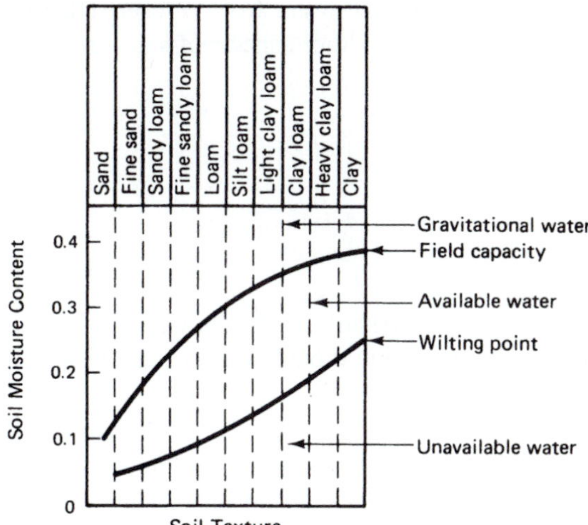

FIGURE 3–11 Soil moisture content as a function of soil texture.

from the ground surface, it is this portion of the zone of aeration that has been the subject of considerable research.

The process of water drainage under the force of gravity from a soil mass was described previously. As the gravitational water discharges from the soil mass, a film of moisture remains around the soil particles. This water film is held in place by surface tension forces, and the force of gravity is not sufficient to overcome the tension forces. This tension force is referred to as soil moisture tension or soil water suction. As the thickness of the film decreases, both the surface tension force and the curvature of the water film increase. The pressure of the air adjacent to the water film is atmospheric, with the pressure in the water film being less than atmospheric. This reduced pressure forms a suction force that prevents the water from draining under the force of gravity.

In a saturated soil, the concepts of hydrostatic forces indicate that the pressure increases with depth, with the pressure being equal to zero at the water table. In the zone of aeration, where the water is held by surface tension forces, the water is at a pressure that is less then atmospheric. If we define our pressure datum to be zero for atmospheric pressure, the pressure of the water in the zone of saturation is positive, and the pressure of the water in the zone of aeration is negative. We know from elementary fluid mechanics that pressure forces represent a potential for doing work. Where the pressure is greater than atmospheric, work can be done by the fluid. Where the pressure is less than atmospheric, work must be done on the fluid. The negative pressure in the zone of aeration is referred to as the matric potential. The matric potential is measured using a tensiometer.

3.4.6 Hydrologic Soil Groups and Soil Cover Condition

In addition to influencing the movement of water through the soil profile, the soil texture influences the rate at which surface water enters the soil profile. This infiltration capacity then affects the amount of water that enters the streams and rivers as direct or surface runoff. Soils that have a low infiltration capacity prevent surface water from entering the soil profile, thus increasing the water available for surface runoff. Other aspects of soils are discussed in Section 3.7.

3.5 CHANNEL GEOMORPHOLOGY

3.5.1 Channel Length

In addition to the drainage area and the watershed length, the channel length is used frequently in hydrologic computations. Two computational schemes are used to compute the channel length:

1. The distance measured along the main channel from the watershed outlet to the end of the channel as indicated on a map, which is denoted as L_c.
2. The distance measured along the main channel between two points located 10 and 85% of the distance along the channel from the outlet, which is denoted as L_{10-85}.

These definitions along with the watershed length are illustrated in Figure 3–12. The watershed length requires the user to extend a line on the map from the end of the main channel to the divide; this requires some subjective assessment, which is often a source of inaccuracy. The definitions for channel length also involve a measure of subjectivity because the endpoint of the channel is dependent on the way that map was drawn; the location of the end of

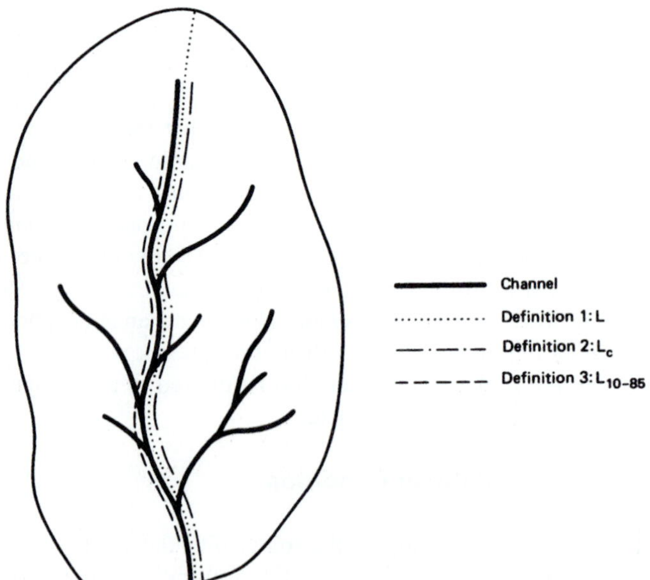

FIGURE 3–12 Delineation of watershed length.

the channel may depend on the level of flow at the time the map was compiled. These subjective assessments can introduce an unknown degree of inaccuracy into the final design. It is important to know with certainty exactly which definition was used in the development of a design aid. For example, if the first definition was used in developing the design method, the use of the second definition in computing the channel length can introduce a bias into the design.

For very small watersheds, the watershed may not include a stream channel; it may consist of a section of overland flow and a section where the flow is in a swale or gully. As the size of the watershed increases, channel flow dominates and the watershed and channel lengths are essentially the same. The design problem of interest will determine which length should be computed.

Example 3–7

The channel system shown in Figure 3–3b can be viewed as consisting of three reaches. The lengths of the reaches in subareas 1, 2, and 3 are 4940, 2440, and 3670 ft, respectively. If the lengths are defined as subwatershed lengths, then the length measurements for subareas 1 and 2 must be extended to the basin divide, which is also shown in Figure 3–3b. The lengths of subwatersheds 1 and 2 are 6810 and 4875 ft, respectively. For subwatershed 3 the watershed length should be defined as the flow path from the most distant point on the watershed divide to the outlet. For subwatershed 3 this point would be the northern most point, which is on the boundary separating subareas 2 and 3. This flow path consists of 2550 ft of channel and 2450 ft of overland and gully flow for a total watershed length of 5000 ft. The importance of distinguishing between the channel length and the length of the subwatershed will be evident when discussing time parameters. If one was not distinguishing between the subwatersheds and was interested in the total watershed length, the length would be the sum of the channel length for the reach through subarea 3 (for example, 3670 ft) and the watershed length of subarea 1 (i.e., 6810 ft), or 10,480 feet.

3.5.2 Channel Slope

The channel slope can be described with any one of a number of computational schemes. The most common is

$$S_c = \frac{\Delta E_c}{L_c} \tag{3–19}$$

in which ΔE_c is the difference in elevation (ft) between the points defining the upper and lower ends of the channel, and L_c is the length of the channel (ft) between the same two points. The 10–85 slope can also be used:

$$S_{10-85} = \frac{\Delta E_{10-85}}{L_{10-85}} \tag{3–20}$$

in which $\Delta E_{10–85}$ is the difference in elevation (ft) between the points defining the channel length $L_{10–85}$.

For cases where the channel slope is not uniform, a weighted slope may provide an index that better reflects the effect of slope on the hydrologic response of the watershed. The following defines a channel slope index:

$$S_i = \left(\frac{n}{k}\right)^2 \tag{3–21a}$$

in which n is the number of segments into which the channel is divided, and k is given by

$$k = \sum_{i=1}^{n} \frac{1}{(\Delta e_i/l_i)^{0.5}} \tag{3–21b}$$

in which Δe_i is the difference in elevation between the endpoints of channel segment i, and l_i is the length of segment i. The channel is divided into segments where the slope is relatively constant over each segment.

Example 3–8

The watershed of Figure 3–3 can be used to illustrate the calculation of the channel and watershed slope. If the watershed were subdivided as shown in Figure 3–3b, the channel reach in each subwatershed must be used for computing the channel slope. Table 3–1 gives the channel and watershed slopes for both the subdivided watershed and the watershed as a whole. In general, the watershed slopes are greater than the channel slopes, which is the rule because the side slopes of the watershed are almost always steeper than the channel.

3.5.3 Drainage Density

The drainage density (D) is the ratio of the total length of streams within a watershed to the total area of the watershed; thus D has units of the reciprocal of length. A high value of the drainage density would indicate a relatively high density of streams and thus a rapid storm response.

Example 3–9

The area and stream lengths for the watershed of Figure 3–13 can be used to illustrate the calculation of the drainage density. The total watershed area is 280 mi^2. Assuming that the only streams are those delineated in Figure 3–13, the total length of streams is 23.07 mi. Thus the drainage density is

$$D = \frac{L_T}{A} = \frac{23.07 \text{ mi}}{280 \text{ mi}^2} = 0.0824 \frac{\text{mi}}{\text{mi}^2} \tag{3–22}$$

This value is especially lower than normal since the channel network was not given in detail. Values typically range from 1.5 to 6 mi/mi^2. The example, however, demonstrates the importance of accurately delineating the channel reaches. Drainage density is not used frequently because of the difficulty in deriving consistent estimates of D due to the map dependency.

3.5.4 Horton's Laws

Horton developed a set of "laws" that are indicators of the geomorphological characteristics of watersheds. The stream order is a measure of the degree of stream branching within a watershed. Each length of stream is indicated by its order (for example, first-order, second-order, etc.). A first-order stream is an unbranched tributary, and a second-order stream is a tributary formed by two or more first-order streams. A third-order stream is formed by two or more second-order streams, and in general, an nth-order stream is a tributary formed by two or more streams of order (n–1) and streams of lower order. For a watershed, the princi-

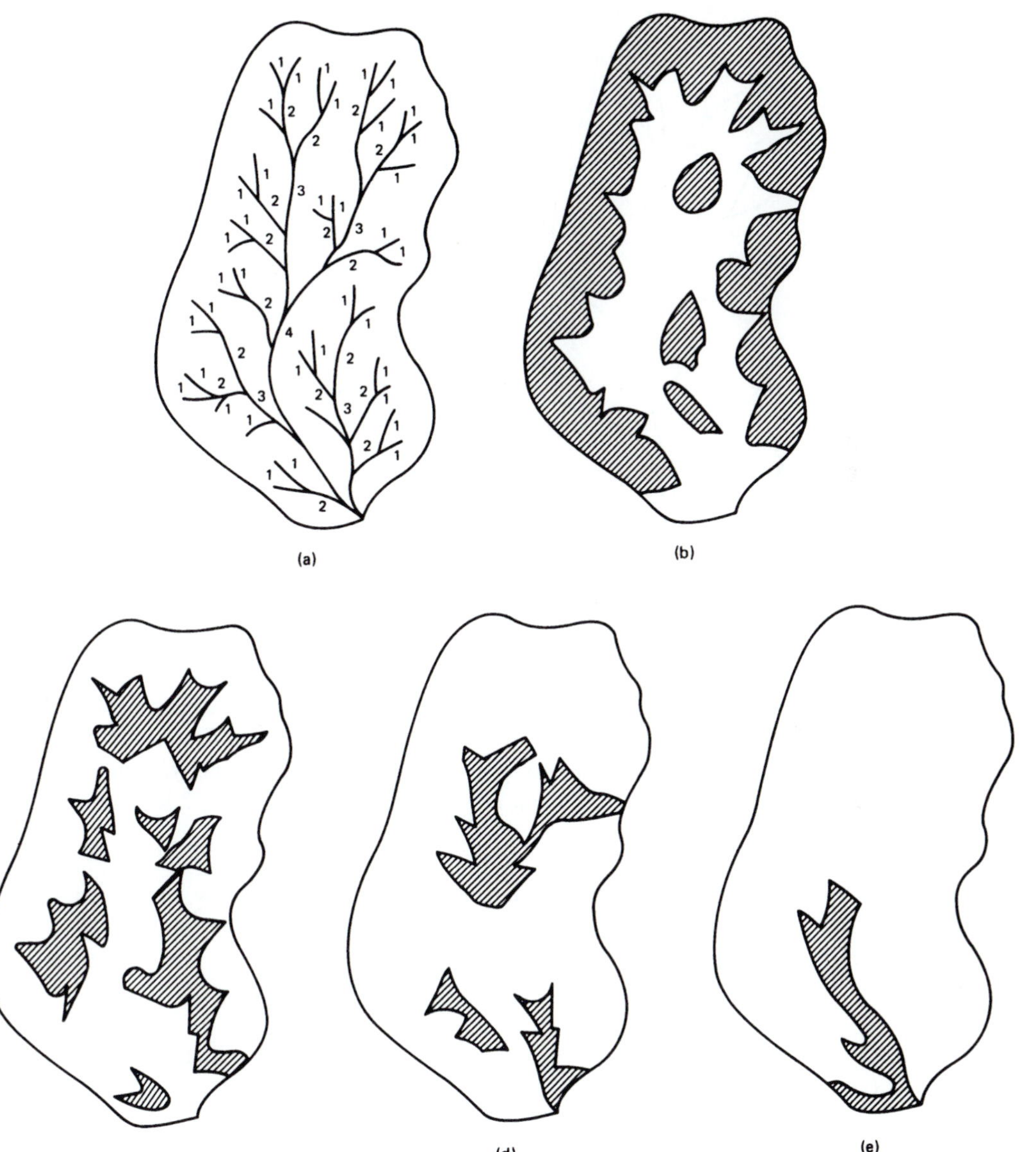

FIGURE 3–13 Identification of stream orders (a) and delineation of stream areas for first-order (b), second-order (c), third-order (d), and fourth-order (e) streams.

pal order is defined as the order of the principal channel (for example, the order of the tributary passing through the watershed outlet). The ordering of streams with a hypothetical watershed is shown in Figure 3–14. For this case, the watershed has a principal order of 4.

The concept of stream order is used to compute other indicators of drainage character. The bifurcation ratio (R_b) is defined as the ratio of the number of streams of any order to the number of streams of the next higher order. Values of R_b typically range from 2 to 4. Figure

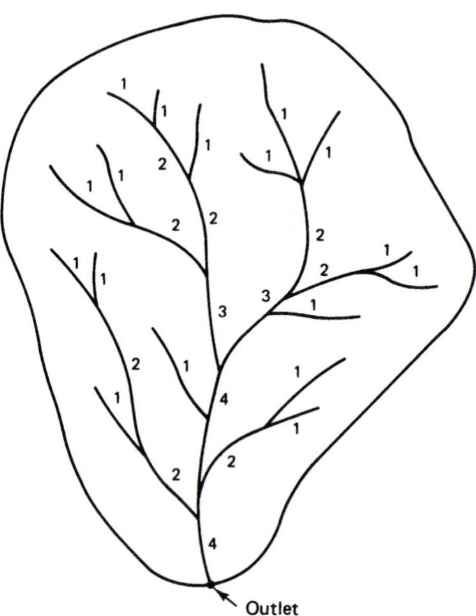

FIGURE 3–14 Ordering of stream of watershed having a principal order of 4.

3–15 displays the same watershed as in Figure 3–14, but the streams of each order have been specifically delineated. Figure 3–15a shows that there are seventeen first-order streams, with Figure 3–15b to d indicating 6, 2, and 1 streams of orders 2, 3, and 4, respectively. This yields bifurcation ratios of 2.83, 3.0, and 2.0 for stream orders 1 to 2, 2 to 3, and 3 to 4, respectively, and an average value of 2.6.

Law of stream numbers. Horton also proposed the law of stream numbers, which relates the number of streams of order i (N_i) to the bifurcation ratio and the principal order (k):

$$N_i = R_b^{k-i} \tag{3–23}$$

For a watershed with a bifurcation ratio of 2.6 and a fourth-order principal stream, Equation 3–23 becomes

$$N_i = 2.6^{4-i} \tag{3–24}$$

Thus the law of stream numbers would predict 18, 7, and 3 streams of order 1, 2, and 3, respectively; these agree closely with the 17, 6, and 2 streams for the watershed of Figures 3–14 and 3–15.

Law of stream lengths. The law of stream lengths relates the average length of streams of order i (L_i) to the stream length ratio (r_L) and the average length of first-order streams ($\bar{L}_1$):

$$L_i = \bar{L}_1 r_L^{i-1} \tag{3–25}$$

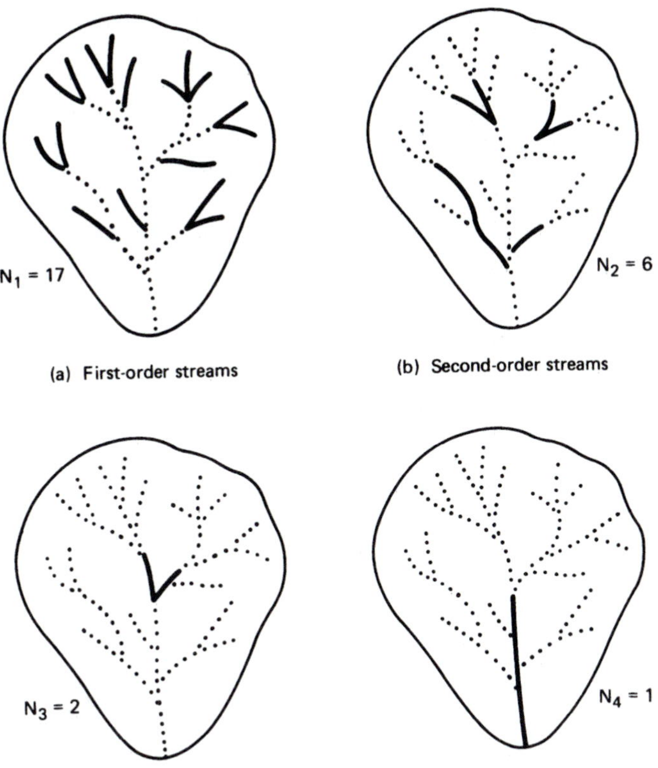

FIGURE 3–15 Stream-order separation for estimating the bifurcation ratio.

where the stream length ratio is defined as the average length of streams of any order to the average length of streams of the next lower order.

Law of stream areas. The law of stream areas is similar to the law of stream lengths. Specifically, the law relates the mean tributary area of streams of order i (A_i) to the mean drainage area of first-order basins ($\overline{A}_1$) and the stream area ratio (r_a):

$$A_i = \overline{A}_1 \, r_a^{i-1} \tag{3–26}$$

where the stream area ratio is the average basin area of streams of one order to the average area of basins of the next lower order. The similarity in Equations 3–25 and 3–26 reflects the high correlation that exists between watershed length and area.

Law of stream slopes. The law of stream slopes relates the average slope of streams of order i (S_i) to the average slope of first-order streams ($\overline{S}_1$) and the stream slope ratio (r_s):

$$S_i = \overline{S}_1 \, r_s^{1-i} \tag{3–27}$$

where the stream slope ratio is the average slope of streams of order j to the average slope of streams of the next higher order, $j + 1$.

Example 3–10

To illustrate the estimation of the three geomorphic parameters R_b, r_L, and r_a, the stream order for each reach of the channels in a 280-mi^2 watershed were delineated. The delineation of Figure 3–13a indicates a fourth-order principal stream. There are 38, 16, 4, and 1 streams of order 1, 2, 3, and 4, respectively; these values can be substituted into Equation 3–10 to get the following:

$$38 = R_b^3 \tag{3–28a}$$

$$16 = R_b^2 \tag{3–28b}$$

$$4 = R_b \tag{3–28c}$$

Equations 3–28 can be solved for R_b, using the principle of least squares. Taking the natural logarithms of both sides of Equation 3–23 yields

$$\ln_e N_i = (k - i)\ln_e R_b \tag{3–29}$$

Using the principle of least squares, it is easily shown that the least squares estimate of $\ln_e R_b$ is given by

$$\ln_e R_b = \frac{\Sigma\,[(\ln_e N_i)\,(k-i)]}{\Sigma\,[(k - i)^2]} \tag{3–30}$$

in which each summation is over the range from $i = 1$ to $i = (k–1)$, where k is the principal order. Thus, applying Equation 3–30 to the values of Equations 3–28 yields an estimate of 3.58 for R_b. Thus the law of stream orders has the following form for the watershed of Figure 3–13:

$$N_i = 3.58^{4-i} \tag{3–31}$$

The length of each stream segment was measured, with average lengths of 1.86, 2.79, 6.42, and 12.0 miles for streams of orders 1, 2, 3, and 4, respectively. The stream length ratio for Equation 3–25 can be determined using the concepts behind the development of Equation 3–30; however, because the form of Equation 3–25 is slightly different from that of Equation 3–23, the resulting equation is different:

$$\ln_e r_L = \frac{\Sigma\,[(\ln_e \bar{L}_i - \ln_e \bar{L}_1)(i - 1)]}{\Sigma\,[(i - 1)^2]} \tag{3–32}$$

where each of the summations is over all values of i from 2 to the principal order for the watershed. For the case of Figure 3–13 the resulting stream length ratio is 1.83, which yields the following law of stream lengths:

$$L_i = \bar{L}_1\,(1.83)^{i-1} \tag{3–33}$$

The area draining each of the subwatersheds for the basin of Figure 3–13 was computed. The average area for streams of each order were computed as 4.1, 4.4, 9.1, and 18.1 mi^2 for stream orders of 1, 2, 3, and 4, respectively. The delineation of these areas is given in Figure 3–13b to e. The least squares estimate of the stream-area ratio (r_a) of Equation 3–26 can be computed from the following:

$$\ln_e r_a = \frac{\Sigma\,[(\ln_e \bar{A}_i - \ln_e \bar{A}_1)(i - 1)]}{\Sigma\,[(i - 1)^2]} \tag{3–34}$$

where each of the summations is over all values of i from 2 to the principal order of the watershed. For this watershed, the stream area ratio is 1.55, which yields the following law of stream areas:

$$\overline{A}_i = \overline{A}_1 (1.55)^{i-1} \tag{3–55}$$

This example illustrates the case of analysis. Where a watershed is not fully defined, regional estimates of the three parameters (R_b, r_L, and r_a) can be obtained using regional averages. For the analysis case, then, it would be necessary to make similar analyses on many watersheds in the region and compute average values of the parameters. The regional averages could then be used for hydrologic synthesis.

3.5.5 Channel Cross Sections

Channel cross sections are a very important part of hydrologic analysis and design for a number of reasons. Many types of design problems require cross-section information, including the cross-sectional area. The wetted perimeter, slope, roughness, and average velocity are other important characteristics.

It is only necessary to walk beside a river or stream to recognize that, even over short distances, stream cross sections can take on a wide variety of shapes and sizes. Small streams in upland areas often have V-shaped cross sections. As the stream drains larger areas, past floods have sculptured out larger, often rectangular or trapezoidal sections. Many rivers have cross sections that consist of a relatively small V- or U-shaped channel that carries the runoff from small storms and the runoff during periods between storm events. This small channel area is often part of a much larger flat area on one or both sides of the channel; this flat area is called the floodplain, which is the area that is covered with water during times of higher discharges. A river cross section is shown in Figure 3–16. The low-flow cross section is approximately trapezoidal; however, it represents a small portion of the total cross section that is used to pass large flood events. The floodplain has low side slopes and could easily be represented as a rectangular section. Because of channel instability and erosional processes cross sections can change shape during flood events, and where such changes take place, they should be accounted for in developing a discharge rating table. Channel cross sections can also change because of development within the floodplain. Figure 3–16 shows the cross section after a bridge was built. While the low-flow section remains unchanged, the width of the section was decreased because of the abutments on each side and the four twin piers that support the bridge. From the continuity equation we know that for a given discharge either the average velocity will have to increase because of the reduced cross-sectional area created

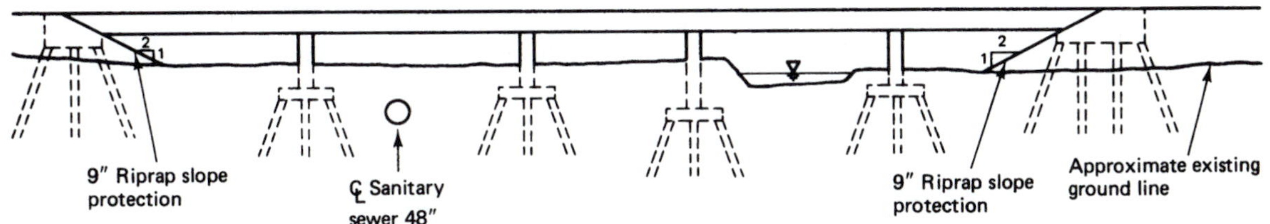

FIGURE 3–16 Elevation view of river cross section at site of proposed rail structure.

by the bridge or the flood surface elevation will increase. The calculations of water surface profiles are usually performed using hydraulic computations in computer programs such as the Corps of Engineers' HEC-2 or the Soil Conservation Service's WSP-2.

In practice, design problems require the calculation of flood surface elevations along the entire section of the channel affected by the structure. Figure 3–17 shows the plan view of the floodplain for the stream reach near the bridge. The change in the cross section at the bridge affects the flood profile both above and below the bridge. The delineation of flood profiles will be discussed later.

3.5.6 Channel Roughness

The roughness of a surface affects the characteristics of runoff, whether the water is on the surface of the watershed or in the channel. With respect to the hydrologic cycle, the roughness of the surface retards the flow. For overland flow, increased roughness delays the runoff and should increase the potential for infiltration. Reduced velocities associated with increased roughness should also decrease the amount of erosion. The general effects of roughness on flow in a channel are similar to those for overland flow.

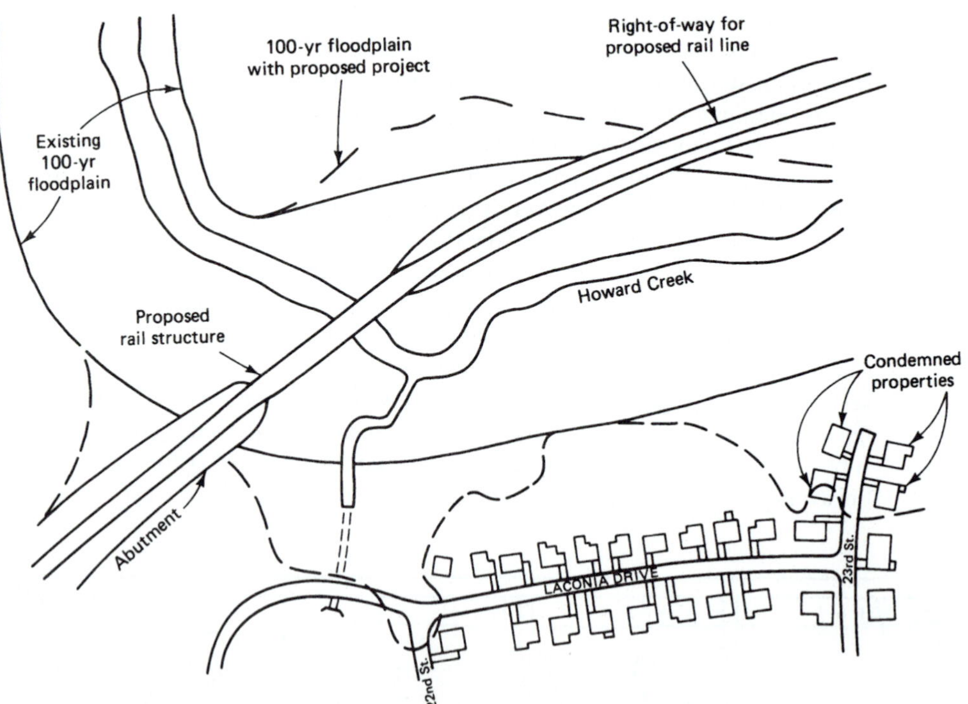

FIGURE 3–17 Plan view of site of proposed rail structure with existing and developed condition 100-year floodplains.

Manning's roughness coefficient (n) is required for a number of hydraulic computations. It is a necessary input in floodplain delineation. Also, a number of methods for estimating the timing of runoff use n as an input. It is also used in the design of stable channel systems. Thus it is an important input.

A number of methods exist for estimating the roughness coefficient. First, books that provide a series of pictures of stream channels with a recommended n value for each picture are available; thus for any natural channel a value of n can be obtained by comparing the roughness characteristics of the channel with the pictures and using the n value from the picture that appears most similar. Second, there are tables (for example, see Table 3–4) that give typical or average values of n for various channel conditions. However, the accuracy of n values from such tables are highly dependent on the degree of homogeneity of channel conditions and the degree of specificity of the table. The picture comparison and tabular look-up methods are used frequently because of their simplicity and because studies have not shown them to be highly inaccurate.

A third method is available that is potentially more accurate than either of the two methods discussed previously. The method involves the selection of a base value of n and then correcting the base value for each of the following five factors:

1. The degree of regularity of the surfaces of the channel cross section.
2. The character of variations in the size and shape of cross sections.
3. The presence and characteristics of obstructions in the channel.
4. The affect of vegetation on flow conditions.
5. The degree of channel meandering. This method is basically a table look-up solution, with the basic n value and each of the corrections obtained from a table. Each of the steps is outlined below. The basic n value represents a minimum of retardance due to friction. The corrections for the five factors attempt to correct a basic value for their affect on the turbulence of flow. The five corrections add to the basic n value, with a larger value of n indicating greater turbulence and retardance.

Step 1: Select Basic Coefficient (n_1). The basic n values of Table 3–5, are assumed to be for a channel having a straight and uniform cross section, with the sides and bottom surfaces cut into the natural material. This "basic" cross section has no vegetation on the bottom or sides.

Step 2: Correction for Channel Irregularity (n_2). The basic n assumes that the channel surfaces are the best (i.e., least retarding) obtainable for the materials involved. As the degree of irregularity increases, the turbulence and retardance is expected to increase. Thus the modifying value (n_2) increases as the degree of irregularity increases (see Table 3–6).

Step 3: Correction for Cross-Section Variation (n_3). The basic roughness value n assumes that the cross section has a relatively constant size and shape throughout the channel reach. As the size and shape of the channel changes, the degree of retardance increases, and

TABLE 3–4 Recommended Design Values of Manning Roughness Coefficients, n[a]

	Manning n Range[b]
I. Unlined open channels[c]	
A. Earth, uniform section	
1. Clean, recently completed	0.016–0.018
2. Clean, after weathering	0.018–0.020
3. With short grass, few weeds	0.022–0.027
4. In graveled soil, uniform section, clean	0.022–0.025
B. Earth, fairly uniform section	
1. No vegetation	0.022–0.025
2. Grass, some weeds	0.025–0.030
3. Dense weeds or aquatic plants in deep channels	0.030–0.035
4. Sides, clean gravel bottom	0.025–0.030
5. Sides, clean, cobble bottom	0.030–0.040
C. Dragline excavated or dredged	
1. No vegetation	0.028–0.033
2. Light brush on banks	0.035–0.050
D. Rock	
1. Based on design section	0.035
2. Based on actual mean section	
a. Smooth and uniform	0.035–0.040
b. Jagged and irregular	0.040–0.045
E. Channels not maintained, weeds and brush uncut	
1. Dense weeds, high as flow depth	0.08–0.12
2. Clean bottom, brush on sides	0.05–0.08
3. Clean bottom, brush on sides, highest stage of flow	0.07–0.11
4. Dense brush, high-stage	0.10–0.14
II. Roadside channels and swales with maintained vegetation[d,e] (values shown are for velocities of 2 and 6 ft/sec):	
A. Depth of flow up to 0.7 ft	
1. Bermuda grass, Kentucky bluegrass, buffalo grass	
a. Mowed to 2 in.	0.07–0.045
b. Length 4 to 6 in.	0.09–0.05
2. Good stand, any grass	
a. Length about 12 in.	0.18–0.09
b. Length about 24 in.	0.30–0.15
3. Fair stand, any grass	
a. Length about 12 in.	0.14–0.08
b. Length about 24 in.	0.25–0.13

[a]Estimates are by Bureau of Public Roads unless otherwise noted and are for straight alignment. A small increase in value of n may be made for channel alignment other than straight.

[b]Ranges of section I are for good to fair construction. For poor-quality construction, use larger values of n.

[c]For important work and where accurate determination of water profiles is necessary, the designer is urged to consult the following references and to select n by comparison of the specific conditions with the channels tested: *Flow of Water in Irrigation and Similar Canals*, by F.C. Scobey, U.S. Department of Agriculture, Technical Bulletin 652, February 1939 and *Flow of Water in Drainage Channels*, by C.E. Ramser, U.S. Department of Agriculture, Technical Bulletin 129, November 1929.

[d]*Handbook of Channel Design for Soil and Water Conservation*, prepared by the Stillwater Outdoor Hydraulic Laboratory in cooperation with the Oklahoma Agricultural Experiment Station, published by the Soil Conservation Service, U.S. Department of Agriculture, Publ. SCS-TP-61, March 1947, rev. June 1954.

[e]*Flow of Water in Channels Protected by Vegetative Linings*, W.O. Ree and V.J. Palmer, Division of Drainage and Water Control Research, Soil Conservation Service, U.S. Dept. of Agriculture Tech. Bull. 967, February, 1949.

TABLE 3–4 Recommended Design Values of Manning Roughness Coefficients, n[a] (*Continued*)

	Manning n Range[b]
B. Depth of flow 0.7–1.5 ft	
1. Bermuda grass, Kentucky bluegrass, buffalo grass	
a. Mowed to 2 in.	0.05–0.035
b. Length 4 to 6 in.	0.06–0.04
2. Good stand, any grass	
a. Length about 12 in.	0.12–0.07
b. Length about 24 in.	0.20–0.10
3. Fair stand, any grass	
a. Length about 12 in.	0.10–0.06
b. Length about 24 in.	0.17–0.09
III. Natural stream channels[f]	
A. Minor streams[g] (surface width at flood stage less than 100 ft)	
1. Fairly regular section	
a. Some grass and weeds, little or no brush	0.030–0.035
b. Dense growth of weeds, depth of flow materially greater than weed height	0.035–0.05
c. Some weeds, light brush on banks	0.04–0.05
d. Some weeds, heavy brush on banks	0.05–0.07
e. Some weeds, dense willows on banks	0.06–0.08
f. For trees within channel, with branches submerged at high stage, increase all above values by	0.01–0.10
2. Irregular sections, with pools, slight channel meander; increase value in la-e by	0.01–0.02
3. Mountain streams, no vegetation in channel, banks usually steep, trees and brush along banks submerged at high stage	
a. Bottom of gravel, cobbles, and few boulders	0.04–0.05
b. Bottom of cobbles, with large boulders	0.05–0.07
B. Floodplains (adjacent to natural streams)	
1. Pasture, no brush	
a. Short grass	0.030–0.035
b. High grass	0.035–0.05
2. Cultivated areas	
a. No crop	0.03–0.04
b. Mature row crops	0.035–0.045
c. Mature field crops	0.04–0.05
3. Heavy weeds, scattered brush	0.05–0.07
4. Light brush and trees[h]	
a. Winter	0.05–0.06
b. Summer	0.06–0.08

[f]For calculations of stage or discharge in natural stream channels, it is recommended that the designer consult the local district office of the Surface Water Branch of the U.S. Geological Survey, to obtain data regarding values of n applicable to streams of any specific locality. Where this procedure is not followed, the table may be used as a guide. The values of n tabulated have been derived from data reported by C.E. Ramser (see footnote c) and for other incomplete data.

[g]The tentative values of n cited are principally derived from measurements made on fairly short but straight reaches of natural streams. Where slopes calculated from flood elevations along a considerable length of channel, involving meanders and bends, are to be used in velocity calculations by the Manning formula, the value of n must be increased to provide for the additional loss of energy caused by bends. The increase may be in the range of perhaps 3% to 15%.

[h] The presence of foliage on trees and brush under flood stage will materially increase the value of n. Therefore, roughness coefficients for vegetation in leaf will be larger than for bare branches. For trees in channels or on banks, and for brush on banks where submergence of branches increases with depth of flow, n will increase with rising stage.

(*continued*)

TABLE 3–4 Recommended Design Values of Manning Roughness Coefficients, n[a] (*Continued*)

	Manning n Range[b]
5. Medium to dense brush[h]	
a. Winter	0.07–0.11
b. Summer	0.10–0.16
6. Dense willows, summer, not bent over by current	0.15–0.20
7. Cleared land with tree stumps, 100-150 per acre	
a. No sprouts	0.04–0.05
b. With heavy growth of sprouts	0.06–0.08
8. Heavy stand of timber, a few down trees, little undergrowth	
a. Flood depth below branches	0.10–0.12
b. Flood depth reaches branches	0.12–0.16
C. Major streams (surface width at flood stage more than 100 ft): Roughness coefficient is usually less than for minor streams of similar description on account of less effective resistance offered by irregular banks or vegetation on banks. Values of n may be somewhat reduced. Follow recommendation of note g if possible. The value of n for larger streams of most regular sections, with no boulders or brush, may be in the range shown.	0.028–0.033

the value of the basic n_1 should be increased. Table 3–7 provides values of n_3 for modifying the value of n_1 for the character of variations in size and shape of the cross section. The largest change reflects frequent alterations in the size and shape of the cross section.

Step 4: Correction for Obstructions (n_4). Obstructions, such as debris deposits, tree stumps and exposed tree roots, and large rocks and boulders, increase turbulence. In assessing the degree to which n_1 should be modified to reflect the effect of obstructions, the following factors should be considered: (1) the degree to which the obstructions occupy or reduce the average cross-sectional area; (2) the character of the obstructions (sharp-edged or angular objects induce greater turbulence than curved, smooth-surfaced objects); and (3) the position and spacing of obstructions transversely and longitudinally in the reach. The basic value n_1 assumes no obstructions, so the correction n_4 is applied to increase the n value for the presence of obstructions. Values of n_4 can be obtained from Table 3–8.

Step 5: Correction for Vegetation (n_5). The basic value n_1 assumes that the channel is devoid of vegetation. Since vegetation increases the surface roughness of the channel, the

TABLE 3–5 Roughness Coefficient Modifier (n_1)

Character of Channel	Basic n
Channels in earth	0.02
Channels cut into rock	0.025
Channels in fine gravel	0.024
Channels in coarse gravel	0.028

Source: Cowan, 1956.

TABLE 3–6 Roughness Coefficient Modifier (n_2)

Degree of Irregularity	Surface Comparable to	Modifying Value
Smooth	The best attainable for the materials involved	0.000
Minor	Good dredged channels; slightly eroded or scoured side slopes of canals or drainage channels	0.005
Moderate	Fair to poor dredged channels; moderately sloughed or eroded side slopes of canals or drainage channels	0.010
Severe	Badly sloughed banks of natural streams; badly eroded or sloughed sides of canals or drainage channels; unshaped, jagged, and irregular surfaces of channels excavated in rock	0.020

Source: Cowan, 1956.

TABLE 3–7 Roughness Coefficient Modifier (n_3)

Character of Variations of Size and Shape of Cross Sections	Modifying Value
Change in size or shape occurring gradually	0.000
Large and small sections alternating occasionally or shape changes causing occasional shifting of main flow from side to side	0.005
Large and small sections alternating frequently or shape changes causing frequent shifting of main flow from side to side	0.010–0.015

Source: Cowan, 1956.

TABLE 3–8 Roughness Coefficient Modifier (n_4)

Relative Effect of Obstructions	Modifying Value
Negligible	0.000
Minor	0.010–0.015
Appreciable	0.020–0.030
Severe	0.040–0.060

Source: Cowan, 1956.

value of n_1 must be increased to reflect the increased roughness due to vegetation. The values of n_5 in Table 3–9 can be used to modify the value of n_1 for the presence and character of vegetation in the channel. In selecting a value of n_5, it is necessary to reflect on the time of year (for example, growing season condition versus dormant season condition). Additionally, the following factors should be considered: (1) the height of the vegetation with respect to the depth of flow; (2) the ability of the vegetation to resist bending under the force of the flowing water; (3) the degree to which the vegetation occupies the flow path within the channel; and (4) the transverse and longitudinal variation of vegetation of different types, densities, and heights.

TABLE 3–9 Roughness Coefficient Modifier (n_5)

Vegetation and Flow Conditions Comparable to:	Degree of Effect on n	Range in Modifying Value
Dense growths of flexible turf grasses or weeds, of which Bermuda grass and bluegrass are examples, where the average depth of flow is two or more times the height of the vegetation. Supple seedling tree switches, such as willow, cottonwood, or salt cedar, where the average depth of flow is three or more times the height of the vegetation.	Low	0.005–0.010
Turf grasses where the average depth of flow is one to two times the height of the vegetation. Stemmy grasses, weeds, or tree seedlings with moderate cover, where the average depth of flow is two to three times the height of the vegetation. Bushy growths, moderately dense, similar to willows one to two years old, dormant season, along side slopes with no significant vegetation along bottom, where the hydraulic radius is greater than 2 ft.	Medium	0.010–0.020
Turf grasses where the average depth of flow is about equal to the height of vegetation. Willow or cottonwood trees 8 to 10 years old intergrown with some weeds and brush, dormant season, where the hydraulic radius is 2 to 4 ft. Bushy willows about 1 year old interwoven with some weeds in full foliage along side slopes, no significant vegetation along channel bottom where hydraulic radius is 2 to 4 ft.	High	0.025–0.050
Turf grasses where the average depth of flow is less than one-half the height of the vegetation. Bushy willows about 1 year old intergrown with weeds along side slopes, dense growth of cattails along channel bottom, all vegetation in full foliage, any value of hydraulic radius up to 10 or 12 ft. Trees intergrown with weeds and brush, all vegetation in full foliage, any value of hydraulic radius up to 10 to 12 ft.	Very high	0.050–0.100

Source: Cowan, 1956.

Step 6: Correction for Channel Meandering (n_6). The basic value n_1 assumes a straight channel reach. The degree of meandering can be measured using the ratio of the total meander length of the channel in the reach (l_m) to the straight length of the reach (l_s). Table 3–10 provides values of n_6 for selected ranges of the ratio l_m/l_s. The value of n_6 is also a function of the value n_s, which is the sum of the values of n_1 to n_5. If values of l_m and l_s are not readily available, the degree of meandering can be judged with the ordinal scale of minor, appreciable, or severe.

Step 7: Computation of Reach n. The value of the reach roughness coefficient equals the sum of the values of the basic value n_1 and the modifying values n_2 to n_6. The calculations can be performed on the computation sheet of Table 3–11.

Further considerations in estimating *n*. Tables 3–7 to 3–9 include ranges of values for modifying n values. For example, Table 3–8 gives a range for n_4 from 0.04 to 0.06

TABLE 3–10 Roughness Coefficient Modifier (n_6)

Ratio of Meander Length to Straight Length	Degree of Meander	Modifying Value[a]
1.0–1.2	Minor	0.000
1.2–1.5	Appreciable	$0.15\ n_s$
1.5 and greater	Severe	$0.30\ n_s$

[a] $n_s = n_1 + n_2 + n_3 + n_4 + n_5$

Source: Cowan, 1956.

for "severe" obstructions. The use of ranges for modifying values reflects both the inaccuracy of the tabled values and recognition that the effect of obstructions will vary from site to site. Recognition of these complicating factors does not simplify the task of selecting a representative value from the given range. Unless one has convincing documentation for selecting another value, the mean of the range is probably the best choice for a value.

The five factors for which corrections are provided are intended to reflect independent effects on flow retardance or roughness. In practice, channel conditions may suggest an in-

TABLE 3–11 Computation Sheet for Manning's Roughness Coefficient

Variable	Description Alternatives	Recommended Value	Actual Value
Basic, n_1	Earth	0.020	$n_1 =$ ________
	Rock	0.025	
	Fine gravel	0.024	
	Coarse gravel	0.028	
Irregularity, n_2	Smooth	0.000	$n_2 =$ ________
	Minor	0.005	
	Moderate	0.010	
	Severe	0.020	
Cross section, n_3	Gradual	0.000	$n_3 =$ ________
	Occasional	0.005	
	Alternating	0.010–0.015	
Obstructions, n_4	Negligible	0.000	$n_4 =$ ________
	Minor	0.010–0.015	
	Appreciable	0.020–0.030	
	Severe	0.040–0.060	
Vegetation, n_5	Low	0.005–0.010	$n_5 =$ ________
	Medium	0.010–0.020	
	High	0.025–0.050	
	Very high	0.050–0.100	
			Subtotal $n_s =$ ________
Meandering, n_6	Minor	0.000	$n_6 =$ ________
	Appreciable	$0.15n_s$	
	Severe	$0.30n_s$	
			Total = Reach $n =$ ________

terdependence of these factors. For example, dense vegetation may be seen as obstructions, and obstructions may contribute to irregularity of the channel cross section. Where interdependencies are evident in evaluating n for a stream reach, every effort should be made to determine the primary factor and use that factor to modify the value of n_1. Double counting will lead to biased estimates of n, with the bias suggesting an overly rough channel.

It should be of interest to examine the sensitivity of the final n value to the various channel conditions that produce the value. In examining the basic value n_1, relatively little scatter exists among the values, with a range of 0.008. Thus variation in the final value of n will not be sensitive to the underlying natural material. A mean value of about 0.025 could be used with little loss of accuracy. However, the actual value from Table 3–3 should be used.

For all of the correction factors except meandering, it should be evident from the values summarized on the worksheet (Table 3–11) that vegetation is the dominant factor. Not only does this factor have the largest possible correction (0.10) but the ranges of the values are relatively large. Thus the final value of n will be most sensitive to the accuracy of the vegetation adjustment. Obstructions have the next greatest potential for effect, with an effect that is approximately one-half the effect of vegetation. The range for the corrections for irregularity (n_2) and cross section (n_3) are relatively small when compared with the ranges provided for obstructions (n_4) and vegetation (n_5); thus for many channels the irregularity and cross-section adjustments will not be dominant factors. The sensitivity of n to meandering can be appreciable, especially if a large value of the subtotal roughness coefficient n_s results because of large values of either n_4 or n_5. Using approximate mean values for both n_4 and n_5 of 0.02 and 0.04, respectively, severe meandering would equal about 0.025 because of n_1, n_4, and n_5 alone. This correction is larger than the maximum values of both n_2 and n_3; thus meandering can be a sensitive factor.

Example 3–11

The example in Table 3–12 was obtained from a U.S. Department of Agriculture publication. It serves not only as an example of the estimation of numbers but also as an illustration of the type of data that must be observed to estimate n. While a value could possibly be estimated from detailed photographs of the channel, a site visit is highly recommended (probably an absolute necessity). The example of Table 3–12 is based on a case where n has been determined so that comparison between the estimated and actual n can be shown. The method of Tables 3–5 to 3–10 suggests a value of 0.11, while computations from actual flow conditions yield values of 0.095 and 0.104, respectively; these represent errors of 15.8% and 5.8%, respectively. These errors are well within the tolerances suggested from the range of n values given for each of the conditions of Table 3–4.

3.5.7 Uniform Flow Computations

Given the complexity of flood runoff processes, both overland and channel flow, it is necessary to make a number of assumptions in hydrologic design work. It is quite common to assume that flood runoff can be described using equations developed for uniform flow conditions. Uniform flow exists when both the depth of flow and the flow rate (discharge) are constant with distance either along the flow path for overland flow or for every section in a channel reach. Although one could argue that this is not a valid assumption during flooding,

TABLE 3–12 Calculation of Manning's Roughness Coefficient: Example 3–11

Channel: Camp Creek dredged channel near Seymour, Illinois.

Description: Course straight; 661 ft long. Cross section, very little variation in shape; variation in size moderate, but changes not abrupt. Side slopes fairly regular; bottom uneven and irregular. Soil, lower part yellowish gray clay; upper part, light gray silty clay loam. Condition, side slopes covered with heavy growth of poplar trees 2 to 3 in. in diameter, large willows and climbing vines; thick growth of water weed on bottom; summer condition with vegetation in full foliage.

Average cross section approximates a trapezoid with side slopes about 1.5 to 1 and bottom width about 10 ft. At bankfull stage, average depth and surface width are about 8.5 and 40 ft, respectively.

Step	Remarks	Modifying Values
1.	Soil materials indicate minimum basic *n*.	0.02
2.	Description indicates moderate irregularity.	0.01
3.	Changes in size and shape judged insignificant	0.00
4.	No obstructions indicated.	0.00
5.	Description indicates very high effect of vegetation.	0.08
6.	Reach described as straight.	0.00
	Total estimated *n*	0.11

USDA Tech. Bull. 129, Table 9, p. 96, gives the following determined values for n for the channel; for average depth of 4.6 ft, $n = 0.095$; for average depth of 7.3 ft, $n = 0.104$.

Source: USDA Tech. Bull. 129, Plate 29-C for photograph and Table 9, p. 86, for data.

the assumption permits necessary simplifications. Also, it has not been shown that the assumption of uniform flow leads to incorrect hydrologic designs.

In uniform flow, the energy grade line, the water surface, and the channel bottom are parallel to each other; that is, the slopes of the three are equal. As indicated, flow characteristics (discharge, flow depth, velocity distribution, and cross-sectional area) are constant along the flow path. If these conditions exist, the resistance of the channel cross section will equal and be opposite to the force component due to gravity.

To develop an equation that can be used to describe uniform flow, a force diagram of an incremental section of the flow regime will be used. Figure 3–18 shows a force diagram for an increment section of length Δx. For the body of fluid in the section, the following forces must be considered: (1) the weight of the water W; (2) the hydrostatic pressure forces, F_1 and F_2, which act on each end of the section; and (3) the friction force F_f of the channel cross section on the water. Summing forces in the x direction yields

$$W \sin\theta + F_1 - F_2 - F_f = 0 \tag{3–36}$$

Recognizing that the hydrostatic forces are equal and that the weight of the water in the section equals $\gamma A \Delta X$ yields

$$\gamma A \Delta X \sin\theta - F_f = 0 \tag{3–37}$$

in which γ is the specific weight of the water and A is the cross-sectional area of the flow. For small slopes $\sin\Theta$ is approximately equal to the bed slope S_o. Since for uniform flow the depth is constant, the slopes of the energy grade line (EGL) and hydraulic grade line (HGL) equal the bed slope; thus

$$F_f = \gamma A \Delta X S_o \tag{3–38}$$

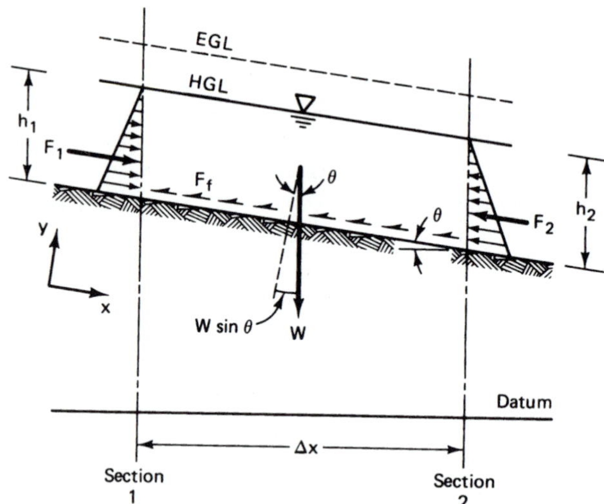

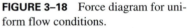

FIGURE 3–18 Force diagram for uniform flow conditions.

The friction force F_f equals the product of the shear stress, τ, and the surface contact area between the flow path and the water. The contact area equals the wetted perimeter P times the length of the flow path ΔX. Chezy showed that the shear stress is proportional to the velocity squared. Thus the friction force is given by

$$F_f = \tau P \Delta X = C_f V^2 P \Delta X \tag{3–39}$$

where C_f is a proportionality constant. Substituting Equation 3–39 into Equation 3–38 and solving for the velocity yields

$$V = C R_h^{0.5} S^{0.5} \tag{3–40}$$

in which R_h is called the hydraulic radius (in ft) and equals A/P, and C is a constant called the Chezy coefficient. In practice, C is not constant but is a function of the hydraulic conditions of the flow. Manning found experimentally that C could be represented by $1.49 R_h^{1/6}/n$, where n is Manning's roughness coefficient that is characteristic of the flow path. Thus Manning's equation for computing the velocity of uniform flow is

$$V = \frac{1.49}{n} R_h^{2/3} S^{1/2} \tag{3–41}$$

Using the continuity equation $Q = VA$, the discharge is given by

$$Q = \left(\frac{1.49}{n}\right) A R_h^{2/3} S^{1/2} = \left(\frac{1.49}{n}\right) \frac{A^{5/3} S^{1/2}}{P^{2/3}} \tag{3–42}$$

3.6 TRAVEL TIME

Most hydrologic designs involve some measure of a flood discharge. Discharges have units of volume per time (L^3/T) such as ft^3/sec, m^3/sec, or acre-ft/day. Various watershed and hydrometeorological characteristics can be used to reflect the volume of flood runoff;

for example, the product of the drainage area and the depth of rainfall gives a volume of water that is potentially available for runoff. But volume alone is not adequate for many design problems. As the dimensions of discharge indicate, time is an important element in hydrologic design. A given volume of water may or may not present a flood hazard; the hazard will depend on the time distribution of the flood runoff. If a significant portion of the total volume passes a given location at about the same time, flood damages will occur. Conversely, if the total volume is distributed over a relatively long period of time, flood damages may be minimal. In some cases, instantaneous inundation causes major damages, such as when an automobile is submerged. Short-duration inundation of an agricultural-crop field does not usually cause major damage. Duration of flooding is an important factor.

Because of the importance of the timing of runoff, most hydrologic models require a watershed characteristic that reflects the timing of runoff. A number of time parameters have been developed. While the timing of runoff is a characteristic of a watershed, time parameters are formulated as a function of other watershed characteristics and, in some cases, rainfall intensity.

A number of time parameters are commonly used with hydrologic and hydraulic models, including the time of concentration, the time lag, and reach travel time. Why is time an important input to hydrologic and hydraulic models? Recognizing that this is a somewhat philosophical question from a hydrologic design standpoint, it may still be worthwhile to consider three types of hydrologic design problems. For designs on small watersheds, such as for the design of either rooftop drains or street-drainage inlets, the time parameter may be an indicator of the intensity and volume of rainfall and the degree to which the rainfall will be attenuated; a short time period would suggest little attenuation of the rainfall intensity because the storages of the hydrologic processes have minimal effects. For designs on larger watersheds, time parameters may be indicators of watershed storage and the effect of storage on the time distribution of runoff. Watershed storage directly affects the shape and the time distribution of the runoff hydrograph. For designs where the time variation of flood runoff is routed through channel reaches, the reach travel time reflects the effect of channel storage on the attenuation of the flood discharge.

Errors in a time parameter will cause errors in designs based on the time parameter. As much as 75% of the total error in an estimate of the peak discharge can result from errors in the estimated value of the time parameter. Recognizing the importance of time parameters in hydrologic design and evaluation, hydrologists have developed numerous methods for estimating the various time parameters. Unfortunately, most of the empirical formulas have been based on very limited data. Their applicability is constrained by the lack of diversity in the data; an empirical estimation method should be used with considerable caution either for watersheds having characteristics different from those of the watersheds used to calibrate the method or for watersheds in other geographic regions. Both uses represent extrapolation of data. Thus a design engineer, when faced with a design problem requiring an estimate of a time parameter, must choose from among many of the alternatives and often apply the time parameter with little or no knowledge of its accuracy.

The most frequently used time parameters are the time of concentration, the lag time, the time to peak, the time to equilibrium, and the time-area curve. Time parameters are usually defined in terms of either the physical characteristics of a watershed or the distribution of rainfall excess and direct runoff. The time of concentration is the most widely used time parameter, so the discussion will be limited at this point to the time of concentration; other time parameters will be discussed in other chapters.

3.6.1 Definitions

There are two commonly accepted definitions of the time of concentration. First, t_c is defined as the time required for a particle of water to flow hydraulically from the most distant point in the watershed to the outlet or design point; methods of estimation based on this definition use watershed characteristics, and sometimes a precipitation index such as the 2-yr, 2-hr rainfall intensity. A number of empirical equations that are based on this definition will be discussed herein.

The second definition is introduced here only for completeness because some of the terms have not been previously introduced. In the second definition, the t_c is based on a rainfall hyetograph (defined in Chapter 4) and the resulting runoff hydrograph (defined in Chapter 9). From the actual hyetograph and hydrograph, the rainfall excess and direct runoff are computed. The time of concentration is the time between the center of mass of rainfall excess and the inflection point on the recession of the direct runoff hydrograph. As an alternative, t_c is sometimes computed as the time difference between the end of rainfall excess and the inflection point. This definition is discussed in more detail in Chapter 9.

It is recognized that both methods of estimating t_c do not provide either the true value or reproducible values of t_c. Methods based on watershed characteristics may assume, for example, that Manning's Equation is always valid and that the roughness coefficient applies throughout the flow regime. Ignoring both the difficulties in selecting a single value of the roughness coefficient for a seemingly homogeneous flow regime and the assumption that the hydraulic radius remains constant, one must still contend with the problem of reproducibility of the input; that is, different users of a method would select different values of the input variables even when applying the method on the same drainage area. In summary, while the time of concentration is an important input to hydrologic design, it is neither a highly accurate input nor highly reproducible.

Obtaining t_c estimates from rainfall and runoff data suffers from some difficulties. Universally accepted methods of separating either base flow from direct runoff or losses from rainfall excess do not exist. Both of these separation requirements can introduce significant variation in estimated t_c values. In using rainfall and runoff data to estimate t_c, one must also recognize and adjust for factors such as antecedent soil moisture, intermittent rainfall patterns, nonlinearities in the convolution process, and variation in the recurrence intervals of the storm. In summary, every designer should recognize that a single correct method for estimating t_c is not possible, and therefore, the true value can never be determined.

3.6.2 Classifying Time Parameters

A system for classifying time parameters is helpful because numerous empirical formulas have been developed. Almost all of the methods are based on four types of input: slope, watershed size, flow resistance, and water input. These characteristics are either for the overland flow portion of the watershed, the pipe system, or the channel system. Methods for overland flow can be further subdivided into sheet flow and concentrated flow methods.

Methods for estimating time parameters can be distinguished on the basis of both their input requirements and dominant flow regime. The classification system presented here (Table 3–13) could be further separated on the basis of the time parameters (for example, time of concentration, time lag, or time to peak); however, this additional separation criterion will not be used because the time of concentration is of primary interest herein.

Methods that are intended to predict the time of concentration for watersheds on which overland flow dominates can include variables to represent each of the four input types. Measures of the overland-flow resistance include Manning's roughness coefficient, n; the runoff coefficient of the Rational Method, C; the SCS runoff curve number, CN; the percentage of imperviousness, I; or a qualitative descriptor of the land cover type. Any one of a number of size and slope parameters can be used to represent the length and slope of overland flow. A rainfall parameter, such as the 2-yr, 24-hr rainfall depth, is sometimes used as input to indicate the effect of the surface runoff availability on the time of travel.

Methods for estimating t_c that include channel characteristics as input variables should be used when channel flow makes a significant contribution to the total travel time of runoff (see Table 3–13). Manning's roughness coefficient, n, is the most widely used index of channel-flow resistance. Espey and Winslow developed a coefficient to indicate the degree of channelization, which could also be an indicator of flow resistance. Several slope and size parameters are used as variables, with the length and slope of the main channel being the most frequently used. The input factor for a channel can be represented by one of several variables, including the precipitation intensity, the volume of surface runoff, and the hydraulic radius for bankfull flow.

For many small urban watersheds, the timing characteristics of the runoff are controlled by a pipe system (see Table 3–13). Thus methods for predicting time parameters on such watersheds should include one or more input variables that reflect the physical characteristics of the pipe system. Manning's roughness coefficient for the pipe could be used to reflect flow resistance. The input could be represented by the maximum flow rate, q_p, the largest or smallest pipe diameter, or the mean hydraulic radius of the pipe system. Any one of a number of size variables could be included in these equations to represent size. The slope of the pipe is obviously the variable used to represent slope.

The classification system given in Table 3–13 can be used for classifying time parameter prediction methods. While some prediction methods will fall into one of the four classes

TABLE 3–13 Criteria for Classifying Time Parameters and Variables[a] Commonly Used to Represent the Input Type

	Input Type			
Flow Regime	Flow Resistance	Watershed Size	Slope	Water Input
Sheet flow	n	L	S	i
Concentrated flow	n, C, CN	L, A	S	i
Channel	n, ϕ	L_c, $L_{10\text{–}85}$, L_{ca}	S_c, $S_{10\text{–}85}$	R_h, i, Q
Pipe	n	L	S	R_h, q_p

[a]Variables are defined in the Notation.

based on flow regime, a significant number of others will have to be identified as a "mixed" method (i.e., one that includes variables reflecting different flow regimes). For example, designs required for urbanized watersheds that include both a significant pipe system and overland flow may require a time-parameter model that includes variables that reflect all four flow regimes: sheet flow, concentrated flow, channel flow, and pipe flow.

In practice, terminology is not always applied consistently; therefore, it is worthwhile clarifying some common terms associated with the time of concentration. The term overland flow could be applied to a number of flow regimes. It is, however, worthwhile separating this into sheet flow and concentrated flow. Sheet flow occurs in the upper reaches of a watershed, usually over very short flow paths. Typically, this is evident on steeply sloped paved surfaces, where it appears as shallow layers, often with small waves. A kinematic wave equation is usually used to compute travel times for sheet flow. After some distance, topography usually causes the flow to concentrate in rills, then swales or gutters, and then gullies. This is referred to as concentrated flow. Either Manning's Equation or the velocity method is commonly used to estimate travel times of concentrated flow.

Manning's Equation is used to estimate velocities of overland flow, both sheet flow and concentrated flow. It can be shown that the kinematic-wave equation for estimating sheet-flow travel time is based on Manning's Equation with the assumption that the hydraulic radius equals the product of the rainfall intensity and the travel time. The velocity method for concentrated flow uses Manning's Equation with an assumed depth and Manning's n. Frequently used curves of velocity versus slope are valid only as long as the assumed depth and n are accurate.

3.6.3 Velocity Method

The velocity method is based on the concept that the travel time (T_t) for a particular flow path is a function of the length of flow (L) and the velocity (V):

$$T_t = \frac{L}{60\ V} \tag{3–43}$$

in which T_t, L, and V have units of min, ft, and ft/sec, respectively. The travel time is computed for the principal flow path. Where the principal flow path consists of segments that have different slopes or land cover, the principal flow path should be divided into segments and Equation 3–43 should be used for each flow segment. The time of concentration is then the sum of the travel times:

$$t_c = \sum_{i=1}^{k} T_{ti} = \sum_{i=1}^{k} \left(\frac{L_i}{60\ V_i} \right) \tag{3–44}$$

in which k is the number of segments and the subscript i refers to the flow segment.

The velocity of Equation 3–43 is a function of the type of flow (sheet, concentrated flow, gully flow, channel flow, pipe flow), the roughness of the flow path, and the slope of the flow path. A number of methods have been developed for estimating the velocity. Flow velocities in pipes and open channels can be computed using Manning's Equation:

$$V = \frac{1.486}{n} R_h^{2/3} S^{1/2} \tag{3–45}$$

in which V is the velocity (ft/sec), n is the roughness coefficient, R_h is the hydraulic radius (ft), and S is the slope (ft/ft). Values of n for overland-flow surfaces are given in Table 3–3. Values of n for channels can be obtained from either Table 3–4 or through the computation of Cowan's method (Table 3–11).

Equation 3–45 can be simplified so that V is only a function of the slope by assuming values for n and R_h. This gives a relationship between the velocity and the average slope of the surface:

$$V = kS^{0.5} \tag{3–46}$$

in which V is the velocity (ft/sec) and S is the slope (ft/ft). Thus, k equals $1.486R_h^{2/3}/n$. For Equation 3–46, the value of k is a function of the land cover with the effect measured by the value of n and R_h. Values for selected land covers are given in Table 3–14. These are shown in Figure 3–19. Equation 3–46 and Figure 3–19 are a simplification of Manning's Equation (Equation 3–45). Values were assumed for n and R_h and used to compute the values for k for Equation 3–46. Other curves could be developed for different flow regimes by selecting values for n and R_h. The assumed values of n and R_h are given in Table 3–14.

After short distances, runoff tends to concentrate in rills and then gullies of increasing proportions. Where roughness coefficients exist for such flow, Manning's Equation can be

TABLE 3–14 Coefficients of Velocity Versus Slope Relationship for Estimating Travel Times with the Velocity Method

Land Use/Flow Regime	n	R_h (ft)	k
Forest			
Dense underbrush	0.8	0.25	0.7
Light underbrush	0.4	0.22	1.4
Heavy ground litter	0.2	0.20	2.5
Grass			
Bermudagrass	0.41	0.15	1.0
Dense	0.24	0.12	1.5
Short	0.15	0.10	2.1
Short grass pasture	0.025	0.04	7.0
Conventional tillage			
With residue	0.19	0.06	1.2
No residue	0.09	0.05	2.2
Agricultural			
Cultivated straight row	0.04	0.12	9.1
Contour or strip cropped	0.05	0.06	4.6
Trash fallow	0.045	0.05	4.5
Rangeland	0.13	0.04	1.3
Alluvial fans	0.017	0.04	10.3
Grassed waterway	0.095	1.0	15.7
Small upland gullies	0.04	0.5	23.5
Paved area (sheet flow)	0.011	0.06	20.8
Paved area (sheet flow)	0.025	0.2	20.4
Paved gutter	0.011	0.2	46.3

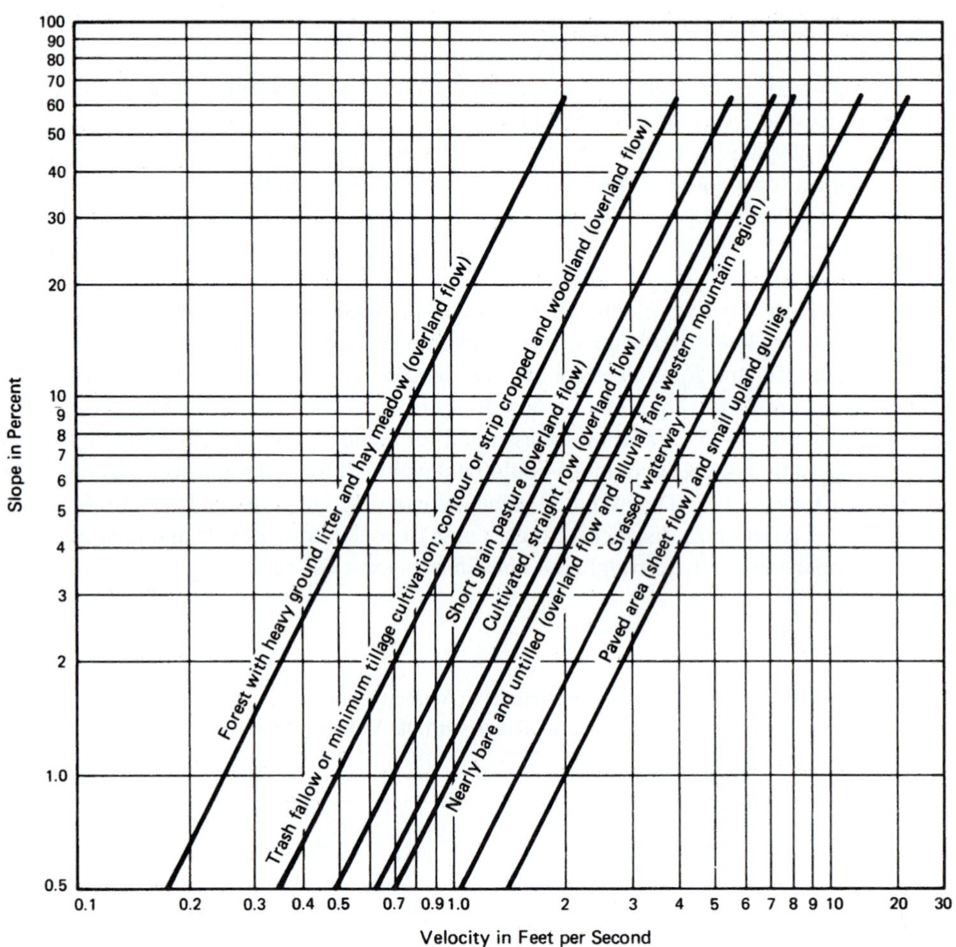

FIGURE 3–19 Velocities for upland method of estimating t_c.

used. Alternatively, the average velocity can be obtained using Equation 3–46 or Figure 3–19, where values are given for grassed waterways, unpaved gullies, and paved areas.

Flow in gullies empties into channels or pipes. Open channels are assumed to begin where either the blue stream shows on USGS quadrangle sheets or the channel is visible on aerial photographs. Cross-section information (for example, depth-area and roughness) should be obtained for all channel reaches in the watershed. Manning's Equation or water surface profile information can be used to estimate average flow velocities. The velocity should be computed for normal depth, which varies with the magnitude of the storm. Flows with return periods from 1.5 to 3 years often represent bankfull conditions, with flows greater than bankfull causing inundation of the floodplain.

3.6.4 Sheet-Flow Travel Time

At the upper reaches of a watershed, runoff does not concentrate into well-defined flow paths, such as gullies or swales. Instead it flows over the surface at reasonably uniform, shallow depths. Hydrologically speaking, this is sheet flow. It is evident on long, sloping streets during rainstorms. After some distance, the flow begins to converge into concentrated flow paths that have depths noticeably greater than that of the shallow flow. The distance from the upper end of the watershed or flow surface to the point where significant concentrated flow begins is termed the sheet-flow length. For steeply sloped, impervious surfaces the sheet-flow length can be several hundred feet. For shallow slopes or for pervious surfaces concentrated flow will begin after relatively short sheet-flow lengths.

In the upper reaches of a watershed, sheet-flow runoff during the intense part of the storm will flow as a shallow layer with a reasonably constant depth. An equation, referred to as the kinematic wave equation for the equilibrium time, can be developed using Manning's equation with the assumption that the hydraulic radius equals the product of the rainfall intensity and the travel time, for example, $R_h = i\,T_t$. Using the velocity equation of Equation 3–43 with the travel time equal to the time of concentration, Manning's Equation becomes

$$V = \frac{L}{T_t/60} = \frac{1.49(60)}{n} R_h^{2/3} S^{1/2} = \frac{1.49(60)}{n} \left(\frac{i\,T_t}{60(12)}\right)^{2/3} S^{1/2}$$

in which i [=] in./hr, T_t [=] min, S [=] ft/ft, and L [=] ft. Solving for the travel time yields

$$T_t = \frac{0.938}{i^{0.4}} \left(\frac{nL}{\sqrt{S}}\right)^{0.6} \tag{3–47}$$

Equation 3–47 requires the rainfall intensity i for the time of concentration. Since T_t is not initially known, it is necessary to assume a value of T_t to obtain i from a rainfall *IDF* curve (see Figure 4–4) and then compute T_t. If the initial assumption for T_t was incorrect, then a new estimate of i is obtained from the *IDF* curve using the computed value of T_t. The iterative process should be repeated until the value of T_t does not change.

To by-pass the necessity to solve Equation 3–47 iteratively, Welle and Woodward (1986) assumed a power-model relationship between rainfall intensity and rainfall duration. Using a return period of two years, they substituted the 2-yr, 24-hr rainfall depth for the rainfall intensity i and derived the following alterative model for Equation 3–47:

$$T_t = \frac{0.42}{P_2^{0.5}} \left(\frac{nL}{S^{0.5}}\right)^{0.8} \tag{3–48}$$

in which L is the flow length (ft), S is the average slope (ft/ft), P_2 is the 2-yr, 24-hr rainfall depth (in.) and T_t [=] min. Equation 3–48 has the important advantage that an iterative solution is not necessary.

The n values for use with Equations 3–47 and 3–48 are given in Table 3–3 and are for very shallow flow depths, 0.1 ft or so; these values reflect the effects of rain drop impact; drag over plane surfaces; obstacles such as litter, crop ridges, and rocks; and erosion and transportation of sediment. The rainfall depth P_2 for Equation 3–48 can be obtained from an *IDF* curve such as Figure 4–4.

In addition to the previously mentioned assumptions, these kinematic equations make the following assumptions: (1) no local inflow; (2) no backwater effects; (3) no storage effects; (4) the discharge is only a function of depth, for example, $q = ay^b$; and (5) planar, non-converging flow. These assumptions became less realistic for any of the following conditions: the slope decreases, the surface roughness increases, or the length of the flow path increases. Recognizing that Equations 3–47 and 3–48 can yield unusually long times of concentration, limits are often, somewhat arbitrarily, placed on flow lengths. Common length limits are from 100 ft to 300 ft. TR-55 (SCS, 1986) places a limit of 300 ft on Equation 3–48; however, many feel that this is too long and a shorter length of 100 ft should be used. This limit would not necessarily apply to Equation 3–47. However, it should be recognized that the five assumptions given above can be violated for the other inputs, n and S. Empirical evidence (McCuen and Speiss, 1995) suggests that Equation 3–47 (but not Equation 3–48) should only be used for flow conditions where the term $nL/\sqrt{S}$ equals approximately 100. For a given n and S, the allowable length can be estimated as $100\ S^{0.5}n^{-1}$.

Example 3–12

Consider a short-grass surface of 120 ft in length and at a slope of 0.2%, or 0.002 ft/ft. Assume that the local drainage policy requires use of a 10-yr return period for design. Manning's n for short grass is 0.15 (see Table 3–3). When using Equation 3–47, an estimate of T_t is used to obtain i from Figure 4–4. In this example, a T_t of 5 min is assumed; thus, the initial estimate of the intensity is 8 in./hr, which gives the following estimate of the sheet-flow travel time:

$$T_t = \frac{0.938}{i^{0.4}}\left[\frac{0.15\ (120)}{(0.002)^{0.5}}\right]^{0.6} = \frac{34.28}{i^{0.4}} = \frac{34.28}{(8)^{0.4}} = 14.9 \text{ min.}$$

Since this is substantially greater than the initial estimate of 5 min, the revised travel time of 15 min is entered into Figure 4–4 to obtain a new estimate of the intensity, which is 5.1 in./hr. This produces a new estimate of T_t:

$$T_t = \frac{34.28}{(5.1)^{0.4}} = 17.9 \text{ min.}$$

Another iteration produces $i = 4.7$ in./hr and $T_t = 18.5$ min. A final iteration yields $i = 4.6$ in./hr and $T_t = 18.6$ min.

As an alternative, the *SCS* kinematic wave equation could be used. From Figure 4–4, the 2-yr, 24-hr intensity is 0.13 in./hr, which yields a P_2 of 3.12 in. Thus, Equation 3–48 yields

$$T_t = \frac{0.42}{(3.12)^{0.5}}\left[\frac{0.15\ (120)}{0.002^{0.5}}\right]^{0.8} = 28.8 \text{ min}$$

This is substantially greater than the 18.6 min obtained with Equation 3–47. The difference between the two estimates is due in part to the return period used and the structural coefficients of the two equations.

Example 3–13

The data of Table 3–15 can be used to illustrate the estimation of the time of concentration with the velocity method. Two watershed conditions are indicated: pre- and post-development. In the pre-development condition, the 4-acre drainage area is primarily forested, with a natural channel having a good stand of high grass. In the post-development condition, the channel has been eliminated and replaced with a 15-in.-diameter pipe.

TABLE 3–15 Characteristics of Principal Flow Path for Time of Concentration Estimation: Example 3–13

Watershed Condition	Flow Segment	Length (ft)	Slope (ft/ft)	Type of Flow	k
Existing	1	140	0.010	Overland (forest)	2.5
	2	260	0.008	Grassed waterway	15.7
	3	480	0.008	Small upland gully	23.5
Developed	1	50	0.010	Overland (short grass)	2.1
	2	50	0.010	Paved	20.8
	3	300	0.008	Grassed waterway	15.7
	4	420	0.009	Pipe-concrete (15 in. dia.)	—

For the pre-development condition, the velocities of flow for the overland and grassed waterway segments can be obtained from either Equation 3–46 and Table 3–14 or Figure 3–19. For the slopes given in Table 3–15, the velocities are 0.25 and 1.40 ft/sec, respectively. For the roadside channel, the velocity can be estimated using the equation for concentrated flow in an unpaved gully, which has a k of 23.5. Thus, the velocity is 2.1 ft/sec. The pre-development time of concentration can be computed with Equation 3–44:

$$t_c = \frac{140}{0.25} + \frac{260}{1.4} + \frac{480}{2.1} = 560 + 186 + 229 = 975\ sec = 16.2 \text{ min}$$

For the post-development conditions, the flow velocities for the first three segments can be determined from Equation 3–46 and Table 3–14. For the slopes given in Table 3–15, the velocities for the first three segments are 0.21, 2.08, and 1.40 ft/sec, respectively. Assuming Manning's coefficient of 0.011 for the concrete pipe, the full-flow velocity is

$$V = \frac{1.486}{0.011}\left(\frac{1.25}{4}\right)^{0.67}(0.009)^{0.5} = 5.9 \text{ ft/sec}$$

A slope of 0.009 ft/ft is used since the meandering roadside channel was replaced with a pipe, which resulted in a shorter length of travel and, therefore, a larger slope. Thus the time of concentration is

$$t_c = \frac{50}{0.21} + \frac{50}{2.08} + \frac{300}{1.40} + \frac{420}{5.9}$$

$$= 238 + 24 + 214 + 71 = 547 \text{ sec} = 9.1 \text{ min}$$

Thus the land development decreased the time of concentration from 16.2 min to 9.1 min.

Example 3–14

Figure 3–20a shows the principal flow path for the existing conditions of a small watershed. The characteristics of each section are given in Table 3–16, including the land use/cover, slope, and length.

Equation 3–46 is used to compute the velocity of flow for section A to B:

$$V = kS^{0.5} = 2.5(0.07)^{0.5} = 0.661 \text{ ft/sec}$$

Thus, the travel time, which is computed with Equation 3–43, is

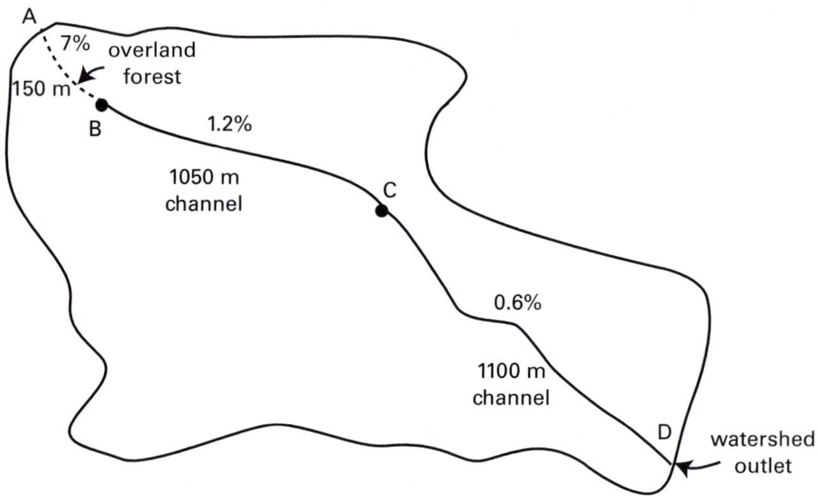

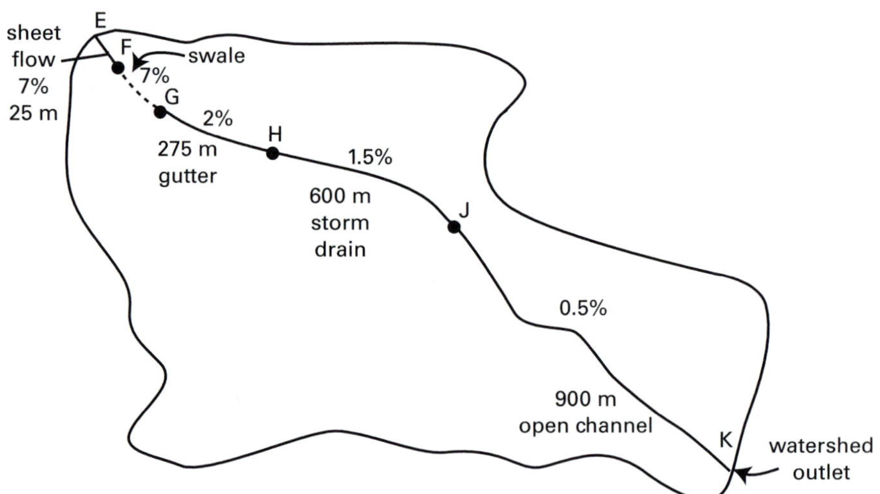

FIGURE 3–20 Time of concentration estimation: (a) principal flow path for existing conditions; (b) principal flow path for developed conditions.

$$T_t = \frac{500 \text{ ft}}{0.661 \text{ ft/sec } (60)} = 12.6 \text{ min}$$

For the section from B to C, Manning's Equation is used. For a trapezoidal channel the hydraulic radius is

$$R_h = \frac{A}{WP} = \frac{dw + zd^2}{w + 2d\sqrt{1 + z^2}} + \frac{1.0\,(2.3) + 2\,(1.0)^2}{2.3 + 2\,(1.0)\sqrt{1 + (2)^2}}$$
$$= 0.635 \text{ ft}$$

TABLE 3–16 Characteristics of Principal Flow Path: Example 3–14

Watershed Condition	Reach	Length (ft)	Slope (%)	n	Land use/cover
Existing	A to B	500	7.0	—	Overland (forest)
	B to C	3450	1.2	0.040	Natural channel (trapezoidal): $w = 2.3$ ft, $d = 1.0$ ft, $z = 2{:}1$
	C to D	3600	0.6	0.030	Natural channel (trapezoidal): $w = 4.1$ ft, $d = 2.3$ ft, $z = 2{:}1$
Developed	E to F	80	7.0	0.013	Sheet flow: $i = 1.85/(0.285 + D)$ where i[=] in./hr, D[=] hr.
	F to G	400	7.0	—	Grassed swale
	G to H	900	2.0	—	Gutter flow (paved)
	H to J	1950	1.5	0.015	Storm drain ($D = 18$ in.)
	J to K	2950	0.5	0.019	Open channel (trapezoidal): $w = 5.2$ ft, $d = 3.3$ ft, $z = 1{:}1$

Thus, Manning's Equation yields a velocity of

$$V = \frac{1.486}{0.040}(0.635)^{0.67}(0.012)^{0.5} = 3.01 \text{ ft/sec}$$

and the travel time is

$$T_t = \frac{3450 \text{ ft}}{3.01 \text{ ft/sec } (60)} = 19.1 \text{ min}$$

For the section from C to D, Manning's Equation is used. The hydraulic radius is

$$R_h = \frac{2.3\,(4.1) + 2\,(2.3)^2}{4.1 + 2\,(2.3)\sqrt{1 + (2)^2}} = 1.39 \text{ ft}$$

Thus, the velocity is

$$V = \frac{1.486}{0.030}(1.39)^{2/3}(0.006)^{0.5} = 4.78 \text{ ft/sec}$$

and the travel time is

$$T_t = \frac{3600 \text{ ft}}{4.78 \text{ ft/s } (60)} = 12.6 \text{ min}$$

Thus, the total travel time is the sum of the travel times for the individual segments (Equation 3–44):

$$t_c = 12.6 + 19.1 + 12.6 = 44.3 \text{ min}$$

For the developed conditions, the principal flow path is segmented into five parts (see Figure 3–20b). For the first part of the overland-flow portion, the runoff is sheet flow; thus, the kinematic wave equation (Equation 3–47) is used. An initial time of concentration of 5 min is assumed and used with the 2-yr *IDF* curve to estimate the intensity:

$$i = \frac{1.85}{0.285 + D} = \frac{1.85}{0.285 + 5/60} = 5.02 \text{ in./hr}$$

Thus, Equation 3–47 yields a revised estimate of the travel time:

$$T_t = \frac{0.938}{5.02^{0.4}}\left(\frac{0.013\ (80)}{\sqrt{0.07}}\right)^{0.6} = 1.12 \text{ min}$$

Since this differs from the initial estimate of 5 min, the intensity is recomputed from the *IDF* relationship. A value of 6.09 in./hr results, which then yields a revised travel time of 1.1 min, which is used as the value for this part of the flow path because it is essentially the same as the travel time of 1.12 min computed after the first iteration.

For the section from F to G, the flow path consists of grass-lined swales. Equation 3–43 can be used to compute the velocity:

$$V = 15.7(0.07)^{0.5} = 4.15 \text{ ft/sec}$$

Thus, the travel time is

$$T_t = \frac{L}{V} = \frac{400 \text{ ft}}{4.15 \text{ ft/sec } (60)} = 1.6 \text{ min}$$

For the segment from G to H, the principal flow path consists of paved surface and gutters. Thus, Equation 3–43 with Table 3–15 is used:

$$V = 46.3\ S^{0.5} = 46.3\ (0.02)^{0.5} = 6.55 \text{ ft/sec}$$

and the travel time is

$$T_t = \frac{900 \text{ ft}}{6.55 \text{ ft/sec } (60)} = 2.3 \text{ min}$$

The segment from H to J is a 18-in. pipe. Thus, Manning's equation is used. The full-flow hydraulic radius is one-fourth the diameter ($D/4$), so the velocity for full flow is

$$V = \frac{1.486}{0.015}\left(\frac{1.5}{4}\right)^{0.67} (0.015)^{0.5} = 6.33 \text{ ft/sec}$$

and the travel time is

$$T_t = \frac{L}{V} = \frac{1950 \text{ ft}}{6.33 \text{ ft/sec } (60)} = 5.1 \text{ min}$$

The final section J to K is an improved trapezoidal channel. The hydraulic radius is

$$R_h = \frac{wd + zd^2}{w + 2d\sqrt{1 + z^2}} = \frac{5.2(3.3) + (3.3)^2}{5.2 + 2(3.3)\sqrt{1 + 1^2}} = 1.93 \text{ ft}$$

Manning's Equation is used to compute the velocity:

$$V = \frac{1.486}{0.019}(1.93)^{0.67}\ (0.005)^{0.5} = 8.57 \text{ ft/sec}$$

and the travel time is

$$T_t = \frac{L}{V} = \frac{2950 \text{ ft}}{8.57 \text{ ft/s } (60)} = 5.7 \text{ min}$$

Thus, the total travel time through the five segments is the post-development time of concentration:

$$t_c = \Sigma\, T_t = 1.1 + 1.6 + 2.3 + 5.1 + 5.7 = 15.8 \text{ min}$$

The t_c for developed conditions is 36 % of the t_c for the existing conditions.

Example 3–15

For large watersheds, the overland flow processes are less important than the channel processes when estimating times of concentration. The contribution of the upland-area overland flow to the overall time of concentration can usually be neglected. Therefore, the principal flow path is usually taken as channel segments that are homogeneous in roughness, slope, and flow depth. Manning's Equation is used to compute the velocities in the segments of the principal flow path. While the roughness and slope are obtained in the conventional way, the hydraulic radius must be estimated. The bankfull discharge rate is usually used with the station rating curve, if available, or a regional relationship between flow depth and discharge, if necessary. The hydraulic radius is approximated by the depth of flow.

Consider the following data for a 52-mi^2 watershed:

Station	A (mi^2)	n	S (ft/ft)	L (ft)	d (ft)
1	10	0.052	0.007	28,400	2.1
2	24	0.047	0.006	19,600	2.7
3	37	0.044	0.005	14,200	3.1
4	52	0.043	0.005	14,100	3.4

The drainage area is the total area upstream of the station. The channel length L is the length between stations. The depth (d), which is used as the hydraulic radius, was estimated from a regional equation that related depth and drainage area:

$$d = 1.1A^{0.287}$$

Manning's equation is used to compute the velocity, and Equation 3–43 is used to compute the reach travel times (T_t):

Station	V (fps)	T_t (hr)
1	3.92	2.01
2	4.75	1.15
3	5.08	0.78
4	5.53	0.71

Therefore, the time of concentration, which equals the sum of the travel times, is 4.65 hours.

3.6.5 Empirical Formulas

Nine empirical methods of estimating t_c will be reviewed. Not all the methods were originally presented as equations for computing the time of concentration; therefore, it was necessary to adjust the empirical equations so that they would compute t_c in hours. For those methods designed to predict the lag time, the computed lag values were multiplied by a constant; the value of the constant depended on the definition of the lag. A value of 1.417 was used for the lag defined as the time difference between the centers of mass of rainfall excess and direct runoff, determined on the basis of the relationship between the time lag and the time of concentration for a SCS triangular hydrograph. This assumption is probably unimportant; that is, the results of comparisons are insensitive to this assumption. A conversion factor of 1.67 was used for methods in which the lag was defined as the time difference between the center of mass rainfall excess and the peak discharge; this constant was also based on analysis of a triangular hydrograph. Again, the use of this conversion factor would not be expected to affect the accuracy of the methods.

The Carter lag equation for partially sewered watersheds. Using data from the Washington, D.C., area, Carter (1961) calibrated an equation for predicting the watershed lag for watersheds that have natural channels and partially sewered land uses. The length (L_m in mi) and slope (S_m in ft/mi) variables in the equation should be measured from the longest channel:

$$t_c = 100\, L_m^{0.6}\, S_m^{-0.3} \tag{3–49}$$

in which t_c is the time of concentration (min). The data that were used to calibrate the equation included watersheds less than 8 mi^2 in area, with channel lengths less than 7 mi and slopes less than 2%. Manning's coefficients for the channels varied between 0.013 and 0.025. While the input variables reflect the channel characteristics, one must assume that the coefficients reflect a significant amount of pipe flow because the watersheds were in urbanized areas and the Manning's n suggest a concrete surface.

The Eagleson lag model. Eagleson (1962) presented an equation for predicting the time between the center of gravity of the rainfall excess and the peak of direct runoff; Equation 3–50 includes a factor for converting the time lag to a time of concentration:

$$t_c = 0.0111\, L_f n R_h^{-2/3}\, S_f^{-1/2} \tag{3–50}$$

in which L_f is the hydraulic length (ft), R_h is the hydraulic radius (ft), S_f is the slope of the principal flowpath (ft/ft), and t_c is the time of concentration (min). The original equation was calibrated from data for watersheds less than 8 mi^2. The variables that were used in calibrating the model were computed using the characteristics of the sewer system. The length, slope, and n value are for the main sewer, while R_h is for the main channel when flowing full. This is a "mixed" method.

The Espey-Winslow equation. Espey and Winslow, as reported by Schultz and Lopez (1974) and Espey et al. (1966), calibrated an equation for predicting the time to peak using data measured in Houston from 1964 to 1967. Six of the seventeen watersheds were predominately rural; the remaining were urbanized. The watersheds ranged in size from 1 to 35 mi^2. The length and slope variables were measured from the channel. The channelization factor ϕ was designed to measure subjectively the hydraulic efficiency of the drainage network. The value of ϕ is the sum of two parts, one indicating the amount of channel vegetation and the other indicating the degree of channel improvement. The impervious area factor represents the resistance of the overland flow portion of the travel time. Equation 3–51 includes a coefficient to convert the model from a time to peak to a time of concentration equation:

$$t_c = 31\phi\, L_c^{0.29}\, S_c^{-0.145}\, I^{-0.6} \qquad (3\text{–}51)$$

in which L_c is the channel length (ft), S_c is the channel slope (ft/ft), I is the percent imperviousness, and t_c is the time of concentration (min). Because it includes variables for both overland and channel flow, it is considered to be a "mixed" method.

Federal Aviation Agency equation. Equation 3–52 was developed from airfield drainage data (FAA, 1970):

$$t_c = 1.8(1.1 - C)\, L^{0.5}\, S^{-0.333} \qquad (3\text{–}52)$$

in which C is the Rational Formula runoff coefficient, L is the flow length (ft), S is the slope (ft/ft), and t_c is the time of concentration (min). Thus it is probably most valid for small watersheds where sheet flow and overland flow dominate. The length, slope, and resistance variables are for the principal flow path.

Kerby-Hathaway formula. Kerby (1959) calibrated Equation 3–53 for computing the time of concentration on very small watersheds in which surface flow dominated:

$$t_c = 0.83\left(\frac{nL}{S^{0.5}}\right)^{0.47} \qquad (3\text{–}53)$$

in which L is the flow length (ft), S_c is the slope (ft/ft), and t_c is the time of concentration (min). The length used in the equation is the straight-line distance from the most distant point of the watershed to the outlet and is measured parallel to the slope until a well-defined channel is reached. Watersheds of less than 10 acres were used to calibrate the model; the slopes were less than 1%, and Manning's n values were 0.8 and less. Consider the similarities between Equations 3–53 and 3–47.

Kirpich's methods. Kirpich (1940) calibrated two equations for computing the time of concentration (min) for small watersheds in Pennsylvania:

$$t_c = 0.0013\, L_c^{0.77}\, S_c^{-0.5} \qquad (3\text{–}54)$$

and Tennessee:

$$t_c = 0.0078\, L_c^{0.77}\, S_c^{-0.385} \qquad (3\text{–}55)$$

The length (ft) and slope (ft/ft) parameters in the equations are for the channel. The Tennessee watersheds ranged in size from 1 to 112 acres, with slopes from 3 to 10%. The computed times of concentration should be multiplied by 0.4 and 0.2 for watersheds where the overland flow path is either concrete or asphalt and the channel is concrete lined, respectively.

The SCS lag formula. The Soil Conservation Service (SCS) provided an equation for estimating the watershed lag, which was defined as the time in hours from the center of mass of the excess rainfall to the peak discharge. They also indicate that the time of concentration equals 1.67 times the lag. Thus, t_c in min is

$$t_c = 0.00526\, L^{0.8} \left(\frac{1000}{CN} - 9\right)^{0.7} S^{-0.5} \tag{3–56}$$

in which L is the watershed length (ft) and S is the watershed slope (ft/ft). Equation 3–56 is intended for use on watersheds where overland flow dominates and was developed for nonurban watersheds. The SCS had recommended that the lag equation be used for homogeneous watersheds 2000 acres and less. The method primarily reflects concentrated flow. While the 1986 TR-55 does not include this formula, it was shown by McCuen et al. (1984) to provide accurate estimates of t_c up to 4000 acres.

The Van Sickle equation. Van Sickle provided a time-to-peak equation calibrated from data collected in Houston, with drainage areas less than 36 mi^2. The equation is based on two length variables; the first, L_t (mi), is the total length of all drainageways and storm sewers greater than 36 in. in diameter, while the second, L_m (mi), is the total basin length. The prediction equation reflects both channel and pipe flow; thus it is a "mixed" method. Equation 3–57 includes a factor for converting the time-to-peak equation to one for predicting the time of concentration:

$$t_c = 0.55 \left(\frac{L_t L_m}{\sqrt{S_f}}\right)^{0.13} \tag{3–57}$$

in which S_f is the slope (ft/ft) and t_c is the time of concentration (min).

3.6.6 Summary

The input parameters of t_c formulas can be separated into four groups: roughness or flow resistance, slope, watershed size, and water input. The roughness or flow resistance effect is most often represented by a parameter such as Manning's roughness coefficient (n) or the SCS runoff curve number (CN). The slope of the watershed reflects the momentum of flood runoff. The watershed length is most often used as the watershed size factor, while a rainfall intensity is usually used to represent the water input effect. The importance of each of these inputs varies with the type of analysis being made. For small watershed analyses, the water input is one of the more important inputs. For large watershed analyses and channel reach routings, the water input is much less important, even to the point that methods for estimating time parameters often do not include a variable to reflect the water input. The variation in importance of these inputs reflects the hydrologic factors for which the inputs are used to indicate. Since rainfall intensity is an important factor in small watershed design, the rainfall intensity is an important input for time parameters calibrated from small watershed data. The slope and roughness input parameters become more important in the time parameters calibrated using data from large watersheds; these inputs reflect the storage effects on runoff. Length is more important for channel time parameters because it is a good indicator of the distance a flood wave must travel and the potential for channel storage to attenuate the flood wave. The point here is not to assess the relative importance of each input, but to indicate that time parameters and the input variables that make up these time parameters should be viewed as indicators of the physical processes that are important in runoff estimation. Such recognition should lead to proper application and, thus, more accurate assessments of runoff rates.

3.7 RUNOFF CURVE NUMBERS

The SCS runoff *curve number* (*CN*) was developed as an index that represents the combination of a hydrologic soil group and a land use and treatment class. Empirical analyses suggested that the *CN* was a function of three factors: soil group, the cover complex, and antecedent moisture conditions.

3.7.1 Soil Group Classification

The soil scientists of the U.S. Soil Conservation Service (SCS) classified more than 4000 soils on the basis of their runoff potential and grouped them into four hydrologic soil groups that are identified by the letters A, B, C, and D. Soil characteristics that are associated with each group are given in Table 3–17.

The SCS soil group can be identified at a site using one of three ways:

1. Soil characteristics
2. County soil surveys
3. Minimum infiltration rate

The soil characteristics associated with each group are given in Table 3–17 and provide one means of identifying the SCS soil group and are discussed in Section 3.7.4. County soil surveys, which are available from Soil Conservation Districts, give detailed descriptions of the soils at locations within a county; these surveys are usually the best means of identifying the soil group and are discussed in Section 3.7.4. As a third method, soil sampling and analysis can also be used to estimate the minimum infiltration rates, which can be used to classify the soil using the following values:

Group	Minimum Infiltration Rate (in./hr)
A	0.30–0.45
B	0.15–0.30
C	0.05–0.15
D	0–0.05

The published soil survey is the most common method for identifying the soil group.

TABLE 3–17 Characteristics of Soils Assigned to Soil Groups

Group A:	Deep sand; deep loess; aggregated silts
Group B:	Shallow loess; sandy loam
Group C:	Clay loams; shallow sandy loam; soils low in organic content; soils usually high in clay
Group D:	Soils that swell significantly when wet; heavy plastic clays; certain saline soils

3.7.2 Cover Complex Classification

The SCS cover complex classification consists of three factors: land use, treatment or practice, and hydrologic condition. Over twenty different land uses are identified for estimating runoff curve numbers. Agricultural land uses are often subdivided by treatment or practice, such as contoured or straight row; this separation reflects the different hydrologic runoff potential associated with variation in land treatment. The hydrologic condition reflects the level of land management. It is separated into three classes: poor, fair, and good. Not all of the land uses are separated by treatment or condition.

3.7.3 Hydrologic Condition

The type of vegetation or ground cover on a watershed, and the quality or density of that cover, have a major impact on the infiltration capacity of a given soil. Further refinement in the cover type is provided by the definition of cover quality as follows:

Poor: Heavily grazed or regularly burned areas. Less than 50% of the ground surface is protected by plant cover or brush and tree canopy.

Fair: Moderate cover with 50 to 75% of the ground surface protected by vegetation.

Good: Heavy or dense cover with more than 75% of the ground surface protected by vegetation.

In most cases, the cover type and quality of a watershed in existing conditions can be readily determined by a field review of a watershed. In ultimate-planned open spaces, the soil cover condition shall be considered as "good."

3.7.4 SCS Soil Survey Reports

The soil scientists of the SCS have classified over 4000 soils. For most areas of the United States, county soil surveys are available from the SCS. These county reports provide detailed information about the soils in the county. All of the soils in the county are shown on a detailed map. The map consists of many sheets that were made from aerial photographs. On each sheet of the detailed map, areas with the same soil characteristics are outlined and identified by symbols. In addition to the soil maps, a guide is provided. The guide lists all of the soils in the county in alphabetical order by map symbol. The hydrologic soil group is also given in the county soil survey. Additional details about soils in the county are provided in the survey.

3.7.5 Curve Number Tables

Table 3–18 shows the *CN* values for the different land uses, treatment, and hydrologic condition; separate values are given for each soil group. For example, the *CN* for a wooded area with good cover and soil group B is 55; for soil group C, the *CN* would increase to 70. If the cover (on soil group B) is poor, the *CN* will be 66.

TABLE 3–18 Runoff Curve Numbers (average watershed condition, $I_a = 0.2S$)

Land Use Description		Curve Numbers for Hydrologic Soil Group A	B	C	D
Fully developed urban areas[a] (vegetation established)					
Lawns, open spaces, parks, golf courses, cemeteries, etc.					
Good condition; grass cover on 75% or more of the area		39	61	74	80
Fair condition; grass cover on 50% to 75% of the area		49	69	79	84
Poor condition; grass cover on 50% or less of the area		68	79	86	89
Paved parking lots, roofs, driveways, etc.		98	98	98	98
Streets and roads					
Paved with curbs and storm sewers		98	98	98	98
Gravel		76	85	89	91
Dirt		72	82	87	89
Paved with open ditches		83	89	92	93
	Average % impervious[b]				
Commercial and business areas	85	89	92	94	95
Industrial districts	72	81	88	91	93
Row houses, town houses, and residential with lots sizes 1/8 acre or less	65	77	85	90	92
Residential: average lot size					
1/4 acre	38	61	75	83	87
1/3 acre	30	57	72	81	86
1/2 acre	25	54	70	80	85
1 acre	20	51	68	79	84
2 acre	12	46	65	77	82
Developing urban areas[c] (no vegetation established)					
Newly graded area		77	86	91	94
Western desert urban areas					
Natural desert landscaping (pervious area only)[f]		63	77	85	88
Artificial desert landscaping		96	96	96	96

Land Use Description	Treatment or Practice[d]	Hydrologic Condition	Curve Numbers for Hydrologic Soil Group A	B	C	D
Cultivated agricultural land						
Fallow	Straight row or bare soil		77	86	91	94
	Conservation tillage	Poor	76	85	90	93
	Conservation tillage	Good	74	83	88	90
Row crops	Straight row	Poor	72	81	88	91
	Straight row	Good	67	78	85	89
	Conservation tillage	Poor	71	80	87	90
	Conservation tillage	Good	64	75	82	85
	Contoured	Poor	70	79	84	88
	Contoured	Good	65	75	82	86
	Contoured and	Poor	69	78	83	87
	conservation tillage	Good	64	74	81	85

(continued)

TABLE 3–18 Runoff Curve Numbers (average watershed condition, $I_a = 0.2S$) (*Continued*)

Land Use Description	Treatment or Practice[d]	Hydrologic Condition	Curve Numbers for Hydrologic Soil Group A	B	C	D
	Contoured and terraces	Poor	66	74	80	82
	Contoured and terraces	Good	62	71	78	81
	Contoured and terraces	Poor	65	73	79	81
	and conservation tillage	Good	61	70	77	80
Small grain	Straight row	Poor	65	76	84	88
	Straight row	Good	63	75	83	87
	Conservation tillage	Poor	64	75	83	86
	Conservation tillage	Good	60	72	80	84
	Contoured	Poor	63	74	82	85
	Contoured	Good	61	73	81	84
	Contoured and	Poor	62	73	81	84
	conservation tillage	Good	60	72	80	83
	Contoured and terraces	Poor	61	72	79	82
	Contoured and terraces	Good	59	70	78	81
	Contoured and terraces	Poor	60	71	78	81
	and conservation tillage	Good	58	69	77	80
Close-seeded	Straight row	Poor	66	77	85	89
legumes	Straight row	Good	58	72	81	85
rotations	Contoured	Poor	64	75	83	85
meadows[e]	Contoured	Good	55	69	78	83
	Contoured and terraces	Poor	63	73	80	83
	Contoured and terraces	Good	51	67	76	80
Noncultivated agricultural land						
Pasture or range	No mechanical treatment	Poor	68	79	86	89
	No mechanical treatment	Fair	49	69	79	84
	No mechanical treatment	Good	39	61	74	80
	Contoured	Poor	47	67	81	88
	Contoured	Fair	25	59	75	83
	Contoured	Good	6	35	70	79
Meadow		—	30	58	71	78
Forestland—grass or		Poor	55	73	82	86
orchards—evergreen		Fair	44	65	76	82
deciduous		Good	32	58	72	79
Brush		Poor	48	67	77	83
		Fair	35	56	70	77
		Good	30	48	65	73
Woods		Poor	45	66	77	83
		Fair	36	60	73	79
		Good	25	55	70	77
Farmsteads		—	59	74	82	86
Forest-range						
Herbaceous		Poor	[g]	80	87	93
		Fair		71	81	89
		Good		62	74	85
Oak-aspen		Poor		66	74	79
		Fair		48	57	63
		Good		30	41	48

TABLE 3–18 Runoff Curve Numbers (average watershed condition, $I_a = 0.2S$) (*Continued*)

Land Use Description	Treatment or Practice[d]	Hydrologic Condition	Curve Numbers for Hydrologic Soil Group A	B	C	D
Juniper		Poor	[g]	75	85	89
		Fair		58	73	80
		Good		41	61	71
Sage-grass		Poor		67	80	85
		Fair		51	63	70
		Good		35	47	55

[a]For land uses with impervious areas, curve numbers are computed assuming that 100% of runoff from impervious areas is directly connected to the drainage system. Pervious areas (lawn) are considered to be equivalent to lawns in good condition. The impervious areas have a *CN* of 98.

[b]Includes paved streets.

[c]Use for the design of temporary measures during grading and construction. Impervious area percent for urban areas under development vary considerably. The user will determine the percent impervious. Then using the newly graded area *CN*, the composite *CN* can be computed for any degree of development.

[d]For conservation tillage poor hydrologic condition, 5 to 20% of the surface is covered with residue (less than 750-lb/acre row crops or 300-lb/acre small grain). For conservation tillage good hydrologic condition, more than 20% of the surface is covered with residue (greater than 750-lb/acre row crops or 300-lb/acre small grain).

[e]Close-drilled or broadcast.

For noncultivated agricultural land:

Poor hydrologic condition has less than 25% ground cover density.

Fair hydrologic condition has between 25 and 50% ground cover density.

Good hydrologic condition has more than 50% ground cover density.

For forest-range:

Poor hydrologic condition has less than 30% ground cover density.

Fair hydrologic condition has between 30 and 70% ground cover density.

Good hydrologic condition has more than 70% ground cover density.

[f]Composite *CN*'s for natural desert landscaping should be computed using Figure 3–21 based on the impervious area percentage (*CN* = 98) and the pervious area *CN*. The pervious area *CN*'s are assumed equivalent to desert shrub in poor hydrologic condition.

[g]Curve numbers for group A have been developed only for desert shrub.

3.7.6 Antecedent Soil Moisture Condition

Antecedent soil moisture is known to have a significant effect on both the volume and rate of runoff. Recognizing that it is a significant factor, SCS developed three antecedent soil moisture conditions, which were labeled I, II, and III. The soil condition for each is as follows:

Condition I: Soils are dry but not to wilting point; satisfactory cultivation has taken place

Condition II: Average conditions

Condition III: Heavy rainfall, or light rainfall and low temperatures have occurred within the last five days; saturated soil

The following tabular summary gives seasonal rainfall limits for the three antecedent soil moisture conditions:

AMC	Total 5-Day Antecedent Rainfall (in.)	
	Dormant Season	Growing Season
I	Less than 0.5	Less than 1.4
II	0.5–1.1	1.4–2.1
III	Over 1.1	Over 2.1

The *CN* values obtained from Table 3–18 are for antecedent soil moisture condition II. If either soil condition I or III is to be used, the *CN* can be adjusted using Table 3–19.

When should antecedent soil moisture analyses be used to adjust *CN* values? How were the values of Table 3–19 derived? Both of these questions are important to the design engineer and to those involved in writing drainage and stormwater control policies. The *CN* values for AMC I and AMC III represent values that defined envelope curves for actual measurements of rainfall (P) and runoff (Q). That is, for any one watershed, similar storm events having nearly equal values of P may result in widely different values of Q. The *CN* values of Table 3–19 reflect this variation in Q, and the variation was assumed to result from variation

TABLE 3–19 Adjustment of Curve Numbers for Dry (condition I) and Wet (condition III) Antecedent Moisture Conditions

CN for condition II	Corresponding *CN* for condition I	Corresponding *CN* for condition III
100	100	100
95	87	99
90	78	98
85	70	97
80	63	94
75	57	91
70	51	87
65	45	83
60	40	79
55	35	75
50	31	70
45	27	65
40	23	60
35	19	55
30	15	50
25	12	45
20	9	39
15	7	33
10	4	26
5	2	17
0	0	0

in antecedent soil moisture conditions. Having answered the second question, the first question can be discussed. The *CN* values for AMC I and III represent ranges of likely values of the *CN*, but for use in design one usually wants a "most likely" value rather than an extreme value that will occur much less frequently. There are better ways of incorporating risk analysis into design than by adjusting the *CN* for antecedent moisture conditions using the transformation of Table 3–19. This is not to imply that the *CN* values of Table 3–18 are the "best" for all locations and design situations. Other site-specific data may suggest a more representative value of the *CN* than the value from Table 3–18. However, in the absence of such analysis, the *CN* values of Table 3–18 for AMC II should be used for all design situations. The adjustment of Table 3–19 should only be used after it has been confirmed from a local or regional analysis.

3.7.7 Estimation of *CN* Values for Urban Land Uses

The *CN* table (Table 3–18) provides *CN* values for a number of urban land uses. For each of these, the *CN* is based on a specific percent of imperviousness. For example, the *CN* values for commercial land use are based on an imperviousness of 85%. For urban land uses with percentages of imperviousness different than those shown in Table 3–18, curve numbers can be computed using a weighted *CN* approach, with a *CN* of 98 used for the impervious areas and the *CN* for open space (good condition) used for the pervious portion of the area. Thus *CN* values of 39, 61, 74, and 80 are used for hydrologic soil groups A, B, C, and D, respectively. These are the same *CN* values for pasture in good condition. The following equation can be used to compute a weighted *CN* (CN_w):

$$CN_w = CN_p(1 - f) + f(98) \tag{3–58}$$

in which f is the fraction (not percentage) of imperviousness and CN_p is the curve number for the pervious portion (39, 61, 74, or 80). To show the use of Equation 3–58, the *CN* values for commercial land use with 85% imperviousness are as follows:

A soil: $39(0.15) + 98(0.85) = 89$
B soil: $61(0.15) + 98(0.85) = 92$
C soil: $74(0.15) + 98(0.85) = 94$
D soil: $80(0.15) + 98(0.85) = 95$

These are the same values shown in Table 3–18.

Equation 3–58 can be placed in graphical form (see Figure 3–21a). By entering with the percentage of imperviousness on the vertical axis at the center of the figure and moving horizontally to the pervious area *CN*, the weighted *CN* can be read. The examples given for commercial land use can be used to illustrate the use of Figure 3–21a for an 85% imperviousness. For a commercial land area with 60% imperviousness on a B soil, the composite *CN* would be

$$CN_w = 61(0.4) + 98(0.6) = 83$$

The same value can be obtained from Figure 3–21a.

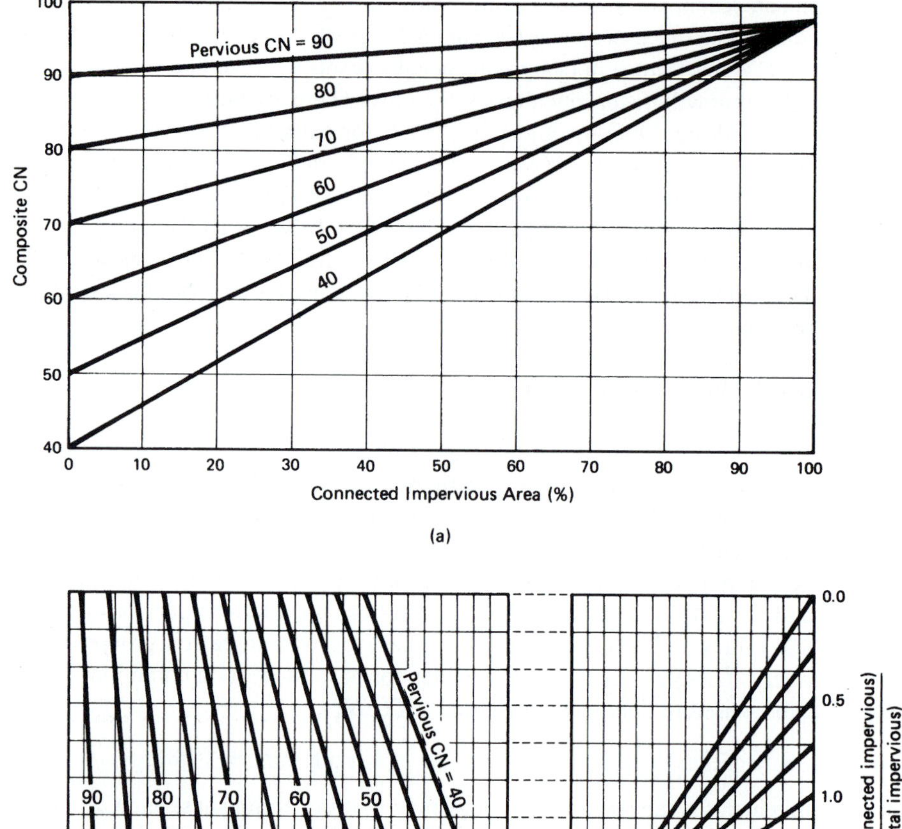

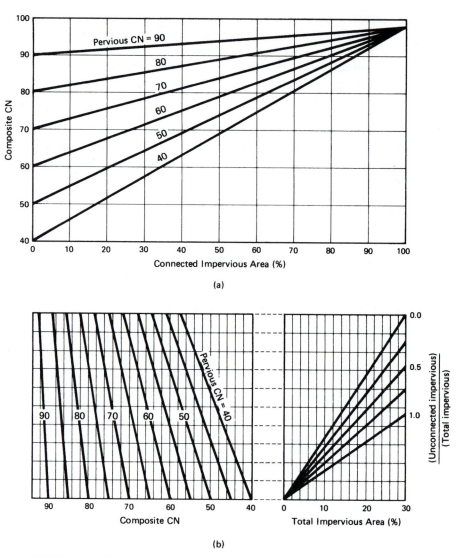

FIGURE 3–21 Graphical computation of composite curve numbers when (a) all impervious surfaces are connected to the storm drain system and (b) less than 30% is not connected.

3.7.8 Effect of Unconnected Impervious Area on Curve Numbers

Many local drainage policies are requiring runoff from certain types of impervious land cover (for example, rooftops, driveways, patios) to be directed to pervious surfaces rather than connected to storm drain systems. Such a policy is based on the belief that disconnecting these impervious areas will require smaller and less costly drainage systems and lead both to increased ground water recharge and to improvements in water quality. In the belief that disconnecting some impervious surfaces will reduce both the peak runoff rates and volumes of direct flood runoff, some developers believe that they should be given credit for the reductions in peak rates and volumes in the design of the drainage system. Thus, a need to account for the effect of disconnecting impervious surfaces on runoff rates and volumes exists.

Estimating *CN*s for areas with some unconnected imperviousness requires three important variables: the pervious area *CN*, the percentage of impervious area, and the percentage of the imperviousness that is unconnected. The existing figure (Figure 3–21a) for computing composite *CN* values is based on the pervious area *CN* and the percentage of imperviousness. A correction factor to the composite *CN* is a function of the percentage of unconnected imperviousness. The method of adjustment is given in Figure 3–21b. The correction can only be applied when the percentage of impervious is less than 30%. A correction is not made if the total imperviousness is greater than 30%. The adjusted *CN* (CN_c), Figure 3–21b, can be computed with the following equation:

$$CN_c = CN_p + I_f (98 - CN_p)(1 - 0.5R) \tag{3–59}$$

in which I_f is the fraction of impervious cover, CN_p is the pervious area *CN*, and *R* is the ratio of the unconnected impervious area to the total impervious area.

Example 3–16

To illustrate the use of Figure 3–21, consider the case of a drainage area having 25% imperviousness, a pervious area *CN* of 61, and 50% of the imperviousness unconnected. Enter the right-hand side of Figure 3–21b using 25% imperviousness and move up the graph to the point on the line corresponding to 50% unconnected. Then move from this point horizontally to the left-hand side of the figure until the horizontal line intersects the interpolated line corresponding to the pervious area *CN* of 61. From this intersection move down to the abscissa and read the value of the composite *CN*. In this case, the composite *CN* would be 68. Using Equation 3–59

$$CN_c = 61 + 0.25(98 - 61)(1 - 0.5(0.5)) = 68$$

If credit were not given for unconnected imperviousness, the percent of imperviousness could be entered on Figure 3–21a and the composite *CN* found using the pervious area *CN*. For 25% imperviousness and a pervious area *CN* of 61 the composite *CN* would equal 70, which means that disconnecting 50% of the impervious cover allows for the *CN* to be reduced by 2. This savings can lead to a reduction in peak discharge and runoff volume.

PROBLEMS

3–1. A rectangular 3-acre lot is being developed into multifamily housing. Can this plot of land be considered a watershed? Why or why not?

3–2. Delineate the boundary of the watershed in Figure 3–3a that has its outlet with east-west coordinate of 0.00 and north-south coordinate of 5.77.

3–3. Delineate the boundary of the watershed in Figure 3–3a that has its outlet with east-west and north-south coordinates of 5.5 and 0.89, respectively. Also, delineate a subwatershed with the outlet at 8.34 and 3.4, respectively.

3–4. Compute the area of the watershed of Problem 3–2 in square feet, acres, and square miles. Provide an estimate of the accuracy of the estimated area.

3–5. Compute the area of both subwatersheds of Problem 3–3 in acres and square miles. Provide an estimate of the accuracy of the estimates.

3–6. Measure the length for the watershed of Problem 3–2. What is the expected accuracy of the estimated length?

3–7. For the watershed of Problem 3–3, measure the watershed length of the upper subarea and the subwatershed length (L) of the lower subarea. Provide an estimate of the error in each of the measurements.

3–8. Based on the area computed in Problem 3–4, estimate the watershed length using the following empirical formula: $L = 209A^{0.6}$, where L is in feet and A is in acres. Compare the estimated length with the measured length (Problem 3–6). What is the error in the predicted value, and what are possible reasons that the predicted value differs from the measured value?

3–9. Using the upper subwatershed for the watershed of Problem 3–3 and its area measured in Problem 3–5, estimate the watershed length using the equation given in Problem 3–8. Comparing the estimated and measured lengths and list possible reasons why the two values differ.

3–10. For the watershed of Problem 3–2, compute the watershed slope. Estimate the accuracy of the computed value.

3–11. Two adjacent subwatersheds have characteristics as follows: Subarea 1: length = 8600 ft; elevation drop = 170 ft. Subarea 2: length = 3300 ft; elevation drop = 250 ft. Compute the slope of each subarea. Can you legitimately conclude that the average watershed slope for the entire area is the average of the two subarea slopes? Explain.

3–12. For the following hypothetical watersheds, compute the hypsometric curve and compute D_m, L_d, F_p, and H_a.

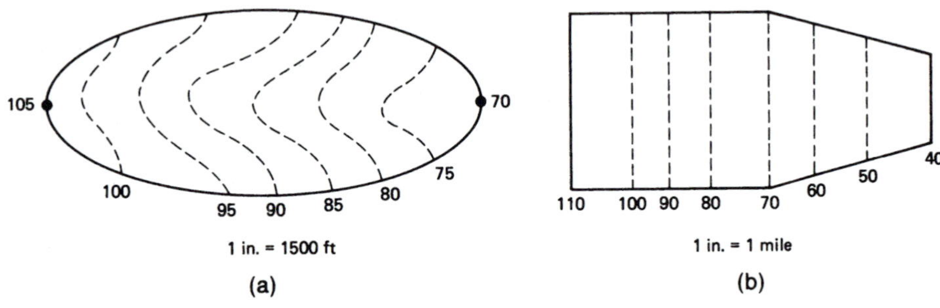

3–13. Derive the hypsometric curve for the watershed shown in Figure 3–3a. Compute D_m, L_d, F_p, and H_a.

3–14. (a) Construct a hypsometric curve in standardized form for a semicircular watershed with the outlet at the center point of the circle. Assume a constant slope over the entire watershed. (b) Form a watershed that is a quarter of a circle that has the same total area as the watershed of part (a). Assume a constant slope over the entire watershed. Construct a hypsometric curve in standardized form and plot it on the same graph as the curve of part (a). (c) Comment on the expected flood runoff suggested by the hypsometric curves of the two watersheds.

3–15. Construct two hypothetical, rectangular watersheds, A and B. The tops and bottoms of the watersheds have elevations of 400 ft and 200 ft, respectively. On watershed A, 50, 75, and 90% of the watershed area lies above elevations of 350, 300, and 250 ft, respectively. On watershed B, 10, 25, and 50% of the watershed area lies above elevations of 350, 300, and 250 ft, respectively. Construct the hypsometric curves and relate the curves to the flood potential of the two watersheds.

3–16. Derive expressions for computing the shape parameters (L_{ca}, L, L_1, R_e) of an elliptical watershed with lengths for the major and minor axes $2a$ and $2b$, respectively. Assume the watershed outlet is located at one end of the major axis.

3–17. For the watershed delineated in Problem 3–2, find the values of the shape parameters of Equations 3–2 to 3–5.

3–18. What factors affect the accuracy of an estimate of L_{ca}?

3–19. Estimate the values of the shape parameters of Equations 3–2 to 3–5 for the total watershed delineated in Problem 3–3.

3–20. Measure the channel length and $L_{10\text{–}85}$ for the watershed of Problem 3–2. Estimate the accuracy of the computed values.

3–21. For the channel of Problem 3–3, measure the channel length (L) of the lower subarea. Provide an estimate of the error in the measurement.

3–22. Using Equation 3–19, compute the channel slope between sections 1 and 6 and for each of the five reaches. Compute the average of the computed slopes for the five reaches and compare it to the estimate for the entire reach. Discuss the results.

Section	Elevation (ft)	Distance from Outlet (ft)
1	134	0
2	141	4300
3	146	5600
4	180	7800
5	235	10,400
6	242	11,300

3–23. Compute the channel slope and the $S_{10\text{–}85}$ slope for the watershed of Problem 3–2. Estimate the accuracy of both values.

3–24. For the entire watershed of Problem 3–3 (EW 5.5, NS 0.89), estimate the channel slope and the $S_{10\text{–}85}$ slope. Estimate the accuracy of both values.

3–25. Using Equations 3–21 compute the average channel slope using the station data from Problem 3–22.

3–26. Find the drainage density for the following watersheds. The reach lengths, in miles, are shown.

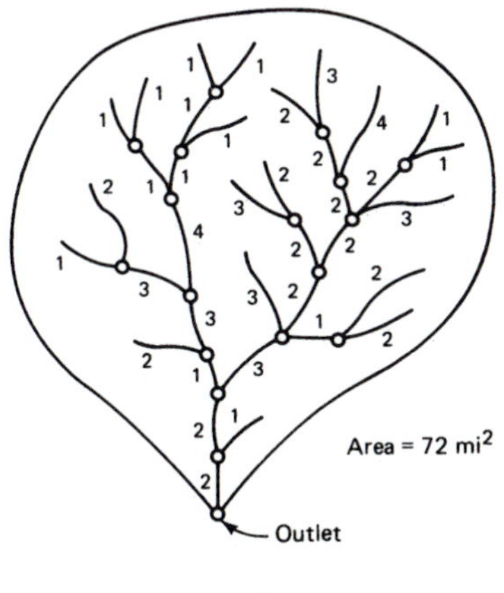

(a)

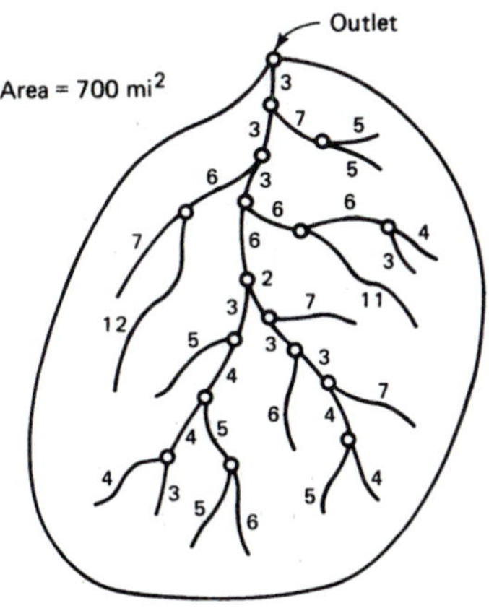

(b)

3–27. For the watershed of Problem 3–26(a), identify the order of each stream segment, find the actual number of streams for each order, and compute the bifurcation ratio. Compute the expected number of streams of each order. For the watershed of Problem 3–26(b), identify the order of each stream segment, find the actual number of streams for each order, and compute the bifurcation ratio. Compute the expected number of streams of each order.

3–28. For the watershed of Problem 3–26(a), determine the average length of streams for each stream order and the stream length ratio. Compute the expected average length for each stream order. Compare the computed and actual values. For the watershed of Problem 3–26(b), determine the average length of streams for each stream order and the stream length ratio. Compute the expected average length for each stream order. Compare the computed and actual values.

3–29. For stream orders 1 to 6, the total watershed areas for an 11.41 mi^2 watershed are 3762, 1927, 853, 408, 256, and 96 acres. The number of streams of orders 1 to 6 are 142, 56, 19, 9, 3, and 1, respectively. Compute both the stream-area ratio and the bifurcation ratio. Compare the actual and estimated values of both $\bar{A}_i$ and N_i.

3–30. For stream orders 1 to 5, the average stream slopes measured on a moderately sloping watershed are 0.025, 0.021, 0.019, 0.017, and 0.016. Compute the stream slope ratio and compare the measured and computed average slopes for each stream order.

3–31. Based on the Manning's n values of Table 3–4, determine the average relative error that should be expected in a value of n for the following conditions: (a) Unlined open channel (earth, uniform section); (b) roadside channel and swales (depth up to 0.7 ft); (c) natural stream channels, minor stream (fairly regular section); (d) natural stream channels, floodplains (cultivated areas). Assuming that the velocity V and roughness are related by $V = 0.2/n$, what would be the mean velocity and relative error in the mean velocity for each of the four flow conditions?

For Problems 3–32 to 3–34, estimate the value of Manning's n *using the method by Cowan (Table 3–11). Compare the estimate with the value from Table 3–4 using the category that best matches the stated description of the channel reach.*

3–32. An earthen channel has moderately eroded side slopes, minor obstructions, short grass with some weeds, and appreciable meandering.

3–33. A coarse gravel channel bed has slightly eroded side slopes, frequent changes in the cross section, negligible obstructions, minimal vegetation, and severe meandering.

3–34. A straight channel has a bed of fine gravel, slightly eroded side slopes, minor obstructions throughout the reach; trees, weeds, and brush throughout the cross section.

3–35. For an unlined open-channel cross section with a width of 10 ft, a depth of 3 ft, and side slopes of 3:1 (h:v), find the flow rate assuming a slope of 0.001 ft/ft and an earthen surface with short grass and a few weeds.

3–36. For the cross section of Problem 3–35, find the flow rate if the channel has clean sides and a gravel bottom and the channel slope is 0.005 ft/ft.

3–37. Assume a slope of 0.01 ft/ft and a hydraulic radius of 3 ft. Show the effect of each of Cowan's adjustments for roughness by computing the velocity for the recommended values of n_1, n_2, n_3, n_4, and n_5. For n_2 through n_5 assume $n_1 = 0.02$. For n_1 through n_4 assume $n_5 = 0.005$. Where ranges are given, use an average value. Plot curves of velocity versus n for each of the five variables. What do the results suggest about the importance of the five variables?

3–38. Find the velocity and flow rate in 10-ft-wide grass-lined swale (Kentucky bluegrass with a height of 2 in.) when the water is flowing at a depth of 3 in. The swale has a slope of 3%.

3–39. Using the estimates of the accuracy of the channel slope from either Problem 3–23 or 3–24 and the accuracy of n from Problem 3–31, evaluate the sensitivity of the computed flow rate in the channel of Problem 3–35 to errors in the slope and n.

3–40. Using the estimates of the accuracy of the channel slope from either Problem 3–23 or 3–24 and the accuracy of n from Problem 3–31, evaluate the sensitivity of the computed flow rate in the channel of Problem 3–36 to errors in the slope and n.

3–41. For a flow of 350 ft^3/sec, find the normal depth for each channel reach between the stations indicated in the following table. Assume that the channel is trapezoidal with the side slopes shown in the table. Use Table 3–4 to estimate the roughness coefficients.

Station	Elevation (ft)	Distance from Outlet (ft)	Bottom Width (ft)	Side Slopes (horiz:vert)	Description of Channel
1	134	0	15	9:2	Earth, uniform section, short grass
2	141	4,300	14	7:2	Earth, uniform section, graveled soil
3	146	5,600	14	3:1	Earth, fairly uniform section, no vegetation
4	180	7,800	12	5:2	Earth, fairly uniform section, grass, some weeds
5	235	10,400	10	2:1	Earth, fairly uniform section, grass, some weeds
6	242	11,300			

3–42. A flow of 74 ft^3/sec is measured at a depth of 1.5 ft in a trapezoidal channel with a bottom width of 16 ft and side slopes of 3:1 (h:v). The channel reach has a slope of 1%. Estimate the channel roughness.

3–43. A flow of 600 ft^3/sec is measured at a depth of 3.4 ft in trapezoidal channel with a bottom width of 25 ft and side slopes of 3:1 (h:v). The channel reach has a slope of 0.002. Estimate the channel roughness.

3–44. The average velocity (V) and depth of flow (d) in a channel reach are measured, as shown below, during four storm events. The trapezoidal reach has a bottom width of 15 ft, side slopes of 2.5:1 (h:v), and a bottom slope of 0.008. Mean velocity measurements are assumed to be accurate to ± 10% and the depth has an expected accuracy of ± 0.3 ft. What is the best estimate of n and what is its expected accuracy?

Storm Event	Velocity (ft/sec)	Depth (ft)
1	4.6	3.2
2	5.7	3.9
3	4.9	3.4
4	6.2	4.1

3–45. A small agricultural watershed (straight row crops) has an elevation drop of 21 ft and a principal flow path of 780 ft. Compute the time of concentration for the watershed using the velocity method.

3–46. A paved highway section has a slope of 4% and a length of 1800 ft. Determine the time of concentration using the velocity method.

3–47. A hydrologist expects to make time-of-concentration estimates for a land use not covered by the coefficients of Table 3–14. The hydrologist makes the following measurements:

Site	Elevation Drop (ft)	Length (ft)	Travel Time (min)
1	6	180	3.9
2	2	230	7.5
3	1	90	3.8
4	22	310	4.7
5	9	150	2.7

Estimate the value of k for Equation 3–46.

3–48. A hydrologist expects to make time-of-concentration estimates for a land use not covered by the coefficients of Table 3–14. The hydrologist makes the following measurements:

Site	Elevation Drop (ft)	Length (ft)	Travel Time (min)
1	4	460	6.2
2	4	390	5.3
3	10	520	4.7
4	12	470	4.1
5	12	360	3.0
6	20	480	2.4
7	39	550	2.8

Estimate the value of k for Equation 3–46.

3–49. A flow path consists of the following sections:

Section	Slope (%)	Length (ft)	Land Use
1	3.5	260	Forest (light underbrush)
2	4.1	490	Short grass
3	3.4	370	Upland gulley
4	3.1	420	Grassed waterway

Estimate the time of concentration using the velocity method.

3–50. A flow path consists of the following sections:

Section	Slope (%)	Length (ft)	Land Use
1	2.8	190	Forest (dense underbrush)
2	2.5	310	Paved gutter
3	1.9	440	Grassed waterway

Estimate the time of concentration using the velocity method.

3–51. The principal flow path for a 95-acre watershed is a 3100-ft stream with a base width of 7 ft, side slopes of 2.5:1 (h:v), and a depth of about 1 ft at bankfull flow. The stream is a regular section with heavy brush on the banks. The elevation drop over the length of the stream is 56 ft. Estimate the time of concentration.

3–52. The principal flow path for a 210-acre watershed is a 5300-ft stream with a base width of 10 ft, side stops of 3:1 (h:v), and a depth of 1.4 ft at bankfull flow. The stream is an unlined open channel, earth, uniform section, with short grass. The elevations of the upper and lower sections of the stream are 258 ft and 146 ft, respectively. Estimate the time of concentration.

3–53. Estimate the time of concentration for shallow sheet flow on a 420-ft section of asphalt roadway at a slope of 8%. Assume a 25-yr design frequency and the *IDF* curve of Figure 4–4.

3–54. Estimate the time of concentration for shallow sheet flow on a 1280-ft section of asphalt roadway at a slope of 2.5%. Assume a 5-yr design frequency and the *IDF* curve of Figure 4–4.

3–55. Estimate the time of concentration using the *SCS* sheet flow equation for a 790 ft section of asphalt pavement at a slope of 0.8 percent. Use the *IDF* curve of Figure 4–4.

REVIEW QUESTIONS

3–1. Which one of the following characteristics is not a measure of basin shape? (a) The length to the center of area; (b) the elongation ratio; (c) the drainage density; (d) the circularity ratio.

3–2. The value of the circularity ratio for a circular watershed is (a) 0; (b) 1; (c) the radius of the watershed; (d) the perimeter of the watershed.

3–3. The hypsometric curve is a description of (a) the relation between stream order and the proportion of the drainage area associated with that stream order; (b) the cumulative relation between elevation and area within (time of travel) isochrones; (c) the cumulative relation between elevation and area within elevation intervals; (d) the relation between elevation and rainfall intensity within elevation intervals.

3–4. Horton's stream order is a measure of (a) the drainage density; (b) the relative stream lengths; (c) the ratio of stream areas; (d) stream branching within a watershed.

3–5. The bifurcation ratio is the ratio of the __________ of streams of any order to the __________ of streams of the next higher order: (a) number; (b) area; (c) length; (d) slope; (e) none of the above.

3–6. If the stream length ratio is 2 and the average length of fourth-order streams is 1.2 mi, the average length in miles of first-order streams is (a) 0.075; (b) 0.15; (c) 0.3; (d) 9.6.

3–7. Which one of the following is not an assumption made in uniform flow computations? (a) The flow depth is constant; (b) the friction force is proportional to the shear stress; (c) the energy grade line is parallel to the channel bottom; (d) the channel has a rectangular cross section; (e) all of the assumptions above apply.

3–8. When using Manning's Equation for open-channel flow, it is common to assume that the hydraulic radius is equal to (a) the channel width; (b) one-fourth of the depth, $D/4$; (c) the square root of the wetted perimeter; (d) the depth of flow.

3–9. Which one of the following is not a factor in estimating Manning's roughness coefficient? (a) The height of channel vegetation; (b) the degree of stream meandering; (c) the regularity of the channel cross sections; (d) the channel slope; (e) all of the above are factors.

3–10. Which one of the following is not a factor in controlling times of concentration of watershed runoff? (a) The drainage density; (b) the roughness of the flow surface; (c) the rainfall intensity; (d) the flow lengths; (e) all of the above are factors.

3–11. Which one of the following is not a factor in the accuracy of a t_c equation for any one watershed? (a) The number of subareas into which the watershed is divided; (b) the location of the watershed relative to the locations of the watersheds used in calibrating the model; (c) the number of watersheds used in calibrating the model; (d) the similarity between the characteristics of the watershed of interest and the characteristics of the watershed used in calibrating the model; (e) all of the factors above affect the accuracy.

3–12. In the soil profile, the horizons vary in (a) composition; (b) structure; (c) texture; (d) color; (e) all of the above.

3–13. Which one of the following does not reflect soil texture? (a) The soil particle diameter; (b) the tendency of soil particles to aggregate into lumps; (c) the particle shape; (d) variation in silt, clay, and sand content; (e) all of the above reflect soil texture.

3–14. The average soil column may have a pore space (percent) of about (a) 15; (b) 30; (c) 45; (d) 60; (e) 75.

3–15. Porosity is the ratio of (a) the volume of water required to saturate the pore space (V_W) to the total volume (V); (b) the volume of solid matter (V_S) to V_W; (c) V_W/V_S; (d) the volume of pore space to V_W.

3–16. The porosity of soil material with mixed diameters is usually (a) less than; (b) the same as; (c) greater than the porosity of a soil column having material of the same diameter.

3–17. The specific yield is the ratio of (a) the volume of water that can drain under the force of gravity (V_g) to V; (b) V_g/V_W; (c) V_W/V; (d) V_g/V_S.

3–18. The specific retention is the ratio of (a) Vg/V; (b) the volume of water retained by surface tension and molecular forces (V_r) to V_W; (c) V_g/V_r; (d) V_r/V.

DISCUSSION QUESTION

The technical content of this chapter is important to the professional hydrologist, but practice is not confined to making technical decisions. The intent of this discussion question is to show that hydrologists must often address situations where value issues intermingle with the technical aspects of a project. In discussing the stated problem, at a minimum include responses to the following questions:

1. What value issues are involved, and how are they in conflict?
2. Are technical issues involved with the value issues? If so, how are they in conflict with the value issues?
3. If the hydrologist attempted to rationalize the situation, what rationalizations might he or she use? Provide arguments to suggest why the excuses represent rationalization.
4. What are the hydrologist's alternative courses of action? Identify all alternatives, regardless of the ethical implications.
5. How should the conflict be resolved?

You may want to review Sections 1.6 to 1.12 in Chapter 1 in responding to the problem statement.

Case. An engineer is designing a debris flow protection facility for use at the outlet of a steep canyon. The method that is used to predict the volume of debris (mud flow) that can be expected at the outlet of the canyon requires values of some labor-intensive watershed characteristics (a hypsometric index and the length to the center of area). To keep project costs at a minimum by avoiding the computations for the complex watershed, the engineer uses average values from past projects on which she has worked rather than making the computations.

4

Precipitation

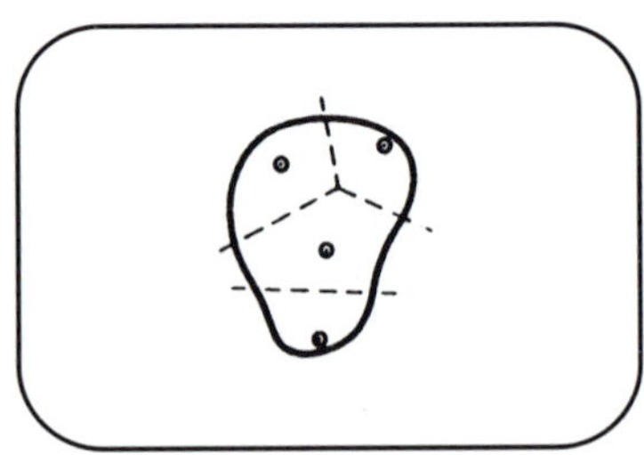

CHAPTER OBJECTIVES

1. Introduce the important characteristics of rainfall.
2. Introduce the concept of a design storm, which is an input to many hydrologic design problems.
3. Provide computational methods for estimating both average areal rainfall amounts and missing rainfall depths.
4. Provide a method for evaluating the consistency of a gaged record of rainfall amounts.
5. Introduce the fundamentals of the probable maximum precipitation.

4.0 NOTATION

A_s = storm area size
A_x = average annual rainfall at gage X
d_i = distance between gage i and regional center point
D = storm duration
i = rainfall intensity (in./hr)
n = number of raingage stations
p = exceedence probability
$\overline{p}$ = average proportion
$\hat{p}$ = estimated mean areal rainfall
P_i = rainfall depth (in.) at gage i
P_x = missing rainfall not measured at gage x
$\overline{P}$ = weighted average rainfall depth
S_i = slope of part i of double mass curve
T = exceedence frequency (return period; years)
w_i = weight applied to rainfall depth at gage i
y_i = adjusted rainfall for section i of double mass curve
ΔX_i = difference in regional catch for section i of double mass curve
ΔY_i = difference in catch at base station for section i of double mass curve

4.1 INTRODUCTION

Precipitation can take many forms, including rain, snow, sleet, hail, and mist. With respect to hydrologic design, only rain and snow are important. This chapter is concerned with rainfall. The problem of snowmelt runoff is addressed in another chapter.

Rainfall is the driving force of most hydrologic designs. Design needs arise when and where rainfall occurs at extreme volumes or rates. High rates of rainfall on small urban watersheds often cause flooding of streets and parking lots because drainage facilities are not usually designed to drain all of the water generated by high rainfall intensities. High rainfall rates can also severely damage crops. The absence of rain over long periods of time reduces the rate of flow in streams and rivers, as well as causing lake levels to decline sharply. Low streamflow rates due to the lack of rain can damage stream habitat as well as reduce the capacity of the stream to assimilate wastes discharged into the stream. The decline of lake levels can reduce the recreational capacity of a lake, as well as reduce the water available for power generation and irrigation. Excessive moisture from low-intensity, long-duration storms can create plant stress, thus limiting the growth of crops. It should be evident that problems can occur from extremes in rainfall, with the extremes in either the rate, the duration, or the time interval between storms.

Some hydrologic planning and design problems only require a volume of rainfall. However, for purposes of hydrologic analysis and design, the distribution of rain with respect to time is usually required; this time distribution of rainfall is called a *hyetograph*. A *hyetograph* is a graph of the rainfall intensity or volume as a function of time. The concept of a hyetograph will be discussed in detail after definitions are provided for precipitation characteristics.

Storm events can be separated into two groups: actual storms and design storms. Rainfall analysis is based on actual storms. Either actual or design storms can be used in design. Measurements during an actual storm event are recorded as a series of rainfall depths that oc-

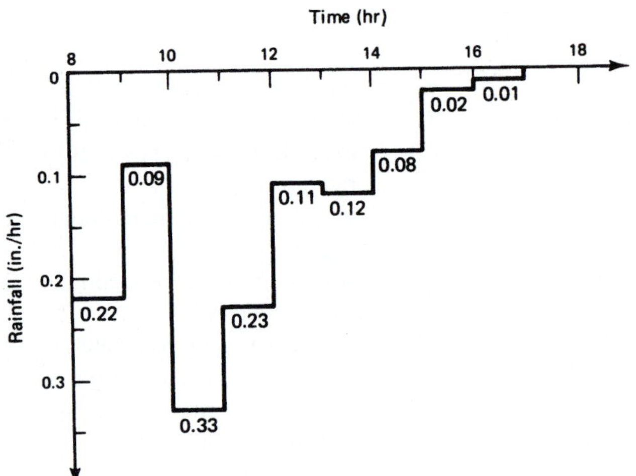

FIGURE 4–1 Hourly precipitation at Washington, D.C., on September 13, 1983.

curred during known time intervals. For example, the precipitation record of the catch at the Washington, D.C., National Airport for September 1983 made at l-hr intervals is shown in Table 4–1. The storm event of September 13, 1983 starting at 8 a.m. had a total volume of 1.21 in. and a duration of 9 hr. The maximum hourly rate was 0.33 in./hr, which occurred between 10 and 11 a.m. Figure 4–1 shows the hyetograph for this rainfall event.

Almost all hydrologic designs are based on what is called the design-storm approach. A design storm is a rainfall hyetograph that has preselected characteristics. A design storm is not an actual measured storm event; in fact, a real storm identical to the design storm has probably never occurred and it is unlikely that it will ever occur. However, most design storms have characteristics that are the average of the characteristics of storms that occurred in the past and, thus, are hypothetical storms that have the average characteristics of storm events that are expected to occur in the future.

4.2 RAINFALL CHARACTERISTICS

Storms differ in a number of characteristics, and the characteristics have a significant affect on hydrologic design.

4.2.1 Volume-Duration-Frequency

What are the characteristics that must be identified in either assessing an actual storm or developing a design storm? The following three characteristics are very important in hydrologic analysis and design:

1. *Duration:* the length of time over which a precipitation event occurs
2. *Volume:* the amount of precipitation occurring over the storm duration
3. *Frequency:* the frequency of occurrence of events having the same volume and duration

TABLE 4–1 Hourly Precipitation Record for Washington, D.C., for September 1983

	HOURLY PRECIPITATION (WATER EQUIVALENT IN INCHES)																								
	A.M. HOUR ENDING AT												P.M. HOUR ENDING AT												
DATE	1	2	3	4	5	6	7	8	9	10	11	12	1	2	3	4	5	6	7	8	9	10	11	12	DATE
1																									1
2																									2
3																									3
4																									4
5																									5
6																									6
7																									7
8																									8
9																									9
10																									10
11																									11
12															T	0.01			T	T	T		0.01	0.03	12
13	0.06	T						T	0.22	0.09	0.33	0.23	0.11	0.12	0.08	0.02	0.01	T							13
14																									14
15																									15
16																									16
17																	1								17
18																									18
19																									19
20																									20
21									T	T	T	T	0.01	0.06	T	0.21	0.17	0.40	0.05	0.02	0.07	0.03	T		21
22																									22
23												T	T												23
24																									24
25																									25
26																									26
27																									27
28																									28
29																									29
30				T	0.04	0.17	0.03	0.11	0.07	0.03	0.01	0.02	0.03	0.03	0.02	T	T								30

Note: T indicates trace amount.

MAXIMUM SHORT DURATION PRECIPITATION

TIME PERIOD (MINUTES)	5	10	15	20	30	45	60	80	100	120	150	180
PRECIPITATION (INCHES)	0.12	0.18	0.21	0.26	0.35	0.40	0.45	0.50	0.58	0.63	0.75	0.81
ENDED: DATE	13	13	13	13	13	13	13	13	13	13	21	21
ENDED: TIME	1053	1058	1058	1100	1110	1115	1122	1142	1208	1222	1748	1818

THE PRECIPITATION AMOUNTS FOR THE INDICATED TIME INTERVALS MAY OCCUR AT ANY TIME DURING THE MONTH. THE TIME INDICATED IS THE ENDING TIME OF THE INTERVAL. DATE AND TIME ARE NOT ENTERED FOR TRACE AMOUNTS.

Closely related to these definitions is the concept of intensity, which equals the volume divided by the duration. For example, a storm having a duration of 2 hr and a volume of 24 acre-in. would have an intensity of 12 acre-in./hour.

Volume and depth. The volume of a storm is most often reported as a depth, with units of length such as inches or centimeters; in such cases, the depth is assumed to occur uniformly over the watershed. Thus the volume equals the depth times the watershed area. For example, if the 2-hr storm with a volume of 24 acre-in. occurred on a 6-acre watershed, the depth of rainfall would be 4 in. and the intensity would be 2 in./hr. This interchanging use of units for storm volume often leads to confusion because the terms depth and volume are applied to a quantity having units of length. One might speak of the rainfall volume, but express it in inches. Such statements imply that the depth occurred uniformly over the entire watershed and the units are "area-inches," with the area of the watershed used to compute a volume in acre-inches or some similar set of units.

Just as each concept is important by itself, it is also important to recognize the interdependence of these terms. A specified depth of rainfall may occur from many different combinations of intensities and durations, and these different combinations of intensities and durations will have a significant effect on both runoff volumes and rates, as well as on engineering designs that require rainfall characteristics as input. For example, 3 in. of precipitation may result from any of the following combinations of intensity and duration:

Intensity (in./hr)	Duration (hr)	Depth (in.)
12	0.25	3
6	0.50	3
3	1.00	3
1.5	2.00	3

Because the rainfall intensity is an important determinant of the hydrologic response, it is important to specify both the depth and duration (or intensity and duration) and not just the total volume (i.e., depth).

Frequency. Just as intensity, duration, and volume are interdependent, the fourth concept, frequency, is also a necessary determinant. Frequency can be discussed in terms of either the exceedence probability or the return period, which are defined as follows:

Exceedence probability: the probability that an event having a specified depth and duration will be exceeded in one time period, which is most often assumed to be one year.

Return period: the average length of time between events having the same depth and duration.

The exceedence probability (p) and return period (T) are related by:

$$p = \frac{1}{T} \tag{4–1}$$

The following tabulation gives selected combinations of Equation 4–1:

T(yr)	p	T(yr)	p
2	0.50	5	0.20
10	0.10	25	0.04
100	0.01	500	0.002

For example, if a storm of a specified duration and depth has a 1% chance of occurring in any one year, it has an exceedence probability of 0.01 and a return period of 100 years. The argument for not using the term return period to interpret the concept of frequency is that it is sometimes improperly interpreted. Specifically, some individuals believe that if a 100-yr rain (or flood) occurs in any one year, it cannot occur for another 100 years; this belief is false because it implies that storm events occur deterministically rather than randomly. Because storm events occur randomly, there is a finite probability that the 100-yr event could occur in two consecutive years, or not at all in a period of 500 years. Thus the exceedence probability concept is preferred by many. However, engineers commonly use the term return period, and its meaning should be properly understood.

***VDF* interdependence.** Events having similar intensities may differ significantly in depth and duration when a difference in frequency occurs. For example, the three storms have similar intensities, but differ significantly in depth, duration, and frequency:

Depth (in.)	Duration (hr)	Frequency (yr)	Intensity (in./hr)
5.8	8	100	0.78
3.3	4	10	0.82
1.6	2	2	0.80

The table illustrates the need to consider the frequency of the event, as well as the depth, intensity, and duration.

It is also important to recognize that the relationship between volume, duration, and frequency (*VDF*) is location dependent; that is, a 3-in. storm for a duration of 3 hr in Miami will occur at a different frequency than the same storm in St. Louis. This implies that, since design storms depend on the *VDF* characteristics, design storms will vary from one location to another. The following tabulation gives the approximate depth of rainfall for the 2-year, 2-hr storm event:

Location	P(in.)
Miami	3.0
Atlanta	2.0
St. Louis	1.75
New York	1.5
Tucson	0.75

Thus a storm of 3 in. over a period of 2 hr can be expected once every 2 yr in Miami, but a lot less frequently in the other cities, especially Tucson. The values above indicate that the volume of a 2-yr, 2-hr design storm will vary with location.

One more tabulation can be used to illustrate further the location dependency of the VDF relationship; the following return periods are for a 4-in. storm event with a duration of 24 hr:

Location	*T*(yr)
Miami	<2
Atlanta	4
St. Louis	5
New York	7
Tucson	>100

In Miami, such a storm occurs more often than once every 2 yr, but in Tucson, the same rainfall would occur less frequently than once every 100 yr.

Relationships between rainfall depth, duration, and frequency can be displayed graphically. The relationship is shown with the depth as the ordinate and the duration on the abscissa, with separate curves given for selected exceedence frequencies. Frequencies of 0.5, 0.2, 0.1, 0.04, 0.02, and 0.01, which correspond to return periods of 2, 5, 10, 25, 50, and 100 years, respectively, are the ones most commonly used. *VDF* curves are developed using a statistical method called frequency analysis, which is the subject of another chapter.

4.2.2 Rainfall Maps

Rainfall depths are required for many design problems. Because of the frequent need for such information, the depth of rainfall for selected durations and frequencies are often provided in the form of maps that show lines of equal rainfall depths; the lines are called isohyets. Maps are available for the U.S. (see Figure 4–2 for the 100-yr, 24-hr map) and for individual states (see Figure 4–3 for the 2-yr, 24-hr map for Maryland).

4.2.3 Intensity-Duration-Frequency

In addition to volume-duration-frequency curves, intensity-duration-frequency curves are readily available because a rainfall intensity is used as input to many hydrologic design methods. Because of the importance of the intensity-duration-frequency (*IDF*) relationship in hydrologic analyses, *IDF* curves have been compiled for most localities; the *IDF* curve for Baltimore, Maryland, is shown in Figure 4–4.

The intensity-duration-frequency (*IDF*) curve is most often used by entering with the duration and frequency to find the intensity. For example, the 10-yr, 2-hr rainfall intensity for Baltimore is found from Figure 4–4 by entering with a duration of 2 hr, moving vertically to the 10-yr frequency curve, and then moving horizontally to the intensity ordinate, which yields $i = 1.3$ in./hour. This corresponds to a storm depth of 2.6 in.

The *IDF* curve could also be used to find the frequency for a measured storm event. The predicted frequency is determined by finding the intersection of the lines defined by the measured intensity and the storm duration. If the volume rather than the intensity is mea-

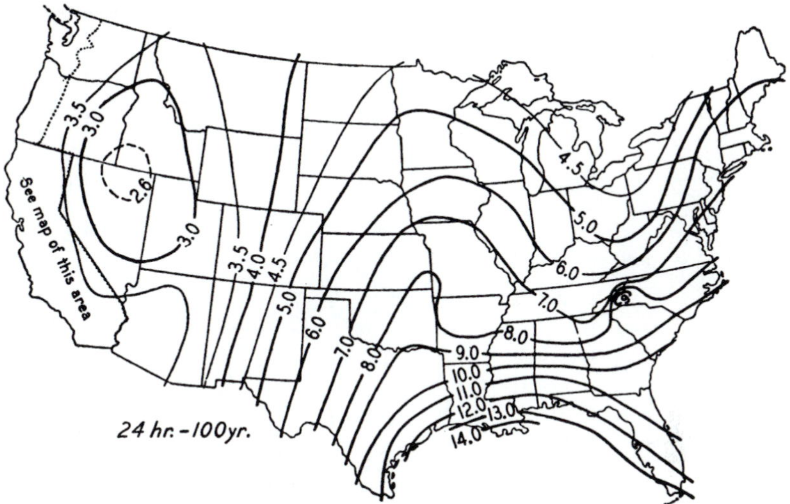

FIGURE 4–2 Twenty-four-hour rainfall, in inches, to be expected once in 100 years. (National Weather Service, 1961.)

sured, the intensity must be determined prior to determining the frequency. For example, if a 3-in. storm occurs in Baltimore during a period of 3 hr, the intensity is 1 in./hr. Using an intensity of 1 in./hr and a duration of 3 hr yields a return period of 10 yr or an exceedence probability of 0.1.

4.2.4 Mathematical Representation of *IDF* Curves

Given the interest in computerizing the elements of hydrologic design, it is worthwhile to present mathematical forms to represent the *IDF* curve. The following equations can provide a reasonably accurate representation of an *IDF* curve:

$$i = \begin{cases} \dfrac{a}{D + b} & \text{for } D \leqslant 2 \text{ hr} \quad (4\text{–}2a) \\ cD^d & \text{for } D > 2 \text{ hr} \quad (4\text{–}2b) \end{cases}$$

in which i is the rainfall intensity (in./hr); D is the duration (hr); and a, b, c, and d are fitting coefficients that vary with the frequency. Values for the coefficients are easily obtained by transforming Equation 4–2 to linear forms and evaluating the coefficients using any two points of the *IDF* curve by solving two simultaneous equations. Equation 4–2a is transformed to linear form as follows:

$$i = \frac{a}{b + D} \tag{4–3a}$$

$$\frac{1}{i} = \frac{b + D}{a} \tag{4–3b}$$

$$y = f + gD \tag{4–3c}$$

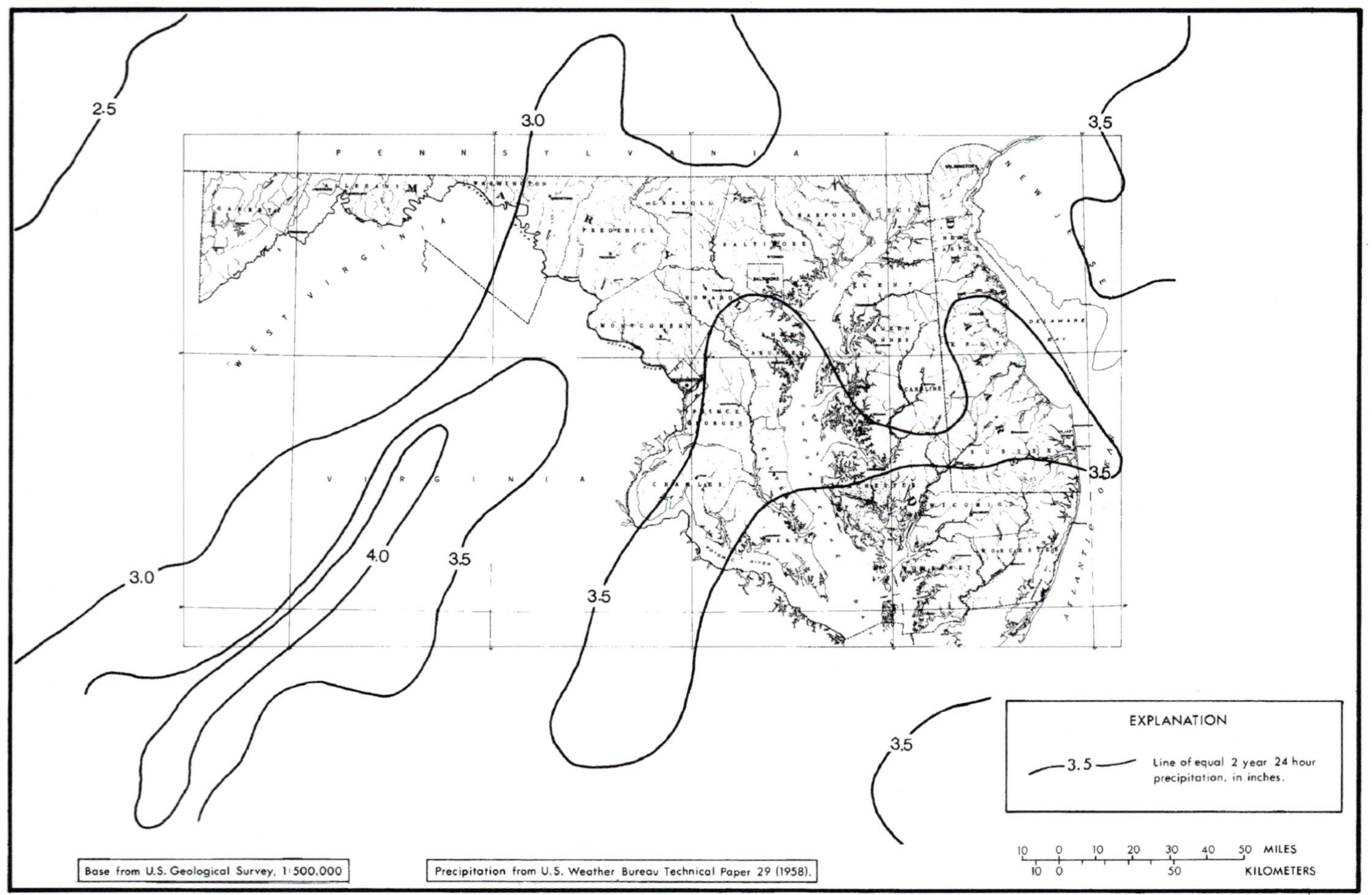

FIGURE 4–3 The 2-year 24-hour precipitation.

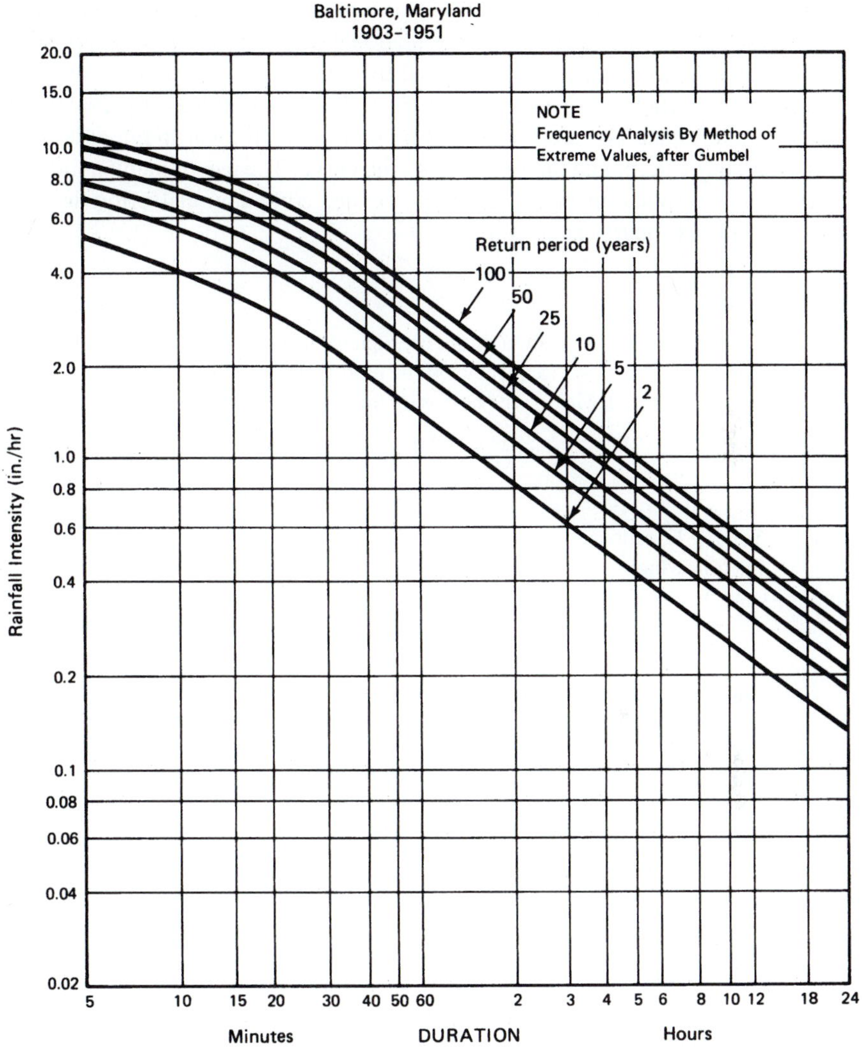

FIGURE 4–4 Rainfall intensity-duration-frequency. (National Weather Service, 1961.)

in which $y = 1/i$, $f = b/a$, and $g = 1/a$. Equation 4–2b is transformed to linear form by making a logarithmic transform:

$$i = cD^d \tag{4–4a}$$

$$\log i = \log c + d \log D \tag{4–4b}$$

$$y = h + dx \tag{4–4c}$$

in which $y = \log i$, $h = \log c$, and $x = \log D$. Either natural or base-10 logarithms can be used. Equations 4–3c and 4–4c are linear and can be solved as two simultaneous equations using any two points from an *IDF* curve. The accuracy depends on the reading accuracy of the two points selected and on the ability of Equation 4–2a to represent the shape of the *IDF* curve.

Greater accuracy can be achieved by using numerous points from the *IDF* curve and fitting the coefficients of Equations 4–3c and 4–4c using least squares (see Chapter 2). Because the ordinate and abscissa of Figure 4–4 are both logarithmic, it is not always possible to accurately interpolate the values from the curve. The use of many points will minimize the effect of interpolation error; however, greater computational effort is obviously required when the least squares approach is used.

Example 4–1

To illustrate the calculation of the coefficients of Equation 4–2a, the following two points from Figure 4–4 for a 2-yr frequency will be used with Equation 4–3c: ($i = 4.1$ in./hr, $D = 1/6$ hr) and ($i = 0.81$, $D = 2$). Making the necessary transformations yields

$$0.2439 = f + g(0.1667) \tag{4–5a}$$

$$1.2346 = f + g(2) \tag{4–5b}$$

Solving Equations 4–5 for f and g yields $f = 0.1538$ and $g = 0.5404$. Solving for a and b yields the following:

$$i = \frac{1.8505}{0.2847 + D} \quad \text{for } D \leqslant 2 \text{ hr} \tag{4–6a}$$

Equation 4–6 should provide intensities that are within 0.01 of values obtained from Figure 4–2.

For durations greater than 2 hr, the following two points can be used to estimate c and d for Equation 4–2b: ($i = 0.5$, $D = 4$) and ($i = 0.3$, $D = 8$). Making natural-logarithm transformations yields $d = -0.7370$ and $h = 0.3286$. Thus Equation 4–2b is

$$i = 1.389D^{-0.7370} \quad \text{for } D > 2 \text{ hr} \tag{4–6b}$$

Using two points to define the function can lead to a discontinuity at the 2-hr point of separation. For a D of 2 hr, Equation 4–6b yields a computed intensity of 0.833 in./hr, which is greater than the value predicted by Equation 4–6a. It is also higher than the value taken from Figure 4–4 for a duration of 2 hr. This problem can be solved either by using a more complex fitting method, such as numerical least squares with a constraint that the lines match at $D = 2$ hrs, or by including a greater number of sample points from the curve near the 2-hr duration. Equations 4–6 can be used to represent the *IDF* curve for the 2-yr event in Baltimore. Similar equations can be developed for the other return periods.

Example 4–2

A set of points, given in Table 4–2, were taken for both segments of the 2-yr *IDF* curve of Figure 4–4. The following equations were fit using least squares:

$$i = \frac{1.804}{D + 0.284} \quad \text{for } D < 2 \text{ hr} \tag{4–7a}$$

$$i = 1.37\, D^{-0.74} \quad \text{for } D \geqslant 2 \text{ hr} \tag{4–7b}$$

The curves agree closely with the measured points except for durations less than 10 min. The two computed curves also show a discontinuity at the point of intersection (i.e., 2 hr).

TABLE 4–2 Fitting the 2-year *IDF* Curve for Baltimore, MD

Model	Duration (hr)	Computed intensity (in./hr)	*IDF* intensity (in./hr)	Error (in./hr)
Equation 4–7a	0.0833	4.91	5.2	–0.29
	0.10	4.70	5.0	–0.30
	0.1667	4.00	4.1	–0.10
	0.25	3.38	3.4	–0.02
	0.3333	2.92	3.0	–0.08
	0.50	2.30	2.3	0.00
	0.75	1.74	1.65	0.09
	1.0	1.40	1.35	0.05
	1.5	1.01	1.00	0.01
	2.0	0.79	0.81	–0.02
Equation 4–7b	2	0.82	0.81	0.01
	4	0.49	0.50	–0.01
	6	0.36	0.36	0.00
	8	0.30	0.30	0.00
	10	0.25	0.25	0.00
	12	0.22	0.22	0.00
	18	0.16	0.16	0.00
	24	0.13	0.13	0.00

4.2.5 Depth-Area Adjustments

The rainfall depths for *IDF* curves, such as those of Figure 4–4, represent estimates for small areas. For designs on areas larger than a few square miles, the point rainfall estimates obtained from *IDF* curves must be adjusted. The point estimates represent extreme values. As the spatial extent of a storm increases, the average depth of rainfall over the watershed decreases; actual storms have a spatial pattern as well as time variation. Figure 4–5 shows the storm pattern for the total storm rainfall (inches) for the August 1, 1985, event in Cheyenne, Wyoming. Figure 4–6 shows a design storm rainfall pattern that is used in estimating probable maximum floods. Both of these figures show that the rainfall depth decreases from the center of the storm as the size of the storm cell increases. When selecting a point rainfall to apply uniformly over a watershed, the point value from an *IDF* curve needs to be reduced to account for the areal extent of the storm. The reduction is made using a depth-area adjustment factor. The factor is a function of the drainage area (square miles) and the rainfall duration. Figure 4–7 shows depth-area adjustment factors based on Weather Bureau Technical Paper No. 40. This set of curves can be used unless specific curves derived from regional analyses are available. Figure 4–7 shows that the adjustment factor decreases from 100% as the watershed area increases and as the storm duration decreases. Beyond a drainage area of 300 mi^2 the adjustment factor shows little change with increasing area.

Example 4–3

To illustrate the use of Figures 4–4 and 4–7, consider the case of a 200-mi^2 watershed near Baltimore, Maryland, where the design standard requires the use of a 24-hr storm duration and a

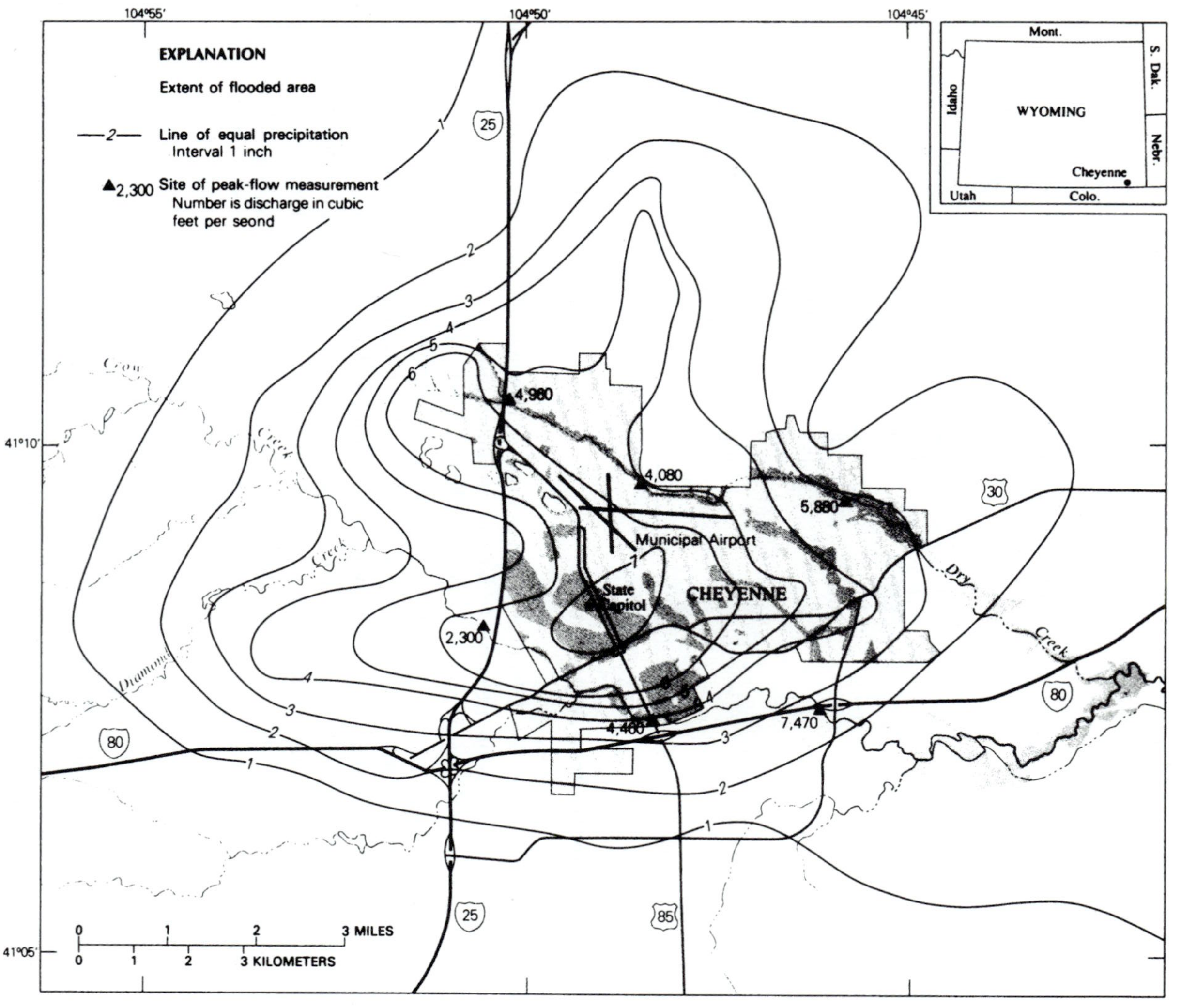

FIGURE 4–5 Storm and flood of August 1, 1985, in Cheyeene, WY.

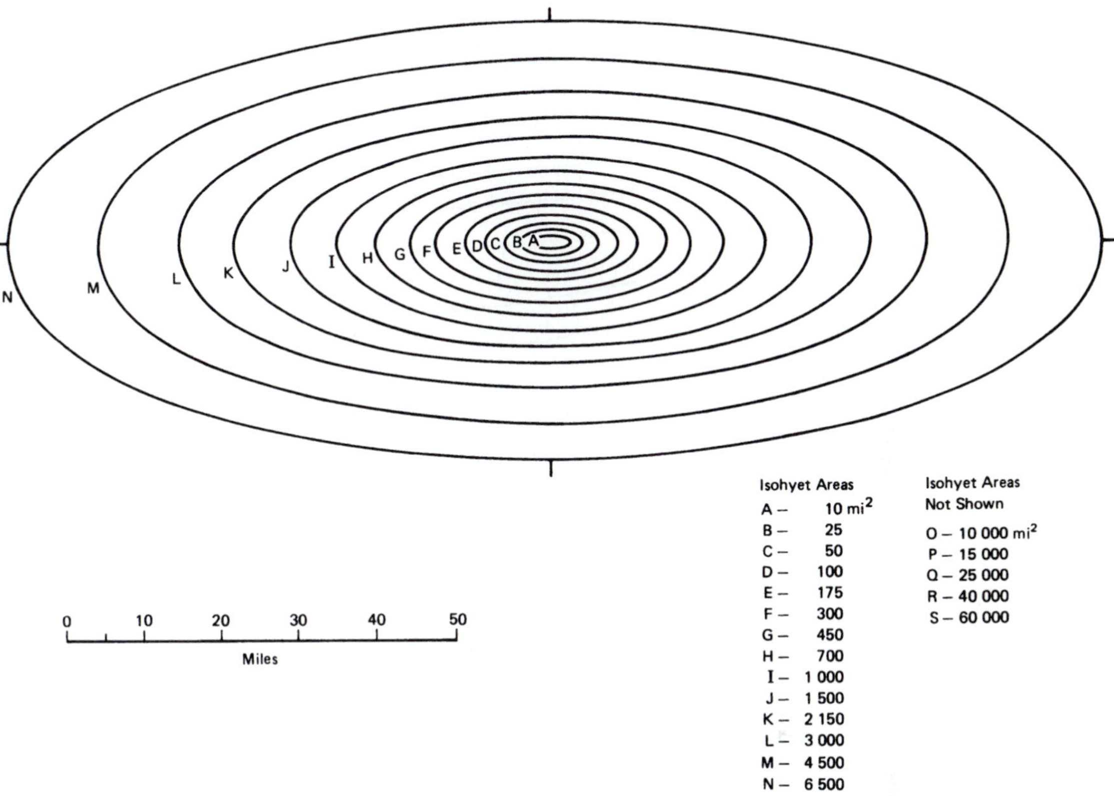

FIGURE 4–6 Standard isohyetal pattern.

100-yr return period. The point rainfall from Figure 4–4 is 0.3 in./hr or a depth of 7.2 in. The depth-area curve of Figure 4–7 indicates an adjustment factor of 91% or 0.91. Therefore, the rainfall depth, which will be assumed constant over the 200-mi^2 watershed, is 6.55 in. The reduction of 0.65 in. from the point rainfall of 7.2 in. accounts for the reduction in rainfall volume due to the variability of rainfall depth over the watershed, as illustrated by the actual and hypothetical rainfall patterns of Figures 4–5 and 4–6, respectively.

4.2.6 Precipitation Extremes and Means

Precipitation, specifically rainfall, is an important element in many engineering designs. Some designs are based on average expected rainfall depths, while high-hazard projects are often based on extreme amounts. Therefore, it is necessary to have some idea of both the typical and the not-so-typical amounts. Table 4–3 provides the recorded maximum point rainfalls at different locations. Since these are maximum recorded values, it would be difficult, maybe impossible, to accurately estimate the return frequency of the values; however, they provide a measure of extreme rainfalls.

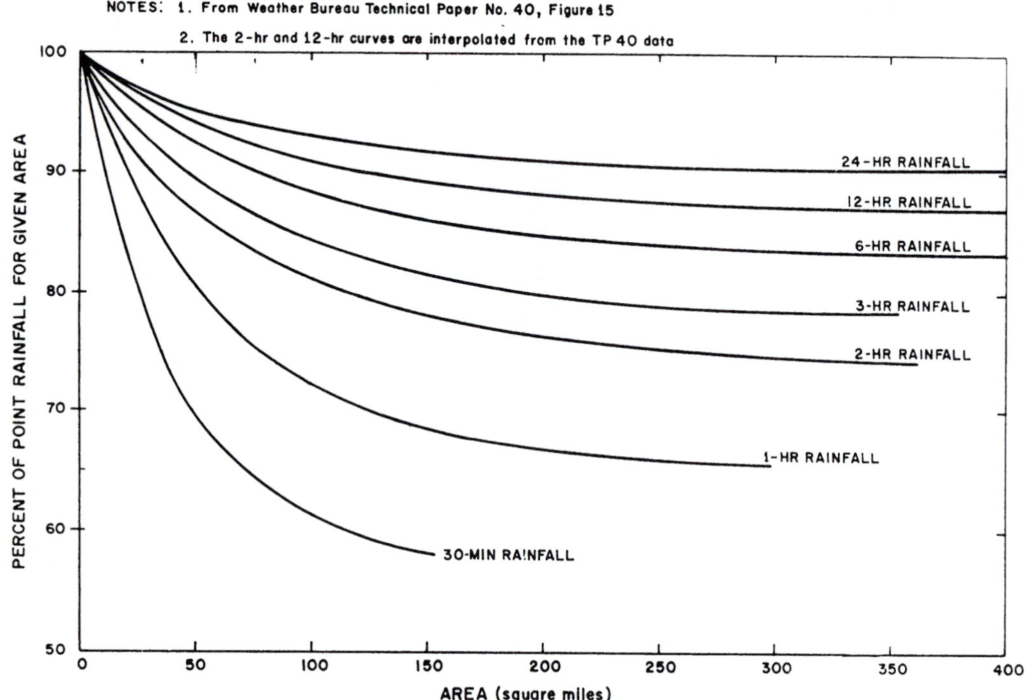

FIGURE 4–7 Depth-area curves for adjusting point rainfalls.

Table 4–4 gives approximate depths for selected durations, return periods, and locations. These values are typical of those used frequently in designs. It is interesting to compare these expected values for selected return periods with the extremes of Table 4–3. For example, the 15-min maximum point rainfall of 7.8 in. is over four times greater than the 100-yr event for New York; it is even larger than the 100-yr event expected for a 24-hr dura-

TABLE 4–3 Recorded Maximum Point Rainfalls

Duration	Depth (in.)	Intensity (in./hr)	Location
1 min	1.23	73.80	Unionville, MD
15 min	7.80	31.20	Jamaica
42 min	12	17.14	Holt, Missouri
130 min	19.00	8.76	Rockport, WV
15 hr	34.50	2.30	Smethport, PA
24 hr	74	3.08	Cilaos, La Reunion I.
48 hr	65.79	1.37	Taiwan
5 day	152	1.27	Cilaos, La Reunion I.
1 month	366	0.51	Cherrapunji, India
1 yr	1042	0.12	India
2 yr	1605	0.09	India

TABLE 4–4 Approximate Expected Rainfall Depths for Selected Durations, Return Periods, and Locations

Return Period (yr)	Duration	Depth (in.) for: Atlanta	New York	Chicago	Los Angeles	Phoenix
2	15 min	1.0	1.0	0.9	0.5	0.5
2	1 hr	1.8	1.4	1.4	0.8	0.9
2	24 hr	4.2	3.5	2.8	3.5	2.0
100	15 min	1.9	1.7	1.5	1.9	1.0
100	1 hr	3.5	3.0	2.7	2.2	3.0
100	24 hr	8.0	6.7	5.7	9.0	5.0
Mean	Annual	48	48	34	16	8

tion in New York. The 24-hr recorded maximum of about 46 in. (Philippines) is almost seven times larger than the 24-hr event that can be expected in New York once every 100 years. The 24-hr recorded maximum in the Philippines is almost the same as the annual average rainfall in New York.

These comparisons show that rainfall depths show considerable variation with duration, frequency, and location. Given the importance of rainfall to hydrologic design, it is important to recognize typical values of point rainfalls and appreciate the variations that can be expected.

4.2.7 Storm-Event Isohyetal Patterns

Where many rain gages are located in a region, the measured depths of rainfall can be used to draw isohyets for total rainfall depths of a storm. Even in cases where a dense network of rain gages exists, considerable skill is required to draw the isohyets. The resulting maps are used frequently in assessing flood damages, especially in litigation following major storms.

Figure 4–5 shows the storm isohyets for the August 1, 1985, event in Cheyenne, Wyoming. The largest depth occurred around the state capitol. The isohyets for actual storms are often characterized by irregular shapes, as evidenced by the pattern of Figure 4–5.

4.3 ESTIMATING MISSING RAINFALL DATA

Measured precipitation data are important to many problems in hydrologic analysis and design. Because of the cost associated with data collection, it is very important to have complete records at every station. Obviously, conditions sometimes prevent this. For gages that require periodic observation, the failure of the observer to make the necessary visit to the gage may result in missing data. Vandalism of recording gages is another problem that results in incomplete data records, and instrument failure because of mechanical or electrical malfunctioning can result in missing data. Any such causes of instrument failure reduce the length and information content of the precipitation record.

Certainly, rainfall records are important. Rainfall data are an important input to hydrologic designs, whether measured storm event data or synthetic data based on characteristics of measured data. A number of federal and state agencies, most notably the National

Weather Service, have extensive data collection networks. They collect and analyze data to provide those who need such information with reasonably complete data records and accurate data summaries, such as *IDF* curves. Those involved in legal cases, such as for flooding that may be the result of human-made watershed modifications, also require accurate and complete rainfall records. Those involved in hydrologic research also require precipitation records to test models and evaluate hydrologic effects. Where parts of records are missing, it may be desirable, even necessary, to estimate the missing part of the record.

A number of methods have been proposed for estimating missing rainfall data. The station-average method is the simplest method. The normal-ratio and quadrant methods provide a weighted mean, with the former basing the weights on the mean annual rainfall at each gage and the latter having weights that depend on the distance between the gages where recorded data are available and the point where a value is required. The isohyetal method is the fourth alternative.

4.3.1 Station-Average Method

The station-average method for estimating missing data uses n gages from a region to estimate the missing point rainfall, $\hat{P}$, at another gage:

$$\hat{P} = \frac{1}{n}\sum_{i=1}^{n} P_i \tag{4–8}$$

in which P_i is the catch at gage i. Equation 4–8 is conceptually simple, but may not be accurate when the total annual catch at any of the n regional gages differ from the annual catch at the point of interest by more than 10%. Equation 4–8 gives equal weight to the catches at each of the regional gages. The value $1/n$ is the weight given to the catch at each gage used to estimate the missing catch.

Example 4–4

As an example, consider the following data:

Gage	Annual P (in.)	Storm Event P (in.)
A	42	2.6
B	41	3.1
C	39	2.3
X	41	?

The storm-event catch at gage X is missing. Ten percent of the annual catch at gage X is 4.1 in., and the average annual catch at each of the three regional gages is within ± 4.1 in.; therefore, the station-average method can be used. The estimated catch at the gage with the missing storm event total is

$$\hat{P} = \frac{1}{3}(2.6 + 3.1 + 2.3) = 2.67 \text{ in.} \tag{4–9}$$

Using this method requires knowledge of the average annual catch, even though this information is not used in computing the estimate $\hat{P}$.

4.3.2 Normal-Ratio Method

The normal-ratio method is conceptually simple, but it differs from the station-average method of Equation 4–8 in that the average annual catch is used to derive weights for the rainfall depths at the individual stations. The general formula for computing is $\hat{P}$:

$$\hat{P} = \sum_{i=1}^{n} w_i P_i \tag{4–10}$$

in which w_i is the weight for the rainfall depth P_i at gage i. The weight for station i is computed by

$$w_i = \frac{A_x}{nA_i} \tag{4–11}$$

in which A_i is the average annual catch at gage i, A_x is the average annual catch at station X, and n is the number of stations. The weights will not sum to 1 unless the mean of the average annual catches for the n stations equals the average annual catch at gage X. The sum of the weights will equal the ratio of the average annual catch at gage X to the mean of the average annual catches for the n gages. This is rational because, if gage X has a lower average annual catch, we would then expect the catch for an individual storm to be less than the mean of the storm event catches for the gages. When the average annual catches for all of the gages are equal, the station-average and normal-ratio methods yield the same estimate. When the variation in the average annual catches is small (less than 10%), the two methods provide estimates that are nearly similar, and the difference in the estimates does not warrant use of the additional information in the form of the average annual catches. When the average annual catches differ by more than 10%, the normal-ratio method is preferable; such differences might occur in regions where there are large differences in elevation (for example, regions where orographic effects are present) or where average annual rainfall is low but has high annual variability.

The conceptual basis of the normal-ratio method may be more apparent if Equations 4–10 and 4–11 are rewritten as

$$\frac{P_x}{A_x} = \frac{1}{n} \sum_{i=1}^{n} \frac{P_i}{A_i} = \bar{p} \tag{4–12}$$

The ratio P_i/A_i is the proportion for gage i of the mean annual catch that occurs in the specific storm; thus the right-hand side of Equation 4–12 is the average proportion. This average proportion is then used as the proportion at gage X where amount P_x was not recorded, and the estimate of P_x is found by multiplying the average proportion, $\bar{p}$, by the average annual catch at gage X, A_x. The value of $\bar{p}$ is considered to be the best estimate of the true proportion p of the catch during the storm for which the recorded value is missing.

Example 4–5

To illustrate the normal-ratio method, consider the following data:

Gage	Average Annual Rainfall (in.)	Total Storm Rainfall (in.)
A	41	2.4
B	37	2.3
C	46	3.1
X	40	?

The storm-event catch for gage X is missing and can be estimated using the data from the table. The average annual rainfall at the four gages is 41 in. Ten percent of this is 4.1 in. Since the annual rainfall at gage C is more than 4.1 in. from the mean of 40 in., the normal-ratio method should be used. The estimate from the normal-ratio method is

$$\begin{aligned}\hat{P} &= w_A P_A + w_B P_B + w_C P_C \\ &= \frac{A_x}{nA_A} P_A + \frac{A_x}{nA_B} P_B + \frac{A_x}{nA_C} P_C \\ &= \left[\frac{40}{3(41)}\right](2.4) + \left[\frac{40}{3(37)}\right](2.3) + \left[\frac{40}{3(46)}\right](3.1) \\ &= 0.325(2.4) + 0.360(2.3) + 0.290(3.1) = 2.51 \text{ in.}\end{aligned} \tag{4–13a}$$

Whereas the station-average method would yield an estimate of 2.60 in., Equation 4–13a indicates a value of 2.51 in. with the normal-ratio method. The latter value is less than the average of the other stations because the mean annual catch at gage X is less than the mean of the average annual catches at gages A, B, and C (i.e., 40 in. versus 41.33 in.).

For the data used with Equation 4–13 the storm rainfall estimated with the normal-ratio method expressed in the form of Equation 4–10 is

$$\begin{aligned}\hat{P} &= 40\left[\frac{1}{3}\left(\frac{2.4}{41} + \frac{2.3}{37} + \frac{3.1}{46}\right)\right] \\ &= 40\left[\frac{1}{3}(0.05854 + 0.06216 + 0.06739\right] \\ &= 40(0.06270) = 2.51 \text{ in.}\end{aligned} \tag{4–14}$$

Thus the proportion at the gages with a known storm catch ranged from 0.059 to 0.067, with a mean proportion of 0.0627; this mean proportion is used as the best estimate of the proportion for the gage with the missing storm event rainfall. The estimate equals the estimate obtained from Equations 4–10 and 4–11.

4.3.3 Isohyetal Method

The isohyetal method is another alternative for estimating missing rainfall data. The procedure is essentially the same as that used for the isohyetal method when it is applied to the problem of estimating mean areal rainfall (see Section 4.5.3). The location and catch for each gage are located on a map and used to draw lines of equal catch (that is, isohyets) for the storm duration of interest. The location of the gage for which data are missing is then plotted on the map and the catch estimated by interpolation within the isohyets. Of course, the accuracy of the estimate will depend on the number of gages used to draw the isohyets, the homogeneity of the meteorological conditions that generated the storm, and, if they exist, orographic effects.

Example 4–6

The isohyetal pattern shown in Figure 4–5 can be used to illustrate the isohyetal method. If a rain gage were located at the confluence of Diamond and Crow Creeks and the data for the August 1, 1985, storm event were missing, the isohyets would suggest a total storm depth of 3.5 in. In some parts of the isohyetal map, interpolation is not wholly objective; for example, in the area about the point defined by a latitude and longitude of 41°10′N and 104°45′W, the isohyet for 2 in. deviates considerably from the isohyet for 3 in. and there is no isohyet shown for 1 in. A similar problem of interpolation occurs near the intersection of Routes 25 and 85.

4.3.4 Quadrant Method

The station-average method does not account for either the closeness of the other gages to the location of the missing measurement or the density of the raingage network. The normal-ratio method requires information in addition to the storm event catches at all gages; namely, the average annual catch at each gage is required. The isohyetal method requires a reasonably dense network of gages in order to accurately construct the isohyets. Each of the methods has disadvantages.

The quadrant method is an alternative to these other methods. Once again, the estimated catch $\hat{P}$ is a weighted average of catches at other gages, and Equation 4–10 is used. The method is based on two assumptions: (1) catches at gages that are located close to each other are not independent estimates of the catch at the unknown point, and therefore, all gages are not necessary; and (2) the weight assigned to a gage used to estimate $\hat{P}$ should decrease as the distance between the gage and the point where an estimate is required increases.

To estimate P, the region is divided into four quadrants using north-south and east-west lines that intersect at the point where the catch is missing (point X in Figure 4–8). The coordinates of each gage with respect to the center location are determined and used to compute the distance between the gages and the center point. The quadrant method uses only the point in each quadrant that is closest to the center location (that is, point X); this attempts to ensure that the gages used to estimate P are somewhat independent. The weights are defined as a function of the reciprocal of the square of the distance between the gage and the center location X, $1/d_i^2$. If there is a gage in each quadrant, the weight for quadrant i is

$$w_i = \frac{1/d_i^2}{\sum_{j=1}^{4}\left(\frac{1}{d_j^2}\right)} \tag{4–15}$$

where d_i is the distance between gage i and the center location.

In some cases, there may not be a gage in one or more quadrants or a gage may be too far from the center point to have a significant weight. In such cases, the summation in the de-

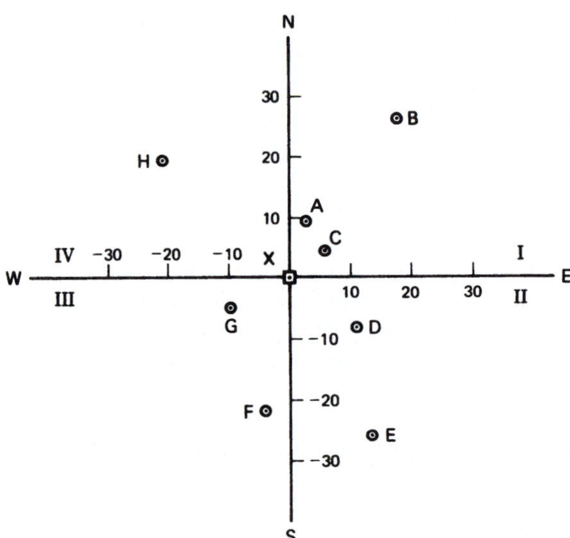

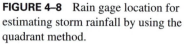

FIGURE 4–8 Rain gage location for estimating storm rainfall by using the quadrant method.

nominator of Equation 4–15 will be over those quadrants that include a gage that yields a significant weight.

The delineation of quadrants does not have to be based on a north-south, east-west axis system. If there are hydrometeorological reasons for a different axis system, such as that caused by orographic effects, a different system can be used. In such cases, the reasons for adopting the alternative system should be clearly stated. There may even be cases where a non-Cartesian coordinate system is used, where justified by hydrometeorological conditions.

Example 4–7

The data of Figure 4–8 and Table 4–5 can be used to illustrate the quadrant method. Eight gages with known catches are located in the region around the point where the catch at a gage is miss-

TABLE 4–5 Calculation of Total Storm Rainfall Using the Quadrant Method[a]

		Coordinates							
Quadrant	Gage	N-S	E-W	d_i^2	d_i	$(1/d_i^2) \times 10^3$	P_i (in.)	w_iP_i	w_i
I	*A*	9	3	90	9.49		3.9		
	B	26	18	1000	31.26		1.4		
	C	4	6	52	7.21*	19.231	3.7	2.102	0.568
II	*D*	−8	11	185	13.60*	5.405	1.6	0.254	0.159
	E	−26	14	872	29.53		0.2		
III	*F*	−22	−4	500	22.36		0.9		
	G	−5	−10	125	11.18*	8.000	3.0	0.708	0.236
IV	*H*	19	−21	802	28.32*	1.247	0.9	0.033	0.037
						33.883		3.097	

*Asterisks indicate gages selected to compute the weights.

ing. The distance is computed from the coordinates and one point from each quadrant selected (points *C*, *D*, *G*, and *H*). Only these four gages are used to compute the weights and estimate *P*. Since gage *C* is closest to point *X*, it has the largest weight. It is of interest to note that gage *A* is not used to estimate *P* even though it is the second closest gage to point *X*; it is not used because gage *C* is in the same quadrant and closer to point *X*. Even though gage *H* is much farther from point *X*, it is used to estimate *P* because it is in a different quadrant and, therefore, assumed to be an independent estimate of the rainfall at point *X*. The weights for gages *C*, *D*, *G*, and *H* are 0.568, 0.159, 0.236, and 0.037, respectively. The sum of the weights equals 1.0. The weighted estimate of the rainfall equals 3.10 in.

4.4 GAGE CONSISTENCY

Estimating missing data is one problem that hydrologists need to address. A second problem occurs when the catch at rain gages is inconsistent over a period of time and adjustment of the measured data is necessary to provide a consistent record. A consistent record is one where the characteristics of the record have not changed with time. Adjusting for gage consistency involves the estimation of an effect rather than a missing value. An inconsistent record may result from any one of a number of events; specifically, adjustment may be necessary due to changes in observation procedures, changes in exposure of the gage, changes in land use that make it impractical to maintain the gage at the old location, and where vandalism frequently occurs.

Double-mass-curve analysis is the method that is used to check for an inconsistency in a gaged record. A *double-mass curve* is a graph of the cumulative catch at the rain gage of interest versus the cumulative catch of one or more gages in the region that has been subjected to similar hydrometeorological occurrences and are known to be consistent. If a rainfall record is a consistent estimator of the hydrometeorological occurrences over the period of record, the double-mass curve will have a constant slope. A change in the slope of the double-mass curve would suggest that an external factor has caused changes in the character of the measured values. If a change in slope is evident, then the record needs to be adjusted, with either the early or later period of record adjusted. Conceptually, adjustment is nothing more than changing the values so that the slope of the resulting double-mass curve is a straight line.

The method of adjustment is easily understood with the schematic of Figure 4–9. The double-mass curve between the cumulative regional catch *X* and the cumulative catch at the gage *Y* where a check for consistency is needed is characterized by two sections, which are denoted in Figure 4–9 by the subscripts 1 and 2. The slopes of the two sections, S_1 and S_2, can be computed from the cumulative catches:

$$S_i = \frac{\Delta Y_i}{\Delta X_i} \tag{4–16}$$

in which S_i is the slope of section *i*, ΔY_i is the change in the cumulative catch for gage *Y* between the endpoints of the section *i*, and ΔX_i is the change in the cumulative catch for the sum of the regional gages between the endpoints of section *i*.

Either section of the double-mass curve can be adjusted for consistency. If the gage has been permanently relocated, it would be of interest to adjust the initial part of the record (that

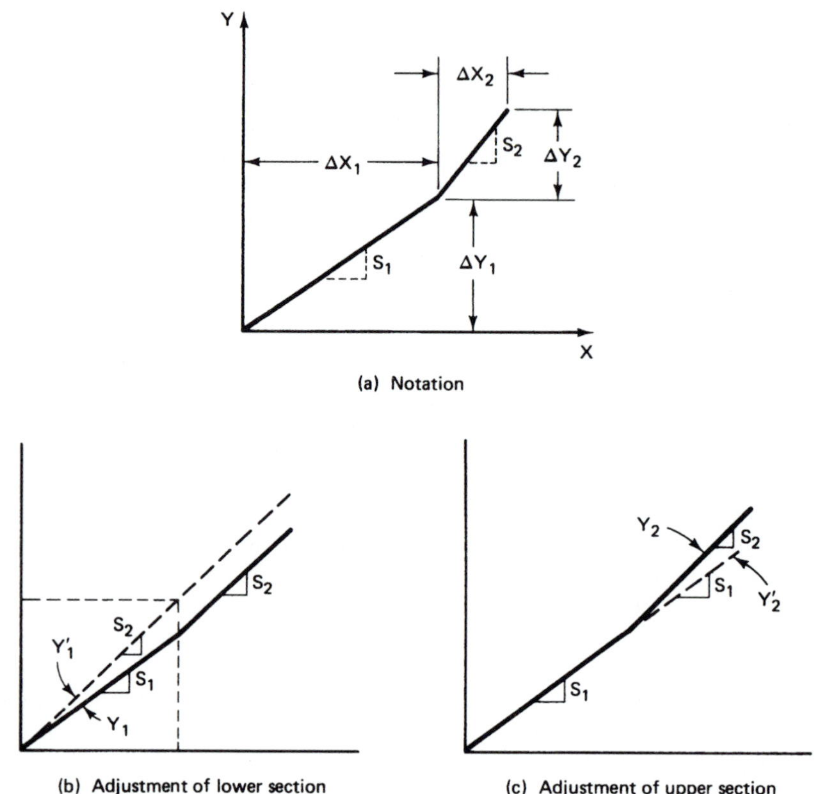

FIGURE 4–9 Adjustment of a double-mass curve for gage consistency.

is, part 1 of Figure 4–9a) so that it is consistent with the latter part of the record (that is, part 2 of Figure 4–9a) and the data that will be collected in the future. If the gage has been relocated only temporarily or if the exposure at the gage adversely affected the catch for a brief period, it would be of interest to adjust the latter part of the record so that it would be consistent with the initial part of the record; this adjustment will make the latter part of the record consistent with both the initial part of the record and the data that will be collected in the future.

If it is of interest to adjust the initial part of the record of station Y, then S_2, which is based on Y_2 and X_2, is correct, and it is necessary to adjust the Y_1 values. The adjustment will provide an adjusted series of data, which will be denoted as y_i. If S_2 is correct, the slope S_1 should be removed from Y_1 and replaced with slope S_2; this adjustment is made by

$$y_1 = \frac{S_2}{S_1} Y_1 \tag{4–17}$$

in which the ratio S_2/S_1 is the adjustment factor. The adjustment is shown in Figure 4–9b. By multiplying each value in the Y_1 series by the adjustment factor of Equation 4–17, a series y_1 that has the same slope as the upper section, Y_2, will be produced.

The procedure for adjusting the upper section of the double-mass curve is similar to the adjustment of the lower section. The intent is to replace the slope of the upper section with the slope of the lower section (see Figure 4–9c). The adjusted values are computed by

$$y_2 = \left(\frac{S_1}{S_2}\right)Y_2 \tag{4–18}$$

Dividing by S_2 removes the effect of the existing slope and replaces it with S_1 through multiplication by S_1. The series y_2 will have slope S_1.

Example 4–8

The data of Table 4–6 can be used to illustrate the use of Equation 4–17 to adjust the lower section of a double-mass curve. Gage *H* was permanently relocated after a period of 3 yr (at the end of 1981); thus it is necessary to adjust the recorded values from 1979 through 1981. The double-mass curve is shown in Figure 4–10, with the cumulative catch at three gages in the region plotted against the cumulative catch at the gage of interest, gage *H*. The slope for the 1979–1981 period is

$$S_1 = \frac{99 - 0}{229 - 0} = 0.4323$$

The slope from 1982 to 1986 is

$$S_2 = \frac{230 - 99}{573 - 229} = 0.3808$$

Using Equation 4–17, the adjusted values from 1979 through 1981 can be computed using

$$h_1 = \left(\frac{0.3808}{0.4323}\right)H_1 = 0.8809H_1 \tag{4–19}$$

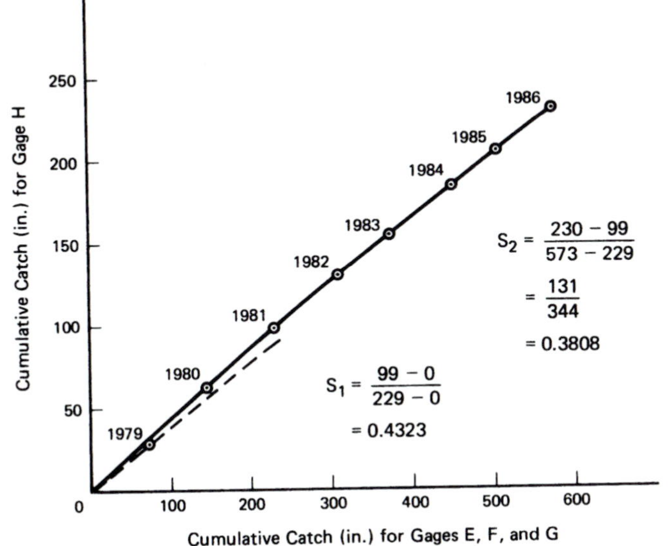

FIGURE 4–10 Adjustment of the lower section of a double-mass curve.

TABLE 4–6 Illustration of Adjustment of the Lower Section of a Double-Mass Curve

	Annual Catch (in.) at Gage					Cumulative Catch (in.) for Gage		
Year	E	F	G	H	TOTAL $E+F+G$	$E+F+G$	H	h_1
1979	22	26	23	28	71	71	28	24.7
1980	21	26	25	33	72	143	61	29.1
1981	27	31	28	38	86	229	99	33.5
1982	25	29	29	31	83	312	130	
1983	19	22	23	24	64	376	154	
1984	24	25	26	28	75	451	182	
1985	17	19	20	22	56	507	204	
1986	21	22	23	26	66	573	230	

Thus the measured data will be reduced by about 12%. The adjusted slope for the lower portion of the double-mass curve is shown in Figure 4–10, and the adjusted catches are given in Table 4–6.

Example 4–9

The data of Table 4–7 can be used to illustrate the use of Equation 4–18 to adjust the upper portion of a double-mass curve. Figure 4–11 shows the double-mass curve of the cumulative catch at gages A, B, and C versus the cumulative catch at gage D. The slopes for the two sections of the curve were computed using 1983 as the midsection point; the slope of the upper portion of the double-mass curve is different for the period from 1984 to 1986. The slope for the 1977–1983 period is

$$S_1 = \frac{205 - 0}{621 - 0} = 0.3301$$

TABLE 4–7 Illustration of Adjustment of the Upper Section of a Double-Mass Curve

	Annual Catch (in.) at Gage					Cumulative Catch (in.)		
Year	A	B	C	D	TOTAL $A+B+C$	$A+B+C$	D	d_2
1977	30	31	27	29	88	88	29	
1978	33	36	32	32	101	189	61	
1979	26	27	24	28	77	266	89	
1980	27	26	27	29	80	346	118	
1981	34	34	30	30	98	444	148	
1982	31	33	31	29	95	539	177	
1983	28	30	24	28	82	621	205	
1984	35	34	33	39	102	723	244	35.0
1985	37	39	36	41	112	835	285	36.8
1986	34	35	35	37	104	939	322	33.2

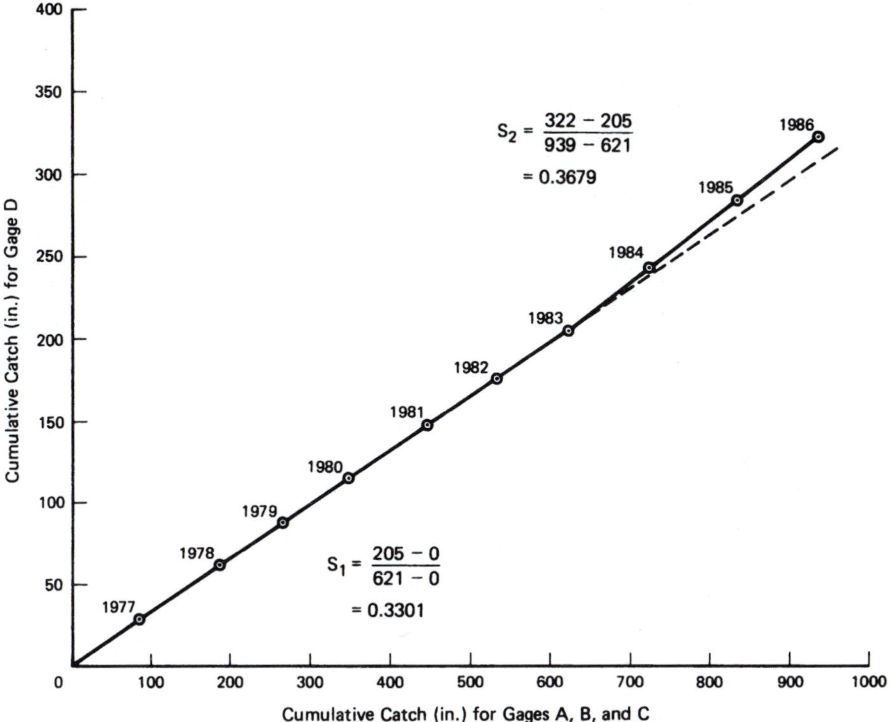

FIGURE 4–11 Adjustment of the upper section of a double-mass curve.

The slope for the 1984–1986 period is

$$S_2 = \frac{322 - 205}{939 - 621} = 0.3679$$

Based on Equation 4–18, the equation for adjusting the upper section of the curve is

$$d_2 = \left(\frac{0.3301}{0.3679}\right) D_2 = 0.8973\, D_2 \tag{4–20}$$

The adjusted values for 1984 to 1986 are given in Table 4–7, and the adjusted double-mass curve for this period is shown as a dashed line in Figure 4–11. The adjusted portion of the curve has the same slope as the lower section of the curve.

4.5 AVERAGE AREAL RAINFALL

For analyses involving areas larger than a few square miles, it may be necessary to make estimates of average rainfall depths over subwatershed areas. Two of the more common problems where this type of analysis could be used is in the computation of the probable maximum flood for larger watersheds and in the analysis of actual storm events in which the rainfall depths were measured at more than one rain gage. The probable maximum flood is used in the design of many large structures. The analysis of actual storm events, such as hurricanes, may be required for legal cases involving flooding or to assess the effect of the larger storms of record on a proposed structure. In these design cases, it is necessary to extend the point rainfall measurements to areal estimates.

Three methods of extending point estimates to areal averages will be presented. The station-average method is the simplest to apply; however, when the rain gages are not uniformly dispersed throughout the watershed, the station-average method may not provide an areal average rainfall that reflects the actual spatial distribution of rainfall. For such cases, a method that can account for the nonhomogeneity of the rain gage locations would provide a more accurate estimate of the actual rainfall depth. Two such methods will be demonstrated. The Theissen Polygon Method assigns weights to the rain gages according to the proportions of the total watershed area that are geographically closest to each of the rain gages. The isohyetal method, which should be more accurate than the Theissen Polygon Method, weights the catch at each of the rain gages according to the watershed area that is physically associated with the catch of the rain gage. Whereas the Theissen Method is easier to apply and is more reproducible (that is, different individuals would be more likely to provide the same estimates) than the isohyetal method, the isohyetal method permits the hydrologist to incorporate orographic effects and storm morphology into the calculation of the weights. In summary, each of the three methods can be viewed as providing a weighted average of measured catches. The station-average method assumes equal weights. The weights for the Theissen Method are proportional to the size of the area geographically closest to each gage. The isohyetal method assigns weights on the basis of storm morphology, the spatial distribution of the rain gages, and orographic effects.

As a general model, the average watershed catch is defined by

$$\overline{P} = \sum_{i=1}^{n} w_i P_i \qquad (4\text{–}21)$$

in which $\overline{P}$ is the average rainfall, P_i is the rainfall measured at station i, n the number of rain gages, and w_i is the weight assigned to the catch at station i. The mean $\overline{P}$ has the same units as the P_i. Equation 4–21 is valid only when the following holds:

$$\sum_{i=1}^{n} w_i = 1 \qquad (4\text{–}22)$$

The general model of Equation 4–21 can be applied with all the methods discussed here. The difference in the application lies solely in the way that the weights are estimated.

4.5.1 Station-Average Method

For the station-average method, each gage is given equal weight; thus w_i equals $1/n$, and Equation 4–21 becomes

$$\overline{P} = \frac{1}{n} \sum_{i=1}^{n} P_i \qquad (4\text{–}23)$$

The use of Equation 4–23 will provide reasonably accurate estimates of $\overline{P}$ when there are no significant orographic effects, the gages are uniformly spaced throughout the watershed, and the rainfall depth over the entire watershed is nearly constant.

Example 4–10

Consider the hypothetical watershed of Figure 4–12a. Five rain gages are located within the vicinity of the watershed, with two of the five within the watershed boundary. For a 24-hr storm the catch at each gage is given in Table 4–8. Using Equation 4–23 the station average rainfall is 3.48 in. In this case, the catch at each gage is given equal weight.

4.5.2 Theissen Polygon Method

For cases where there are large differences in the catches at the rain gages and/or the rain gages are not uniformly distributed throughout the watershed, the Theissen Polygon Method is a good alternative to the station-average method. Polygons are formed about each gage by constructing perpendicular bisectors between each pair of nearby gages. The resulting polygons indicate the areal extent assigned to the gage; only the area within the watershed is used to develop the weights of Equation 4–21 even though gages outside the watershed boundary can be used to compute the Theissen weighted estimate of the rainfall. The weights equal the fraction of the total watershed area within the polygons.

Example 4–11

Using the data of Table 4–8 and the rain gage distribution shown in Figure 4–12a, perpendicular bisectors were constructed between pairs of gages. The resulting layout of Theissen polygons is

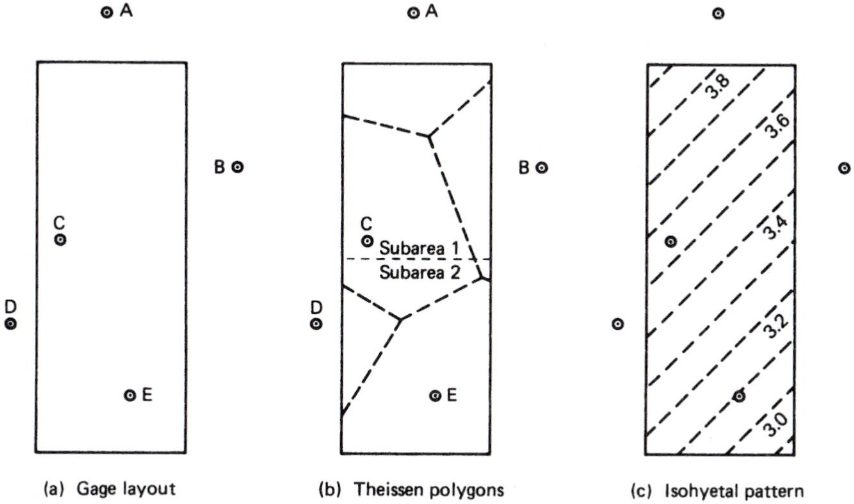

FIGURE 4–12 Hypothetical watershed for calculating weighted average rainfall.

TABLE 4–8 Calculation of Average Rainfall for the Station-Average, Theissen Polygon, and Isohyetal Methods

Gage	Catch (in.)	Theissen Weight	Isohyetal Cell, j	Average Catch, P_j	Area Fraction, w_j	w_jP_j
A	3.90	0.143	1	2.95	2/90	5.90/90
B	3.35	0.119	2	3.05	6/90	18.30/90
C	3.55	0.350	3	3.15	10/90	31.50/90
D	3.50	0.063	4	3.25	12/90	39.00/90
E	3.10	0.325	5	3.35	12/90	40.20/90
Sum	17.40	1.000	6	3.45	12/90	41.40/90
Average	3.48		7	3.55	12/90	42.60/90
			8	3.65	11.5/90	41.98/90
			9	3.75	8/90	30.00/90
			10	3.85	4/90	15.40/90
			11	3.925	0.5/90	1.96/90
					1.000	3.42

shown in Figure 4–12b. The fractions of the total area are also given in Table 4–8. Thus, the Theissen weighted rainfall is

$$\begin{aligned}\bar{P} &= 0.143(3.90) + 0.119(3.35) + 0.350(3.55)\\ &\quad + 0.063(3.50) + 0.325(3.10)\\ &= 3.43 \text{ in.}\end{aligned} \tag{4–24}$$

In this case, the two gages within the watershed boundary represent 67.5% of the total area. While gage D is closer to the boundary than either gage A or gage B, it reflects the smallest area because gage D is closer to gages C and E.

The station-average estimate of 3.48 in. agrees closely with the Theissen average because there is little variation in the catches at the gages and the distribution of the gages is reasonably uniform.

4.5.3 Isohyetal Method

The Theissen Method is considered to be more accurate than the station-average method. But the Theissen Method does not allow the hydrologist to consider factors, such as topography, that can affect the catch at a gage in delineating areas in which the catch at the gage is representative. The isohyetal method is designed to overcome this deficiency. Specifically, the isohyetal method requires the hydrologist to delineate the boundaries of the region associated with the catch at each gage by drawing isohyets (that is, lines of equal rainfall depths). The area within each pair of adjacent isohyets can then be used to weight the average rainfall associated with the adjacent isohyets. Mathematically, the estimated mean areal precipitation for a watershed of area A is

$$\bar{P} = \sum_{i=1}^{n} \left(\frac{A_i}{A}\right) \bar{P}_i \tag{4–25}$$

in which A_i is the area of the watershed between isohyets i and $i + 1$, $\bar{P}_i$ is the average precipitation for isohyets i and $i + 1$, and n the number of isohyetal intervals. The ratio A_i/A is the weight applied to the particular precipitation range.

While the opportunity to incorporate knowledge of the factors affecting the catch at each gage into making a weighted estimate is an advantage of the isohyetal method, it is also its disadvantage. Specifically, the drawing of the isohyets is somewhat subjective when the network of rain gages is not dense. Thus two or more hydrologists will not necessarily arrive at the same rainfall pattern, which will result in unequal estimates of the weighted average rainfall. Thus the accuracy of this method is dependent on the knowledge and skill of the hydrologist. Additionally, the accuracy is dependent on the density of the rain gage network. It is difficult to accurately delineate the isohyets when there are only a few rain gages.

Example 4–12

To illustrate the use of Equation 4–25, consider the following data for a 30-mi^2 watershed:

Isohyetal Range (in.)	Mean Rainfall (in.)	Area (mi^2)
0–1	0.5	5.7
1–2	1.5	11.4
2–3	2.5	9.3
3–4	3.5	3.6
		30.0 = sum

The mean areal rainfall is

$$\hat{P} = \frac{5.7}{30}(0.5) + \frac{11.4}{30}(1.5) + \frac{9.3}{30}(2.5) + \frac{3.6}{30}(3.5)$$

$$= 0.19(0.5) + 0.38(1.5) + 0.31(2.5) + 0.12(3.5) \tag{4–26}$$

$$= 1.86 \text{ in.}$$

In this case, the sum of the weights equals 1. The values of $\overline{P}_i$ are average values between isohyets and within the watershed; they are not values recorded at individual gages, as is the case for the station-average and Theissen Methods.

Example 4–13

The gage shown in Figure 4–12a can be used to illustrate the isohyetal method. It should be obvious that there are a large number of possible isohyetal patterns that could be constructed from the data for the five gages. For this hypothetical example, the catches given in Table 4–8 were actually obtained from the isohyetal pattern shown in Figure 4–12c. The isohyets of Figure 4–12c can be used to estimate the average rainfall; the calculations are shown in Table 4–8, with the drainage area divided into eleven subareas. For each pair of isohyets, the average of the rainfall amounts is used to represent the $\overline{P}_i$ of Equation 4–25. The weight of w_i of Equation 4–21 is taken to be the fraction of the area between the isohyets. For this hypothetical example, the weighted average rainfall is 3.42 in., which agrees closely with the estimates obtained with the station-average and Theissen Methods.

4.5.4 Average Rainfall for Subwatersheds

In making a watershed analysis, it is often necessary to divide the watershed into subareas to account for nonhomogeneity of land use or soil characteristics. Also, some hydrologic models place limits on the size of subareas; in such cases, watershed subdivision is also necessary. In these cases, the average rainfall may have to be estimated for each subarea, and one of the methods above for computing average rainfalls will have to be applied to each subarea separately.

When using the station-average method, a decision rule must be developed for assigning gages to specific subareas. For example, for the subdivided watershed of Figure 4–12b, would one use gages *A*, *B*, and *C* for computing the average for subarea 1 and gages *D* and *E* for subarea 2? This may seem rational, but the Theissen weight for Theissen section *C* in subarea 2 is still larger than the weight for section *D* (that is, 0.222 versus 0.126); this suggests that rainfall for gage *C* is possibly more indicative of the rainfall on subarea 2 than the catch at gage *D*. In any case, the decision rule is not as clear with the station-average method as it is with the Theissen Polygon Method.

The isohyetal method can be used with a subdivided watershed by computing the weights according to the parts of the isohyets that are located within each subarea. The isohyets should be developed in the same manner regardless of the way the watershed is subdivided.

Example 4–14

Consider the hypothetical watershed of Figure 4–12. If we assume that the watershed is divided into two equal subareas, as shown in Figure 4–12b, the weights for Equation 4–21 must now be calculated for each subwatershed. For the Theissen Method these are shown in Table 4–9. Using Equation 4–21, we get the following weighted average rainfalls for the subareas:

$$\bar{P}_1 = 3.90(0.286) + 3.35(0.236) + 3.55(0.478) = 3.60 \text{ in.} \tag{4–27a}$$

$$\bar{P}_2 = 3.35(0.002) + 3.55(0.222) + 3.50(0.126) + 3.10(0.650) = 3.25 \text{ in.} \tag{4–27b}$$

Within the limits of roundoff accuracy, the mean watershed rainfall for the subdivided watershed equals the mean obtained by not subdividing:

$$\bar{P} = 0.5(3.60) + 0.5(3.25) = 3.425 \text{ in.} \tag{4–28}$$

which agrees with Equation 4–24.

TABLE 4–9 Theissen Weights for Subdivided Watershed

Subarea	Gage	Catch (in.)	Area Fraction
1	*A*	3.90	0.286
	B	3.35	0.236
	C	3.55	0.478
2	*B*	3.35	0.002
	C	3.55	0.222
	D	3.50	0.126
	E	3.10	0.650

4.6 DEVELOPMENT OF A DESIGN STORM

To this point, the following characteristics of storms have been discussed: volume, duration, frequency, and intensity. Some design problems only require the total volume of rainfall for a specified duration and frequency. However, for many problems in hydrologic design, it is necessary to show the variation of the rainfall volume with time. That is, some hydrologic design problems require the storm input to the design method to be expressed as a hyetograph and not just as a total volume for the storm. Characteristics of a hyetograph that are important are the peak, the time to peak, and the distribution, as well as the volume, duration, and frequency. In developing a design storm for any region, empirical analyses of measured rainfall records are made to determine the most likely arrangement of the ordinates of the hyetograph. For example, some storm events will have an early peak (front loaded), some a late peak (rear loaded), some will peak in the center of the storm (center loaded), and some will have more than one peak. The empirical analysis of measured rainfall hyetographs at a location will show the most likely of these possibilities, and this finding can be used to develop the design storm.

4.6.1 Constant-Intensity Design Storm

A design storm that is used frequently for hydrologic designs on very small urban watersheds is the constant-intensity storm. It is quite common to assume that the critical cause of flooding is the short-duration, high-intensity storm. Therefore, it is assumed that for the critical storm duration, the rainfall intensity will be constant. It is common to assume that the largest peak runoff rate occurs when the entire drainage area is contributing, which leads to the assumption that the duration of the design storm equals the time of concentration of the watershed. The intensity of the storm is obtained from an intensity-duration-frequency curve for the location, often using the time of concentration as the duration and the frequency specified by the design standard; the storm volume is the intensity multiplied by the time of concentration. Uses of this design storm are discussed in other chapters.

Example 4–15

To illustrate the constant intensity design storm, we will assume the following condition: (l) the design standard specifies a 10-yr return period for design; (2) the watershed time of concentration is 15 min; and (3) the watershed is located in Baltimore, Maryland. From Figure 4–4 the rainfall intensity for a duration of 15 min is 5.5 in./hr, which yields a storm volume of 1.375 in. The resulting design storm is shown in Figure 4–13.

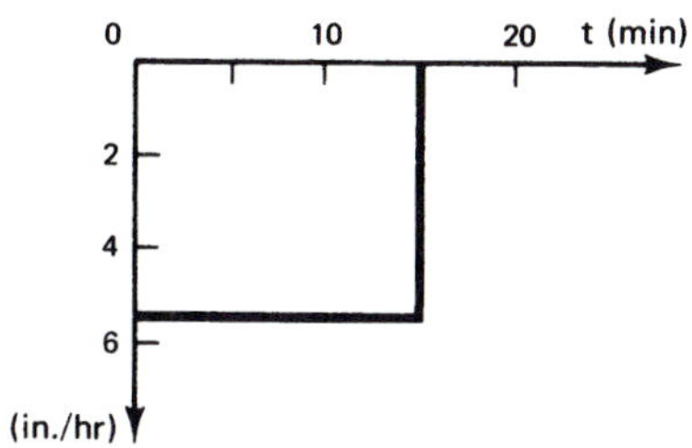

FIGURE 4–13 Constant-intensity design storm for a 15-minute time of concentration and a 10-year return period (Baltimore, MD).

4.6.2 The SCS 24-Hour Storm Distributions

The SCS developed four dimensionless rainfall distributions using the Weather Bureau's Rainfall Frequency Atlases (NWS, 1961). The rainfall-frequency data for areas less than 400 mi^2, for durations to 24 hr, and for frequencies from 1 yr to 100 yr were used. Data analyses indicated four major regions, and the resulting rainfall distributions were labeled type I, IA, II, and III. The locations where these design storms are applicable are shown in Figure 4–14.

The distributions are based on the generalized rainfall volume-duration-frequency relationships shown in technical publications of the Weather Bureau. Rainfall depths for durations from 6 min to 24-hr were obtained from the volume-duration-frequency information in these publications and used to derive the storm distributions. Using increments of 6 min, incremental rainfall depths were determined. For example, the maximum 6-min depth was subtracted from the maximum 12-min depth and this 12-min depth was subtracted from the maximum 18-min depth, and so on to 24 hours. The distributions were formed by arranging these 6-min incremental depths such that for any duration from 6 min to 24 hr, the rainfall depth for that duration and frequency is represented as a continuous sequence of 6-min depths. The location of the peak was found from the analysis of measured storm events to be location dependent. For the regions with type I and IA storms, the peak intensity occurred at a storm time of about 8 hr, while for the regions with type II and III storms, the peak was found to occur at the center of the storm. Therefore for type II and III storm events, the greatest 6-min depth is assumed to occur at about the middle of the 24-hr period, the second largest 6-min incremental depth during the next 6 min and the third largest in the 6-min inter-

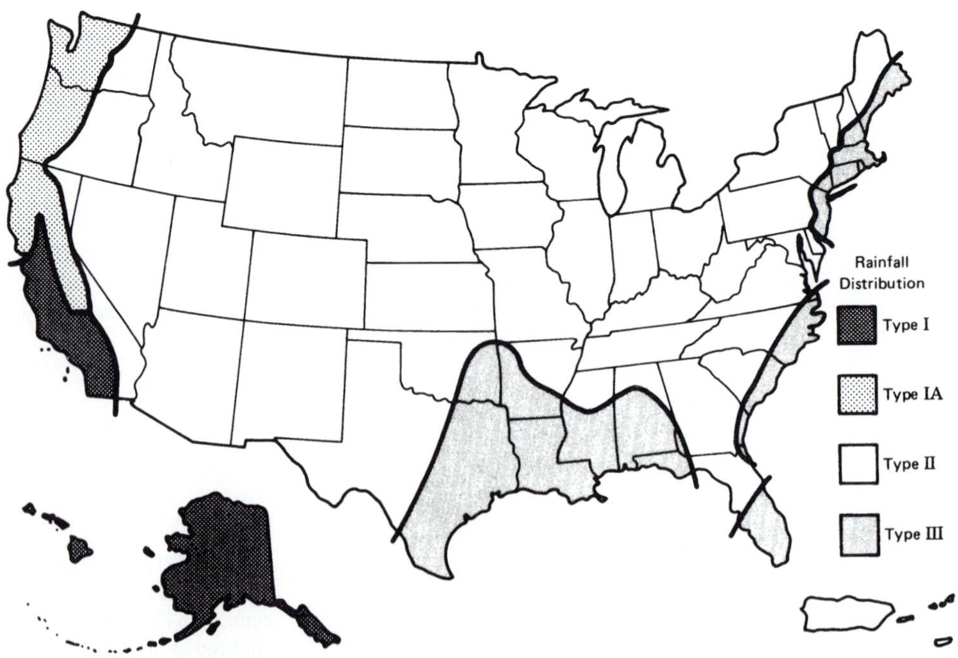

FIGURE 4–14 Approximate geographic areas for SCS rainfall distributions. (SCS, 1986.)

val preceding the maximum intensity. This continues with each incremental rainfall depth in decreasing order of magnitude. Thus the smaller increments fall at the beginning and end of the 24-hr storm. This procedure results in the maximum 6-min depth being contained within the maximum l-hr depth, the maximum l-hr depth is contained within the maximum 6-hr depth, and so on. Because all of the critical storm depths are contained within the storm distributions, the distributions are appropriate for designs on both small and large watersheds.

This procedure can be used to show the values of the cumulative rainfall distribution of the SCS type II storm for a time increment of 0.25 hr; the values are given in Table 4–10 for hyetograph times from 10.00 to 14.00 hr. The largest incremental rainfall (27.6%) occurred between hyetograph times 11.75 and 12.00 hr. The next largest (10.4%) was placed just prior to the largest incremental amount. The third largest (4.4%) was placed just after the largest incremental amount. This procedure is followed approximately for the remainder of the storm.

The resulting distributions (Figure 4–15) are most often presented with the ordinates given on a dimensionless scale. The SCS type I and type II dimensionless distributions plot as a straight line on log-log paper. Although they do not agree exactly with *IDF* values from all locations in the region for which they are intended, the differences are within the accuracy of the rainfall depths read from the Weather Bureau atlases. Figure 4–15 shows the distributions, and the ordinates are given in Table 4–11 for all four SCS synthetic design storms.

Example 4–16

The procedure of forming a design storm can be illustrated with a simplified example. The design storm will have the following characteristics: duration, 6 hr; frequency, 50 yr; time increment, 1 hr; location, Baltimore. The intensity-duration-frequency curve of Figure 4–4 can be

TABLE 4–10 SCS 24-hour Type II Design Storm for Hyetograph Times from 10 to 14 Hours and an Increment of 0.25 Hour

Time (hr)	Cumulative Rainfall (in.)	Incremental Rainfall (in.)
10.00	0.181	0.010
10.25	0.191	0.012
10.50	0.203	0.015
10.75	0.218	0.018
11.00	0.236	0.021
11.25	0.257	0.026
11.50	0.283	0.104
11.75	0.387	0.276
12.00	0.663	0.044
12.25	0.707	0.028
12.50	0.735	0.023
12.75	0.758	0.018
13.00	0.776	0.015
13.25	0.791	0.013
13.50	0.804	0.011
13.75	0.815	0.010
14.00	0.825	

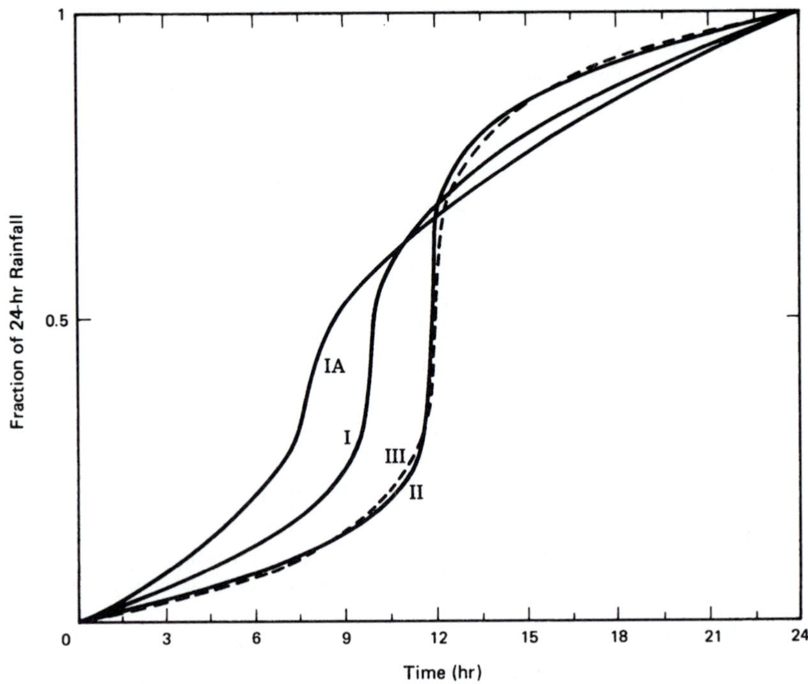

FIGURE 4–15 SCS 24-hour rainfall distributions (not to scale). (SCS, 1984.)

used to obtain the rainfall intensities for durations of 1 to 6 hr in increments of 1 hr; these intensities are shown in column 3 of Table 4–12. The depth (that is, duration times intensity) is shown in column 4, with the incremental depth (column 5) taken as the difference between the depths for durations 1 hr apart. The incremental depths were used to form the 50-yr design storm, which is shown in column 6, by placing the largest incremental depth in hour 3 and the second largest incremental depth in hour 4. The remaining incremental depths were placed by alternating their location before and after the maximum incremental depth. It is important to notice that the maximum three hours of the design storm has a volume of 3.90 in., which is the depth for a 3-hr duration from Figure 4–4; this will be true for any storm duration from 1 to 6 hr. The cumulative form of the design storm is given in column 7 of Table 4–12.

A dimensionless design storm can be developed by transforming the cumulative design storm of column 7 of Table 4–12 by dividing each ordinate by the total depth of 4.62 in. The dimensionless cumulative design storm derived from the 50-yr intensities is shown in column 8 of Table 4–12. The calculation of a dimensionless design storm for a 2-yr frequency is shown in the lower part of Table 4–12. In comparing the 50-yr and 2-yr dimensionless design storms, it should be apparent that a cumulative design storm could be developed for any design frequency by multiplying the 6-hr rainfall depth for that frequency by the average ordinates of the dimensionless cumulative design storms of Table 4–12, which is approximately [0.05, 0.14, 0.77, 0.89, 0.97, 1.00]. Based on this dimensionless cumulative design storm (see Figure 4–16a), the 10-yr cumulative design storm, which has a 6-hr depth of 3.48 in. (0.58 in./hr from Figure 4–4 multiplied by 6 hr), would be [0.17, 0.49, 2.68, 3.10, 3.38, 3.48 in.], which is shown using the ordinate on the right-hand side in Figure 4–16a. Thus the 10-yr design storm would be [0.17, 0.32,

TABLE 4–11 SCS Cumulative, Dimensionless One-Day Storms

Time (hrs)	Type I Storm	Type IA Storm	Type II Storm	Type III Storm
0	0	0	0	0
0.5	0.008	0.010	0.0053	0.0050
1.0	0.017	0.020	0.0108	0.0100
1.5	0.026	0.035	0.0164	0.0150
2.0	0.035	0.050	0.0223	0.0200
2.5	0.045	0.067	0.0284	0.0252
3.0	0.055	0.082	0.0347	0.0308
3.5	0.065	0.098	0.0414	0.0367
4.0	0.076	0.116	0.0483	0.0430
4.5	0.087	0.135	0.0555	0.0497
5.0	0.099	0.156	0.0632	0.0568
5.5	0.112	0.180	0.0712	0.0642
6.0	0.126	0.206	0.0797	0.0720
6.5	0.140	0.237	0.0887	0.0806
7.0	0.156	0.268	0.0984	0.0905
7.5	0.174	0.310	0.1089	0.1016
8.0	0.194	0.425	0.1203	0.1140
8.5	0.219	0.480	0.1328	0.1284
9.0	0.254	0.520	0.1467	0.1458
9.5	0.303	0.550	0.1625	0.1659
10.0	0.515	0.577	0.1808	0.1890
10.5	0.583	0.601	0.2042	0.2165
11.0	0.624	0.624	0.2351	0.2500
11.5	0.655	0.645	0.2833	0.2980
12.0	0.682	0.664	0.6632	0.5000
12.5	0.706	0.683	0.7351	0.7020
13.0	0.728	0.701	0.7724	0.7500
13.5	0.748	0.719	0.7989	0.7835
14.0	0.766	0.736	0.8197	0.8110
14.5	0.783	0.753	0.8380	0.8341
15.0	0.799	0.769	0.8538	0.8542
15.5	0.815	0.785	0.8676	0.8716
16.0	0.830	0.800	0.8801	0.8860
16.5	0.844	0.815	0.8914	0.8984
17.0	0.857	0.830	0.9019	0.9095
17.5	0.870	0.844	0.9115	0.9194
18.0	0.882	0.858	0.9206	0.9280
18.5	0.893	0.871	0.9291	0.9358
19.0	0.905	0.844	0.9371	0.9432
19.5	0.916	0.896	0.9446	0.9503
20.0	0.926	0.908	0.9519	0.9570
20.5	0.936	0.920	0.9588	0.9634
21.0	0.946	0.932	0.9653	0.9694
21.5	0.956	0.944	0.9717	0.9752
22.0	0.965	0.956	0.9777	0.9808
22.5	0.974	0.967	0.9836	0.9860
23.0	0.983	0.978	0.9892	0.9909
23.5	0.992	0.989	0.9947	0.9956
24.0	1.000	1.000	1.0000	1.0000

TABLE 4–12 Development of 6-Hour Dimensionless Cumulative Design Storms for Baltimore

T (yr)	Duration (hr)	Intensity (in./hr)	Depth (in.)	Incremental Depth (in.)	Design Storm (in.)	Cumulative Design Storm (in.)	Dimensionless Cumulative Design Storm
50	1	3.00	3.00	3.00	0.25	0.25	0.054
	2	1.75	3.50	0.50	0.40	0.65	0.141
	3	1.30	3.90	0.40	3.00	3.65	0.790
	4	1.05	4.20	0.30	0.50	4.15	0.898
	5	0.89	4.45	0.25	0.30	4.45	0.963
	6	0.77	4.62	0.17	0.17	4.62	1.000
2	1	1.35	1.35	1.35	0.10	0.10	0.046
	2	0.82	1.64	0.29	0.19	0.29	0.134
	3	0.61	1.83	0.19	1.35	1.64	0.759
	4	0.50	2.00	0.17	0.29	1.93	0.894
	5	0.42	2.10	0.10	0.17	2.10	0.972
	6	0.36	2.16	0.06	0.06	2.16	1.000

2.19, 0.42, 0.28, 0.10 in.]. The 10-yr, 6-hr design storm is shown in Figure 4–16b with ordinates expressed as an intensity.

4.6.3 Comparison of Design Storms

Concern about the 24-hr duration of the SCS design storms has been expressed. The argument is that flooding on small watersheds results from short duration storms, not 24-hr storms. Many prefer the constant-intensity design storm even though such a storm is unlikely to occur because storms usually begin and end with a period of low intensity. In this regard, the SCS design storms are rational.

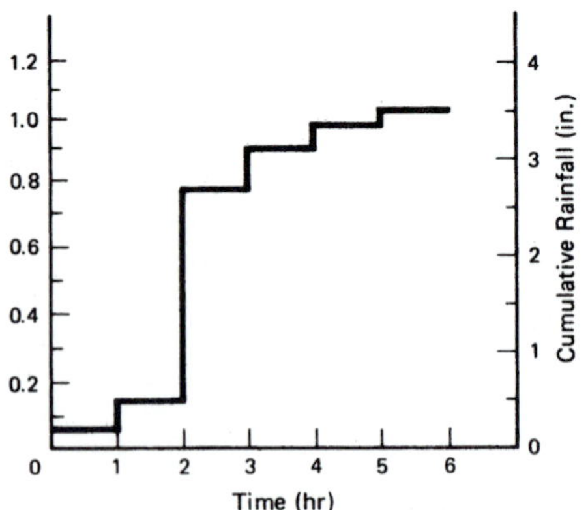

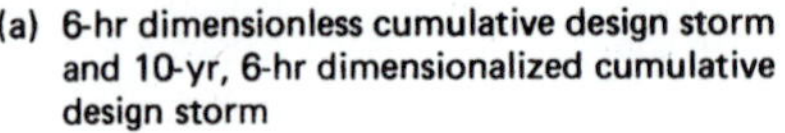

(a) 6-hr dimensionless cumulative design storm and 10-yr, 6-hr dimensionalized cumulative design storm

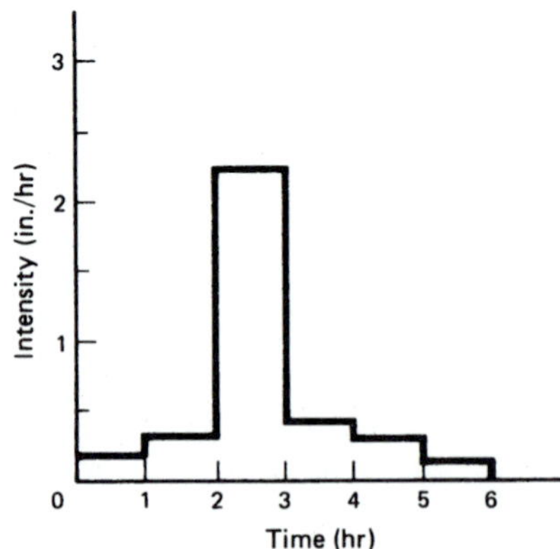

(b) 10-yr, 6-hr design storm for Baltimore

FIGURE 4–16 Formation of dimensionless and dimensionalized design storms.

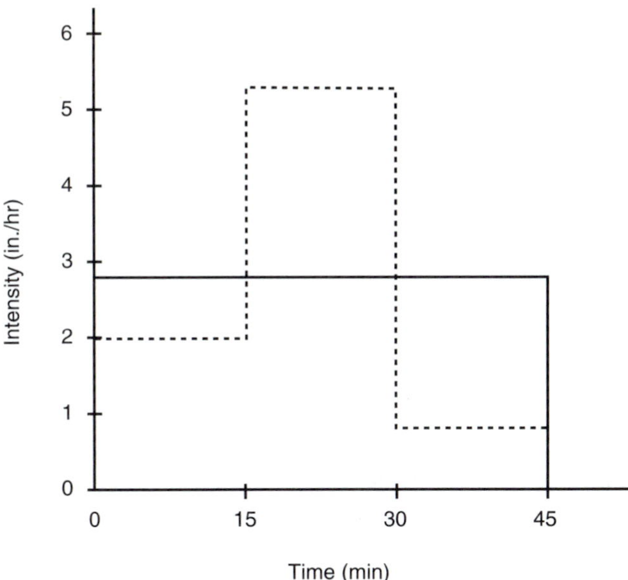

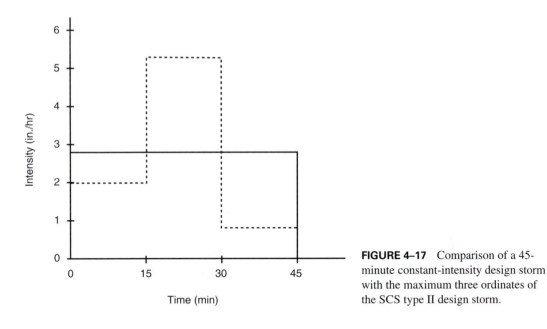

FIGURE 4–17 Comparison of a 45-minute constant-intensity design storm with the maximum three ordinates of the SCS type II design storm.

Example 4–17

Assume that a small watershed has a time of concentration of 45 min and that a 10-yr return period is used. From Figure 4–4, the design intensity is 2.8 in./hr, which yields a depth of 2.1 in. Using the maximum 45-min period of the SCS 24-hr design storm (see Table 4–11), the corresponding proportions are 0.104, 0.276, and 0.044. Multiplying these by the 24-hr depth of 4.8 in. for a 10-yr storm yields ordinates of 0.50, 1.32, and 0.21 in., respectively, for a total depth of 2.03 in. This is close to the 2.1-in. depth for the constant-intensity storm. The intensity hyetographs for the two storms are shown in Figure 4–17. The three intensities are 2.00, 5.30, and 0.84 in./hr. This example illustrates that the 24-hr design storm is capable of reflecting intensities of small-duration storms.

4.6.4 HEC-1 Storm Distribution Options

The HEC-1 program provides for three input options of rainfall: (1) values of a historical storm event; (2) the depth of rainfall for each time interval; and (3) a dimensionless hyetograph and the total storm depth P. The program also permits rainfall information to be varied for different subareas. The third option could be used to input one of the SCS distributions with the total storm depth.

4.7 PROBABLE MAXIMUM PRECIPITATION (PMP)

The procedure given herein is based on that recommended in HMR No. 52. There are six major elements, with specific computations in each:

1. Determination of incremental PMP
2. Placement of the isohyetal pattern
3. Determination of maximum precipitation volume
4. Distribution of the storm-area averaged PMP
5. Determination of the temporal distribution of the PMS
6. Determination of temporal distributions for subwatersheds.

An outline of each of these major elements follows.

Determination of 6-Hour incremental rainfall. To obtain the 6-hr incremental PMP, depth-area-duration (DAD) data are obtained from HMR No. 51 (Figures 18 to 47) for the location of the watershed. The geographic center of the watershed is usually used in obtaining the DAD data. The center is determined using the method outlined in Section 3.3.5.

Because of the necessity to interpolate for standard isohyet areas, relationships between the DAD data and storm-area size (A_s) will be necessary. For graphical relationships, the storm-area size should be plotted on the log scale of semilog graph paper. The curves should be plotted so that they are parallel or converging with increasing size; the convergence should be such that the lines will not cross. Alternatively, semilog relationships can be developed for all durations (6, 12, 24, 48, and 72 hours):

$$P = a + b \log A_s \tag{4–29}$$

After obtaining values for a and b for each duration, the coefficients should be adjusted to ensure rational estimates.

Using either the graphical or semilog equations, DAD values should be obtained for standard isohyet-area sizes, with at least four values smaller and four values larger than the storm-area size corresponding to the maximum precipitation volume.

To obtain DAD estimates for the 18-hr duration, relationships between the DAD data and the duration are obtained for the storm-area sizes from the previous step. Graphical relationships on linear paper can be used or polynomial equations can be fitted. In either case, adjusting may be necessary to ensure reasonable estimates of the 18-hr depth.

Placement of the isohyetal pattern. Figure 4–6 shows the isohyetal pattern used to distribute the rainfall over the watershed. The spatial concentration of rainfall is determined largely by the location of the storm center and the orientation of the storm pattern. HMR No. 52 recommends centering the isohyetal pattern over the drainage area such that the runoff volume will be the greatest. This recommendation is based on the assumption of high correlation between the peak discharge and the maximum volume of runoff. While this correlation is usually high, it is good practice to move the storm center slightly closer to the watershed outlet since the upper reaches of a watershed are often less important in determining the maximum discharge. The greatest peak flow is more likely to correlate with the greatest volume when the watershed is not subdivided for analysis. This occurs because the channel routing scheme assumes less storage than that assumed by the unit hydrograph. Irregularly shaped watersheds may require more analyses than a regularly shaped watershed.

The location of the storm center is fairly straightforward for homogeneous watersheds. Where there are significant variations in land cover/use, soils, or channel characteristics, finding the optimum storm center location may be more involved. The existence of storage reservoirs also can be an important factor. In the SCS TR-55, a peak discharge correction factor, which reduces the design peak discharge, is used to adjust for the presence of small storage volumes. The correction factor depends on the location of the storage, with a larger correction factor being used when the storage is near the watershed outlet. While the TR-55 procedure is intended for use on small watersheds, the basic concept is still valid.

The locational effects of detention storage have also been widely discussed with reference to detention ponds in urban areas. Studies have shown that, when the detention basins are located near the outlet of the basin, the effect can be to increase the peak discharge over that which would have occurred even without the detention storage. The argument is as follows: Storage basins tend to delay the time of the peak of the hydrograph out of the detention basin, thus giving the upstream flow the time to drain to the area of the downstream drainage facility. Thus, the peak of the upstream hydrograph occurs at about the time of the peak flow out of the detention facility. The point is, storage can alter the runoff characteristics for a watershed and it is necessary to perform more analyses to identify the maximum peak discharge.

There is a need for guidelines on locating the storm center for a developed watershed that includes significant amounts of detention storage. If the storage facilities are distributed over the entire watershed, then the storm center will need to be positioned near the center of the watershed. However, additional analyses need to be made. The storm center should first be moved closer to the watershed outlet. Several runs can be made with the storm center moved successively farther from the center of the watershed until the optimum is found. The extent of the storage will be a determining factor in finding the optimum storm positioning. Also, the interdependence of storage structures on the same tributary is a factor. In general, the effect of a storage facility is to delay the time of the peak of the outflow and reduce the peak. However, structures don't have a large effect on the volume of runoff, which itself is a determinant of the peak discharge at the watershed outlet.

If the detention structures are located in one portion of the watershed rather than being uniformly distributed over the entire watershed, the effects are easier to describe. When the structures are predominantly in the headwaters of the watershed, the delay in the flow through the structures essentially reduces the contributing effects of the upper portion of the watershed. Thus, the storm center will be closer to the watershed outlet. Conversely, if the structures are concentrated near the outlet, then the delay of the flow through the structures can cause relatively large increases in peak discharges because the delay enables the water from the upstream portion of the watershed to "catch up" with the flow going through the structures. The elongated, flat peak of the hydrograph out of the structures combines with the peak of the flow from upstream to produce the peak of the total flow. In this case, the storm center should be located more in the portion of the watershed away from the structures, which means the storm center is above the center of the watershed. This will give a larger peak from the upstream portion of the watershed, which will combine with the volume and flat peak of the portion of the watershed that is controlled by the structure.

The problem of locating the storm center is compounded by the need to provide equal protection at every hydrologic facility in the watershed. Assume that the 100-yr storm is used for all design work. This implies that the runoff will have a 100-yr exceedence frequency. But the 100-yr storm for the entire watershed will not provide a 100-yr runoff at every point within the watershed. It will only provide 100-yr runoff at the outlet. Designs at other points in the watershed will require separate analyses including the proper positioning of the design isohyets to achieve the maximum peak at the interior design point. Use of the 100-yr rainfall pattern for design within the entire watershed will lead to underdesigned facilities. The largest peak discharge must be considered as the 100-yr peak, so this will occur only when the optimum design storm is evaluated separately for each site within the watershed.

There are several factors that can make the problem more complex. First, if there are points within a watershed that act to control runoff rates, such as pipes designed for a small return period, these can cause surcharges. In such cases, the hydrologic facilities act like storage structures. Second, assumed antecedent moisture conditions are important when the watershed does not have uniform land cover. The positioning of the storm center will be more sensitive to the location of the undeveloped land and less to the urbanized land, especially if dry initial conditions are used. Third, the shape of the watershed must also be considered in positioning the storm center. Small parcels of land that are at the extreme headwater of the watershed may not contribute to the peak so they can be ignored when positioning the storm isohyets.

Determination of maximum precipitation volume. The objective of this step is to determine the maximum precipitation volume for the three largest 6-hr time increments. This requires a somewhat subjective step of moving the isohyetal pattern over the drainage area and assessing the volume at each trial location. The area in each isohyet can be computed graphically or planimetered and used to compute the rainfall volume. The watershed should be completely covered by the isohyetal pattern and all partial areas used to compute the volume.

Distribute the storm-area averaged PMP. The 6-hr incremental depths equal the incremental PMP divided by the drainage area. For cases where increments less than 6 hr are needed, the depth-duration curve should be used. HMR 52 provides a figure for interpolating ratios of the 1-hr to 6-hr rainfall.

Determine the temporal distribution of the PMS. The temporal distribution is formed in much the same way as the SCS 24-hr design storms (see Section 4.6.2). The individual 6-hr ordinates are arranged in decreasing proportion about the maximum 6-hr ordinate. HMR 52 indicates that large storms are usually rear loaded, so the maximum 6-hr ordinate should not be placed within the first 24 hr of the 72-hr storm. HMR 52 does not state the specific sequence. It is usually better to analyze rainfall records from nearby gages and determine the temporal distribution of recent large events. Such information can be used in developing an appropriate temporal pattern.

Determine temporal distributions for subwatersheds. Large watersheds will most often be subdivided into subwatersheds that are relatively homogeneous in hydrometeorological characteristics. The volumes for each subwatershed need to be consistent with the isohyetal pattern for the entire watershed. Subareas not near the location of the center of the isohyetal pattern will have smaller volumes. However, the same temporal pattern of rainfall must be used for each subarea.

4.8 SUMMARY

Watershed characteristics and precipitation are necessary inputs to many hydrologic design methods. For designs based on rainfall, the rainfall represents an upper limit on the amount of water that must be considered in the design. For a small asphalt parking lot in an urban area, almost all of the rainfall will run off immediately, so the success of the outlet design will depend largely on the accuracy of the design rainfall estimate. For a large forested watershed of several hundred square miles, a design may be less sensitive to the accuracy of the rainfall input; however, the rainfall characteristics will still be important to the design. The degree of importance of the rainfall input also depends on the effect of watershed and channel storage. In the small parking lot, there is very little storage, and thus the outflow is much more sensitive to the input (the rainfall). In the case of the large forested watershed, large amounts of storage in the form of interception, depression, subsurface, and channel storage decrease the sensitivity of the outflow from the watershed to the rainfall input. Thus, the importance of rainfall and watershed characteristics varies with the design problem but both remain important determinants of the final design.

4.8.1 Analysis and Synthesis

Most of the concepts discussed in this chapter relate to the design or synthesis phase; little attention has been given to the analysis problem. In terms of the systems framework introduced in Chapter 1, the measured rainfall record is the input, and rainfall intensity-duration-frequency characteristics are one type of output. An important task in rainfall analysis is to transform measured rainfall data into *IDF* characteristic curves such as those displayed in Figure 4–4. The method of making such a transformation was not introduced in this chapter since it is a task that is more involved than would be appropriate for an introductory text; however, the general approach is based on frequency analysis (see Chapter 5).

The topics discussed in this chapter represent problems in synthesis. Using an *IDF* curve to develop a synthetic hyetograph uses the *IDF* curve as input and a procedure or set of decision rules for setting up the output, which is the design storm. Depth-area-duration curves, which have been derived from the analysis of many measured storm events, serve as a model for converting a point rainfall depth to an area averaged rainfall depth; thus figures such as Figure 4–7 serve as the transformation tool. The methods of estimating areal averages of rainfall volumes are models that synthesize areal averages from point rainfall estimates. Analysis of actual storm event data have shown that the methods provide reasonable approximations. The same is true for the methods that are used to estimate missing rainfall data, although these methods may also be used as part of a method in precipitation analysis.

In summary, this chapter has centered on discussions of methods used for design purposes. It has been assumed that the products of rainfall analyses are available for solving design problems and that problems in rainfall analysis are more involved than would be appropriate to outline here.

PROBLEMS

4–1. Using the rainfall summary for September 1983 (Table 4–1) at Washington, D.C., graph the hourly hyetograph (in./hr) for the storm of September 21. Determine the total depth (in.). Also, determine the peak hourly rate (in./hr) and 2-hr maximum rate of rainfall for the storm. What volume of rain (in acre-ft) would fall on a 10-mi^2 watershed for this storm?

4–2. Using the rainfall summary for September 1983 (Table 4–1) at Washington, D.C., graph the hourly hyetograph (in./hr) for the storm of September 30. Determine the total depth (in.). Also, determine the peak hourly rate (in./hr) and the maximum 3-hr rate of rainfall for the storm. What volume of rain (in acre-ft) would fall on a 50-acre watershed for this storm event?

4–3. Given the following depths of rainfall for a 50-min storm, graph the storm hyetograph, with the ordinates in in./hr.

Storm time (min)	10	20	30	40	50
Rainfall depth (in.)	0.05	0.20	0.30	0.25	0.10

What is the maximum 30-min intensity (in./hr)?

4–4. Given the following depths of rainfall for an 8-hr storm event, graph the storm hyetograph, with the ordinates in in./hr.

Storm time (min)	2	4	6	8
Rainfall depth (in.)	0.7	1.5	0.9	0.5

What is the maximum 4-hr intensity (in./hr)?

4–5. Given the following depths of rainfall for a 45-min storm, graph the storm hyetograph, with the ordinates in in./hr.

Storm time (min)	0		6		18		21		30		36		45
Rainfall depth (in.)		0.06		0.24		0.18		0.54		0.30		0.18	

What is the total depth of rainfall for the storm? What is the maximum intensity?

4–6. Compute the average rainfall intensity (in./hr) for both of the following storms that occur on a 0.5-mi^2 watershed: (a) A depth of 1.5 in. in 30 min; (b) a volume of 45 acre-ft during 40 min.

4–7. Compute the total volume of rainfall (acre-ft) for each of the storms and watersheds: (a) Depth of rainfall = 3 in., drainage area = 0.25 mi^2; (b) rainfall intensity = 0.6 in./hr, storm duration = 90 min, drainage area = 10 acres; (c) depth of rainfall = 0.9 in., storm duration = 45 min, drainage area = 1 mi^2.

4–8. (a) Using a climatic atlas, find the 10-yr, 2-hr rainfall depths (in.) and intensities (in./hr) for the following cities: New York; Atlanta; Chicago; Dallas; Denver; Seattle. (b) Using Figure 4–2, find the 100-yr, 24-hr rainfall depths (inches) and intensities (in./hr) for the following cities: Boston; Cleveland; New Orleans; Miami; Houston; Phoenix.

4–9. Using the rainfall map of Figure 4–3, find the 2-yr, 24-hr depth (in.) and rainfall intensity (in./hr) for each of the following cities: Frostburg (Allegheny County), Washington, D.C.; Dover, Delaware.

4–10. From the intensity-duration-frequency curve of Figure 4–4 determine the missing values of intensity, depth, duration, or frequency for the following table:

Intensity (in./hr)	Depth (in.)	Duration (hr)	Frequency (yr)
—	—	0.75	5
—	—	6	25
8	—	0.25	—
—	2.5	0.5	—
0.3	—	24	—
4.0	—	—	50

4–11. From the intensity-duration-frequency curve of Figure 4–4 determine the missing values of intensity, depth, duration, or frequency for the following table:

Intensity (in./hr)	Depth (in.)	Duration (hr)	Frequency (yr)
2.0	4.0	—	—
—	4.8	24	—
—	—	6	75
—	0.58	0.083	—
0.3	—	12	—

4–12. Derive equations (use Equations 4–2) to represent the *IDF* curve for Baltimore for the 10-yr return period. Use the transformations of Equations 4–3 and 4–4.

4–13. Derive equations (use Equations 4–2) to represent the *IDF* curve for Baltimore for the 100-yr return period. Use the transformations of Equations 4–3 and 4–4.

4–14. Using an *IDF* curve for your locality compute equations of the form of Equations 4–2 to represent the *IDF* curves for the 2-, 10-, and 100-yr return periods.

4–15. Determine the average depth of rainfall for a 10-yr, 6-hr storm over a 150-mi^2 watershed in Baltimore (Figure 4–4). Also, compute the total storm volume in acre-ft.

4–16. Determine the average depth of rainfall for a 50-yr, 12-hr storm over a 250-mi^2 watershed in Baltimore (Figure 4–4). Also, compute the total storm volume in acre-ft.

4–17. Using a log-log plot, show the variation of the depth-area adjustment factor of Figure 4–7 with storm duration for a 100-mi^2 watershed. Plot the adjustment factor as the ordinate and the storm duration as the abscissa.

4–18. Using a log-log plot, show the variation of the depth-area adjustment factor of Figure 4–7 with storm duration for a 50 m^2 watershed. Plot the adjustment factor as the ordinate and the storm duration as the abscissa.

4–19. On log-log paper, graph the intensity versus duration using the values of the data of Table 4–3. Also include the 2-, 10-, and 100-yr curves from Figure 4–4 on the same graph. Comment on the return period of the envelope curve of the extreme rainfalls.

4–20. Using the station-average method, compute the missing rainfall for a storm event in which the rainfall at four gages in the region had the following catches: 2.1, 2.7, 3.4, and 2.5 in.

4–21. Using the station-average method, compute the missing rainfall for a storm event in which the rainfall at five gages in the region had the following catches: 5.1, 5.4, 5.3, 5.7, and 5.2 in.

4–22. A newscaster reported a rainfall of 6.0 in. during the event that produced the measurements of Problem 4–21. Can we conclude that the newscaster overestimated the rainfall?

4–23. The long-term average annual rainfall (in.) and storm event total (in.) for each of four gages are given below. Determine the missing rainfall for the storm event if the annual average rainfall at the gage with the missing total is 32 in.

Annual	29.1	34.4	30.9	30.2
Storm event	1.3	2.7	1.8	1.9

4–24. Using the data of Problem 4–23, compute the rainfall at the gage where the catch was not recorded with the normal-ratio method. Compare the result with the estimate obtained with the station-average method.

4–25. Using the data from problem 4–23 and assume the annual catch is 38 in., compute the storm event rainfall at the gage where the catch was not recorded using both the normal-ratio and station-average methods. Which method should provide the more accurate estimate? Why?

4–26. The long-term average annual rainfall (in.) and storm event totals (in.) for five gages are given below. Determine the missing storm event catch at a gage with a mean annual catch of 24 in. using both the station-average and the normal-ratio methods. Discuss the difference. Which method should provide the more accurate result? Why?

Annual	17.5	26.8	23.2	31.1	27.4
Storm event	0.6	1.2	1.1	1.5	0.8

4–27. The long-term average annual rainfall (in.) and storm event totals (in.) for three gages are given below. Determine the missing storm event catch at a gage with a mean annual catch of 23.0 in. using both the station-average and the normal-ratio methods. Discuss the difference. Which method should provide the more accurate result? Why?

Annual	33.7	36.4	40.8
Storm Event	4.3	5.1	5.5

4–28. A study of long-term regional rainfall records indicates that the ratio of the January total rainfall in the region is 13% of the average annual catch in the region. Using the normal-ratio method, estimate the catch at a gage for a particular January in which the gage was inoperative when the average annual rainfall is 36 in.

4–29. Show the develoment of Equation 4–12.

4–30. Assuming that inoperative gages exist at two locations for the isohyetal pattern shown in Figure 4–5, estimate the missing point rainfall catch at two locations: (41°9.30′, 104°51.44′) and (41°11.27', 104°49.53'). Assess the accuracy of each estimate and discuss the factors that contributed to your assessment of the accuracy.

4–31. Assume that inoperative gages exist at the intersection of routes 80 and 25 and at the point of trisection of the runways of the Cheyenne Municipal Airport (see Figure 4–5). Estimate the missing point rainfall catch at the two locations. Assess the accuracy of each estimate, and discuss the factors that contributed to your assessment of the accuracy.

4–32. The following tabular data give the catches (in.), for a storm event and the coordinates of the gages. Estimate the missing catch at a gage located at (40°07.4′, 103°46.9′) using the isohyetal method. Estimate and discuss the accuracy of the estimated value.

Latitude	Longitude	Catch (in.)	Latitude	Longitude	Catch (in.)
40°07.6′	103°51.8′	1.4	40°07.5′	103°53.1′	1.0
40°05.0′	103°52′	0.7	40°09.5′	103°44.1′	3.2
40°05.1′	103°50.2′	1.2	40°11.9′	103°49.6′	2.8
40°06.8′	103°45.8′	2.1	40°09.6′	103°54.2′	1.4
40°07.1′	103°43.6′	2.4	40°10.2′	103°53.8′	1.9
40°05.5′	103°47.3′	1.6	40°07.0′	103°54.5′	0.3
40°09.3′	103°49.6′	2.4			

4–33. Using the catches given in Problem 4–32, estimate the missing catch at a gage located at (40°10.5′, 103°46.5′) using the isohyetal method. Estimate and discuss the accuracy of the estimated value.

4–34. The following data are the catch (in.) during a month at the nearest gage in each of four quadrants and the distance (mi) between the gage and the point of the missing monthly catch. Estimate the missing catch using the quadrant method.

Quadrant	Catch (in.)	Distance (mi)
I	3.6	12.3
II	5.2	7.9
III	4.4	10.3
IV	4.7	8.6

4–35. The following data are the catch (in.) during a month at the nearest gage in each of four quadrants and the distance (mi) between the gage and the point of the missing monthly catch. Estimate the missing catch using the quadrant method. Compare the estimate with the station average method. Which do you expect is more accurate? Why?

Quadrant	Catch (in.)	Distance (mi)
I	1.9	14.7
II	2.2	9.2
III	1.4	5.4
IV	2.5	18.6

4–36. The following data are the catch (in.) for a storm event at the gages in the region and the coordinates of the gages relative to the location of the gage where the catch was not recorded. Estimate the missing catch using the quadrant method.

Coordinates		Catch	Coordinates		Catch
N-S	E-W	(in.)	N-S	E-W	(in.)
12	5	2.9	−19	6	2.3
16	17	2.8	−14	−16	1.9
−5	22	2.0	13	−6	2.1
−6	8	3.3	9	−13	1.8

4–37. The following data are the catch (in.) at the gages in a region and the coordinates of the gages relative to the location of the gage where the monthly catch was not recorded. Estimate the missing catch using the quadrant method.

Coordinates		Catch	Coordinates		Catch
N-S	E-W	(in.)	N-S	E-W	(in.)
15	6	4.8	10	−9	5.6
8	14	4.5	7	−18	4.2
−8	7	3.9	19	−15	3.9
−8	14	4.0			

4–38. Using the data given in Problem 4–32, estimate the missing catch at a gage located at (40°07.4′, 103°46.9′). Use the quadrant method.

4–39. If all of the eight gages in Problem 4–36 were to be used to compute the weighted catch, what weight would be applied to each catch? Plot the weight versus the distance between the gage at the location of the missing storm total to the gage where the catch was recorded. Is the estimate based on all eight gages more accurate than the estimate based on the four gages?

4–40. Using the following data, check the catch at gage A for consistency using the records at gages W, X, Y, and Z, which have consistent records. Gage A was temporarily relocated in January 1983 and will be returned to its original location in January 1987. Adjust the record for the period from 1983 to 1986 inclusive.

	Annual Catch (in.) at Gage				
Year	*W*	*X*	*Y*	*Z*	*A*
1977	21	19	20	23	17
1978	27	24	27	29	24
1979	29	28	27	29	25
1980	25	23	23	26	22
1981	19	22	17	23	16
1982	21	20	18	22	20
1983	23	20	22	25	24
1984	17	16	18	20	20
1985	18	16	18	20	22
1986	22	19	20	25	25

4-41. Using the following data, check the catch at gage X for consistency using the records at gages, A, B, and C, which have consistent records. Gage X was subjected to a change in exposure in early 1980. Adjust the catch at gage X for the period from 1980 to 1985.

	Annual Catch (in.) at Gage					Annual Catch (in.) at Gage			
Year	A	B	C	X	Year	A	B	C	X
1970	41	37	42	43	1978	36	30	38	37
1971	44	41	44	46	1979	42	37	43	46
1972	36	33	40	40	1980	43	39	42	40
1973	37	36	40	42	1981	38	36	40	36
1974	40	38	43	45	1982	43	39	46	39
1975	39	39	44	43	1983	45	40	45	42
1976	35	31	36	37	1984	42	41	43	38
1977	34	31	36	38	1985	38	35	37	34

4-42. Gage T was installed in January 1973 and relocated in December 1977. Using the records for gages F, G, and H, adjust the 1973–1977 record of gage T.

	Annual Catch (in.) at Gage			
Year	F	G	H	T
1973	27	25	28	25
1974	31	30	33	29
1975	32	29	36	30
1976	34	33	34	31
1977	29	29	30	26
1978	28	26	27	30
1979	32	34	34	36
1980	36	34	37	38
1981	31	29	34	36
1982	28	27	29	31
1983	25	22	25	28
1984	28	25	29	30
1985	33	31	34	37
1986	34	33	36	37

4-43. (a) A double-mass curve between one station of interest Y and the sum of n gages, which is denoted as X, is defined by:

$$Y = \begin{cases} b_0 X & \text{for } X \leqslant X_0 \\ b_1 + b_2 X & \text{for } X > X_0 \end{cases}$$

in which Y is the cumulative catch for gages Y; X is the cumulative of the total of the n gages; X_0 is the value of X where the catch at gage Y becomes inconsistent with the past record; Y_i is the catch at Y during any year i; X_i is the sum of the catches at the n gages in year i; b_0, b_1, and b_2 are constants. Determine the adjustment factor and develop an expression for computing the adjusted values of Y_i for $X > X_0$.

(b) Apply the results of part (a) to the data of Problem 4–40.

4–44. (a) Using the notation of Problem 4–34, derive expressions for the adjustment factor and for adjusting the lower portion of a double-mass curve defined by the form of Problem 4–43.

(b) Apply the results of part (a) to the data of Problem 4–42.

4–45. Assume the 1973–1986 catches at gages *F*, *G*, and *H* of Problem 4–42 are consistent. The values at gage *P* for the same years are, respectively, 24, 28, 29, 30, 26, 28, 35, 37, 33, 25, 22, 25, 29, 31. The rain gage was later found to be biased during the period from January 1978 through December 1981. Adjust the annual catches at gage *P* for that period such that they are consistent with the record for 1973–1977 and 1982–1986.

4–46. Four rain gages are located within a 15-mi^2 area. For a particular storm event the following depths were recorded: 1.2, 0.9, 1.3, and 1.7 in. Using the station-average method, find the mean rainfall depth; also find the total volume of rain (acre-ft) on the watershed.

4–47. Seven rain gages are located within a 240-mi^2 area. For a particular storm event the following depths were recorded: 0.8, 1.8, 2.2, 1.4, 3.1, 2.5, and 1.6 in. Using the station-average method, find the mean rainfall depth; also find the total volume of rain (acre-ft) on the watershed.

4–48. For the watershed shown below, compute the average depth of storm rainfall using the Theissen Method.

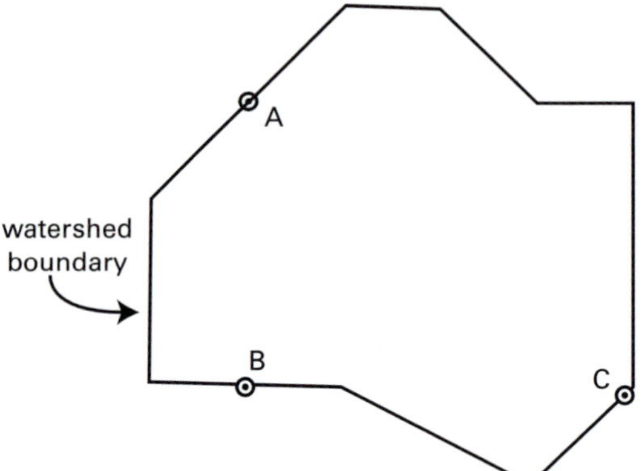

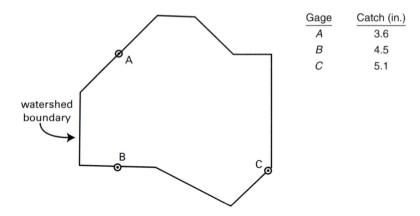

Gage	Catch (in.)
A	3.6
B	4.5
C	5.1

4–49. For the watershed shown below, compute the average depth of storm rainfall using the Theissen Method.

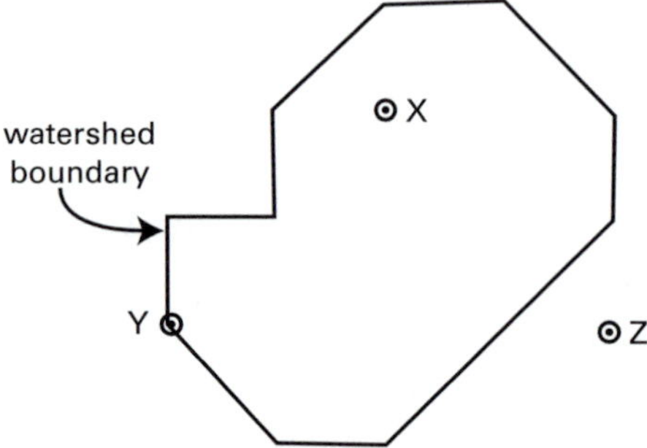

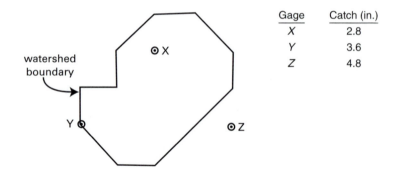

Gage	Catch (in.)
X	2.8
Y	3.6
Z	4.8

4–50. Erosion rates (tons/acre/yr) are estimated at three points in a large field. Using the Theissen Weighting Method, estimate the annual erosion from the 1000 ft * 700 ft plot within the field.

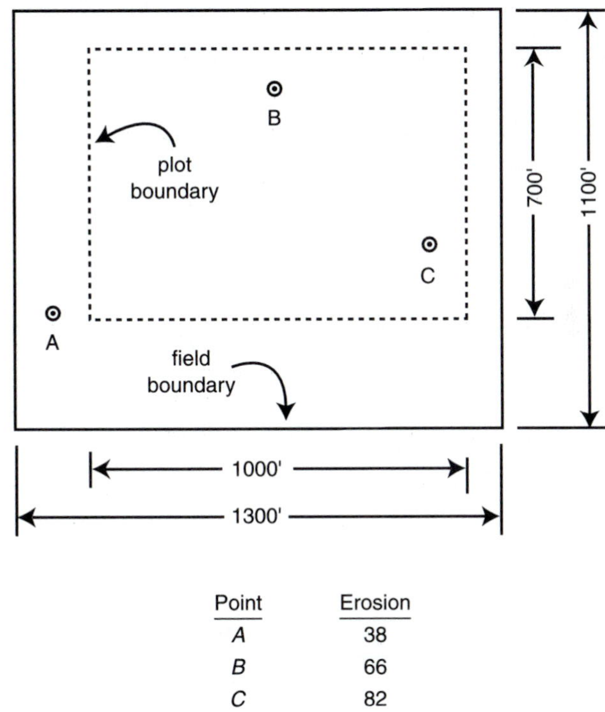

Point	Erosion
A	38
B	66
C	82

4–51. Using the isohyetal method, compute the mean storm rainfall (in.).

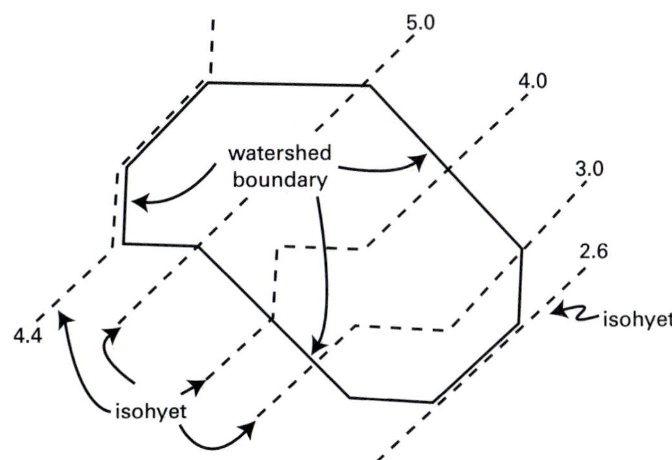

4–52. Using the isohyetal method, compute the mean storm rainfall (in.).

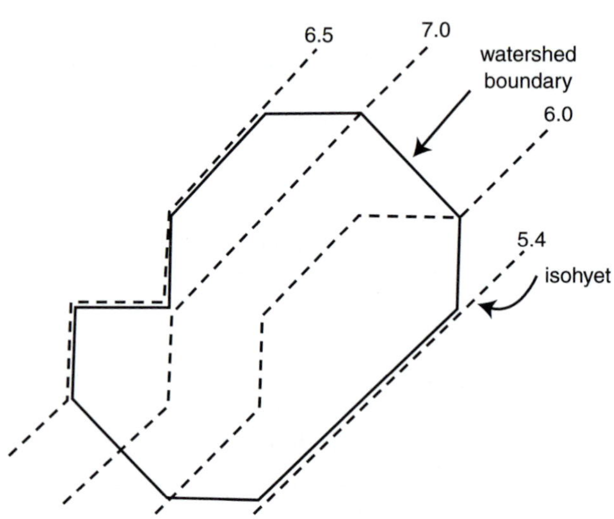

4–53. (a) Estimate the mean areal rainfall using the isohyetal method if an analysis provides the following data:

Rainfall range (in.)	0–0.5	0.5–1	1–1.5	1.5–2	2–2.5	2.5–3
Area enclosed (acres)	500	720	2100	1150	1080	250

(b) Recompute the mean areal rainfall for the data of part (a) if the rainfall ranges are computed on a range of 1 in. (that is, 0–1, 1–2, 2–3). Discuss the results of parts (a) and (b).

4–54. Estimate the mean areal rainfall using the isohyetal method if an analysis provides the following data:

Rainfall range (in.)	0–1	1–2	2.–2.5	2.5–3	3–3.5	3.5–5
Area enclosed (acre)	12	29	36	26	14	8

4–55. For the watershed shown below, find the average rainfall depth using the station-average, Theissen, and isohyetal methods.

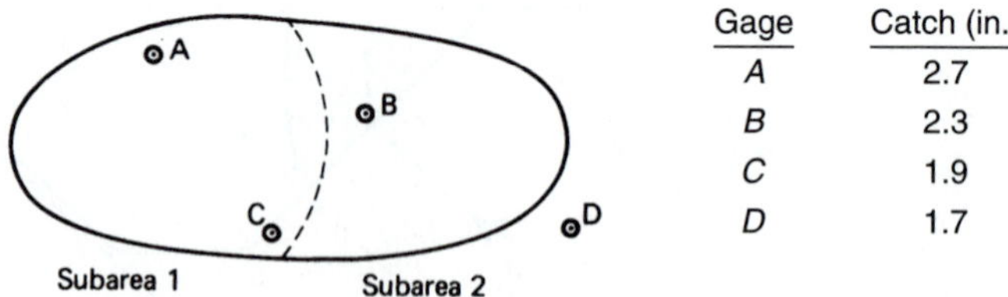

Gage	Catch (in.)
A	2.7
B	2.3
C	1.9
D	1.7

4–56. For the watershed shown below, find the average rainfall depth using the station-average, Theissen, and isohyetal methods.

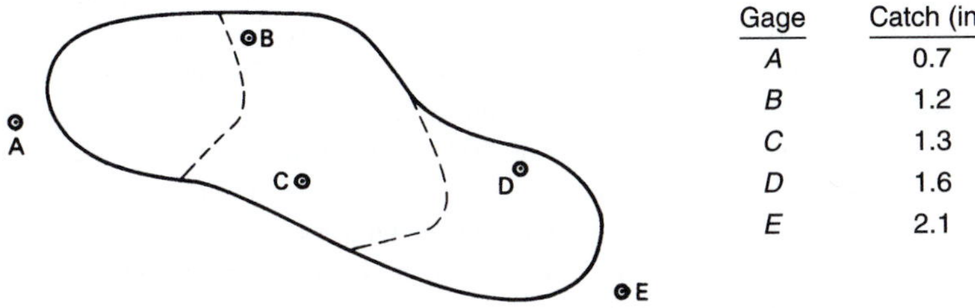

Gage	Catch (in.)
A	0.7
B	1.2
C	1.3
D	1.6
E	2.1

4–57. If the watershed of Problem 4–55 is subdivided as shown by the dashed line, determine the mean rainfall depth for the two subwatersheds using the station-average, Theissen, and isohyetal methods.

4–58. If the watershed of Problem 4–56 is subdivided as shown by the dashed lines, determine the mean rainfall depth for each subwatershed using the station-average, Theissen, and isohyetal methods.

4–59. Show the 50-yr, constant-intensity design storm for a watershed in Baltimore with a time of concentration of 30 min.

4–60. Show the 10-yr, 15-min, constant-intensity design storm for a watershed in Baltimore.

4–61. Develop a 2-yr, 1-hr, constant-intensity design storm for your locality.

4–62. Develop a 100-yr, 30-min, constant-intensity design storm for your locality.

4–63. Using the procedure described in Section 4.6.2 develop a 10-yr, 1-hr design storm for Baltimore using increments of 10 min. Also, place the design storm in dimensionless form.

4–64. Using the procedure described in Section 4.6.2 develop a 100-yr, 24-hr design storm for Baltimore using a time increment of 3 hr. Also, place the design storm in dimensionless form.

4–65. Using the procedure described in Section 4.6.2 develop a 50-yr, 12-hr design storm for your locality using a 2-hr time increment. Also, place the design storm in demensionless form.

4–66. Using the 25-yr, 24-hr rainfall volume for Baltimore (Figure 4–4), convert the dimensionless SCS type II storm (Table 4–11) to a dimensioned design storm. Use a 3-hr time increment.

4–67. Using the 100-yr, 24-hr rainfall volume for your locality and the appropriate dimensionless SCS design storm (Table 4–11) form a dimensioned design storm. Use a 2-hr time increment.

4–68. For the design storm of Problem 4–63, adjust the ordinates to represent the areal rainfall over a 200-mi^2 watershed.

4–69. For the design storm of Problem 4–64, adjust the ordinates to represent the areal rainfall over a 225-mi^2 watershed.

4–70. For the design storm of Problem 4–67, adjust the ordinates to represent the areal rainfall over a 225-mi^2 watershed.

REVIEW QUESTIONS

4–1. A rainfall hyetograph shows (a) the total volume of rainfall for a storm event; (b) the cumulative volume of rainfall with time over the duration of a storm event; (c) the variation of storm event rainfall over a watershed; (d) the intensity or volume of rainfall as a function of time; (e) none of the above.

4–2. A 15-yr rainfall (a) will occur six times in a period of 90 years; (b) has a probability of occurring of 0.0667 in a 15-yr period; (c) can occur only once in any period of 15 yr; (d) will occur more frequently than a 10-yr rainfall; (e) none of the above.

4–3. A rainfall with an exceedence probability of 0.025 (a) will occur on the average once every 40 years; (b) has a probability of 0.025 of occurring in any one year; (c) will have a larger magnitude than a 25-yr rainfall; (d) is a 40-yr event; (e) all of the above.

4–4. What is the area in acres of a watershed if a 100-acre-in. rainfall occurs during a 3-hr storm with a depth of 4 in.? (a) 8.33; (b) 25; (c) 75; (d) 100; (e) none of the above.

4–5. For a watershed in Baltimore (where Figure 4–4 applies), the duration (min) of a 50-yr storm that has the same volume as the 5-yr, 10-min storm is (a) 6; (b) 10; (c) 25; (d) 60; (e) none of the above.

4–6. For a 10-acre watershed in Baltimore (where Figure 4–4 applies) with a 25-min time of concentration, the depth (in.) of a 5-yr, constant-intensity design storm is (a) 1.46; (b) 3.50; (c) 14.6; (d) 35.0; (e) none of the above.

4–7. For a 25-acre watershed in Baltimore (where Figure 4–4 applies) with a 45-min time of concentration, the total volume of rainfall (acre-ft) of a 25-yr, constant-intensity design storm is (a) 0.206; (b) 5.16; (c) 61.9; (d) 309.4; (e) none of the above.

4–8. The SCS dimensionless design storm (a) should only be used for designs on watersheds with times of concentrations of about 24 hr; (b) varies in shape for different return periods; (c) enables the use of rainfall depths from the *IDF* curve for all storm durations up to 24 hr; (d) is dimensionalized using the 100-yr storm rainfall depth; (e) none of the above.

4–9. The purpose of the depth-area adjustment is to correct (a) rainfall depths for storm duration; (b) rainfall intensities for watershed size; (c) the distribution of rainfall for storm location; (d) the point rainfall depth for storm orientation.

4–10. The depth-area adjustment factor is a function of which of the following? (a) The rainfall frequency; (b) the total depth of storm rainfall; (c) the orientation of the watershed; (d) the duration of the rainfall; (e) none of the above.

4–11. The depth-area adjustment factor (a) decreases as the storm duration decreases; (b) increases with increases in the drainage area; (c) is not a function of storm duration; (d) can be ignored for drainage areas greater than about 350 mi^2; (e) none of the above.

4–12. Which one of the following is not a method for computing the average areal rainfall? (a) The Theissen Polygon Method; (b) the isohyetal method; (c) the normal-ratio method; (d) the station-average method.

4–13. Weights for the Theissen Polygon Method are a function of (a) the number of rain gages only; (b) the proportions of the total watershed area that are geographically closest to each of the rain gages; (c) the relative distribution throughout the watershed of the average annual catch at each rain gage; (d) the proportions of the area that are physically associated with the catch of the rain gage.

4–14. Which one of the following methods assumes that the weights applied to the catch of each of the rain gages are equal? (a) The Theissen Polygon Method; (b) the isohyetal method; (c) the normal-ratio method; (d) the station-average method.

4–15. Which one of the following methods accounts for storm morphology in determining weights to apply to the rain gage catches? (a) The quadrant method; (b) the Theissen Polygon Method; (c) the station-average method; (d) the isohyetal method.

4–16. Which one of the following methods does not account for nonhomogeneity in the spatial distribution of rain gages? (a) The station-average method; (b) the Theissen polygon method; (c) the isohyetal method; (d) the quadrant method.

4–17. When estimating a missing rainfall volume with the normal-ratio method, the weights sum to (a) 1; (b) the mean proportion of the storm rainfall to the average annual rainfall; (c) the ratio of the average annual catch at the gage with the missing storm total to the mean of the average annual catches for the other gages; (d) the number of regional gages used to estimate the missing rainfall depth.

4–18. When using the isohyetal method to estimate missing rainfall volumes, the weights given to the individual rain gages (a) are embedded within the isohyets; (b) sum to 1; (c) depend on the average minimal catch at each gage; (d) are equal for each rain gage; (e) none of the above.

4–19. For the isohyetal pattern of Figure 4–5 the isohyetal method of estimating the rainfall (in.) at the intersection of Highways 80 and 85 is (a) 2.8; (b) 3.0; (c) 3.3; (d) 3.8.

4–20. When using the quadrant method for estimating missing rainfall, which one of the following is not an assumption? (a) Catches at gages that are located close to each other are not independent estimates of the catch at the unknown point, and therefore all gages are not necessary; (b) the weight assigned to a gage should decrease as the distance between the gage and the point where an estimate is required increases; (c) gages within a quadrant represent similar regional information concerning storm morphology; (d) gages in different quadrants contain independent information about regional storm rainfall; (e) all of the above are assumptions of the quadrant method.

4–21. Which of the following is not a reason for making a double mass curve analysis for gage consistency? (a) To adjust data because of frequent vandalism at the gage; (b) to estimate the areal average rainfall; (c) to adjust for a change in exposure of the gage; (d) to adjust for change in observational procedures.

4–22. Which one of the following is not true for adjustment of a rainfall record using double-mass curve analysis? (a) It is more accurate for long rainfall records; (b) it is applicable only for cases where the cause of gage inconsistency is abrupt; (c) it is based on the change in the slope of a plot of cumulative gage records; (d) it assumes that the regional records used to form the multisite cumulative total rainfall represent homogeneous records.

DISCUSSION QUESTION

The technical content of this chapter is important to the professional hydrologist, but practice is not confined to making technical decisions. The intent of this discussion question is to show that hydrologists must often address situations where value issues intermingle with the technical aspects of a project. In discussing the stated problem, at a minimum include responses to the following questions:

1. What value issues are involved, and how are they in conflict?
2. Are technical issues involved with the value issues? If so, how are they in conflict with the value issues?
3. If the hydrologist attempted to rationalize the situation, what rationalizations might he or she use? Provide arguments to suggest why the excuses represent rationalization.
4. What are the hydrologist's alternative courses of action? Identify all alternatives, regardless of the ethical implications.
5. How should the conflict be resolved?

You may want to review Sections 1.6 to 1.12 in Chapter 1 in responding to the problem statement.

Case. A hydrologist is hired as a consultant on a legal case. A small shopping mall, behind which is a stream, was flooded. The owners of the stores in the mall sue the owners of a large arena that was built five years earlier upstream of the mall. The arena has a large paved parking lot, which the store owners argue caused the flood levels to increase. They indicate that they had never experienced flooding in the twenty years prior to the construction of the arena. The hydrologist is hired by the owners of the arena. In making his analysis, the hydrologist collects data for the storm event from the four rain gages in the region. None of the gages are located in the 32-mi^2 watershed that drains to the location of the mall. The hydrologist wants to show that the flooding occurred because the storm was unusually large, not because of the impervious area on the arena owner's land. In computing a weighted average rainfall, the hydrologist elects not to include one gage in the analysis because it has a much lower total storm rainfall than the other three gages. Including the rainfall from the one station would reduce the strength of the hydrologist's argument that the arena was not a factor in the flooding. The hydrologist omits any reference to the one gage in his report.

5

Frequency Analysis

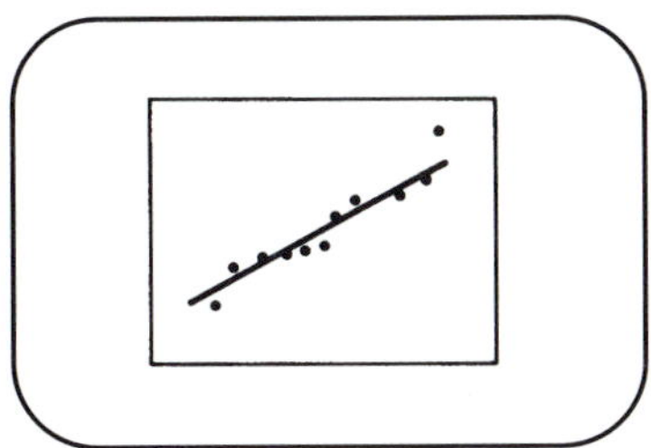

CHAPTER OBJECTIVES

1. Introduce models that relate the magnitude and the uncertainty of occurrence of values of a single random variable.
2. Introduce concepts related to the accuracy of a frequency curve, including standard errors and confidence intervals.
3. Discuss basic concepts of regionalizing and weighting flood skew.
4. Introduce probabilistic concepts associated with uncertainty over the design life of a project.
5. Provide a method for adjusting a nonhomogeneous flood record for the effects of urbanization.

5.0 NOTATION

a = coefficient of the gamma distribution
A = drainage area
b = coefficient
d_i = difference in ranks for Spearman test
f = relative increase in peak discharge for a percent imperviousness U
f_1, f_2 = peak adjustment factors
g = standardized skew
g_h = historically adjusted logarithmic skew
g_{hw} = historically adjusted logarithmic weighted skew
g_w = weighted skew
h = depth of flow
H = historic record length
i = rank of an event
I = likelihood
K = frequency factor
K_L = distribution factor for lower confidence limit
K_o = frequency factor for outlier detection
K_U = distribution factor for upper confidence limit
L = logarithm of the likelihood function
L = number of low extreme events
m_{hi} = historically adjusted rank
MSE_g = mean square error for station skew
$MSE_{\bar{g}}$ = mean square error for generalized skew
MSE_{gh} = mean square error for historically adjusted skew
n = sample size
n_z = number of zero-flood years
N = sample size
N_t = total record length
p = exceedence probability
p_a = adjustment probability
p_c = conditional probability
P_c = probability for Cunnane plotting position
P_e = volume of rainfall excess distribution
P_h = probability for Hazen plotting position
P_w = probability for Weibull plotting position
q = discharge rate
Q = discharge rate
Q_a = adjusted discharge rate
Q_o = critical flow rate
Q_p = peak discharge rate
$\hat{Q}_7$ = estimated 7-day low flow
R_s = Spearman correlation coefficient
S = sample standard deviation
S_g = standard error of the standardized skew
S_R = relative sensitivity
S_s = standard error of the standard deviation
$S_{\bar{x}}$ = standard error of the mean
S_y = standard deviation of random variable y
S_{yh} = historically adjusted logarithmic standard deviation
T = return period (years)
T_R = duration of rainfall excess
U = percentage of imperviousness
V = average flow velocity
W = historic adjustment weight
X = random variable
$\bar{X}$ = sample mean of random variable X
$\bar{X}_{hi}$ = historically adjusted flood estimate
Y = random variable
$\bar{Y}$ = sample mean of random variable Y
$\hat{Y}$ = estimated value of Y
$\bar{Y}_h$ = historically adjusted logarithmic mean
Y_L = lower confidence limit on Y

Y_{oh} = detection criterion for high outliers
Y_{ol} = detection criterion for low outliers
Y_u = upper confidence limit on Y
z = frequency factor for normal distribution (standard normal deviate)
Z = number of historic events
α = location parameter for uniform distribution
β = scale parameter for uniform distribution
γ = confidence coefficient
μ = population mean
σ = population standard deviation

5.1 INTRODUCTION

Hydrologic design centers around runoff, not the rainfall that generates the runoff or the watershed processes that transform the rainfall into runoff. Like rainfall, runoff is a function of time and space. Imagine standing near a stream during a very heavy rainfall. At the start of the storm the water level in the stream would probably be relatively low; the stream bed might even be dry. After a period of time the water level would begin to rise, especially shortly after a heavy burst of rainfall. At some point in time, the rainfall rate would taper off and shortly thereafter, the level in the stream would reach its highest level. The rainfall might then stop and the level of the water in the stream would start to decrease. The time that it takes to decrease from the time of the highest level to the level that existed prior to the start of rainfall is usually much longer than the time that it took to reach the maximum level. This brief description of a rainfall-runoff event can be used to illustrate several points.

5.1.1 Hydrographs and Discharge

In Chapter 4 the time sequence of rainfall was called a *hyetograph*. The time sequence of runoff is called a *hydrograph*. If you had taken measurements of the level of water in the stream above some datum, such as the bottom of the channel, at regular intervals of time, say every minute or every five minutes, a stage hydrograph of water level versus time could be plotted. The variation in the stage hydrograph would probably show some degree of association with the variation in the rainfall hyetograph, although the watershed processes usually smooth out much of the variation in the hyetograph. Table 5–1 gives the daily discharge hydrograph for the 1984 water year for the Northeast Branch of the Anacostia River at Riverdale, MD. The minimum occurred over a four-day period in September. The maximum mean daily discharge occurred on March 29. The maximum of 2190 ft^3/sec was much less than the maximum record of 12,000 ft^3/sec (June 22, 1972).

While a stage hydrograph is important, hydrographs are usually reported in terms of the discharge rate, which has units of length cubed per unit time; for example, the discharge might be reported in cubic feet per second, cubic meters per second, or acre-feet per day. These units reflect the relationship between the cross-sectional area of the stream (A) and the average velocity of flow (V), which are related by the continuity equation, $q = AV$, in which q is the discharge rate. It is important to note that A in the continuity equation is the cross-sectional area of the stream and not the drainage area of the watershed, which is also com-

TABLE 5–1 Mean Daily Discharge (ft^3/sec) Record for the 1984 Water Year (Oct. 1983 to Sept. 1984) for Gage 01649500 Northeast Branch Anacostia River at Riverdale, MD

Day	Oct	Nov	Dec	Jan	Feb	Mar	Apr	May	Jun	Jul	Aug	Sep
1	30	32	59	75	61	103	139	132	69	800	36	15
2	20	32	48	67	55	84	118	77	53	400	71	14
3	17	36	48	60	74	75	109	232	44	70	51	90
4	15	51	668	62	160	69	281	817	40	40	38	30
5	15	40	180	60	180	160	596	169	37	60	33	16
6	16	36	403	61	117	138	280	157	35	75	31	14
7	13	32	251	59	74	90	154	173	35	40	42	13
8	12	30	103	52	61	80	121	339	37	30	40	12
9	12	30	74	48	56	106	106	253	33	28	84	12
10	12	240	67	110	54	75	100	118	31	38	76	11
11	16	248	56	162	79	80	95	89	31	30	40	11
12	320	78	237	76	74	82	90	82	29	28	160	11
13	127	40	1920	60	67	296	91	70	29	26	380	12
14	103	31	607	79	410	261	106	60	29	24	60	20
15	33	80	168	71	723	139	188	55	28	22	32	13
16	23	173	108	59	382	98	356	52	27	22	28	11
17	20	57	82	53	158	80	169	49	33	40	26	13
18	20	35	69	49	158	70	117	47	77	100	26	11
19	35	32	63	48	106	64	106	50	69	44	30	10
20	32	164	57	48	89	59	91	48	32	34	70	9.5
21	52	459	51	48	76	250	85	43	28	220	24	9.0
22	28	84	770	50	68	141	102	41	28	90	22	9.0
23	693	47	244	55	257	82	220	168	26	55	20	9.0
24	429	82	115	90	299	66	136	101	37	39	19	9.0
25	163	937	80	190	134	171	102	55	32	32	18	12
26	86	181	60	200	92	572	83	294	30	31	17	13
27	56	80	100	169	79	180	80	256	30	60	17	15
28	47	237	532	120	263	608	75	74	30	35	16	85
29	40	206	262	82	194	2190	115	344	34	31	16	31
30	34	82	119	74	—	399	104	383	100	31	16	22
31	32	—	83	75	—	186	—	125	—	31	18	—
MEAN	82.3	130	248	81.0	159	228	151	160	39.1	84.1	50.2	18.8
MIN	12	30	48	48	54	59	75	41	26	22	16	9.0
IN.	1.30	1.99	3.93	1.28	2.35	3.60	2.31	2.53	.60	1.33	.80	.29

Source: U.S. Geological Survey, 1984.

monly denoted by A. As the depth of flow during a storm increases, both the cross-sectional area of the stream and the average velocity of flow will increase. The relationship between the depth of flow and the discharge is often assumed to have the following form:

$$q = ah^b \tag{5–1}$$

in which h is the depth of flow and a and b are constants for a particular stream section. The depth of flow and the area of the cross section are related geometrically. If we measured the velocity of flow in the stream and simultaneously measured the depth of flow, q could be determined from the continuity equation and used with the measurements of depth and Equation 5–1 to evaluate the values of a and b. Equation 5–1 could be fitted using regression analysis after taking the logarithms of both h and q. The relationship of Equation 5–1 is called the rating curve. Figure 5–1 is an example of a rating curve. In practice, values for h and V are measured for many storm events before estimating a and b of Equation 5–1. The cross-sectional area A is computed from the flow depth h and the discharge is obtained with the continuity equation. In some cases, considerable scatter about the curve of best fit is apparent.

5.1.2 Rainfall-Runoff Association

As suggested above, the rates of rainfall and runoff are correlated. A sudden burst of rainfall can noticeably change the trend of a hydrograph. However, the watershed and channel processes that affect the transformation of rainfall to runoff smooth out many of the extreme variations in the rainfall; this is especially true of large watersheds and watersheds where there is a lot of storage, such as in a heavily forested basin. For small urban watersheds, such as an asphalt-covered parking lot, the opportunity for smoothing by watershed processes is less, so the common variations in rainfall and runoff would be more evident.

Figure 5–2 shows the hyetograph and four hydrographs for a storm in Los Angeles. The rainfall was measured in the Eaton Wash watershed, and the rain gage is the only gage in the vicinity of the four urban watersheds, which are adjacent to each other. From Figure 5–2, the association between rainfall and runoff is evident; however, the association is differ-

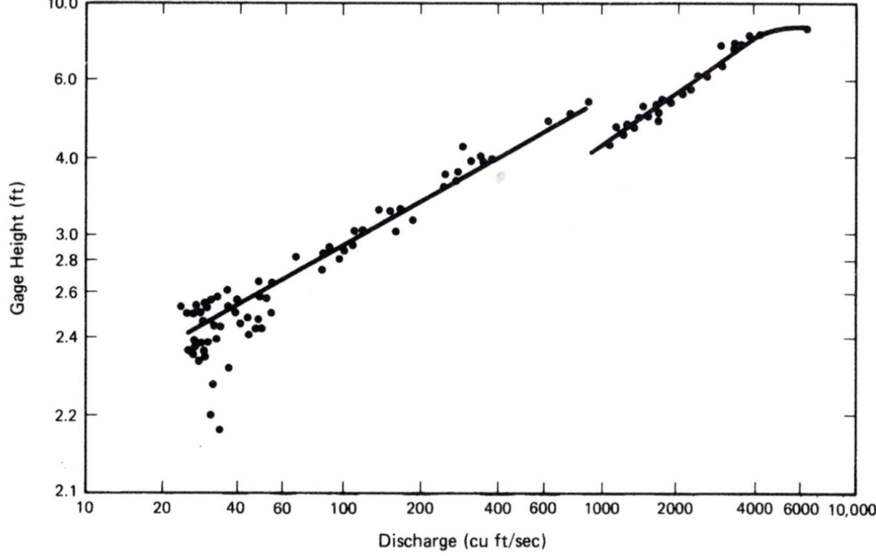

FIGURE 5–1 Stage-discharge relation for station 34 on Pigeon Roost Creek, Miss.

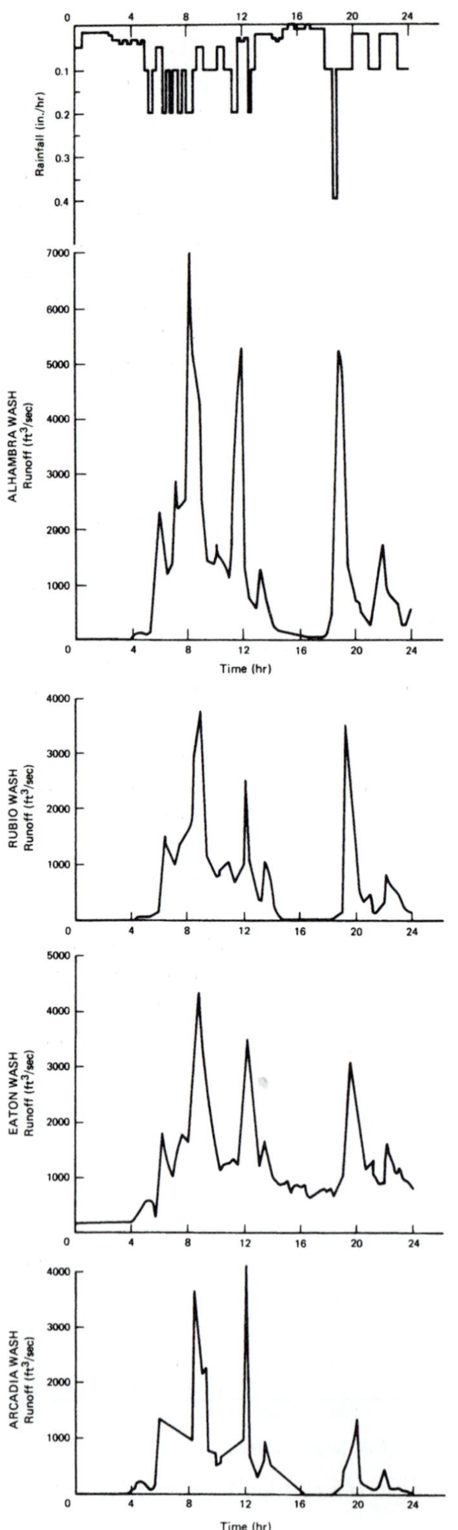

FIGURE 5–2 Rainfall hyetograph and runoff hydrographs for March 1, 1983, Los Angeles, CA.

ent for each of the four watersheds. The difference in the association may be due to the spatial variation in the rainfall on the four watersheds or because of differences in the watershed processes of the four watersheds. The large burst of rainfall at about 19 hr has a different effect on each of the four watersheds. On Alhambra and Rubio Washes the rainfall results in a spike in the hydrographs. For Eaton and Arcadia Washes the effect is less peaked and suggests a greater degree of smoothing by the watershed processes. For the burst of rain at about 12 hr, the runoff on Arcadia Wash is the largest for the storm, while for the other three watersheds the peak runoff at about 12 hr is less than the peaks at about 9 hr. This may be due to either spatial variability in rainfall that cannot be assessed because data for only one rain gage is available or because of differences in watershed processes over the four watersheds.

5.1.3 Runoff Frequency

The concept of frequency was introduced in Chapter 4 as a characteristic of precipitation. Precipitation was shown to be a random variable, for which the underlying concepts were introduced in Chapter 2. Since the future occurrences of a random variable cannot be predicted exactly, the concepts of probability were used to describe its expected behavior.

Runoff can also be viewed as a random variable, so the concept of frequency applies to runoff characteristics as well as to rainfall characteristics. The peak of the discharge hydrograph is an important design variable, so the frequency of a peak discharge plays a central role in hydrology. Just as we talked of the 100-yr rainfall intensity or depth, engineers commonly estimate the 100-yr peak discharge in their design work. The frequency concept for runoff can be discussed in terms of either the return period or the exceedence probability.

Since discharge rates are central to many design problems in hydrology, the process of analysis as it applies to discharge warrants special treatment. While runoff frequency will be the primary illustration used in this chapter, we must continually recognize that frequency analysis is a statistical tool that can be applied to any random variable, not just runoff. We could, for example, make a frequency analysis of rainfall characteristics, water quality or toxic waste parameters, or the magnitude of earthquakes. The topic could have been introduced in Chapter 2 as a statistical tool, but because of its importance a separate discussion is warranted.

5.2 FREQUENCY ANALYSIS AND SYNTHESIS

Design problems such as the delineation of flood profiles requires estimates of discharge rates. A number of methods of estimating peak discharge rates will be discussed throughout this book. The methods can be divided into two basic groups: those intended for use at sites where gaged stream flow records are available and those intended for use at sites where such records are not available; these two groups will be referred to as methods for gaged and ungaged sites, respectively.

Statistical frequency analysis is the most commonly used procedure for the analysis of flood data at a gaged location. Actually, statistically frequency analysis is a general procedure that can be applied to any type of data. Because it is so widely used with flood data, the term "flood frequency analysis" is common. However, statistical frequency analysis can also be applied to other hydrologic variables such as rainfall data for the development of intensity-duration-frequency curves and low-flow discharges for use in water quality control. The variable could also be the mean annual rainfall, the peak discharge, the 7-day low flow, or a water quality parameter. Therefore, the topic will be treated in both general and specific terms.

5.2.1 Population versus Sample

In frequency modeling, it is important to distinguish between the population and the sample. Frequency modeling is a statistical method that deals with a single random variable and is, thus, classified as a univariate method. The goal of univariate prediction is to make estimates of either probabilities or magnitudes of random variables. A first step in achieving this goal is to identify the population. The objective of univariate data analysis is to use sample information to determine the appropriate population density function, with the probability density function (PDF) being the univariate model from which probability statements can be made. The input requirements for frequency modeling include a data series and a probability distribution that is assumed to describe the occurrence of the random variable. The data series could include the largest instantaneous peak discharge to occur each year of the record. The probability distribution could be the normal distribution. Analysis is the process of using the sample information to estimate the population. The population consists of a mathematical model that is a function of one or more parameters. For example, the normal distribution is a function of two parameters: the mean μ and standard deviation σ. In addition to identifying the correct PDF, it is necessary to quantify the parameters of the PDF. The population consists of both the probability distribution function and the parameters.

A frequently used procedure called the method of moments equates characteristics of the sample (for example, sample moments) to characteristics of the population (for example, population parameters). It is important to note that estimates of probability and magnitudes are made using the assumed population and not the sample of data; the sample is used only in identifying the population.

5.2.2 Analysis versus Synthesis

As with many hydrologic methods that have a statistical basis, the terms analysis and synthesis apply to the statistical frequency method. Frequency analysis is the process of "breaking down" data in a way that leads to either a mathematical or graphical model of the relationship between the flood magnitude and its probability of occurrence. Conversely, synthesis refers to the estimation of either (1) a value of the random variable X for some selected exceedence probability, or (2) the exceedence probability for a selected value of the random variable X. In other words, analysis is the derivation of a model that can represent the relation between the random variable and its likelihood of occurrence, while synthesis refers to the use of the resulting relation for the purposes of estimation.

It is important to point out that frequency analysis may actually be part of a more elaborate problem of synthesis. Specifically, separate frequency analyses can be performed at a large number of sites within a region and the value of the random variable X for a selected exceedence probability determined for each site; these values can then be used to develop a regression model using the random variable X as the criterion or dependent variable. As an example, regression equations relating a peak discharge of a selected exceedence probability to watershed characteristics are widely used in hydrologic design. These equations are derived by (1) making a frequency analysis of annual maximum discharges at a number (n) of stream gage stations in a region, (2) selecting the value of the peak discharge from each of the n frequency curves for a selected exceedence probability, say the 100-yr flood, and (3) developing the regression equation relating the n values of peak discharge to watershed characteristics for the same n watersheds. This process of regionalizing flood magnitudes for specific exceedence probabilities is discussed in detail in Chapter 7, where the material of this chapter is combined with the material of Chapter 2.

5.2.3 Probability Paper

Frequency analysis is a common task in hydrologic studies. A frequency analysis usually produces a graph of the value of a single hydrologic variable versus the probability of its occurrence. The computed graph represents the best estimate of the statistical population from which the sample of data were drawn.

Since frequency analyses are often presented graphically, a special type of graph paper, which is called probability paper, is required. The paper has two axes, with the ordinate being used to plot the value of the random variable, and the probability of its occurrence is given on the abscissa. The probability scale will vary depending on the probability distribution being used. In hydrology, the normal and Gumbel extreme-value distributions are the two PDFs used most frequently to define the probability scale. The example of Figure 5–3 uses normal probability paper. The probability scale represents the cumulative normal distribution. The scale at the top of the graph is the exceedence probability; the probability that the random variable will be equaled or exceeded in one time period. It changes from 99.99% to 0.01%. The lower scale is the nonexceedence probability, which is the probability that the corresponding value of the random variable will not be exceeded in any one time period. This scale extends from 0.01% to 99.99% The ordinate of probability paper is used for the random variable, such as the peak discharge. The example shown in Figure 5–3 has an arithmetic scale. Log-normal probability paper is also available, with the scale for the random variable in logarithmic form. Gumbel and log-Gumbel papers can also be obtained and used to describe the probabilistic behavior of random variables that follow these probability distributions.

A frequency curve provides a probabilistic description of the likelihood of occurrence or nonoccurrence of the variable. Figure 5–3 provides an example of a frequency curve, with the value of the random variable X versus its probability of occurrence. The upper probability scale gives the probability of X being exceeded in one time period, while the lower probability scale gives the probability of X not being exceeded. For example, for the frequency curve of Figure 5–3, the probability of X being greater than 7 in one time period

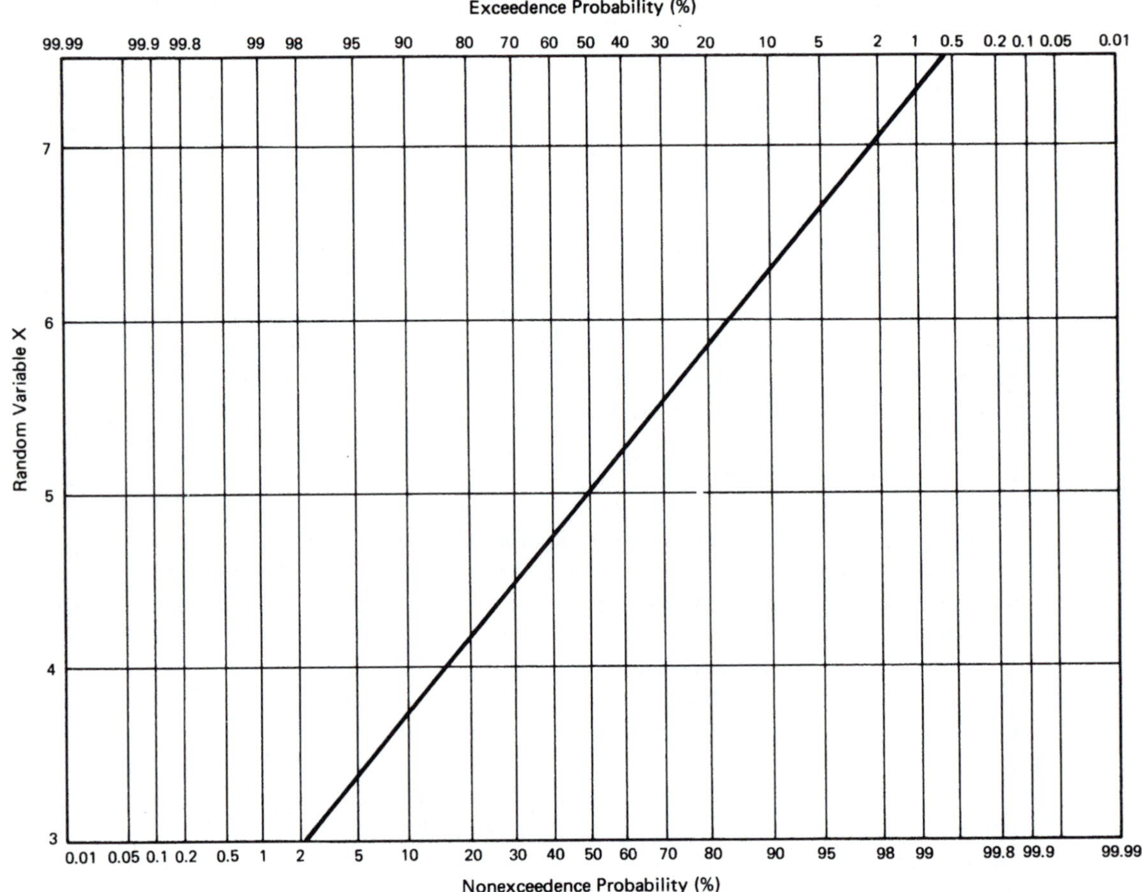

FIGURE 5–3 Frequency curve for a normal population with $\mu = 5$ and $\sigma = 1$.

is 0.023; therefore, there is a probability of 0.977 of X not being greater than 7 in one time period.

Although a unique probability plotting paper could be developed for each probability distribution, probability papers for the normal and extreme-value distributions are the most frequently used probability papers. The probability paper is presented as a cumulative distribution function. If the sample of data is from the distribution function that was used to scale the probability paper, the data will follow the pattern of the population line when properly plotted on the paper. If the data do not follow the population line, then either (1) the sample is from a different population or (2) sampling variation has produced a nonrepresentative sample. In most cases, the former reason is assumed to be the cause, especially when the sample size is reasonably large.

5.2.4 Mathematical Model

As an alternative to a graphical solution using probability paper, a frequency analysis may be conducted using a mathematical model. A model that is commonly used in hydrology for normal, log-normal, and log-Pearson Type III analyses has the form

$$X = \overline{X} + KS \tag{5–2}$$

in which X is the value of the random variable having mean $\overline{X}$ and standard deviation S, and K is a frequency factor. Depending on the underlying population, the specific value of K reflects the probability of occurrence of the value X. Equation 5–2 can be rearranged to solve for K when X, $\overline{X}$, and S are known and we wish to estimate the probability of X occurring:

$$K = \frac{X - \overline{X}}{S} \tag{5–3}$$

In summary, Equation 5–2 is used for the synthesis case when the probability is known and an estimation of the magnitude is needed, while Equation 5–3 is used for the synthesis case where the magnitude is known and the probability is needed.

5.2.5 Procedure

In a broad sense, frequency analysis can be divided into two phases: deriving the population curve and plotting the data to evaluate the goodness of fit. The following procedure is often used to derive the frequency curve to represent the population:

1. Hypothesize the underlying density function.
2. Obtain a sample and compute the sample moments.
3. Equate the sample moments and the parameters of the proposed density function.
4. Construct a frequency curve that represents the underlying population.

This procedure is referred to as method-of-moments estimation because the sample moments are used to provide numerical values for the parameters of the assumed population. The computed frequency curve, which represents the population, can then be used for estimation of magnitudes for a given return period or probabilities for specified values of the random variable.

It is important to recognize that it is not necessary to plot the data points in order to make probability statements about the random variable. While the four steps listed above lead to an estimate of the population frequency curve, the data should be plotted, as described in Section 5.2.7, to ensure that the population curve is a good representation of the data. The plotting of the data is a somewhat separate part of a frequency analysis; its purpose is to assess the quality of the fit rather than act as a part of the estimation process.

5.2.6 Sample Moments

The formulas for computing the moments were given in Chapter 2; however, they are worth repeating here. For the random variable X, the sample mean ($\overline{X}$), standard deviation (S), and standardized skew (g) are, respectively, computed by

$$\overline{X} = \frac{1}{n}\sum_{i=1}^{n} X_i \tag{5–4a}$$

$$S = \left[\frac{1}{n-1}\sum_{i=1}^{n}(X_i - \overline{X})^2\right]^{0.5} \tag{5–4b}$$

$$g = \frac{n\sum_{i=1}^{n}(X_i - \overline{X})^3}{(n-1)(n-2)S^3} \tag{5–4c}$$

For use in frequency analyses where the skew is used, Equation 5–4c represents a standardized value of the skew. Equations 5–4 can also be used when the data are transformed by taking the logarithms; in this case, the log transformation should be made prior to computing the moments.

5.2.7 Plotting Position Formulas

It is important to note from the discussed steps that it is not necessary to plot the data before probability statements can be made using the frequency curve; however, the data should be plotted to determine how well the measured data agree with the fitted curve of the assumed population. A rank-order method is used to plot the data. This involves ordering the data from the largest event to the smallest event, assigning a rank of 1 to the largest event and a rank of n to the smallest event, and using the rank (i) of the event to obtain a probability plotting position; numerous plotting position formulas are available. Bulletin 17B (Interagency Advisory Committee on Water Data, 1982) provides the following generalized equation for computing plotting position probabilities:

$$P_i = \frac{i - a}{n - a - b + 1} \tag{5–5a}$$

where a and b are constants that depend on the probability distribution. For example, $a = b = 0$ for the uniform distribution. Numerous formulas have been proposed, including the following:

Weibull:
$$P_i = \frac{i}{n+1} \tag{5–5b}$$

Hazen:
$$P_i = \frac{2i-1}{2n} = \frac{i - 0.5}{n} \tag{5–5c}$$

Cunnane:
$$P_i = \frac{i - 0.4}{n + 0.2} \tag{5–5d}$$

in which i is the rank of the event, n is the sample size, and P_i values give the exceedence probabilities for an event with rank i. The data are plotted by placing a point for each value of the random variable at the intersection of the value of the random variable and the value of the exceedence probability at the top of the graph. The plotted data should approximate the population line if the assumed population model is a reasonable assumption.

The different plotting position formulas provide different probability estimates, especially in the tails of the distributions. The following tabular summary gives the computed probabilities for each rank for a sample of 9 using three plotting position formulas:

$n = 9$				$n = 99$			
Rank	p_w	p_h	p_c	Rank	p_w	p_h	p_c
1	0.1	0.05	0.065	1	0.01	0.005	0.006
2	0.2	0.15	0.174	2	0.02	0.015	0.016
3	0.3	0.25	0.283	•			
4	0.4	0.35	0.391	•			
5	0.5	0.45	0.500	•			
6	0.6	0.55	0.609	98	0.98	0.985	0.984
7	0.7	0.65	0.717	99	0.99	0.995	0.994
8	0.8	0.75	0.826				
9	0.9	0.85	0.935				

The Hazen formula gives smaller probabilities for all ranks than either the Weibull or Cunnane formulas. The probabilities for the Cunnane formula are more dispersed than either of the other two formulas. For a sample size of 99 the same trends exist.

5.2.8 Return Period

The concept of return period was introduced in Chapter 4 with respect to rainfall. It is also used to describe the likelihood of flood magnitudes. As Equation 4–1 indicates, the return period is the reciprocal of the exceedence probability. Just as a 25-yr rainfall has a probability of 0.04 of occurring in any one year, a 25-yr flood has a probability of 0.04 of occurring in any one year. As indicated in Chapter 4, it is incorrect to believe that once a 25-yr event occurs it will not occur again for another 25 years. Two 25-yr events can occur in consecutive years. Then again, a period of 100 years may pass before a second 25-yr event occurs following the passage of a 25-yr event.

Does the 25-yr rainfall cause the 25-yr flood magnitude? Some hydrologic models make this assumption; however, it is unlikely to be the case in actuality. It is a reasonable assumption for modeling because models are based on the average of expectation or "on the average" behavior. In actual occurrence of storms, the 25-yr flood magnitude will not occur if the 25-yr rainfall occurs on a dry watershed. Similarly, a 50-yr flood could occur from a 25-yr rainfall if the watershed was saturated. Modeling often assumes that a T-yr rainfall on a watershed that exists in a T-yr hydrologic condition will produce a T-yr flood.

For some engineering problems, discharge rates are needed for short recurrence intervals, such as six or three months. These magnitudes cannot be treated in the same way as the less frequent events, such as a 50-yr event. The best way to estimate discharge magnitudes for short recurrence intervals is with the partial-duration series, which is discussed in Section 5.6.

5.3 POPULATION MODELS

Step 1 of the procedure outlined in Section 5.2.5 indicates that frequency analysis requires the selection of a model to represent the population. Any probability distribution can be used as the model; however, the log-normal and log-Pearson Type III distribution are the most widely used distributions in hydrologic analysis. These two distributions will be introduced along with the normal distribution, which is a basic model.

5.3.1 The Normal Distribution

Normal probability paper, which is commercially available, is commonly used in hydrology. Following the general procedure outlined above, the specific steps used to develop a curve for a normal population are as follows:

1. Assume that the random variable has a normal distribution with population parameters μ and σ.
2. Compute the sample moments $\bar{X}$ and S (the skew is not needed).
3. For the normal distribution, the parameters and sample moments are related by $\mu = \bar{X}$ and $\sigma = S$.
4. A curve is fitted as a straight line with $(\bar{X} - S)$ plotted at an exceedence probability of 0.8413 and $(\bar{X} + S)$ at an exceedence probability of 0.1587.

The frequency curve of Figure 5–3 is an example for a normal distribution with a mean of 5 and a standard deviation of 1. It is important to note that the curve passes through the two points: $(\bar{X} - S, 0.8413)$ and $(\bar{X} + S, 0.1587)$. It also passes through the point defined by the mean and a probability of 0.5. Two other points that could be used are $(\bar{X} + 2S, 0.0228)$ and $(\bar{X} - 2S, 0.9772)$. Using the points farther removed from the mean has the advantage that inaccuracies in the line drawn to represent the population will be smaller than when using more interior points.

In practice, the sample values should then be plotted (see Section 5.2.7) to decide whether or not the measured values closely approximate the population. If the data provide a reasonable fit to the line, one can assume that the underlying population is the normal distribution and the sample mean and standard deviation are reasonable estimates of the location and scale parameters, respectively. A poor fit would indicate that either the normal distribution is not the population or that the sample statistics are not good estimators of the population parameters, or both.

When using a frequency curve, it is common to discuss the likelihood of events in terms of either the exceedency frequency, the exceedence probability, or a return period. The concept of exceedence frequency was discussed in Chapter 4 with respect to rainfall and in Section 5.1.3 with respect to runoff. The return period (T) is related to the exceedence probability (p) by $p = 1/T$, or $T = 1/p$. Thus an event with an exceedence probability of 0.01 should be expected to occur 1 time in 100. In many cases, a time unit is attached to the return period. For example, if the data represent annual floods at a location, the basic time unit is 1 yr. In such a case, the return period for an event with an exceedence probability of 0.01 would be the 100-yr event (that is, $T = 1/0.01 = 100$); similarly, the 25-yr event has an exceedence probability of 0.04 (that is, $p = 1/25 = 0.04$). It is important to emphasize that two T-yr events will not necessarily occur exactly T-yr apart. Two T-yr events could occur in successive years; then again, two T-yr events may be spaced three times T-yr apart. But on the average the events will be spaced T-yr apart. Thus, in a long period of time, say 10,000-yr, we would expect $10{,}000/T$ such events to occur. In any one 10,000-yr period, we may have more or less occurrences than the mean number of occurrences (for example, $10{,}000/T$).

Estimation with a normal frequency curve. For the normal distribution, estimation may involve either finding a probability corresponding to a specified value of the random variable or finding the value of the random variable for a given probability. Both estimation problems can be made using either a graphical analysis or the mathematical models of Equations 5–2 and 5–3.

In a graphical analysis, estimation involves simply either entering with the probability and finding the corresponding value of the random variable or entering with the value of the random variable and finding the corresponding exceedence probability. In both cases, the fitted line (that is, the population) is used.

Example 5–1

Figure 5–4 shows a frequency histogram for the data of Table 5–2. The sample consists of 58 annual maximum instantaneous discharges, with a mean of 8620 ft^3/sec, a standard deviation of 4128 ft^3/sec, and a standardized skew of 1.14. In spite of the large skew, the normal frequency curve was fitted using the procedure of the preceding section. Figure 5–5 shows the cumulative normal distribution using the sample mean and the standard deviation as estimates of the location and scale parameters. The population line was drawn by plotting $\overline{X} + S = 12{,}748$ at $p = 15.87\%$ and $\overline{X} - S = 4492$ at $p = 84.13\%$, using the upper scale for the probabilities. The data were plotted using the Weibull plotting position formula (Equation 5–5a). It is evident that the data do not provide a reasonable fit to the population; the data show a significant skew with an especially poor fit to the tails of the distribution (that is, high and low exceedence probabilities). Because of the poor fit, the line shown in Figure 5–5 should not be used to make probability statements about the future occurrences of floods; for example, the normal distribution (that is, the line) would suggest a 1% chance flood magnitude of slightly more than 18,000 ft^3/sec. However, if a line was drawn subjectively through the trend of the points, the 1% chance flood would be considerably larger, say about 23,000 ft^3/sec.

The 100-yr flood is estimated by entering with a probability of 0.01 and finding the corresponding flood magnitude. We could also estimate probabilities. For example, if a levee system at this site would be overtopped at a magnitude of 16,000 ft^3/sec, the curve indicates a corresponding probability of about 4%, which is the 25-yr flood.

To estimate either probabilities or flood magnitudes using the mathematical model, Equation 5–2 becomes $X = \overline{X} + zS$ because the frequency factor K of Equation 5–2 becomes the standard normal deviate z for a normal distribution, where values of z are from Table 2–2. To find the value of the random variable X, values for $\overline{X}$ and S must be known and the value of z obtained from Table 2–2 for any probability. To find the probability for a given value of the ran-

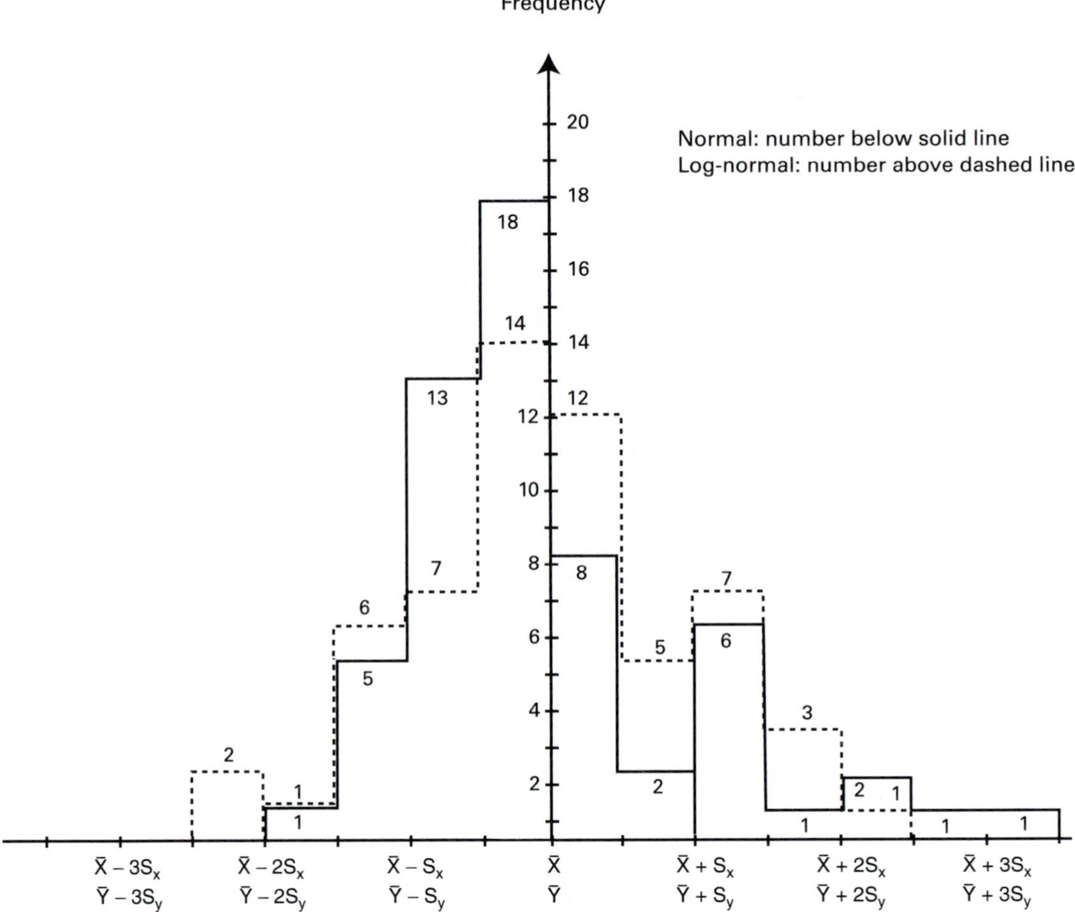

FIGURE 5–4 Frequency histograms of the annual maximum flood series (solid line) and logarithms (dashed line) based on means ($\bar{X}$ and, for logarithms, $\bar{Y}$) and standard deviations (S_x and, for logarithms, S_y): Piscataquis River near Dover-Foxcroft, Maine.

dom variable X, Equation 5–3 is used to solve for the frequency factor z (which is K in Equation 5–3); the probability is then obtained from Table 2–2 using the computed value of z. For example, the value of z from Table 2–2 for a probability of 0.01 (that is, the 100-yr event) is 2.327; thus the flood magnitude for the case of Example 5–1 is

$$X = \bar{X} + zS = 8620 + 2.327(4128) = 18{,}226 \text{ ft}^3/\text{sec}$$

which agrees with the value obtained from the graphical analysis. For a discharge of 16,000 ft^3/sec, the corresponding z value is

$$z = \frac{X - \bar{X}}{S} = \frac{16{,}000 - 8{,}620}{4{,}128} = 1.788$$

From Table 2–2 the probability is 0.0377, which agrees with the graphical estimate of about 4%.

TABLE 5–2 Frequency Analysis of Peak Discharge Data: Piscataquis River

Rank	Weibull Probability	Random Variable	Logarithm of Variable
1	.0169	21500	4.332438
2	.0339	19300	4.285557
3	.0508	17400	4.240549
4	.0678	17400	4.240549
5	.0847	15200	4.181844
6	.1017	14600	4.164353
7	.1186	13700	4.136721
8	.1356	13500	4.130334
9	.1525	13300	4.123852
10	.1695	13200	4.120574
11	.1864	12900	4.110590
12	.2034	11600	4.064458
13	.2203	11100	4.045323
14	.2373	10400	4.017034
15	.2542	10400	4.017034
16	.2712	10100	4.004322
17	.2881	9640	3.984077
18	.3051	9560	3.980458
19	.3220	9310	3.968950
20	.3390	8850	3.946943
21	.3559	8690	3.939020
22	.3729	8600	3.934499
23	.3898	8350	3.921686
24	.4068	8110	3.909021
25	.4237	8040	3.905256
26	.4407	8040	3.905256
27	.4576	8040	3.905256
28	.4746	8040	3.905256
29	.4915	7780	3.890980
30	.5085	7600	3.880814
31	.5254	7420	3.870404
32	.5424	7380	3.868056
33	.5593	7190	3.856729
34	.5763	7190	3.856729
35	.5932	7130	3.853090
36	.6102	6970	3.843233
37	.6271	6930	3.840733
38	.6441	6870	3.836957
39	.6610	6750	3.829304
40	.6780	6350	3.802774
41	.6949	6240	3.795185
42	.7119	6200	3.792392
43	.7288	6100	3.785330
44	.7458	5960	3.775246
45	.7627	5590	3.747412
46	.7797	5300	3.724276
47	.7966	5250	3.720159

(continued)

TABLE 5–2 Frequency Analysis of Peak Discharge Data: Piscataquis River (*Continued*)

Rank	Weibull Probability	Random Variable	Logarithm of Variable
48	.8136	5150	3.711807
49	.8305	5140	3.710963
50	.8475	4710	3.673021
51	.8644	4680	3.670246
52	.8814	4570	3.659916
53	.8983	4110	3.613842
54	.9153	4010	3.603144
55	.9322	4010	3.603144
56	.9492	3100	3.491362
57	.9661	2990	3.475671
58	.9831	2410	3.382017
Mean		8620	3.889416
Standard deviation		4128	0.203080
Standardized skew		1.14	–0.066

5.3.2 The Log-Normal Distribution

When a poor fit to observed data is obtained, a different distribution function should be considered. For example, when the data demonstrate a concave-upward curve, as in Figure 5–5, it is reasonable to try a log-normal distribution or an extreme-value distribution. In some cases, it may be preferable to fit with a distribution that requires an estimate of the skew coefficient, such as a log-Pearson Type III distribution. However, sample estimates of the skew coefficient may be inaccurate for small samples, and thus they should be used with caution.

The same procedure that is used for fitting the normal distribution can be used to fit the log-normal distribution. The underlying population is assumed to be log normal. The data must first be transformed to logarithms, $Y = \log X$. This transformation creates a new random variable Y. The mean and standard deviation of the logarithms are computed and used as the parameters of the population; it is important to recognize that the logarithm of the mean does not equal the mean of the logarithms, which is also true for the standard deviation. Thus one should not use the logarithms of the mean and standard deviation as parameters; the mean and standard deviation of the logarithms should be computed and used as the parameters. Either natural or base 10 logarithms may be used, although the latter is more common in hydrology. The population line is defined by plotting the straight line on arithmetic probability paper between the points $(\bar{Y} + S_y, 0.1587)$ and $(\bar{Y} - S_y, 0.8413)$, where $\bar{Y}$ and S_y are the mean and standard deviation of the logarithms, respectively. In plotting the data, either the logarithms can be plotted on an arithmetic scale or the untransformed data can be plotted on a logarithmic scale.

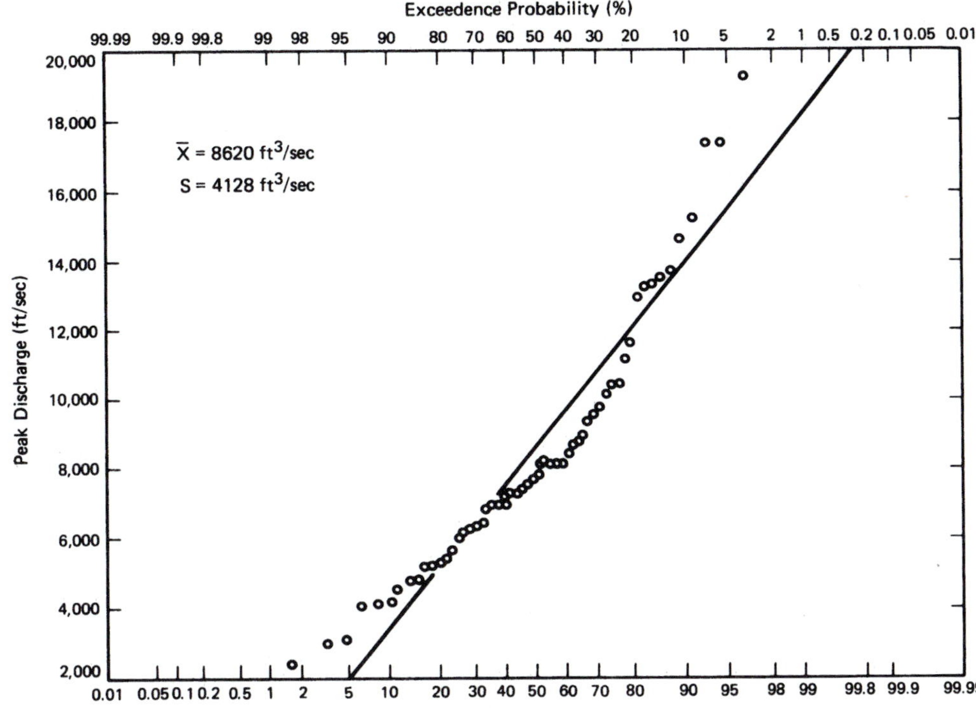

FIGURE 5–5 Piscataquis River near Dover-Foxcroft, Maine.

When using a frequency curve for a log-normal distribution, the value of the random variable Y and the moments of the logarithms ($\overline{Y}$ and S_y) are related by the equation

$$Y = \overline{Y} + zS_y \tag{5–6}$$

in which z is the value of the standardized normal variate; values of z and the corresponding probabilities can be taken from Table 2–2. Equation 5–6 can be used to estimate either flood magnitudes for a given exceedence probability or an exceedence probability for a specific discharge. To find a discharge for a specific exceedence probability the standard normal deviate z is obtained from Table 2–2 and used in Equation 5–6 to compute the discharge. To find the exceedence probability for a given discharge Y, Equation 5–6 is rearranged by solving for z. With values of Y, $\overline{Y}$, and S_y, a value of z is computed and used with Table 2–2 to compute the exceedence probability. Of course, the same values of both Y and the probability can be obtained directly from the frequency curve.

Example 5–2

The peak discharge data for the Piscataquis River were transformed by taking the logarithm of each of the 58 values. The moments of the logarithms are as follows: $\overline{Y} = 3.8894$, $S_y = 0.20308$, and $g = -0.07$. A histogram of the logarithms is shown in Figure 5–6. In comparison to the histogram of Figure 5–4, the logarithms of the sample data are less skewed. While a skew of -0.07 would usually be rounded to -0.1, this is sufficiently close to 0 such that the discharges can be

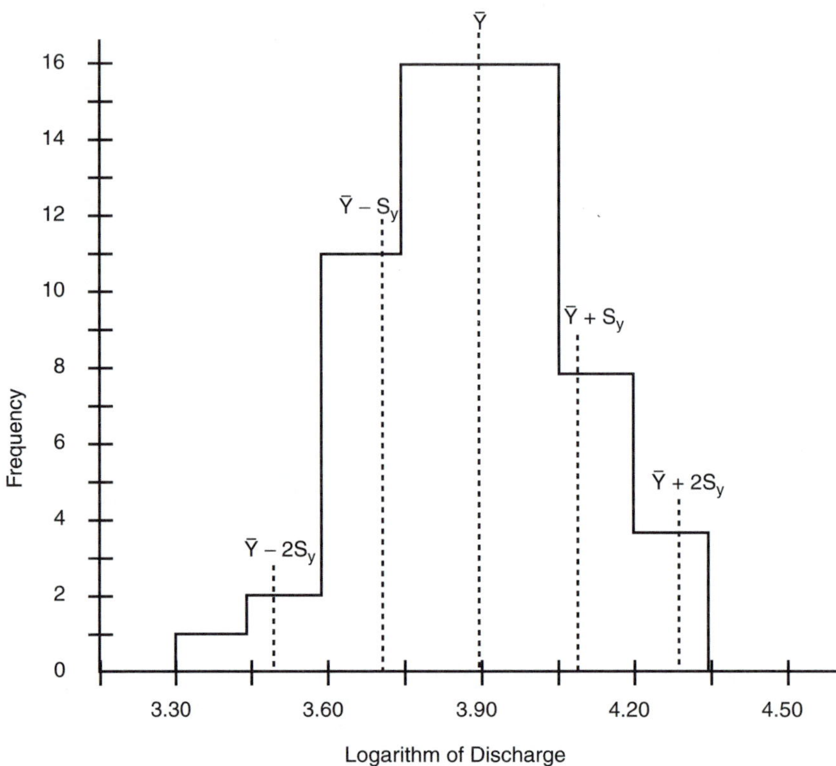

FIGURE 5–6 Histogram of logarithms of annual maximum series: Piscataquis River.

represented with a log-normal distribution. The frequency curve is shown in Figure 5–7. To plot the log-normal population curve, the following two points were used: $\overline{Y} - S_y = 3.686$ at $p = 84.13\%$ and $\overline{Y} + S_y = 4.092$ at $p = 15.87\%$. The data points were plotted on Figure 5–7 using the Weibull formula and show a much closer agreement with the population line in comparison to the points for the normal distribution in Figure 5–5. Therefore, it is reasonable to assume that the measured peak discharge rates can be represented by a log-normal distribution and that the future flood behavior of the watershed can be described statistically using a log-normal distribution.

If one were interested in the probability that a flood of 20,000 ft^3/sec would be exceeded in any one year, the logarithm of 20,000, which is 4.301, would be entered on the discharge axis and followed to the assumed population line. Reading the exceedence probability corresponding to that point on the frequency curve, one concludes that there is about a 1.7% chance of a flood of 20,000 ft^3/sec being exceeded in any one year. It can also be interpreted that over the span of 1000 years, one could expect the flood discharge to exceed 20,000 ft^3/sec in 17 of those years; it is important to understand that this is an average value. In any period of 1000 years a value of 20,000 ft^3/sec may be exceeded more or less than 17, but on the average there would be 17 exceedences in a period of 1000 years.

The probability of a discharge of 20,000 ft^3/sec can also be estimated mathematically. The standard normal deviate is

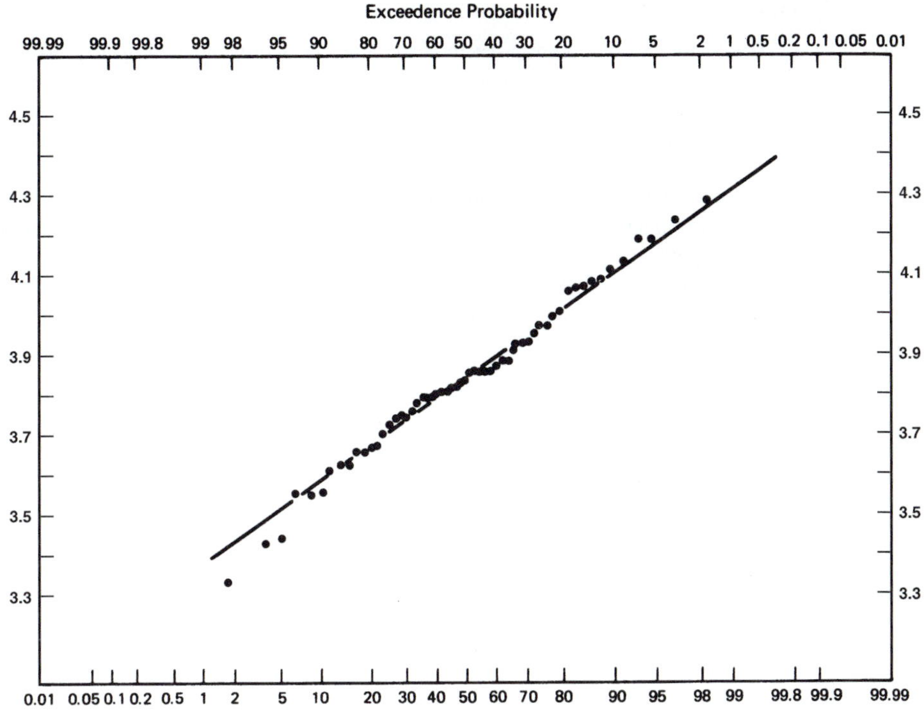

FIGURE 5–7 Frequency curve for the logarithms of the annual maximum discharge.

$$z = \frac{\log(20000) - \overline{Y}}{S_y} = \frac{4.301 - 3.8894}{0.20308} = 2.027$$

The value of z is entered into Table 2–2, which yields a probability of 0.9786. Since the exceedence probability is of interest, this is subtracted from 1, which yields a value of 0.0214. This corresponds to a 47-yr flood. The difference between the mathematical estimate of 2.1% and the graphical estimate of 1.7% is due to the error in the graph. The computed value of 2.1% should be used.

The frequency curve can also be used to estimate flood magnitudes for selected probabilities. The flood magnitude is found by entering the figure with the exceedence probability, moving vertically to the frequency curve, and finally moving horizontally to the flood magnitude. For example, the 100-yr flood for the Piscataquis River can be found by entering with an exceedence probability of 1%, moving to the curve of Figure 5–7 and then to the ordinate, which indicates a logarithm of about 4.3586 or a discharge of 22,800 ft^3/sec. Discharges for other exceedence probabilities can be found in the same way or by using the mathematical model of Equation 5–2.

In addition to the graphical estimate, Equation 5–6 can be used to obtain a more exact estimate. For an exceedence probability of 0.01, a z value of 2.327 is obtained from Table 2–2. Thus, the logarithm is

$$Y = \overline{Y} + zS_y = 3.8894 + 2.327\,(0.20308) = 4.3620$$

Taking the anti-logarithm yields a discharge of 23013 ft^3/sec.

Example 5–3

The histogram of Figure 5–8a indicates that the sediment yield data of Table 5–3 are from a highly skewed probability density function. Figure 5–8b shows the histogram of the logarithms. The shape of the histogram suggests that the logarithms could be normally distributed. The data were subjected to a frequency analysis for a log-normal population. The mean, standard deviation, and standardized skew of the logarithms are –0.5506, 0.6154, and –0.0334, respectively. The data and logarithms in ranked order and Weibull plotting positions are given in Table 5–3, and the frequency curve is shown in Figure 5–9. As expected, the points follow the straight line for a log-normal population.

Estimates of sediment yield or probabilities can be obtained graphically from Figure 5–9 or mathematically using Equations 5–2 and 5–3. For example, the probability of a sediment

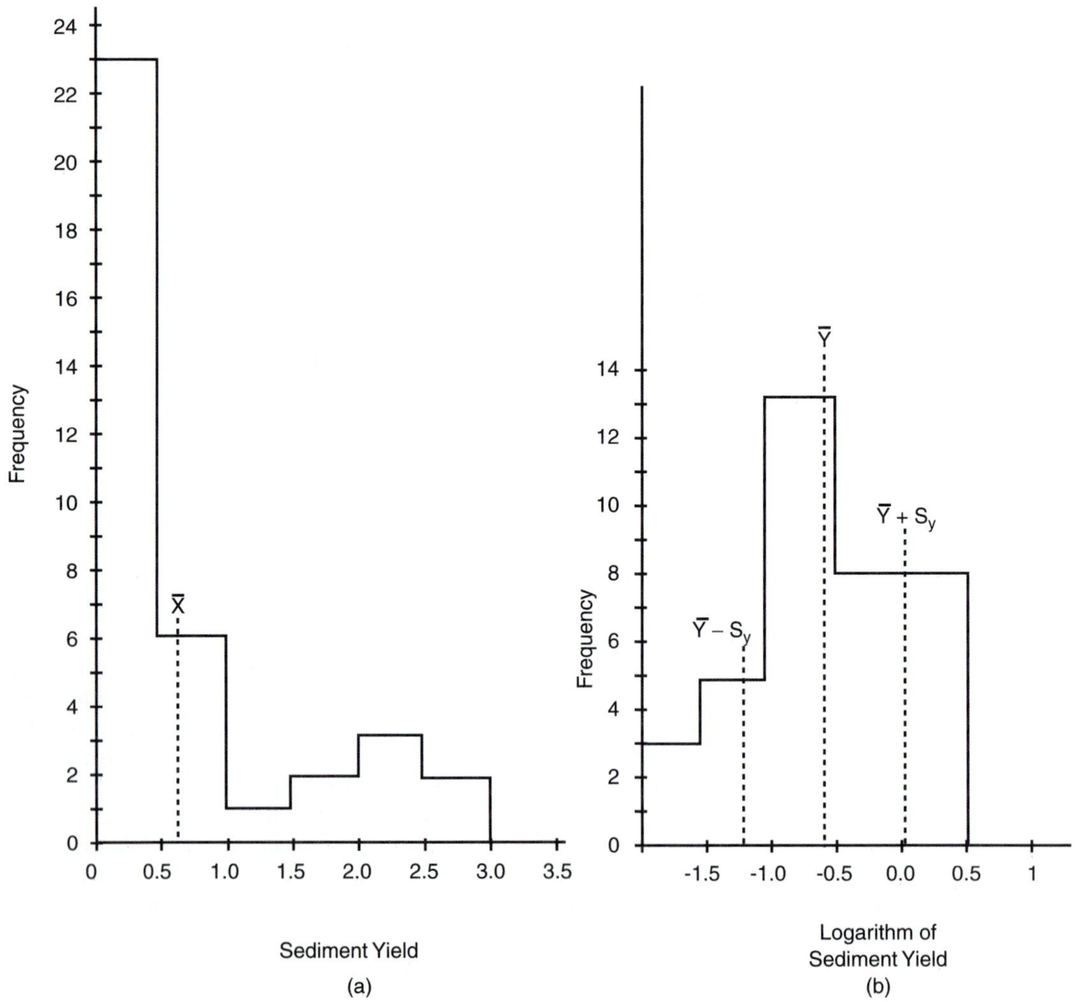

FIGURE 5–8 Histograms of (a) sediment yield and (b) logarithms of the sediment yield.

TABLE 5–3 Frequency Analysis of Sediment Yield Data

Rank	Weibull Probability	Random Variable	Logarithm of Variable
1	.0263	2.67	.426511
2	.0526	2.65	.423246
3	.0789	2.37	.374748
4	.1053	2.31	.363612
5	.1316	2.20	.342423
6	.1579	1.65	.217484
7	.1842	1.55	.190332
8	.2105	1.42	.152288
9	.2368	.99	−.004365
10	.2632	.69	−.161151
11	.2895	.66	−.180456
12	.3158	.64	−.193820
13	.3421	.61	−.214670
14	.3684	.51	−.292430
15	.3947	.37	−.431798
16	.4211	.35	−.455932
17	.4474	.25	−.602060
18	.4737	.22	−.657577
19	.5000	.21	−.677781
20	.5263	.21	−.677781
21	.5526	.20	−.698970
22	.5789	.18	−.744727
23	.6053	.17	−.769551
24	.6316	.17	−.769551
25	.6579	.16	−.795880
26	.6842	.15	−.823909
27	.7105	.14	−.853872
28	.7368	.14	−.853872
29	.7632	.12	−.920819
30	.7895	.09	−1.045758
31	.8158	.076	−1.119186
32	.8421	.07	−1.154902
33	.8684	.04	−1.397940
34	.8947	.04	−1.443698
35	.9211	.036	−1.522879
36	.9474	.02	−1.698970
37	.9737	.02	−1.698970
Mean		0.6579	−0.550612
Standard deviation		0.8286	0.615371
Skew		1.4489	−0.033371

Source: Flakman, 1972.

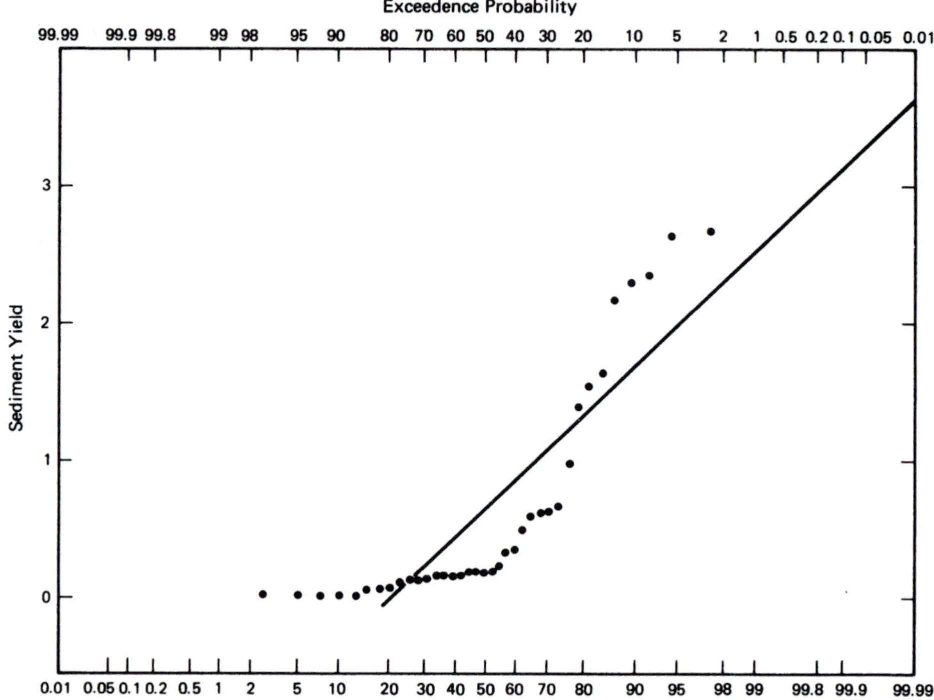

FIGURE 5–9a Frequency analysis of sediment yield data for a normal population.

yield greater than 2 is estimated by computing the value of the standard normal deviate and finding the corresponding nonexceedence probability from Table 2–2:

$$z = \frac{\log(2) - \overline{Y}}{S_y} = \frac{0.3010 - (-0.5506)}{0.6154} = 1.3839$$

From Table 2–2, the corresponding probability in 0.9168. Therefore, the exceedence probability is 0.0832. The graphical estimate agrees with the computed probability.

The magnitude corresponding to an exceedence probability of 2% is computed using a z of 2.054 (from Table 2–2):

$$Y = \overline{Y} + zS_y = -0.5506 + 2.054\,(0.6154) = 0.7134$$

Therefore, the sediment yield is $10^{0.7134} = 5.17$. This is larger than any of the measured values. The histogram of Figure 5–8b indicates there were no large values in the upper tail of the sample distribution of the logarithms. Thus, the computed 50-yr value is beyond the upper ordinate of the measured frequencies, which would be expected given the small sample size.

Example 5–4

As indicated, frequency analysis can be used in the development of intensity-duration-frequency curves. A separate frequency analysis would be made for each duration and the results combined to form an *IDF* curve such as the one in Figure 4–4. In practice, the rainfall records would be an-

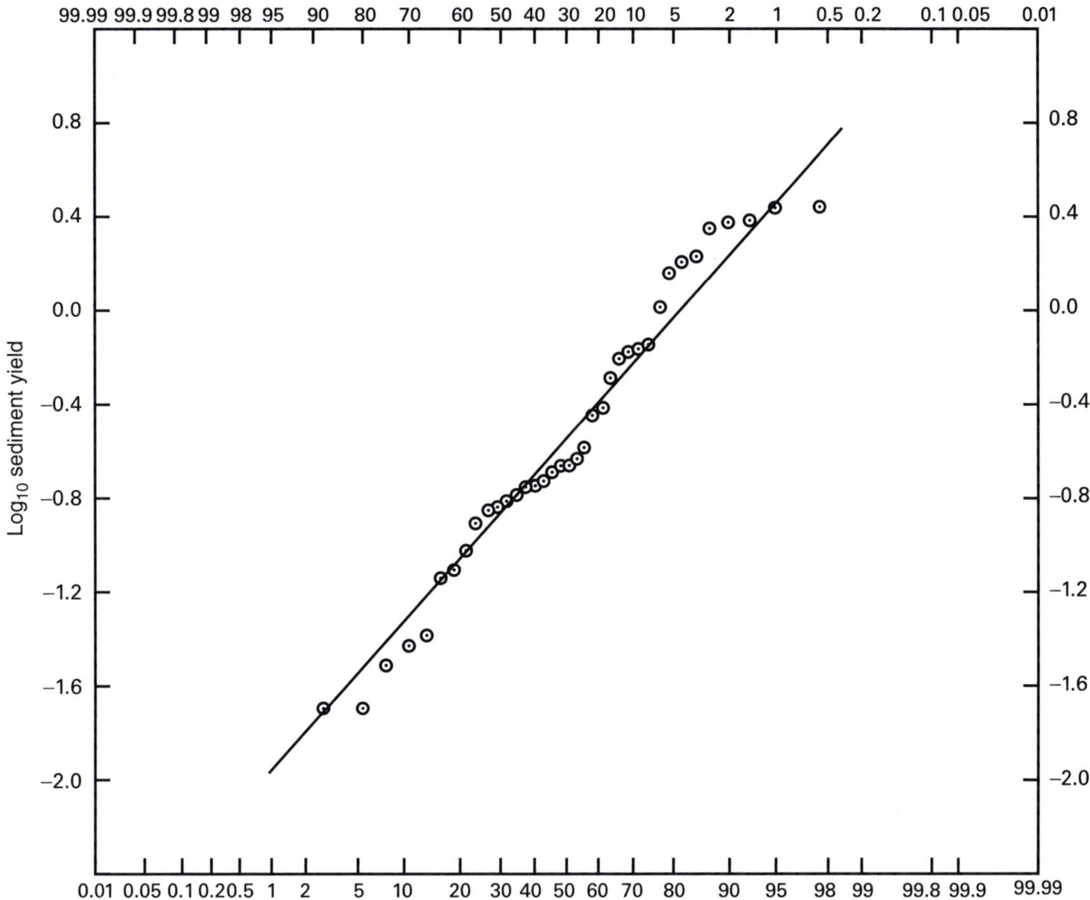

FIGURE 5–9b Frequency analysis of sediment yield data for a log-normal population.

alyzed and the largest intensity in each year for the specified duration would be obtained from each year of record. For a record of n years, the data base would then consist of n intensities that are the highest intensity for a given duration during any storm event in the year. The n values would then be subjected to a frequency analysis.

The data in Table 5–4 are the annual peak intensities for 2-hr duration storms in a 19-yr record. Past analyses have suggested that peak intensities can be represented with a log-normal distribution; thus each intensity was transformed using base 10 logarithms, with the resulting values given in Table 5–4. The rank and plotting probability are also given in Table 5–4. The mean and standard deviation of the logarithms were computed as $\overline{Y} = -0.081256$ and $S_y = 0.14698$, respectively. Thus the assumed model for the population is

$$Y = \overline{Y} + zS_y = -0.081256 + 0.14698z \tag{5–7}$$

The data are plotted in Figure 5–10 and suggest that the log-normal distribution is a reasonable representation of the population.

TABLE 5–4 Log-Normal Frequency Analysis of Annual Peak 2-Hour Rainfall Intensities

Year	i(in./hr)	log i	Rank	Probability	Return Period	Plotting Position	Normal deviate z	y	$x = 10^y$
1966	1.04	0.0170	5	0.25	2	0.50	0	−0.08126	0.83
1967	0.72	−0.1427	12	0.60	5	0.20	0.8418	0.04247	1.10
1968	0.58	−0.2366	17	0.85	10	0.10	1.2817	0.10713	1.28
1969	1.41	0.1492	2	0.10	25	0.04	1.7511	0.17612	1.50
1970	0.70	−0.1549	14	0.70	50	0.02	2.0540	0.22064	1.66
1971	1.00	0.0000	6	0.30	100	0.01	2.3267	0.26072	1.82
1972	0.76	−0.1192	11	0.55					
1973	1.26	0.1004	3	0.15					
1974	0.42	−0.3768	19	0.95					
1975	0.95	−0.0223	8	0.40					
1976	0.70	−0.1549	15	0.75					
1977	1.43	0.1553	1	0.05					
1978	0.71	−0.1000	13	0.65					
1979	0.46	−0.3372	18	0.90					
1980	0.99	−0.0044	7	0.35					
1981	0.81	−0.0915	10	0.50					
1982	1.10	0.0414	4	0.20					
1983	0.68	−0.1675	16	0.80					
1984	0.89	−0.0506	9	0.45					

The assumed model of Equation 5–7 was used to compute the 2-hr rainfall intensities for return periods of 2, 5, 10, 25, 50, and 100 years, which are also given in Table 5–4. The values can also be obtained by entering Figure 5–10 with the corresponding exceedence probability and finding the corresponding logarithm. For example, for the 100-yr intensity the graph is entered with an exceedence probability of 1% and the population line yields a logarithm of about 0.261, which agrees closely with the value obtained with Equation 5–7.

The population line could also be used to find the probability of a specified intensity. For example, to find the probability of a rainfall depth of 5 in. in 2 hr, or an intensity of 2.5 in./hr, the logarithm of 2.5 is 0.3979. Entering Figure 5–10 with an ordinate of 0.398 yields an exceedence probability of 0.056% or 0.00056, which corresponds to a return period of 1800 years.

5.3.3 The Log-Pearson Type III Distribution

Normal and log-normal frequency analyses were introduced because they are easy to understand and have a variety of uses in hydrology. However, the statistical distribution most commonly used in hydrology in the United States of America is the log-Pearson Type III (LP3) distribution because it was recommended by the U.S. Water Resources Council in Bulletin 17B. The Pearson Type III distribution is a probability density function. It is widely accepted because it is easy to apply when the parameters are estimated using the method of moments and because it usually provides a good fit to measured data. An LP3 analysis requires a logarithmic transformation of the data; specifically, the common logarithms of the data are used as the variate, and the Pearson Type III distribution is used as the PDF. This results in an LP3 analysis.

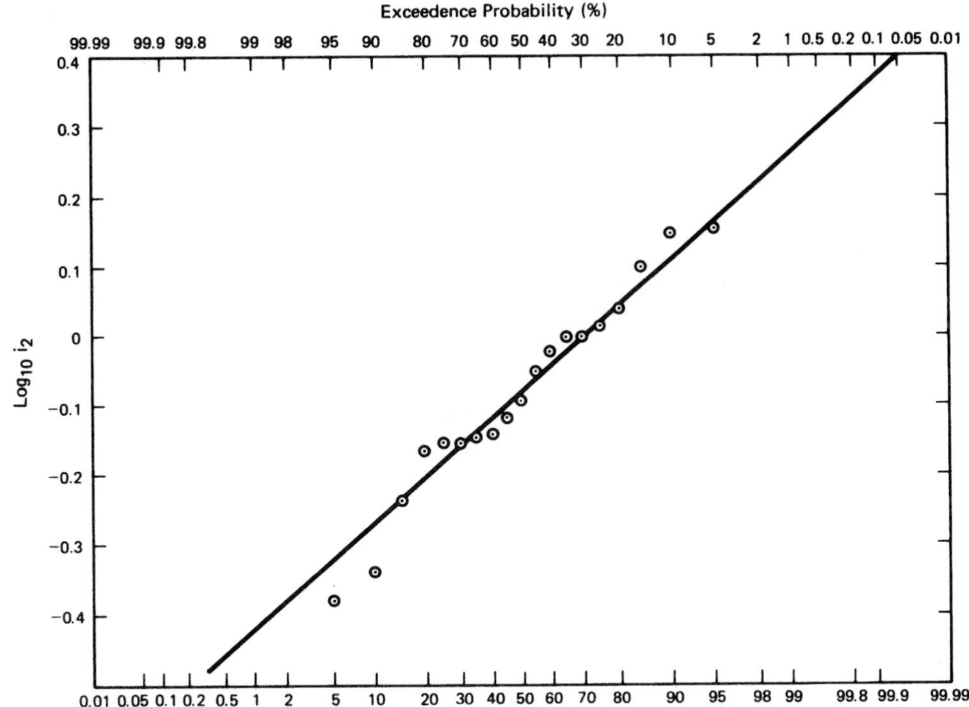

FIGURE 5–10 Log-normal frequency analysis of 2-hr rainfall intensity (i_2).

While LP3 analyses have been made for most stream gage sites in the United States and can be obtained from the U.S. Geological Survey, a brief description of the analysis procedure will be given here. While the method of analysis presented here will follow the procedure recommended by the Water Resources Council in Bulletin 17B, when performing an analysis, Bulletin 17B should be consulted because a number of options and adjustments that are included in Bulletin 17B are not discussed here. The intent here is only to provide sufficient detail so that a basic frequency analysis can be made and properly interpreted.

Bulletin 17B provides details for the analysis of three types of data: a systematic gage record, regional data, and historic information for the site. The systematic record consists of the annual maximum flood record. It is not necessary for the record to be a continuous record as long as the missing part of the record is not the result of flood experience, such as the destruction of the stream gage during a large flood. Regional information includes a generalized skew coefficient, a weighting procedure for handling independent estimates, and a means of correlating a short systematic record with a longer systematic record from a nearby stream gaging station. Historic evidence, such as high-water marks and newspaper accounts of flooding that occurred prior to the installation of the gage, can be used to augment information at the site.

The procedure for fitting an LP3 curve with a measured systematic record is similar to the procedure used for the normal and log-normal analyses described previously. The Bulletin 17B procedure for analyzing a systematic record is based on a method-of-moments analysis and is as follows:

1. Create a series that consists of the logarithms Y_i of the annual maximum flood series.
2. Using Equations 5–4 compute the sample mean, $\overline{Y}$, standard deviation, S_y, and standardized skew, g_s, of the logarithms of Step 1.
3. For selected values of the exceedence probability (p) obtain values of the standardized variate K from Appendix 5–1 (round the skew to the nearest tenth).
4. Determine the values of the LP3 curve for the exceedence probabilities selected in Step 3 using the equation

$$Y = \overline{Y} + KS_y \tag{5–8}$$

   in which Y is the logarithmic value of the LP3 curve.
5. Take the antilogarithm of the Y_j values and use them to plot the LP3 frequency curve.

Having determined the LP3 population curve, the data can be plotted to determine the adequacy of the curve. Confidence limits can also be placed on the curve; the procedure for computing confidence intervals is discussed in Bulletin 17B.

In Step 3 it is necessary to select two or more points to compute and plot the LP3 curve. If the absolute value of the skew is small, the line will be nearly straight and only a few points are necessary to draw the line accurately. When the absolute value of the skew is large, more points must be used because of the greater curvature of the LP3 curve. When selecting exceedence probabilities, it is common to include 0.5, 0.2, 0.1, 0.04, 0.02, 0.01, and 0.002 because these correspond to return periods that are usually of interest.

Example 5–5

Data for the Back Creek near Jones Springs, West Virginia (USGS gaging station 01-6140), is given in Table 5–5. Based on the 38 years of record (1929–1931, 1939–1973), the mean, standard deviation, and skew of the common logarithms are 3.722, 0.2804, and –0.731, respectively; the skew will be rounded to –0.7. Table 5–6 shows the K values of Equation 5–8 for selected values of the exceedence probability (p); these values were obtained from Appendix 5–1 using p and the sample skew of –0.7. Equation 5–8 was used to compute the logarithms of the LP3 discharges for the selected exceedence probabilities (Table 5–6).

The logarithms of the discharges were then plotted versus the exceedence probabilities, as shown in Figure 5–11. The rank of each event is also shown in Table 5–5 and was used to compute the exceedence probability using the Weibull plotting position formula (Equation 5–5a). The logarithms of the measured data were plotted versus the exceedence probability (Figure 5–11). The data show a reasonable fit to the frequency curve, although the fit is not especially good for the few highest and the lowest measured discharges; the points with exceedence probabilities between 80 and 92% suggest a poor fit. Given that we can only make subjective assessments of the goodness of fit, the computed frequency curve appears to be a reasonable estimate of the population.

The curve of Figure 5–11 could be used to estimate either flood discharges for selected exceedence probabilities or exceedence probabilities for selected discharges. For example, the 100-yr flood ($p = 0.01$) equals 16,992 ft^3/sec (that is, $\log Q = 4.2285$ from Table 5–6). The exceedence probability for a discharge of 10,000 ft^3/sec (that is, $\log Q = 4$) is approximately 0.155 from Figure 5–11; this would correspond to the 6-yr event.

TABLE 5–5 Annual Maximum Floods for Back Creek

Year	Q	log Q	Rank	p
1929	8,750	3.9420	7	0.179
1930	15,500	4.1903	3	0.077
1931	4,060	3.6085	27	0.692
1939	6,300	3.7993	14	0.359
1940	3,130	3.4955	35	0.897
1941	4,160	3.6191	26	0.667
1942	6,700	3.8261	12	0.308
1943	22,400	4.3502	1	0.026
1944	3,880	3.5888	30	0.769
1945	8,050	3.9058	9	0.231
1946	4,020	3.6042	28	0.718
1947	1,600	3.2041	37	0.949
1948	4,460	3.6493	23	0.590
1949	4,230	3.6263	25	0.641
1950	3,010	3.4786	36	0.923
1951	9,150	3.9614	6	0.154
1952	5,100	3.7076	19	0.487
1953	9,820	3.9921	5	0.128
1954	6,200	3.7924	15	0.385
1955	10,700	4.0294	4	0.103
1956	3,880	3.5888	31	0.795
1957	3,420	3.5340	33	0.846
1958	3,240	3.5105	34	0.872
1959	6,800	3.8325	11	0.282
1960	3,740	3.5729	32	0.821
1961	4,700	3.6721	20	0.513
1962	4,380	3.6415	24	0.615
1963	5,190	3.7152	18	0.462
1964	3,960	3.5977	29	0.744
1965	5,600	3.7482	16	0.410
1966	4,670	3.6693	21	0.538
1967	7,080	3.8500	10	0.256
1968	4,640	3.6665	22	0.564
1969	536	2.7292	38	0.974
1970	6,680	3.8248	13	0.333
1971	8,360	3.9222	8	0.205
1972	18,700	4.2718	2	0.051
1973	5,210	3.7168	17	0.436

Source: Bulletin 17B, 1982.

5.3.4 Low-Flow Frequency Analysis

Partially-treated wastewater is commonly discharged into streams and rivers where it mixes with the existing flow. Natural processes improve the overall quality of the total flow. During periods of low flow the volume of wastewater may be too large to be safely discharged without reducing the quality of the water below established water-quality standards. When evaluating sites for suitability for a manufacturing or commercial business, it is important to

TABLE 5–6 Computation of Log Pearson Type III Frequency Curve for Back Creek Near Jones Springs, West Virginia

p	K	$\overline{Y} + KS_y$	Q(ft^3/sec)
0.99	−2.82359	2.9303	852
0.90	−1.33294	3.3482	2,230
0.70	−0.42851	3.6018	3,998
0.50	0.11578	3.7545	5,682
0.20	0.85703	3.9623	9,169
0.10	1.18347	4.0538	11,320
0.04	1.48852	4.1394	13,784
0.02	1.66325	4.1884	15,430
0.01	1.80621	4.2285	16,922

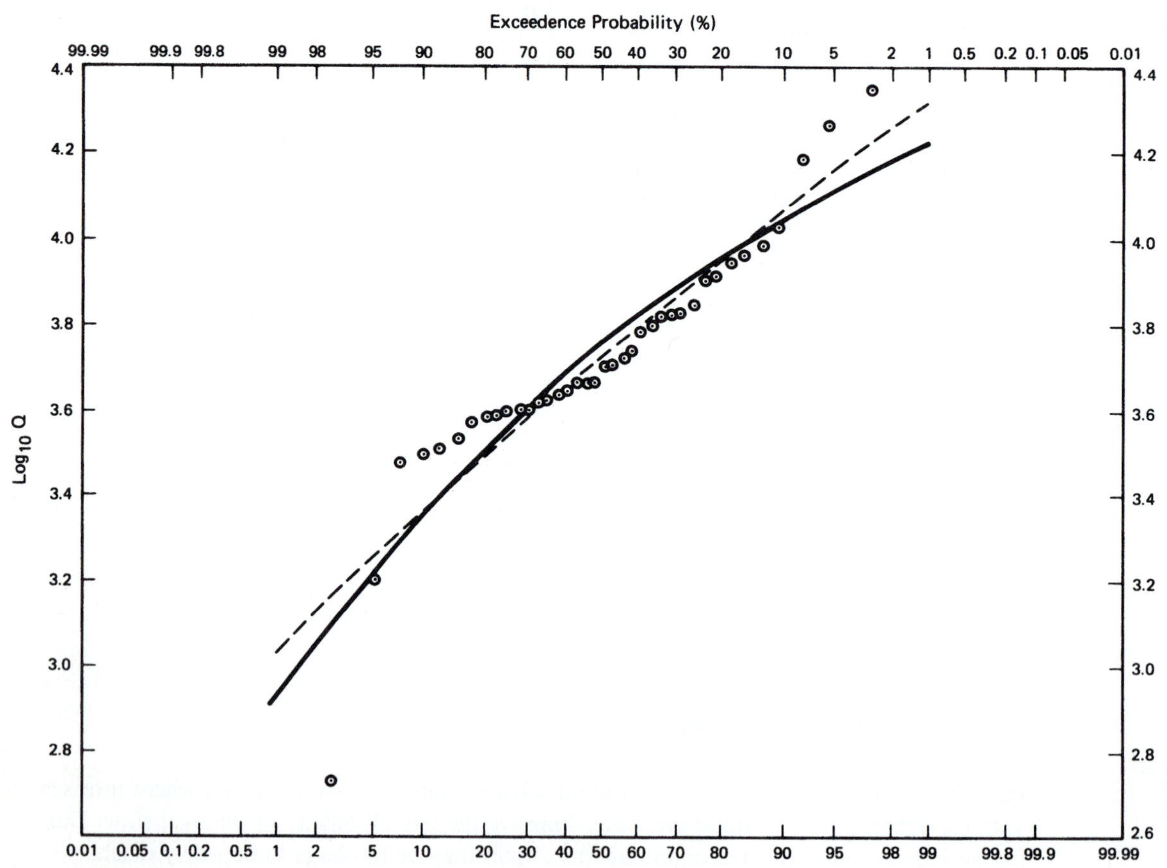

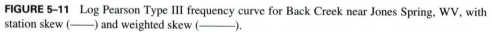

FIGURE 5–11 Log Pearson Type III frequency curve for Back Creek near Jones Spring, WV, with station skew (——) and weighted skew (———).

assess the probability that the stream will almost always have sufficient flow to meet the need for discharging wastewater. Such probabilities are estimated using a low-flow frequency analysis at the site.

While instantaneous maximum discharges are used for flood frequency analyses, low-flow frequency analyses usually specify a flow duration (for example, 7-day). The instantaneous discharge is used with high flows because damage often occurs even if the site is inundated only for a very short period of time. This may not be true for low flows because high pollution concentrations over very short periods of time may not be damaging to the aquatic life of the stream. Thus, the duration, such as seven days or one month, is specified in establishing the policy.

One difference between low-flow and flood frequency analyses is that the data for low-flow analyses consist of the annual events that have the lowest average flow of the required duration D during each water year of record. Thus, the records of flow for each water year are evaluated to find the period of D days during which the average flow was the lowest; these annual values are used as the sample data. The record of n years is then evaluated using frequency analysis. The log-normal distribution is commonly selected for low-flow frequency analyses, however an LP3 analysis could be used.

Once the data record has been collected, the procedure for making a low-flow frequency analysis is quite similar to that used for flood frequency analysis. The major differences are listed here:

1. Instead of using the exceedence probability scale, the nonexceedence scale (that is, the scale at the bottom of the probability paper) is used to obtain probabilities. The nonexceedence scale is important because the T-yr event is the value that will not be exceeded.
2. The data are ranked from low to high, with the smallest sample magnitude associated with a Weibull probability of $1/(n+1)$ and the largest magnitude associated with a probability of $n/(n+1)$; any other plotting position formula could be used in place of the Weibull. However, plotting position probabilities are nonexceedence probabilities.
3. The mathematical model for predicting magnitudes remains

$$Y = \overline{Y} + zS_y \tag{5–9}$$

but since nonexceedence is of interest, the T-yr values are from the lower tail of the distribution; thus, values obtained from a normal distribution table should be from the lower tail (that is, negative values of z).

In obtaining estimates from either the graph or Equation 5–9, the low magnitudes are the extreme events.

Example 5–6

Table 5–7 includes the 7-day, low-flow series for a 19-yr record. The data were subjected to a log-normal, low-flow frequency analysis, with a mean and standard deviation of the logarithms equal to 1.4028 and 0.1181, respectively. The population curve is plotted using two points: $\overline{Y} - S_y = 1.2847$ plotted at a nonexceedence probability of 0.1587 and $\overline{Y} + S_y = 1.5209$ plotted at

TABLE 5–7 Data for Example 5–6

Rank	Plotting Probability	(ft**3/s)	Discharge Logarithm
1	.05	14.8	1.17026
2	.10	17.4	1.24055
3	.15	20.0	1.30103
4	.20	20.2	1.30535
5	.25	20.4	1.30963
6	.30	20.7	1.31597
7	.35	21.4	1.33041
8	.40	24.0	1.38021
9	.45	25.7	1.40993
10	.50	26.0	1.41497
11	.55	26.3	1.41996
12	.60	26.6	1.42488
13	.65	27.5	1.43933
14	.70	28.2	1.45025
15	.75	29.5	1.46982
16	.80	31.3	1.49554
17	.85	34.7	1.54033
18	.90	40.7	1.60959
19	.95	42.2	1.62531

a nonexceedence probability of 0.8413. The data are also plotted in Figure 5–12, with the Weibull equation used to compute the plotting position probabilities. The population curve agrees with the measured data, which suggests that the log-normal distribution is a good representation of the population.

Estimates of 7-day low flows can be obtained from the frequency plot of Figure 5–12 or from the mathematical model

$$\hat{Y} = 1.4028 + 0.1181\, z$$

in which z is the standard normal deviate. The following gives the estimated 7-day low flow ($\hat{Q}_7$) for selected exceedence frequencies (T):

T (yrs)	z	$\hat{Y}$	$\hat{Q}_7$
2	0.0000	1.403	25.3
5	−0.8418	1.303	20.1
10	−1.2817	1.251	17.8
25	−1.7511	1.196	15.7
50	−2.0540	1.160	14.5
100	−2.3267	1.128	13.4

It is evident that, on the average, once every 50 years, the lowest 7-day average flow will fall to 14.5 ft^3/sec.

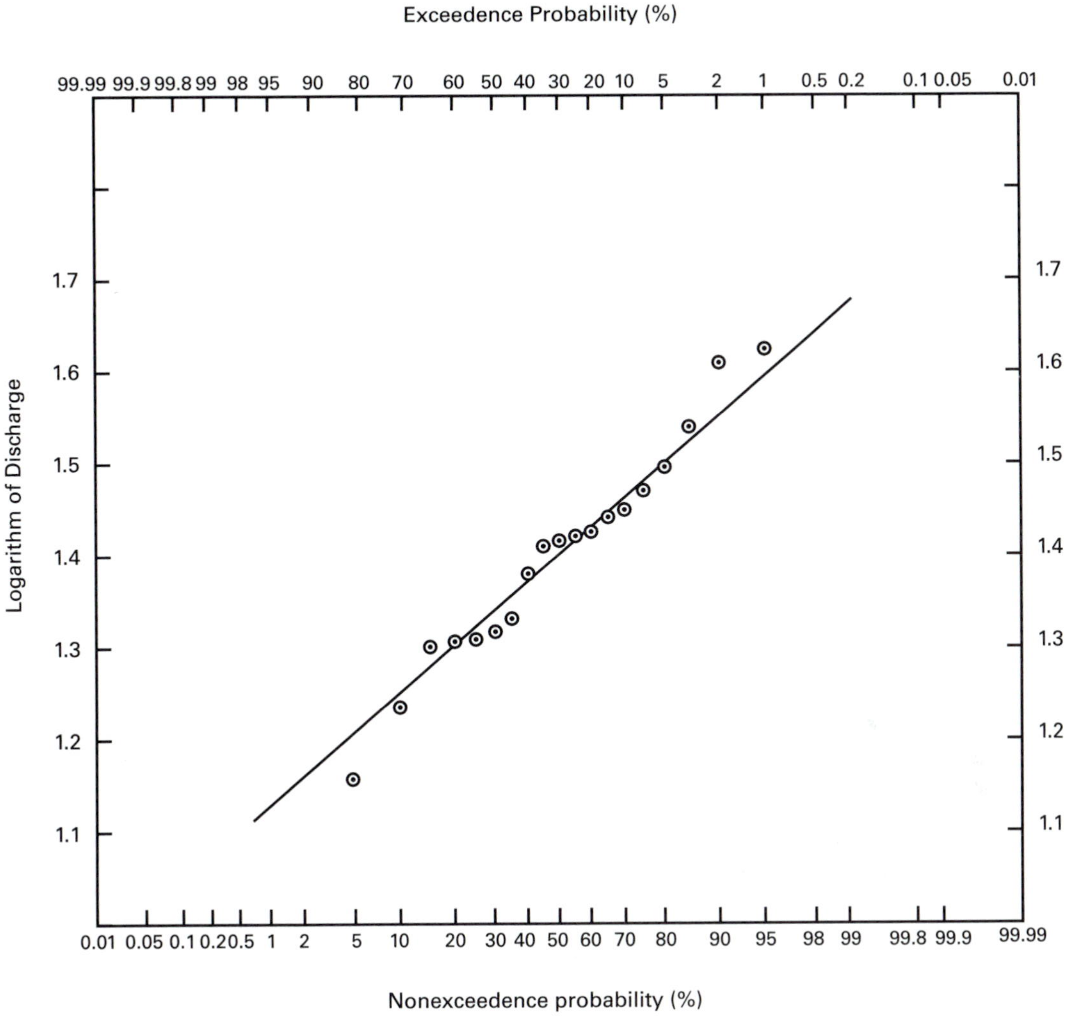

FIGURE 5–12 Low-flow frequency curve: Example 5–6.

5.4 ADJUSTMENTS OF THE FREQUENCY CURVE

The methods of Section 5.3 provide a basic frequency curve. The accuracy of a computed frequency curve can be improved when other information is available. For small samples, the expected accuracy of a computed station skew is relatively poor; thus, for small samples, accuracy can be improved if a weighted skew coefficient is used in place of the computed station skew. If historic data are available, the accuracy of a computed curve can be improved by incorporating the data into the analysis. Outliers, both high and low, can significantly affect the magnitude of either computed floods or exceedence probabilities. Thus, outliers in a sample should be detected and proper adjustments made in order to attain the best accuracy possible. Finally, frequency analysis assumes that the sample is from a homogeneous population; if it is not, then the data should be adjusted so that it has the characteristics of a homogeneous population. Watershed changes, such as urbanization or deforestation, can cause a flood record to be nonhomogeneous. These topics (standard errors of the moments, weighted skew, historic data, outliers, and watershed change) are discussed in this section.

5.4.1 Standard Errors of the Moments

The accuracy of statistical moments (that is, the mean, variance, and skew) is a function of the sample size, with the accuracy improving as the sample size increases. The standard errors for the mean ($S_{\bar{x}}$), standard deviation (S_s), and standardized skew (S_g) are as follows:

$$S_{\bar{x}} = \frac{S}{\sqrt{n}} \tag{5–10a}$$

$$S_s = S\sqrt{\frac{1 + 0.75\,g^2}{2n}} \tag{5–10b}$$

$$S_g = [10^{A - B\,log_{10}(n/10)}]^{0.5} \tag{5–10c}$$

in which A and B are given by

$$A = \begin{cases} -0.33 + 0.08|g| & if\,|g| \leq 0.90 \\ -0.52 + 0.30|g| & if\,|g| > 0.90 \end{cases} \tag{5–10d}$$

and

$$B = \begin{cases} 0.94 - 0.26|g| & if\,|g| \leq 1.50 \\ 0.55 & if\,|g| > 1.50 \end{cases} \tag{5–10e}$$

in which $|g|$ is the absolute value of the station skew, which is used as an estimate of the population skew; n is the record length. The standard errors are shown in Figure 5–13. Figure 5–13 suggests that for a given sample size, the relative accuracy of the skew is much less than that of both the mean and standard deviation, especially for small sample sizes.

Example 5–7

It is important to demonstrate that even an error of one standard error can cause significant variation in computed discharges. Data for the Back Creek near Jones Springs, West Virginia (USGS gaging station 01-6140), can be used to illustrate the effect of error. The standard errors would be computed from Equations 5–10:

$$S_{\bar{x}} = \frac{0.280}{\sqrt{38}} = 0.04542$$

$$S_s = 0.280\sqrt{\frac{1 + 0.75(-0.731)^2}{2(38)}} = 0.03801$$

$$A = -0.33 + 0.08(0.731) = -0.2715$$

$$B = 0.94 - 0.26(0.731) = 0.7499$$

$$S_g = [10^{-0.2715 - 0.7499\,log\,(38/10)}]^{0.5} = 0.4434\ (use\ 0.4)$$

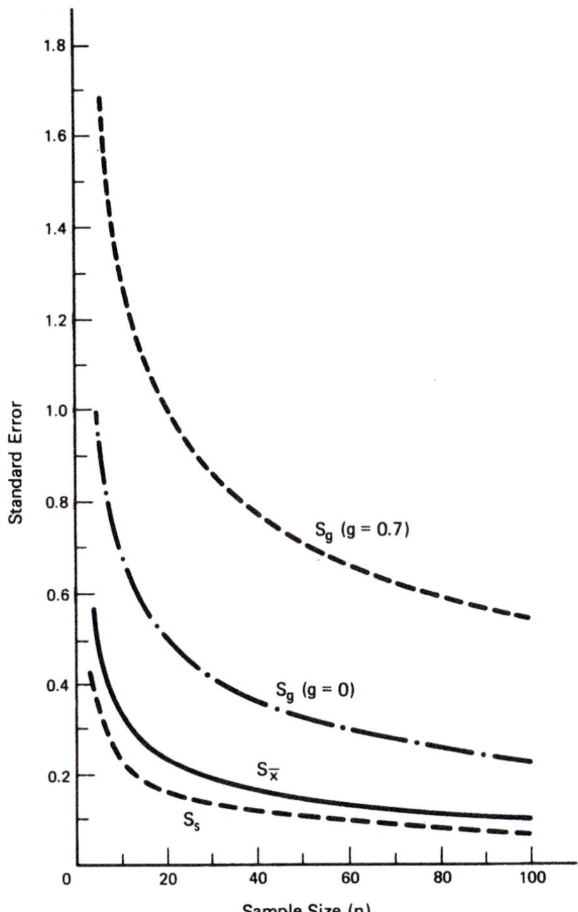

FIGURE 5–13 Standard errors of the mean ($S_{\bar{x}}$), standard deviation (S_s), and skew (S_g).

If the statistics from above ($\bar{Y} = 3.722$, $S_y = 0.280$, $g = -0.731$) are used as the baseline frequency curve and the statistics are increased successively by one standard error, three curves can be computed to show the effects of errors in the moments. The combination of parameters are ($\bar{Y} = 3.722 + 0.045$, $S_y = 0.280$, $g = -0.7$), ($\bar{Y} = 3.722$, $S_y = 0.280 + 0.038$, $g = -0.7$), and ($\bar{Y} = 3.722$, $S_y = 0.280$, $g = -0.7 + 0.4$). The discharges for exceedence probabilities of 0.5, 0.2, 0.1, 0.04, 0.02, and 0.01 for these parameters are given in Table 5–8. As expected, the change in the mean has the greatest relative effect at the mean (that is, exceedence probability = 0.5) but a much smaller relative effect at the extreme frequencies. Such changes in the discharge would have significant effects on design, such as the design of a spillway. There is a need to have accurate estimates of the moments of the data.

5.4.2 Weighted Skew

The skew is the moment most sensitive to sampling variation, especially for small samples, which are common in hydrology. Recognizing this, the Bulletin 17B recommends using a weighted skew coefficient rather than the sample skew. The weighted skew weights the sta-

TABLE 5–8 Sensitivity to the Flood Frequency Curve for Back Creek Near Jones Springs, West Virginia, for a Change of One Standard Error in the Mean (Q_1), Standard Deviation (Q_2), and the Standardized Skew (Q_3)

p	K for $g = -0.7$	K_1 for $g_1 = -0.3$	$X = \bar{X} + KS$	Q	$X_1 = \bar{X}_1 + KS$	Q_1	$X_2 = \bar{X} + KS_1$	Q_2	$X_3 = \bar{X} + KS$	Q_3
0.5	0.11578	0.04993	3.754	5,681	3.799	6,301	3.759	5,739	3.736	5,445
0.2	0.85703	0.85285	3.962	9,162	4.007	10,162	3.995	9,875	3.961	9,137
0.1	1.18347	1.24516	4.053	11,308	4.098	12,542	4.098	12,541	4.071	11,766
0.04	1.48852	1.64329	4.139	13,765	4.184	15,268	4.195	15,680	4.182	15,210
0.02	1.66325	1.88959	4.188	15,407	4.233	17,089	4.251	17,820	4.251	17,827
0.01	1.80621	2.10394	4.228	16,894	4.273	18,739	4.296	19,787	4.311	20,469

tion skew and a generalized skew using the mean square errors of the skew estimates. The generalized skew is a regional estimate of the skew; it is based on regional trends in the skew coefficients from nearby gaging stations. Bulletin 17B includes a map of the generalized skew for the United States, as well as a general procedure for deriving estimates of generalized skew. For small samples the weighting should significantly improve the accuracy of the skew. The two parts of the process will be introduced separately.

Estimating generalized skew. Generalized skew is a skew value obtained through regionalization, that is, from pooling information about skew from all sites within a homogeneous region. Three methods can be used: (l) the construction of skew isolines on a map of a region, where the isolines are based on the station (sample) skew values for all sites in the region; (2) the development of prediction equations for estimating the skew using watershed characteristics as predictor variables; and (3) the mean of the skew coefficients (g) for all stations in the region. Although these methods have been studied, their success has been limited.

Isoline maps of the standardized skew of Equation 5–4c can be constructed by plotting the station values of g for all gages in the region and drawing lines of equal skew, or isolines. Geographic and topographic trends should be considered in mapping the isolines. The value from the resulting map for the point of interest is referred to as the generalized skew and is denoted by $\bar{g}$. The mean square error (MSE) should be computed and used to assess the accu racy of the isoline map. The accuracy of the map is significant only if the MSE of the map is less than the MSE of the mean skew for the stations used to construct the map. The MSE of the map is considered to be the accuracy of generalized skew estimates.

The map of Figure 5–14 is a generalized skew map for the United States; the map is from the Bulletin No. 17B. This map can be used where data for a more complete map are not available. The MSE for the map of Figure 5–14 is 0.302.

As an alternative to mapping the skew coefficient, a prediction equation can be developed using the skew as the criterion variable and watershed and meteorologic characteristics as predictor variables. The MSE of the regression equation can be used as a measure of the accuracy of the equation. Like the isoline map, this procedure has not had much success. It appears that our lack of understanding of the factors that cause the skew coefficient to vary has limited the success of both the mapping and the prediction equation approaches. While

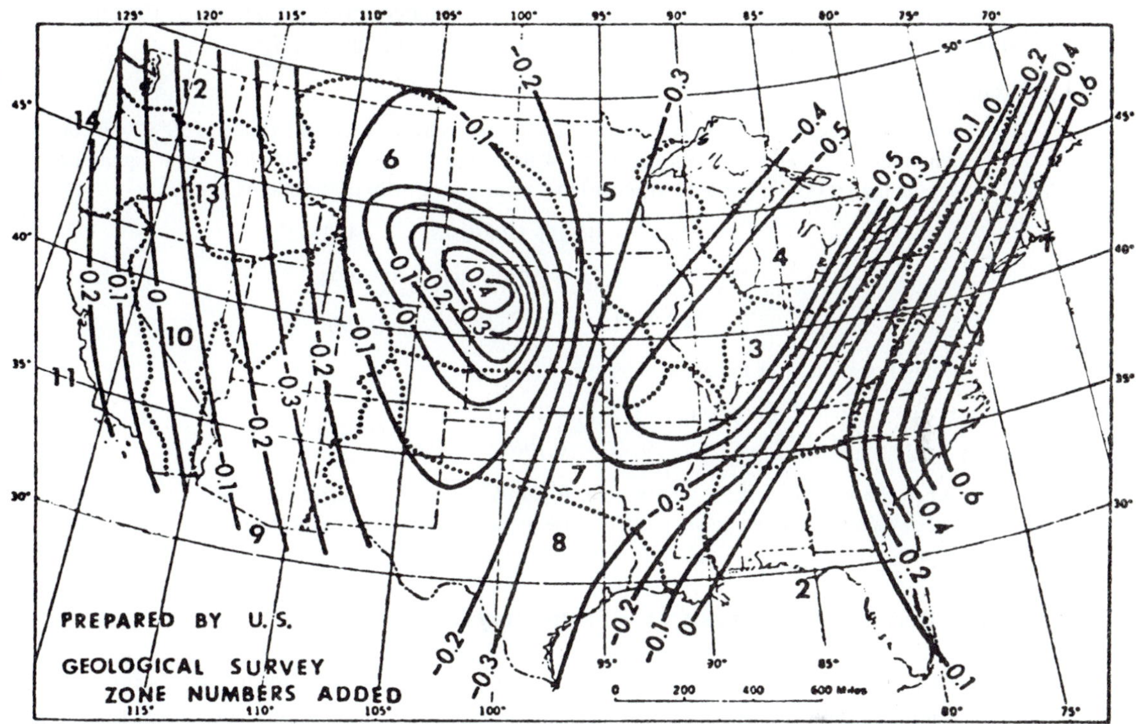

FIGURE 5–14 Generalized skew coefficients of annual maximum streamflow logarithms. (Bulletin 17B, 1982.)

watershed storage is probably the hydrologic reason for variation in skew, most watershed characteristics do not directly reflect storage.

As an alternative to the mapping and prediction-equation approaches, the mean of station skews in the region can be computed and used as the generalized estimate of skew. The MSE of the mean skew should be compared with the mean square errors of the isoline map and the prediction equation. The estimate with the smallest MSE should be used as the estimate of generalized skew.

Weighting of station and generalized skew. Given estimates of station skew g and a generalized skew $\bar{g}$, a weighted skew coefficient can be computed using the following equation:

$$g_w = \frac{MSE_{\bar{g}}(g) + MSE_g\,(\bar{g})}{MSE_{\bar{g}} + MSE_g} \tag{5–11}$$

in which $MSE_{\bar{g}}$ and MSE_g are the mean square errors for generalized skew and station skew, respectively. Equation 5–11 provides an estimate with the minimum mean square error, assuming that the generalized skew is unbiased and independent of station skew. The mean square error of the station skew estimate can be estimated from Equations 5–10c; that is, MSE_g of Equation 5–11 equals the square of Equation 5–10c, that is, S_g^2.

Example 5–8

The Back Creek near Jones Springs, West Virginia, is located with a latitude of 39°30'43" and a longitude of 78°02'15". From Figure 5–14 the generalized skew is 0.5. The mean square error for the map is 0.302. For a 38-yr record the mean square error of the station skew is

$$A = -0.33 + 0.08(0.731) = -0.27152$$

$$B = 0.94 - 0.26(0.731) = 0.74994$$

$$MSE_g = 10^{-0.272 - 0.750\, log\,(38/10)} = 0.196$$

Using Equation 5–11 the weighted skew is

$$g_w = \frac{0.302(-0.731) + 0.196(0.5)}{0.302 + 0.196} = -0.247 \; (use \; -0.2)$$

When the weighted skew is available, it should be used to derive the flood frequency curve rather than the station skew. The flood frequency curve using weighted skew is also shown in Figure 5–11.

5.4.3 Confidence Intervals on a Frequency Curve

A frequency curve developed from sample data is the best estimate of the population curve. If we knew the data that would occur over the next two decades or any such short period, a different frequency record would result; thus we would have a different estimate of the *T*-yr event. If we could collect data for many time periods of *n* years, we would have an array of frequency curves and thus an array of estimates of the *T*-yr event. The variation of these *T*-yr events represents sampling error, or the variation that can be expected from taking a sample from a population. If we were to use this distribution to hypothesize about the true value, we could select a level of confidence about our statement and determine limits between which we would expect the true value to lie with γ percent confidence. The interval between these limits is termed a confidence interval.

The sample flood frequency curve represents some average estimate of the true population curve. When the flood estimate obtained from a frequency curve is used for design, we recognize that there is a high probability, possibly 50%, that the true value actually lies above the sample estimate of the *T*-yr event. This implies that there is a good likelihood that locations will be flooded by a *T*-yr event even when the design is based on the *T*-yr estimate obtained from a flood frequency analysis. Due to the possibility for flooding in areas where public facilities have been designed to provide flood protection at a *T*-yr level of protection, public agencies may wish to reduce the average flooding due to the *T*-yr event to some percentage well below that provided by the *T*-yr frequency curve estimate. This is one possible motivation for computing and using a confidence interval on the frequency curve estimate.

Either one-sided or two-sided confidence intervals can be developed. A two-sided confidence interval provides both an upper and lower limit. For one-sided confidence intervals, either a lower or an upper limit, but not both, can be constructed. The type of problem will determine which form is used.

Factors affecting a confidence interval. The width of a confidence interval is affected by a number of factors. First is the degree of confidence, which is indicated by the confidence coefficient, γ, that one wishes to have in the decision. While a 95% confidence coefficient is commonly used, 75%, 90%, and 99% are also used. The larger the level of confidence, the greater the confidence one has in a decision; thus the width of the interval also increases as γ increases.

Second, the sample size used to compute the frequency curve affects the width of the confidence interval. A larger sample suggests greater accuracy in the statistics used to define the curve, so the width of the confidence interval decreases as the sample size increases.

Third, the exceedence probability for which the confidence interval is needed affects the width of the interval. The interval will be smallest for the mean flood, which is the 2-yr event, or the event with an exceedence probability of 0.5. Other factors (for example, n, γ) being the same, the width of the confidence interval will increase as the exceedence frequency deviates from a value of 0.5.

The width of the confidence interval is also affected by the skew coefficient. As is evident from the standard errors of Section 5.4.1, the skew affects the standard error, with the width of the interval increasing as the absolute value of the skew increases. This reflects the greater error in the computed value of a standard deviation or skew coefficient as the magnitude of the skew increases. This must be reflected in the width of the confidence interval computed for a frequency curve based on sample statistics.

Bulletin 17B procedure for confidence intervals. Bulletin 17B provides an approximate solution for developing confidence intervals on a frequency curve. The general forms of the upper and lower bounds on the γ percent confidence interval are

$$Y_u = \overline{Y} + K_u S_y \tag{5–12a}$$

$$Y_L = \overline{Y} - K_L S_y \tag{5–12b}$$

in which $\overline{Y}$ and S_y are the mean and standard deviation of the logarithms, respectively; K_U and K_L are the upper and lower distribution factors; and Y_u and Y_L are the upper and lower limits of the confidence interval. If a one-sided interval is computed, only one of the limits is computed. The distribution factors are computed by

$$K_u = \frac{K + \sqrt{K^2 - ab}}{a} \tag{5–13a}$$

$$K_L = \frac{K - \sqrt{K^2 - ab}}{a} \tag{5–13b}$$

in which

$$a = 1 - \frac{z^2}{2(n-1)} \tag{5–14a}$$

$$b = K^2 - \frac{z^2}{n} \tag{5–14b}$$

in which K is the frequency factor of Appendix 5–1, n is the sample size, and z is the confidence limit deviate values for a normal distribution from Table 2–2.

In summary, the following inputs are required to compute confidence intervals: the level of confidence, the return period or exceedence probability, the sample moments $(\overline{Y}, S_y, g)$ for the logarithms of the flood series, and the sample size.

Example 5–9

The data for the Back Creek watershed can be used to illustrate the computation of confidence intervals for frequency curves. The 95% level of confidence is selected because of convention, and a one-sided upper limit will be computed. The upper limit will be computed for return periods of 2, 5, 10, 25, 50, and 100 years since these return periods are commonly used in hydrologic design. The sample mean and standard deviation of the common logarithms were given as 3.722 and 0.2804, respectively. The weighted skew was computed in Equation 5–11 as –0.2. The sample size was 38.

The calculations for the 100-yr event will be shown here, with the results for the other return periods given in Table 5–9. To apply Equations 5–12 to 5–14, the standard normal deviate for a 95% limit is required; from Table 2–2 z equals 1.645. From Appendix 5–1 the frequency factor for $T = 100$ yrs and $g = -0.2$ is 2.17840; therefore, Equations 5–14 yield

$$a = 1 - \frac{(1.645)^2}{2(38 - 1)} = 0.96343$$

$$b = (2.17840)^2 - \frac{(1.645)^2}{38} = 4.67422$$

Since only an upper limit is needed, Equation 5–13a can be used to compute the distribution factor:

$$K_u = \frac{2.17840 + \sqrt{(2.17840)^2 - (0.96343)(4.67)}}{0.96343}$$

$$= 2.77185$$

Equation 5–12a is used to compute the upper confidence limit:

$$Y_u = 3.722 + 0.2804(2.77185) = 4.4992$$

Taking the antilog yields an upper limit of 31,567 ft^3/sec, which can be compared with the 100-yr flood estimate of

$$\hat{Y} = 3.722 + 0.2804(2.17840) = 4.33282$$

TABLE 5–9 Computation of 95% Upper Confidence Limits for Back Creek Watershed

T (yr)	a	K	b	K_u	Y_u	Q_u	Y_T	Q_T
2	0.96343	0.03325	–0.07011	0.30646	3.80793	6,426	3.73132	5,387
5	0.96343	0.84986	0.65105	1.20207	4.05906	11,457	3.96030	9,126
10	0.96343	1.25824	1.51196	1.67517	4.19172	15,550	4.07481	11,880
25	0.96343	1.67999	2.75116	2.17401	4.33159	21,458	4.19307	15,598
50	0.96343	1.94499	3.71177	2.49100	4.42048	26,332	4.26738	18,509
100	0.96343	2.17840	4.67422	2.77185	4.49923	31,567	4.33282	21,519

which has an antilog of 21,519 ft^3/sec. The values of Table 5–9 indicate that the difference between the upper limit and the frequency curve increases as the return period increases, which reflects the decrease in accuracy of the sample statistics and therefore the frequency curve.

The confidence limit on the frequency curve defines a one-sided interval, and we can conclude that we are 95% certain that true T-yr flood is less than the confidence limit. Both the points that describe the frequency curve and the confidence limits could be plotted on probability paper, with intermediate points then available by interpolation. For example, if either the 20-yr or 75-yr event at the gaged location were of interest, their values could be obtained from the plotted frequency line. However, it is important to emphasize that the resulting upper confidence interval defines an upper limit on the T-yr event and not an interval on the entire frequency curve. That is, confidence intervals computed using the procedure of Equations 5–12 to 5–14 is a method of estimating the confidence limit on the mean T-yr event. It should only be interpreted in terms of the mean T-yr event.

5.4.4 Adjusting for Historic Floods

In addition to a systematic record of N years, historic flood information, such as high-water marks, is sometimes available at or near the gaging station. When such information can be used to provide a reliable estimate of a flood magnitude, the computed value can be used to adjust the frequency curve based on the systematic record.

Procedure. Bulletin 17B provides the following procedure for adjusting a frequency curve for historic information:

1. Obtain data.
   (a) Identify the N annual maximum flows in the systematic record.
   (b) Identify the L low values to be excluded; this includes the number of zero flows or flows below a measurable base and the number of low outliers (outliers are discussed in Section 5.4.5).
   (c) Identify the length of time (H years) over which the historic flood information is known to represent (note: the historic period includes the length of the systematic record).
   (d) Identify the number (Z) of historic peaks (including high outliers); the historic peaks and high outliers must be larger than the peaks in the systematic record.
2. Compute weights.
   (a) Assign a weight of 1 to each of the Z historic events in the historic period H and to any high outliers.
   (b) Compute a weight W to assign to each of the N events in the systematic record:

   $$W = \frac{H - Z}{N + L} \tag{5–15}$$

   (Note: If high outliers are removed from the systematic record for treatment as historic peaks, the value of N should be adjusted accordingly.)
3. Compute moments.
   (a) If values for the mean, standard deviation, and skew have not previously been computed, compute the historically adjusted statistics $\overline{Y}_h$, S_{yh}, g_h:

$$\overline{Y_h} = \frac{W \sum_{i=1}^{n} Y_i + \sum_{i=1}^{Z} Z_i}{H - WL} \tag{5–16a}$$

$$S_{yh}^2 = \frac{W \sum_{i=1}^{N} (Y_i - \overline{Y_h})^2 + \sum_{i=1}^{Z} (Z_i - \overline{Y_h})^2}{(H - W * L - 1)} \tag{5–16b}$$

$$g_h = \frac{H - W * L}{(H - W * L - 1)(H - W * L - 2)} \cdot \left[\frac{W \sum_{i=1}^{N} (Y_i - \overline{Y_h})^3 + \sum_{i=1}^{Z} (Z_i - \overline{Y_h})^3}{S_{yh}^3} \right] \tag{5–16c}$$

(b) If values for the mean $(\overline{Y})$, standard deviation (S_y), and skew (g) have previously been computed, the historically adjusted statistics $(\overline{Y_h}, S_{yh}, g_h)$ can be computed by

$$\overline{Y_h} = \frac{W * N * \overline{Y} + \sum_{i=1}^{Z} Y_i}{H - W * L} \tag{5–17a}$$

$$S_h^2 = \frac{W(N - 1)S_y^2 + W * N(\overline{Y} - \overline{Y_h})^2 + \sum_{i=1}^{Z} (Z_i - \overline{Y_h})^2}{H - W * L - 1} \tag{5–17b}$$

$$g_h = \left[\frac{H - W * L}{(H - W * L - 1)(H - W * L - 2)S_h^3} \right] * \tag{5–17c}$$

$$\left[\frac{W(N - 1)(N - 2)gS_y^3}{N} + 3W(N - 1)(\overline{Y} - \overline{Y_h})S_y^2 + W{*}N{*}(\overline{Y} - \overline{Y_h})^3 + \sum_{i=1}^{Z} (Z_i - \overline{Y_h})^3 \right]$$

(c) If warranted, compute a weighted skew using g_h and generalized skew $(\overline{g})$:

$$g_{hw} = \frac{MSE_{\overline{g}}(g_h) + MSE_{gh}(\overline{g})}{MSE_{\overline{g}} + MSE_{gh}} \tag{5–18}$$

The mean square error for the historically weighted skew is computed from Equation 5–10 using H in place of N. If generalized skew is obtained from the Bulletin 17B map, the mean square error $MSE_{\overline{g}}$ equals 0.302.

4. Compute the historically adjusted frequency curve.

(a) Using the historically adjusted (g_h) or historically weighted (g_{hw}) skew, obtain LP3 deviates K from Appendix 5–1 for selected exceedence probabilities.

(b) Using the historically adjusted moments $(\overline{Y_h}$ and $S_{yh})$ compute the logarithms of the historically adjusted flood estimates:

$$\hat{Y}_{hi} = \overline{Y_h} + KS_{yh} \tag{5–19}$$

(c) Transform the logarithms:

$$\hat{X}_{hi} = 10^{\hat{Y}_{hi}} \tag{5–20}$$

5. Plot the measured flows using historically weighted plotting probabilities.

(a) Compute the historically adjusted rank (m_h) for each flood magnitude:

$$m_{hi} = \begin{cases} i & for\ 1 \leqslant i \leqslant Z \\ W * i - (W-1)(Z+0.5) & for\ (Z+1) \leqslant i \leqslant (Z+N+L) \end{cases} \tag{5–21b}$$

(b) Compute the historically weighted plotting position for each event:

$$P_i = \frac{m_{hi} - a}{H + 1 - 2a} * 100 \tag{5–22}$$

in which a is a constant that is characteristic of the plotting position formula ($a = 0$ for Weibull, $a = 0.5$ for Hazen, and $a = 0.4$ for Cunnane).

(c) Plot the magnitude versus the probability P_i.

Example 5–10

The flood record in Bulletin 17B from the Big Sandy River at Bruceton, TN, will be used to illustrate the steps for the historical adjustment. The record consists of a systematic record of 44 years (1930–1973), 3 historic events (1897, 1919, 1927) and a historic period of 77 years (1897–1973). The flood peaks are given in Table 5–10a (column 2) and the logarithms are given in column 3. The weight is computed with Equation 5–15 :

$$W = \frac{H - Z}{N + L} = \frac{77 - 3}{44 + 0} = 1.6818$$

The historically adjusted moments are

$$\overline{Y}_h = \frac{1.6818(162.40) + 12.987}{77} = 3.71581$$

$$S_h^2 = \frac{1.6818(3.09755) + 1.13705}{77 - 1} = 0.08351$$

$$S_h = 0.28898$$

$$g_h = \frac{77[1.6818(-0.37648) + 0.70802]}{76(75)(0.28898)^3} = 0.042$$

Since generalized skew is available, a weighted skew can be computed. The generalized skew of –0.2 is obtained from the Bulletin 17B map (see Figure 5–14), which has a mean square error, $MSE_{\bar{g}}$, of 0.302. The mean square error, MSE_{gh}, of the historically weighted skew is computed from Equation 5–10c using the historic length H in place of N:

$$A = -0.33 + 0.08(0.042) = -0.3266$$

$$B = 0.94 - 0.26(0.042) = 0.9291$$

$$MSE_{gh} = 10^{A - B * \log(77/10)} = 0.0708$$

TABLE 5–10a Historically Weighted Log Pearson Type III—Annual Peaks (from Bulletin 17B)

Station: 3-6065, Big Sandy River at Bruceton, TN. D. A. 205 square miles; Record: 1897, 1919, 1927, 1930–1973 (47 years)
Historical period: 1897–1973 (77 years) $N = 44$; $Z = 3$; $H = 77$

Year	$Q(\text{ft}^3/\text{s}) = Y$	Log $Y = X$	Departure From Mean log $X = (X–M)$	Weight = W	Event Number = E	Weighted order Number = m	Plotting Position (Weibull) PP
1897	25,000	4.39794	0.68212	1.00	1	1.00	1.28
1919	21,000	4.32222	0.60640	1.00	2	2.00	2.56
1927	18,500	4.26717	0.55136	1.00	3	3.00	3.85
1935	17,000	4.23045	0.51464	1.68182	4	4.34	5.56
1937	13,800	4.13988	0.42407		5	6.02	7.72
1946	12,000	4.07918	0.36337		6	7.71	9.88
1972	12,000	4.07918	0.36337		7	9.39	12.04
1956	11,800	4.07188	0.35607		8	11.07	14.19
1942	10,100	4.00432	0.28851		9	12.75	16.35
1950	9,880	3.99475	0.27895		10	14.43	18.50
1930	9,100	3.95904	0.24323		11	16.12	20.67
1967	9,060	3.95713	0.24132		12	17.80	22.82
1932	7,820	3.89321	0.17740		13	19.48	24.97
1973	7,640	3.88309	0.16728		14	21.16	27.13
1962	7,480	3.87390	0.15809		15	22.84	29.28
1965	7,180	3.85612	0.14031		16	24.53	31.45
1936	6,740	3.82866	0.11285		17	26.21	33.60
1948	6,130	3.78746	0.07165		18	27.89	35.76
1939	5,940	3.77379	0.05798		19	29.57	37.91
1945	5,630	3.75051	0.03470		20	31.25	40.06
1934	5,580	3.74663	0.03082	$W = (H - Z)N$	21	32.94	42.23
1955	5,480	3.73878	0.02297	$= (77 - 3)/(44)$	22	34.62	44.38
1944	5,340	3.72754	0.01173	$= 1.68182$	23	36.30	46.54
1951	5,230	3.71850	0.00269		24	37.98	48.69
1957	5,150	3.71181	–0.00400		25	39.66	50.85
1971	5,080	3.70586	–0.00995		26	41.35	53.01
1953	5,000	3.69897	–0.01684		27	43.03	55.17
1949	4,740	3.67578	–0.04003		28	44.71	57.32
1970	4,330	3.63649	–0.07932		29	46.39	59.47
1938	4,270	3.63043	–0.08538		30	48.07	61.62
1952	4,260	3.62941	–0.08640		31	49.76	63.79
1947	3,980	3.59988	–0.11593		32	51.44	65.95
1943	3,780	3.57749	–0.13832		33	53.12	68.10
1961	3,770	3.57634	–0.13947		34	54.80	70.25
1958	3,350	3.52504	–0.19077		35	56.49	72.42
1954	3,320	3.52114	–0.19467		36	58.17	74.58
1933	3,220	3.50786	–0.20795		37	59.85	76.73
1964	3,100	3.49136	–0.22445		38	61.53	78.88
1968	3,080	3.48855	–0.22725		39	63.21	81.04
1969	2,800	3.44716	–0.26865		40	64.90	83.21
1963	2,740	3.43775	–0.27806		41	66.58	85.36
1959	2,400	3.38021	–0.33560		42	68.26	87.51
1931	2,060	3.31387	–0.40194		43	69.94	89.67
1966	1,920	3.28330	–0.43251		44	71.62	91.82
1940	1,680	3.22531	–0.49050		45	73.31	93.99
1960	1,460	3.16435	–0.55146		46	74.99	96.14
1941	1,200	3.07918	–0.63663	1.68182	47	76.67	98.29

TABLE 5–10b Historically Weighted Frequency Curve

Prob. (%)	K	$S*K$	Log (Q)	Q (ft^3/s)
95	−1.64599	−0.47566	3.24014	1,738
50	0.00067	0.00019	3.71600	5,200
20	0.84180	0.24326	3.95907	9,100
10	1.28110	0.37021	4.08602	12,190
4	1.74929	0.50551	4.22132	16,646
2	2.05159	0.59289	4.30868	20,355
1	2.32340	0.67142	4.38723	24,391

Thus, the historically weighted skew from Equation 5–11 is

$$g_{hw} = \frac{0.302(0.042) + 0.0708(-0.2)}{0.302 + 0.0708} = -0.004$$

Therefore, a historically weighted skew is 0, which indicates log-normal distribution.

The historically adjusted frequency curve is computed using a mean of 3.71581, a standard deviation of 0.28898, and a skew of 0. Equations 5–19 and 5–20 are used; the results are given in column 5 of Table 5–10b.

To plot the data, Equations 5–21 are used to compute the historically adjusted order number. Values of 1, 2, and 3 are used for the three historic values. Values for the events in the systematic record are computed with Equation 5–21b:

$$\begin{aligned} m_{hi} &= 1.6818\, i - 0.6818\,(3.5) \\ &= 1.6818\, i - 2.3863 \qquad \mathit{for}\ 4 \leq i \leq 47 \end{aligned}$$

The values are given in column 7 of Table 5–10a. The historically adjusted frequency curve and the measured data are shown in Figure 5–15. The close fit of the data indicates that the assumed population is a reasonable estimate of the population.

5.4.5 Outliers

When the procedures of Sections 5.2 and 5.3 are used to compute a frequency curve and plot the data, it is not uncommon for the data points to depart from the trend of the line in the tails of the distribution. Those points that do not appear to follow the trend of the data are considered to be extreme events. Figure 5–16 provides an example of a frequency plot that shows a 25-yr record on the Republican River near Hardy, Nebraska. The largest flood in the 25-yr period of record before the 1935 event (1910–1934) is 38,000 ft^3/sec. In 1935, a flood of 225,000 ft^3/sec occurred. For the Hazen plotting position formula, such an event would plot with an exceedence probability of 1/52 or 0.019, which would obviously place the flood high above the trend of the other 25 flood magnitudes. If such an extreme event is included in the standard analysis, the data points will not follow the trend of the assumed population regardless of the probability distribution. Several options are available: (1) use the standard analysis regardless of the apparent disparity; (2) make a standard analysis after censoring (that is, deleting) the extreme event from the record; and (3) adjust the plotting position of the extreme event. Each of these options has advantages and disadvantages. Since all extreme

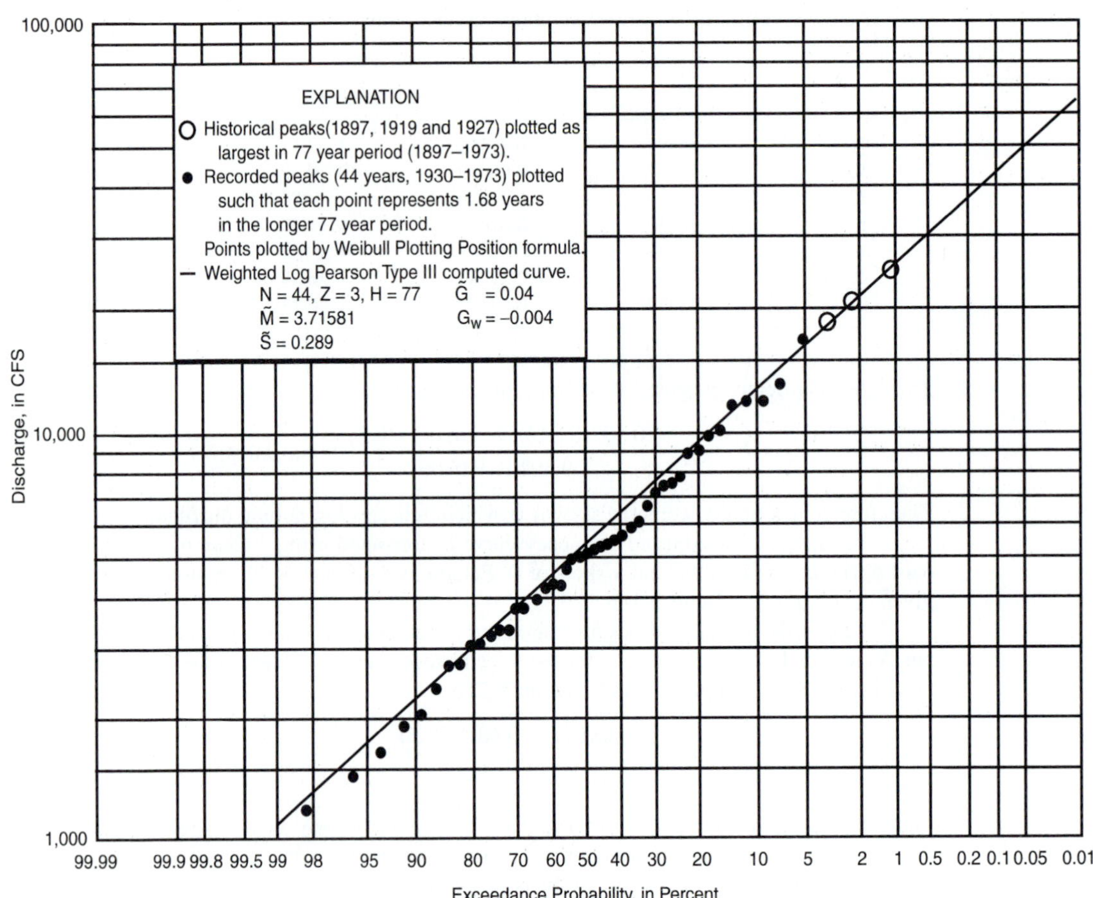

FIGURE 5–15 Historically weighted log-Pearson Type III: Big Sandy River at Bruceton, TN (from Bulletin 17B).

points do not deviate as much as that apparent in Figure 5–16, it is necessary to have a systematic criterion for deciding how to handle extreme events.

An extreme event can be tested to determine whether or not it can be declared to be an outlier. A systematic test is needed. If the test shows that the extreme event is unlikely to be from the same population as the remainder of the data, then it can be called an outlier and handled appropriately. Outliers can occur in either, or both, the upper and lower tails, which are referred to as high and low outliers, respectively.

Bulletin 17B detection procedure. Extreme events that appear to be high outliers can be tested with the following procedure:

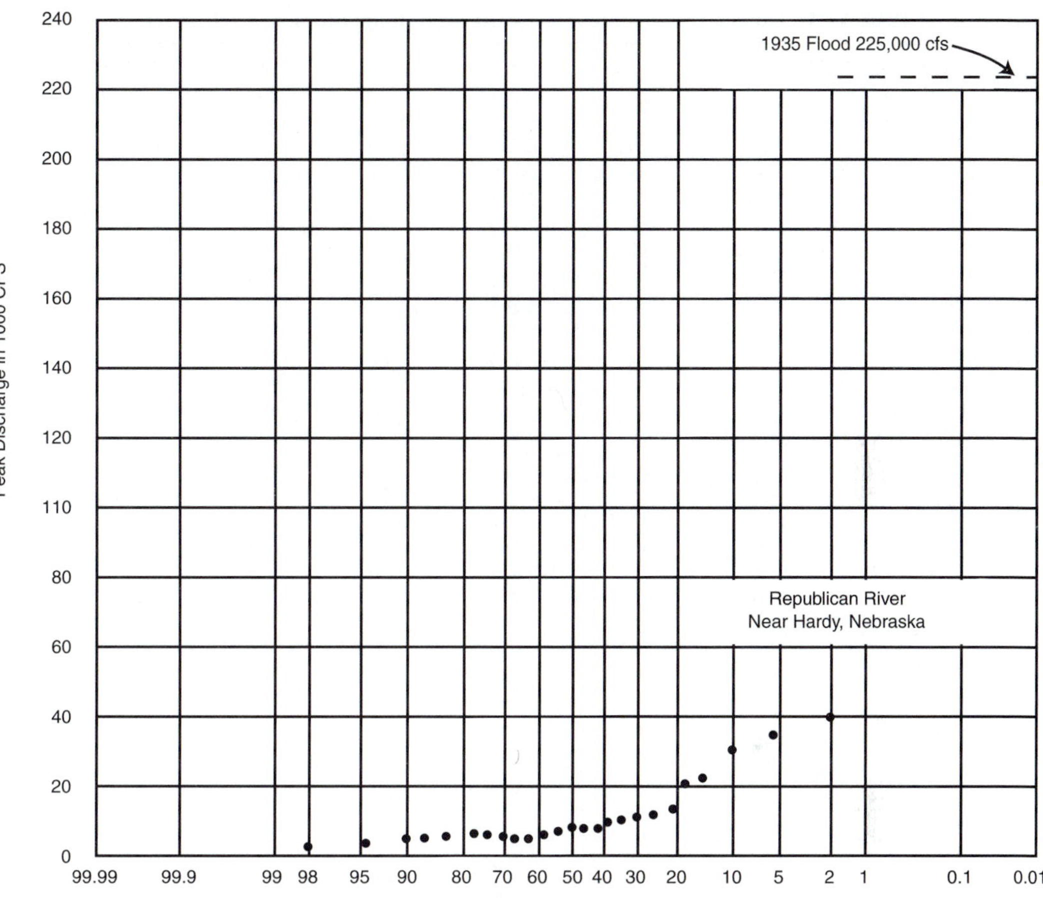

FIGURE 5–16 Example of probability distribution of annual peak discharge observations for the Republican River near Hardy, Nebraska. (Source: Office of Water Data Coordination, *Feasibility of Assigning a Probability of the Probable Maximum Flood.* Washington, D.C., 1986.)

1. Using the sample size (n) obtain the value of the detection deviate (K_o) from Table 5–11.
2. Compute the mean ($\overline{Y}$) and standard deviation (S_y) of the logarithms of the series including the extreme event being considered but excluding zero-flood events, peaks below the gage base, and outliers previously detected.
3. Compute the value of the detection criterion for high outliers (Y_{oh}):

$$Y_{oh} = \overline{Y} + K_o S_y \tag{5–23}$$

4. Compare the logarithm of the extreme event being considered (Y_h) with the criterion (Y_{oh}); if $Y_h > Y_{oh}$, then the event can be considered a high outlier.

TABLE 5–11 Outlier Test Deviates (K_O) at 10 Percent Significance Level

Sample Size	K_O Value	Sample Size	K_O Value	Sample Size	K_O Value	Sample Size	K_O Value
10	2.036	45	2.727	80	2.940	115	3.064
11	2.088	46	2.736	81	2.945	116	3.067
12	2.134	47	2.744	82	2.949	117	3.070
13	2.165	48	2.753	83	2.953	118	3.073
14	2.213	49	2.760	84	2.957	119	3.075
15	2.247	50	2.768	85	2.961	120	3.078
16	2.279	51	2.775	86	2.966	121	3.081
17	2.309	52	2.783	87	2.970	122	3.083
18	2.335	53	2.790	88	2.973	123	3.086
19	2.361	54	2.798	89	2.977	124	3.089
20	2.385	55	2.804	90	2.981	125	3.092
21	2.408	56	2.811	91	2.984	126	3.095
22	2.429	57	2.818	92	2.989	127	3.097
23	2.448	58	2.824	93	2.993	128	3.100
24	2.467	59	2.831	94	2.996	129	3.102
25	2.487	60	2.837	95	3.000	130	3.104
26	2.502	61	2.842	96	3.003	131	3.107
27	2.510	62	2.849	97	3.006	132	3.109
28	2.534	63	2.854	98	3.011	133	3.112
29	2.549	64	2.860	99	3.014	134	3.114
30	2.563	65	2.866	100	3.017	135	3.116
31	2.577	66	2.871	101	3.021	136	3.119
32	2.591	67	2.877	102	3.024	137	3.122
33	2.604	68	2.883	103	3.027	138	3.124
34	2.616	69	2.888	104	3.030	139	3.126
35	2.628	70	2.893	105	3.033	140	3.129
36	2.639	71	2.897	106	3.037	141	3.131
37	2.650	72	2.903	107	3.040	142	3.133
38	2.661	73	2.908	108	3.043	143	3.135
39	2.671	74	2.912	109	3.046	144	3.138
40	2.682	75	2.917	110	3.049	145	3.140
41	2.692	76	2.922	111	3.052	146	3.142
42	2.700	77	2.927	112	3.055	147	3.144
43	2.710	78	2.931	113	3.058	148	3.146
44	2.720	79	2.935	114	3.061	149	3.148

Source: Bulletin 17B

The procedure for low outliers is similar to that for high outliers. Specifically, Steps 1 and 2 are the same, with Steps 3 and 4 as follows:

3. Compute the value of the detection criterion for low outliers (Y_{ol}):

$$Y_{o1} = \overline{Y} - K_o S_y \tag{5–24}$$

4. Compare the logarithm of the extreme event being considered (Y_l) with the criterion (Y_{ol}); if $Y_l < Y_{ol}$, then the event can be considered a low outlier.

Bulletin 17B treatment procedure. The flowchart of Figure 5–17 shows the procedure for treating outliers recommended in Bulletin 17B. If the station skew is greater

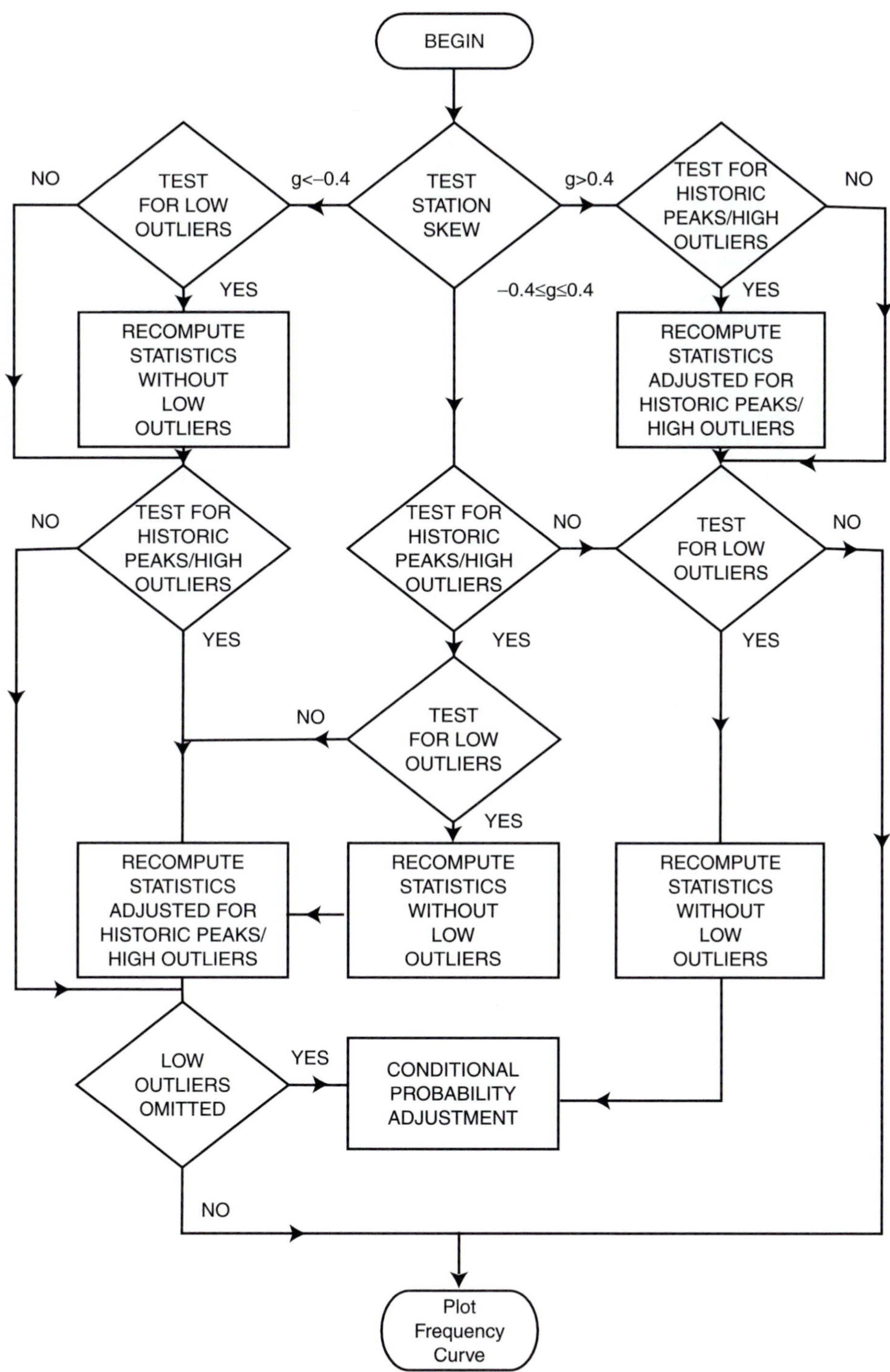

FIGURE 5–17 Flow diagram for historic and outlier adjustment.

than 0.4, extreme events that may be high outliers are tested first. If the station skew is less than –0.4, extreme events that may be low outliers are tested first. If the station skew is between –0.4 and 0.4, then tests should be made for both the high and low outliers before eliminating any outliers from the flood series.

The rationale behind the 0.4 limit for high outliers and –0.4 for low outliers is as follows. High outliers, such as the 1935 magnitude in Figure 5–16, will cause a significant increase in the computed value of the station skew. Similarly, a very low flood magnitude will cause a more negatively skewed sample. Since the average skew for the United States is approximately 0, then a reasonable likelihood exists that computed skews above 0.4 are due to the presence of a high outlier and that computed skews more negative than –0.4 are due to the presence of a low outlier. If both a high outlier and a low outlier exist within a flood record, then the skew may be near zero because of the offsetting effects on the summation in Equation 5–4c; thus, they should be tested simultaneously before censoring or making an adjustment.

If a high outlier is detected, then it is necessary to have historic flood data or flood information at nearby sites in order to make an adjustment. A flood shown to be a high outlier can be treated as a historic flood using the procedure of Section 5.4.4. If the other flood information is not available, then the event should be retained with the remainder of the systematic flood record.

If a low outlier is detected, then the value is censored and the conditional probability adjustment is applied. For this adjustment, the adjustment probability p_a is computed:

$$p_a = \frac{N}{n} \tag{5–25}$$

in which N is the number of events after outliers have been censored and n is the systematic record length. The conditional exceedence probabilities p_m are computed using the plotting probabilities from one of the formulas of Equation 5–5:

$$p_m = p_a * p_i \tag{5–26}$$

in which p_i is p_w, p_h, or p_c from Equations 5–5. More details of the conditional probability adjustment are given in Section 5.4.8 in the discussion of zero-flood records.

5.4.6 Adjusting a Flood Record for Urbanization

A statistical flood frequency analysis is based on the assumption of a homogeneous annual flood record. Significant changes in land use is a source of nonhomogeneity of flood characteristics, thus violating the assumptions that underline frequency analysis. A flood frequency analysis based on a nonhomogeneous record will result in inaccurate estimates of flood estimates for any return period. Therefore, the affect of nonhomogeneity must be estimated prior to making a frequency analysis so that the flood record can be adjusted.

Urbanization is a primary cause of nonhomogeneity of flood records. Although this problem has been recognized for decades, there have been few attempts at developing a systematic procedure for making the necessary adjustment of flood records. Multiparameter watershed models have been used for this purpose; however, a single model or procedure for adjustment has not been widely accepted by the professional community. Comparisons of methods for adjusting records have not been made.

The affect of urbanization. A number of models that enable the effect of urbanization on peak discharges to be assessed are available. Some models provide a basis for accounting for urbanization, but it is difficult to develop a general statement of the affect of urbanization from these models. For example, with the Rational method, urban development would have an affect on both the runoff coefficient and the time of concentration. Thus it is not possible to make a general statement that a 5% increase in imperviousness will cause an x% increase in the peak discharge for a specific return period. Other models are not so constrained.

A number of regression equations are available that include the percentage of imperviousness as a predictor variable. With such models it is possible to develop a general statement on the effect of urbanization. Sarma et al. (1969) provided one such example:

$$q_p = 484.1\, A^{0.723}(1 + U)^{1.516} P_E^{1.113} T_R^{-0.403} \tag{5–27}$$

in which A is the drainage area (mi^2), U is the fraction of imperviousness, P_E is the volume of excess rainfall (in.), T_R is the duration of rainfall excess (hours), and q_p the peak discharge (ft^3/sec). Since the model has the power-model form, the specific effect of urbanization depends on the values of the other predictor variables (that is, A, P_E, and T_R). However, the relative sensitivity of Equation 5–27 can be used as a measure of the effect of urbanization. The relative sensitivity (S_R) is given by

$$S_R = \frac{\partial q_p}{\partial U} \cdot \frac{U}{q_p} \tag{5–28}$$

Evaluation of Equation 5–28 yields a relative sensitivity of 1.516. Thus a 1% change in U will cause a change of 1.516% in the peak discharge. This estimate is an average effect since it is independent of both the value of U and the return period.

Based on the work of Carter (1961) and Anderson (1970), Dunne and Leopold (1978) provided the following equation for estimating the effect of urbanization:

$$f = 1 + 0.015U \tag{5–29}$$

in which f is a factor that gives the relative increase in peak discharge for a percent imperviousness of U. The following is a summary of the effect of urbanization based on the model of Equation 5–29:

U	0	10	20	30	40	50	100
f	1	1.15	1.3	1.45	1.6	1.75	2.5

Thus a 1% increase in U will increase the peak discharge by 1.5%, which is the same effect shown by the model of Equation 5–28.

The SCS provided an adjustment for urbanization for the TR-55 (1975) chart method. The adjustment depended on the percentages of imperviousness and the hydraulic length modified (see Figure 5–18), as well as the runoff curve number (CN). Although the adjustment does not specifically include the return period as a factor, the chart method incorporates

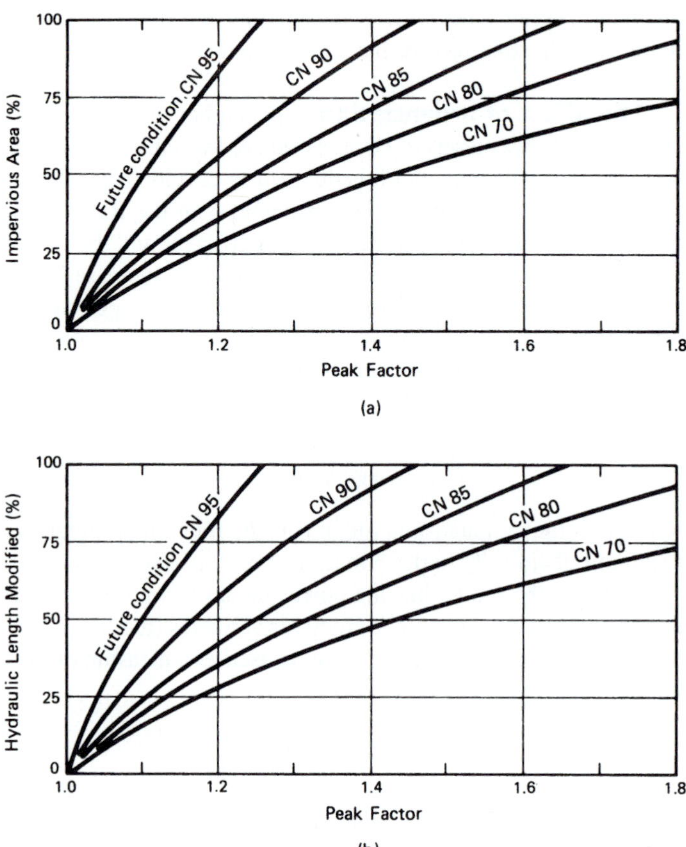

FIGURE 5–18 Factors for adjusting peak discharges for a given future-condition runoff curve number based on (a) the percentage of impervious area in the watershed and (b) the percentage of the hydraulic length modified.

the return period through the rainfall input. Table 5–12 provides the adjustment factors for imperviousness and the hydraulic length modified. Assuming that these changes occur in the same direct proportion, the effect of urbanization on peak discharges would be the square of the factor. Approximate measures of the effect of changes in f_2 from change in U are also shown in Table 5–12 (R_s); these values of R_s represent the change in peak discharge due to the peak factors provided in TR-55 (1975). Additional effects of urban development on the peak discharge would be reflected in change in the *CN*. However, the relative sensitivities of the SCS chart method suggest a change in peak discharge of 1.3 to 2.6% for a 1% change in urbanization, which represents the combined effect of changes in imperviousness and modifications of the hydraulic length.

The USGS urban peak discharge equations provide another alternative for assessing the effects of urbanization. The equations are given in Table 7–5. Figures 5–19 and 5–20

TABLE 5–12 Adjustment Factors for Urbanization

SCS Chart Method					USGS Urban Equations			
CN	*U*	*f*	f^2	R_s	*T*	*U*	f_1	R_S
70	20	1.13	1.28	0.018	2 yr	20	1.70	0.016
	25	1.17	1.37	0.019		25	1.78	0.016
	30	1.21	1.46	0.025		30	1.86	0.018
	35	1.26	1.59	0.026		35	1.95	0.020
	40	1.31	1.72	—		40	2.05	—
80	20	1.10	1.21	0.013	100 yr	20	1.23	0.010
	25	1.13	1.28	0.014		25	1.28	0.008
	30	1.16	1.35	0.019		30	1.32	0.008
	35	1.20	1.44	0.015		35	1.36	0.010
	40	1.23	1.51	—		40	1.41	—

show the ratio of the urban to rural discharge as a function of the percentage of imperviousness and a basin development factor. For the 2-yr event, the ratio ranges from 1 to 4.5, with the latter value for complete development. For the 100-yr event the ratio has a maximum value of 2.7. For purposes of illustration and assuming that basin development occurs in direct proportion to changes in imperviousness, the values of Table 5–12 (R_S) show the effect of urbanization on peak discharge. The average change in peak discharge due to a 1% change in urbanization is 1.75 and 0.9% for the 2- and 100-yr events, respectively. While the methods discussed previously provided an effect of about 1.5%, the USGS equations suggest that

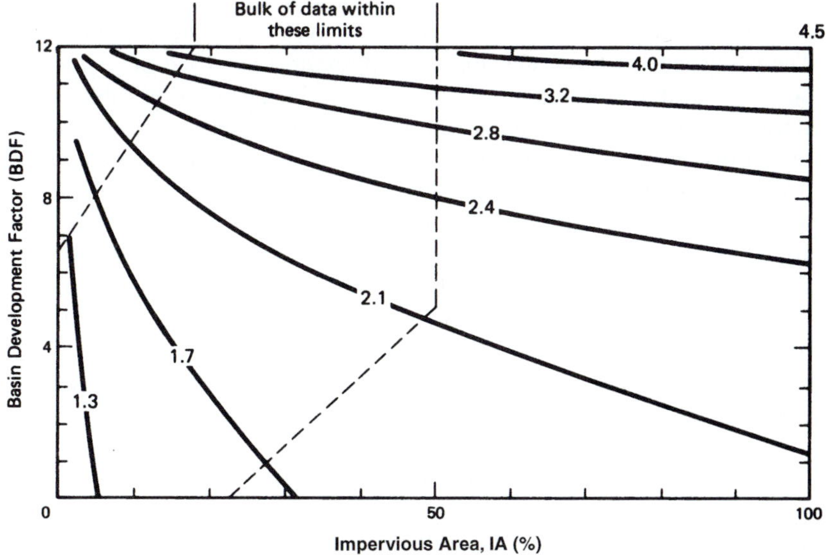

FIGURE 5–19 Ratio of the urban to rural 2-year peak discharge as a function of basin development factor (BDF) and impervious area (IA).

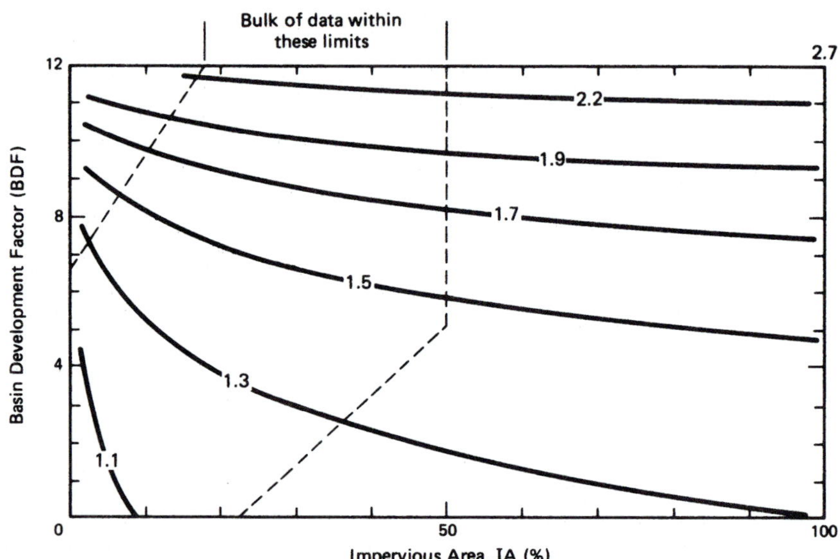

FIGURE 5–20 Ratio of the urban to rural 100-year peak discharge as a function of basin development factor and impervious area.

the effect is slightly higher for the more frequent storm events and slightly lower for the less frequent storm events.

Rantz (1971) provided a method for assessing the effect of urbanization on peak discharges using simulated data of James (1965) for the San Francisco Bay area. Urbanization is characterized by two variables: the percentages of channels sewered and basin developed. The percentage of basin developed is approximately twice the percentage of imperviousness. The peak factors are shown in Figure 5–21. The data of Table 5–13 show the relative sensitivity of the peak discharge to (a) the percent imperviousness, and (b) the combined effect of two variables: the percentages of channels sewered and basin developed. For urbanization as measured by the percentage change in imperviousness, the mean relative sensitivities are 2.6%, 1.7%, and 1.2% for the 2-, 10-, and 100-yr events, respectively. These values are larger by about 30 to 50% than the values computed from the USGS urban equations. When both the percentages of channels sewered and basin developed are used as indices of development, the relative sensitivities are considerably higher. The mean relative sensitivities are 7.1, 5.1, and 3.5% for the 2-, 10-, and 100-yr events, respectively. These values are much larger than the values suggested by the other methods discussed in the preceding paragraphs.

A method for adjusting a flood record. The literature does not identify a single method that is considered to be the best method for adjusting a flood record. Each method depends on the data used to calibrate the prediction process, and the data base used to calibrate the methods are very sparse. However, the sensitivities suggest that a 1% increase in imperviousness causes an increase in peak discharge of about 1 to 2.5% with the former

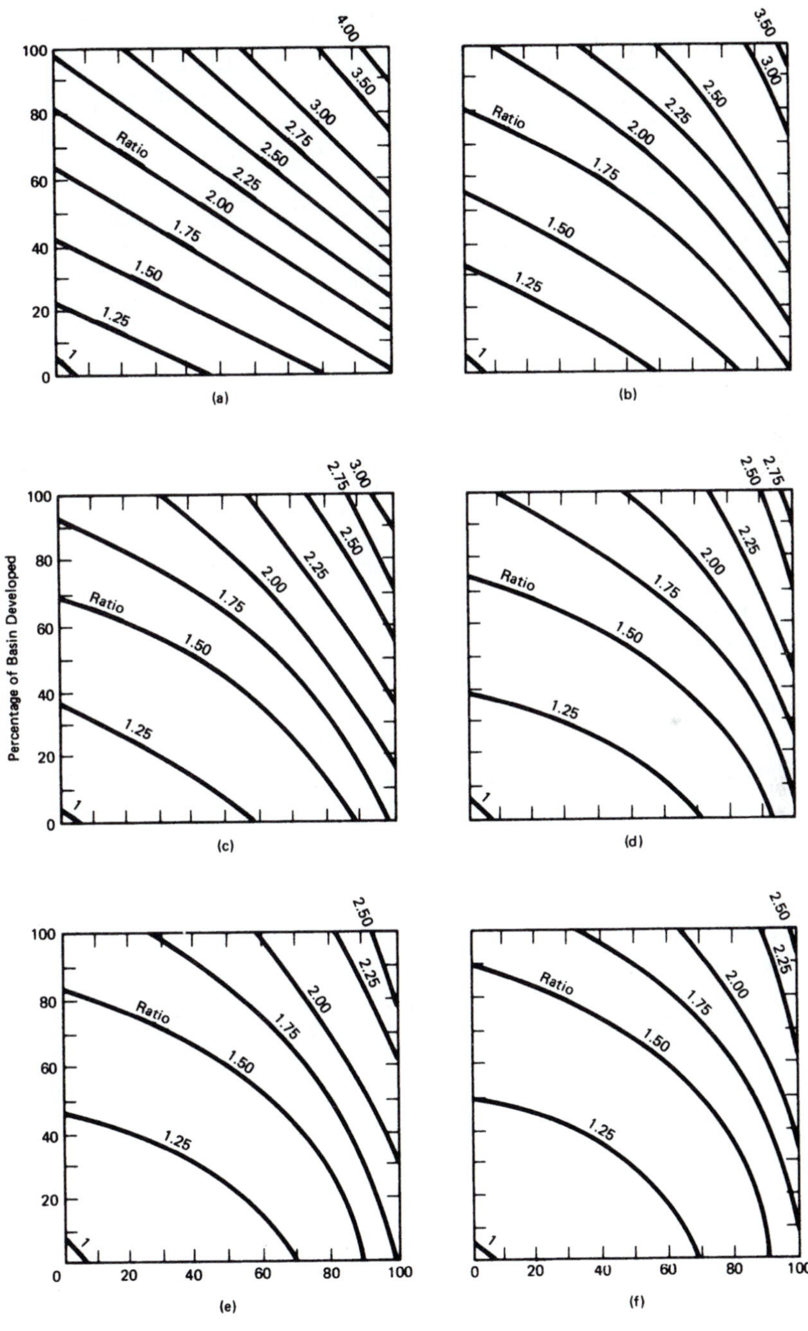

FIGURE 5–21 Ratio of post-development to pre-development peak discharge rates as a function of the percentages of basin developed and channels sewered for recurrence intervals of: (a) 2-yr; (b) 5-yr; (c) 10-yr; (d) 25-yr; (e) 50-yr; and (f) 100-yr.

TABLE 5–13 Effect on Peak Discharge of (a) the Percentage of Imperviousness (U) and (b) the Combined Effect of Urban Development (D)

		T = 2 yr		T = 10 yr		T = 100 yr	
(a)	U (%)	f	R_S	f	R_S	f	R_S
	10	1.22	0.025	1.13	0.015	1.08	0.011
	20	1.47	0.025	1.28	0.017	1.19	0.012
	30	1.72	0.026	1.45	0.018	1.31	0.013
	40	1.98	0.029	1.63	0.018	1.44	0.012
	50	2.27		1.81		1.56	
(b)	D (%)						
	10	1.35	0.040	1.18	0.022	1.15	0.010
	20	1.75	0.060	1.40	0.040	1.25	0.025
	30	2.35	0.085	1.80	0.050	1.50	0.050
	40	3.20	0.100	2.30	0.092	2.00	0.055
	50	4.20		3.22		2.55	

value for the 100-yr event and the latter for the 2-yr event. However, there was considerable variation at any return period.

Based on the general trends of the data, a method of adjusting a flood record was developed. Figure 5–22 shows the peak adjustment factor as a function of the exceedence probability for percentages of imperviousness up to 60%. The greatest effect is for the more frequent events and the highest percentage of imperviousness. Given the return period of a flood peak for a nonurbanized watershed, the effect of an increase in imperviousness can be assessed by multiplying the discharge by the peak adjustment factor for the return period and percentage of imperviousness. Where it is necessary to adjust a discharge from a partially urbanized watershed to a discharge for another watershed condition, the discharge can be divided by the peak adjustment factor for the existing condition and then the resulting "rural" discharge multiplied by the peak adjustment factor for the second watershed condition. The first operation (that is, division) adjusts the discharge to a magnitude representative of a nonurbanized condition. The adjustment method of Figure 5–22 requires an exceedence probability. For a flood record, the best estimate of the probability is obtained from a plotting position formula.

The following procedure can be used to adjust a flood record for which the individual flood events have occurred on a watershed that is undergoing a continuous change in the level of urbanization:

1. Identify both the percentage of imperviousness for each event in the flood record and the percentage of imperviousness for which an adjusted flood record is needed.
2. Compute the rank (i) and exceedence probability (p) for each event in the flood record (a plotting position formula can be used to compute the probability).
3. Using the exceedence probability and the actual percentage of imperviousness, find from Figure 5–22 the peak adjustment factor (f_1) to transform the measured peak from the actual level of imperviousness to a nonurbanized condition.

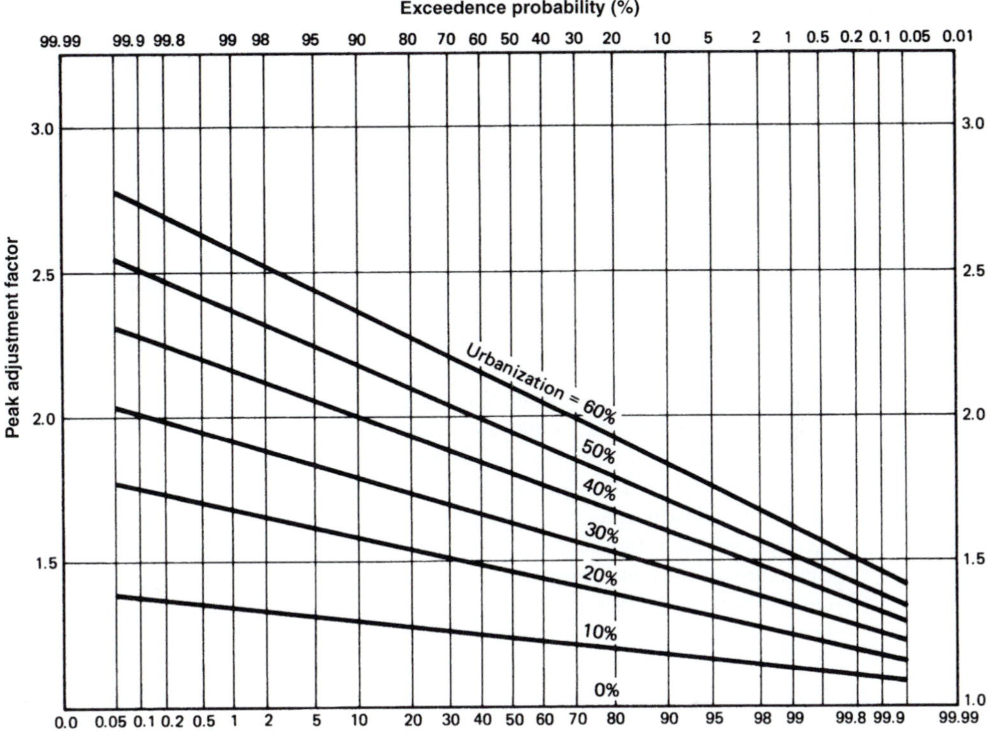

FIGURE 5–22 Peak adjustment factors for urbanizing watersheds.

4. Using the exceedence probability and the percentage of imperviousness for which a flood series is needed, find from Figure 5–22 the peak adjustment factor (f_2) that is necessary to transform the nonurbanized peak to a discharge for the desired level of imperviousness.
5. Compute the adjusted discharge (Q_a) by

$$Q_a = \frac{f_2}{f_1} Q \tag{5–30}$$

in which Q is the measured discharge.

6. Repeat Steps 3, 4, and 5 for each event in the flood record and rank the adjusted series.
7. If there are significant changes in the ranks of the measured (Q) and adjusted (Q_a) flood series, repeat Steps 2 through 6 until the ranks do not change.

It is important to note that this procedure should be applied to the flood peaks and not to the logarithms of the flood peaks. This is true even when the adjusted series will be used to compute a log-normal or log-Pearson Type III frequency curve. The peak adjustment factors of Figure 5–22 are based on methods that show the effect of imperviousness on flood peaks, not on the logarithms. The following example illustrates the application of the method.

Example 5–11

Table 5–14 contains the 48-yr record of annual maximum peak discharges for the Rubio Wash watershed in Los Angeles. Between 1929 and 1964 the percent of impervious cover, which is also given in Table 5–17, increased from 18 to 40%. The mean and standard deviation of the logarithms of the record are 3.2517 and 0.1910, respectively. The station skew was –0.53, and the map skew was –0.45. Therefore, a weighted skew of -0.5 was used.

The procedure was used to adjust the flood record from actual levels of imperviousness for the period from 1929 to 1963 to current impervious cover conditions. For example, while the peak discharges for 1931 and 1945 occurred when the percent cover was 19 and 34%, respectively, the values were adjusted to a common percentage of 40%, which is the watershed state after 1964. Three iterations of adjustments were required. The iterative process is required because the return period for some of the earlier events changed considerably from the measured record; for example, the rank of the 1930 peak changed from 30 to 22 on the first trial, and the rank of the 1933 event went from 20 to 14. Because of such changes in the rank, the exceedence probabilities change and thus the adjustment factors, which depend on the exceedence probabilities, change. After the second adjustment is made, the rank of the events did not change, so the process is complete. The adjusted series is given in the last part of Table 5–14.

The adjusted series has a mean and standard deviation of 3.2800 and 0.1785, respectively. As expected, the mean increased and the standard deviation decreased. The mean increased because the earlier events occurred when less impervious cover existed. The standard deviation decreased because the measured data include both natural variation and variation due to different levels of imperviousness. The adjustment corrected for the latter source of variation. The adjusted flood frequency curve will, in general, be higher than the curve for the measured series but will have a smaller slope. The higher curve reflects the effect of the higher amount of imperviousness (40%). The lower slope reflects the fact that the adjusted series is for a single level of imperviousness. The computations for the adjusted and unadjusted flood frequency curves are given in Table 5–15. The percent increase in the 2-, 5-, 10-, 15-, 50-, and 100-yr flood magnitudes are also given in Table 5–15. The change is relatively minor because the imperviousness did not change after 1964 and the change was only minor (that is, 10%) from 1942 to 1964; also most of the larger storm events occurred after the watershed had reached the developed condition. The adjusted series would represent the annual flood series for a constant urbanization condition (for example, 40% imperviousness). Of course, the adjusted series is not a measured series, and its accuracy depends on the representativeness of Figure 5–22 for measuring the effects of urbanization.

5.4.7 Testing the Significance of Urbanization

A basic assumption of frequency analysis is that each of the values in the record were sampled from the same population. In hydrologic terms, this assumption implies that the watershed has not undergone a systematic change. Obviously, the state of a watershed is continually changing. Seasonal variations of land cover and soil moisture conditions are important hydrologically. These sources of variation along with the variation introduced into floods by the storm-to-storm variation in rainfall cause the variation in the floods in the annual maximum flood series. These factors are usually considered as sources of random variation and do not violate the assumption of watershed homogeneity.

Major land use changes, including urban development, afforestation, and deforestation, introduce a source of variation into a flood record that violates the assumption of homogene-

TABLE 5–14 Adjustment of Rubio Wash Annual Flood Record for Urbanization

Measured Series					Ordered Data			
Year	Urbanization (%)	Annual Peak	Rank	Exceed. Prob.	Rank	Annual Peak	Year	Exceed. Prob.
1929	18.0	661.	47	.9592	1	3700.0	1970	.0204
1930	18.0	1690.	30	.6122	2	3180.0	1974	.0408
1931	19.0	798.	46	.9388	3	3166.0	1972	.0612
1932	20.0	1510.	34	.6939	4	3020.0	1951	.0816
1933	20.0	2071.	20	.4082	5	2980.0	1956	.1020
1934	21.0	1680.	31	.6327	6	2890.0	1968	.1224
1935	21.0	1370.	35	.7143	7	2781.0	1958	.1429
1936	22.0	1181.	40	.8163	8	2780.0	1942	.1633
1937	23.0	2400.	14	.2857	9	2740.0	1957	.1837
1938	25.0	1720.	29	.5918	10	2650.0	1946	.2041
1939	26.0	1000.	43	.8776	11	2610.0	1976	.2245
1940	28.0	1940.	26	.5306	12	2540.0	1969	.2449
1941	29.0	1201.	38	.7755	13	2460.0	1967	.2653
1942	30.0	2780.	8	.1633	14	2400.0	1937	.2857
1943	31.0	1930.	27	.5510	15	2310.0	1953	.3061
1944	33.0	1780.	28	.5714	16	2300.0	1965	.3265
1945	34.0	1630.	32	.6531	17	2290.0	1950	.3469
1946	34.0	2650.	10	.2041	18	2200.0	1952	.3673
1947	35.0	2090.	19	.3878	19	2090.0	1947	.3878
1948	36.0	530.	48	.9796	20	2071.0	1933	.4082
1949	37.0	1060.	42	.8571	21	2070.0	1975	.4286
1950	38.0	2290.	17	.3469	22	2041.0	1966	.4490
1951	38.0	3020.	4	.0816	23	2040.0	1964	.4694
1952	39.0	2200.	18	.3673	24	1985.0	1973	.4898
1953	39.0	2310.	15	.3061	25	1970.0	1955	.5102
1954	39.0	1290.	36	.7347	26	1940.0	1940	.5306
1955	39.0	1970.	25	.5102	27	1930.0	1943	.5510
1956	39.0	2980.	5	.1020	28	1780.0	1944	.5714
1957	39.0	2740.	9	.1837	29	1720.0	1938	.5918
1958	39.0	2781.	7	.1429	30	1690.0	1930	.6122
1959	39.0	985.	44	.8980	31	1680.0	1934	.6327
1960	39.0	902.	45	.9184	32	1630.0	1945	.6531
1961	39.0	1200.	39	.7959	33	1570.0	1963	.6735
1962	39.0	1180.	41	.8367	34	1510.0	1932	.6939
1963	39.0	1570.	33	.6735	35	1370.0	1935	.7143
1964	40.0	2040.	23	.4694	36	1290.0	1954	.7347
1965	40.0	2300.	16	.3265	37	1240.0	1971	.7551
1966	40.0	2041.	22	.4490	38	1201.0	1941	.7755
1967	40.0	2460.	13	.2653	39	1200.0	1961	.7959
1968	40.0	2890.	6	.1224	40	1181.0	1936	.8163
1969	40.0	2540.	12	.2449	41	1180.0	1962	.8367
1970	40.0	3700.	1	.0204	42	1060.0	1949	.8571
1971	40.0	1240.	37	.7551	43	1000.0	1939	.8776
1972	40.0	3166.	3	.0612	44	985.0	1959	.8980
1973	40.0	1985.	24	.4898	45	902.0	1960	.9184
1974	40.0	3180.	2	.0408	46	798.0	1931	.9388
1975	40.0	2070.	21	.4286	47	661.0	1929	.9592
1976	40.0	2610.	11	.2245	48	530.0	1948	.9796

(continued)

TABLE 5–14 Iteration 1 (*Continued*)

Year	Urbanization (%)	Measured Peak	Correction Factor Exist.	Correction Factor Ultim.	Correction Factor Peak	Adjusted Series Rank	Adjusted Series Exceed. Prob.
1929	18.0	661.	1.560	2.075	879.3	47	.9592
1930	18.0	1690.	1.434	1.846	2175.6	22	.4490
1931	19.0	798.	1.573	2.044	1037.1	44	.8980
1932	20.0	1510.	1.503	1.881	1889.3	32	.6531
1933	20.0	2071.	1.433	1.765	2551.4	13	.2653
1934	21.0	1680.	1.506	1.855	2069.4	25	.5102
1935	21.0	1370.	1.528	1.890	1694.8	34	.6939
1936	22.0	1181.	1.581	1.943	1451.4	36	.7347
1937	23.0	2400.	1.448	1.713	2838.1	8	.1633
1938	25.0	1720.	1.568	1.838	2016.7	28	.5714
1939	26.0	1000.	1.690	1.984	1173.8	42	.8571
1940	28.0	1940.	1.603	1.814	2194.3	20	.4082
1941	29.0	1201.	1.703	1.920	1354.0	37	.7551
1942	30.0	2780.	1.508	1.648	3037.7	5	.1020
1943	31.0	1930.	1.663	1.822	2114.5	23	.4694
1944	33.0	1780.	1.705	1.830	1910.3	31	.6327
1945	34.0	1630.	1.752	1.863	1733.5	33	.6735
1946	34.0	2650.	1.585	1.672	2795.3	10	.2041
1947	35.0	2090.	1.675	1.757	2191.6	21	.4286
1948	36.0	530.	2.027	2.123	555.1	48	.9796
1949	37.0	1060.	1.907	1.969	1094.7	43	.8776
1950	38.0	2290.	1.708	1.740	2332.7	16	.3265
1951	38.0	3020.	1.557	1.583	3068.9	4	.0816
1952	39.0	2200.	1.732	1.748	2220.5	19	.3878
1953	39.0	2310.	1.706	1.722	2331.0	17	.3469
1954	39.0	1290.	1.881	1.900	1303.2	38	.7755
1955	39.0	1970.	1.788	1.806	1989.1	29	.5918
1956	39.0	2980.	1.589	1.602	3004.4	6	.1224
1957	39.0	2740.	1.646	1.660	2763.7	11	.2245
1958	39.0	2781.	1.620	1.634	2804.5	9	.1837
1959	39.0	985.	1.979	2.001	995.7	45	.9184
1960	39.0	902.	1.999	2.020	911.9	46	.9388
1961	39.0	1200.	1.911	1.931	1212.5	40	.8163
1962	39.0	1180.	1.935	1.956	1192.5	41	.8367
1963	39.0	1570.	1.853	1.872	1585.9	35	.7143
1964	40.0	2040.	1.790	1.790	2040.0	27	.5510
1965	40.0	2300.	1.731	1.731	2300.0	18	.3673
1966	40.0	2041.	1.781	1.781	2041.0	26	.5306
1967	40.0	2460.	1.703	1.703	2460.0	15	.3061
1968	40.0	2890.	1.619	1.619	2890.0	7	.1429
1969	40.0	2540.	1.693	1.693	2540.0	14	.2857
1970	40.0	3700.	1.480	1.480	3700.0	1	.0204
1971	40.0	1240.	1.910	1.910	1240.0	39	.7959
1972	40.0	3166.	1.559	1.559	3166.0	3	.0612
1973	40.0	1985.	1.798	1.798	1985.0	30	.6122
1974	40.0	3180.	1.528	1.528	3180.0	2	.0408
1975	40.0	2070.	1.773	1.773	2070.0	24	.4898
1976	40.0	2610.	1.683	1.683	2610.0	12	.2449

TABLE 5–14 Iteration 2 (*Continued*)

Year	Urbanization (%)	Measured Peak	Correction Factor Exist.	Correction Factor Ultim.	Correction Factor Peak	Adjusted Series Rank	Adjusted Series Exceed. Prob.
1929	18.0	661.	1.560	2.075	879.3	47	.9592
1930	18.0	1690.	1.399	1.781	2152.6	22	.4490
1931	19.0	798.	1.548	2.001	1031.6	44	.8980
1932	20.0	1510.	1.493	1.863	1885.1	32	.6531
1933	20.0	2071.	1.395	1.703	2528.5	14	.2857
1934	21.0	1680.	1.475	1.806	2056.7	25	.5102
1935	21.0	1370.	1.522	1.881	1692.9	34	.6939
1936	22.0	1181.	1.553	1.900	1444.6	36	.7347
1937	23.0	2400.	1.405	1.648	2814.1	8	.1633
1938	25.0	1720.	1.562	1.830	2015.1	28	.5714
1939	26.0	1000.	1.680	1.969	1172.4	42	.8571
1940	28.0	1940.	1.567	1.765	2185.4	21	.4286
1941	29.0	1201.	1.695	1.910	1353.1	37	.7551
1942	30.0	2780.	1.472	1.602	3025.8	5	.1020
1943	31.0	1930.	1.637	1.790	2110.2	23	.4694
1944	33.0	1780.	1.726	1.855	1912.5	31	.6327
1945	34.0	1630.	1.760	1.872	1734.1	33	.6735
1946	34.0	2650.	1.585	1.672	2795.3	10	.2041
1947	35.0	2090.	1.690	1.773	2192.9	20	.4082
1948	36.0	530.	2.027	2.123	555.1	48	.9796
1949	37.0	1060.	1.921	1.984	1094.9	43	.8776
1950	38.0	2290.	1.699	1.731	2332.4	16	.3265
1951	38.0	3020.	1.557	1.583	3068.9	4	.0816
1952	39.0	2200.	1.741	1.757	2220.6	19	.3878
1953	39.0	2310.	1.724	1.740	2331.3	17	.3469
1954	39.0	1290.	1.901	1.920	1303.4	38	.7755
1955	39.0	1970.	1.820	1.838	1989.5	29	.5918
1956	39.0	2980.	1.606	1.619	3004.8	6	.1224
1957	39.0	2740.	1.668	1.683	2764.2	11	.2245
1958	39.0	2781.	1.646	1.660	2805.1	9	.1837
1959	39.0	985.	1.999	2.020	995.8	45	.9184
1960	39.0	902.	2.022	2.044	912.0	46	.9388
1961	39.0	1200.	1.923	1.943	1212.6	40	.8163
1962	39.0	1180.	1.935	1.956	1192.5	41	.8367
1963	39.0	1570.	1.871	1.890	1586.0	35	.7143
1964	40.0	2040.	1.822	1.822	2040.0	27	.5510
1965	40.0	2300.	1.748	1.748	2300.0	18	.3673
1966	40.0	2041.	1.814	1.814	2041.0	26	.5306
1967	40.0	2460.	1.722	1.722	2460.0	15	.3061
1968	40.0	2890.	1.634	1.634	2890.0	7	.1429
1969	40.0	2540.	1.713	1.713	2540.0	13	.2653
1970	40.0	3700.	1.480	1.480	3700.0	1	.0204
1971	40.0	1240.	1.931	1.931	1240.0	39	.7959
1972	40.0	3166.	1.559	1.559	3166.0	3	.0612
1973	40.0	1985.	1.846	1.846	1985.0	30	.6122
1974	40.0	3180.	1.528	1.528	3180.0	2	.0408
1975	40.0	2070.	1.798	1.798	2070.0	24	.4898
1976	40.0	2610.	1.693	1.693	2610.0	12	.2449

(*continued*)

TABLE 5–14 Iteration 3 (*Continued*)

Year	Urbanization (%)	Measured Peak	Correction Factor Exist.	Correction Factor Ultim.	Correction Factor Peak	Adjusted Series Rank	Adjusted Series Exceed. Prob.
1929	18.0	661.	1.560	2.075	879.3	47	.9592
1930	18.0	1690.	1.399	1.781	2152.6	22	.4490
1931	19.0	798.	1.548	2.001	1031.6	44	.8980
1932	20.0	1510.	1.493	1.863	1885.1	32	.6531
1933	20.0	2071.	1.401	1.713	2532.2	14	.2857
1934	21.0	1680.	1.475	1.806	2056.7	25	.5102
1935	21.0	1370.	1.522	1.881	1692.9	34	.6939
1936	22.0	1181.	1.553	1.900	1444.6	36	.7347
1937	23.0	2400.	1.405	1.648	2814.1	8	.1633
1938	25.0	1720.	1.562	1.830	2015.1	28	.5714
1939	26.0	1000.	1.680	1.969	1172.4	42	.8571
1940	28.0	1940.	1.573	1.773	2186.9	21	.4286
1941	29.0	1201.	1.695	1.910	1353.1	37	.7551
1942	30.0	2780.	1.472	1.602	3025.8	5	.1020
1943	31.0	1930.	1.637	1.790	2110.2	23	.4694
1944	33.0	1780.	1.726	1.855	1912.5	31	.6327
1945	34.0	1630.	1.760	1.872	1734.1	33	.6735
1946	34.0	2650.	1.585	1.672	2795.3	10	.2041
1947	35.0	2090.	1.683	1.765	2192.2	20	.4082
1948	36.0	530.	2.027	2.123	555.1	48	.9796
1949	37.0	1060.	1.921	1.984	1094.9	43	.8776
1950	38.0	2290.	1.699	1.731	2332.4	16	.3265
1951	38.0	3020.	1.557	1.583	3068.9	4	.0816
1952	39.0	2200.	1.741	1.757	2220.6	19	.3878
1953	39.0	2310.	1.724	1.740	2331.3	17	.3469
1954	39.0	1290.	1.901	1.920	1303.4	38	.7755
1955	39.0	1970.	1.820	1.838	1989.5	29	.5918
1956	39.0	2980.	1.606	1.619	3004.8	6	.1224
1957	39.0	2740.	1.668	1.683	2764.2	11	.2245
1958	39.0	2781.	1.646	1.660	2805.1	9	.1837
1959	39.0	985.	1.999	2.020	995.8	45	.9184
1960	39.0	902.	2.022	2.044	912.0	46	.9388
1961	39.0	1200.	1.923	1.943	1212.6	40	.8163
1962	39.0	1180.	1.935	1.956	1192.5	41	.8367
1963	39.0	1570.	1.871	1.890	1586.0	35	.7143
1964	40.0	2040.	1.822	1.822	2040.0	27	.5510
1965	40.0	2300.	1.748	1.748	2300.0	18	.3673
1966	40.0	2041.	1.814	1.814	2041.0	26	.5306
1967	40.0	2460.	1.722	1.722	2460.0	15	.3061
1968	40.0	2890.	1.634	1.634	2890.0	7	.1429
1969	40.0	2540.	1.703	1.703	2540.0	13	.2653
1970	40.0	3700.	1.480	1.480	3700.0	1	.0204
1971	40.0	1240.	1.931	1.931	1240.0	39	.7959
1972	40.0	3166.	1.559	1.559	3166.0	3	.0612
1973	40.0	1985.	1.846	1.846	1985.0	30	.6122
1974	40.0	3180.	1.528	1.528	3180.0	2	.0408
1975	40.0	2070.	1.798	1.798	2070.0	24	.4898
1976	40.0	2610.	1.693	1.693	2610.0	12	.2449

TABLE 5–15 Computation of Flood Frequency Curves for Rubio Wash Watershed in Both Actual (Q) and Ultimate Development (Q_a) Conditions[a]

(1) p	(2) K	(3) $\log_{10}Q$	(4) Q (ft³/sec)	(5) $\log_{10}Q_a$	(6) Q_a (ft³/sec)	(7)
0.99	−2.68572	2.7386	548	2.8006	632	
0.90	−1.32309	2.9989	998	3.0439	1106	
0.70	−0.45812	3.1642	1459	3.1983	1579	
0.50	0.08302	3.2676	1852	3.2949	1972	0.065
0.20	0.85653	3.4153	2602	3.4329	2710	0.041
0.10	1.21618	3.4840	3048	3.4971	3142	0.031
0.04	1.56740	3.5511	3557	3.5598	3629	0.020
0.02	1.77716	3.5912	3901	3.5973	3956	0.014
0.01	1.95472	3.6251	4218	3.6290	4256	0.009

[a] p, exceedence probability; K, LP3 variate for $g_w = -0.5$.

(3) $\log_{10}Q = \bar{Y} + KS_y$

(4) $Q = 10^{**}\log_{10}Q$

(5) $\log_{10}Q_a = \bar{Y}_a + KS_{ya}$

(6) $Q_a = 10^{**}\log_{10}Q_a$

ity. Where these systematic variations are hydrologically significant, they can introduce considerable variation into a flood record such that flood magnitudes or probabilities estimated from a flood frequency analysis will not be accurate indicators of flooding for the changed watershed condition. For example, if a watershed had undergone urban development from 0% to 50% imperviousness over the duration of the flood record and flood magnitudes will be computed from a flood frequency analysis for the current state of 50% imperviousness, then estimates from the frequency curve will not be representative of this condition because many of the floods used to derive the frequency curve were for a condition of much less urban development.

Where a flood record is suspected of lacking homogeneity, the data should be tested prior to making a flood frequency analysis. Statistical tests are intended to be used for this purpose, although the result of a statistical test should not be accepted unless a hydrologic basis for the result is strongly suspected. That is, before a flood record is adjusted, the results of both hydrologic and statistical assessments should suggest the need for adjustment.

Several statistical methods are available to test for nonhomogeneity. The selection of a test is not arbitrary. The type of nonhomogeneity should be considered in selecting the test to use. Urban development can take place abruptly or gradually. In hydrologic terms, an abrupt change is one that occurs over a small portion of the length of record. For example, a 40-yr flood record collected from 1950 to 1989 on a small watershed that urbanized from 1962 to 1966 would be suitable for analysis with a statistical method appropriate for an abrupt change. Conversely, if the urbanization occurred from 1955 to 1982, then a statistical method sensitive to gradual change would be more appropriate. A test designed for one type of nonhomogeneity but used with another type may not detect the hydrologic effect.

If a flood record has been analyzed and found to lack homogeneity, then the flood magnitudes should be adjusted prior to making the frequency analysis. Flood magnitudes computed from an adjusted series can be significantly different from magnitudes computed from the unadjusted series.

Two statistical tests will be introduced here to test for the nonhomogeneity of a flood record. Both are appropriate for records that have undergone a gradual change. Others are available for detecting change in a flood record due to an abrupt watershed change.

Spearman test. The Spearman test is appropriate for cases where the watershed change is gradual and annual measures of the watershed change are available for each year of the flood record. In this sense, it is a bivariate test. As an example, if a 20-yr flood record is available and the percentage of impervious cover in the watershed is available for each year of record, then the Spearman test could be used.

The Spearman test follows the six general steps of a hypothesis test, which are discussed in Chapter 2:

1. State the hypotheses. With respect to watershed change, the null (H_O) and alternative (H_A) hypotheses are as follows:

   H_O: The magnitude of the flood peaks are independent of the urbanization.

   H_A: Urbanization increases the magnitudes of the flood peaks.

   The statement of the alternative hypothesis indicates that the test is applied as a one-tailed test, with a positive correlation expected.

2. Specify the test statistic. The Spearman test uses the following test statistic, which is referred to as the Spearman correlation coefficient (R_s):

$$R_s = 1 - \frac{6\sum_{i=1}^{n} d_i^2}{n^3 - n} \tag{5–31}$$

   where n is the sample size, and d_i is the ith difference in the ranks of the two series. In this case, the two series would be the series of annual peaks and the series of the average annual percentage of urbanization (that is, imperviousness). The measurements of each series are ranked from largest to smallest and the difference in ranks d_i computed.

3. Set the level of significance. While the value of 5% is used most frequently, values of 1% and 10% are also used. The decision is most often made from convention rather than as a rational analysis of the situation.

4. Compute the sample value of the test statistic. The sample value of R_s is found using the following steps:
   (a) While keeping the two series in chronological order, find the rank of each event in each series.
   (b) Compute the difference in ranks for each i.
   (c) Compute the value of the Spearman test statistic R_s using Equation 5–31.

5. Obtain the critical value of the test statistic. For this test, the critical value is obtained from Table 5–16. The values in the table are for the upper tail. Since the distribution of R_s is symmetric, critical values for the lower tail, if needed, are obtained by placing a negative sign in front of the value from Table 5–16.

6. Make a decision. For an upper one-tailed test, the null hypothesis is rejected if the computed value (Step 4) is larger than the critical value (Step 5). For a lower one-tailed

TABLE 5–16 Critical Values for the Spearman Correlation Coefficient for the Null Hypothesis H_o: $|\rho_s| = 0$ and both the One-Tailed Alternative H_a: $|\rho_s| > 0$ and the Two-Tailed Alternative H_a: $\rho_s \neq 0$

	Level of Significance for a One-Tailed Test			
	0.050	0.025	0.010	0.005
	Level of Significance for a Two-Tailed Test			
Sample size	0.100	0.050	0.020	0.010
5	.900	1.000	1.000	1.000
6	.829	.886	.943	1.000
7	.714	.786	.893	.929
8	.643	.738	.833	.881
9	.600	.700	.783	.833
10	.564	.648	.745	.794
11	.536	.618	.709	.818
12	.497	.591	.703	.780
13	.475	.566	.673	.745
14	.457	.545	.646	.716
15	.441	.525	.623	.689
16	.425	.507	.601	.666
17	.412	.490	.582	.645
18	.399	.476	.564	.625
19	.388	.462	.549	.608
20	.377	.450	.534	.591
21	.368	.438	.521	.576
22	.359	.428	.508	.562
23	.351	.418	.496	.549
24	.343	.409	.485	.537
25	.336	.400	.475	.526
26	.329	.392	.465	.515
27	.323	.385	.456	.505
28	.317	.377	.448	.496
29	.311	.370	.440	.487
30	.305	.364	.432	.478

test, the null hypothesis is rejected if the computed value is more negative than the critical value. For a two-tailed test, the null hypothesis is rejected if the computed value lies in either tail, but the level of significance is twice that of the one-tailed test. If the null hypothesis is rejected, one can conclude that the watershed change has caused a significant hydrologic change in the annual maximum flood series.

Example 5–12

Table 5–17 gives nine annual maximum floods for the Compton Creek watershed at Greenleaf Drive in Los Angeles. The corresponding imperviousness is also given. Since the temporal trend is gradual, rather than abrupt, the Spearman test is appropriate.

The ranks of the series are given in columns 4 and 5 of Table 5–17, along with the differences in column 6. The computed value of R_s is

TABLE 5–17 Spearman and Spearman-Conley Tests of Compton Creek Flood Record

Year	Annual Maximum Discharge Y_i	Average Imperv-iousness $x_i(\%)$	Rank of Y_i r_{yi}	Rank of X_i r_{xi}	Difference $d_i = r_{yi} - r_{xi}$
1949	425	40	9	9	0
1950	900	42	7	8	−1
1951	700	44	8	7	1
1952	1250	45	3	6	−3
1953	925	47	6	5	1
1954	1200	48	4	4	0
1955	950	49	5	3	2
1956	1325	51	2	2	0
1957	1950	52	1	1	0
					$\Sigma d_i^2 = 16$

Annual Maximum Discharge Y_t(cfs)	X_t = offset Y_t (cfs)	Rank of Y_i r_{yi}	Rank of X_i r_{xi}	Difference $d_i = r_{yi} - r_{xi}$
900	425	7	8	−1
700	900	8	6	2
1250	700	3	7	−4
925	1250	6	2	4
1200	925	4	5	−1
950	1200	5	3	2
1325	950	2	4	−2
1950	1325	1	1	0
				$\Sigma d_i^2 = 44$

$$R_s = 1 - \frac{6(16)}{9^3 - 9} = 0.867$$

From Table 5–16 the critical value is 0.600. Therefore, the null hypothesis is rejected, which leads to the conclusion that the urbanization has had a significant effect on the annual maximum series.

Spearman-Conley test. In many cases, the record for the land-use-change variable is incomplete. Typically, records of the imperviousness are not available on a year-to-year basis. In such cases, the Spearman test cannot be used.

The Spearman-Conley test (Conley and McCuen, 1997) is an alternative test that can be used to test for serial correlation in cases where the values of the land-use-change variable are incomplete. The steps for applying the Spearman-Conley test are as follows:

1. State the hypotheses. For this test, the hypotheses are as follows:

   H_O: The annual flood peaks are serially independent.

   H_A: Adjacent values of the annual flood series are correlated.

For a flood series suspected of being influenced by urbanization the alternative hypothesis could be expressed as a one-tailed test with an indication of positive correlation. Significant urbanization would cause the peaks to increase, which would produce a positive correlation coefficient.

2. Specify the test statistic. Equation 5–31 can also be used as the test statistic for the Spearman-Conley test. However, it will be denoted as R_{sc}. In applying it, the value of n is the number of pairs, which is 1 less than the number of annual maximum flood magnitudes in the record. To compute the value of R_{sc}, a second series X_t is formed, where $X_t = Y_{t-1}$. The computed value of R_{sc} ranks the values of the two series, each with one fewer value than the original flood series, and uses Equation 5–31.
3. Set the level of significance. Again, this is usually set by convention, with a 5% value common.
4. Compute the sample value of the test statistic. The sample value of R_{sc} is computed using the following steps:
   (a) Create a second series of flood magnitudes (X_t) by offsetting the actual series (Y_{t-1}).
   (b) While keeping the two series in chronological order, identify the rank of each event in each series using a rank of 1 for the largest value and successively larger ranks for events in decreasing order.
   (c) Compute the difference in ranks for each i.
   (d) Compute the value of the Spearman-Conley test statistic R_{sc} using Equation 5–31.
5. Obtain the critical value of the test statistic. Unlike the Spearman test, the distribution of R_{sc} is not symmetric and is different from that of R_s. Table 5–18 gives the critical values for the upper and lower tails. Enter Table 5–18 with the number of pairs of values used to compute R_{sc}.
6. Make a decision. For a one-tailed test, reject the null hypothesis if the computed R_{sc} is greater in magnitude than the value of Table 5–18. If the null hypothesis is rejected, one can conclude that the annual maximum floods are serially correlated. The hydrologic engineer can then draw the conclusion that the correlation reflects the effect of urbanization.

Example 5–13

To demonstrate the Spearman-Conley test, the annual flood series of Compton Creek will be used; however, the test will be made with the assumption that estimates of imperviousness (column 3 of Table 5–17) are not available.

The flood series, except for the 1949 event, are listed chronologically in column 7 of Table 5–17. The offset values are given in column 8. While the record has nine floods, only eight pairs are listed in columns 7 and 8. One value is lost because the record must be offset. Thus, $n = 8$ for this test. The ranks are given in columns 9 and 10, and the difference in ranks in column 11. The sum of squares of the d_i values equals 44. Thus, the computed value of the test statistic is

$$R_{sc} = 1 - \frac{6(44)}{8^3 - 8} = 0.476$$

For a 5% level of significance and a one-tailed upper test, the critical value (Table 5–18) is 0.464. Therefore, the null hypothesis can be rejected. The values in the flood series are serially correlated. The serial correlation is assumed to be the result of urbanization.

TABLE 5–18 Critical Values for Univariate Analyses with the Spearman-Conley One-Tailed Serial Correlation Test

Sample Size	Lower-Tail Level of Significance (%)					Upper-Tail Level of Significance (%)				
	10	5	1	0.5	0.1	10	5	1	0.5	0.1
5	−0.8	−0.8	−1.0	−1.0	−1.0	0.4	0.6	1.0	1.0	1.0
6	−0.7	−0.8	−0.9	−1.0	−1.0	0.3	0.6	0.8	1.0	1.0
7	−0.657	−0.771	−0.886	−0.943	−0.943	0.371	0.486	0.714	0.771	0.943
8	−0.607	−0.714	−0.857	−0.893	−0.929	0.357	0.464	0.679	0.714	0.857
9	−0.572	−0.667	−0.810	−0.857	−0.905	0.333	0.452	0.643	0.691	0.809
10	−0.533	−0.633	−0.783	−0.817	−0.883	0.317	0.433	0.617	0.667	0.767
11	−0.499	−0.600	−0.746	−0.794	−0.867	0.297	0.406	0.588	0.648	0.745
12	−0.473	−0.564	−0.718	−0.764	−0.836	0.291	0.400	0.573	0.627	0.727
13	−0.448	−0.538	−0.692	−0.734	−0.815	0.287	0.385	0.552	0.601	0.706
14	−0.429	−0.516	−0.665	−0.714	−0.791	0.275	0.369	0.533	0.588	0.687
15	−0.410	−0.495	−0.644	−0.692	−0.771	0.270	0.358	0.521	0.574	0.671
16	−0.393	−0.479	−0.621	−0.671	−0.754	0.261	0.350	0.507	0.557	0.656
17	−0.379	−0.462	−0.603	−0.650	−0.735	0.256	0.341	0.491	0.544	0.641
18	−0.365	−0.446	−0.586	−0.633	−0.718	0.250	0.333	0.480	0.530	0.627
19	−0.354	−0.433	−0.569	−0.616	−0.703	0.245	0.325	0.470	0.518	0.613
20	−0.344	−0.421	−0.554	−0.600	−0.687	0.240	0.319	0.460	0.508	0.601
21	−0.334	−0.409	−0.542	−0.586	−0.672	0.235	0.313	0.451	0.499	0.590
22	−0.325	−0.399	−0.530	−0.573	−0.657	0.230	0.307	0.442	0.489	0.580
23	−0.316	−0.389	−0.518	−0.560	−0.643	0.226	0.301	0.434	0.479	0.570
24	−0.307	−0.379	−0.506	−0.549	−0.632	0.222	0.295	0.426	0.471	0.560
25	−0.301	−0.370	−0.496	−0.537	−0.621	0.218	0.290	0.419	0.463	0.550
26	−0.294	−0.362	−0.486	−0.526	−0.610	0.214	0.285	0.412	0.455	0.540
27	−0.288	−0.354	−0.476	−0.516	−0.600	0.211	0.280	0.405	0.448	0.531
28	−0.282	−0.347	−0.467	−0.507	−0.590	0.208	0.275	0.398	0.441	0.523
29	−0.276	−0.341	−0.458	−0.499	−0.580	0.205	0.271	0.392	0.434	0.515
30	−0.270	−0.335	−0.450	−0.491	−0.570	0.202	0.267	0.386	0.428	0.507

5.4.8 Zero-flood Records

Many parts of the western United States are classified as arid or semiarid. The classification is based, in part, on the rainfall. However, vegetation and soils are also factors in classification. Generally speaking, arid lands are those where natural rainfall is inadequate to support crop growth. Semiarid lands are those where rainfall is only sufficient to support short-season crops.

From an engineering hydrology standpoint, arid and semiarid lands are characterized by little rainfall, which, when it does occur, is usually of an intense nature with runoff having a rapid response. Flash flooding is a major concern in such areas.

Hydrologic data are typically not available in arid and semiarid areas, at least in significant quantities. Where gages have been installed, the records often include years in which little or no rainfall occurred and thus, no significant runoff. In other years, intense rainfalls of short duration produce high peak discharges relative to the total volume of runoff. These factors make it comparatively difficult to provide estimates of flood magnitudes or probabilities.

Annual maximum flood records that include values of zero are not uncommon in arid regions. Thus, a frequency curve based on logarithms, such as the log-Pearson Type III, cannot be developed because the logarithm of zero is minus infinity. Furthermore, very small runoff magnitudes that are not associated with out-of-bank flow are not really floods and should not be included in flood analyses, even when they are the largest flow for the year. In such cases, Bulletin 17B provides a method for computing a frequency curves for records that include zero-flood years. The method involves the conditional probability adjustment. When this method is applied, three frequency curves are computed; these are referred to as the initial or unadjusted curve, the conditional frequency curve, and the synthetic frequency curve. The selection from among the curves to make estimates of flood magnitudes is based on the assessment of the hydrologist.

The procedure to follow when analyzing records that include zero-flood years consists of the following six steps:

1. Preliminary analysis
2. Check for outliers
3. Compute unadjusted frequency curve
4. Compute conditional frequency curve
5. Compute synthetic frequency curve
6. Select a curve to make estimates

These steps are discussed in detail.

Preliminary analysis. The first step in the analysis is to separate the record into two parts, the non-zero floods and the zero floods. The procedures can only be applied when the number of zero-flood years does not exceed 25% of the record; thus,

$$\frac{n_z}{N_t} \leqslant 0.25 \tag{5–32}$$

in which n_z is the number of zero-flood years, and N_t is the total record length including those years with zero floods. After eliminating the zero-flood years, the mean, standard deviation, and standardized skew coefficient of the logarithms are computed with the remainder of the data. The skew should be rounded to the nearest tenth.

Check for outliers. The test for outliers given in Bulletin 17B should be applied (see Section 5.4.5). While low outliers are more common than high outliers in flood records from arid regions, tests should be made for both. The procedure depends on the station skew. If station skew is less than –0.4, check for low outliers first. If station skew is greater than +0.4, check for high outliers first. If the skew is between –0.4 and +0.4, a check for both high and low outliers should be made simultaneously. If low outliers are identified, then they are censored (for example, deleted from the flood record) and the moments recomputed. When high outliers are identified, the moments must be recomputed using the historic-peak adjustment; this requires historic flood information. If historic information is not available, then the high outlier must be retained in the record.

Compute unadjusted frequency curve. The moments of the logarithms from Step 1, or from Step 2 if outliers were identified, are used to compute the unadjusted frequency curve. For this step, station skew rather than weighted skew should be used. For selected exceedence probabilities, values of the LP3 deviates (K) are obtained from Appendix 5–1 for the station skew. The deviates are then used with the log mean ($\bar{Y}$) and log standard deviation (S_y) to compute the logarithm of the discharge:

$$Y = \bar{Y} + KS_y \tag{5–33}$$

The computed discharges are then obtained by $Q = 10^Y$. The frequency curve can be plotted using the Q values and the exceedence probabilities used to obtain the corresponding values of K. The data points can be plotted using a plotting position formula.

Compute the conditional frequency curve. To derive the conditional frequency curve, the conditional probability (p_c) is computed as:

$$p_c = \frac{N_t - n_z}{N_t} = \frac{\textit{number of floods excluding zero values}}{\textit{total number of years in record}} \tag{5–34}$$

If historic information is available, then the conditional probability should be computed by Equation 5–35 rather than Equation 5–34:

$$p_c = \frac{H - WL}{H} \tag{5–35}$$

in which H is the historic record length, L is the number of peaks truncated, and W is the systematic record weight. The probability p_c is then multiplied by each probability used to obtain the K values and plot the unadjusted frequency curve in Step 3. The adjusted probabilities and the discharges computed with Equation 5–33 are plotted on frequency paper to form the conditional frequency curve. If the curve is plotted on the same paper as the unadjusted curve from Step 3, then the conditional frequency curve can be compared to the measured data points.

Compute the synthetic frequency curve. The conditional frequency of Step 4 does not have known moments. Approximate values, which are referred to as synthetic statistics, can be computed. Since there are three moments, three points on the conditional frequency curve will be used to fit the synthetic frequency curve. Specifically, the discharge values for exceedence probabilities of 0.01, 0.1, and 0.50 are used. The discharges are obtained from the conditional frequency curve. The following equations are used to compute the synthetic statistics:

$$G_s = -2.50 + 3.12 \log (Q_{0.01}/Q_{0.10}) \log(Q_{0.10}/Q_{0.50}) \tag{5–36}$$

$$S_s = [\log (Q_{0.01}/Q_{0.50})]/(K_{0.01} - K_{0.50}) \tag{5–37}$$

$$\bar{X}_s = \log (Q_{0.50}) - K_{0.50} (S_s) \tag{5–38}$$

in which $K_{0.50}$ and $K_{0.01}$ are the LP3 deviates obtained from Appendix 5–1 for the synthetic skew G_s. Equation 5–36 for the synthetic skew is an approximation for use between skew values of −2.0 and +2.5. If appropriate, the synthetic skew can be used to compute a weighted skew, which would be used in place of the synthetic skew.

The synthetic statistics can then be used to compute the synthetic frequency curve. When verifying the synthetic curve, the plotting positions for the synthetic curve should be based on either the total number of years of record or the historic record length H, if the historic adjustment is used.

Select a frequency curve. The first five steps have resulted in three frequency curves: the unadjusted, the conditional, and the synthetic curves. All three are of potential value for making flood estimates. Each should be compared to the measured data and the goodness of fit assessed. The disadvantages of the unadjusted curve are that it uses station skew, which can be highly variable for small record lengths, and that it does not account for the zero years in the record, which can be significant if the number of zero-flood years is relatively large. The adjustment with Equation 5–34 or Equation 5–35 is an attempt to overcome the lack of accountability for zero-flood years, but since the zero-flood years are essentially years of low rainfall, applying the adjustment of Equation 5–34 to the high-flow years may produce a distortion on the high end of the curve. The disadvantages of the synthetic curve are that it depends on three exceedence probabilities, which have been selected conceptually, and that the form of Equations 5–36, 5–37, and 5–38 are subjective. These disadvantages should be considered when selecting one of the curves to make estimates.

Example 5–14

Table 5–19 contains the annual maximum discharge record (1932–1973) for Orestimba Creek near Newman, CA (USGS Gaging Station 11-2745). This record was analyzed in Bulletin 17B. The record includes six years in which there was no discharge. To ensure that the adjustment method is applicable, the ratio of the number of zero-flood years to the total record length must be less than or equal to 0.25. In this case, the method can be applied since 6/42 equals 0.143.

The six zero values are dropped from the record, which gives $n = 36$, and the moments of the logarithms computed:

3.0786 = Mean of the logarithms

0.6443 = Standard deviation of the logarithms

−0.8360 = Standardized skew coefficient of the logarithms

The skew is rounded to the nearest tenth, which in this case is −0.8.

The remaining record ($n = 36$) should be checked for outliers. Since the skew is less than −0.4, the record is first checked for low outliers. For a 36-yr record length, an outlier deviate (K_o) of 2.639 is obtained from Table 5–11. The logarithm of the critical flow for low outliers is computed as follows:

$$\begin{aligned}\log Q_o &= \overline{X} - K_o S \\ &= 3.0786 - 2.639(0.6443) \\ &= 1.3783\end{aligned}$$

Therefore, the critical flow (Q_0) is 23.9 ft^3/sec. Since one of the flows in Table 5–19 is less than this critical flow, the 1955 flow of 16 cfs is considered a low outlier. The value is censored and the remaining 35 values are used to compute the following moments of the logarithms:

TABLE 5–19 Annual Maximum Flood Series: Orestimba Creek, CA (Station 11-2745)

Year	Flow (cms)	Log of Flow	Plotting Probability
1932	120.630	2.081	0.222
1933	9.769	0.990	0.806
1934	14.611	1.165	0.750
1935	37.378	1.573	0.556
1936	33.980	1.531	0.611
1937	61.731	1.791	0.417
1938	91.463	1.961	0.333
1939	3.256	0.513	0.972
1940	97.410	1.989	0.306
1941	86.933	1.939	0.361
1942	53.236	1.726	0.444
1943	182.644	2.262	0.083
1944	36.529	1.563	0.583
1945	169.052	2.228	0.111
1946	22.144	1.345	0.667
1947	0.000	*	*
1948	0.000	*	*
1949	9.486	0.977	0.833
1950	4.955	0.695	0.861
1951	82.685	1.917	0.389
1952	103.640	2.016	0.278
1953	4.163	0.619	0.917
1954	0.000	*	*
1955	0.453	−0.344	+
1956	159.141	2.202	0.139
1957	40.776	1.610	0.528
1958	288.832	2.461	0.028
1959	152.345	2.183	0.167
1960	12.686	1.103	0.778
1961	0.000	*	*
1962	49.271	1.693	0.472
1963	235.030	2.371	0.056
1964	4.417	0.645	0.889
1965	15.857	1.200	0.722
1966	3.625	0.559	0.944
1967	118.931	2.075	0.250
1968	0.000	*	*
1969	143.850	2.158	0.194
1970	28.600	1.456	0.639
1971	16.537	1.218	0.694
1972	0.000	*	*
1973	42.758	1.631	0.500

*Zero-flow year

[+]Low outlier

$$3.1321 = \text{Mean of the logarithms}$$

$$0.5665 = \text{Standard deviation of the logarithms}$$

$$-0.4396 = \text{Standardized skew of the logarithms}$$

It is next necessary to check for high outliers. For a sample size of 35, the outlier deviate (K_o) from Table 5–11 is 2.628. Thus, the logarithm of the critical flow for high outliers is

$$\begin{aligned}\log Q_0 &= \bar{X} + K_o S \\ &= 3.1321 + 2.628(0.5665) \\ &= 4.6209\end{aligned}$$

Therefore, the critical flow (Q_0) is 41,770 ft^3/sec. None of the flows in the record exceeded this; thus, there are no high outliers.

The unadjusted curve is computed using the 35 values. The mean, standard deviation, and skew of the logarithms are given above. The skew is rounded to –0.4. The computations of the unadjusted curve are given in Table 5–20 and the curve is shown in Figure 5–23, with the values of column 4 of Table 5–20 plotted versus the exceedence probabilities of column 1.

Using the statistics for the censored series with $n = 35$, the conditional frequency curve is computed using the conditional probability adjustment. Log-Pearson Type III deviates are obtained from Appendix 5–1 for a skew of –0.4 and selected exceedence probabilities (see Table 5–20). Since there are 35 events remaining after removing the zero flows and the outlier, the expected probability of Equation 5–33 is 35/42 = 0.8333.

The frequency curves with and without the conditional probability adjustment are shown in Figure 5–23. The conditional curve graphs the flow of column 4 of Table 5–20 versus the probability of column 5. The measured data ($n = 35$) are also plotted in Figure 5–23. Neither curve provides a good representation of the data in the lower tail. While the low outlier is not included with the plotted data, it is of interest to plot the point.

The synthetic statistics can be computed using Equations 5–36, 5–37, and 5–38. These require values of discharges from the adjusted frequency curve for exceedence probabilities of 0.01, 0.1, and 0.5, which are denoted as $Q_{0.01}$, $Q_{0.10}$, and $Q_{0.50}$, respectively. These three values must be estimated graphically because the probabilities do not specifically appear in the computations (column 5) of Table 5–20. There is no mathematical equation that represents the adjusted curve. The values from the adjusted curve of Figure 5–23 are as follows: $Q_{0.01}$ = 17,940 cfs, $Q_{0.10}$ = 6,000 cfs, and $Q_{0.50}$ = 1060 cfs, respectively. Thus, the synthetic skew of Equation 5–36 is

$$G_s = -2.50 + 3.12\left[\frac{\log(17{,}940/6000)}{\log(6000/1060)}\right] = -0.529$$

The computed value of –0.529 should be rounded to the nearest tenth; thus, $G_s = -0.5$. The synthetic standard deviation is

$$S_s = \frac{\log(17{,}940/1060)}{1.95472 - 0.08302} = 0.6564$$

where the values of $K_{0.01}$ and $K_{0.50}$ are obtained from Appendix 5–1 using the synthetic skew of –0.5. The synthetic mean is

$$\begin{aligned}\bar{X}_s &= \log(1060) - 0.08302(0.6564) \\ &= 2.971\end{aligned}$$

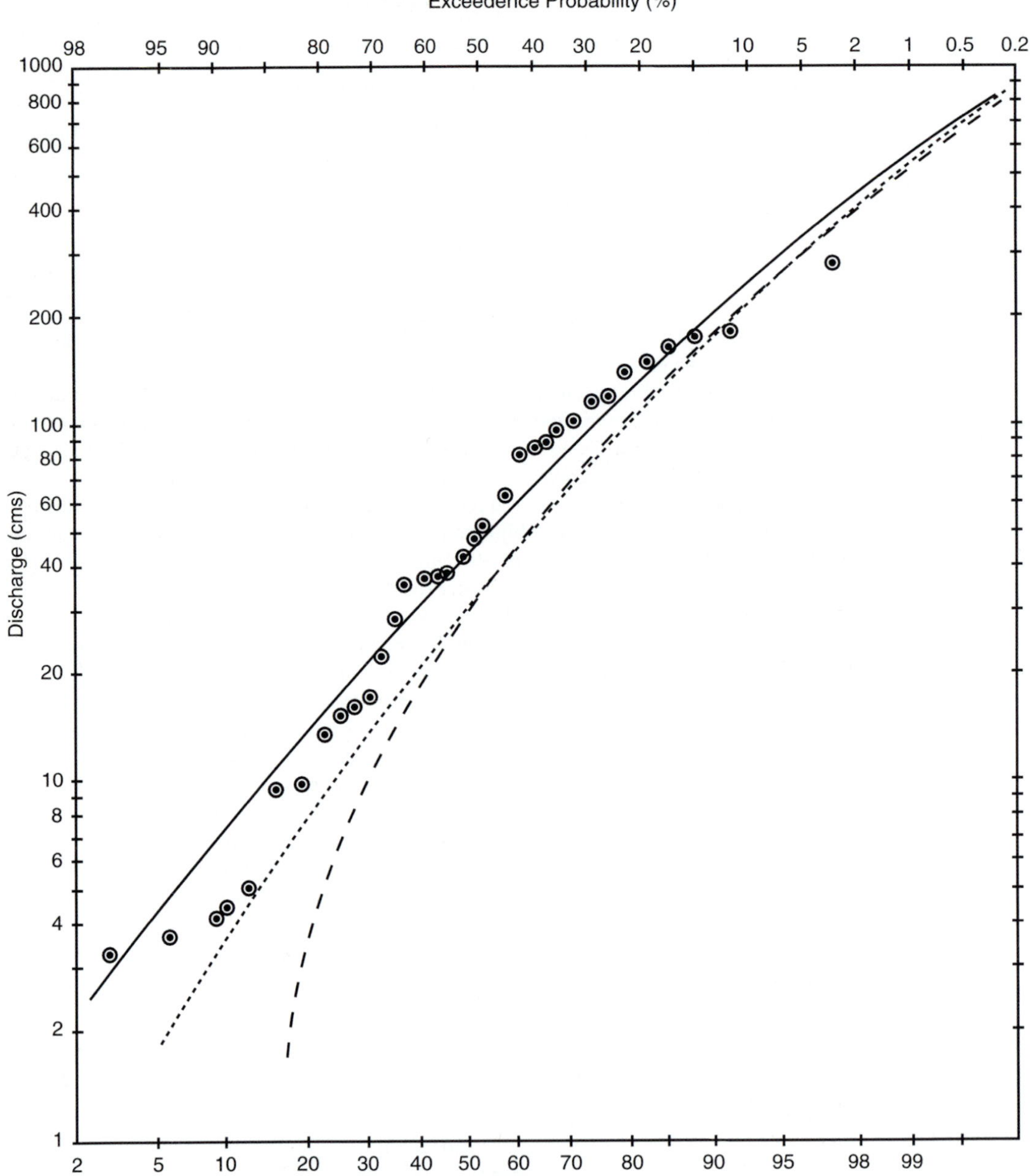

FIGURE 5–23 Unadjusted (——), conditional (— —), and synthetic (– – – –) frequency curves for Orestimba Creek, CA.

TABLE 5–20 Computation of Unadjusted and Conditional Frequency Curves

(1) Exceedence Probability P_d	(2) LP3 Deviate (K) for $g = -0.4$	(3) log Q	(4) Q (cms)	(5) Adjusted Exceedence Probability
0.99	−2.61539	0.1030	1.268	0.825
0.90	−1.31671	0.8387	6.897	0.750
0.70	−0.47228	1.3171	20.75	0.583
0.50	0.06651	1.6223	41.91	0.417
0.20	0.85508	2.0690	117.2	0.167
0.10	1.23114	2.2820	191.4	0.083
0.04	1.60574	2.4943	312.1	0.033
0.02	1.83361	2.6233	420.1	0.017
0.01	2.02933	2.7342	542.3	0.0083
0.002	2.39942	2.9439	878.8	0.0017

(3) $\log Q = \overline{X} + KS$
$= 1.5846 + 0.5665\,K$

(5) $(35/42) * P_d$

A weighted skew should be used with the synthetic mean and synthetic standard deviation to compute the final frequency curve. The generalized skew coefficient for the location of the gage is –0.3, with a mean square error of 0.302. The mean square error for the synthetic skew is 0.163. Thus, the weighted skew is

$$G_w = \frac{0.302\,(-0.529) + 0.163\,(-0.3)}{0.302 + 0.163} = -0.449$$

This can be rounded to the nearest tenth; thus, $G_w = -0.4$, which is used to obtain the deviate K values from Appendix 5–1.

The synthetic curve is computed using the following equation:

$$\log Q = 2.971 + K\,(0.6564)$$

where the K values are obtained from Appendix 5–1. The computations are provided in Table 5–21. The synthetic curve is plotted in Figure 5–23.

None of the three curves closely follow the trend in the measured data, especially in the lower tail. The synthetic curve is based, in part, on the generalized skew, which is the result of regionalization of values from watersheds that may have different hydrologic characteristics than those of Orestimba Creek. In order to make estimates of flood magnitudes, one of the curves must be selected. This would require knowledge of the watershed and sound judgment of the hydrologist responsible for the analysis.

5.5 RISK ASSESSMENT

Three probability distributions have been used to construct flood frequency curves: the normal, the log-normal, and the log-Pearson Type III. Other probability distributions, including the extreme value and the Wakeby, have been proposed and used as alternatives to the above

TABLE 5–21 Computation of the Synthetic Frequency Curve

(1) Exceedence Probability P_d	(2) LP3 Deviate (K) for $g = -0.4$	(3) log Q	(4) Q (cms)
0.99	−2.61539	−0.2969	0.5
0.90	−1.31671	0.5569	3.6
0.70	−0.47228	1.1120	12.9
0.50	0.06651	1.4662	29.3
0.20	0.85508	1.9846	96.5
0.10	1.23114	2.2319	170.5
0.04	1.60574	2.4781	300.7
0.02	1.83361	2.6279	424.5
0.01	2.02933	2.7566	570.9
0.002	2.39942	2.9999	999.7

(3)$\log Q = \overline{X}_s + KS_s$
$= 1.4225 + 0.6574\,K$

three distributions. This suggests that the true distribution is unknown. Our lack of knowledge about the true distribution is one source of uncertainty with computed flood frequency curves. Random sampling is a second source of uncertainty that would exist even if we knew the correct distribution function to use in representing the population. Statistical confidence intervals are the means of placing bounds on the uncertainty due to statistical sampling.

Even if we know the true or correct probability distribution to use in computing a flood frequency curve and also the correct parameter values, we would not be certain about the occurrence of floods over the design life of an engineering structure. A culvert might be designed to pass the 10-yr flood (that is, the flood that has an exceedence probability of 0.1), but over any period of 10 years, the capacity may be reached as many as 10 times or not at all. A cofferdam constructed to withstand up to the 50-yr flood may be exceeded shortly after being constructed, even though the dam will only be in place for one year. These are chance occurrences that are independent of the lack of knowledge of the true probability distribution. That is, the risk would exist even if we knew the true population. Such risk of failure or design uncertainty can be estimated using the concept of binomial risk.

The binomial distribution is a probability mass function that provides a means of estimating probabilities of discrete random variables that meet the following four conditions: (1) there are n trials in the experiment; (2) there are only two possible outcomes for each trial; (3) the probability of occurrence is constant from trial to trial; and (4) the trials are independent. If we define our random variable to be the occurrence of a flood in any one year and assume that the probability defined by the flood frequency curve is constant from year to year, then the occurrence of the random variable can be described by the binomial probability distribution. Letting n and p represent the project life and probability of occurrence of a specific flood magnitude, respectively, the probability that a flood of a certain magnitude will occur exactly x times in the n years is given by

$$p\,(x; n, p) = \binom{n}{x} p^x (1 - p)^{n-x} \tag{5–39}$$

The probability of the 25-yr flood (that is, $p = 0.04$) not occurring (that is, $x = 0$) in a period of 5 years (that is, $n = 5$) is

$$p\,(0; 5, 0.04) = \binom{5}{0} (0.04)^0 (0.96)^{50} = 0.815 \tag{5–40}$$

Therefore, there is an 18.5% chance of getting one or more (up to five) such floods in 5 years. The probability of exactly one occurrence of the 25-yr flood is

$$p\,(1; 5, 0.04) = \binom{5}{1}(0.04)^1(0.96)^4 = 0.170 \tag{5–41}$$

Table 5–22 provides some additional examples of binomial risk for the 10-yr and 100-yr events.

Example 5–15

A bridge with a design life of 40-yr is designed to pass the 100-yr flood. The design engineer is interested in the probability that the bridge will be submerged during its design life.

For a 100-yr design, the exceedence probability is 0.01. For a design life of 40-yr, the probability of the bridge being submerged exactly i times in 40-yr is

$$\binom{40}{i} p^i (1 - p)^{40-i} = \binom{40}{i} (0.01)^i (0.99)^{40-i}$$

where $i = 0, 1, 2, ..., 40$. The probability of exactly two submergences in 40-yr is:

$$\binom{40}{2} (0.01)^2 (0.99)^{38} = \frac{40(39)}{2(1)} (0.01)^2 (0.99)^{38} = 0.0532$$

TABLE 5–22 Binomial Risk Probabilities ($P(i)$) of i Occurrences for Design Life of n Years and an Exceedence Probability p

(a) $n = 10$ years, $p = 0.10$, $T = 10$ years		(b) $n = 100$ years, $p = 0.01$, $T = 100$ years		(c) $n = 50$ years, $p = 0.10$, $T = 10$ years	
i	$P(i)$	i	$P(i)$	i	$P(i)$
0	0.35	0	0.37	0	0.01
1	0.39	1	0.37	1	0.03
2	0.19	2	0.18	2	0.08
3	0.06	3	0.06	3	0.14
4	0.01	4	0.01	4	0.18
5+	<0.01	5+	<0.01	5	0.18
				6	0.15
				7	0.11
				8	0.06
				9	0.03
				10	0.02
				11+	0.01

The following summarizes the probabilities:

i	$p(i)$
0	0.6690
1	0.2703
2	0.0532
3	0.0067
4	0.0006
	0.9998

Thus, the probability that the bridge will be submerged in a 40-yr period is 1-0.6690 = 0.331. The probability of 5 or more submergences is 0.0002.

5.6 PARTIAL-DURATION SERIES

For typical engineering design problems, the annual-maximum series analysis (see Section 5.3) provides the necessary discharge magnitude and exceedence probability. Generally, these design problems require a discharge magnitude for a return period of five years or more. Some types of problems require discharge magnitudes that are much smaller than the 5-yr flood. Flow rates that occur more frequently than once a year are needed in many types of problems, such as in nonpoint source pollution analyses, aquatic habitat analyses, constructed wetland design, small watershed design where damages are not expected, stream restoration projects, and channel erosion analyses. In cases such as these, the return period concept of Equation 4–1 is not appropriate because it does not apply to events with less than the 1-yr magnitude.

Langbein (1949) indicates that the lowest of the annual floods is considered to be the 1-yr flood. The Weibull plotting position formula would suggest a return period of $(N+1)/N$, where N is the record length. Where smaller magnitudes are of interest, a partial-duration series analysis can be undertaken.

A partial-duration-series analysis is a frequency analysis that uses all measured discharges (or stages) above a threshold, which can be either the bankfull discharge, the lowest value of record, or some other recognizable criterion. The threshold may be set so that the number of events in the series equals an average of three to four flows per year. Thus, for a record of 30 years, the highest 90 to 120 magnitudes would be used. Each flood magnitude should be hydrologically independent of the other events in the series. This is difficult to assess. Langbein (1949) indicates that every pair of discharge values in a partial series should be separated by a substantial recession in stage and discharge. A more specific criterion is needed to actually assemble a partial duration series from a flood record. As an example, the criterion might be that the discharge on the recession between the two peaks must fall to at least 25% of the first discharge. The specific criterion should be set prior to analyzing the event. It may also be realistic to set a time period, say one week, that is used to identify independent events.

The concept of a return period, at least as it is applied to an annual-maximum series, cannot be rigorously applied to partial-duration series. The return period from an annual-maximum series represents an average interval in which a flood magnitude will recur as an annual-maximum (Langbein, 1949). The recurrence interval from a partial-duration series is the average interval between floods of a specified magnitude; this does not reflect a time period, whereas the annual-maximum series specifies a time period of one year. Langbein (1949) provided the correspondence between return periods of an annual-maximum series and the recurrence interval for partial-duration series (see Table 5–23).

Example 5–16

A flood record for the Whitemarsh Run at Whitemarsh, MD (USGS Gage No. 01585100) for the period of record available (Feb. 1959 to Sept. 1989) is given in Table 5–24. The threshold discharge is 282 ft^3/sec; this was used because it is the smallest discharge in the annual maximum record and it provided a partial-series record length of 121 events, which represents an average of 4.2 storms per year. Since the largest measured discharge (that is, 8000 ft^3/s) was over twice the second annual maximum (that is, 3880 ft^3/s), the largest value was tested as an outlier using the method of Section 5.4.5. The computed deviate is (Equation 5–23)

$$K = \frac{\log(8000) - 3.129}{0.3166} = 2.445$$

The detection deviate K_0 from Table 5–11 for a sample size of 29 is 2.549. Since the computed value is less than the critical value, the largest event is not detected as an outlier.

The partial-duration series is given in Table 5–24. The annual-maximum events are noted. The return periods for the annual-maximum events and the recurrence intervals for the partial-duration series are expressed using the Weibull plotting position formula.

TABLE 5–23 Relationship between the Return Period of an Annual-Maximum Series and the Recurrence Interval of a Partial-Duration Series (Langbein, 1949)

Partial-Duration Series (years)	Annual Series (years)	Partial-Duration Series (years)	Annual Series (years)
0.25	1.02	5.0	5.52
0.5	1.16	10	10.5
1.0	1.58	20	20.5
1.45	2.00	50	50.5
2.0	2.54	100	100.5

TABLE 5–24 Annual-Maximum and Partial-Duration Series: White Marsh Run (USGS Gage No. 01585100)

				Partial-Duration		Annual-Maximum	
Water Year	Date	Peak Discharge (ft^3/sec)	Gage Height (feet)	Rank	Exceedence Interval (years)	Rank	Return Period (years)
1960	09/12/60	1260	6.60	48	0.62	16	1.9
1961	01/01/61	442	3.39	119	0.25	27	1.11
1962	02/26/62	526	3.63	118	0.25	26	1.15
1963	08/01/63	396	3.26	120	0.25	28	1.07
1964	01/09/64	686	4.09	117	0.26	25	1.20
1965	02/07/65	830	4.60	103	0.29		
1965	08/05/65	902	4.91	89	0.34	22	1.36
1966	02/13/66	794	4.48	108	0.28	23	1.30
1967	03/07/67	978	5.21	73	0.41		
1967	07/03/67	998	5.29	69	0.43		
1967	08/25/67	1140	6.00	53	0.57	19	1.6
1967	08/27/67	1060*	5.60	61	0.49		
1968	01/14/68	872	4.79	94	0.32		
1968	05/28/68	767	4.39	112	0.27		
1968	09/10/68	1060	5.62	63	0.47	21	1.4
1969	06/19/69	282	2.93	121	0.25	29	1.03
1970	12/10/69	752	4.34	116	0.26	24	1.25
1971	02/07/71	767	4.39	111	0.27		
1971	06/14/71	1090	5.73	58	0.52		
1971	08/01/71	8000	14.05	1	30	1	30
1971	08/04/71	1610*	7.73	19	1.6		
1971	08/06/71	1090*	5.75	57	0.53		
1971	08/27/71	1580	7.64	22	1.4		
1971	09/11/71	3760	11.46	3	10		
1972	10/10/71	1180	6.22	52	0.58		
1972	10/26/71	1540	8.00	25	1.2		
1972	11/25/71	918	4.97	86	0.35		
1972	06/22/72	2170	10.04	9	3.3	8	3.7
1972	06/29/72	965	5.19	75	0.40		
1972	07/03/72	868	4.77	96	0.31		
1973	11/08/72	1000	5.31	67	0.45		
1973	11/14/72	980	5.22	71	0.42		
1973	11/20/72	812	4.55	104	0.29		
1973	11/26/72	948	5.09	79	0.38		
1973	12/08/72	1120	5.92	54	0.56		
1973	04/04/73	998	5.29	68	0.44		
1973	05/28/73	808	4.53	106	0.28		
1973	06/22/73	761	4.37	113	0.27		
1973	07/03/73	1670	8.63	17	1.8	11	2.7
1974	12/20/73	1040	5.48	65	0.46		
1974	03/30/74	958	5.11	78	0.38		
1974	05/12/74	860	4.74	97	0.31		
1974	07/06/74	1060	5.61	62	0.48	20	1.5
1975	12/02/74	940	5.07	81	0.37		
1975	12/16/74	1190	6.23	51	0.59		

TABLE 5–24 Annual-Maximum and Partial-Duration Series: White Marsh Run (USGS Gage No. 01585100) (*Continued*)

				Partial-Duration		Annual-Maximum	
Water Year	Date	Peak Discharge (ft³/sec)	Gage Height (feet)	Rank	Exceedence Interval (years)	Rank	Return Period (years)
1975	03/19/75	1460	7.59	29	1.0		
1975	04/25/75	844	4.68	101	0.30		
1975	06/16/75	807	4.53	107	0.28		
1975	06/28/75	1480	7.68	27	1.1		
1975	07/10/75	965*	5.19	76	0.39		
1975	07/13/75	1670	8.66	16	1.9	10	3.0
1975	07/13/75	1460*	7.62	30	1.0		
1975	09/25/75	837*	4.65	102	0.29		
1975	09/26/75	1280	6.70	43	0.70		
1976	01/01/76	1350	7.06	38	0.79		
1976	08/06/76	857	4.73	99	0.30		
1976	09/16/76	3460	11.40	4	7.5	3	10
1977	10/20/76	1100	5.80	56	0.54		
1977	10/31/76	1000	5.36	66	0.45		
1977	03/13/77	1090	5.76	60	0.50		
1977	03/22/77	936	5.05	83	0.36		
1977	05/28/77	1600	8.29	20	1.5	12	2.5
1977	07/25/77	859	4.74	98	0.31		
1977	08/31/77	1570	8.17	23	1.3		
1978	12/18/77	1930	9.57	10	3.0		
1978	01/09/78	1360	7.11	35	0.86		
1978	01/17/78	980	5.26	72	0.42		
1978	01/26/78	905	4.92	88	0.34		
1978	03/26/78	1330	6.95	40	0.75		
1978	06/21/78	902	4.91	90	0.33		
1978	08/01/78	1330	6.95	39	0.77		
1978	08/11/78	1360	7.09	36	0.83		
1978	08/27/78	3880	11.73	2	15	2	15
1979	01/02/79	900	4.90	91	0.33		
1979	01/24/79	849	4.70	100	0.30		
1979	02/24/79	1260	6.58	46	0.65		
1979	02/26/79	1060*	5.59	64	0.47		
1979	04/27/79	1440	7.52	32	0.94		
1979	07/17/79	910	4.94	87	0.34		
1979	07/29/79	1650*	8.57	18	1.7		
1979	08/02/79	1700	8.81	14	2.1		
1979	08/03/79	953*	5.18	77	0.39		
1979	09/05/79	1700*	8.79	15	2.0		
1979	09/06/79	2640	10.65	6	5.0	5	6.0
1979	09/21/79	1580	8.19	21	1.4		
1980	10/01/79	965	5.19	74	0.41		
1980	10/05/79	872	4.79	93	0.32		
1980	11/26/79	936	5.05	82	0.37		
1980	03/13/80	869	4.78	95	0.32		
1980	03/21/80	754	4.35	115	0.26		

(*continued*)

TABLE 5–24 Annual-Maximum and Partial-Duration Series: White Marsh Run (USGS Gage No. 01585100) (*Continued*)

				Partial-Duration		Annual-Maximum	
Water Year	Date	Peak Discharge (ft^3/sec)	Gage Height (feet)	Rank	Exceedence Interval (years)	Rank	Return Period (years)
1980	04/27/80	1350	7.07	37	0.81	15	2.0
1981	06/14/81	1420	7.40	33	0.91	14	2.1
1981	06/25/81	1110	5.86	55	0.55		
1981	07/04/81	1270	6.67	45	0.67		
1981	08/11/81	1270	6.68	44	0.68		
1982	04/26/82	931	5.03	84	0.36		
1982	05/13/82	1210	6.37	49	0.61	17	1.8
1983		2300	10.20	8	3.7	7	4.3
1984		1500	7.80	26	1.2	13	2.3
1985	08/21/85	1.190	6.26	50	0.60	18	1.7
1986	11/20/86	925	5.00	85	0.35		
1986	12/02/86	767	4.40	110	0.27		
1986	12/24/86	1880	9.37	11	2.7	9	3.3
1987	09/08/87	2300	10.21	7	4.3	6	5.0
1987	11/29/87	944	5.09	80	0.37		
1987	05/17/88	882	4.84	92	0.33		
1987	05/23/88	1320*	6.90	42	0.71		
1987	05/24/88	1450	7.53	31	0.97		
1987	07/23/88	1330	6.97	41	0.73		
1987	07/28/88	1260	6.63	47	0.64		
1988	08/29/88	984	5.28	70	0.43		
1988	05/05/89	759*	4.37	114	0.26		
1988	05/06/89	1710	8.81	13	2.3		
1988	05/23/89	1090	5.77	59	0.51		
1988	05/27/89	780	4.45	109	0.28		
1988	06/09/89	1480	7.69	28	1.1		
1988	06/23/89	1570	8.14	24	1.2		
1988	07/05/89	811*	4.57	105	0.29		
1988	07/06/89	1390	7.25	34	0.88		
1988	07/16/89	3140	11.11	5	6.0	4	7.5
1988	07/20/89	1880*	9.43	12	2.5		
$n = 121$							

(1) USGS Flow Gaging Station No. 01585100, now inactive

(2) Period of Record: 1960–1988 Water Years

(3) Partial Duration Series — all floods > selected base flood peak discharge

(4) Recurrence interval from this series is the average frequency of occurrence between floods of a given size irrespective of their relation to the year.

(5) * indicates an event that may not be independent of another event in the series

PROBLEMS

5–1. Simultaneous measurement of depth and flow rate for seven storm events yielded the following data. Using regression analysis with the model of Equation 5–1, find the coefficients of the discharge rating curve. Plot the curve and the data points.

h(ft)	0.6	0.7	0.9	1.1	1.1	1.5	2.2
Q(ft^3/sec)	3	5	15	19	23	94	451

5–2. Eight measurements of depth and velocity during a storm event result in the following data. Using regression analysis with the model of Equation 5–1, find the coefficients of the discharge rating curve. Plot the curve and show the data points.

h(ft)	0.5	0.7	0.7	0.8	1.1	1.3	1.9	2.8
A (ft^2)	10.1	14.3	14.3	17.2	20.8	26.6	40.3	57.3
V (ft/sec)	2.4	3.0	3.3	3.4	4.1	4.3	5.1	6.1

5–3. (a) Determine the cross-sectional area of the natural channel section shown below. If the average velocity is 3.8 ft/sec, compute the discharge (ft^3/sec).

(b) If the cross section is modified for a bridge including abutments and two piers, as shown by the dashed lines, compute the depth versus cross-sectional area relationship; also compute the depth of flow through the cross section if both the velocity and discharge were to be the same as for the natural cross section.

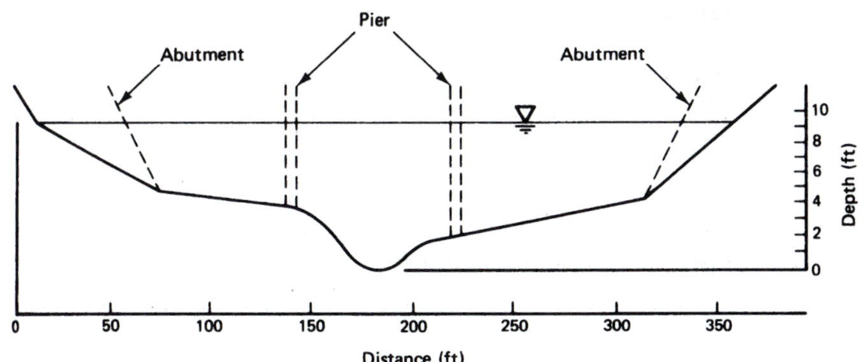

5–4. For the problem of flood frequency analysis, what are the knowns and the unknowns? For synthesis with a flood frequency curve, what are the knowns and the unknowns?

5–5. Draw a vertical axis of about 6 inches. Perpendicular to this, draw two horizontal axes, one connecting at the top and one connecting at the bottom of the vertical axis. On the upper side of the lower horizontal axis, make a linear scale with values of z from −2 to +2 in increments of 0.5. Make the same values on the lower side of the upper horizontal axis. Using Table 2–2, find the probabilities that correspond to the z values on the lower axis and insert the probabilities on the lower side of the lower horizontal axis. Insert 1 minus the probability on the upper side of the upper horizontal axis. Compare the resulting graph to the probability paper of Figure 5–3. Comment on the similarities.

5–6. Given a mean and standard deviation of 3500 and 2000 ft^3/sec, respectively, find the 2-, 10-, and 100-yr peak floods for a normal distribution.

5–7. Given a mean and standard deviation of 1000 and 600 ft^3/sec, respectively, find the 5- and 25-yr peak floods for a normal distribution.

5–8. Assuming a normal distribution with a mean and standard deviation of 2400 and 1200 ft^3/sec, find the exceedence probability and return period for flood magnitudes of 3600 and 5200 ft^3/sec.

5–9. Assuming a normal distribution with a mean and standard deviation of 5700 and 1300 ft^3/sec, respectively, find the exceedence probability and return period for flood magnitudes of 7400 and 8700 ft^3/sec.

5–10. For a sample of 60 compute the Weibull, Hazen, and Cunnane plotting probabilities and the corresponding return periods for the two largest and two smallest ranks. Comment on the results.

5–11. For sample sizes of 5, 20, and 50 compute the Weibull, Hazen, and Cunnane plotting probabilities and the corresponding return periods for the largest and smallest ranks. Discuss the magnitude of the differences as a function of sample size.

5–12. For the random variable X of Figure 5–3, find the following probabilities: (a) $P(X > 4.5)$; (b) $P(X < 6.2)$; (c) $P(X > 6.7)$; (d) $P(X = 7.1)$.

5–13. For the random variable X of Figure 5–3, find the following probabilities: (a) $P(X > 5.3)$; (b) $P(X < 6.5)$; (c) $P(X > 7.1)$; (d) $P(4.3 < X < 6.6)$.

5–14. For the random variable X of Figure 5–3, find X_O for each of the following probabilities: (a) $P(X > X_o) = 0.1$; (b) $P(X > X_o) = 0.8$; (c) $P(X < X_O) = 0.05$; (d) $P(X < X_O) = 0.98$.

5–15. If the random variable X of Figure 5–3 is the peak rate of runoff in a stream (in $ft^3/sec \times 10^3$), find the following: (a) $P(X > 7040\ ft^3/sec)$; (b) $P(X > 3270\ ft^3/sec)$; (c) $P(X < 3360\ ft^3/sec)$; (d) $P(X < 6540\ ft^3/sec)$.

5–16. If the random variable X of Figure 5–3 is the annual sediment yield (in tons/acre $\times 10^1$), find the following: (a) $P(X > 40$ tons/acre/year); (b) $P(X > 67.3$ tons/acre/year); (c) $P(X < 32.7$ tons/acre/year).

5–17. If the random variable X of Figure 5–3 is the peak flow rate (in $ft^3/sec \times 10$) from a 20-acre urban watershed, what is (a) the 10-yr flood; (b) the 25-yr flood; (c) the 100-yr flood?

5–18. Assuming a normal distribution, make a frequency analysis of the January rainfall (P) on Wildcat Creek for the period 1954–1971. Plot the data using the Weibull plotting position formula. Based on the frequency curve, estimate the following: (a) the 100-yr January rainfall; (b) the probability that the January rainfall in any one year will exceed 7 in.

Year	*P*	*Year*	*P*	*Year*	*P*	*Year*	*P*
1954	6.43	1958	3.59	1962	4.56	1967	5.19
1955	5.01	1959	3.99	1963	5.26	1968	6.55
1956	2.13	1960	8.62	1964	8.51	1969	5.19
1957	4.49	1961	2.55	1965	3.93	1970	4.29
				1966	6.20	1971	5.43

5–19. Assuming a normal distribution, make a frequency analysis of the July evaporation (E) on Wildcat Creek for the period 1954-1971. Plot the data using the Weibull plotting position formula.

Based on the frequency curve, estimate the following: (a) the 50-yr July evaporation; (b) the probability that the July evaporation in any one year will exceed 10 in.

Year	*E*	*Year*	*E*	*Year*	*E*	*Year*	*E*
1954	8.30	1958	8.39	1962	8.05	1967	6.16
1955	7.74	1959	7.94	1963	8.03	1968	7.44
1956	6.84	1960	9.24	1964	6.91	1969	9.19
1957	8.45	1961	8.79	1965	7.50	1970	9.24
				1966	6.20	1971	5.43

5–20. Assuming a normal distribution, make a frequency analysis of the annual maximum flood series on the Grey River at Reservoir near Alpine, WY. Plot the data using the Weibull plotting position formula. Estimate the following: (a) The 20-yr annual maximum; (b) the exceedence probability and the return period for an event of 5500 cfs; (c) the index ratio of the 100-yr flood to the 2-yr flood.

4210	3720	3110	3150	2550
2010	2920	2420	4050	5220
5010	2500	4280	3260	3650
4290	2110	3860	7230	3590
			5170	650

5–21. Given a mean and standard deviation of the base 10 logarithms of 0.36 and 0.15, respectively, find the 10-, 50-, and 100-yr rainfall intensities assuming a log-normal distribution.

5–22. Given a mean and standard deviation of the base 10 logarithms of –0.43 and 0.12, respectively, find the 5-, 25-, and 100-yr rainfall intensities assuming a log-normal distribution.

5–23. Assuming a log-normal distribution with a mean and standard deviation of –0.89 and 0.10, respectively, find the exceedence probability and return period for rainfall intensities of 0.22 in./hr and 0.28 in./hr.

5–24. The following data are annual 15-min peak rainfall intensities I (in./hr) for 9 years of record. Compute and plot the $\log_{10}$-normal frequency curve and the data. Use the Weibull plotting position formula. Using both the curve and the mathematical equation estimate (a) the 25-yr, 15-min peak rainfall intensity; (b) the return period for an intensity of 7 in./hr; (c) the probability that the annual maximum 15-min rainfall intensity will be between 4 and 6 in./hr.

Year	*I*	*Year*	*I*
1972	3.16	1977	2.24
1973	2.29	1978	4.37
1974	4.07	1979	6.03
1975	4.57	1980	2.75
1976	2.82		

5–25. Assuming a log-normal distribution, make a frequency analysis of the total annual runoff Q (in.) from Wildcat Creek for the period 1954–1970. Plot the data using the Weibull plotting position formula. Based on the frequency curve, estimate the following: (a) The 50-yr annual runoff; (b) the probability that the total annual runoff in any one year will be less than 5 in.; (c) the probability that the total annual runoff will exceed 15 in.

Year	*Q*	*Year*	*Q*	*Year*	*Q*	*Year*	*Q*
1954	8.56	1959	7.72	1964	22.73	1969	15.77
1955	6.86	1960	11.48	1965	13.28	1970	9.51
1956	14.13	1961	16.76	1966	18.42		
1957	11.76	1962	13.68	1967	19.49		
1958	10.74	1963	15.74	1968	13.09		

5–26. Assuming a log-normal distribution, make a frequency analysis of the mean August temperature (T) at Wildcat Creek for the 1954–1971 period. Plot the data using the Weibull plotting position formula. Based on the frequency curve, estimate the following: (a) The August temperature that can be expected to be exceeded once in 10 years; (b) the probability that the mean August temperature will exceed 83°F in any one year; (c) the probability that the mean August temperature will not exceed 72°F.

Year	*T*	*Year*	*T*	*Year*	*T*	*Year*	*T*
1954	82.5	1959	80.6	1964	76.1	1968	79.6
1955	80.1	1960	78.9	1965	77.7	1969	76.2
1956	80.4	1961	74.1	1966	76.6	1970	79.1
1957	79.5	1962	75.7	1967	74.6	1971	76.7
1958	78.9	1963	78.0				

5–27. Using the data of Problem 5–20, perform a log-normal frequency analysis. Make the three estimations indicated in Problem 5–20 and compare the results for the normal and log-normal analyses.

5–28. Assuming a log-Pearson Type III distribution, make a frequency analysis of the total annual rainfall P at Wildcat Creek for the 1954–1970 period. Plot the data using the Weibull plotting position formula. Based on the frequency curve, estimate the following: (a) The 100-yr annual rainfall; (b) the probability that the total annual rainfall in any one year will be greater than 50 in.

Year	*P*	*Year*	*P*	*Year*	*P*	*Year*	*P*
1954	34.67	1958	42.39	1962	56.22	1966	57.60
1955	36.46	1959	42.57	1963	56.00	1967	64.17
1956	50.27	1960	43.81	1964	69.18	1968	50.28
1957	44.10	1961	58.06	1965	49.25	1969	51.91
						1970	38.23

5–29. Compute a log-Pearson Type III frequency analysis for the data of Problem 5–25. Compare the fit with the log-normal and the estimates of parts (a), (b), and (c).

5–30. Using the data of Problem 5–20, perform a log-Pearson Type III analysis. Make the three estimations indicated in Problem 5–20, and compare the results for the normal, log-normal (Problem 5–27), and LP3 analyses.

5–31. Obtain precipitation data for a local weather bureau station for a period of at least 20 years. Determine the maximum 2-hr rainfall for each year of record. Perform a frequency analysis for a log-Pearson Type III distribution and estimate the 2-yr, 2-hr rainfall, the 10-yr, 2-hr rainfall, and the 100-yr, 2-hr rainfall. Compare these with the values from the local *IDF* curve.

5–32. Obtain a streamflow record of at least 20 years from the U.S. Geological Survey Water Supply Papers. Find the maximum instantaneous peak discharge for each year of record. Perform a frequency analysis for a log-Pearson Type III distribution and estimate the 2-yr (Q_2), 10-yr (Q_{10}), and 100-yr (Q_{100}) peak discharges. Also compute the index ratios of the Q_{10}/Q_2 and Q_{100}/Q_2.

5–33. Make log-Pearson Type III analysis of the data of Problem 5–19. Based on the frequency curve, estimate the following: (a) The 50-yr July evaporation; (b) the probability that the July evaporation in any one year will exceed 10 in.

5–34. The following table gives the annual minimum 7-day average low flows Q from Shoal Creek, Tennessee, for the period from 1926 to 1959. Compute a $\log_{10}$ normal frequency curve and estimate the 25- and 100-yr minimum 7-day average flows. Plot both the frequency curve and the data.

Year	*Q*	*Year*	*Q*	*Year*	*Q*	*Year*	*Q*
1926	99	1936	83	1946	120	1956	80
1927	90	1937	111	1947	116	1957	97
1928	116	1938	112	1948	98	1958	125
1929	142	1939	127	1949	145	1959	124
1930	99	1940	97	1950	202		
1931	63	1941	71	1951	133		
1932	128	1942	84	1952	111		
1933	126	1943	56	1953	101		
1934	93	1944	108	1954	72		
1935	83	1945	123	1955	80		

5–35. The following table gives the annual minimum 7-day average low flows Q for the Buffalo River in Tennessee for the period from 1926 to 1959. Compute a $\log_{10}$-normal frequency curve and estimate the 10- and 50-yr minimum 7-day average low flows. Plot both the frequency curve and the data.

Year	*Q*	*Year*	*Q*	*Year*	*Q*	*Year*	*Q*
1926	99	1936	115	1946	182	1956	120
1927	145	1937	146	1947	146	1957	156
1928	130	1938	155	1948	120	1958	166
1929	159	1939	149	1949	200	1959	165
1930	110	1940	124	1950	226		
1931	96	1941	93	1951	208		
1932	168	1942	102	1952	164		
1933	149	1943	97	1953	146		
1934	125	1944	131	1954	112		
1935	122	1945	170	1955	112		

5–36. Using the data from Problem 5–35, compute a log-Pearson Type III frequency curve and estimate the 10- and 50-yr minimum 7-day low flows. Plot both the frequency curve and the data.

5–37. For the data of Problem 5–18, compute the standard errors of the mean, standard deviation, and skew of the logarithms.

5–38. For the data of Problem 5–20, compute the standard errors of the mean, standard deviation, and skew of the logarithms.

5–39. Assuming that the generalized skew has a mean square error of 0.302, show the variation of weighted skew g_w for sample sizes of 10, 20, and 30; a generalized (map) skew of 0.5; and station skews of 0.1, 0.4, and 0.7. Comment on the effect of both sample size and station skew on the value of weighted skew.

5–40. Assuming that the generalized skew has a mean square error of 0.302, show the variation of g_w for sample sizes of 5, 25, and 50; a generalized (map) skew of –0.5; and station skew of 0, –0.25, and –0.75. Comment on the effect of both sample size and station skew on the value of weighted skew.

5–41. The following data are the annual maximum series q_p for the Anacostia River at Hyattsville, MD (Gage No. 01651000) (1939–1988). Perform a log-Pearson Type III for the cases with station skew, map skew (Figure 5–14), and weighted skew. The station is near Washington, D.C. Compute and compare the 10- and 100-yr flood magnitudes for the three curves. Also compute the return period for a flood magnitude of 20,000 ft^3/sec for each of the three skews. Discuss differences in the results.

Year	q_p	*Year*	q_p	*Year*	q_p	*Year*	q_p
1939	1880	1951	2130	1963	3200	1976	5540
1940	1750	1952	3360	1964	1700	1977	3620
1941	1050	1953	2710	1965	2390	1978	4440
1942	2180	1954	2980	1966	7000	1979	12000
1943	2280	1955	2930	1967	5030	1980	3780
1944	2000	1956	3010	1968	1900	1981	2060
1945	2300	1957	1550	1969	6050	1982	2470
1946	1300	1958	3590	1970	3220	1983	2230
1947	1300	1959	4170	1971	2510	1984	1930
1948	1900	1960	3180	1972	18000	1985	3520
1949	1650	1961	2210	1973	3200	1986	7390
1950	2280	1962	1810	1974	2640	1987	4030
				1975	14800	1988	2190

5–42. The following data are the annual maximum series for the Elizabeth River at Elizabeth, N.J. (1924–1988). Perform a log-Pearson Type III analysis for the cases of station skew, map skew (Figure 5–14), and weighted skew. Compute and compare the 20- and 75-yr flood magnitudes for the three curves. Also compute the return period for a flood magnitude of 4000 ft^3/sec for each of the three skews. Discuss differences in the results.

Year	q_p	*Year*	q_p	*Year*	q_p	*Year*	q_p
1924	1280	1940	1380	1956	1530	1972	2240
1925	980	1941	1030	1957	795	1973	3210
1926	741	1942	820	1958	1760	1974	2940
1927	1630	1943	1020	1959	806	1975	2720
1928	829	1944	998	1960	1190	1976	2440
1929	903	1945	3500	1961	952	1977	3130

(*continued*)

Year	q_p	Year	q_p	Year	q_p	Year	q_p
1930	418	1946	1100	1962	1670	1978	4500
1931	549	1947	1010	1963	824	1979	2890
1932	686	1948	830	1964	702	1980	2470
1933	1320	1949	1030	1965	1490	1981	1900
1934	850	1950	452	1966	1600	1982	1980
1935	614	1951	2530	1967	800	1983	2550
1936	1720	1952	1740	1968	3330	1984	3350
1937	1060	1953	1860	1969	1540	1985	2120
1938	1680	1954	1270	1970	2130	1986	1850
1939	760	1955	2200	1971	3770	1987	2320
						1988	1630

5–43. Compute an upper 95% confidence interval for the 2-, 10-, and 100-yr points on the frequency curve of Problem 5–28. Use the Bulletin 17B procedure.

5–44. Compute an upper 95% confidence interval for the 2-, 10-, and 100-yr points on the frequency curve of Problem 5–41. Use the Bulletin 17B procedure and a weighted skew.

5–45. Assume that floods of 25,000 and 22,000 ft^3/sec occurred between 1880 and 1938 for the gage of Problem 5–41. Use the historic information to make a LP3 frequency analysis for the site. Use weighted skew. Estimate the historically adjusted 100-yr flood magnitude and compare it with the value based on the 1939–1988 record.

5–46. Assume that between 1895 and 1923, floods of 6130 and 5460 ft^3/sec occurred for the gage of Problem 5–42. Use the historic information to make a LP3 frequency analysis for the site. Estimate the 50-yr flood for the site. Use weighted skew. Estimate the 50-yr flood magnitude and compare it with the value based on the systematic record.

5–47. Test the data of Problem 5–41 for two high and two low outliers using the Bulletin 17B procedure.

5–48. Test the data of Problem 5–42 for three high and three low outliers using the Bulletin 17B procedure.

5–49. Test the lowest value of the data from Problem 5–20 as an outlier. Use the Bulletin 17B procedure.

5–50. The following data are the annual maximum series (Q_p) and the percentage of impervious area (I) for the Alhambra Wash watershed for the period from 1930 through 1977. Using the method of Figure 5–22 adjust the flood series to ultimate development of 50%. Perform LP3 analyses on both the adjusted and unadjusted series and evaluate the effect on the estimated 2-, 10-, 25-, and 100-yr floods.

Year	I %	Q_p (cfs)	Year	I %	Q_p (cfs)	Year	I %	Q_p (cfs)
1930	21	1870	1946	35	1600	1962	45	2560
1931	21	1530	1947	37	3810	1963	45	2215
1932	22	1120	1948	39	2670	1964	45	2210
1933	22	1850	1949	41	758	1965	45	3730
1934	23	4890	1950	43	1630	1966	45	3520
1935	23	2280	1951	45	1620	1967	45	3550
1936	24	1700	1952	45	3811	1968	45	3480
1937	24	2470	1953	45	3140	1969	45	3980

(continued)

Year	I %	Q_p (cfs)	Year	I %	Q_p (cfs)	Year	I %	Q_p (cfs)
1938	25	5010	1954	45	2410	1970	45	3430
1939	25	2480	1955	45	1890	1971	45	4040
1940	26	1280	1956	45	4550	1972	46	2000
1941	27	2080	1957	45	3090	1973	46	4450
1942	29	2320	1958	45	4830	1974	46	4330
1943	30	4480	1959	45	3170	1975	46	6000
1944	32	1860	1960	45	1710	1976	46	1820
1945	33	2220	1961	45	1480	1977	46	1770

5–51. Residents in a community at the discharge point of a 240-mi^2 watershed believe that recent increase in peak discharge rates is due to deforestation by a logging company that has been occurring in recent years. Use the Spearman test to analyze the following data: the annual maximum discharges (q_p) and an average fraction of forest cover (f) for the watershed.

Year	q_p	f	Year	q_p	f	Year	q_p	f
1982	8000	53	1987	12200	54	1992	5800	46
1983	8800	56	1988	5700	51	1993	14300	44
1984	7400	57	1989	9400	50	1994	11600	43
1985	6700	58	1990	14200	49	1995	10400	42
1986	11100	55	1991	7600	47			

5–52. Analyze the data of Problem 5–50 to assess whether or not the increase in urbanization has been accompanied by an increase in the annual maximum discharge. Apply the Spearman test with both a 1% and 5% level of significance. Discuss the results.

5–53. Assume that reasonable estimates of the fraction of forested area were not available for the flood record of Problem 5–51. Use the Spearman-Conley test to assess whether or not the discharges show a significant increasing trend.

5–54. Use the Spearman-Conley test and the flood record of Problem 5–50 to assess whether or not the discharges show a significant increasing trend.

5–55. Using the following annual maximum-series for Oak Creek near Mojave, CA (1958–1984) perform a zero-flood year analysis for a LP3 population:

1740, 750, 235, 230, 165, 97, 82, 44, 36, 29, 22, 12, 11,
9.8, 9.2, 8.2, 6.9, 6.8, 6.3, 4.0, 3.6, 0, 0, 0, 0, 0, 0

5–56. Using the following annual-maximum series for San Emigdio Creek, CA, perform a zero-flood year analysis for a LP3 population:

6690, 538, 500, 340, 262, 234, 212, 208, 171, 159, 150, 145,
118, 94, 92, 60, 58, 54, 40, 38, 28, 27, 26, 25, 0, 0, 0, 0, 0

5–57. Compute the probability of the following: (a) exactly two 10-yr floods in a 25-yr period; (b) three or more 5-yr floods in a period of 10 years.

5–58. Compute the probability of the following: (a) exactly two 5-yr floods in a 15-yr period; (b) two or fewer 20-yr floods in a period of 10 years.

5–59. Compute the probabilities of the following: (a) two 5-yr floods in two years; (b) two 5-yr floods in a period of 10 years.

5–60. Compute and graph the probability mass function of the number of 2-yr floods in 10 years.

5–61. Compute and graph the probability mass function of the number of 5-yr floods in 5 years.

5–62. For the problem statement of Example 5–15, assume that upstream land development causes an increase in the flood frequency curve such that now the bridge will only pass the 80-yr discharge rate. Compute the mass function and compare it with that given in Example 5–15.

REVIEW QUESTIONS

5–1. A frequency analysis relates (a) the logarithm of a discharge versus the return period; (b) the value of a random variable versus the probability of being exceeded; (c) the value of a random variable versus the rank of the event; (d) the return period of an event versus the exceedence probability; (e) all of the above.

5–2. In frequency analyses, the population consists of (a) a probability density function and its parameters; (b) a straight line relationship between the logarithm of the event and the probability of exceedence; (c) a set of measured points versus the probability of occurrence of the points; (d) a linear relationship defined by the mean and standard deviation of the sample points; (e) all of the above.

5–3. Which one of the following is not provided by a histogram analysis? (a) A means of identifying the presence of extreme events; (b) a quantitative understanding of the central tendency, the spread, and the symmetry of the underlying population; (c) the degree to which the sample points deviate from the frequency curve; (d) a qualitative understanding of the distribution of the data.

5–4. Which one of the following is required to compute the best estimate of the frequency curve for a log-normal distribution? (a) The mean and standard deviation of the logarithms of the sample points; (b) the probability function for a normal distribution; (c) log probability paper; (d) the sample skew; (e) none of the above.

5–5. Plotting measured data on probability paper using a plotting position formula is used to (a) estimate the probability of observed events; (b) verify the assumed population; (c) determine if there are extreme events in the data; (d) identify the population of the data; (e) none of the above.

5–6. The fact that the sample data points do not fall near a computed frequency curve indicates that (a) the sample moments are not accurate indicators of the population parameters; (b) the incorrect density function was assumed; (c) the data include extreme events; (d) any of the above; (e) none of the above.

5–7. The 2500-yr flood has an exceedence probability of (a) 0.004; (b) 4×10^{-4}; (c) 0.00025; (d) 2.5×10^{-3}.

5–8. If the $\log_{10}$ mean and $\log_{10}$ standard deviation are 3.21 and 0.45, respectively, and the population of peak discharges Q is assumed to be a log-normal distribution, the 2-yr peak discharge is (a) 3.21; (b) 3.66; (c) 1622; (d) 4571.

5–9. If the $\log_{10}$ mean and $\log_{10}$ standard deviation are 4.06 and 0.375, respectively, and the population of peak discharges is assumed to be a log-normal distribution, a discharge of 79,433 ft^3/sec has a return period of (a) 24 years; (b) 55 years; (c) 80 years; (d) 100 years; (e) none of the above.

5–10. Which of the following is not a method of estimating the skew coefficient for a flood frequency analysis? (a) A mean value of station skews in a region; (b) a prediction equation using watershed characteristics; (c) station skew; (d) a regional map of skew; (e) all of the above could be used.

5–11. Which one of the following factors would not affect a skew coefficient and should not be considered in estimating skew for a flood frequency analysis? (a) Channel constrictions at bridges; (b) meteorological characteristics; (c) topographic and geographic characteristics; (d) antecedent soil moisture.

5–12. The method of moments is (a) a method of estimating the parameters of a probability density function; (b) a method for weighting station skew and a regional skew map estimate; (c) a method for identifying the correct probability density function; (d) a method of estimating the skew coefficient from the sample moments.

5–13. When using the binomial probability function to estimate flood risk, which one of the following is not a necessary assumption? (a) The floods follow either a log-normal or log-Pearson Type 111 distribution; (b) floods in adjacent years are independent; (c) the probability of flooding is constant from year to year; (d) there are only two possible outcomes (flood or no flood) in any one year; (e) all of the above are assumptions.

5–14. The probability that a 50-yr flood will occur in two consecutive years is (a) 0.01; (b) 0.02; (c) 0.04; (d) 0.0004.

5–15. Which one of the following factors would probably not be important for correcting an annual flood record for continuous urbanization over the period of record? (a) The percentage of imperviousness; (b) the rainfall characteristics; (c) the length of the flood record; (d) the percentage of area sewered.

DISCUSSION QUESTION

The technical content of this chapter is important to the professional hydrologist, but practice is not confined to making technical decisions. The intent of this discussion question is to show that hydrologists must often address situations where value issues intermingle with the technical aspects of a project. In discussing the stated problem, at a minimum include responses to the following questions:

1. What value issues are involved, and how are they in conflict?
2. Are technical issues involved with the value issues? If so, how are they in conflict with the value issues?
3. If the hydrologist attempted to rationalize the situation, what rationalizations might he or she use? Provide arguments to suggest why the excuses represent rationalization.
4. What are the hydrologist's alternative courses of action? Identify all alternatives, regardless of the ethical implications.
5. How should the conflict be resolved?

You may want to review Sections 1.6 to 1.12 in Chapter 1 in responding to the problem statement.

Case. A state agency issues a request for proposals on a floodplain study. Because of the special nature of the specific project, a hydrologist recognizes that only three companies will be qualified to bid on the project; the firm that he works for and two other companies. Given these circumstances, he recognizes that his company could make a substantial profit by rigging the bid. Specifically, he approaches the other two firms with the proposal that they submit proposals with highly inflated price tags. The hydrologist will then submit a proposal with a cost figure that is below the dollar amount submitted by the other companies but still highly inflated. The hydrologist indicates to the two other companies that he will work with them on future efforts where the other companies will be awarded highly inflated contracts. The other firms agree to participate in the bid rigging scheme.

APPENDIX 5–1 Exceedence Probability Coefficients

p	$g = 0.0$	$g = 0.1$	$g = 0.2$	$g = 0.3$	$g = 0.4$	$g = 0.5$	$g = 0.6$
0.9999	−3.71902	−3.50703	−3.29921	−3.09631	−2.89907	−2.70836	−2.52507
0.9995	−3.29053	−3.12767	−2.96698	−2.80889	−2.65390	−2.50257	−2.35549
0.9990	−3.09023	−2.94834	−2.80786	−2.66915	−2.53261	−2.39867	−2.26780
0.9980	−2.87816	−2.75706	−2.63672	−2.51741	−2.39942	−2.28311	−2.16884
0.9950	−2.57583	−2.48187	−2.38795	−2.29423	−2.20092	−2.10825	−2.01644
0.9900	−2.32635	−2.25258	−2.17840	−2.10394	−2.02933	−1.95472	−1.88029
0.9800	−2.05375	−1.99973	−1.94499	−1.88959	−1.83361	−1.77716	−1.72033
0.9750	−1.95996	−1.91219	−1.86360	−1.81427	−1.76427	−1.71366	−1.66253
0.9600	−1.75069	−1.71580	−1.67999	−1.64329	−1.60574	−1.56740	−1.52830
0.9500	−1.64485	−1.61594	−1.58607	−1.55527	−1.52357	−1.49101	−1.45762
0.9000	−1.28155	−1.27037	−1.25824	−1.24516	−1.23114	−1.21618	−1.20028
0.8000	−0.84162	−0.84611	−0.84986	−0.85285	−0.85508	−0.85653	−0.85718
0.7000	−0.52440	−0.53624	−0.54757	−0.55839	−0.56867	−0.57840	−0.58757
0.6000	−0.25335	−0.26882	−0.28403	−0.29897	−0.31362	−0.32796	−0.34198
0.5704	−0.17733	−0.19339	−0.20925	−0.22492	−0.24037	−0.25558	−0.27047
0.5000	0.0	−0.01662	−0.03325	−0.04993	−0.06651	−0.08302	−0.09945
0.4296	0.17733	0.16111	0.14472	0.12820	0.11154	0.09478	0.07791
0.4000	0.25335	0.23763	0.22168	0.20552	0.18916	0.17261	0.15589
0.3000	0.52440	0.51207	0.49927	0.48600	0.47228	0.45812	0.44352
0.2000	0.84162	0.83639	0.83044	0.82377	0.81638	0.80829	0.79950
0.1000	1.28155	1.29178	1.30105	1.30936	1.31671	1.32309	1.32850
0.0500	1.64485	1.67279	1.69971	1.72562	1.75048	1.77428	1.79701
0.0400	1.75069	1.78462	1.81756	1.84949	1.88039	1.91022	1.93896
0.0250	1.95996	2.00688	2.05290	2.09795	2.14202	2.18505	2.22702
0.0200	2.05375	2.10697	2.15935	2.21081	2.26133	2.31084	2.35931
0.0100	2.32635	2.39961	2.47226	2.54421	2.61539	2.68572	2.75514
0.0050	2.57583	2.66965	2.76321	2.85636	2.94900	3.04102	3.13232
0.0020	2.87816	2.99978	3.12169	3.24371	3.36566	3.48737	3.60872
0.0010	3.09023	3.23322	3.37703	3.52139	3.66608	3.81090	3.95567
0.0005	3.29053	3.45513	3.62113	3.78820	3.95605	4.12443	4.29311
0.0001	3.71902	3.93453	4.15301	4.37394	4.59687	4.82141	5.04718

APPENDIX 5–1 Exceedence Probability Coefficients *(Continued)*

p	$g = 0.7$	$g = 0.8$	$g = 0.9$	$g = 1.0$	$g = 1.1$	$g = 1.2$	$g = 1.3$
0.9999	−2.35015	−2.18448	−2.02891	−1.88410	−1.75053	−1.62838	−1.51752
0.9995	−2.21328	−2.07661	−1.94611	−1.82241	−1.70603	−1.59738	−1.49673
0.9990	−2.14053	−2.01739	−1.89894	−1.78572	−1.67825	−1.57695	−1.48216
0.9980	−2.05701	−1.94806	−1.84244	−1.74062	−1.64305	−1.55016	−1.46232
0.9950	−1.92580	−1.83660	−1.74919	−1.66390	−1.58110	−1.50114	−1.42439
0.9900	−1.80621	−1.73271	−1.66001	−1.58838	−1.51808	−1.44942	−1.38267
0.9800	−1.66325	−1.60604	−1.54886	−1.49188	−1.43529	−1.37929	−1.32412
0.9750	−1.61099	−1.55914	−1.50712	−1.45507	−1.40314	−1.35153	−1.30042
0.9600	−1.48852	−1.44813	−1.40720	−1.36584	−1.32414	−1.28225	−1.24028
0.9500	−1.42345	−1.38855	−1.35299	−1.31684	−1.28019	−1.24313	−1.20578
0.9000	−1.18347	−1.16574	−1.14712	−1.12762	−1.10726	−1.08608	−1.06413
0.8000	−0.85703	−0.85607	−0.85426	−0.85161	−0.84809	−0.84369	−0.83841
0.7000	−0.59615	−0.60412	−0.61146	−0.61815	−0.62415	−0.62944	−0.63400
0.6000	−0.35565	−0.36889	−0.38186	−0.39434	−0.40638	−0.41794	−0.42899
0.5704	−0.28516	−0.29961	−0.31368	−0.32740	−0.34075	−0.35370	−0.36620
0.5000	−0.11578	−0.13199	−0.14807	−0.16397	−0.17968	−0.19517	−0.21040
0.4296	0.06097	0.04397	0.02693	0.00987	−0.00719	−0.02421	−0.04116
0.4000	0.13901	0.12199	0.10486	0.08763	0.07032	0.05297	0.03560
0.3000	0.42851	0.41309	0.39729	0.38111	0.36458	0.34772	0.30054
0.2000	0.79002	0.77986	0.76902	0.75752	0.74537	0.73257	0.71915
0.1000	1.33294	1.33640	1.33889	1.34039	1.34092	1.34047	1.33904
0.0500	1.81864	1.83916	1.85856	1.87683	1.89395	1.90992	1.92472
0.0400	1.96660	1.99311	2.01848	2.04269	2.06573	2.08758	2.10823
0.0250	2.26790	2.30764	2.34623	2.38364	2.41984	2.45482	2.48855
0.0200	2.40670	2.45298	2.49811	2.54206	2.58480	2.62631	2.66657
0.0100	2.82359	2.89101	2.95735	3.02256	3.08660	3.14944	3.21103
0.0050	3.22281	3.31243	3.40109	3.48874	3.57530	3.66073	3.74497
0.0020	3.72957	3.84981	3.96932	4.08802	4.20582	4.32263	4.43839
0.0010	4.10022	4.24439	4.38807	4.53112	4.67344	4.81492	4.95549
0.0005	4.46189	4.63057	4.79899	4.96701	5.13449	5.30130	5.46735
0.0001	5.27389	5.50124	5.72899	5.95691	6.18480	6.41249	6.63980

(continued)

APPENDIX 5–1 Exceedence Probability Coefficients *(Continued)*

p	$g = 1.4$	$g = 1.5$	$g = 1.6$	$g = 1.7$	$g = 1.8$	$g = 1.9$	$g = 2.0$
09.999	−1.41753	−1.32774	−1.24728	−1.17520	−1.11054	−1.05239	−0.99990
0.9995	−1.40413	−1.31944	−1.24235	−1.17240	−1.10901	−1.05159	−0.99950
0.9990	−1.39408	−1.31275	−1.23805	−1.16974	−1.10743	−1.05068	−0.99900
0.9980	−1.37981	−1.30279	−1.23132	−1.16534	−1.10465	−1.04898	−0.99800
0.9950	−1.35114	−1.28167	−1.21618	−1.15477	−1.09749	−1.04427	−0.99499
0.9900	−1.31815	−1.25611	−1.19680	−1.14042	−1.08711	−1.03695	−0.98995
0.9800	−1.26999	−1.21716	−1.16584	−1.11628	−1.06864	−1.02311	−0.97980
0.9750	−1.25004	−1.20059	−1.15229	−1.10537	−1.06001	−1.01640	−0.97468
0.9600	−1.19842	−1.15682	−1.11566	−1.07513	−1.03543	−0.99672	−0.95918
0.9500	−1.16827	−1.13075	−1.09338	−1.05631	−1.01973	−0.98381	−0.94871
0.9000	−1.04144	−1.01810	−0.99418	−0.96977	−0.94496	−0.91988	−0.89464
0.8000	−0.83223	−0.82516	−0.81720	−0.80837	−0.79868	−0.78816	−0.77686
0.7000	−0.63779	−0.64080	−0.64300	−0.64436	−0.64488	−0.64453	−0.64333
0.6000	−0.43949	−0.44942	−0.45873	−0.46739	−0.47538	−0.48265	−0.48917
0.5704	−0.37824	−0.38977	−0.40075	−0.41116	−0.42095	−0.43008	−0.43854
0.5000	−0.22535	−0.23996	−0.25422	−0.26808	−0.28150	−0.29443	−0.30685
0.4296	−0.05803	−0.07476	−0.09132	−0.10769	−0.12381	−0.13964	−0.15516
0.4000	0.01824	0.00092	−0.01631	−0.03344	−0.05040	−0.06718	−0.08371
0.3000	0.31307	0.29535	0.27740	0.25925	0.24094	0.22250	0.20397
0.2000	0.70512	0.69050	0.67532	0.65959	0.64335	0.62662	0.60944
0.1000	1.33665	1.33330	1.32900	1.32376	1.31760	1.31054	1.30259
0.0500	1.93836	1.95083	1.96213	1.97227	1.98124	1.98906	1.99573
0.0400	2.12768	2.14591	2.16293	2.17873	2.19332	2.20670	2.21888
0.0250	2.52102	2.55222	2.58214	2.61076	2.63810	2.66413	2.68888
0.0200	2.70556	2.74325	2.77964	2.81472	2.84848	2.88091	2.91202
0.0100	3.27134	3.33035	3.38804	3.44438	3.49935	3.55295	3.60517
0.0050	3.82798	3.90973	3.99016	4.06926	4.14700	4.22336	4.29832
0.0020	4.55304	4.66651	4.77875	4.88971	4.99937	5.10768	5.21461
0.0010	5.09505	5.23353	5.37087	5.50701	5.64190	5.77549	5.90776
0.0005	5.63252	5.79673	5.95990	6.12196	6.28285	6.44251	6.60090
0.0001	6.86661	7.09277	7.31818	7.54272	7.76632	7.98888	8.21034

APPENDIX 5–1 Exceedence Probability Coefficients *(Continued)*

p	$g = -0.0$	$g = -0.1$	$g = -0.2$	$g = -0.3$	$g = -0.4$	$g = -0.5$	$g = -0.6$
0.9999	−3.71902	−3.93453	−4.15301	−4.37394	−4.59687	−4.82141	−5.04718
0.9995	−3.29053	−3.45513	−3.62113	−3.78820	−3.95605	−4.12443	−4.29311
0.9990	−3.09023	−3.23322	−3.37703	−3.52139	−3.66608	−3.81090	−3.95567
0.9980	−2.87816	−2.99978	−3.12169	−3.24371	−3.36566	−3.48737	−3.60872
0.9950	−2.57583	−2.66965	−2.76321	−2.85636	−2.94900	−3.04102	−3.13232
0.9900	−2.32635	−2.39951	−2.47226	−2.54421	−2.61539	−2.68572	−2.75514
0.9800	−2.05375	−2.10697	−2.15935	−2.21081	−2.26133	−2.31084	−2.35931
0.9750	−1.95996	−2.00688	−2.05290	−2.09795	−2.14202	−2.18505	−2.22702
0.9600	−1.75069	−1.78462	−1.81756	−1.84949	−1.88039	−1.91022	−1.93896
0.9500	−1.64485	−1.67279	−1.69971	−1.72562	−1.75048	−1.77428	−1.79701
0.9000	−1.28155	−1.29178	−1.30105	−1.30936	−1.31671	−1.32309	−1.32850
0.8000	−0.84162	−0.83639	−0.83044	−0.82377	−0.81638	−0.80829	−0.79950
0.7000	−0.52440	−0.51207	−0.49927	−0.48600	−0.47228	−0.45812	−0.44352
0.6000	−0.25335	−0.23763	−0.22168	−0.20552	−0.18916	−0.17261	−0.15589
0.5704	−0.17733	−0.16111	−0.14472	−0.12820	−0.11154	−0.09478	−0.07791
0.5000	0.0	0.01662	0.03325	0.04993	0.06651	0.08302	0.09945
0.4296	0.17733	0.19339	0.20925	0.22492	0.24037	0.25558	0.27047
0.4000	0.25335	0.26882	0.28403	0.29897	0.31362	0.32796	0.34198
0.3000	0.52440	0.53624	0.54757	0.55839	0.56867	0.57840	0.58757
0.2000	0.84162	0.84611	0.84986	0.85285	0.85508	0.85653	0.85718
0.1000	1.28155	1.27037	1.25824	1.24516	1.23114	1.21618	1.20028
0.0500	1.64485	1.61594	1.58607	1.55527	1.52357	1.49101	1.45762
0.0400	1.75069	1.71580	1.67999	1.64329	1.60574	1.56740	1.52830
0.0250	1.95996	1.91219	1.86360	1.81427	1.76427	1.71366	1.66253
0.0200	2.05375	1.99973	1.94499	1.88959	1.83361	1.77716	1.72033
0.0100	2.32635	2.25258	2.17840	2.10394	2.02933	1.95472	1.88029
0.0050	2.57583	2.48187	2.38795	2.29423	2.20092	2.10825	2.01644
0.0020	2.87816	2.75706	2.63672	2.51741	2.39942	2.28311	2.16884
0.0010	3.09023	2.94834	2.80786	2.66915	2.53261	2.39867	2.26780
0.0005	3.29053	3.12767	2.96698	2.80889	2.65390	2.50257	2.35549
0.0001	3.71902	3.50703	3.29921	3.09631	2.89907	2.70836	2.52507

(continued)

APPENDIX 5–1 Exceedence Probability Coefficients *(Continued)*

p	$g = -0.7$	$g = -0.8$	$g = -0.9$	$g = -1.0$	$g = -1.1$	$g = -1.2$	$g = -1.3$
0.9999	−5.27389	−5.50124	−5.72899	−5.95691	−6.18480	−6.41249	−6.63980
0.9995	−4.46189	−4.63057	−4.79899	−4.96701	−5.13449	−5.30130	−5.46735
0.9990	−4.10022	−4.24439	−4.38807	−4.53112	−4.67344	−4.81492	−4.95549
0.9980	−3.72957	−3.84981	−3.96932	−4.08802	−4.20582	−4.32263	−4.43839
0.9950	−3.22281	−3.31243	−3.40109	−3.48874	−3.57530	−3.66073	−3.74497
0.9900	−2.82359	−2.89101	−2.95735	−3.02256	−3.08660	−3.14944	−3.21103
0.9800	−2.40670	−2.45298	−2.49811	−2.54206	−2.58480	−2.62631	−2.66657
0.9750	−2.26790	−2.30764	−2.34623	−2.38364	−2.41984	−2.45482	−2.48855
0.9600	−1.96660	−1.99311	−2.01848	−2.04269	−2.06573	−2.08758	−2.10823
0.9500	−1.81864	−1.83916	−1.85856	−1.87683	−1.89395	−1.90992	−1.92472
0.9000	−1.33294	−1.33640	−1.33889	−1.34039	−1.34092	−1.34047	−1.33904
0.8000	−0.79002	−0.77986	−0.76902	−0.75752	−0.74537	−0.73257	−0.71915
0.7000	−0.42851	−0.41309	−0.39729	−0.38111	−0.36458	−0.34772	−0.33054
0.6000	−0.13901	−0.12199	−0.10486	−0.08763	−0.07032	−0.05297	−0.03560
0.5704	−0.06097	−0.04397	−0.02693	−0.00987	0.00719	0.02421	0.04116
0.5000	0.11578	0.13199	0.14807	0.16397	0.17968	0.19517	0.21040
0.4296	0.28516	0.29961	0.31368	0.32740	0.34075	0.35370	0.36620
0.4000	0.35565	0.36889	0.38186	0.39434	0.40638	0.41794	0.42899
0.3000	0.59615	0.60412	0.61146	0.61815	0.62415	0.62944	0.62400
0.2000	0.85703	0.85607	0.85426	0.85161	0.84809	0.84369	0.83841
0.1000	1.18347	1.16574	1.14712	1.12762	1.10726	1.08608	1.06413
0.0500	1.42345	1.38855	1.35299	1.31684	1.28019	1.24313	1.20578
0.0400	1.48852	1.44813	1.40720	1.36584	1.32414	1.28225	1.24028
0.0250	1.61099	1.55914	1.50712	1.45507	1.40314	1.35153	1.30042
0.0200	1.66325	1.60604	1.54886	1.49188	1.43529	1.37929	1.32412
0.0100	1.80621	1.73271	1.66001	1.58838	1.51808	1.44942	1.38267
0.0050	1.92580	1.83660	1.74919	1.66390	1.58110	1.50114	1.42439
0.0020	2.05701	1.94806	1.84244	1.74062	1.64305	1.55016	1.46232
0.0010	2.14053	2.01739	1.89894	1.78572	1.67825	1.57695	1.48216
0.0005	2.21328	2.07661	1.94611	1.82241	1.70603	1.59738	1.49673
0.0001	2.35015	2.18448	2.02891	1.88410	1.75053	1.62838	1.51752

APPENDIX 5–1 Exceedence Probability Coefficients *(Continued)*

p	$g = -1.4$	$g = -1.5$	$g = -1.6$	$g = -1.7$	$g = -1.8$	$g = -1.9$	$g = -2.0$
0.9999	−6.86661	−7.09277	−7.31818	−7.54272	−7.76632	−7.98888	−8.21034
0.9995	−5.63252	−5.79673	−5.95990	−6.12196	−6.28285	−6.44251	−6.60090
0.9990	−5.09505	−5.23353	−5.37087	−5.50701	−5.64190	−5.77549	−5.90776
0.9980	−4.55304	−4.66651	−4.77875	−4.88971	−4.99937	−5.10768	−5.21461
0.9950	−3.82798	−3.90973	−3.99016	−4.06926	−4.14700	−4.22336	−4.29832
0.9900	−3.27134	−3.33035	−3.38804	−3.44438	−3.49935	−3.55295	−3.60517
0.9800	−2.70556	−2.74325	−2.77964	−2.81472	−2.84848	−2.88091	−2.91202
0.9750	−2.52102	−2.55222	−2.58214	−2.61076	−2.63810	−2.66413	−2.68888
0.9600	−2.12768	−2.14591	−2.16293	−2.17873	−2.19332	−2.20670	−2.21888
0.9500	−1.93836	−1.95083	−1.96213	−1.97227	−1.98124	−1.98906	−1.99573
0.9000	−1.33665	−1.33330	−1.32900	−1.32376	−1.31760	−1.31054	−1.30259
0.8000	−0.70512	−0.69050	−0.67532	−0.65959	−0.64335	−0.62662	−0.60944
0.7000	−0.31307	−0.29535	−0.27740	−0.25925	−0.24094	−0.22250	−0.20397
0.6000	−0.01824	−0.00092	0.01631	0.03344	0.05040	0.06718	0.08371
0.5704	0.05803	0.07476	0.09132	0.10769	0.12381	0.13964	0.15516
0.5000	0.22535	0.23996	0.25422	0.26808	0.28150	0.29443	0.30685
0.4296	0.37824	0.38977	0.40075	0.41116	0.42095	0.43008	0.43854
0.4000	0.43949	0.44942	0.45873	0.46739	0.47538	0.48265	0.48917
0.3000	0.63779	0.64080	0.64300	0.64436	0.64488	0.64453	0.64333
0.2000	0.83223	0.82516	0.81720	0.80837	0.79868	0.78816	0.77686
0.1000	1.04144	1.01810	0.99418	0.96977	0.94496	0.91988	0.89464
0.0500	1.16827	1.13075	1.09338	1.05631	1.01973	0.98381	0.94871
0.0400	1.19842	1.15682	1.11566	1.07513	1.03543	0.99672	0.95918
0.0250	1.25004	1.20059	1.15229	1.10537	1.06001	1.01640	0.97468
0.0200	1.26999	1.21716	1.16584	1.11628	1.06864	1.02311	0.97980
0.0100	1.31815	1.25611	1.19680	1.14042	1.08711	1.03695	0.98995
0.0050	1.35114	1.28167	1.21618	1.15477	1.09749	1.04427	0.99499
0.0020	1.37981	1.30279	1.23132	1.16534	1.10465	1.04898	0.99800
0.0010	1.39408	1.31275	1.23805	1.16974	1.10743	1.05068	0.99900
0.0005	1.40413	1.31944	1.24235	1.17240	1.10901	1.05159	0.99950
0.0001	1.41753	1.32774	1.24728	1.17520	1.11054	1.05239	0.99990

6

Subsurface Hydrology

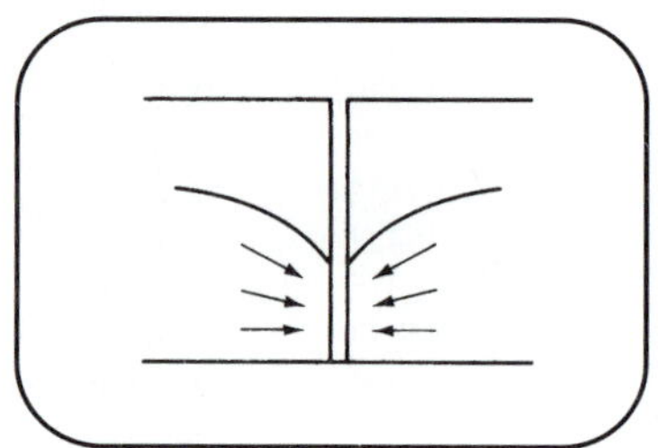

CHAPTER OBJECTIVES

1. Discuss the importance of the subsurface storage of a watershed to problems in engineering design and provide a discussion of soil characteristics.
2. Introduce the basics of water movement through the soil profile.
3. Provide methods for estimating the relationship between discharge from a well and the characteristics of the soil profile of the watershed.
4. Introduce methods for the proper long-term management of water stored in the soil profile.

6.0 NOTATION

A = cross-sectional area
b = depth of confined aquifer
d = depth of aquifer
d = drawdown
d = particle diameter
d = thickness of aquifer
D = annual draft
$\hat{D}$ = predicted annual draft
E_j = total energy at section j
F = dimension of force
h = head or height of the piezometric surface
h_L = head loss
h_o = depth of water table
h_w = height of water in well
i = hydraulic gradient
I = inflow
k = specific permeability
K = hydraulic conductivity
L = a characteristic length for Reynold's number
L = dimension of length
L = length of a section
O = output
p_j = pressure at section j
P = annual precipitation
q = discharge per unit width
Q = discharge rate
Q = surface runoff
Q_p = annual draft
r = radial distance
r_o = radial distance from pumping well to observation well
r_w = radius of well
$\boldsymbol{R}$ = Reynold's Number
s = storage
S = storativity
S_S = specific storage
t = time
t_o = time when the drawdown is zero
T = dimension of time
T = transmissivity
T_t = travel time
u = variable for well function $W(u)$
V = Darcy's velocity
V = flow velocity
w = width of aquifer
$W(u)$ = well function
x = horizontal distance along the aquifer
z_j = elevation head at section j
γ = specific weight of the fluid
Δe = change in the depth to the water table
μ = absolute viscosity
ν = kinematic viscosity
ρ = mass density of the fluid

6.1 INTRODUCTION

A brief look at a list of job openings in the field of hydrology is one way to develop an awareness of the importance of knowledge of ground water hydrology. It is probably the area of specialty in the field of hydrology where the greatest demand for employment currently exists. In many locations, water stored in geologic formations is the primary source of water for personal, municipal, commercial, industrial, and agricultural uses. In these cases, surface water supply sources are nonexistent or too costly to use. In many agricultural communities, irrigation is required, and knowledge of soil moisture and soil-water-plant interactions is necessary to maintain productivity.

In the past, hazardous chemicals were disposed of by burying tanks filled with the waste chemicals; at the time this was the least costly alternative for disposal. In time, corrosion of the tanks has allowed the wastes to leak into adjacent geologic formations, which represents a significant public health hazard. Many job openings are available for those with experience and/or knowledge of the transport of hazardous wastes through geologic formations.

An understanding of the flow of water into and through the soil system is important to many engineering design problems. The characteristics of soil and their interaction with water is important in the design of building foundations, roadways and other transportation facilities, sewage lagoons, sanitary landfills, septic tanks, dams, bridge piers, and levees. Soil characteristics are an important factor in estimating flood magnitudes, which is a primary topic of this book.

Knowledge of ground water hydrology is also important to those who work in the field of surface water flood hydrology. While maximum flow rates (that is, volumes per unit time) and velocities may cause flood damage and be the primary design variables, this should not imply that soil properties are not important to surface runoff. Soil properties control the rate at which water infiltrates into the soil, percolates through the subsurface geologic formations, and travels through the soil to seep into surface water bodies from below the water surface. This rate affects the proportions of rainfall that appear as surface runoff and as ground water losses. Thus the soil properties affect surface water hydrologic designs.

Where designs or analyses require knowledge of ground water, factors that can be important include the characteristics of the geologic formation, properties of the water and chemicals if appropriate, the extent of the geologic formation, and both the temporal and spatial variation of the water. The objective of this chapter is to introduce the basic concepts and variables related to the geologic formation and to provide examples of how these concepts are important in solving problems related to the movement of water in geologic formations.

6.1.1 Ground Water: A Component of the Hydrologic Cycle

The schematic of the hydrologic cycle in Figure 1–1 shows that some of the surface water passes into the ground; this water represents soil moisture in the unsaturated zone, and it is called ground water once it enters the zone of saturation. As was indicated in Chapter 1, urban development reduces the potential for water to enter into the ground, both through reductions in the permeability of the ground surface due to land disturbance and the covering of the ground surface with impervious cover materials.

The processes of water in the ground are more complex than suggested by the schematic of Figure 1–1. Figure 6–1 shows a schematic of the hydrologic cycle that emphasizes subsurface water. Water enters (infiltrates) the upper layers of the soil and percolates downward toward the water table. The area above the water table is called the unsaturated zone or zone of aeration, while the zone of saturation is the area below the water table.

The soil profile can be divided into two major sections, the zones of aeration and saturation. The zone of aeration can be further divided into three subzones: the root zone, the gravitational zone, and the capillary zone. The root zone is also called the *soil water zone*. In the zone of saturation, water occupies all of the pore space and is under hydrostatic pressure; this water is referred to as ground water. The upper edge of the zone of saturation is called

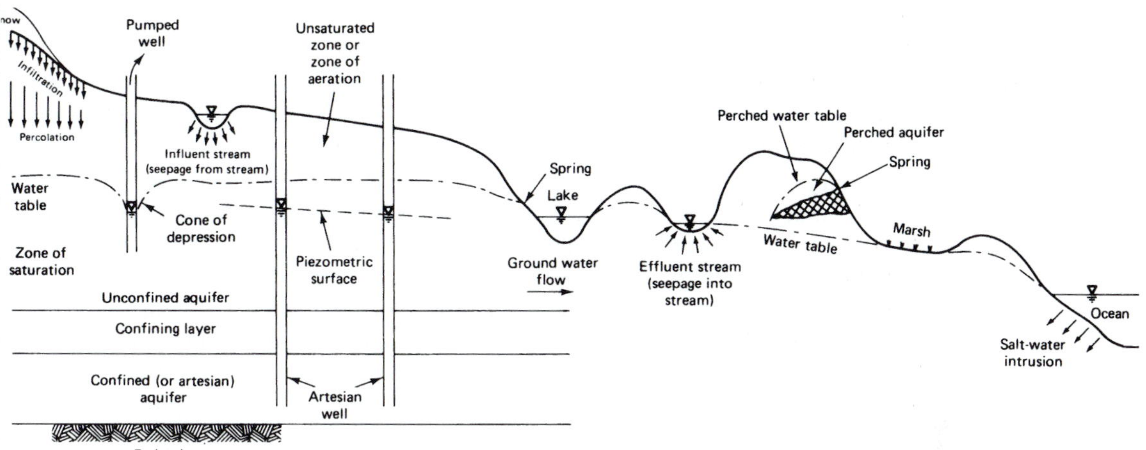

FIGURE 6–1 Hydrologic cycle: groundwater perspective.

the water table. At the water table, the water is at atmospheric pressure. The hydrostatic pressure increases with depth below the water table. Above the water table, the pores in the soil structure contain both air and water. The root zone extends from the ground surface to the bottom of the root systems of the surface vegetation; this zone is saturated only for brief periods following substantial rain. It is of special importance to those involved in agricultural uses of the land. The gravitational zone extends from the bottom of the root zone to the top of the capillary zone. Water moves through this zone under gravitational forces, with some water held in place by hygroscopic and capillary forces. Water under gravitational forces in the gravitational zone moves into the capillary zone, which extends from the bottom of the gravitational zone to the water table. Capillary forces control the water content of the capillary zone.

Very often, confined or artisan aquifers are formed where an impermeable layer creates a confined ground water storage area. A confined aquifer is recharged from surface waters at a distant location where the aquifer intersects the ground surface. An *unconfined aquifer* is a permeable underground formation having a surface at atmospheric pressure; that is, it is a formation having the water table as its upper layer. Flow in underground formations depends on the slope of the water table (ground water flow in an unconfined aquifer), or the piezometric surface, in a confined aquifer.

A considerable volume of water is extracted from underground formations for agricultural, municipal, commercial, and industrial water uses. Water pumped from unconfined aquifers can cause significant decreases in the water table, especially in areas close to the well. The depression of the water table near pumped wells is called the *zone of depression.* As the slope of the water table increases toward the well, the velocity of flow into the well increases.

Wells drilled into confined aquifers are called *artisan wells*. The water will rise in the well to a level that defines the piezometric surface. If the piezometric surface is above the ground level, water will flow freely from the well.

6.1.2 Hydrologic Classification

Given the importance of ground water, it is useful to classify regions of a soil column on the basis of the water content. Two zones can be specified. Where all voids within a column are filled with water, the column is said to be saturated. The water table is the upper limit of the saturated zone. Very often, the term *ground water* is used to refer to water in the zone of saturation. Above the water table, the pore space in the soil column may have both water and air; this zone is referred to as the zone of aeration, or the unsaturated zone. The terms *suspended water* or *vadose water* are used to refer to water in the zone of aeration; in practice, this water is simply referred to as soil moisture. Separating the soil column into these two zones is necessary because the physical processes and methods of estimation are different in the two zones.

The zone of aeration can be further divided into three subzones; the soil-water or root zone, the intermediate zone, and the capillary zone, which is sometimes referred to as the capillary fringe. Water in the zone of aeration may exist as gravity water (that is, water that can drain under the force of gravity), capillary water (that is, water retained by capillary forces), and hygroscopic moisture (that is, water that adheres to the surface of the soil particles as a thin film).

6.2 DARCY'S LAW

To illustrate factors that affect the flow of a fluid through a soil column, consider the apparatus of Figure 6–2. The cylinder contains a saturated porous medium. From Bernoulli's equation the total head at both cross sections consists of the elevation head, z_j, which is the distance from the datum to the center of area of the section; the pressure head, p_j/γ; and the velocity head, $V_j^2/2g$, where the subscript j refers to the upper ($j = 1$) or lower ($j = 2$) cross section. Each of these elements of the total energy at section j have units of length (for example, ft). The fluid that travels through the cylinder moves from the section with higher total

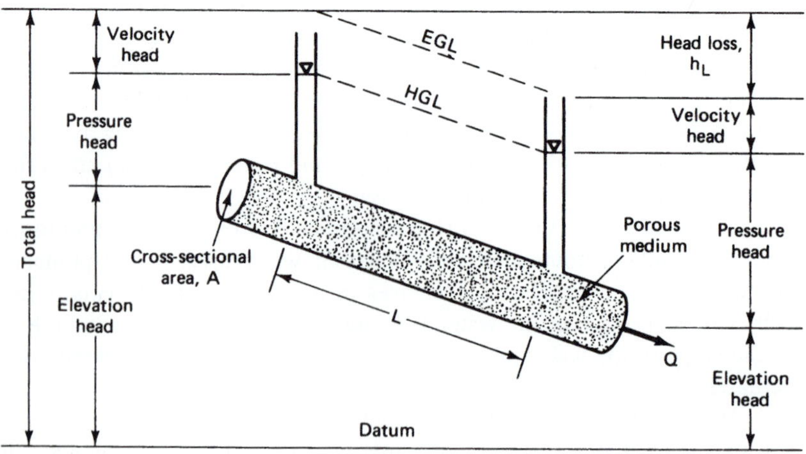

FIGURE 6–2 Apparatus for experiments on Darcy's Law.

energy to the section with the lower energy; in Figure 6–2 this is from section 1 to section 2. The difference in the total energy, $E_1 - E_2$, is the head loss, h_L, between the two sections. The head loss is the energy lost to friction and other minor losses. A standpipe is inserted into the cylinder at both cross sections. The water rises in each standpipe to a level equal to the sum of the elevation and pressure heads. A line connecting the water elevations in the stand pipes is referred to as the hydraulic grade line (HGL). The energy grade line (EGL) is the line connecting the points of total energy E_1 and E_2.

Using the apparatus of Figure 6–2, a number of experiments can be conducted. If the flow rate Q is increased, the head loss increases; a direct relationship between Q and h_L is evident from the Darcy-Weisbach equation, although the Darcy-Weisbach equation does not apply here. As a second experiment, the length of the cylinder could be doubled while holding the head loss constant. In this case the discharge would be halved; thus the discharge is inversely proportional to the length. Based on these two experiments, the discharge has been found to be proportional to the ratio of the head loss to the length (for example, $Q \sim h_L/L$). The ratio h_L/L equals the slope of the hydraulic grade line and will be denoted here as i; it will be referred to as the hydraulic gradient.

A third experiment can be performed by replacing the cylinder of Figure 6–2 with a similar cylinder having an area $2A$. The experiment would show that the discharge would double; thus $Q \sim A$. The three experiments show that the discharge is related to the area and hydraulic gradient by

$$Q = KiA \tag{6–1}$$

where K is a proportionality constant. Values for K could be determined by performing a number of experiments in which both the fluid passing through the cylinder and the characteristics of the soil medium are varied. Because of this dependency, K is called the hydraulic conductivity.

The velocity of the flow through the cylinder can be determined using the equation of continuity:

$$V = \frac{Q}{A} = Ki \tag{6–2}$$

The velocity of Equation 6–2 is not the actual velocity of the fluid flowing through the porous medium because the area A of Equation 6–1 is the total cross-sectional area and not the area through which the fluid is flowing. The actual velocity would be larger because the actual area of the pores would be smaller by about 50%.

Studies have shown that Darcy's Law is valid for Reynold's numbers less than 1, which covers almost all cases of flow in natural porous media. However, turbulent flow can occur in isolated pores of a porous medium, and Darcy's Law is not valid where such conditions dominate. The Reynold's Number (**R**), which is the ratio of the inertia forces to the viscous forces, is given by

$$\mathbf{R} = \frac{\rho V L}{\mu} = \frac{VL}{\nu} \tag{6–3}$$

in which ρ is the mass density of the fluid, μ is the absolute or dynamic viscosity of the fluid, ν is the kinematic viscosity, V is the flow velocity, and L is a characteristic length. The

Reynold's number is dimensionless and can be used with any system of dimensions. The characteristic length L can be a dimension of either the flow path or an obstacle in the flow path. For porous media flow, the mean grain diameter is often used.

6.2.1 Hydraulic Conductivity

Solving Equation 6–1 for the coefficient K yields

$$K = \frac{Q}{iA} = \frac{Q}{(dh_L/dL)A} \tag{6–4}$$

Since i is dimensionless, K has units of length per unit time, which is the units of a velocity. It reflects the velocity of flow through the porous medium. The proportionality constant K is called either the hydraulic conductivity or the coefficient of permeability.

Experiments have shown that K is a function of both the fluid and the medium through which the fluid is passing. Specifically, K is a function of the specific weight of the fluid (γ), the viscosity of the fluid (μ), and the size of the pores in the medium. Since the pore size cannot be measured directly, it can be replaced by a representative grain diameter, d. Thus K is functionally related by

$$K = C\gamma^{a_1}\mu^{a_2}d^{a_3} \tag{6–5}$$

where a_1, a_2, and a_3 are constants and C is a dimensionless proportionality constant. The Buckingham pi theorem (that is, dimensional analysis) yields

$$LT^{-1}[=](FL^{-3})^{a_1}(FL^{-2}T)^{a_2}(L)^{a_3} \tag{6–6}$$

Equating the exponents of Equation 6–6 for F, L, and T yields

$$\textit{For } F\text{:}\quad 0 = a_1 + a_2 \tag{6–7a}$$

$$\textit{For } L\text{:}\quad 1 = -3a_1 - 2a_2 + a_3 \tag{6–7b}$$

$$\textit{For } T\text{:}\quad -1 = a_2 \tag{6–7c}$$

Solving Equations 6–7 yields $a_1 = 1$, $a_2 = -1$, and $a_3 = 2$. Thus Equation 6–5 becomes

$$K = \frac{C\,d^2\gamma}{\mu} \tag{6–8}$$

Recognizing that γ and μ are properties of the fluid and that d^2 is of the porous medium, the term Cd^2 can be presented as a separate term and is called the *specific permeability, k*:

$$k = Cd^2 \tag{6–9}$$

which, when substituted into Darcy's Law of Equation 6–1, yields

$$Q = \frac{k\gamma}{\mu}iA \tag{6–10}$$

From Equation 6–9 it is evident that k has dimensions of L^2, or area, which can be viewed as a measure of the area of the pore spaces. The constant C reflects other characteristics of the porous medium, such as the distribution and shape of the particles and the packing and structure of the soil mass. Instead of using square feet or square inches as units for the specific permeability k, it is usually expressed in darcys, where 1 darcy equals 1.062×10^{-11} ft^2. Table 6–1a provides typical ranges of k for different soil classes.

Example 6–1

A pump is discharging 70 gal/min from a well in a sandy aquifer. Two observation wells located in a radial line from the pumping well have a 1-ft difference in the water surface elevations; the two observation wells are 50 ft apart. Thus the hydraulic gradient is 2%. At the observation well closer to the pumping well, which is at a distance of 300 ft, the depth from the water table to bedrock is 80 ft. Thus the area through which the 70 gal/min passes at a distance of 300 ft from

TABLE 6–1 Various Soil Class Parameters

(a) Magnitude of Specific Permeability k for Different Soil Classes

Soil Class	Approximate Range of k (darcys)
Clean gravel	10^3–10^5
Clean sands	1–10^3
Very fine sands and silts	10^{-4}–1
Unweathered clays	$<10^{-4}$

(b) Approximate Velocities, V, for Different Soil Classes and Hydraulic Gradients, i

Soil Class	K (darcys)	i (ft/ft)	V (ft/yr)
Clean gravel	10^4	0.001	8,860
		0.05	443,000
Clean sand	10^2	0.001	89
		0.05	4,430
Very fine sand	10^{-2}	0.001	0.0089
		0.05	0.44
Unweathered clay	10^{-5}	0.001	8.86×10^{-6}
		0.05	4.43×10^{-4}

(c) Approximate Conductivity, K (ft/day), for Selected Materials

Material	K
Coarse sand	150
Medium sand	40
Fine sand	8
Silt	0.6
Clay	0.0007
Loess	0.3

the pumping well is $2\pi rb = 2\pi(300)(80) = 48{,}000\ \pi\ \text{ft}^2$. Thus the hydraulic conductivity of the aquifer is

$$K = \frac{Q}{iA} = \frac{(70\ \text{gal/min})[2.228 \times 10^{-3}\ \text{ft}^3/\text{sec}/(\text{gal/min})]}{(0.02)(48{,}000\pi\ \text{ft}^2)}$$

$$= 0.5171 \times 10^{-4}\ \text{ft/sec}$$

The specific permeability is

$$k = \frac{K\mu}{\gamma} = \frac{(0.5171 \times 10^{-4}\ \text{ft/sec})(2.735 \times 10^{-5}\ \text{lb-sec/ft}^2)}{62.41\ \text{lb/ft}^3}$$

$$= 2.266 \times 10^{-11}\ \text{ft}^2 = 2.134\ \text{darcy}$$

Example 6–2

Assume that a soil analysis indicates a hydraulic conductivity of 0.0007 ft/sec. Two observation wells located 75 ft apart and in line with the flow have water surface elevations that differ by 2.2 ft. The geologic formation has a depth of 60 ft and a width of 540 ft. The flow through the cross section is

$$Q = 0.0007\ \frac{\text{ft}}{\text{sec}}\left(\frac{2.2\ \text{ft}}{75\ \text{ft}}\right)(60\ \text{ft})\ (540\ \text{ft}) = 0.665\ \text{ft}^3/\text{sec}$$

$$= 298.6\ \text{gal/min}$$

6.2.2 Ground Water Velocities

Using Equations 6–8 and 6–9, the flow velocity is given by

$$V = \frac{k\gamma}{\mu} i \tag{6–11}$$

For a temperature of 60°F and the appropriate conversions the velocity in ft/yr can be estimated for different soil classes and hydraulic gradients. Table 6–1b shows the velocities for four soil classes and hydraulic gradients of 0.001 ft/ft and 0.05 ft/ft. The velocity ranges from virtually 0 to about 1200 ft/day.

Example 6–3

Figure 6–3 shows a confined aquifer that is 2.4 mi wide and 65 ft thick. The hydraulic conductivity for the sandy soil is 0.035 ft/sec, and the porosity is 0.35. The water surface elevation in two observation wells, which are located 800 ft apart (see Figure 6–3), is 7.2 ft; therefore, the hydraulic gradient i is 0.009 ft/ft. The rate of flow is

$$Q = 0.035\ \frac{\text{ft}}{\text{sec}}\left(0.009\ \frac{\text{ft}}{\text{ft}}\right)(65\ \text{ft})(2.4\ \text{mi})(5280\ \text{ft/mi})$$

$$= 259.5\ \text{ft}^3/\text{sec}$$

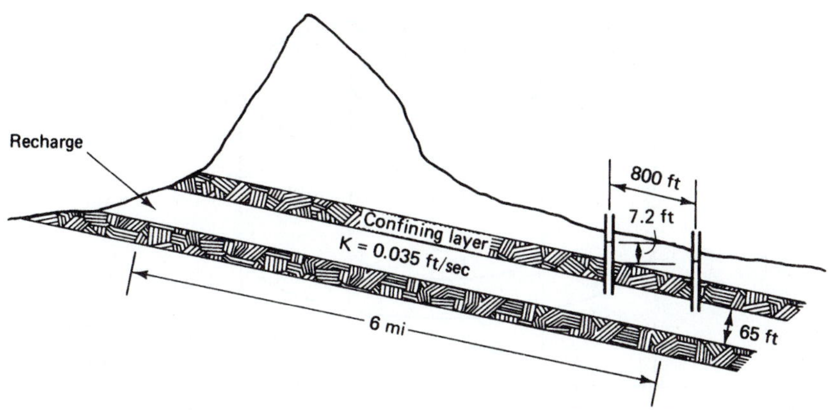

FIGURE 6–3 Schematic of a confined aquifer for Example 6–3.

The Darcy velocity is

$$V = \frac{Q}{A} = Ki = 0.035\,\frac{\text{ft}}{\text{sec}}\left(0.009\,\frac{\text{ft}}{\text{ft}}\right) = 0.000315 \text{ ft/sec}$$

and the seepage velocity is

$$V_s = \frac{V}{P} = \frac{0.000315 \text{ ft/sec}}{0.35} = 0.0009 \text{ ft/sec}$$

Therefore, the travel time from the recharge area to a point of interest 6 mi from the recharge area is

$$T_t = \frac{L}{V} = \frac{(6 \text{ mi})\,(5280 \text{ ft/mi})}{(0.0009 \text{ ft/sec})\,(3600 \text{ sec/hr})\,(24 \text{ hr/day})} = 407 \text{ days}$$

6.3 HYDRAULICS OF WELLS: STEADY STATE

While Darcy's Law is a simple relationship that is used to represent flow in a complex physical system, it is still a very useful model. It is commonly used in describing flow from an aquifer to a well. To introduce the conceptual framework the case of a confined aquifer with a full penetrating well can be used. We will make the assumption that the water level in the aquifer is initially horizontal and the water is not moving. When the pump in the well is started, the pressure around the intake of the pump is reduced, which creates a pressure difference between the water in the surrounding aquifer and the pump intake; the water in the aquifer then begins movement to the pump intake where the pressure is lower. This causes the level of water in the aquifer to decrease from its initial horizontal position. Initially, unsteady flow conditions will exist and the level of the water will continue to decrease. At some point in time, the drawdown in the well will cease and steady-state conditions will

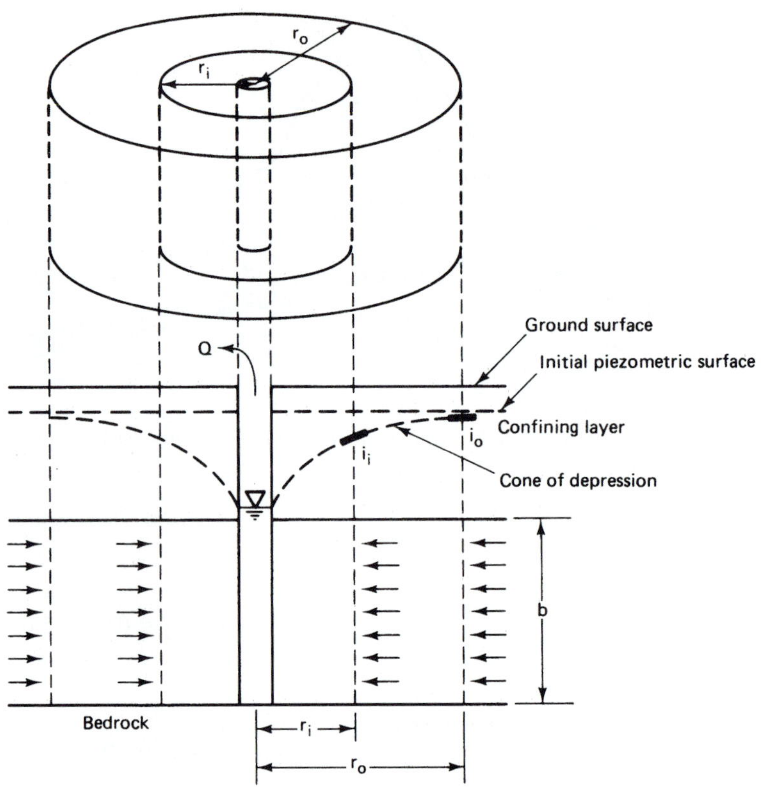

FIGURE 6–4 Illustration of pumping from a well in a confined aquifer.

exist. The water level in the aquifer is called the cone of depression because it has a characteristic conic shape. The shape of cone can be evaluated by considering flow through two imaginary concentric cylinders toward a pumping well located at the center (see Figure 6–4). The radii of the inner and outer cylinders are r_i and r_o, respectively. Since the flow rate is constant across the surface of each of the cylinders and assuming Darcy's Law applies, we have

$$Q = Ki_i(2\pi r_i b) = Ki_o(2\pi r_o b)$$

With $2\pi bK$ constant, we get the equality $i_i r_i = i_o r_o$. Since $r_o > r_i$, then $i_o < i_i$, which means that as the distance from the center of the well decreases, the hydraulic gradient increases. The curvature is defined by the product ir = constant; however, as r decreases, Darcy's Law becomes less valid and the relationship becomes less valid.

6.3.1 Unidirectional Flow: Confined Aquifer

For the case of a confined aquifer with a uniform thickness and where Darcy's Law is valid, steady flow is

$$\frac{\partial^2 h}{\partial x^2} = 0 \tag{6–12}$$

in which h is the head measured from a datum at the top of the aquifer and x is the horizontal distance along the aquifer measured from some boundary point, at which $h = 0$ when $x = 0$. From differential calculus, the solution of Equation 6–12 has the form

$$h = c_1 x + c_2 \tag{6–13}$$

in which c_1 and c_2 are constants that depend on the boundary conditions. Since $h = 0$ at $x = 0$, then c_2 of Equation 6–13 is 0. Darcy's Law can be used to define the second boundary condition:

$$\frac{dh}{dx} = \frac{V}{K} \tag{6–14}$$

From Equation 6–13, $dh/dx = c_1$; therefore, from Equation 6–14, we get $c_1 = V/K$. Substituting this into Equation 6–13 gives the solution

$$h = \frac{V}{K}x \tag{6–15}$$

which states that h varies linearly with x. This is shown in Figure 6–5.

Example 6–4

Consider the case of an aquifer with a depth of 12 ft and a hydraulic conductivity of 125,000 ft/yr. If measurements at two observation wells 2500 ft apart indicate a drop in the piezometric surface of 8 ft, then Equation 6–15 would indicate a velocity of

$$V = \frac{hK}{x} = \frac{8(125,000)}{2500} = 400 \text{ ft/yr} \tag{6–16}$$

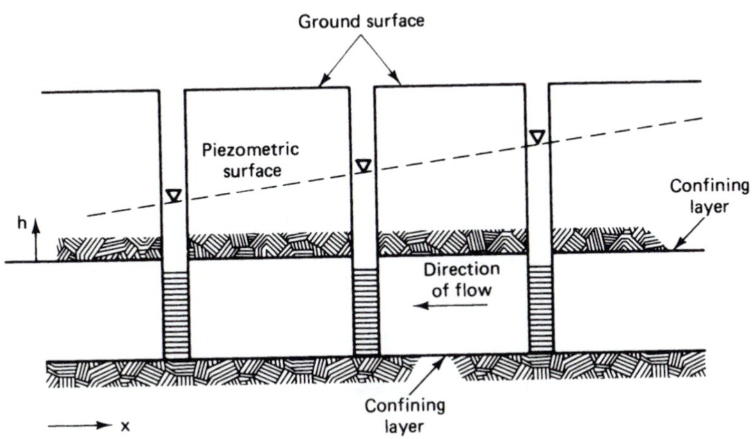

FIGURE 6–5 Unidirectional flow in a confined aquifer.

Using the continuity equation for a unit width of 1 ft yields a flow rate of

$$Q = AV = 12\ \text{ft}^2(400\ \text{ft/yr}) = 0.088\ \frac{\text{ft}^3/\text{sec}}{\text{ft width}} \tag{6–17}$$

6.3.2 Unidirectional Flow: Unconfined Aquifer

For an unconfined aquifer, the solution is not as simple as the solution for a confined aquifer. To make the problem manageable, we can consider the case of water flowing between two water bodies, with the water source for the water body with the higher water level sufficient to maintain steady-state conditions. The system is shown in Figure 6–6. Assuming that the aquifer is sufficiently wide so that the flow is only in the x direction, the discharge per unit width q is obtained from Darcy's Law as follows:

$$\frac{Q}{w} = q = Ki\frac{A}{w} = K\frac{dh}{dx}h \tag{6–18}$$

in which h is the height of the water table above the bottom of the aquifer, which is assumed to be an impermeable boundary. At the point where $x = 0$, we will define the water table depth as h_0. At any point x, the depth is h. Equation 6–18 can then be rearranged to

$$\int_0^x q\,dx = \int_{h_0}^h Kh\ dh \tag{6–19}$$

Integration yields

$$q = \frac{1}{2}\frac{K}{x}(h^2 - h_0^2) \tag{6–20}$$

Equation 6–20 defines the water surface elevation h for any distance x from the upstream face of the higher water body. Equation 6–20 indicates that the water table has a parabolic shape.

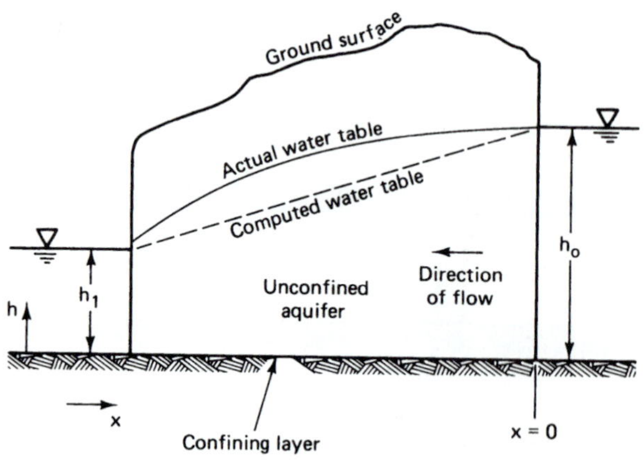

FIGURE 6–6 Steady flow between two water bodies: unconfined aquifer.

The model of Equation 6–20 requires certain assumptions, which are not entirely valid. Although it is beyond the intentions of this discussion to detail the specific effects of the violations of the assumptions, let it suffice to say that for given values of h_1, h_0, and x, Equation 6–20 should provide reasonable estimates of either q or K. Equation 6–20 will also provide reasonable estimates of the water table when the hydraulic gradient is relatively small.

Example 6–5

To illustrate Equation 6–20 consider the case where measurements of h_0 and h_1 of 9 and 6 ft, respectively, are made when the two water bodies are 1500 ft apart. A laboratory test of soil samples indicates that K is 0.0027 ft/sec. Therefore, the steady-state flow rate is

$$q = \frac{(0.0027\text{ ft/sec})(6^2 - 9^2)\text{ ft}}{2(1500\text{ ft})} = -0.405 \times 10^{-4}(\text{ft/sec})/\text{ft}$$

where the minus sign indicates the direction of flow.

6.3.3 Radial Flow: Confined Aquifer

To develop a solution that is manageable, several assumptions must be made: (1) the well completely penetrates the infinite, confined aquifer; (2) the flow is two-dimensional; (3) the aquifer is homogeneous and isotropic; (4) the flow is laminar; and (5) the flow is horizontal. Based on these assumptions and the assumption that Darcy's Law is valid, we can represent Darcy's Law, with the continuity equation, in a polar coordinate system:

$$Q = KiA = K\frac{dh}{dr}(2\pi rb) \tag{6–21}$$

in which b is the depth of the aquifer and r is the distance measured radially from the center of the well. Equation 6–21 can be rearranged to

$$\int Q\frac{dr}{r} = 2\pi bK\,dh \tag{6–22}$$

At the well, which has a radius r_w, the height of the water is h_w. At any other point in the aquifer a distance r from the well, the height of the piezometric surface is h. Therefore, integration of Equation 6–22 yields

$$Q = \frac{2\pi bK(h - h_w)}{\ln_e(r/r_w)} \tag{6–23}$$

Equation 6–22 could be generalized for any two points in the system, where subscripts 1 and 2 can be used to indicate the location of the two observation wells:

$$Q = \frac{2\pi bK(h_1 - h_2)}{\ln_e(r_1/r_2)} \tag{6–24}$$

Given the locations of and depths in two observation wells, either Q or K could be determined with Equation 6–24.

Example 6–6

Consider the case of a well that penetrates a confined aquifer that has a depth of 20 ft and a hydraulic conductivity of 0.0032 ft/sec. Two observation wells are located at 1000 ft and 2700 ft, with water surface elevations of 2.8 and 4.2 ft, respectively.

The discharge can then be computed with Equation 6–24:

$$Q = \frac{2\pi(20\text{ ft})(0.0032\text{ ft/sec})(4.2 - 2.8)\text{ ft}}{\ln_e(2700/1000)} = 0.567\text{ ft}^3/\text{sec}$$

6.3.4 Radial Flow: Unconfined Aquifer

Darcy's Law in polar coordinate form can be used to describe steady radial flow to a well in an unconfined aquifer. Assuming that the well completely penetrates the aquifer, the flow is steady and laminar, the aquifer is homogeneous and isotropic, and the flow is horizontal, Darcy's Law is

$$Q = KiA = K\frac{dh}{dr}(2\pi rh) \tag{6–25}$$

Assuming that $h = h_0$ at $r = r_0$ and $h = h_w$ at $r = r_w$, the integrals are

$$\int_{r_w}^{r_0} \frac{Q}{r}\,dr = \int_{h_w}^{h_0} K2\pi h\,dh \tag{6–26}$$

Therefore, the drawdown curve is described approximately by

$$Q = \frac{\pi K(h_0^2 - h_w^2)}{\ln_e(r_0/r_w)} \tag{6–27}$$

Example 6–7

To illustrate Equation 6–27 for estimating the sustainable flow rate, assume that an aquifer has a K of 0.004 ft/sec. If the depths of the water surface are 8 ft and 10.3 ft, respectively, in the well being pumped ($r_w = 2$ in.), and an observation well is located 750 ft from the well, the flow rate is

$$Q = \frac{\pi(0.004)[(10.3)^2 - (8)^2]}{\ln_e(750/0.167)} = 0.0629\text{ ft}^3/\text{sec}$$

6.4 HYDRAULICS OF WELLS: UNSTEADY FLOW

Before installing a permanent facility for pumping water from an aquifer it is practical to obtain estimates of important characteristics of the aquifer. These will be useful in evaluating the yield that can be sustained once a steady-state pumping rate is achieved. Characteristics of interest are the transmissivity (T) and the storativity (S). The storativity, which is dimensionless, equals the product of the thickness (d) of the aquifer and the specific storage (S_s):

$$S = dS_s \tag{6–28}$$

A number of methods have been proposed for estimating T and S from measurements of the time rate of change in the level of piezometric surface as a volume rate Q is pumped from the aquifer. These methods, which are intended for use with fully penetrating wells, are based on the following assumptions: Darcy's Law is applicable; the aquifer is isotropic and homogeneous, and has a constant thickness and negligible slope. Two methods will be introduced, the Theis method and Jacob's Straight-line method. The latter is an approximation of the former.

6.4.1 Theis Method

C.V. Theis developed a method for describing unsteady flow in a confined aquifer using concepts of heat flow in a thermal conductor. Specifically, the drawdown of the piezometric surface is a function of the constant flow rate (Q), the transmissivity of the aquifer (T), the radial distance from the pumping well to the observation well (r), the aquifer storativity (S), and the time from the start of pumping (t). Theis used the following differential equation, which is expressed in plane polar coordinates, to describe the hydraulics of the problem:

$$\frac{\partial^2 h}{\partial r^2} + \frac{1}{r}\frac{\partial h}{\partial r} = \frac{S}{T}\frac{\partial h}{\partial t} \tag{6–29}$$

in which h is the height of the piezometric surface. Using the boundary conditions that $h = h_0$ before the start of pumping and that h approaches h_0 for very large values of r, where h_0 is the initial height of the piezometric surface, Theis provided the following solution to Equation 6–29:

$$h_0 - h = \frac{Q}{4\pi T}\int_u^\infty \frac{e^{-u}}{u}\,du \tag{6–30a}$$

where

$$u = \frac{r^2 S}{4Tt} \tag{6–30b}$$

The integral of Equation 6–30a is known as the well function, is often denoted as $W(u)$, and is defined by the following series:

$$W(u) = -0.5772 - \ln_e u + u - \frac{u^2}{2(2!)} + \frac{u^3}{3(3!)} - \frac{u^4}{4(4!)} + .. \tag{6–31a}$$

or

$$W(u) = -0.5772 - \ln u + u + \sum_{i=2}^{\infty}\frac{(-1)^{i-1}u^i}{i(i!)} \tag{6–31b}$$

When Q is expressed in gal/min, T in gal/day/foot, r in ft, t in min, $(h_0 - h)$ in ft, and both $W(u)$ and S are dimensionless, the drawdown of Equation 6–30 can be computed by

$$h_0 - h = \frac{114.6\,Q}{T}\,W(u) \tag{6–32}$$

and

$$u = \frac{2693\,r^2 S}{Tt} \tag{6–33}$$

When t is expressed in days, the constant of Equation 6–33 becomes 1.87. The value of $W(u)$ can be obtained from Table 6–2.

In practice, values for S and T are needed, which represent the two unknowns for the two equations, Equations 6–32 and 6–33. Theis provided a graphical solution using log-log plots of $W(u)$ versus u and $(h_0 - h)$ versus r^2/t, with the plots made on log-log paper having the same scale. The curve $W(u)$ versus u is known as the *type curve*. The plot of $(h_0 - h)$ versus r^2/t will be referred to as the *data curve*. The graphical solution is as follows:

1. Plot the type curve.
2. Using the measured data, plot the data curve.
3. Superimposing the data curve onto the type curve while maintaining the coordinate axes of the two curves parallel, move the data curve until the trend of the data curve is approximately coincident with the type curve.
4. Select a point arbitrarily as the match point and record the values of $(h_0 - h)$ and r^2/t from the data curve and the values of $W(u)$ and u from the type curve.
5. Using the values of $(h_0 - h)$ and $W(u)$ from Step 4, along with the value of Q, find T by rearranging Equation 6–32:

$$T = \frac{114.6\,Q}{h_0 - h} W(u) \tag{6–34}$$

6. Compute the estimate of S by rearranging Equation 6–33:

$$S = \frac{Ttu}{2693r^2} = \frac{Tu}{2693(r^2/t)} \tag{6–35}$$

TABLE 6–2 Values of $W(u)$ for Values of u

u	1.0	2.0	3.0	4.0	5.0	6.0	7.0	8.0	9.0
×1	0.219	0.049	0.013	0.0038	0.0011	0.00036	0.00013	0.000051	0.000018
$\times 10^{-1}$	1.82	1.22	0.91	0.70	0.56	0.45	0.37	0.31	0.26
$\times 10^{-2}$	4.04	3.35	2.96	2.68	2.47	2.30	2.15	2.03	1.92
$\times 10^{-3}$	6.33	5.64	5.23	4.95	4.73	4.54	4.39	4.26	4.14
$\times 10^{-4}$	8.63	7.94	7.53	7.25	7.02	6.84	6.69	6.55	6.44
$\times 10^{-5}$	10.94	10.24	9.84	9.55	9.33	9.14	8.99	8.86	8.74
$\times 10^{-6}$	13.24	12.55	12.14	11.85	11.63	11.45	11.29	11.16	11.04
$\times 10^{-7}$	15.54	14.85	14.44	14.15	13.93	13.75	13.60	13.46	13.34
$\times 10^{-8}$	17.84	17.15	16.74	16.46	16.23	16.05	15.90	15.76	15.65
$\times 10^{-9}$	20.15	19.45	19.05	18.76	18.54	18.35	18.20	18.07	17.95
$\times 10^{-10}$	22.45	21.76	21.35	21.06	20.84	20.66	20.50	20.37	20.25
$\times 10^{-11}$	24.75	24.06	23.65	23.36	23.14	22.96	22.81	22.67	22.55
$\times 10^{-12}$	27.05	26.36	25.96	25.67	25.44	25.26	25.11	24.97	24.86
$\times 10^{-13}$	29.36	28.66	28.26	27.97	27.75	27.56	27.41	27.28	27.16
$\times 10^{-14}$	31.66	30.97	30.56	30.27	30.05	29.87	29.71	29.58	29.46
$\times 10^{-15}$	33.96	33.27	32.86	32.58	32.35	32.17	32.02	31.88	31.76

TABLE 6–3 Drawdown Data ($h_0 - h$) for Example 6–8 of the Theis Method

Time (min)	r^2/t	$h_0 - h$ (ft)
15	24,000	1.941
30	12,000	2.672
45	8,000	3.115
60	6,000	3.434
75	4,800	3.683
90	4,000	3.888
105	3,429	4.061
120	3,000	4.212
150	2,400	4.464
180	2,000	4.671
225	1,600	4.925
240	1,500	4.998
300	1,200	5.252

Example 6–8

Data of drawdown at an observation well located 600 ft from a well that fully penetrates a confined aquifer is given in Table 6–3. The well delivers 150 gal/min. The type curve is shown in Figure 6–7, and the drawdown curve is shown in Figure 6–8. The data curve was drawn on tracing paper and superimposed on the type curve (see Figure 6–7); while several positions appeared reasonable, the position shown in Figure 6–9 appears to give the best fit. The data point is with r^2/t equal to 12,000 and ($h_0 - h$) equal to 2.672 feet. This point matched a point on the type curve with $u = 0.058$ and $W(u) = 2.33$. Solving Equation 6–34 yields

$$T = \frac{114.6\ (150)\ (2.33)}{2.672} = 15,000\ \text{gal/day/ft}$$

A value of S can be obtained from Equation 6–35:

$$S = \frac{15{,}000\ (0.058)}{2693\ (12{,}000)} = 0.000027$$

6.4.2 Jacob's Straight-Line Method

Recognizing the effort required by Theis's method, C.E. Jacob simplified the fitting process by truncating the expansion of Equation 6–31. Recognizing that for small values of u, the series terms of Equation 6–31 become negligible, Jacob expressed the drawdown by

$$h_0 - h = \frac{Q}{4\pi T}\left(-0.5772 - \ln_e \frac{r^2 S}{4tT}\right) \tag{6–36a}$$

$$= \frac{Q}{4\pi T}\left(\ln_e \frac{4tT}{r^2 S} - 0.5772\right) \tag{6–36b}$$

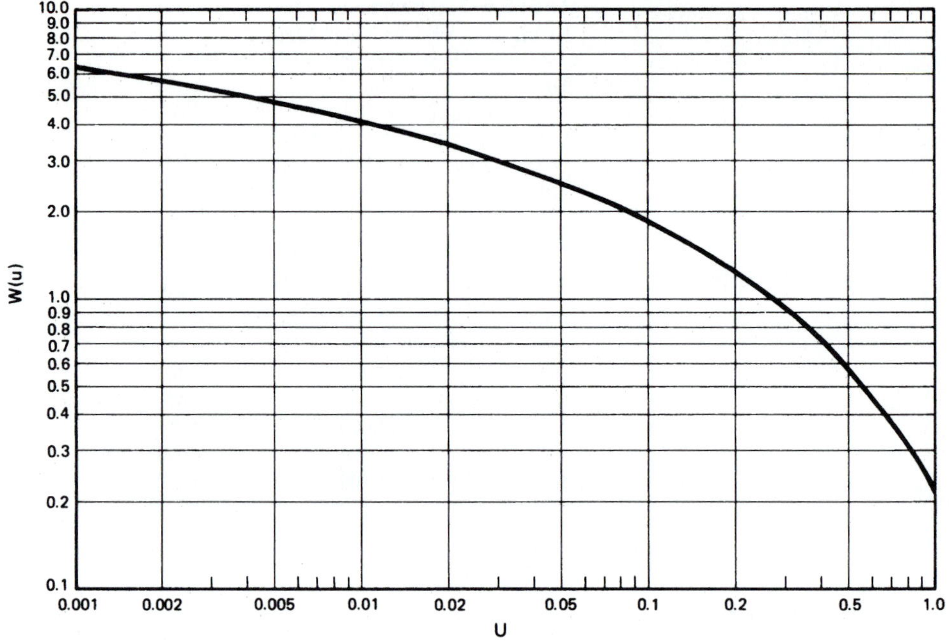

FIGURE 6–7 Curve type for the Theis Method: $W(u)$ versus u.

which reduces to

$$h_0 - h = \frac{2.30\,Q}{4\pi T} \log_{10} \frac{2.25\,tT}{r^2 S} \tag{6–37}$$

If $(h_0 - h)$ is plotted versus $\log_{10} t$, Equation 6–37 indicates that the data should plot as a straight line. Letting $d = h_0 - h$, two points can be selected to determine the slope of the line;

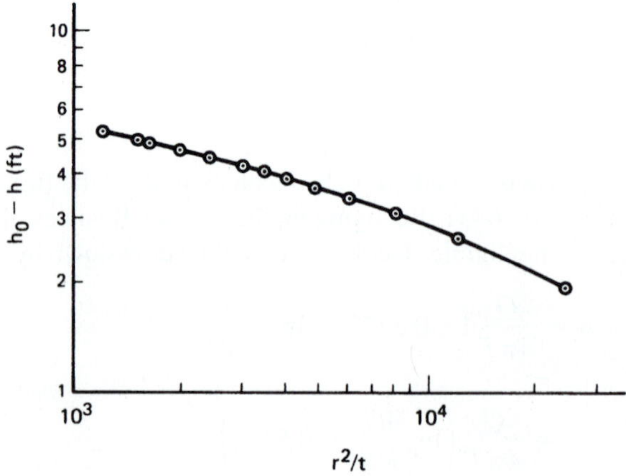

FIGURE 6–8 Data curve for Example 6–8 for the Theis Method.

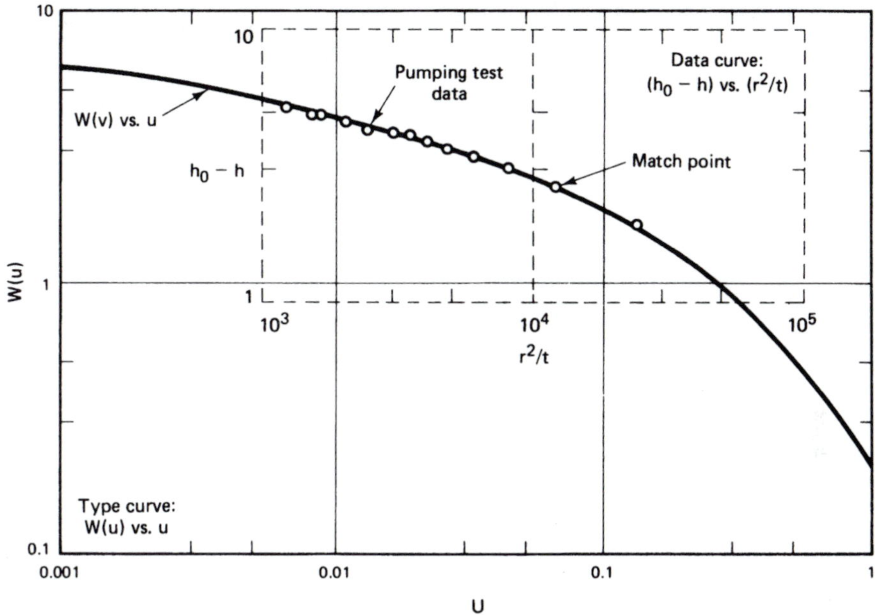

FIGURE 6–9 Data curve overlaid on the type curve for Example 6–8.

for convenience, it is best to choose two points that are exactly one log cycle apart on the time axis. Then letting d_1 and d_2 be the drawdowns from the straight line at the two points one log cycle apart, the $\log_{10}$ term of Equation 6–37 will equal 1 for times t one log cycle apart. Thus the difference in the two drawdowns $d_1 - d_2$, where $(d_1 - d_2) > 0$, is

$$d_1 - d_2 = \frac{2.30\,Q}{4\pi T} \tag{6–38}$$

Solving for T gives

$$T = \frac{2.30\,Q}{4\pi(d_1 - d_2)} \tag{6–39}$$

A value for the storativity can be obtained using the time from the straight-line relationship at which the drawdown is zero; this time will be denoted as t_0. Substituting into Equation 6–37 yields

$$0 = \frac{2.3Q}{4\,\pi\,T}\log_{10}\frac{2.25Tt_0}{r^2S} \tag{6–40a}$$

$$= \log_{10}\frac{2.25Tt_0}{r^2S} \tag{6–40b}$$

Taking the inverse log yields

$$1 = \frac{2.25Tt_0}{r^2S} \tag{6–41}$$

and solving for S yields

$$S = \frac{2.25Tt_0}{r^2} \tag{6–42}$$

For the units used previously with Equations 6–32 and 6–34, Equations 6–39 and 6–42 become

$$T = \frac{264Q}{d_1 - d_2} \tag{6–43}$$

and

$$S = \frac{Tt_0}{4790r^2} \tag{6–44}$$

To summarize the procedure of Jacob's straight-line method, estimates of the aquifer characteristics T and S can be obtained by following these steps

1. On semilog paper plot the drawdown versus the time from the start of pumping with the drawdown on the axis with an arithmetic scale and the time plotted on the axis having a log scale.
2. Draw a straight line through the points from the measured data.
3. Determine the change in the drawdown (in feet) per log cycle; this is denoted as $d_1 - d_2$ in Equation 6–38.
4. Using Equation 6–43, compute the transmissivity.
5. From the straight line on the plot (not the data points) estimate the time at which the drawdown equals zero (the time is denoted as t_0 in Equation 6–44).
6. Compute the storativity using Equation 6–44.

It is important to note that Jacob's Straight-line method uses semilog paper and in application requires just one graphical plot. On the other hand, the Theis Method uses log-log paper and requires plotting both the type curve and the data curve.

The Jacob Straight-line method is an approximation of the Theis method. The accuracy of Jacob's method depends on the importance of terms of Equation 6–31 that are truncated in deriving Equation 6–37. Several rules of thumb have been proposed for limiting the use of Jacob's method as an approximation of the Theis method. One rule states that u should be less than 0.01. Others state that t must be greater than $5r^2S/T$, which occurs for large values of t and small values of r. Table 6–4 provides the value of $W(u)$ for selected values of u as the terms of the expansion of Equation 6–31 are added. For u equal to 0.1 truncation of terms beyond $-\ln_e u$ changes the value of $W(u)$ by about 5.5%; for u equal to 0.01, the error is about 0.2%.

TABLE 6–4 Accuracy of Value of $W(u)$ for Selected Values of u Based on Expansion of Equation 6–31

Term	1	0.1	0.01	0.0001
–0.5772	–0.5772	–0.5772	–0.5772	–0.5772
$-\ln_e u$	–0.5772	1.7254	4.0280	6.3306
u	0.4228	1.8254	4.0380	6.3316
$-u^2/4$	0.1728	1.8229	4.0380	6.3316
$u^3/18$	0.2284	1.8229	4.0380	6.3316
$-u^4/96$	0.2179	1.8229	4.0380	6.3316
$u^5/600$	0.2196	1.8229	4.0380	6.3316
$-u^6/4320$	0.2194	1.8229	4.0380	6.3316
Table Value	0.219	1.82	4.04	6.33

Example 6–9

The drawdown data in Table 6–5 are from an observation well located 475 ft from a pumping well that fully penetrates a confined aquifer. A discharge of 525 gal/min is pumped from the well. The drawdown data are plotted on semilog paper (see Figure 6–10), and a straight line is drawn through the data and extended back to the time t_0 where the drawdown equals 0 ft. The change in drawdown over the log cycle from 10 to 100 min is 1.51 ft. Thus the transmissivity is

$$T = \frac{264Q}{d_1 - d_2} = \frac{264(525)}{1.51} = 91{,}788 \text{ gal/day/ft} \tag{6–45}$$

The value of t_0 obtained graphically from Figure 6–10 is 0.88 min, which is used to compute a value of the storativity:

$$S = \frac{Tt_0}{4790r^2} = \frac{91{,}788\ (0.88)}{4790\ (475)^2} = 0.0000747 \tag{6–46}$$

TABLE 6–5 Drawdown ($h_0 - h$) Versus Time for Example 6–9 of Jacob's Method

Time (min)	Drawdown (ft)	$u \times 10^{-3}$
15	1.86	36.0
30	2.31	18.0
45	2.58	12.0
60	2.77	9.0
75	2.92	7.2
90	3.04	6.0
105	3.14	5.1
120	3.23	4.5
150	3.38	3.6
180	3.50	3.0
210	3.60	2.6
240	3.69	2.3
270	3.77	2.0
300	3.84	1.8

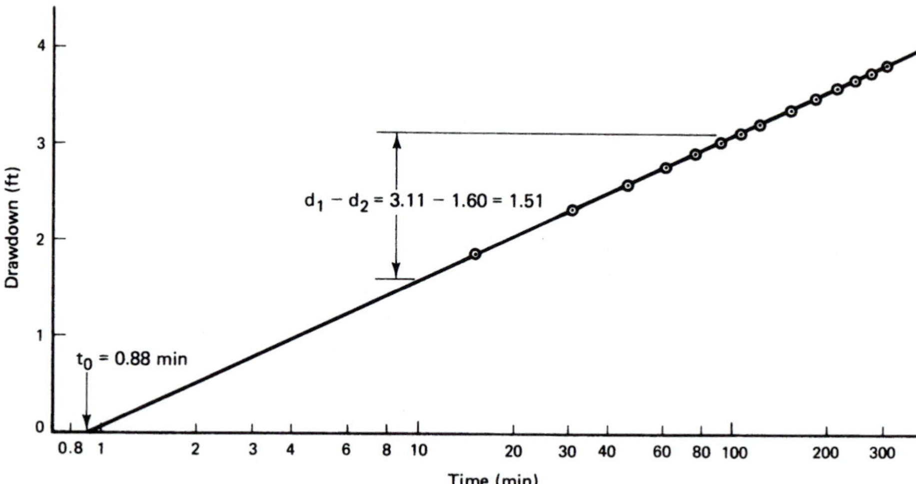

FIGURE 6–10 Drawdown ($h_0 - h$) versus time for Example 6–9 on Jacob's Method.

In addition to computing the aquifer characteristics, it is necessary to test the validity of Jacob's method as an approximation of the Theis method. Equation 6–30b was used to compute the value of u for each value of t; the values of u are given in Table 6–5. Since all the values are smaller than 0.01, Jacob's method is acceptable. Also, the value of $5r^2S/T$ is 0.001, so Jacob's method is valid for all t greater than 0.001. Thus the values of T and S from Equations 6–45 and 6–46 are acceptable.

6.5 GROUND WATER MANAGEMENT

In many localities ground water is a primary source of water supply. Ground water is extracted through wells (see Figure 6–1) as a source of water supply for municipal, industrial, and agricultural uses. Water serves as a necessary resource. Since ground water supplies are recharged naturally, it is a renewable resource. However, like any renewable resource, it can be depleted if it is not properly managed. The most noticeable visual effect of ground water depletion is the depression of the land surface. Depressions occur because the pore water pressure of the ground water is partly responsible for balancing the pressure force of any external burden and the overlying soil. When the ground water supply is severely depleted, the soil structure of the aquifer must support the overlying burden by itself. If it is structurally inadequate, the aquifer consolidates with the depression of the ground surface as a result.

Another important effect of the depletion of ground water supplies is the reduction of the water available for pumping. It is more difficult to extract water from a nearly depleted aquifer and this reduces the supply of water for others, as well as for future generations. The quality of the water may also deteriorate when an aquifer is over-harvested.

6.5.1 Safe Yield: Definition

When water is extracted at a rate that exceeds the recharge of the aquifer, the water table is lowered. Therefore, the rate of water extraction that can be safely harvested from an aquifer is an important characteristic for ground water management. This rate is called the safe yield; specifically, the *safe yield* is the volume of water that can be extracted from an aquifer during any time period without adversely affecting the supply. In practice, the time period used to compute estimates of safe yield is 1 year. However, recognizing that precipitation, surface runoff, and evapotranspiration rates vary from year to year, and therefore the recharge of ground water supplies varies naturally, a period other than 1 year could be used. Water removed from an aquifer in excess of the safe yield is termed *overdraft.*

6.5.2 Determination of Safe Yield

A number of methods have been proposed for estimating the safe yield of an aquifer. The methods reflect differences in the types and quality of data available. The methods are based on a water balance that reflects a water budget of the ground water aquifer. The conservation of mass expressed in Equation 1–1 is repeated here but with a slight change in notation (for example, s for S):

$$I - O = \frac{ds}{dt} \tag{6–47}$$

For ground water storage, the input (I) consists of the infiltration and percolation of precipitation such as seepage from streams, rivers, and lakes as well as water recharged through artificial means such as water spreading or recharge through wells and pits. The output (O) of Equation 6–47 includes evapotranspiration such as seepage to effluent streams and rivers, lakes, and marshes as well as water pumped from the aquifer. The change in the storage (ds/dt) of Equation 6–47 can be measured by either (1) the product of the storage coefficient and the change in the piezometric head, or (2) the product of the specific yield and the change in the elevation of the water table.

If measurements of the ground water table at observation wells over long periods of time are available, the safe yield can be determined as the volume rate of water that can be extracted such that the water surface elevation at some current time is the same as at some point in time in the distant past. If significant volumes of water were not being extracted from the aquifer during this time period, the record of water table fluctuations represents the long-term record of natural draft and recharge. Where significant volumes of water have been extracted from the aquifer over the period of record, these volumes must be considered in estimating the safe yield.

Three methods of estimating the safe water yield at a site will be presented; other methods are available and are described in the literature. The methods are (1) the zero fluctuation method, (2) the average draft method, and (3) the water balance method. The water balance method requires estimates of each of the terms in Equation 6–47; therefore, it is the most data intensive and should be the most accurate. The zero fluctuation method requires only annual measurements of the water table elevation, so it is the least data intensive and probably the least accurate. The average draft method requires both annual measurements of the water table elevation and a record of annual pumpage from the aquifer. Where the pumpage is the most significant output and variations in precipitation are small, the average draft method should be an accurate approximation of the water balance method.

6.5.3 Zero Fluctuation Method

For the zero fluctuation method the data required to estimate the safe yield include the total draft over a period of record when the water table elevation at the beginning and end of the period is the same. The longer the period of record, the more accurate the estimate of the safe yield should be. In practice, the data will probably consist of annual records of water table elevations and the annual draft.

Example 6–10

The data of Table 6–6 can be used to illustrate the zero fluctuation method. The annual draft for a 10-yr period shows a total draft of 42.1×10^2 acre-ft. Since the water table elevation is 38 ft below the surface at the beginning and end of the 10-yr period, an estimate of the safe yield is 4.21×10^2 acre-ft.

6.5.4 The Average Draft Method

The average draft method uses the same type of data as the zero fluctuation method; however, instead of requiring just the total draft for a period during which the water table elevation is the same at the beginning and end of the period, annual values of the change in the water table elevation and the annual draft are required. To compute the safe yield, the annual drafts should be plotted as the abscissa and the change in the annual water table elevation as the ordinate. A line that best represents the data can be fit by either eye or regression; if regression analysis is used as the fitting method, the annual draft should be used as the criterion (dependent) variable. The safe yield is the value of the annual draft estimated from the line at a point corresponding to a zero elevation change.

TABLE 6–6 Example 6–10: Application of the Zero Fluctuation Method of Estimating Safe Field

Year	Depth Water Table (ft)	Annual Draft (acre-ft $\times 10^2$)
1962	38	—
1963	36	3.4
1964	33	6.5
1965	27	7.1
1966	30	3.2
1967	31	3.9
1968	28	6.0
1969	28	3.7
1970	32	2.8
1971	36	2.5
1972	38	3.0

TABLE 6–7 Example 6–11: Average Draft Method

Year	Depth to Water Table (ft)	Change in Depth to Water Table Δe (ft)	Annual Draft, D (acre-ft $\times 10^3$)
1962	42	—	18.9
1963	45	−3	23.0
1964	49	−4	25.5
1965	48	1	17.3
1966	46	2	15.1
1967	48	−2	21.2
1968	54	−6	28.0
1969	53	1	16.9
1970	53	0	18.7
1971	50	3	13.4
1972	45	5	10.7
1973	43	2	15.7
1974	43	0	18.3
1975	47	−4	24.5
1976	49	−2	21.5
1977	52	−3	23.6
1978	57	−5	26.7
1979	56	1	16.4
1980	51	5	11.0
1981	47	4	12.0
1982	42	5	10.5

Example 6–11

The data of Table 6–7 can be used to illustrate the average draft method. The annual data are for a 20-yr period. The change in the depth to the water table (Δe) is plotted against the annual draft (D) in Figure 6–11. The line was fitted using regression analysis:

$$\hat{D} = 18.5 - 1.59\,\Delta e \tag{6–48}$$

in which $\hat{D}$ is the predicted annual draft. Thus for a Δe of 0, the predicted annual draft is 18.5×10^3 acre-ft, which is the best estimate of the safe yield.

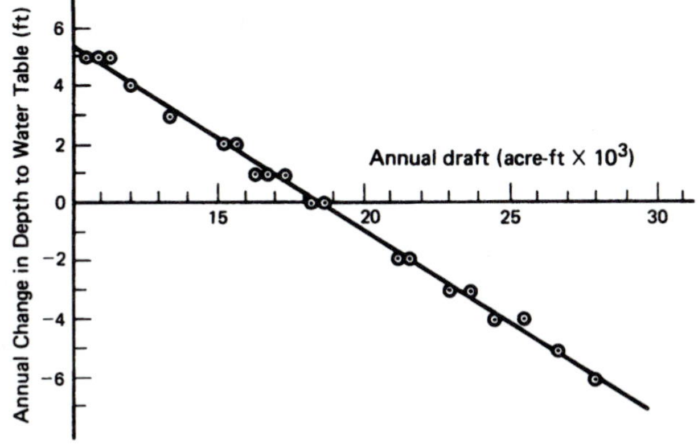

FIGURE 6–11 Annual draft versus the change in the depth of the water table.

6.5.5 The Simplified Water Balance Method

The simplified water balance method uses Equation 6–47 to compute the safe yield. In many cases, annual precipitation and streamflow are the only data available to represent the inflow and outflow of Equation 6–47. Observation wells can be used to estimate the annual changes in the water table elevation. In this case, the other inputs and outputs are assumed to be relatively constant with time and the water balance equation can be reduced to

$$P - (Q + Q_p) = \Delta e \tag{6–49}$$

in which the annual change in the water table elevation Δe represents the change in storage, the annual precipitation is assumed to be the input, and the sum of the surface runoff Q and annual draft Q_p is the output. If values of the annual draft are not available, $P - Q$ can be plotted against the change in the water table elevation to estimate the safe yield, with the value of Δe used as the ordinate. A line can be fitted to the data, and the safe yield can be estimated from the value of $(P - Q)$ corresponding to a value of Δe of 0.

Example 6–12

Table 6–8 gives annual rainfall and runoff volumes, as well as the annual change in the water table elevation. The annual net rainfall excess, $P - Q$, is plotted against the annual values of Δe in Figure 6–12. A line was fit using regression analysis:

$$(P - Q) = 5.36 + 1.25\,\Delta e \tag{6–50}$$

Therefore, the safe yield is estimated to be 5.36×10^3 acre-ft.

TABLE 6–8 Example 6–12: Simplified Water Balance Method

Year	Depth to Water Table (ft)	Change to Water Table (ft)	Annual Precipitation (acre-ft × 10^3)	Annual Streamflow (acre-ft × 10^3)	Annual Surface Inflow – Outflow (acre-ft × 10^3)
1966	53	—	—	—	—
1967	51	2	12.3	3.1	9.2
1968	47	4	19.2	4.2	15.0
1969	46	1	14.9	4.8	10.1
1970	44	2	20.2	7.6	12.6
1971	45	−1	10.4	6.5	3.9
1972	48	−3	3.6	2.2	1.4
1973	53	−5	0.0	0.0	0.0
1974	54	−1	6.0	2.3	3.7
1975	56	−2	9.7	6.4	3.3
1976	57	−1	11.8	5.2	6.6
1977	56	1	0.2	0.0	0.2
1978	56	0	14.4	6.3	8.1
1979	53	3	0.0	0.0	0.0
1980	55	−2	5.9	4.9	1.0
1981	56	−1	8.0	6.5	1.5

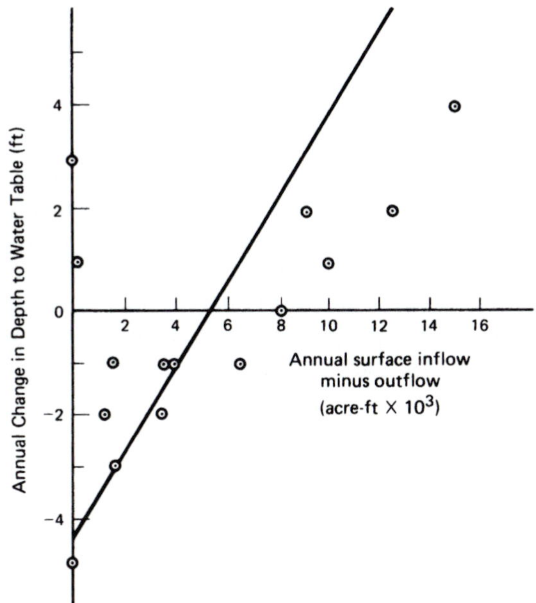

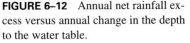

FIGURE 6–12 Annual net rainfall excess versus annual change in the depth to the water table.

6.6 ANALYSIS VERSUS SYNTHESIS

The concepts of analysis and synthesis apply to ground water hydrology in the same way that they apply to other areas of hydrology. The specific elements of the system of Figure 1–3 depend on the particular ground water problem. The input is usually some measure of the available water supply, and the output consists of the supply of water that can be extracted from the ground water system; the system is represented by characteristics of the geologic formation.

For Darcy's Law the slope of the hydraulic grade line reflects the potential of the water supply to move from one point to another point in the geologic formation; thus it is the input function. The velocity of the flow or the flow rate represents the output from the system. The system is characterized by the hydraulic conductivity and the dimensions or size of the aquifer. For a given aquifer, the objective of analysis would be to use measurements of the slope of the hydraulic grade line and the flow rate to estimate the hydraulic conductivity of the aquifer. In the synthesis case, estimates of the hydraulic grade line and the hydraulic conductivity would be used to estimate the flow through the aquifer.

For the methods introduced to describe the hydraulics of wells for steady-state conditions, the variables used to define the piezometric surface represent the system input. The flow rate represents the system output, and the hydraulic conductivity and aquifer characteristics define the system transfer function. For unsteady flow conditions the transfer function is represented by the transmissivity and the storativity, the aquifer thickness, and the location of the observation well with respect to the location of the pumping well. The drawdown curve and the pumping rate represent the input and output functions, respectively. In the analysis case, the objective would be to estimate the aquifer characteristics, the hydraulic conductivity for steady flow conditions, and the transmissivity and storativity for unsteady flow conditions. In the synthesis case, the objective would be to estimate the discharge rate from the pumping well.

In the problems on ground water management, the elevation of the water table reflects the supply of water and is therefore, the input function. The annual draft is the system output. While in the analysis case, measures of the annual draft are available, estimates of the annual draft are made in the synthesis case. The coefficients of the relationship between the change in the water table elevation and the annual draft reflect the characteristics of the aquifer and thus represent the transfer function. In the analysis case, the coefficients must be estimated from measurements of Δe and the annual draft; in the synthesis case, the coefficients and measurements of Δe are used to predict the annual draft.

PROBLEMS

6–1. Compute the Reynold's number for flow in an aquifer having a mean grain diameter of 0.055 in. when the temperature is 60°F and the water has a velocity of 0.02 ft/hr.

6–2. Compute the Reynold's number for flow in an aquifer for the following conditions: temperature = 50°F; velocity = 0.0001 mi/day; mean grain diameter = 0.008 ft.

6–3. What is the maximum velocity for Darcy's Law to be valid when the mean grain diameter is 0.006 ft and the temperature is 70°F?

6–4. What is the maximum velocity for Darcy's Law to be valid given a temperature of 60°F and a mean grain diameter of 0.04 in.?

6–5. What would be the limiting mean grain diameter (in.) for Darcy's Law to be valid when the temperature equals 50°F and the flow velocity is 0.75 ft/day?

6–6. For a clean sand with a specific permeability of 10 darcys, find the velocity of the flow when the difference in the elevation of water in two observation wells 300 ft apart is 1.2 ft. For an aquifer 35 ft thick find the discharge rate in ft^3/sec/ft. Assume a temperature of 50°F.

6–7. Find the velocity of flow in a silt having a specific permeability of 0.02 darcy when the difference in water surface elevations of two observation wells 225 ft apart is 0.85 ft. Assume a temperature of 50°F. Find the discharge rate (ft^3/sec/ft) if the aquifer is 25 ft thick.

6–8. Two observation wells are 85 ft apart, with the difference in water surface elevations being 0.2 ft. A tracer dye is injected into the well that has the higher water level, and a mean travel time of 6.5 days is measured. Find the hydraulic conductivity in both ft/day and ft/yr.

6–9. Two observation wells are 120 ft apart and have a difference in water surface elevations of 0.15 ft. A tracer is injected into the well that has the higher water level, and a mean travel time of 3.75 days is measured. Find the hydraulic conductivity in mi/yr.

6–10. An aquifer is 60 ft thick, with laboratory tests indicating a hydraulic conductivity of 450 gal/day/ft^2. Test wells located 50 and 125 ft from the pumping well have a difference in water surface elevations of 3.5 ft. Find the flow rate from the well.

6–11. Two water bodies, 850 ft apart, are separated by an unconfined aquifer with a hydraulic conductivity of 2000 ft/yr. Assuming steady flow conditions and a water surface elevation difference of 6 ft, compute the discharge rate (ft^3/sec/ft) into the water body that has the lower water surface elevation of 5 ft.

6–12. An unconfined aquifer separates two water bodies 675 ft apart. The aquifer has a hydraulic conductivity of 3.5 ft/day. If the water surface elevation difference is 2.5 ft, compute the steady-state discharge rate (ft^3/sec/ft). Assume a 2 ft elevation for the lower water surface.

6–13. An unconfined aquifer separates two water bodies, 550 ft apart. The two water surfaces have an elevation difference of 0.8 ft and the steady-state discharge rate is 0.2 ft^3/sec/ft. Compute the hydraulic conductivity of the aquifer. Assume a 3 ft elevation for the lower water surface.

6–14. The water surface elevations of two water bodies located 375 ft apart differ by 0.5 ft. For a steady-state discharge of 40 gal/day/ft, compute the hydraulic conductivity of the unconfined aquifer. The elevation of the lower water surface is 4.5 ft. Graph the water table across the aquifer.

6–15. Two water bodies 625 ft apart separated by an unconfined aquifer have water surface elevations that differ by 1.1 ft. For a steady-state discharge of 125 gal/day/ft, compute the hydraulic conductivity of the aquifer. The elevation of the lower water surface is 3.4 ft. Graph the water table across the aquifer.

6–16. For a confined aquifer 65 ft thick, find the discharge if the aquifer has a hydraulic conductivity of 500 gal/day/ft^2, and an observation well located 150 ft from the pumping well has a water surface elevation 1.5 ft above the water surface elevation in the pumping well, which has radius of 6 in.

6–17. For a confined aquifer 85 ft thick, find the discharge (gal/day) if the aquifer has a hydraulic conductivity of 65,000 ft/yr. An observation well located 325 ft from the pumping well has a water surface elevation 1.8 ft above the water surface elevation in the pumping well, which has a radius of 3 in.

REVIEW QUESTIONS

6–1. Confined aquifers are recharged naturally by (a) percolation through the overlying strata; (b) saltwater intrusion; (c) infiltration of surface water; (d) penetrating recharge wells.

6–2. When a well is drilled into a confined aquifer, the water level will rise in the well to the (a) water table; (b) piezometric surface; (c) ground surface; (d) capillary fringe; (e) none of the above.

6–3. The zone of aeration is (a) the soil zone below the water table; (b) another name for the zone of saturation; (c) the soil layers consisting of the A- and E-horizons; (d) composed of the root, intermediate, and capillary zones; (e) none of the above.

6–4. Which one of the following pairs of terms is not identical? (a) Hydraulic conductivity and permeability; (b) effective porosity and specific retention; (c) effective porosity and specific yield; (d) zone of aeration and unsaturated zone.

6–5. Soil moisture is (a) water remaining after field capacity has been reached; (b) water in the zone of saturation; (c) water taken up by plants and transpired to the atmosphere; (d) water in the zone of aeration.

6–6. Which one of the following does not apply to the permanent wilting point (PWP)? (a) The moisture level where the pressure just becomes greater than atmospheric pressure; (b) the PWP is the level of soil moisture where plants wilt; (c) once the PWP is reached, only hygroscopic moisture is present; (d) water below the PWP is referred to as unavailable water.

6–7. The matric potential is (a) measured using a hygroscope; (b) described by Darcy's Law; (c) an indicator of the potential of pressure forces to do work; (d) negative pressure in the zone of aeration.

6–8. For the flow rate Q computed by Darcy's Law, which one of the following is not directly related to Q? (a) The hydraulic conductivity; (b) the cross-sectional area of the flow path; (c) the hydraulic gradient; (d) the flow length.

6–9. Darcy's law is valid (a) when the inertial forces are greater than the viscous forces; (b) when the Reynold's number < 1; (c) for all cases of flow in porous media; (d) when the Reynold's number < 2000.

6–10. Which one of the following is the hydraulic conductivity (K) *not* a function of? (a) The water temperature; (b) the specific weight of water; (c) a representative grain size; (d) the viscosity of the water; (e) all of the above affect K.

6–11. A darcy is a unit of measurement for (a) the hydraulic conductivity; (b) the specific yield; (c) the specific permeability; (d) the coefficient of permeability.

6–12. For unidirectional flow in a confined aquifer, the flow velocity is *not* a function of (a) the depth of the aquifer; (2) the hydraulic conductivity; (c) the drop in the piezometric surface; (d) the soil type.

6–13. For unidirectional flow in an unconfined aquifer that separates two water bodies, the slope of the water table is (a) logarithmic; (b) linear; (c) parabolic; (d) exponential decay.

6–14. Which one of the following is *not* an assumption for defining the discharge rate of radial flow in a confined aquifer? (a) Laminar flow; (b) an isotropic aquifer; (c) a completely penetrating well; (d) three-dimensional flow; (e) horizontal flow.

6–15. Which one of the following is *not* valid for unsteady flow? (a) Darcy's Law; (b) the Theis Method; (c) Jacob's Method; (d) all of the methods are valid for unsteady flow.

6–16. Which one of the following is *not* necessary to compute the transmissivity and storativity using the Theis method? (a) Measured drawdown data; (b) aquifer thickness; (c) radial distance between the observation and pumping wells; (d) the discharge rate; (e) all of the above are required.

6–17. The safe yield is (a) the change in the water table when a well is pumped at a constant rate; (b) the rate at which water can be pumped from an aquifer without lowering the water table; (c) the product of the specific yield and the change in the elevation of the water table; (d) the volume of water that can be safely harvested from an aquifer.

6–18. Which one of the following is *not* a method of estimating the safe yield of an aquifer? (a) The simplified water balance method; (b) the average draft method; (c) the energy balance method; (d) the zero fluctuation method.

DISCUSSION QUESTION

The technical content of this chapter is important to the professional hydrologist, but practice is not confined to making technical decisions. The intent of this discussion question is to show that hydrologists must often address situations where value issues intermingle with the technical aspects of a project. In discussing the stated problem, at a minimum include responses to the following questions:

1. What value issues are involved, and how are they in conflict?
2. Are technical issues involved with the value issues? If so, how are they in conflict with the value issues?

3. If the hydrologist attempted to rationalize the situation, what rationalizations might he or she use? Provide arguments to suggest why the excuses represent rationalization.
4. What are the hydrologist's alternative courses of action? Identify all alternatives, regardless of the ethical implications.
5. How should the conflict be resolved?

You may want to review Sections 1.6 to 1.12 in Chapter 1 in responding to the problem statement.

Case: An entry-level engineer, who specializes in GIS, overlays several surface and subsurface files and recognizes that the design location of a pipeline that will carry hazardous waste material from a manufacturing plant to a special remediation facility transects a porous formation that is used as part of the local water supply. The young engineer recognizes that, if the pipe leaks, the water supply will be contaminated. When discussing the issue with the project manager, the manager insists that it is not a matter with which he needs to be concerned. The original contract that the engineering firm signed with the manufacturer specifically required that location and a change in location might prevent the project from being undertaken.

7

Peak Discharge Estimation

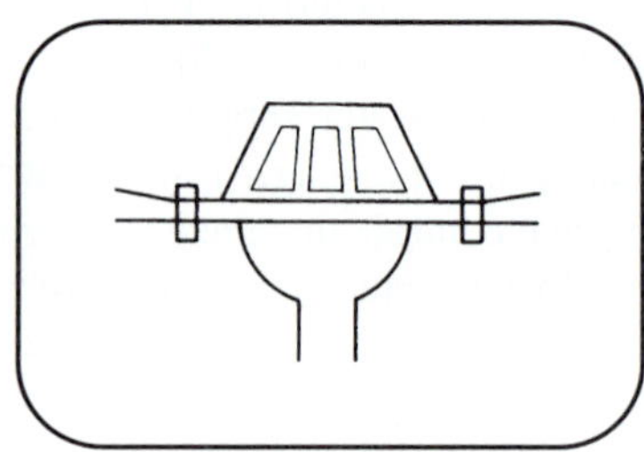

CHAPTER OBJECTIVES

1. Outline the data and analyses required to develop regional peak discharge design methods.
2. Introduce commonly used methods of making peak discharge estimates on ungaged watersheds.
3. Present peak discharge envelope curves.

7.0 NOTATION

a = cross-sectional area of waterway
A = drainage area
A_c = area of waterway
A_m = drainage area (sq. miles)
b_j = j^{th} regression coefficient
BDF = basin development factor
C = runoff coefficient for the Rational method
C_m = coefficient
C_t = Talbot coefficient
C_w = weighted runoff coefficient
CN = runoff curve number
CN_p = pervious area CN
CN_w = weighted CN
D = drainage density
DH = elevation difference
f = fraction of imperviousness
F = actual retention in SCS method
F = percent forest cover
F = value of F statistic
F_p = pond and swamp adjustment factor
i = rainfall intensity
I_a = initial abstraction
IA = impervious area
k_j = index flood ratio for j^{th} return period
K = frequency factor
n = number of watersheds
n = roughness coefficient
p = number of predictor variables
P = depth of rainfall
P = mean annual rainfall
PPS = percent of total watershed in ponds and swamps
q = discharge rate
q_a = adjusted peak discharge
q_a = unit discharge (ft^3/sec/acre)
q_p = peak discharge
q_u = unit peak discharge
q_{um} = unit peak discharge ($ft^3/sec/mi^2/in.$)
Q = depth of runoff
R = isoerodent factor
R = relief ratio
R_h = hydraulic radius
RQ_i = rural peak discharge for return period i
$RI2$ = basin rainfall intensity
S = channel slope
S = ground slope
S = maximum retention in SCS method
S = standard deviation of flows
S_e = standard error of estimate
S_y = standard deviation of y
SL = 10–85 channel slope
ST = basin storage
t_c = time of concentration
T = return period
UQ_i = urban peak discharge for return period i
V = velocity of flow
X_i = i^{th} predictor variable
$\overline{X}$ = mean discharge
$\overline{X}$ = sample mean
Y = peak discharge in a regression equation

7.1 INTRODUCTION

A primary design variable in hydrology is the peak discharge, which corresponds to the maximum water surface elevation during a storm event. Specifically, a peak discharge is the maximum volume flow rate passing a particular location during a storm event; it has units of volume per time, such as cubic feet per second, cubic meters per second, or acre-feet per hour.

TABLE 7–1 Classification of Hydrologic Models

- I. Peak discharge models
  - A. Calibrated models
    1. Single-return-period equations
    2. Index flood estimation
    3. Moment estimation
  - B. Uncalibrated models
- II. Single-event hydrograph models
  - A. Design storm
  - B. Actual storm
- III. Watershed (continuous) multiple-event models
  - A. Actual record
  - B. Synthetic record

The peak discharge is a primary variable for the design of stormwater runoff facilities such as pipe systems, storm inlets and culverts, and small open channels. It is also used for some hydrologic planning such as small detention facilities in urban areas. It is an acceptable design variable for designs where the time variation of storage is not a primary factor in the runoff process. Where variation of storage is important, the design should be based on a hydrograph procedure (see Chapter 9).

As suggested, peak discharge models are just one group of hydrologic models. Other types of hydrologic models are used for other types of hydrologic design problems. Table 7–1 provides a classification system for hydrologic models, with the peak discharge models discussed in this chapter. Other types of hydrologic models of Table 7–1 will be discussed in other chapters.

As shown in Table 7–1, peak discharge models can be separated into two groups: calibrated and uncalibrated models. In a sense, all models are calibrated, so the difference suggested by the two groups of peak discharge models of Table 7–1 lies within the level of formality in the calibration. As used herein, the term calibrated models will be applied to models based on regressions involving peak discharges obtained from flood frequency analyses at gaged sites or the statistical moments of flood records. The term uncalibrated model will refer to models that have not undergone a formal fitting to gaged data. While the line between the two is somewhat fuzzy, the distinction should be clear after examples are introduced in this chapter.

7.2 HISTORICAL REVIEW

Designs based on peak discharges have a long historical basis. In a historical review of the literature on engineering studies of hydrologic determination of waterway areas, Ven Te Chow (1962) included the following events:

1852: preparation of a table expressing the relation between the diameter and slope of a circular outlet sewer and the size of the drainage area.
1879: preparation of the Myers Formula; Major E.T.C. Myers was chief engineer of the Richmond, Fredericksburg, and Potomac Railway in Virginia.
1880: publication of the formula by the Swiss hydraulic engineer A. Burkli-Ziegler.
1887: publication of the Talbot Formula; Professor A.N. Talbot was from the University of Illinois.
1889: publication by Emil Kuichling of the Rational Method.

These are just a few of the historical events relating to design methods using peak discharges. It is also of interest to note that J.C.I. Dooge (1973) credits T.J. Mulvaney with the development of the Rational method as early as 1851 in Ireland.

While the Rational Method and some of the more recent methods will be discussed in detail, it may be of interest to review some of these older formulas. The change in the form of design methods over the last 100 years may seem surprisingly small.

In 1879, Myers presented the following formula for computing the area of a waterway (A_c, in square feet):

$$A_c = C_m A^{0.5} \tag{7–1}$$

in which C_m is a coefficient that reflects the slope of the drainage area and A is the drainage area in acres. Use of the Myers Formula should be limited to small drainage areas. The area A_c appears to be independent of land use and rainfall, although the rainfall intensity may be inherent in the coefficient C_m. It was probably developed independent of land use since it was intended to be used by railroad engineers, and thus the land use would be typical of railway right-of-way areas.

The Burkli-Ziegler Formula computes the unit discharge (ft^3/sec/acre), q_a, using

$$q_a = C_b^i \left(\frac{S}{A}\right)^{0.25} \tag{7–2}$$

in which C_b is a runoff coefficient for Equation 7–2; i is the average rainfall rate (in./hr) during the heaviest storm, S is the average ground slope (ft/1000 ft), and A is the drainage area in acres. The runoff coefficient, C_b, depended on the land cover, with special emphasis on the relative imperviousness.

The Talbot Formula was intended for computing the waterway cross-sectional area (a, ft^2). Talbot based his formula on the Burkli-Ziegler Formula, with the following result:

$$a = C_t A^{0.75} \tag{7–3}$$

Chow (1962) provided criteria for selecting values of the coefficient C_t.

These early attempts at hydrologic design suggest that the dominant factors in peak discharge estimation were recognized as the drainage area, land use, rainfall intensity, and slope. These same factors are at the heart of many of the equations used today. Current methods attempt to develop more precise criteria for selecting values of the empirical coefficients and include additional variables as input. Although these additional variables are probably not as important as drainage area, land use, and rainfall, they should nevertheless improve the accuracy of prediction.

7.3 SINGLE-RETURN-PERIOD EQUATIONS

Hydrologic design engineers would like to have a 50-yr flood record available at every site where a peak discharge estimate is needed for design work. If such data were always available, then a frequency analysis of the flood record (see Chapter 5) could be used to characterize the flood potential at the site of the design work. Unfortunately, flood records are rarely available where peak discharge estimates are needed for design work. Therefore, it is necessary to use either a prediction method that was developed from flood frequency analyses of gaged data in the region or an uncalibrated prediction equation that was designed for use at ungaged sites.

Stream gages are usually located at locations that receive runoff from large watersheds. Therefore, prediction equations based on the regional analysis of flood frequency results are most appropriate for larger drainage areas. Their use for design on small areas of about 100 acres or less may represent extrapolation rather than interpolation, and thus, the accuracy of estimated values would seem to be less than that for the watersheds similar in size and physiohydrologic characteristics to the watersheds used in developing the equations. For the smaller watersheds, one of the uncalibrated methods is usually applied primarily because of their simplicity in contrast to their expected accuracy.

The classification scheme of Table 7–1 includes three types of peak discharge models under the heading "calibrated models." For lack of a better description, the first of the three was called single-return-period equations. This type of model relates the peak discharge of a specific return period to one or more watershed and/or meteorological characteristics. It is quite common to calibrate these using a multiple linear regression analysis after making a logarithmic transform of all the variables. Stepwise regression is frequently used to decide which predictor or independent variables to use. Before providing examples of the single-return-period equations, it may be instructive to discuss the problem in terms of the concepts of analysis and synthesis.

7.3.1 Analysis

Since the form of Equation 2–113 is used most often, the objective of the analysis process is to select the predictor variables to use in the following equation and calculate the values of the regression coefficients b_j ($j = 0, 1, 2, ..., p$):

$$q_p = b_o X_1^{b_1} X_2^{b_2 \cdots} X_p^{b_p} \tag{7–4}$$

in which q_p is the peak discharge and X_j ($j = 1, 2, ..., p$) are the watershed characteristics. Note that the subscript p on q indicates 'peak' while the random variable p denotes the number of predictor variables in the equation. The model is calibrated using data from watersheds in a reasonably homogeneous region. The steps of the analysis procedure are as follows:

1. Obtain the annual maximum flood series for each of the n gaged sites in the region.
2. Perform a separate flood frequency analysis (for example, log-Pearson Type III) on each of the flood series of Step 1 and determine the peak discharges for selected return

periods (for example, the 2-, 5-, 10-, 25-, 50-, and 100-yr discharges are commonly selected).

3. Determine the values of watershed and meteorological characteristics for each watershed for which a flood series was collected in Step 1. Form an $(n \times p)$ data matrix of all the data, where n is the number of flood series (watersheds) of Step 1, and p is the number of watershed characteristics obtained.
4. Form a one-dimensional array of the n peak discharges for the specific return period.
5. Regress the array of n peak discharges of Step 4 on the data matrix of Step 3 to obtain the prediction equation. Peak discharge estimates made with the resulting equation are for the return period selected in Step 2.

If more than one return period is of interest, the procedure can be repeated for each return period, with a separate equation developed for each. In this case, it is also important to review closely the resulting coefficients to ensure that they are consistent across the various return periods. Because of sampling variation it is possible for the regression analyses to produce a set of coefficients that, under certain sets of values for the predictor variables, the 10-yr estimated discharge may, for example, be greater than the computed 25-yr discharge. The coefficients should be adjusted to ensure that this does not occur. One method of smoothing the coefficients is to plot them versus return period and draw a smooth curve through the points. Then, instead of using the actual coefficient, the value for each return period should be taken from the fitted curve and used for the final equations.

Example 7–1

Table 7–2 contains the data matrix for developing a single-return-period peak discharge equation. Annual maximum flood series were obtained for 58 gaged sites in Indiana (Davis, 1974). Each series was subjected to a flood frequency analysis from which the 10-yr peak discharge was obtained; the 10-yr peak for each gaged site is also given in Table 7–2 and denoted as q_p. Seven watershed characteristics were measured for each of the 58 sites; the data matrix, 58 by 7, is also given in Table 7–2. This matrix and the vector of peak discharge values represents the first four steps of the analysis process described previously. The vector of peaks was regressed on the matrix of watershed characteristics after taking the natural logarithms of the data. The following equation resulted:

$$\begin{aligned} ln\, q_p &= 1.6931 + 0.6795\, ln\, A + 0.5002\, ln\, S \\ &+ 0.1683\, ln\, L + 0.7009\, ln\, P + 0.1497\, ln\, R \\ &+ 0.1652\, ln\, D + 1.2606\, ln\, C \end{aligned} \tag{7–5}$$

Equation 7–5 resulted in a correlation coefficient of 0.955 and a sum of the errors equal to zero, both of which apply only to the accuracy of estimating $ln\, q_p$. Equation 7–5 was transformed to the following form:

$$\hat{q}_p = 5.4365 A^{0.6795} S^{0.5002} L^{0.1683} P^{0.7009} R^{0.1497} D^{0.1652} C^{1.2606} \tag{7–6}$$

The average error (that is, $\hat{q}_p - q_p$) for the 58 peak discharges for Equation 7–6 is –215 ft^3/sec, which indicates a small bias. The correlation coefficient computed for Equation 7–5 cannot be used as a measure of accuracy of estimates made with Equation 7–6. The development of Equation 7–6 represents the last step of the analysis process.

TABLE 7–2 Data Matrix for Indiana Flood Data

X_1	X_2	X_3	X_4	X_5	X_6	X_7	Y
121.0	12.8	19.5	11.0	341.0	9.5	.80	13300
38.2	26.0	16.6	13.0	449.0	11.0	.90	12600
188.0	5.5	25.1	15.0	265.0	10.5	.90	14100
129.0	6.3	33.2	16.0	438.0	8.7	.80	13100
41.9	15.4	14.0	17.5	435.0	12.0	.80	5500
85.6	4.7	15.6	9.5	182.0	5.8	.50	3200
133.0	4.6	20.1	10.0	150.0	7.5	.80	7700
113.0	3.6	22.7	7.0	95.0	3.4	.30	600
35.0	5.5	9.6	8.5	77.0	2.2	.30	460
152.0	2.0	19.1	8.0	82.0	2.3	.70	2350
146.0	3.3	24.1	9.0	100.0	2.8	.80	4100
162.0	3.3	9.6	9.0	111.0	3.0	.80	4800
24.7	4.5	12.7	9.5	72.0	4.5	.70	740
132.0	7.2	35.9	12.0	252.0	5.4	.70	13100
133.0	11.4	29.1	13.0	344.0	5.4	.70	16800
35.5	10.2	12.5	10.5	165.0	5.7	.70	1580
131.0	4.0	27.1	10.0	121.0	5.4	.70	3540
40.4	6.2	15.0	9.5	110.0	4.6	.70	2400
18.5	18.7	6.4	9.5	110.0	6.6	.70	3750
169.0	7.2	31.8	11.5	273.0	5.3	.70	5800
42.4	6.7	19.3	10.0	127.0	4.5	.70	1600
103.0	15.2	17.4	10.5	153.0	7.5	.70	9000
174.0	6.8	35.1	10.5	264.0	7.9	.70	11200
23.9	18.8	9.2	11.0	189.0	4.5	.70	1820
28.8	10.6	11.5	11.0	131.0	6.0	.70	3900
14.6	19.8	7.6	13.0	291.0	11.5	.80	4150
59.0	12.6	8.3	12.5	283.0	7.0	.70	9600
184.0	5.8	30.8	12.0	251.0	5.2	.70	7800
107.0	4.3	30.1	12.0	195.0	7.0	.70	8200
91.4	10.3	33.7	12.5	401.0	7.4	.70	8200
155.0	8.9	42.3	13.5	410.0	9.8	.90	14200
77.2	9.4	32.6	14.0	318.0	1.5	1.00	14000
85.9	12.2	29.1	13.5	391.0	9.0	1.00	13400
198.0	9.2	43.2	14.0	448.0	9.1	1.00	29000
38.2	13.0	10.6	14.0	312.0	10.0	.80	5700
76.1	11.6	17.7	13.5	400.0	8.0	.70	7400
120.0	9.0	34.7	13.5	436.0	9.5	.70	11700
48.8	19.1	13.7	14.5	383.0	7.5	.80	8400
60.7	12.7	16.5	15.0	390.0	7.0	.80	6300
171.0	2.6	52.8	15.5	466.0	13.7	.80	5600
69.2	7.4	22.6	6.5	159.0	4.0	.50	2300
125.0	3.6	29.8	7.0	131.0	3.2	.50	2700
160.0	3.2	36.0	8.0	173.0	3.1	.40	2400
62.9	6.2	14.8	12.0	262.0	8.0	.40	2080
78.7	4.7	22.5	8.0	206.0	6.6	.40	1800
80.5	5.2	22.9	4.5	120.0	3.2	.30	530
142.0	3.9	26.6	5.5	130.0	3.5	.30	650
87.3	8.0	20.1	4.5	193.0	5.1	.50	1270
116.0	1.2	23.2	5.5	100.0	3.0	.30	640

X_1	X_2	X_3	X_4	X_5	X_6	X_7	Y
132.0	5.0	12.9	7.0	65.0	2.5	.50	1420
123.0	3.2	22.2	7.5	137.0	2.5	.40	1150
54.7	2.3	21.1	7.0	102.0	6.5	.40	1600
35.6	2.5	8.9	7.5	44.0	2.5	.40	340
144.0	2.9	15.7	8.0	58.0	3.5	.40	1220
203.0	2.5	18.6	8.0	63.0	4.5	.40	1670
21.8	6.4	10.4	8.5	82.0	3.4	.70	750
83.7	2.2	13.2	8.5	95.0	2.5	.70	2100
44.8	6.4	21.4	8.5	193.0	3.0	.70	1970

X_1: drainage area (mi^2)

X_2: channel slope (ft/mi); the difference in elevation at points 10% and 85% of the distance along the channel from a gaging station to the watershed divide, divided by the distance between the two points

X_3: channel length (mi); distance along a stream from a point of discharge to the watershed divide

X_4: precipitation index (in.); mean annual precipitation minus the sum of average annual evapotranspiration and mean annual snowfall (water equivalent)

X_5: basin relief (ft.); the difference in elevation between the highest point on the watershed perimeter and the stream at the point of discharge

X_6: drainage density (mi/mi^2); total length of streams in a watershed divided by the drainage area

X_7: soil runoff coefficient; ratio of the volume of rainfall to the total volume of runoff occurring after the beginning of runoff

Y: instantaneous peak discharge (ft^3/sec) for a 10-yr exceedence frequency

Source: L. G. Davis, *Floods in Indiana: Technical Manual for Estimating Their Magnitude and Frequency*, Geological Survey Circular 710, USGS, Reston, VA, 1974.

Example 7–2

Table 7–3 contains an abbreviated data set from a larger data base. The data matrix includes four predictor variables and three criterion variables, with data for fifteen gaging stations. A correlation analysis between the logarithms of the predictor variables and the logarithms of each of the criterion variables is given in Table 7–4a. Except for the correlation between slope and rainfall, –0.80, the correlations between the predictor variables are small; this suggests that three of the four variables will be needed. For all three criterion variables, the drainage area has the highest correlation, followed by rainfall, forest cover, and slope. Very little variation in the rainfall depths exists (column 4 of Table 7–3), with values ranging from 2.75 to 3.75 in. It is important to note that the correlations between rainfall depth and the three peak discharges are all negative, which is irrational; one would expect a positive correlation between rainfall and runoff. The irrational correlation is due to the small variation in rainfall, which causes it to appear as an unimportant variable.

A stepwise regression analysis (see Chapter 2) was performed for each of the three criterion variables on all four of the predictor variables. The model of Equation 7–4 was fitted after taking logarithms of the data. The results are given in Table 7–4b. In each case, rainfall was not significant, statistically or physically, and so it does not appear in the final equations. All of the equations produce accurate predictions with R^2 values above 0.8 and S_e/S_y values less than 0.5. All of the equations are biased, although the biases are less than 10 percent of the mean values of the peak discharges. All of the total F values are statistically significant.

TABLE 7–3 Data Matrix for Fitting Peak Discharge Equations

Gage Number	Area (sq. mi.)	Channel Slope (ft/mi)	2-yr, 24-hr Rainfall (inches)	Forest Cover (%)	Peak Discharge (ft^3/sec) for Return Period (years)		
					2	10	100
01472174	5.98	39.6	3.25	14	519	1,620	5,240
01478000	20.5	22.7	3.30	19	1,700	2,900	4,990
01479000	87.8	15.0	3.25	23	3,780	6,410	10,900
01480000	47.0	16.6	3.25	18	2,200	3,690	6,270
01480500	45.8	22.9	3.20	25	1,910	5,920	18,500
01480680	17.8	32.9	3.25	30	657	968	1,430
01480700	60.6	24.3	3.25	19	2,890	6,250	13,600
01481000	287.	14.5	3.25	45	6,810	13,200	25,300
01483700	31.9	3.86	3.55	46	500	1,110	2,540
01484000	13.6	6.26	3.55	35	284	828	2,340
01484300	7.08	7.89	3.70	54	37	77	165
01559700	5.28	148.	2.80	69	267	682	1,620
01594600	3.85	22.8	3.75	46	141	538	2,140
01598000	115.	69.8	2.75	83	3,510	7,970	17,000
01668800	15.5	12.5	3.55	75	137	502	1,910

TABLE 7–4 Results of Correlation and Power-Model Regression Analyses for Data Matrix of Table 7–3

(a) Correlation Matrix

	X_1	X_2	X_3	X_4	Y_2*	Y_{10}*	Y_{100}*
X_1: Drainage area	1.00	–.12	–.35	–.07	0.858	0.829	0.757
X_2: Channel slope		1.00	–.80	0.07	0.200	0.222	0.225
X_3: Rainfall			1.00	–.09	–.548	–.540	–.490
X_4: Forest cover				1.00	–.390	–.359	–.320

*Y_T = Peak discharge for return period of T years

(b) Stepwise Regression Results⁺

Return Period (yrs)	Regression Coefficients*										
	b_0	b_1	b_2	b_4	R	R^2	S_e (cfs)	S_e/S_y	$\bar{e}$	$\bar{e}/\bar{Y}$	F_{TOT}
2	139.8	1.049	0.5371	–0.9255	0.986	0.973	353	0.185	6	0.004	133
10	295.3	0.9386	0.5214	–0.7847	0.979	0.959	858	0.229	–60	–.017	85
100	741.3	0.8186	0.4838	–0.6650	0.912	0.831	3561	0.464	–639	–.084	18

*b_0 is the intercept; b_1, b_2, and b_4 are the coefficients for drainage area, channel slope, and forest cover, respectively.

⁺R = Correlation Coefficient; R^2 = fraction of explained variance; S_e = standard error of estimate. S_y = Standard deviation of Y_T; $\bar{e}$ = bias = average error; $\bar{e}/\bar{Y}$ = relative bias; F_{TOT} = total F statistic.

It is of interest to evaluate the values of the regression coefficients across return periods. As the values of Table 7–4b show, the values decrease in magnitude as the return period increases. Problems can sometimes occur when the coefficients do not show consistent trends; in such cases, it may be necessary to smooth the coefficients so that the computed peak discharges are consistent across return periods.

7.3.2 Synthesis

Once the regression equation has been calibrated, it can be used to predict peak discharges for the specified return period. Values for the watershed characteristics used as predictor variables must be obtained for the watershed where a design is to be made. This location will be an ungaged location since the annual maximum flood series does not exist to perform a flood frequency analysis. It is extremely important for anyone who uses a prediction equation calibrated by someone else to know the range of each of the variables used in calibrating the equation. Such equations should not be used outside the range of the data used for calibration. The accuracy of such "extrapolated" values would not be high.

Example 7–3

To use Equation 7–6 at an ungaged site, values for the seven predictor variables would have to be obtained. Consider the case where A = 102 mi^2, S = 14 ft/ mi, L = 21.7 mi, P = 11.4 in., R = 192 ft, D = 8.1 mi/mi^2, and C = 0.6. The computed peak discharge would be

$$\begin{aligned} q_p = {} & 5.4365(102)^{0.6795}(14)^{0.5002}(21.7)^{0.1683}(11.4)^{0.7009} \\ & (192)^{0.1497}(8.1)^{0.1652}(0.6)^{1.2606} = 7102 \text{ ft}^3/\text{sec} \end{aligned} \tag{7–7}$$

If equations for return periods other than the 10-yr event were available, they could be used to obtain the computed data necessary to plot a flood frequency curve for the ungaged site.

7.3.3 USGS Urban Peak Discharge Formulas

Sauer et al. (1981) used 269 urban watersheds to develop regression equations for watersheds having at least 15% of the drainage area covered by residential, commercial, or industrial development. None of the gaged sites had undergone a change in development of more than 50% during the period of record that was used to develop the set of single-return-period regression equations (Table 7–5). The rural discharge can be obtained from USGS reports for the location where the urban peak discharge estimate is required. An equation such as Equation 7–6 is available for most locations in the United States. Sauer et al. (1981) define the basin development factor as follows:

> The most significant index of urbanization was a basin development factor (BDF) which provides a measure of the efficiency of the drainage system. This parameter, which proved to be highly significant in the regression equations, can be easily determined from drainage maps and field inspections of the drainage basin. The basin is first subdivided into thirds and within each third, four aspects of the drainage system are evaluated and assigned a code as follows:
>
> 1. **Channel improvements.** If channel improvements such as straightening, enlarging, deepening, and clearing are prevalent for the main drainage channel and principal tributaries (those that drain directly into the main channel), then a code of one (1) is assigned. Any

TABLE 7–5 Nationwide Urban Flood Frequency Regression Equations

Regression Equations	Log R^2	Standard Error of Regression: Average Units	Standard Error of Regression: Percent
$UQ2 = 2.35A^{0.41}SL^{0.17}(RI2 + 3)^{2.04}(ST + 8)^{-0.65}(13 - BDF)^{-0.32}IA^{0.15}RQ2^{0.47}$	0.93	0.1630	±38
$UQ5 = 2.70A^{0.35}SL^{0.16}(RI2 + 3)^{1.86}(ST + 8)^{-0.59}(13 - BDF)^{-0.31}IA^{0.11}RQ5^{0.54}$	0.93	0.1584	37
$UQ10 = 2.95A^{0.32}SL^{0.15}(RI2 + 3)^{1.75}(ST + 8)^{-0.57}(13 - BDF)^{-0.30}IA^{0.09}RQ10^{0.58}$	0.93	0.1618	38
$UQ25 = 2.78A^{0.31}SL^{0.15}(RI2 + 3)^{1.76}(ST + 8)^{-0.55}(13 - BDF)^{-0.29}IA^{0.07}RQ25^{0.60}$	0.93	0.1705	40
$UQ50 = 2.67A^{0.29}SL^{0.15}(RI2 + 3)^{1.74}(ST + 8)^{-0.53}(13 - BDF)^{-0.28}IA^{0.06}RQ50^{0.62}$	0.92	0.1774	42
$UQ100 = 2.50A^{0.29}SL^{0.15}(RI2 + 3)^{1.76}(ST + 8)^{-0.52}(13 - BDF)^{-0.28}IA^{0.06}RQ100^{0.63}$	0.92	0.1860	44
$UQ500 = 2.27A^{0.29}SL^{0.16}(RI2 + 3)^{1.86}(ST + 8)^{-0.54}(13 - BDF)^{-0.27}IA^{0.05}RQ500^{0.63}$	0.90	0.2071	49
Three-parameter equations			
$UQ2 = 13.2A^{0.21}(13 - BDF)^{-0.43}RQ2^{0.73}$	0.91	0.1797	±43
$UQ5 = 10.6A^{0.17}(13 - BDF)^{-0.39}RQ5^{0.78}$	0.92	0.1705	40
$UQ10 = 9.51A^{0.16}(13 - BDF)^{-0.36}RQ10^{0.79}$	0.92	0.1720	41
$UQ25 = 8.68A^{0.15}(13 - BDF)^{-0.34}RQ25^{0.80}$	0.92	0.1802	43
$UQ50 = 8.04A^{0.15}(13 - BDF)^{-0.32}RQ50^{0.81}$	0.91	0.1865	44
$UQ100 = 7.70A^{0.15}(13 - BDF)^{-0.32}RQ100^{0.82}$	0.91	0.1949	46
$UQ500 = 7.47A^{0.16}(13 - BDF)^{-0.30}RQ500^{0.82}$	0.89	0.2170	52

The input requirements for these equations are as follows:

A = drainage area (mi^2);

$RI2$ = basin rainfall intensity (in.) for a duration of 2 hr and a return period of 2 yr

SL = the 10-85 channel slope (ft/mi) with a maximum value of 70

ST = basin storage (%) in lakes and reservoirs

BDF = basin development factor

IA = impervious cover (%)

RQ_i = rural peak discharge (ft^3/sec) for return period i, which is the same return period for the urban equation

UQ_i = urban peak discharge (ft^3/sec) for return period i.

one, or all, of these improvements would qualify for a code of one (1). To be considered prevalent, at least 50 percent of the main drainage channel and principal tributaries must be improved to some extent over natural conditions. If channel improvements are not prevalent, then a code of zero (0) is assigned.

2. **Channel linings.** If more than 50 percent of the main drainage channel and principal tributaries have been lined with an impervious material, such as concrete, then a code of one (1) is assigned. If less than 50 percent of these channels are lined, then a code of zero (0) is assigned. The presence of channel linings would probably indicate the presence of channel improvements as well. Therefore, this is an added factor and indicates a more highly developed drainage system.
3. **Storm drains or storm sewers.** Storm drains are defined as enclosed drainage structures (usually pipes), frequently used on the secondary tributaries where the drainage is received directly from streets or parking lots. Quite often these drains empty into the main tributaries and channel that are either open channels, or in some basins may be enclosed as box or pipe culverts. When more than 50 percent of the secondary tributaries within a section consists of storm drains, then a code of one (1) is assigned, and conversely if less than 50 percent of the secondary tributaries consists of storm drains, then a code of

TABLE 7–6 Minimum and Maximum Values of Predictor Variables Used to Calibrate Equations of Table 7–5

Variable	Minimum	Maximum	Units
A	0.2	100	square miles
SL	3.0	70[a]	feet per mile
RI2	0.2	2.8	inches
ST	0	11	percent
BDF	0	12	____
IA	3.0	50	percent
LT	0.2	45	hours

[a]Maximum value of slope for use in equations is 70 ft/mi, although numerous watersheds used in this study had *SL* values up to 500 ft/mi.

zero (0) is assigned. It should be noted that if 50 percent or more of the main drainage channels and principal tributaries are enclosed, then the aspects of channel improvements and channel linings would also be assigned a code of one (1).

4. **Curb and gutter streets.** If more than 50 percent of a subarea is urbanized (covered by residential, commercial, and/or industrial development), and if more than 50 percent of the streets and highways in the subarea is constructed with curbs and gutters, then a code of one (1) should be assigned. Otherwise, a code of zero (0) is assigned. Frequently, drainage from curb-and-gutter streets will empty into storm drains.

The above guidelines for determining the various drainage system codes are not intended to be precise measurements. A certain amount of subjectivity is involved. It is recommended that field checking be performed to obtain the best estimate. The basin development factor (*BDF*) is computed as the sum of the assigned codes. Obviously, with three subareas per basin, and four drainage aspects to which codes are assigned in each subarea, the maximum value for a fully developed drainage system would be 12. Conversely, if the drainage system has not been developed, then a *BDF* of zero (0) would result. Such a condition does not necessarily mean that the basin is unaffected by urbanization. In fact, a basin could be partially urbanized, have some impervious area, and have some improvements to secondary tributaries, and still have an assigned *BDF* of zero (0). The range of the data used to develop the urban equations of Table 7–5 are given in Table 7–6.

TABLE 7–7 Peak Discharges for Rural Conditions, Current Urban Conditions, and Future Urban Conditions

Return Period (yr)	Rural Peaks (ft^3/sec)	Urban Peaks 1986 (ft^3/sec)	Urban Peaks 2011 (ft^3/sec)
2	737	911	1350
5	1350	1550	2000
10	1910	2160	2740
25	2880	3020	3760
50	3830	3910	4820
100	5020	4980	6140

Example 7–4

To illustrate the use of the single-return-period equations of Table 7–5, assume that the rural peak discharges (RQ in ft^3/sec) were computed from the USGS rural area equations for the design location; the values of RQ are given in Table 7–7. The watershed has an area of 12 mi^2, a slope of 42 ft/mi, and a 2-yr, 2-hr rainfall of 1.6 in. In its current state (1986) the watershed storage is 3%, the basin development factor equals 3, and the imperviousness is 13%. It is also of interest to project the effects of planned developments to the year 2011. In the 25-yr zoning plan, the storage will increase to 5%, the imperviousness to 25%, and the basin development factor to 9 (see Table 7–8). Based on these characteristics and the rural peak discharges, the seven-parameter urban peak discharge equations of Table 7–5 were used to compute the peak discharges for the current levels of urban development and the levels that will exist in 2011. These discharges are given in Table 7–7. For the smaller return periods, the urban peaks are much greater than the rural peak discharges; the current 2-yr urban peak is 24% greater than the 2-yr rural peak discharge. For the larger return periods, the differences in percent are smaller, and the computed 100-yr urban peak for current conditions is actually smaller than the rural peak. Although this is not rational, the difference of 40 ft^3/sec is certainly not significant and is only indicative of the problem in using regression equations. Since all regression equations are less than 100% accurate, values that are somewhat irrational should be expected on occasion. The fact that the upper end of the rural flood frequency curve approaches the upper end of the urban curve is rational because we would expect the watershed to be saturated during large storms such as the 100-yr event; thus the saturated rural watershed would respond in much the same way as the impervious watershed. This is especially true for watersheds that have small percentages of imperviousness, such as the watershed of this example in the current urban condition with IA equal to 13%.

7.4 INDEX-FLOOD ESTIMATION

Two potential problems arise when adopting the single-return-period equation. First, calibration is quite involved because a separate equation must be fit for each return period. This requires evaluation of six equations when developing a frequency curve. Second, because of random variation, the fitted coefficients may show irrational variation that potentially could yield irrational estimates. This problem is avoided by proper smoothing of the coefficients. In spite of these problems, the single-return-period method is widely used.

The index-flood method avoids these problems associated with the single-return-period-equations method. Simply stated, the index-flood approach has two basic components: a prediction equation for one return period and a set of multipliers to be used for the other return periods. For example, we may define the 2-yr flood to be the index flood, with the Rational Method used to compute the flood. To find the discharges for other return periods, the computed 2-yr flood can be multiplied indices such as the following set:

Return Period (yr)	Index Ratio
5	1.3
10	1.6
25	1.9
50	2.2
100	2.5

TABLE 7–8 Basin Development Factor for 25-Year-Plan Development

Subarea	Area (mi²)	Main Channel Length (mi)	Length of Secondary Tributaries (mi)	Road Miles (mi)	Length of Channel Improved (mi)	Length of Channel Lined (mi)	Storm Drains (mi)	Curb and Gutter (mi)
Upper	4.3	3.93	6.84	4.52	0.84	0	3.63	2.37
Middle	3.8	3.62	4.57	4.06	2.02	0	2.54	2.32
Lower	3.9	3.75	3.89	3.92	1.94	2.01	1.96	2.08
Total	12.0							

	Code
Channel Improvements	
Upper third: 0.84 mi. have been straightened and deepened [0.84/3.93 < 50%]	0
Middle third: 2.02 mi. have been straightened and deepened [2.02/3.62 > 50%]	1
Lower third: 1.94 mi. have been straightened and deepened [1.94/3.75 > 50%]	1
Channel Linings	
Upper third: 0 mi. of channel lined [0/3.93 < 50%]	0
Middle third: 0 mi. of channel lined [0/3.62 < 50%]	0
Lower third: 2.01 mi. of channel lined [2.01/3.75 > 50%]	1
Storm Drains on Secondary Tributaries	
Upper third: 3.63 mi. have been converted to drains [3.63/6.84 > 50%]	1
Middle third: 2.54 mi. have been converted to drains [2.54/4.57 > 50%]	1
Lower third: 1.96 mi. have been converted to drains [1.96/3.89 > 50%]	1
Curb and Gutter Streets	
Upper third: 2.37 mi. installed [2.37/4.52 > 50%]	1
Middle third: 2.32 mi. installed [2.32/4.06 > 50%]	1
Lower third: 2.08 mi. installed [2.08/3.92 > 50%]	1
Total *BDF* =	9

The index-flood method has several advantages. First, the method only requires calibration of one index equation. Second, the approach is computationally simple to apply. Third, the index ratios assure consistency across return periods. However, evidence also indicates that the index-flood approach is not as accurate as the single-return-period approach because the fitting of a larger number of coefficients with the single-return-period method almost always yields greater accuracy in calibration.

7.4.1 Analysis

Analysis in the index-flood approach involves calibrating both the index-flood equation and the index ratios. While the Rational formula was used as the index equation in the example above, most commonly, an equation in the form of Equation 7–4 is used. The index equation is calibrated in the same way as any one of the single-return-period equations. After performing flood frequency analyses on the n annual maximum flood series, the vector of n discharges for the index return period are computed from the flood frequency curves; these are used as the values of the dependent variable in the regression. The watershed characteristics are used as the predictor variables. For each other return period T of interest, the n discharges for the T-yr flood are also obtained from the flood frequency curves. A bivariate regression analysis with a zero intercept is then conducted to determine the index ratio k_T for the Tth return period:

$$q_T = k_T q_i \tag{7–8}$$

in which q_i is the discharge for the index flood obtained from the flood frequency curve, and q_T is the discharge for return period T obtained from the flood frequency curve. A separate analysis is made for each return period. It may be necessary to smooth the coefficients.

7.4.2 Synthesis

To apply the index-flood method, the watershed characteristic data needed to compute the index flood are obtained and used to estimate the index flood. The entire frequency curve can then be computed by multiplying the computed index flood by the index ratios. The accuracy of the flood frequency curve depends on the accuracy of both the index-flood equation and the index ratios.

Example 7–5

The data of Table 7–3 can be used to illustrate the fitting and use of the index-flood method. The 2-yr event will be used as the index flood. The index equation will be the same as that reported for the peak-discharge equation method, with the results given in Table 7–4b. The index ratios were computed by regressing the 10- and 100-yr peaks on the 2-yr peaks, with coefficients of 1.994 and 4.013 for the 10- and 100-yr peak discharges, respectively. The zero-intercept regression produced a S_e/S_y of 0.207 and 0.467 for the 10- and 100-yr events, respectively.

To use the index-flood method for an ungaged watershed with a drainage area of 45 sq. mi, a channel slope of 62 ft/mile, and a 31% forest cover, the index equation yields a 2-yr peak discharge of

$$\begin{aligned} q_2 &= 139.8A^{1.049}\,S^{0.5371}\,F^{-0.9255} \\ &= 139.8\,(45)^{1.049}\,(62)^{0.5371}\,(31)^{-0.9255} = 2898\ \text{ft}^3/\text{sec} \end{aligned}$$

Therefore, the 10- and 100-yr discharges are

$$\begin{aligned} q_{10} &= 1.994q_2 = 5,779\ \text{ft}^3/\text{sec} \\ q_{100} &= 4.013q_2 = 11,630\ \text{ft}^3/\text{sec} \end{aligned}$$

These values could be used to plot a flood frequency curve.

Example 7–6

Trent (1978) used the 10-yr peak discharge as the index flood from small rural watersheds:

$$q_p = b_o A^{b_1} R^{b_2} DH^{b_3} \tag{7–9}$$

in which A is the drainage area, R is an isoerodent factor, defined as the mean annual rainfall kinetic energy times the annual maximum 30-min rainfall intensity (in., hr), and DH is the difference (ft) of the elevation of the main channel between the most distant point on the watershed boundary and the design point. The coefficients b_i ($i = 1, 2, 3$) are a function of the hydrophysiographic zone. The estimated peak discharge must be modified when the surface water storage in lakes, swamps, and ponds exceeds 4%. The 2-yr peak discharge is estimated by multiplying the 10-yr peak (q_{10}) by the index ratio of 0.41. The 100-yr peak (q_{100}) can be estimated by

$$q_{100} = 1.64\, q_{10}^{1.029} \tag{7–10}$$

In this case, the form of Equation 7–8 was used for finding the 2-yr event from the index flood, which had a return period of 10 yr, but the 100-yr event used a slightly more complex form. Equation 7–10 could be calibrated using least squares with a logarithmic transformation of the data.

Example 7–7

Golden (1973) used the index-flood approach for estimating the magnitude of floods in Georgia. A preliminary analysis suggested that the state had to be divided into three regions:

Region 1: The valley and ridge province, Blue Ridge province, and a small part of the Piedmont province
Region 2: Piedmont province except for that small part in region 1
Region 3: Coastal plain

Golden selected the 2-yr event as the index flood and developed the following index-flood equations:

$$\textit{Region 1:}\ q_p = 1180A^{0.72}S^{-1.26} \tag{7–11}$$

$$\textit{Region 2:}\ q_p = 202A^{0.6} \tag{7–12}$$

$$\textit{Region 3:}\ q_p = 2.65A^{0.63}P^{2.54} \tag{7–13}$$

in which q_p is the 2-yr discharge (ft^3/sec), A is the drainage area (mi^2), S is the soil infiltration capacity from the SCS method (in.), and P is the 2-yr, 24-hr rainfall (in.). The following set of index ratios were calibrated from the 10-, 25-, and 50-yr flood events:

Region	k_{10}	k_{25}	k_{50}
1	1.95	2.40	2.75
2	2.05	2.60	3.00
3	2.50	3.25	3.90

The flood frequency curve for a 10-mi^2 watershed in region 2 can be determined from Equation 7–12:

$$q_2 = 202(10)^{0.6} = 804 \text{ ft}^3/\text{sec} \tag{7–14}$$

Thus, the 10-, 25-, and 50-yr peak discharges are 1649, 2131, and 2413 ft^3/sec, respectively:

$$q_{10} = 2.05q_2 = 2.05(804) = 1649 \text{ ft}^3/\text{sec}$$
$$q_{25} = 2.65q_2 = 2.65(804) = 2131 \text{ ft}^3/\text{sec}$$
$$q_{50} = 3.00q_2 = 3.00(804) = 2413 \text{ ft}^3/\text{sec}$$

7.5 MOMENT ESTIMATION

The idea behind the moment estimation method is to develop prediction equations for estimating the mean, standard deviation, and skew. The computed values can then be used to compute a flood frequency curve using a method from Chapter 5 (see Equation 5–2). Studies for using watershed characteristics to estimate the skew coefficient have met with little success, giving some disadvantage to this method.

7.5.1 Analysis

The development of the equations necessary to use the moment estimation method can be summarized by the following steps:

1. Obtain the annual-maximum flood series for each stream gage station in the region and perform a flood frequency analysis for each of the n records.
2. Compute the mean, standard deviation, and skew for each series.
3. Determine watershed characteristics for each of the n watersheds for which a flood frequency analysis was made in Step 1, and form an $(n \times p)$ matrix of data to be used as the predictor variables.
4. Using each of the moments (mean, standard deviation, and skew) as the dependent variable, develop a separate regression equation using the watershed characteristics as the predictor variables.

If the regression equation for predicting the skew coefficient is not statistically significant, the mean of the skew coefficients can be assumed. If the logarithms of the data are taken prior to computing the moments, the log-Pearson Type III distribution is used, or the lognormal if the skew is not significant.

7.5.2 Synthesis

For use at an ungaged site, the three equations are used with the watershed characteristics to predict the moments. The frequency curve can be estimated using the equation

$$q = \overline{X} + KS \tag{7–15}$$

where $\overline{X}$ and S are the mean and standard deviation computed with the regression equations and K is the standardized variate that is a function of the computed skew and the exceedence probability.

Example 7–8

Thomas and Benson (1975) derived the empirical coefficients of equations for predicting the mean and standard deviation of the logarithms of the annual peak-flow series; the regression equations for the skew coefficients were not statistically significant. Equations were derived for four regions of the United States. As an example, the regression equations for the mean ($\overline{X}$) and standard deviation (S) for rural watersheds in the eastern region are

$$\overline{X} = 0.00264A^{1.01}P^{1.58} \tag{7–16}$$

$$S = 0.0142A^{0.99}P^{0.85} \tag{7–17}$$

in which A is the drainage area in square miles and P is the mean annual precipitation in inches.

For given values of A and P, Equations 7–16 and 7–17 can be used to estimate the mean and standard deviation of the logarithms and used to compute a log-normal frequency curve. Since the skew is 0, the values of K in Equation 7–15 become the standard normal deviate.

For a 100-mi^2 watershed in the eastern United States where the mean annual precipitation is 42 in., Equations 7–16 and 7–17 yield

$$\overline{X} = 0.00264(100)^{1.01}(42)^{1.58} = 101 \text{ ft}^3/\text{sec} \tag{7–18}$$

$$S = 0.0142(100)^{0.99}(42)^{0.85} = 32.5 \text{ ft}^3/\text{sec} \tag{7–19}$$

The log-normal flood frequency curve is computed using the values of Table 2–2.

7.6 RATIONAL METHOD

The methods outlined in the three previous sections are based on the analysis of stream gage data. For small watersheds, especially those undergoing urban/suburban development, regional equations that are appropriate for assessing the impact of development on peak discharges are not available, with the possible exception of the USGS regression equations of Section 7.3.3. However, these are not widely used because they do not include variables that are typically used to reflect changes in watershed conditions. Thus, there is a demand for methods that provide peak discharge estimates that use readily available input data, such as watershed and design-storm rainfall characteristics. The remainder of this chapter introduces a few of these methods.

7.6.1 Procedure

The most widely used uncalibrated equation is the Rational Method. Mathematically, the Rational Method relates the peak discharge (q_p, ft^3/sec) to the drainage area (A, acres), the rainfall intensity (i, in./hr), and the runoff coefficient (C):

$$q_p = CiA \tag{7–20}$$

The rainfall intensity is obtained from an intensity-duration-frequency curve (see Figure 4–4) using both the return period and a duration equal to the time of concentration as input. The value of the runoff coefficient is a function of the land use, cover condition, soil group, and watershed slope. Table 7–9 is an example of a table of C values; however, many other tables are in use. For example, Table 7–10 is a commonly used summary of C values. A

TABLE 7–9 Runoff Coefficients for the Rational Formula versus Hydrologic Soil Group (A, B, C, D) and Slope Range

	A			B			C			D		
Land Use	0–2%	2–6%	6%$^+$	0–2%	2–6%	6%$^+$	0–2%	2–6%	6%$^+$	0–2%	2–6%	6%$^+$
Cultivated land	0.08[a]	0.13	0.16	0.11	0.15	0.21	0.14	0.19	0.26	0.18	0.23	0.31
	0.14[b]	0.18	0.22	0.16	0.21	0.28	0.20	0.25	0.34	0.24	0.29	0.41
Pasture	0.12	0.20	0.30	0.18	0.28	0.37	0.24	0.34	0.44	0.30	0.40	0.50
	0.15	0.25	0.37	0.23	0.34	0.45	0.30	0.42	0.52	0.37	0.50	0.62
Meadow	0.10	0.16	0.25	0.14	0.22	0.30	0.20	0.28	0.36	0.24	0.30	0.40
	0.14	0.22	0.30	0.20	0.28	0.37	0.26	0.35	0.44	0.30	0.40	0.50
Forest	0.05	0.08	0.11	0.08	0.11	0.14	0.10	0.13	0.16	0.12	0.16	0.20
	0.08	0.11	0.14	0.10	0.14	0.18	0.12	0.16	0.20	0.15	0.20	0.25
Residential lot	0.25	0.28	0.31	0.27	0.30	0.35	0.30	0.33	0.38	0.33	0.36	0.42
size 1/8 acre	0.33	0.37	0.40	0.35	0.39	0.44	0.38	0.42	0.49	0.41	0.45	0.54
Residential lot	0.22	0.26	0.29	0.24	0.29	0.33	0.27	0.31	0.36	0.30	0.34	0.40
size 1/4 acre	0.30	0.34	0.37	0.33	0.37	0.42	0.36	0.40	0.47	0.38	0.42	0.52
Residential lot	0.19	0.23	0.26	0.22	0.26	0.30	0.25	0.29	0.34	0.28	0.32	0.39
size 1/3 acre	0.28	0.32	0.35	0.30	0.35	0.39	0.33	0.38	0.45	0.36	0.40	0.50
Residential lot	0.16	0.20	0.24	0.19	0.23	0.28	0.22	0.27	0.32	0.26	0.30	0.37
size 1/2 acre	0.25	0.29	0.32	0.28	0.32	0.36	0.31	0.35	0.42	0.34	0.38	0.48
Residential lot	0.14	0.19	0.22	0.17	0.21	0.26	0.20	0.25	0.31	0.24	0.29	0.35
size 1 acre	0.22	0.26	0.29	0.24	0.28	0.34	0.28	0.32	0.40	0.31	0.35	0.46
Industrial	0.67	0.68	0.68	0.68	0.68	0.69	0.68	0.69	0.69	0.69	0.69	0.70
	0.85	0.85	0.86	0.85	0.86	0.86	0.86	0.86	0.87	0.86	0.86	0.88
Commercial	0.71	0.71	0.72	0.71	0.72	0.72	0.72	0.72	0.72	0.72	0.72	0.72
	0.88	0.88	0.89	0.89	0.89	0.89	0.89	0.89	0.90	0.89	0.89	0.90
Streets	0.70	0.71	0.72	0.71	0.72	0.74	0.72	0.73	0.76	0.73	0.75	0.78
	0.76	0.77	0.79	0.80	0.82	0.84	0.84	0.85	0.89	0.89	0.91	0.95
Open space	0.05	0.10	0.14	0.08	0.13	0.19	0.12	0.17	0.24	0.16	0.21	0.28
	0.11	0.16	0.20	0.14	0.19	0.26	0.18	0.23	0.32	0.22	0.27	0.39
Parking	0.85	0.86	0.87	0.85	0.86	0.87	0.85	0.86	0.87	0.85	0.86	0.87
	0.95	0.96	0.97	0.95	0.96	0.97	0.95	0.96	0.97	0.95	0.96	0.97

[a] Runoff coefficients for storm recurrence intervals less than 25 years.

[b] Runoff coefficients for storm recurrence intervals of 25 years or longer.

problem with tables such as Table 7–10 is that for each land use a range of values is provided. This can lead to inconsistency in application. As a general rule, the mean of the range should be used unless a different value can be fully justified. It would be improper for a low value to be selected to reduce the size and therefore the cost of the drainage system.

A primary use of the Rational Method has been for design problems for small urban areas such as the sizing of inlets and culverts, which are characterized by small drainage areas and short times of concentration. For such designs, short-duration storms are critical, which is why the time of concentration is used as the input duration for obtaining i from the intensity-duration-frequency curve. If the storm duration occurs at a constant rate i and uniformly over the entire watershed, the volume of rainfall would equal iAt_c, which would have units of acre-inches

TABLE 7–10 Runoff Coefficients for the Rational Method

Description of Area	Range of Runoff Coefficients	Recommended Value*
Business		
Downtown	0.70–0.95	0.85
Neighborhood	0.50–0.70	0.60
Residential		
Single-family	0.30–0.50	0.40
Multiunits, detached	0.40–0.60	0.50
Multiunits, attached	0.60–0.75	0.70
Residential (suburban)	0.25–0.40	0.35
Apartment	0.50–0.70	0.60
Industrial		
Light	0.50–0.80	0.65
Heavy	0.60–0.90	0.75
Parks, cemeteries	0.10–0.25	0.20
Playgrounds	0.20–0.35	0.30
Railroad yard	0.20–0.35	0.30
Unimproved	0.10–0.30	0.20

It is often desirable to develop a composite runoff coefficient based on the percentage of different types of surface in the drainage area. This procedure often is applied to typical "sample" block as a guide to selection of reasonable values of the coefficient for an entire area. Coefficients with respect to surface type currently in use are listed below.

Character of Surface	Range of Runoff Coefficients	Recommended Value*
Pavement		
Asphaltic and Concrete	0.70–0.95	0.85
Brick	0.75–0.85	0.80
Roofs	0.75–0.95	0.85
Lawns, sandy soil		
Flat, 2%	0.05–0.10	0.08
Average, 2 to 7%	0.10–0.15	0.13
Steep, 7%	0.15–0.20	0.18
Lawns, heavy soil		
Flat, 2%	0.13–0.17	0.15
Average, 2 to 7%	0.18–0.22	0.20
Steep, 7%	0.25–0.35	0.30

The coefficients in these two tabulations are applicable for storms of 5- to 10-year frequencies. Less frequent, higher intensity storms will require the use of higher coefficients because infiltration and other losses have a proportionally smaller effect on runoff. The coefficients are based on the assumption that the design storm does not occur when the ground surface is frozen.

*Recommended value not included in original source.

Source: Design and Construction of Sanitary and Storm Sewers, American Society of Civil Engineers, New York, p. 332, 1969.

when the t_c is expressed in hours. The runoff coefficient then becomes a scaling factor that converts the volume rate (that is, iA in acre-in./hr) of rainfall to a peak discharge. A more detailed discussion of the conceptual basis of the Rational Method will be given in Chapter 9.

Example 7–9

Consider the design problem where a peak discharge is required to size a storm drain inlet for a 2.4-acre parking area in Baltimore, with a time of concentration of 0.1 hr and a slope of 1.5%. For a 25-yr design return period, the rainfall intensity (see Figure 4–4) is 8.6 in./hr and the runoff coefficient (see Table 7–9) is 0.95. Therefore, the design discharge is

$$q_p = 0.95\,(8.6)\,(2.4) = 20\text{ ft}^3/\text{sec} \tag{7–21}$$

Some drainage policies provide for a minimum time of concentration with 15 to 20 min often being specified. If the design above were for a project where the minimum t_c was 15 min, the design intensity would be 6.5 in./hr and the peak discharge would be 15 ft^3/sec.

7.6.2 Runoff Coefficients for Nonhomogeneous Areas

The runoff coefficients of Table 7–9 reflect the effects of land use, soil, and slope on runoff potential. The use of Equation 7–20 assumes that the watershed is homogeneous in these characteristics so that the runoff coefficient used provides unbiased estimates. Where a drainage area is characterized by distinct subareas that have different runoff potentials, the watershed should be subdivided and the equation applied separately to each area; the procedure for this is discussed in Section 7.6.3. Where a watershed is not homogeneous but is characterized by highly dispersed areas that can be characterized by different runoff coefficients, a weighted runoff coefficient should be determined. The weighting is based on the area of each land use and is found by the following equation:

$$C_w = \frac{\sum_{j=1}^{n} C_j A_j}{\sum_{j=i}^{n} A_j} \tag{7–22}$$

in which A_j is the area for land cover j, C_j is the runoff coefficient for area j, n is the number of distinct land covers within the watershed, and C_w is the weighted runoff coefficient. The weighted coefficient can be used with Equation 7–20. The denominator of Equation 7–22 equals the total drainage area, so Equation 7–22 can be substituted into Equation 7–20, which yields the following:

$$q_p = i \sum_{j=1}^{n} C_j A_j \tag{7–23}$$

Example 7–10

Equation 7–23 will be illustrated using the data of Table 7–11. It is assumed that the different land uses are scattered throughout the watershed, and therefore it is impractical to subdivide the watershed. Equation 7–22 can be used to compute a weighted runoff coefficient:

$$\begin{aligned} C_w &= \frac{0.19(14.2) + 0.14(11.6) + 0.32(8.9) + 0.89(4.3) + 0.82(3.9)}{42.9} \\ &= 0.33 \end{aligned} \tag{7–24}$$

TABLE 7–11 Example: Calculation of Weighted Runoff Coefficients

Land Use	C_i	A_i (acres)
Open space	0.19	14.2
Forest	0.14	11.6
Residential (1/2 acre)	0.32	8.9
Light commercial	0.89	4.3
Streets	0.82	3.9
		42.9

For a 25-yr rainfall intensity of 3.6 in./hr, the peak discharge would be

$$q_p = 0.33\ (3.6)\ (42.6) = 51\ \text{ft}^3/\text{sec} \tag{7–25}$$

Equation 7–23 will provide the same estimate of q_p.

7.6.3 Designs on Subdivided Watersheds

The discussion to this point concerning the Rational Method has used the method only to compute the peak discharge for a contributing area. The method can also be used for nonhomogeneous watersheds in which the watershed is divided into homogeneous subareas and where multiple inlets and pipe systems are involved. Where a watershed has distinct areas of nonhomogeneity, every attempt should be made to subdivide the watershed into homogeneous subareas and then use the Rational Method for each subarea or group of subareas.

A number of methods have been proposed for solving such problems. One method will be described here and examples provided for illustration. A second method is given in Chapter 9.

The method described here is an attempt to provide an equal level of protection to each structural element of the total drainage area. It is based on the following two rules for using Equation 7–20:

1. For each inlet area at the headwater of a drainage area the Rational Method (Equation 7–20) is used to compute the peak discharge.
2. For locations where drainage is arriving from two or more inlet areas, the longest time of concentration is used to find the design intensity, a weighted runoff coefficient is computed, and the total drainage area to that point is used with Equation 7–20.

It is important to emphasize that Equation 7–20 is not used to compute the discharge from each inlet area and the discharges summed; this would ignore the differences in timing of runoff that exist for the different subareas. The procedure behind these two steps will be illustrated with two examples.

Example 7–11

Figure 7–1 shows the schematic of a drainage area that has been divided into three subareas, with the characteristics of each shown. Beginning with the upstream subarea, the discharge into inlet 1 can be determined. Since the time of concentration is less than 15 min, a duration of

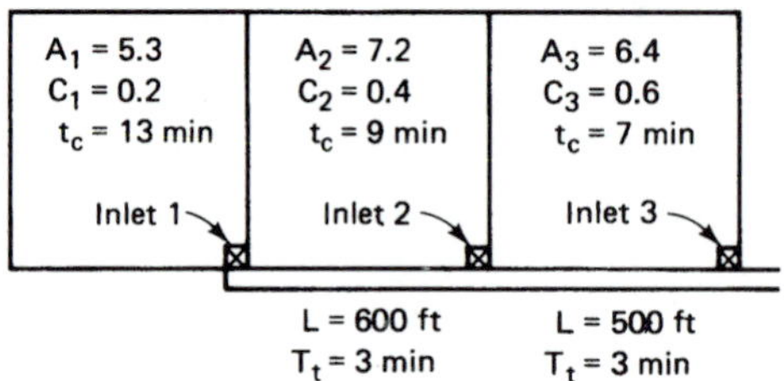

FIGURE 7–1 Schematic diagram of drainage area.

15 min will be used to obtain the rainfall intensity from Figure 4–4; the intensity for a 10-yr event is 5.4 in./hr. Therefore, the peak discharge into inlet 1 is

$$q_{p1} = 0.2\ (5.4\ \text{in./hr})\ (5.3\ \text{acres}) = 5.7\ \text{ft}^3/\text{s} \tag{7–26}$$

The runoff into inlet 1 flows through a pipe, which is 600 ft in length and has a travel time of 3 min. The peak discharge to the inlet from subarea 2 can also be computed from the Rational Formula:

$$q_{p2} = 0.4(5.4\ \text{in./hr})(7.2\ \text{acres}) = 15.6\ \text{ft}^3/\text{sec} \tag{7–27}$$

However, the pipe between inlets 2 and 3 should not necessarily be designed to carry the sum of these subarea peak discharges (that is, 21.3 ft^3/sec). Subareas 1 and 2 have different times of concentration and the flow from subarea 1 must travel through the 600 ft of pipe before arriving at inlet 2. Therefore, the pipe between inlets 2 and 3 will not be subjected to the sum of the two. Instead, it is common practice to recompute the discharge for the total area using a weighted runoff coefficient and a rainfall intensity based on the longest time of concentration. For the drainage area of Figure 7–1, the weighted runoff coefficient for subareas 1 and 2 is

$$C_w = \frac{0.2\ (5.3) + 0.4\ (7.2)}{5.3 + 7.2} = 0.315 \tag{7–28}$$

The longest time of concentration for the two subareas would be the sum of the drainage time from subarea 1 and the travel time in the pipe between inlets 1 and 2, which is 13 + 3 = 16 min. From Figure 4–4 the 10-yr intensity is 5.3 in./hr, which yields a peak discharge of

$$q_p = 0.315\ (5.3)\ (5.3 + 7.2) = 20.9\ \text{ft}^3/\text{sec} \tag{7–29}$$

The discharge of Equation 7–29 should be used to size the pipe from inlet 2 to inlet 3.

The data for subarea 3 could be used to size the inlet for that subarea:

$$q_{p3} = 0.6\ (5.4\ \text{in./hr})\ (6.4\ \text{acres}) = 20.7\ \text{ft}^3/\text{sec} \tag{7–30}$$

However, the size of the pipe draining the three subareas can be determined using a discharge estimate obtained using a weighted runoff coefficient, which is

$$C_w = \frac{0.2\ (5.3) + 0.4\ (7.2) + 0.6\ (6.4)}{5.3 + 7.2 + 6.4} = 0.412 \tag{7–31}$$

The longest time of concentration is 19 min, which includes 6 min of travel time in the pipe from inlet 1 to inlet 3. From Figure 4–4 a rainfall intensity of 4.7 in./hr is obtained and used to compute the peak discharge for the entire 18.9 acres:

$$q_p = 0.412\ (4.8\ \text{in./hr})\ (18.9\ \text{acres}) = 37.4\ \text{ft}^3/\text{sec} \tag{7–32}$$

While the sum of the discharges from the individual subareas is greater than the discharge computed in Equation 7–32 (41.8 ft^3/sec versus 37.4 ft^3/sec), the value computed using the approach

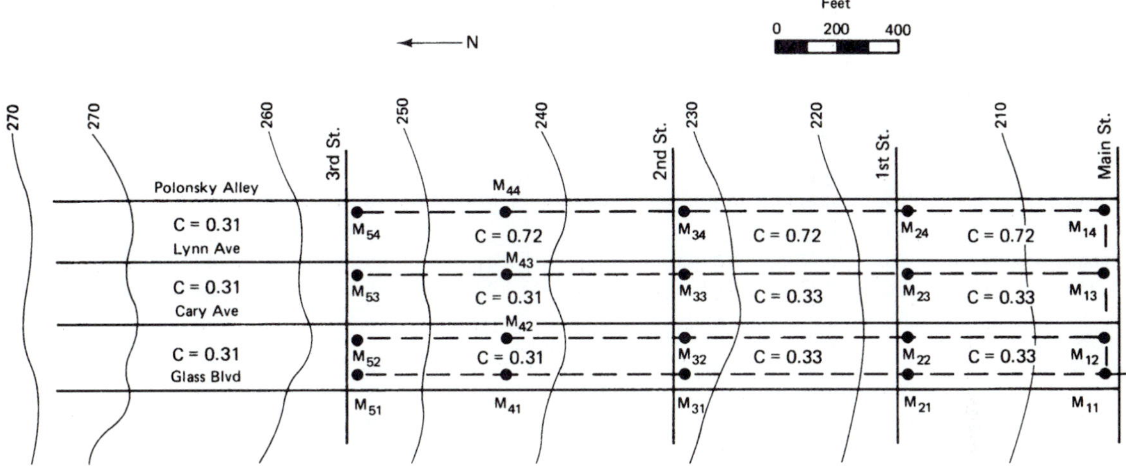

FIGURE 7–2 Watershed layout for Example 7–12.

with the weighted runoff coefficient and the longest time of concentration is an accepted method. It is believed that this approach provides the same level of protection with respect to flood risk at each design point. That is, the inlets and pipe segments would each be designed to pass the flood runoff for the same exceedence frequency, which was 10 years in the example of Figure 7–1. This is probably a reasonable assumption, although it may not be entirely accurate. A method based on hydrographs will be presented in Chapter 9; this method should be more accurate for the given assumptions.

Example 7–12

Using another hypothetical case, Figure 7–2 shows a 43.8-acre drainage area that includes approximately 11 acres of commercial property, with the remainder in ¼-acre and ⅛-acre parcels with residential land use. The existing slope and grading during development has resulted in an area in which all drainage is contained between Glass Boulevard and Polonsky Alley. The upper end of the drainage area is approximately 800 ft north of 3rd Street. The location of the proposed manholes and storm drainage system are shown in Figure 7–2. The outlet will be a single pipe that will run under Main Street at the intersection of Glass Boulevard. The drainage policy requires design on a 10-yr exceedence frequency.

The computations for the peak discharges are given in Table 7–12. The computations begin for flow into manhole M_{54}, which is at the intersection of 3rd Street and Polonsky Alley. Calculations are provided for the design discharges of both the inlets and the pipes. For some locations there may be more than one inlet at the intersection and so the inlet discharges would have to be divided accordingly. Computations proceed down Polonsky Alley to Main Street. At each manhole, a weighted runoff coefficient is computed, with the weights depending on the drainage area in the different land uses. For example, the weighted runoff coefficient for manhole M_{24} is computed by

$$C_w = \frac{1.8\,(0.31) + 1.1\,(0.72) + 1.1\,(0.72) + 1.6\,(0.72)}{1.8 + 1.1 + 1.1 + 1.6} = 0.58 \tag{7–33}$$

Since all of the inlet times of concentration were less than 15 min, a minimum of 15 min was used. The total times of concentration for the downstream manholes are the sum of the inlet t_c for

TABLE 7–12 Peak Discharge Computations for Drainage System of Figure 7–2

	Area (acres)		Runoff Coefficient					Intensity (in./hr)		Peak, q_p (ft^3/sec)	
Manhole	Inlet	Total	Inlet	Weighted	Inlet T_c(min)	Travel Time(min)	Total T_c(min)	Inlet	Total	Inlet	Total
M_{54}	1.8		0.31		15			5.4		3.0	
M_{44}	1.1	2.9	0.72	0.47	15	3	18	5.4	5.0	4.3	6.9
M_{34}	1.1	4.0	0.72	0.54	15	3	21	5.4	4.6	4.3	10.0
M_{24}	1.6	5.6	0.72	0.59	15	4	25	5.4	4.1	6.2	13.6
M_{14}	1.6	7.2	0.72	0.62	15	4	29	5.4	3.9	6.2	17.5
M_{53}	3.7		0.31		15			5.4		6.2	
M_{43}	2.3	6.0	0.31	0.31	15	3	18	5.4	5.0	3.8	9.3
M_{33}	2.3	8.3	0.31	0.31	15	3	21	5.4	4.6	3.8	11.8
M_{23}	3.2	11.5	0.33	0.32	15	4	25	5.4	4.1	5.7	14.9
M_{13}	3.2	21.9	0.33	0.42	15	1	30	5.4	3.8	5.7	34.8
M_{52}	3.7		0.31		15			5.4		6.2	
M_{42}	2.3	6.0	0.31	0.31	15	3	18	5.4	5.0	3.8	9.3
M_{32}	2.3	8.3	0.31	0.31	15	3	21	5.4	4.6	3.8	11.8
M_{22}	3.2	11.5	0.33	0.32	15	4	25	5.4	4.1	5.7	14.9
M_{12}	3.2	36.6	0.33	0.38	15	1	31	5.4	3.6	5.7	49.9
M_{51}	1.8		0.31		15			5.4		3.0	
M_{41}	1.1	2.9	0.31	0.31	15	3	18	5.4	5.0	1.8	4.5
M_{31}	1.1	4.0	0.31	0.31	15	3	21	5.4	4.6	1.8	5.7
M_{21}	1.6	5.6	0.33	0.32	15	4	25	5.4	4.1	2.9	7.2
M_{11}	1.6	43.8	0.33	0.37	15	1	32	5.4	3.5	2.9	56.5

the upper subarea and the travel times through the pipe systems. Independent calculations can be made for the pipes located on the four north-south roadways. Rainfall intensities are obtained from Figure 4–4 for the 10-yr return period and a duration equal to the time of concentration.

The discharges from manholes M_{13}, M_{12}, and M_{11} must be determined using summation of flows from the four feeder lines. The area into manhole M_{13} consists of the 7.2 acres along Polonsky Alley and the 14.7 acres draining into the inlets along Lynn Avenue. The area into manholes M_{12} consists of the 21.9 acres draining into manholes M_{14} and M_{13} and the 14.7 acres draining into inlets along Cary Avenue. The total area draining into a manhole, the weighted runoff coefficient, and an intensity based on a duration equal to the longest time of concentration/travel time combination are used to compute the discharge.

The drainage system of Figure 7–2 has a peak discharge from manhole M_{11} of 56.5 ft^3/sec, which is approximately 1.3 ft^3/sec/acre. The sizing of pipes is discussed in Chapter 8.

7.7 THE SCS RAINFALL-RUNOFF DEPTH RELATION

The volume of storm runoff can depend on a number of factors. Certainly, the volume of rainfall will be an important factor. For very large watersheds, the volume of runoff from one storm event may depend on rainfall that occurred during previous storm events. However, for smaller watersheds, design hydrologists usually assume that runoff for the current storm event is independent of the rainfall of previous storm events. When using the design-storm approach, the assumption of storm independence is quite common.

In addition to rainfall, other factors affect the volume of runoff. A common assumption in hydrologic modeling is that the rainfall available for runoff is separated into three parts: direct (or storm) runoff, initial abstraction, and losses. Factors that affect the split between losses and runoff include the volume of rainfall, land cover and use, soil type, and antecedent moisture conditions. Land cover and land use will determine the amount of depression and interception storage. Because of the large number of factors that affect the separation of rainfall into direct runoff and losses, the process of hydrologic modeling involves the acceptance of a number of simplifying assumptions. The actual method of separating storm runoff is discussed in Chapter 9.

7.7.1 Conceptual Model

In developing the SCS rainfall-runoff relationship, the total rainfall was separated into three components: direct runoff (Q), actual retention (F), and the initial abstraction (I_a). This is shown schematically in Figure 7–3. Conceptually, the following relationship between P, Q, I_a, and F was assumed:

$$\frac{F}{S} = \frac{Q}{P - I_a} \tag{7–34}$$

in which S is the potential maximum retention. The actual retention is

$$F = (P - I_a) - Q \tag{7–35}$$

Substituting Equation 7–35 into Equation 7–34 yields the following:

$$\frac{(P - I_a) - Q}{S} = \frac{Q}{P - I_a} \tag{7–36}$$

Rearranging Equation 7–36 to solve for Q yields

$$Q = \frac{(P - I_a)^2}{(P - I_a) + S} \tag{7–37}$$

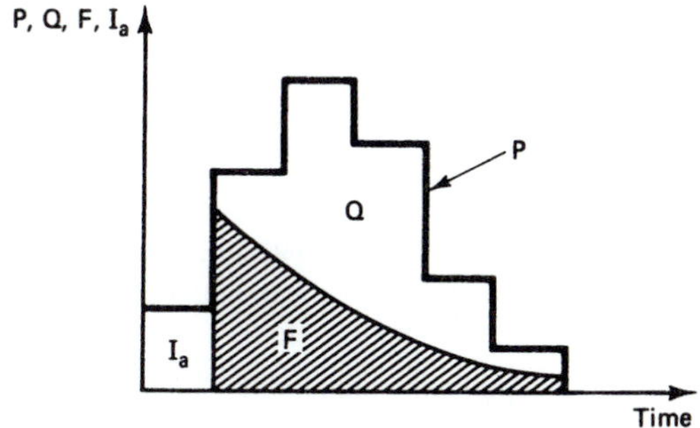

FIGURE 7–3 Separation of rainfall into direct runoff (Q), initial abstraction (I_a), and actual retention (F).

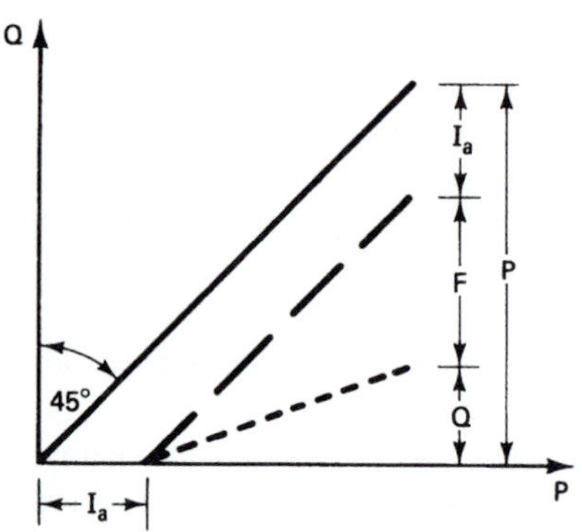

FIGURE 7–4 Schematic diagram of the mass curve relationship between rainfall (P), total runoff depth (Q), initial abstraction (I_a), and actual retention.

Equation 7–37 contains one known, P, and two unknowns, I_a and S. The variables of Equation 7–37 are best understood when placed in the form of a mass curve. Figure 7–4 shows a schematic of the mass curve of Q versus P. The depth (or volume) of rainfall, P, is separated into the initial abstraction, the retention, and the runoff.

Before putting Equation 7–37 in a form that can be used to solve for Q, it may be worthwhile to examine the rationality of the underlying model of Equation 7–34. The initial abstraction is the amount of rainfall at the beginning of a storm that is not available for runoff; therefore, $(P\text{-}I_a)$ is the rainfall that is available after the initial abstraction has been satisfied. S is the amount of storage (that is, depression, interception, subsurface) available to hold rainfall. Letting K_1 equal the ratio of $(P\text{-}I_a)$ to S, K_1 represents the proportion of water available for runoff per unit of available storage. Setting $K_2 = Q/F$, then K_2 is the proportion of water that runs off per unit of water retained. Transforming Equation 7–34 indicates that $K_1 = K_2$. This implies that the proportion of runoff relative to the amount that is retained equals the proportion of the available rain per unit of available storage. Two examples will illustrate the rationale of this equality.

Example 7–13

For the first example, assume that a storm with $P = 4$ in. occurs on two watersheds with different capacities for storage, such as with sandy and clayey soils. We will also assume here that $I_a =$ 0.5 in., and for watersheds 1 and 2 we will assume $S = 3$ and 2 in., respectively. This is shown schematically in Figure 7–5. Rearranging Equations 7–34 and 7–35 provides the following two equations, with two unknowns:

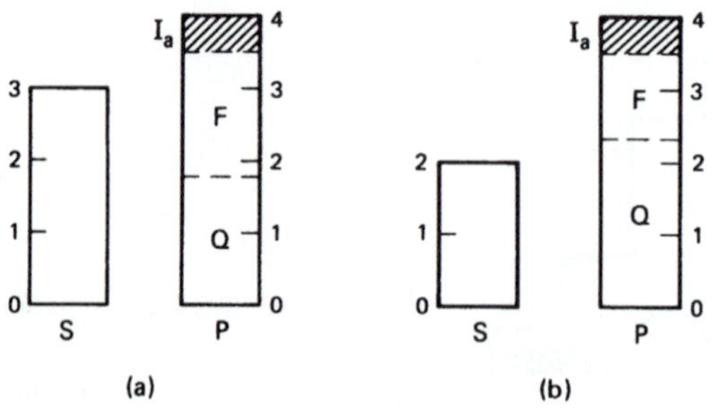

FIGURE 7–5 Separation of rainfall (P) into initial abstraction (I_a), actual retention (F), and direct runoff (Q) for two watersheds with different minimum detention: (a) $S = 3$ in.; (b) $S = 2$ in.

$$0 = F - \frac{S}{P - I_a} Q \tag{7–38a}$$

and

$$P - I_a = F + Q \tag{7–38b}$$

For the case of Figure 7–5a we have

$$0 = F - \left(\frac{3}{3.5}\right) Q \tag{7–39a}$$

and

$$3.5 = F + Q \tag{7–39b}$$

Solving the two linear simultaneous equations of Equations 7–39 yields $Q = 1.88$ in. and $F = 1.62$ in.; therefore, $Q/P = 0.47$. For the case of Figure 7–5b, we have

$$0 = F\left(\frac{2}{3.5}\right) Q \tag{7–40a}$$

and

$$3.5 = F + Q \tag{7–40b}$$

Solving Equations 7–40 yields $Q = 2.33$ in. and $F = 1.27$ in.; therefore, $Q/P = 0.56$. The results show that the watershed having the greater storage (S) has a smaller proportion (Q/P) of surface runoff. For watershed 2 the value of F is 0.35 in. less than for watershed 1, and Q is 0.35 in. greater.

Example 7–14

For the second example, we will consider a single watershed that is subjected to two storm events. For storm events 1 and 2, rainfall depths of 3 and 4 in., respectively, will be used. The watershed will be assumed to have $S = 2$ in. and $I_a = 0.5$ in. These are shown schematically in Figure 7–6. Solving Equations 7–38 for Q and F of storm event 1 yields $Q = 1.39$ in. and $F = 1.11$ in.; therefore, $Q/P = 0.46$. For storm event 2, Equations 7–38 yield $Q = 2.23$ in. and $F = 1.27$ in.; therefore, $Q/P = 0.56$. The results indicate that for the same storage, a greater proportion of rainfall appears as runoff for the larger storm event. From Figure 7–6 it is evident that while F increased by only 0.16 in., Q increased by 0.84 in.; that is, of the last 1 in. of rainfall for storm event 2, 84% went to direct runoff while only 16% went to watershed storage.

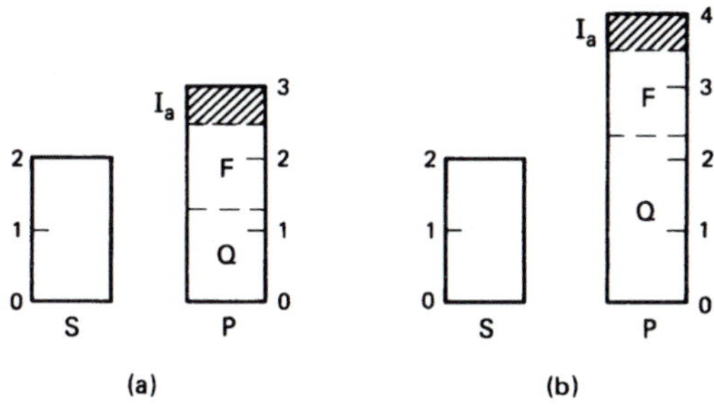

FIGURE 7–6 Separation of rainfall (P) into initial abstractions (I_a), actual retention (F), and direct runoff (Q) for two storm events: (a) $P = 3$ in.; (b) $P = 4$ in.

7.7.2 Runoff Depth Estimation

The results of Examples 7–13 and 7–14 suggest that the model of Equation 7–34 provides rational results. Given Equation 7–37, two unknowns must be estimated, S and I_a. The retention S should be a function of the following five factors: land use, interception, infiltration, depression storage, and antecedent moisture. The initial abstraction will also depend on these elements of the hydrologic cycle. Empirical evidence indicated that they were related by the following equation:

$$I_a = 0.2\,S \tag{7–41}$$

If the five factors above affect S, they also affect I_a. Substituting Equation 7–41 into Equation 7–37 yields the following equation, which contains the single unknown, S:

$$Q = \frac{(P - 0.2\,S)^2}{P + 0.8\,S} \tag{7–42}$$

Equation 7–42 represents the basic equation for computing the runoff depth, Q, for a given rainfall depth, P. It is worthwhile noting that while Q and P have dimensions of depth (that is, inches), Q and P reflect volumes and are often referred to as volumes because we usually assume that the same depths occurred over the entire watershed.

The SCS runoff *CN* was developed for use with Equation 7–42. Thus there was a need to relate S, which was the unknown of Equation 7–42, and the runoff *CN*. An empirical analysis led to the following relationship:

$$S = \frac{1000}{CN} - 10 \tag{7–43}$$

Equations 7–42 and 7–43 can be used to estimate Q when the values of P and *CN* are available. It is important to note the following constraint on Equation 7–42:

$$P \geqslant 0.2\,S \tag{7–44}$$

When $P < 0.2\,S$, it is necessary to assume that $Q = 0$.

In addition to Equations 7–42, 7–43, and 7–44, the runoff depth can be determined using either Figure 7–7 or Table 7–13. The use of Figure 7–7 suffers from the problem of interpolation; that is, the accuracy of Q is limited by the degree to which the figure can be interpolated.

Example 7–15

Determine the runoff volume for the 24-hr, 100-yr rainfall of 7 in. when the soil is of group B and the watershed is in row crop, contoured in good hydrologic condition. Table 3–18 yields a *CN* of 75. Equation 7–43 gives

$$S = \frac{1000}{75} - 10 = 3.333 \text{ in.}$$

Therefore, the initial abstraction I_a from Equation 7–41 is

$$I_a = 0.2S = 0.6667 \text{ in.}$$

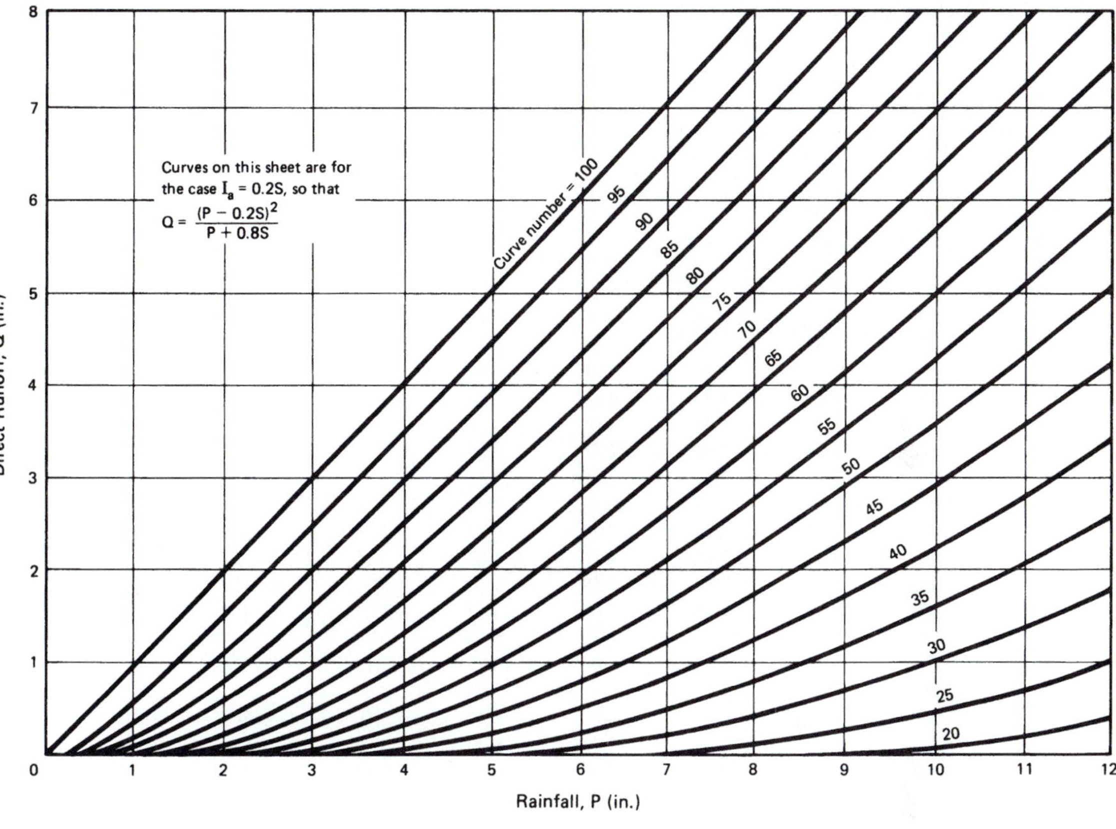

FIGURE 7–7 Solution of runoff equation.

The depth of direct runoff is computed by Equation 7–42:

$$Q = \frac{(7 - 0.6667)^2}{7 + 0.8\,(3.333)} = 4.15 \text{ in.}$$

The mass curve of Figure 7–7, or Table 7–13, also provides a runoff depth of 4.15 in.

Example 7–16

Determine the runoff depth for a 24-hr, 100-yr rainfall of 7 in. for antecedent soil moisture condition II, with the following land uses and soil groups:

Area Fraction	Land Use/Condition	Soil Group
0.40	Meadow: good condition	D
0.25	Wooded: poor cover	C
0.20	Open space: good condition	D
0.15	Residential (¼-acre lots)	C

TABLE 7–13 Runoff Depth in Inches for Selected *CN* Values and Rainfall Amounts[a]

Rainfall (in)	Curve Number, *CN*								
	60	65	70	75	80	85	90	95	98
1.0	.00	.00	.00	.03	.08	.17	.32	.56	.79
1.2	.00	.00	.03	.07	.15	.27	.46	.74	.99
1.4	.00	.02	.06	.13	.24	.39	.61	.92	1.18
1.6	.01	.05	.11	.20	.34	.52	.76	1.11	1.38
1.8	.03	.09	.17	.29	.44	.65	.93	1.29	1.58
2.0	.06	.14	.24	.38	.56	.80	1.09	1.48	1.77
2.5	.17	.30	.46	.65	.89	1.18	1.53	1.96	2.27
3.0	.33	.51	.71	.96	1.25	1.59	1.98	2.45	2.77
3.5	.53	.75	1.01	1.30	1.64	2.02	2.45	2.94	3.27
4.0	.76	1.03	1.33	1.67	2.04	2.46	2.92	3.43	3.77
4.5	1.02	1.33	1.67	2.05	2.46	2.91	3.40	3.92	4.26
5.0	1.30	1.65	2.04	2.45	2.89	3.37	3.88	4.42	4.76
5.5	1.60	1.99	2.41	2.86	3.33	3.83	4.36	4.92	5.26
6.0	1.92	2.35	2.81	3.28	3.78	4.30	4.85	5.41	5.76
6.5	2.26	2.72	3.21	3.71	4.24	4.78	5.33	5.91	6.26
7.0	2.60	3.10	3.62	4.15	4.69	5.25	5.82	6.41	6.76
7.5	2.96	3.49	4.04	4.59	5.16	5.73	6.31	6.90	7.26
8.0	3.33	3.89	4.46	5.04	5.63	6.21	6.81	7.40	7.76
8.5	3.71	4.30	4.90	5.50	6.10	6.70	7.30	7.90	8.26
9.0	4.10	4.72	5.33	5.95	6.57	7.18	7.79	8.40	8.76
9.5	4.50	5.14	5.78	6.41	7.04	7.67	8.29	8.90	9.26
10.0	4.90	5.56	6.22	6.88	7.52	8.16	8.78	9.40	9.76
10.5	5.31	6.00	6.68	7.34	8.00	8.64	9.28	9.89	10.26
11.0	5.72	6.43	7.13	7.81	8.48	9.13	9.77	10.39	10.76
11.5	6.14	6.87	7.59	8.28	8.96	9.62	10.27	10.89	11.26
12.0	6.56	7.32	8.05	8.76	9.45	10.11	10.76	11.39	11.76
12.5	6.99	7.76	8.51	9.23	9.93	10.61	11.26	11.89	12.26
13.0	7.42	8.21	8.98	9.71	10.42	11.10	11.76	12.39	12.76
13.5	7.86	8.67	9.44	10.19	10.90	11.59	12.25	12.89	13.26
14.0	8.30	9.12	9.91	10.67	11.39	12.08	12.75	13.39	13.76
14.5	8.74	9.58	10.38	11.15	11.88	12.58	13.25	13.89	14.26
15.0	9.19	10.04	10.85	11.63	12.37	13.07	13.74	14.39	14.76
15.5	9.63	10.50	11.33	12.11	12.86	13.57	14.24	14.89	15.26

[a]To obtain runoff depths for *CN* values and other rainfall amounts not shown in this table, use an arithmetic interpolation.

Table 3–18 provides the following *CN*s for the four land uses: 78, 77, 80, 83, respectively. Therefore, the weighted *CN* is

$$\begin{aligned} CN &= 0.40\,(78) + 0.25\,(77) + 0.20\,(80) + 0.15\,(83) \\ &= 78.9 \text{ (use 79)} \end{aligned}$$

For a *CN* of 79 and a rainfall of 7 in., the runoff depth is 4.58 in.

In this case, the *CN* values for the subareas are not dissimilar. In such a case, the weighted *CN* is appropriate. For cases where the differences in the CN are greater than 5, it is preferable to weight the *Q* rather than the *CN*. For the data above, the following shows the *Q* for each subarea:

Area Fraction	CN	Q (in.)
0.40	78	4.47
0.25	77	4.37
0.20	80	4.69
0.15	83	5.03

This yields the following weighted Q:

$$\begin{aligned} Q &= 0.40\,(4.47) + 0.25\,(4.37) + 0.20\,(4.69) + 0.15\,(5.03) \\ &= 4.57 \text{ in.} \end{aligned}$$

The weighted Q is only 0.01 in. different from the Q computed with the weighted CN. When the CN variation is greater than in this example (that is, 78 to 83), the two estimates of Q will show a greater difference than the difference of 0.01 in. shown in this example. Example 7–17 will illustrate this point.

Example 7–17

Consider the case of a watershed that is 25% residential (¼-acre lots) on an A soil and 75% open space (fair condition) on an A soil. The CNs are 77 and 49, respectively, so the weighted CN would be

$$CN = 0.25\,(77) + 0.75\,(49) = 56$$

For a rainfall of 3.2 in., Q equals 0.28. The runoff depths for the residential and open space land uses are 1.21 and 0.11 in., respectively. Thus the weighted Q would be

$$Q = 0.25(1.21) + 0.75(0.11) = 0.385 \text{ in.}$$

Thus the two methods provide estimates of Q that differ by more than 30%.

This example was used to illustrate the weighted CN concept. Using a weighted CN to estimate Q will not be significantly in error when P is large and when the CN values are nearly equal. The relative error increases as P decreases and as the differences in CNs increase. It is best to compute the individual Q values and then compute the weighted value of Q than to use a weighted CN to compute Q.

Example 7–18

Figure 7–8 shows a schematic of a watershed that is 40% wooded (good condition) and 60% residential (¼-acre lots). The watershed has 75% soil group B and 25% soil group C. It is necessary to determine the runoff depth for a rainfall of 7 in. The fraction of the watershed in each land use/soil group combination is computed as follows:

Land Use	Soil Group	Area	CN	Q
Wooded	B	0.4(0.75) = 0.30	55	2.12
	C	0.4(0.25) = 0.10	70	3.62
Residential	B	0.6(0.75) = 0.45	75	4.15
	C	0.6(0.25) = 0.15	83	5.03

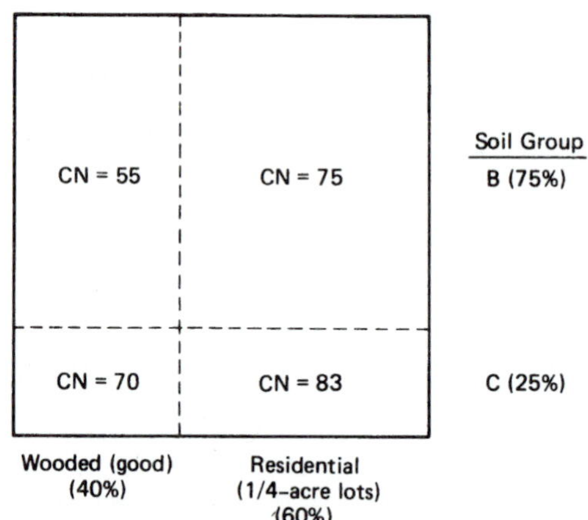

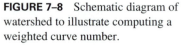

FIGURE 7–8 Schematic diagram of watershed to illustrate computing a weighted curve number.

The weighted Q is

$$Q = 0.30\,(2.12) + 0.10\,(3.62) + 0.45\,(4.15) + 0.15\,(5.03) = 3.62 \text{ in.}$$

Using the weighted CN approach, the weighted CN would be 69.7 (use 70), which yields a Q of 3.62 in. In this case the precipitation is large, so the two methods provide the same estimate of Q.

7.8 SCS GRAPHICAL PEAK DISCHARGE METHOD

For many peak discharge estimation methods, the input includes variables to reflect the size of the contributing area, the amount of rainfall, the potential watershed storage, and the time-area distribution of the watershed. These are often translated into input variables such as the drainage area, the depth of rainfall, an index reflecting land use and soil type, and the time of concentration. The SCS Graphical Method is typical of many peak discharge methods that are based on input such as that described. The Rational Method was based on these same four inputs.

7.8.1 The I_a/P Parameter

I_a/P is a new parameter that has been added to the methods for estimating peak discharge. I_a denotes the initial abstraction, and P is the 24-hr rainfall depth. The I_a/P value can be obtained from Table 7–14 for a given CN and P. If the computed value of I_a/P is less than 0.1, 0.1 should be used; if it is greater than 0.5, then 0.5 should be used. Because SCS procedures for hydrograph generation (see Chapter 9) are based on unit hydrograph theory, classifying hydrographs by unit runoff response is desirable. Using I_a/P will make this possible.

For a given 24-hr rainfall distribution, I_a/P represents the fraction of rainfall that must occur before runoff begins. Vary the I_a/P and the runoff response time changes. Hold I_a/P constant and the unit runoff response is identical no matter what combination of CN and P is used to form the I_a/P ratio. Table 7–15 shows the beginning runoff times and 0.1-hr peak runoff for a range of I_a/P values. This will be more relevant when hydrographs are introduced in Chapter 9.

WORKSHEET 7–1 SCS Graphical Method

Variable	Pre-development	Post-development
Time of concentration, t_c (hrs)		
Drainage area, A (ac)		
Runoff curve number, CN		
$S = (1000/CN) - 10$		
$I_a = 0.2*S$		

Predevelopment Conditions

Rainfall frequency (years)	2	5	10	25	50	100
24-hour intensity, i (in./hr)						
Rainfall depth, P (in.)						
I_a/P						
Runoff depth, Q (in.)						
Unit peak, q_u (ft^3/sec/mi^2/in.)						
Peak discharge, q_p (ft^3/sec)						

Post-development Conditions

Rainfall frequency (years)	2	5	10	25	50	100
24-hour intensity, i (in./hr)						
Rainfall depth, P (in.)						
I_a/P						
Runoff depth, Q (in.)						
Unit peak, q_u (ft^3/sec/mi^2/in.)						
Peak discharge, q_p (ft^3/sec)						

7.8.2 Peak Discharge Computation

The following equation can be used to compute a peak discharge with the SCS Graphical Method:

$$q_p = q_{um} A_m Q \tag{7–45}$$

in which q_p is the peak discharge in ft^3/sec, q_{um} is the unit peak discharge in ft^3/sec/mi^2/in. of runoff, A_m is the drainage area in square miles, and Q is the depth of runoff in inches. The

TABLE 7–14 I_a/P for Selected Runoff Curve Numbers and Rainfall Depths

Rainfall (in.)	Runoff Curve Number 55	60	65	70	75	80	85	90	95
1.0						0.50	0.35	0.22	0.11
1.5	Use 0.5				0.44	0.33	0.24	0.15	
2.0				0.43	0.33	0.25	0.18	0.11	
2.5			0.43	0.34	0.27	0.20	0.14		
3.0		0.44	0.36	0.29	0.22	0.17	0.12		
3.5	0.47	0.38	0.31	0.24	0.19	0.14	0.10		
4.0	0.41	0.33	0.27	0.21	0.17	0.13			
4.5	0.36	0.30	0.24	0.19	0.15	0.11			
5.0	0.33	0.27	0.22	0.17	0.13	0.10			
6.0	0.27	0.22	0.18	0.14	0.11				
7.0	0.23	0.19	0.15	0.12	0.10				
8.0	0.20	0.17	0.13	0.11					
9.0	0.18	0.15	0.12	0.10			Use 0.1		
10.0	0.16	0.13	0.11						
11.0	0.15	0.12	0.10						
12.0	0.14	0.11							
13.0	0.13	0.10							
14.0	0.12	0.10							
15.0	0.11								

TABLE 7–15 Effect of I_a/P on Runoff Response

	Type I Rainfall Distribution		Type II Rainfall Distribution	
	Time Runoff Begins	Maximum 0.1-hr Runoff	Time Runoff Begins	Maximum 0.1-hr Runoff
I_a/P	(hrs)	(%)	(hrs)	(%)
0.1	5.0	8.2	7.1	16.3
0.2	8.1	7.1	10.4	15.7
0.3	9.5	5.0	11.6	13.5
0.4	9.8	3.1	11.8	12.7
0.5	10.0	0.9	11.9	9.8

unit peak discharge is obtained from Figure 7–9, which requires the time of concentration (t_c) in hours and the initial abstraction/rainfall (I_a/P) ratio as input. The runoff depth (Q) is obtained from Equation 7–42, Table 7–13, or Figure 7–7 and is a function of the depth of rainfall and the runoff *CN*. The I_a/P ratio is obtained either directly by $I_a = 0.2S$ or from Table 7–14; the ratio is a function of the *CN* and the depth of rainfall. Computations can be summarized on Worksheet 7–1.

7.8.3 Pond and Swamp Adjustment

The peak discharge obtained from Equation 7–45 assumes that the topography is such that surface flow into ditches, drains, and streams is relatively unimpeded. Where ponding or swampy areas occur in the watershed, a considerable amount of the surface runoff may be retained in temporary storage. The peak discharge rate should be reduced to reflect this condition of increased storage. Values of the pond and swamp adjustment factor (F_p) are provided in Table 7–16. The adjustment factor values in Table 7–16 are a function of the storm exceedence frequency (T), the percent of the total watershed area in ponds and swamps (*PPS*), and the primary location of the pond and swamp areas. Table 7–16 should be used when either the pond and swamp storage is located only in the center part of the watershed (but not within principal drainageways) or the storage is spread uniformly throughout the watershed. The value of F_p is dependent on *PPS* rather than the volume of pond and swamp storage. Given the dependency of peak discharges on the volume of runoff, it is only reasonable that the adjustment factor F_p is an approximation of the true effect of ponds and swamp storage on peak discharge. The accuracy of the adjustment factor F_p depends on the degree to which the actual volume of pond and swamp storage agrees with the "average" volume of storage indicated by *PPS* and used to determine the values of F_p in Table 7–16.

If the watershed includes significant portions of pond and swamp storage, then the peak discharge of Equation 7–45 can be adjusted using the following:

$$q_a = q_p F_p \tag{7–46}$$

in which q_a is the adjusted peak discharge in ft^3/sec.

TABLE 7–16 Adjustment Factor (F_p) for Pond and Swamp Areas that are Spread Throughout the Watershed

Percentage of Pond and Swamp Areas	F_p
0	1.00
0.2	0.97
1.0	0.87
3.0	0.75
5.0	0.72

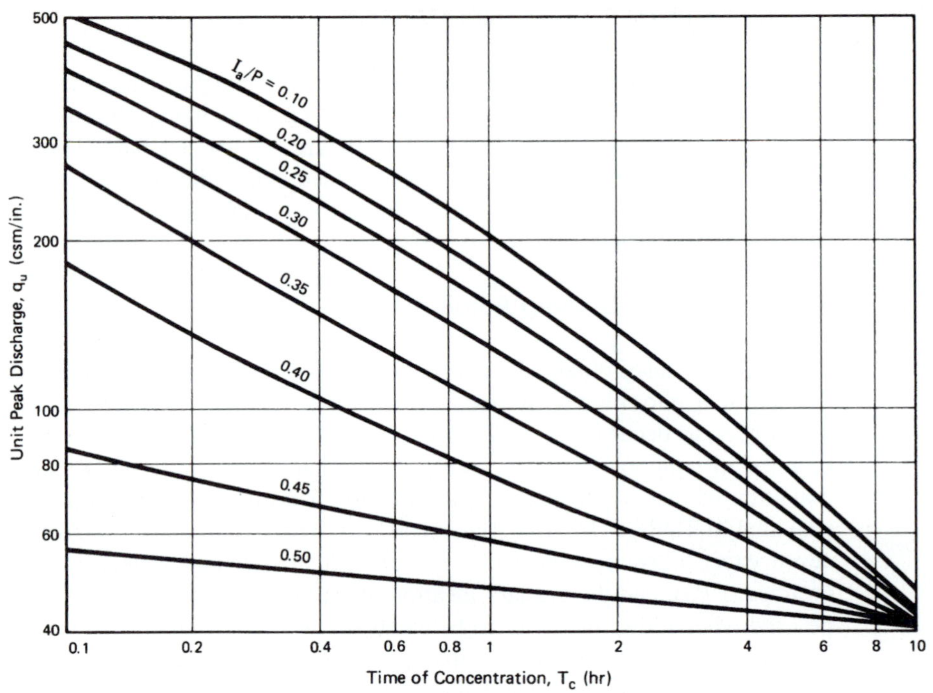

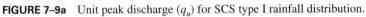

FIGURE 7–9a Unit peak discharge (q_u) for SCS type I rainfall distribution.

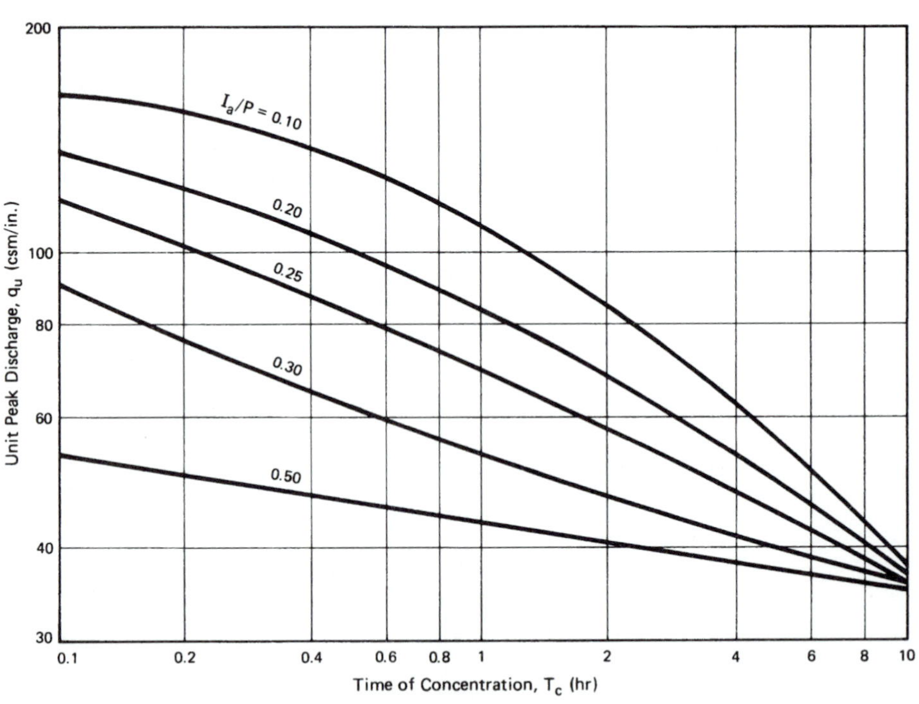

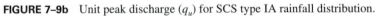

FIGURE 7–9b Unit peak discharge (q_u) for SCS type IA rainfall distribution.

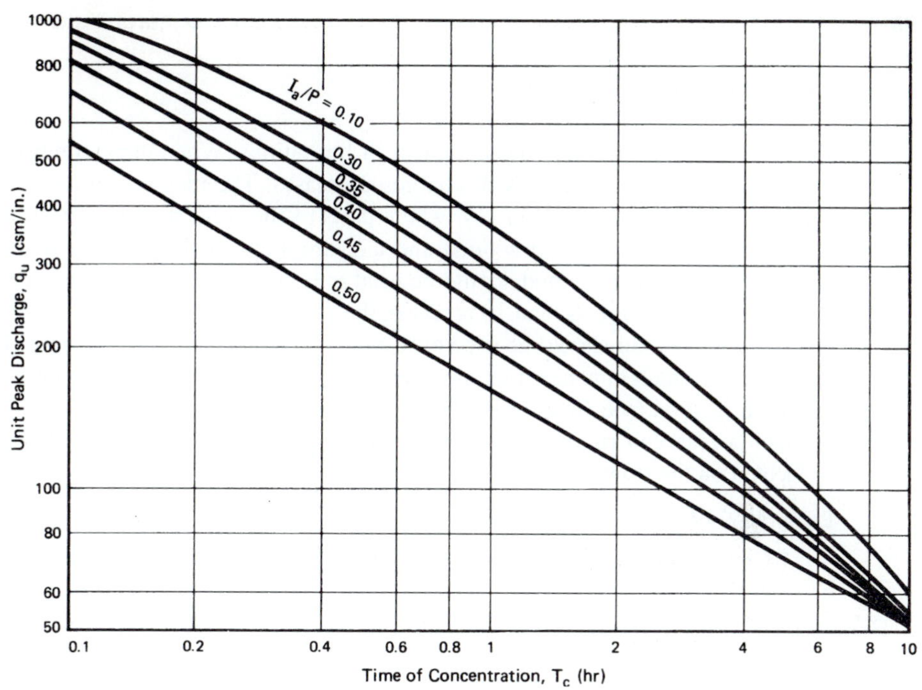

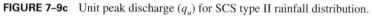

FIGURE 7–9c Unit peak discharge (q_u) for SCS type II rainfall distribution.

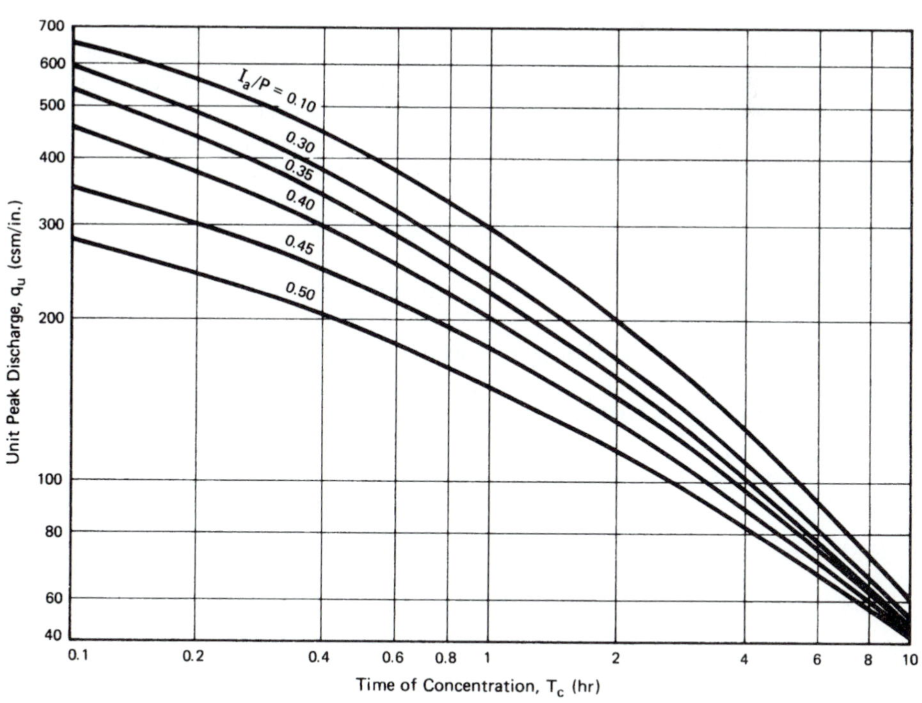

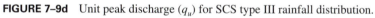

FIGURE 7–9d Unit peak discharge (q_u) for SCS type III rainfall distribution.

7.8.4 Limitations

The SCS Graphical Method has a number of limitations. When these conditions are not met, the accuracy of estimated peak discharges decreases. The method should be used on watersheds that are homogeneous in *CN*; where parts of the watershed have *CN*s that differ by 5, the watershed should be subdivided and a hydrograph method should be used (see Chapter 9). The Graphical Method should be used only when the *CN* is 50 or greater and the t_c is greater than 0.1 hr or less than 10 hr. Also, the computed value of I_a/P should be between 0.1 and 0.5. The method should be used only when the watershed has one main channel or when there are two main channels that have nearly equal times of concentration; otherwise, a hydrograph method (see Chapter 9) should be used. Other methods should also be used when channel or reservoir routing is required, or where watershed storage is either greater than 5% or located on the flow path used to compute the t_c.

Example 7–19

Figure 7–10 shows a strip development on a small upland watershed. The developer is interested in estimating the peak discharge for the contributing area after development. The developed portion of the area is 15.75 acres, with ⅓- and ½-acre lots. The developed area is graded so that runoff will collect in grass-lined swales at the front of the lot and drain to a paved swale that flows along the side of the main road. Flow from the paved swales passes through a pipe culvert to the upper end of the stream channel. The upper portion of the watershed, with is bounded by the contour for 164 ft, is forested on B soil ($CN = 60$) and has an area of 7.25 acres.

To use the Graphical Method, an estimate of the time of concentration is required; the flow path used to compute t_c is shown in Figure 7–10 by a series of four arrows, with each arrow representing a different flow regime. The runoff flows overland in the forest, and then enters a grass-lined swale. Computations of travel times are given in Table 7–17. A total travel time of 2960 sec represents a time of concentration of 0.82 hr.

The Graphical Method also requires the depth of runoff. Assuming the local drainage policy requires 10-yr control, Figure 4–4 indicates a 24-hr rainfall depth of 4.8 in. (0.2 in./hr * 24 hr). A detailed site layout indicates that 21% of the developed area will be impervious, with all of the area connected to the primary drainage system. Thus for a B soil the weighted *CN* is

$$CN = \frac{15.75}{23}[0.21(98) + 0.79(61)] + \frac{7.25}{23}(60) = 66$$

For a *CN* of 66 and a 4.8-in. rainfall, Equation 7–42 yields a runoff depth of 1.59 in. Table 7–14 gives a value of 0.21 for I_a/P. From Figure 7–9c, with a $t_c = 0.82$ hr and $I_a/P = 0.2$, the unit peak discharge is 370 ft^3/sec/mi^2/in.

TABLE 7–17 Calculation of Travel Times Along Principal Flow Path

Path	Flow Regime	Length (ft)	Slope (ft/ft)	Velocity (ft/sec)	Travel Time (sec)
1	Forest	340	0.007	0.2	1700
2	Grass swale	480	0.001	0.5	960
3	Paved swale	350	0.004	1.3	270
4	Pipe	60	0.008	2.0	30
					2960

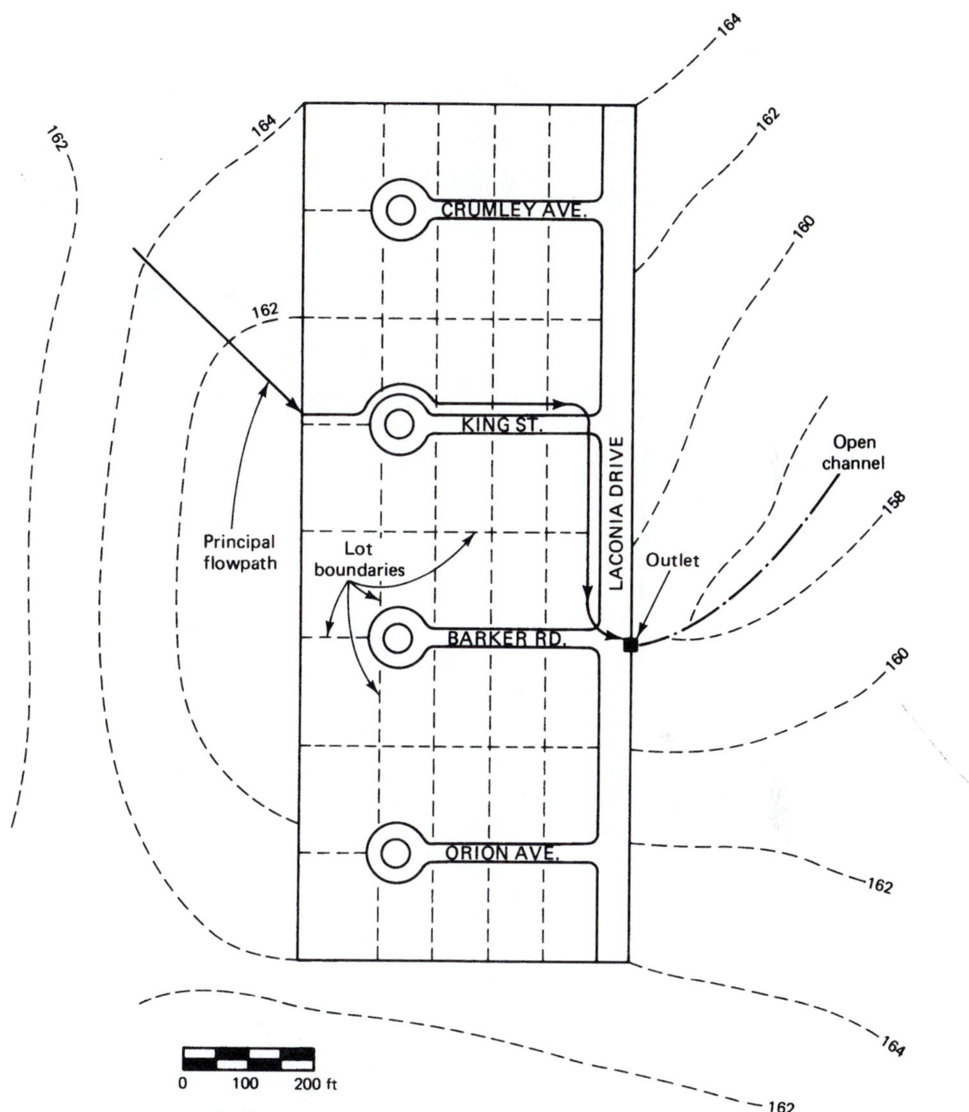

FIGURE 7–10 Topographic and land use map for Example 7–19.

The peak discharge for the total 23-acre watershed is computed with Equation 7–45:

$$q_p = q_u A Q = 370\left(\frac{23}{640}\right)(1.59) = 21\ \text{ft}^3/\text{sec}$$

Since there are no ponds or swampy areas within the watershed, the pond and swamp adjustment is not made.

Example 7–20

A developer is performing a hydrologic design for a small community shopping center adjacent to Duffryn Avenue. The parcel of undeveloped land, which is forested in fair condition, is shown in Figure 7–11a. Superimposing the perimeter of the parcel of land to be developed, the low point of the watershed is the northwest corner of the parcel. The watershed boundary for the post-development conditions is also shown in Figure 7–11a; the boundary is defined in part by the western perimeter line of the developed area, the topographic line for an elevation of 180 ft,

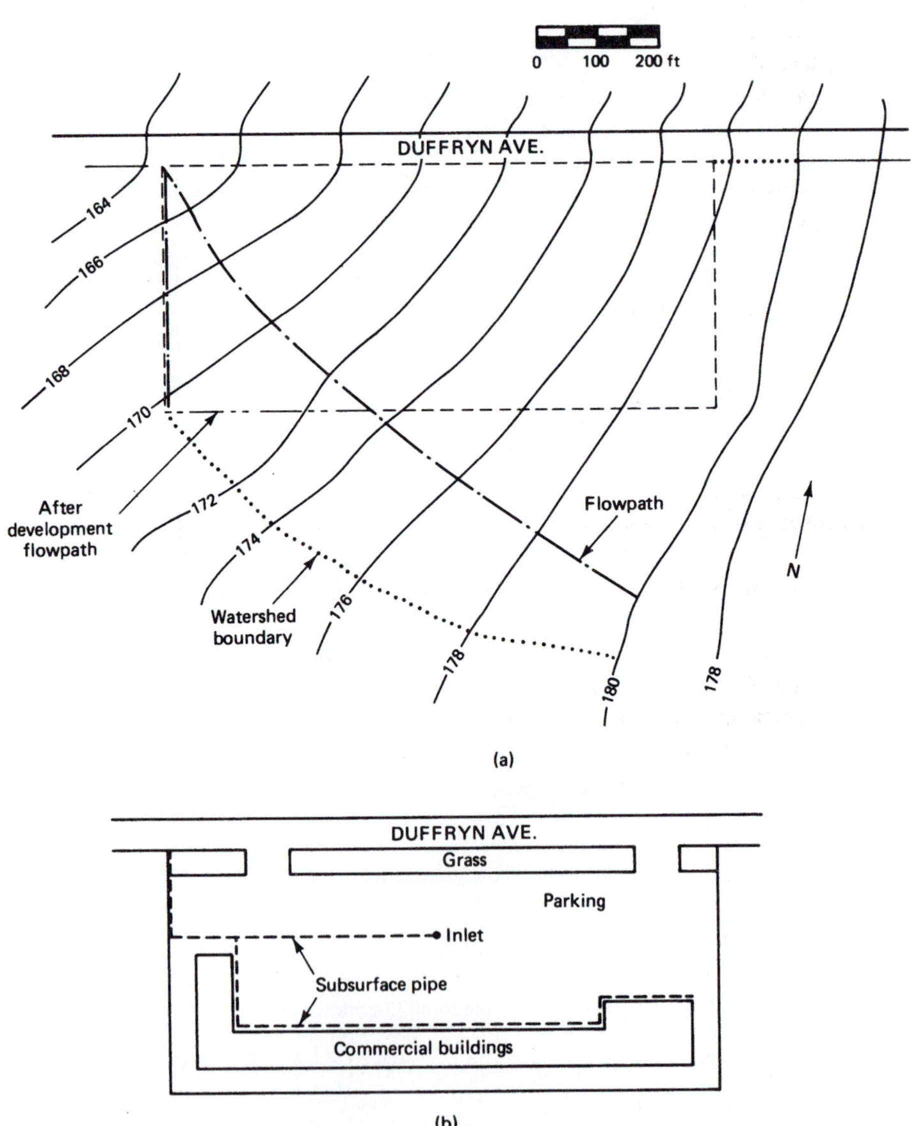

FIGURE 7–11 (a) Topographic and (b) land use maps for Example 7–20 (1 in. = 200 ft).

and the south side of Duffryn Avenue. The drainage area encloses 636,800 ft^2, 14.62 acres, or 0.02284 mi^2. The principal flow path from the outlet to the drainage divide has a length of 1050 ft. A pre- and post-development evaluation will be made, with the design based on a return period of 25 yr.

The basic inputs to use the Graphical Method are the 24-hr rainfall, the *CN*, the drainage area, and the time of concentration. For a 25-yr, 24-hr storm event, the intensity (from Figure 4–4) is 0.245 in./hr, which yields a depth of 5.9 in. For a forested (fair condition) watershed over a C soil, the *CN* (from Table 3–18) is 73. For a drop in elevation of less than 20 ft in 1000 ft, the watershed has a slope of 1.7%. Using a rainfall of 5.9 in. and a *CN* of 73, Equation 7–42 yields a runoff depth of 3.01 in. A value of 0.12 for I_a/P is obtained from Table 7–14. For a slope of 1.7% on a forested watershed, Figure 3–19 indicates a velocity of 0.32 ft/sec; this yields a t_c of 0.911 hr. From Figure 7–9c, we obatin a unit peak discharge of 360 ft^3/sec/mi^2/inch. Thus the pre-development peak discharge computed with Equation 7–42 is

$$\begin{aligned} q_p &= (360 \text{ ft}^3/\text{sec/mi}^2/\text{in.})\,(0.02284 \text{ mi}^2)\,(3.01 \text{ in.}) \\ &= 25 \text{ ft}^3/\text{sec} \end{aligned}$$

The layout for the proposed shopping center is shown in Figure 7–11b. A total of 360,000 ft^2 will be developed with 328,800 ft^2 covered with impervious land covers and 31,200 ft^2 as grass cover in good condition. From Table 3–18, *CN* values of 98 and 74 are obtained for the impervious and grass covers, respectively. Thus the weighted *CN* is

$$CN = \frac{328{,}800}{636{,}800}(98) + \frac{31{,}200}{636{,}800}(74) + \frac{276{,}800}{636{,}800}(73) = 86$$

The increase in *CN* from the pre-development *CN* of 73 results in an increase in the runoff depth. For a rainfall depth of 5.9 in, Equation 7–42 yields a runoff depth of 4.31 in. For an increase in *CN*, I_a will decrease; from Table 7–14, the I_a/P ratio decreases to 0.1. The shopping center alters the longest flow path, as shown in Figure 7–11. The post-development condition has 525 ft of flow in the forested area; for a drop in elevation of 6.5 ft, the slope is 1.2%, which yields a flow velocity of 0.27 ft/sec (from Figure 3–19) and thus a travel time of 0.540 hr. For the flow around the perimeter of the developed shopping area, the flow length is 700 ft, with a slope of 1.3%. From Figure 3–19 we get a flow velocity of 2.3 ft/sec. This yields a travel time of 0.0845 hr. Thus the time of concentration, which is the sum of the travel times, equals 0.625 hr. From Figure 7–9c we get a unit peak discharge of 465 ft^3/sec/mi^2/in. Thus the post-development peak discharge is

$$\begin{aligned} q_p &= (465 \text{ ft}^3/\text{sec/mi}^2/\text{in.})\,(0.02284 \text{ mi}^2)\,(4.31 \text{ in.}) \\ &= 46 \text{ ft}^3/\text{sec} \end{aligned}$$

This represents an increase of 84%.

7.9 SLOPE-AREA METHOD OF DISCHARGE ESTIMATION

Where a stream gage exists, the frequency analysis methods of Chapter 5 can be used to estimate peak discharge rates at gaged locations. The slope-area method is for use at ungaged locations. The following information is necessary to use the slope-area method: (1) cross-section information at both the upstream and downstream cross sections; (2) Manning's roughness coefficient for the channel section; (3) the length of the channel section; (4) the flow depth at both sections for which a discharge is needed (this might be high-water marks for a past flood event); (5) velocity head coefficients at both cross sections. With this information, the following procedure is used to estimate the discharge rate:

1. Using the high-water marks and the cross-section measurements, compute the cross-sectional areas and the wetted perimeters of the upstream and downstream sections.
2. Using Manning's roughness n, calculate the conveyance (K) at both the upstream (K_u) and downstream (K_d) sections:

$$K = 1.49\, AR_h^{2/3}/n \tag{7–47}$$

If the cross section has an irregular shape, then it should be divided into subsections and the conveyance computed for each subsection. The conveyance for the section equals the sum of the conveyances for the subsections:

$$K = \sum_{i=1}^{n_s} K_i \tag{7–48}$$

where n_s is the number of subsections.
3. Calculate the reach conveyance (K):

$$K = (K_u K_d)^{0.5} \tag{7–49}$$

4. Estimate the energy gradient (S) using the difference in elevations of the high-water marks at the two sections (ΔE) divided by the reach length (L) as measured along the thalweg of the channel:

$$S = \frac{\Delta E}{L} \tag{7–50}$$

5. Estimate the peak discharge (q_p):

$$q_p = KS^{0.5} \tag{7–51}$$

6. Calculate the difference in the velocity heads at the two sections:

$$\Delta h_v = \frac{q_p^2}{2g}\left(\frac{\alpha_u}{A_u^2} - \frac{\alpha_d}{A_d^2}\right) \tag{7–52}$$

in which α is the velocity distribution coefficient at the cross section. For simple sections, use $\alpha = 1$. For compound sections, use

$$\alpha = \frac{\sum_{i=1}^{n_s} (K_i^3/a_i^2)}{\left(\sum_{i=1}^{n_s} K_i\right)^3 / A} \tag{7–53}$$

in which a_i is the cross-sectional area of subsection i.

7. Compute a revised estimate of the energy gradient:

$$S = \frac{\Delta E + \Delta h_v}{L} \tag{7–54}$$

8. Recompute the peak discharge using Equation 7–51, and repeat Steps 6 to 8 until the computed discharge rate does not change.

Use of this procedure assumes the following conditions: (1) the channel reach should be relatively uniform; (2) the difference in elevation ΔE should be at least 0.5 ft; (3) the section should not meander significantly and it should be free of obstructions (bridges, culverts, etc.).

Example 7–21

After a flood, a hydrologic engineer estimates the discharge rate at a section where the flood overtopped the banks. At the time when the flow at the upstream cross section just overtopped its banks, the elevation at the downstream section was 3.6 ft. lower. A surveyor obtains cross-section information at both sections. The following is a tabular summary of the input and preliminary calculations:

	Upstream Section	Downstream Section
Depth (ft)	3.0	3.2
Width (ft)	51	54
Area (ft^2)	153	172.8
Wetted perimeter (ft)	57	60.4
Hydraulic radius (ft)	2.684	2.861
Manning's n	0.043	0.046
Conveyance	10,239	11,280

An average conveyance of 10,747 is obtained using Equation 7–49. The initial estimate of the energy gradient is obtained with Equation 7–50:

$$S = \frac{\Delta E}{L} = \frac{3.6}{940} = 0.00383 \text{ ft/ft}$$

Thus, the initial estimate of the discharge is

$$q_p = KS^{0.5} = 10,747\,(0.00383)^{0.5} = 665 \text{ ft}^3\text{/sec}$$

The difference in velocity heads is computed using Equation 7–52:

$$\Delta h_v = \frac{(665)^2}{2\,(32.2)}\left[\frac{1}{(153)^2} - \frac{1}{(172.8)^2}\right] = 0.06339$$

The revised estimate of the energy gradient (Equation 7–54) is

$$S = \frac{3.6 + 0.0634}{940} = 0.00390$$

This is used to get a revised estimate of the peak discharge:

$$q_p = 10{,}747\ (0.00390) = 671\ \text{ft}^3/\text{sec}$$

One more iteration produces the same discharge rate.

7.10 PEAK DISCHARGE ENVELOPE CURVES

Design storms are hypothetical constructs and have never occurred. Many design engineers like to have some assurance that a design peak discharge is unlikely to occur over the design life of a project. This creates an interest in comparing the design peak to actual peaks of record.

Crippen and Bue (1977) developed envelope curves for the conterminous United States, with seventeen regions delineated (see Figure 7–12). Maximum floodflow data from 883 sites that have drainage areas less than 10,000 sq. mi (25,900 sq. km) were graphed versus drainage area. Figure 7–13 shows the nationwide envelope curve. The curves for the seventeen regions were fit to the following logarithmic polynomial model:

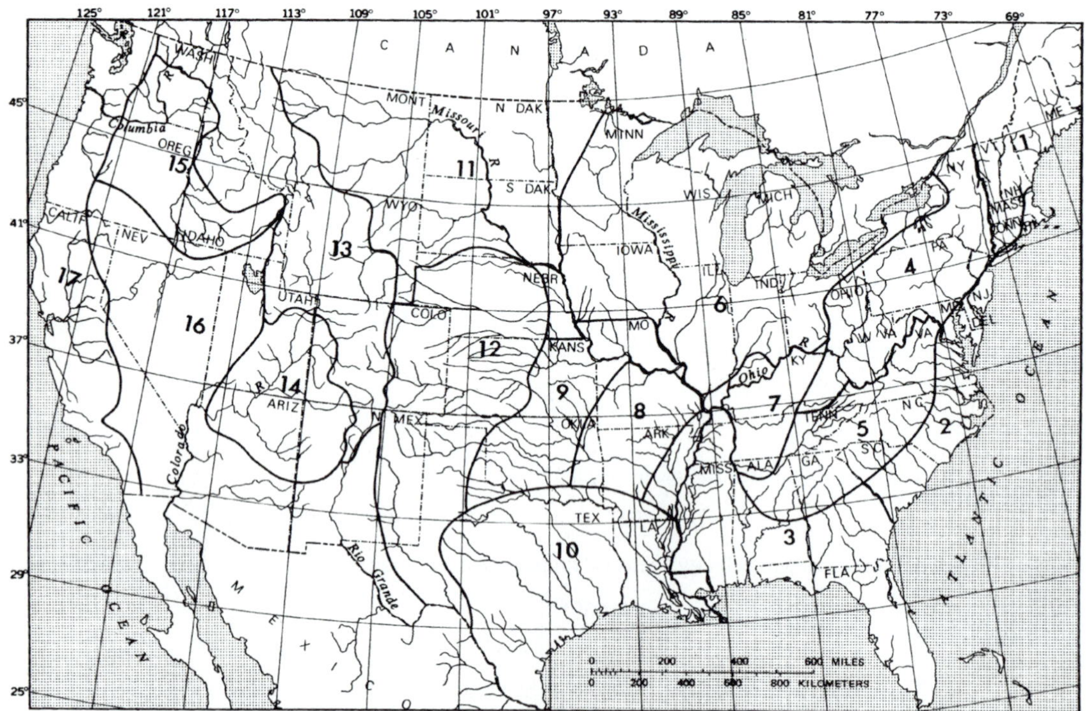

FIGURE 7–12 Map of the conterminous United States showing flood-region boundaries (from Crippen and Bue, 1977).

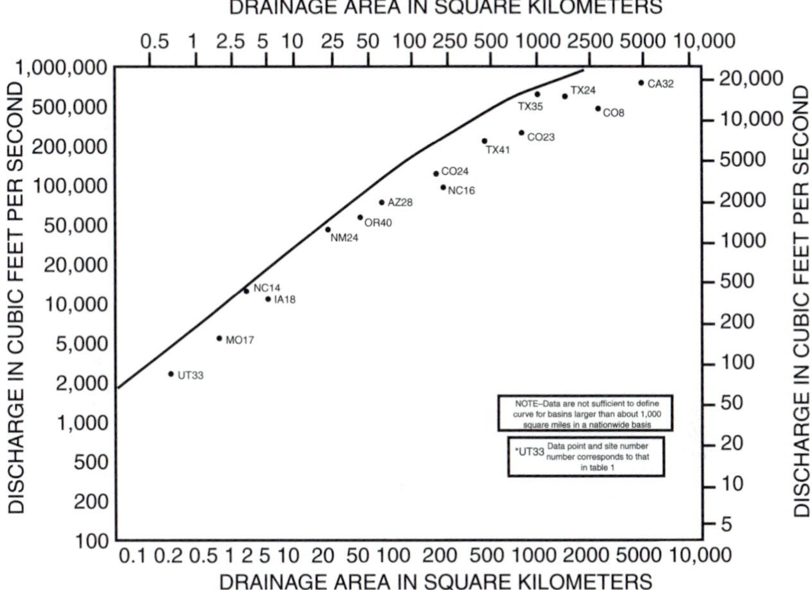

FIGURE 7–13 Selected peak discharges versus drainage areas, and nationwide envelope curve (from Crippen and Bue, 1977).

$$q_p = 10^{b_o + b_1 \log A + b_2 (\log A)^2 + b^3 (\log A)^3} \tag{7–55}$$

in which A is the drainage area (sq. mi) and q_p is the maximum floodflow (ft^3/sec).

Table 7–18 gives the values of the coefficients (b_0, b_1, b_2, and b_3 of Equation 7–55) and the upper limit on the drainage area for each region. The curves are valid for drainage areas greater than 0.1 sq. mi; the smallest area used for the nationwide curve was 0.3 sq. mi. Crippen and Bue did not assign an exceedence probability to the floodflows used to fit the curves, so a probability cannot be given to values estimated from the curves.

Example 7–22

Table 7–19 contains the annual maximum flood record for Oak Creek near Mojave, CA. The record includes a historic flood of 43,500 ft^3/sec that occurred in 1932. The historic event is 25 times larger than the largest flood in the systematic record. The watershed area is 15.8 sq. mi. The watershed is located in region 16 of Figure 7–12. Using the coefficients from Table 7–18, the maximum floodflow for a watershed of that size in the region would be

$$\begin{aligned} q_{\max} &= 10^{3.9549 + 0.91449 \log(15.8) - 0.06926[\log(15.8)]^2 - 0.0034776[\log(15.8)]^3} \\ &= 88,218 \text{ ft}^3/\text{sec} \end{aligned} \tag{7–56}$$

The envelope discharge is approximately twice the magnitude of the historic peak.

TABLE 7–18 Coefficients for Equations of Crippen and Bue (1977) Envelope Curves

Region	Upper Limit (sq. mi.)	Coefficient			
		b_0	b_1	b_2	b_3
1	10000	3.529445	.770808	−.0431261	−.00297574
2	3000	3.770214	.701652	−.0552976	−.0000964921
3	10000	3.668434	.790170	−.0668536	−.00213616
4	10000	3.611647	.819885	−.0842699	.00228029
5	10000	4.040262	.723359	−.0870813	.00227439
6	10000	3.860461	.831197	−.0951892	.00496138
7	10000	3.660894	.770034	−.0946535	.00421293
8	10000	3.600008	.842559	−.0910894	.00294855
9	10000	3.821433	.732911	−.0866673	.00189613
10	1000	3.733268	.987093	−.0650187	−.00425421
11	10000	3.580618	.804757	−.0964107	.00455308
12	7000	3.947249	.818865	−.0747427	.0000138097
13	10000	3.795052	.764519	−.10214	.0058948
14	10000	3.336318	.624470	−.0299691	−.00382852
15	20	3.842281	.917891	−.0688274	−.00571096
16	1000	3.954898	.914490	−.0692602	−.00347756
17	10000	3.767455	.887077	−.0712395	−.00276473
Nationwide	1000	4.072985	.802254	.000592407	−.0192899

7.11 REGIONALIZATION OF HYDROLOGIC VARIABLES

Design methods for specific localities or regions are widely used. The skew map of Chapter 5 and the soil maps discussed in Chapter 3 are examples of important hydrologic design elements that show regional variation. The four SCS rainfall distributions of Chapter 4 indicate that rainfall hyetograph characteristics vary spatially. It was necessary to regionalize the peak discharge envelope curves of Section 7.10 because maximum floodflows are quite different in various parts of the country. When hydrologic variables (such as maximum floodflows, minimum flows, rainfall, or water quality characteristics) vary spatially, an underlying cause related to the physical processes is involved. Regionalization should be based on knowledge of the physical processes; however, statistical analyses can assist in delineating regional boundaries.

TABLE 7–19 *Annual Maximum Flood Series (1958–1984) and a Historical Flood for Oak Creek near Mojave, CA (Area = 15.8 sq. mi.)*

43,000*	97	12	6.9	3.1
1,740	82	11	6.3	3.0
750	44	9.8	6.2	2.0
235	36	9.2	4.0	0.8
230	29	8.2	3.6	0.43
165	22			0.4

*Historic event, 1932

Regionalization of a design method is undertaken because it can improve the accuracy of predictions. For example, if a state could be divided into two parts that indicate where hurricanes are common and where they are rare, the hurricane part would have higher floodflows for a given exceedence probability. If one equation were developed using all flood records in the state, then it would probably tend to underpredict in the hurricane-prone region and overpredict in the nonhurricane-prone region. Such an equation would be biased. By separating the state into two regions and developing separate equations from the flood records in the two regions, we would expect the two equations to provide more accurate predictions in their respective regions. These equations would be unbiased. Greater prediction accuracy should be reflected in the goodness-of-fit statistics (for example, correlation coefficient and standard error of estimate) of the regionalized equations compared to those of the single statewide equation.

The two most difficult aspects of the regionalization process are the decision that regionalization needs to be considered and then the identification of the boundary lines between the regions. Typically, the hydrologist has unique knowledge of the hydrologic characteristics of the area; this will be very useful in interpreting any statistical analyses used to delineate the boundary. Regionalization is one option to consider when the prediction accuracy of a design method is believed to be less than attainable. If the goodness-of-fit statistics for a statistical analysis of regional data are considered low, then analyses should be made to determine whether or not the prediction accuracy can be improved by partitioning the area into regions.

Once a decision has been made to regionalize the design method, it is necessary to determine the number of regions and the boundaries of each region. A common procedure for regionalizing a design method is to develop a prediction equation for the entire area and make a spatial graph of the residuals (for example, differences between the predicted and measured values of the criterion variable). The value of each residual is plotted on a map of the area at the location of the corresponding gage. An attempt is then made to draw isolines of the residuals. If a spatial pattern is apparent in the map of the residuals, then two or more regions may be needed. The presence of a pattern suggests that the design equation for the entire area produced biased estimates in parts of the area. The intent of regionalization will be to eliminate the spatial bias. The hydrologist will then attempt to identify an association between the pattern in the residuals and an important hydrologic characteristic(s).

The boundaries of each region should be established on the basis of the underlying hydrologic processes. The residuals may suggest approximate locations for the boundaries, but they need to be justified using hydrologic criteria. Very often, political boundaries are used within a state, however individual counties often have different policies and design practices, so in some instances the regional boundaries may be made to coincide with political boundaries.

PROBLEMS

7–1. Using the first seven observations on the drainage area (X_1) and the 10-yr peak discharge (Y) data of Table 7–2, calibrate a regression equation of the form

$$Y = b_0 X_1^{b_1}$$

Compute the standard error of estimate. Use the fitted equation to predict Y for observations 8–12 of Table 7–2, and comment on the accuracy of the model.

7–2. Using the nine observations shown below on the drainage area (X_1), the precipitation index (X_4), and the 10-yr peak discharge (Y) data (from Table 7–2), calibrate a regression equation of the form

$$\hat{Y} = b_0 X_1^{b_1} X_4^{b_2}$$

Compute the standard error of estimate. Use the fitted equation to predict Y for observations 10–14 of Table 7–2, and comment on the accuracy of the model.

A	P	Q_p
133	13	16800
103	10.5	9000
91.4	12.5	8200
76.1	13.5	7400
60.7	15	6300
54.7	7	1600
44.8	8.5	1970
24.7	9.5	740
23.9	11.0	1820

7–3. Using all stations that have drainage areas (X_1) less than 100 sq. mi, perform a stepwise regression analysis for the seven predictor variables and the peak discharge (Y) of Table 7–2. Fit the power model of Equation 7–4. A stepwise regression computer program will be necessary. What model would be selected using a partial F test with a 5% level of significance? What model would be selected if predictor variables are inserted as long as the change in R^2 is greater than 5% at any step? Compare the two models.

7–4. Using Equation 7–6, estimate the 10-yr peak discharge for a watershed with the following characteristics: $A = 85$ mi^2; $S = 12.7$ ft/mi; $L = 28$ mi; $P = 14$ in.; $R = 202$ ft; $D = 7.7$ mi/mi^2; $C = 0.6$. Compute the effect on the 10-yr peak discharge if the runoff coefficient were changed to 0.75.

7–5. Using Equation 7–6, estimate the 10-yr peak discharge for a watershed with the following characteristics: $A = 14.6$ mi^2; $S = 1.2$ ft/mi; $L = 6.4$ mi; $P = 4.5$ in.; $R = 44$ ft; $D = 1.5$ mi/mi^2; $C = 0.3$. Compare the computed peak discharge with the measured discharge of Table 7–2 and discuss the findings.

7–6. Use the regression equation of Problem 7–2 to plot a discharge-area curve for drainage areas from 1 to 100 sq. mi. Plot the curve on log-log paper. Show the effect of neglecting the rainfall index.

7–7. Regression analyses of gaged data in a region produced the following single-return-period equations:

$$q_2 = 50\, A^{0.91}$$

$$q_5 = 58\, A^{0.87}$$

$$q_{10} = 71\, A^{0.89}$$

in which A is the drainage area (mi^2) and q_T is the T-yr peak discharge (ft^3/sec). Show the problem that could result from their use in the region from which the data were obtained. Recommend a solution to the problem.

7–8. If the peak discharge computed in Problem 7–4 is the rural peak discharge $RQ10$ of Table 7–5, show the effect of urban development on the peak discharge $UQ10$ by varying BDF from 0 to 6 in the three-parameter equations.

7–9. A 20.7-mi^2 watershed has a rural 25-yr peak discharge of 2420 ft^3/sec. Show the effect of the *BDF* increasing from 0 to 12 in. increments of 1 on the peak discharge. Compute the ratio of *UQ*25 for any *BDF* with the value for *BDF* = 0 and compare the ratios with the values of Figure 5–21(d).

7–10. A 14.1-mi^2 watershed has rural peak discharges for the return periods given below. Compute the peak discharge rates for urban conditions with a *BDF* of 7. Plot the logarithms of both the urban and rural flood frequency curves. Estimate the effect of urbanization on the 40-yr peak discharge.

T (yr)	2	5	10	25	50	100
q_p(ft^3/sec)	450	680	840	1080	1260	1450

7–11. The rural peak discharge rates for selected return periods are tabularized below for a 5.1-mi^2 watershed. Show the effect of urbanization (*BDF* = 8) on the peak discharges using Table 7–5. Plot the logarithms of both the rural and urban frequency curves. Estimate the effect on the 75-yr peak discharge.

T (yr)	2	5	10	25	50	100
q_p(ft^3/sec)	220	290	350	420	480	550

7–12. The USGS Urban Peak Discharge equations (Table 7–5) are based on data from watersheds in the U.S. Would you expect them to be applicable in other parts of the world? Explain.

7–13. The following data include the drainage area and the 2-, 10-, and 50-yr peak discharges for five watersheds. Calibrate the index-flood equation by regressing the 2-yr flood on the drainage area using a power model of Equation 7–4. Also, calibrate the index ratios for the 10- and 50-yr floods. Then compute and plot a flood frequency curve for a 12.8 mi^2 watershed.

T	Peak Discharges (ft^3/sec) for Watershed				
(yr)	1	2	3	4	5
2	245	310	380	485	615
10	490	565	750	920	1170
50	770	850	1215	1420	1790
Area (mi^2)	8.8	11.4	14.6	17.6	25.6

7–14. The following data include the drainage area and the 10-, 25-, and 100-yr peak discharges for five watersheds. Calibrate the index-flood equation by regressing the 10-yr flood on the drainage area using a power model of Equation 7–4. Also, calibrate the index ratios for the 25- and 100-yr floods. Then compute and plot the flood frequency curve for a 17.6 mi^2 watershed.

T	Peak Discharges (ft^3/sec) for Watershed				
(yr)	1	2	3	4	5
10	470	540	710	880	1125
25	570	630	835	1030	1350
100	700	740	990	1230	1645
Area (mi^2)	8.2	10.3	13.6	16.2	19.5

7–15. Using the three-parameter equations for $UQ5$ and $UQ2$ of Table 7–5, develop an urban watershed index-flood ratio for the 5-yr event. What is the relative importance of each of the factors that influence the variation of the index ratio? Reduce it to a constant and compare it to the value from Section 7.4 for rural watersheds.

7–16. Using the index ratios at the beginning of Section 7.4, create a set of index ratios that could be applied with an index-flood equation calibrated with the 10-yr storm being the index return period.

7–17. Using the index ratios of Section 7.4 and the following 2-yr peak discharge equation, compute and plot a flood frequency curve for a 32 sq. mi watershed:

$$q_2 = 55\, A^{0.67}$$

where A is in square miles.

7–18. The following data are the drainage area A (mi^2) and the mean ($\overline{X}$) and standard deviation (S) of the logarithms of the annual flood series for five watersheds. Calibrate two power-model equations regressing both $\overline{X}$ and S on the drainage area. Plot the relationships between $\overline{X}$ and S versus A for $1.0 < A < 100$ mi^2. Also, compute and plot a log-normal flood frequency curve for a 22-mi^2 watershed.

	Watershed				
	1	2	3	4	5
Area	5.1	8.2	12.7	32.6	39.3
Log mean	2.02	2.59	2.49	2.72	3.06
Log-standard deviation	0.25	0.21	0.23	0.19	0.20

7–19. The following data are the drainage area (mi^2) and the mean ($\overline{X}$) and standard deviation (S) of the logarithms of the annual flood series for six watersheds. Calibrate power-model equations regressing both $\overline{X}$ and S on the drainage area. Plot the two relationships between $\overline{X}$ and S versus A for $1.0 < A < 100$ mi^2. Also, compute and plot a log-normal flood frequency curve for a 74.3-mi^2 watershed.

	Watershed					
	1	2	3	4	5	6
Area	36	49	56	63	81	94
Log mean	2.99	3.03	3.07	3.18	3.22	3.37
Log-standard deviation	0.19	0.20	0.18	0.17	0.15	0.15

7–20. For the analysis of Problem 7–17, construct and plot a log-Pearson Type III frequency curve for a 68-mi^2 watershed. Assume a skew coefficient of 0.5.

7–21. Assume that the peak discharge q_p of the Rational Method results from a storm of constant intensity i over a storm duration equal to the time of concentration t_c. Runoff is produced by a proportion of the total volume of rainfall equal to the runoff coefficient C. Assume that the runoff has a peak q_p, a time base equal to 2 t_c, and the shape of an isosceles triangle with a time-to-peak equal to t_c. Based on these assumptions derive the Rational Formula. Show that it would be incorrect to use a time base other than 2 t_c.

7–22. A design engineer makes numerous peak discharge estimates for small watersheds (2 to 10 acres) for 5-yr return periods. Most are made for single-family residential developments in areas where Figure 4–4 would be applicable. Develop a simplified model that would only require the drainage area to compute an estimate of the peak discharge.

7–23. An old-timer tells an entry-level engineer that he has always used the rule of thumb that indicates peak discharge rates of 1 ft^3/sec/acre. Under what conditions might this rule of thumb be valid?

7–24. A nonhomogeneous parcel of land contributes runoff to a proposed inlet. Estimate the weighted runoff coefficient for the following land cover distribution: 10% roof areas, 20% pavement/roadways, 40% lawns (heavy soil, 3% slope), 30% wooded.

7–25. A nonhomogeneous watershed has the following distribution of land cover:

Fraction	Land Cover	Soil Group	Slope (%)
0.25	Cultivated land	C	2.3
0.30	Forest	D	2.2
0.20	Pasture	C	1.9
0.10	Open space	B	1.8
0.15	Open space	C	2.1

Estimate the runoff coefficient for use in predicting a 10-yr peak discharge.

7–26. Data from eight storm events were collected at a 12.7–acre watershed. Based on the data, estimate the runoff coefficient.

i (in./hr)	2.3	3.7	4.1	2.8	3.2	2.5	4.3	3.6
q_p (ft^3/sec)	10.1	18.1	18.6	13.2	14.4	11.2	19.9	16.5

7–27. Using the Rational Method, estimate the 10-yr peak discharge from a 35-acre watershed using the *IDF* curve of Figure 4–4. Assume light industrial development with the principal flow path being paved surfaces including gutter flow. The principal flow path has a length of 1800 ft and a slope of 2%.

7–28. Using the Rational Method, estimate the 25-yr peak discharge into a culvert from an unimproved 17-acre watershed with the primary flow path being short grass. The drainage area has a slope of 3.5% and a length of 1150 ft. Use the *IDF* curve of Figure 4–4.

7–29. Using the Rational Method, estimate the 10-yr peak discharge from a 45-acre watershed using the *IDF* curve of Figure 4–4. The principal flow path consists of 100 ft of sheet flow over grass (3% slope), 450 ft of flow in a grass-lined swale (2.5% slope), and 1500 ft of flow in a shallow concrete channel ($n = 0.035$, $S = 2\%$, $R_h = 0.6$ ft). The land cover consists of residential development with 0.5-acre lots on C soils.

7–30. Using the Rational Method, estimate the 50-yr peak discharge from a 12-acre drainage area that is forested, has C soils, and a slope of 4%. The principal flow path is 900 ft in length with 75 ft of overland flow (heavy ground litter), 175 ft of flow in a vegetated gully, and 650 ft of flow in a shallow channel ($n = 0.025$, $R_h = 0.5$ ft). Use the *IDF* curve of Figure 4–4.

7–31. For the watershed schematic shown below, find the peak discharge into each inlet and the required discharge capacity for each pipe using the Rational Method. Use a 50-yr return period and the *IDF* curve of Figure 4–4.

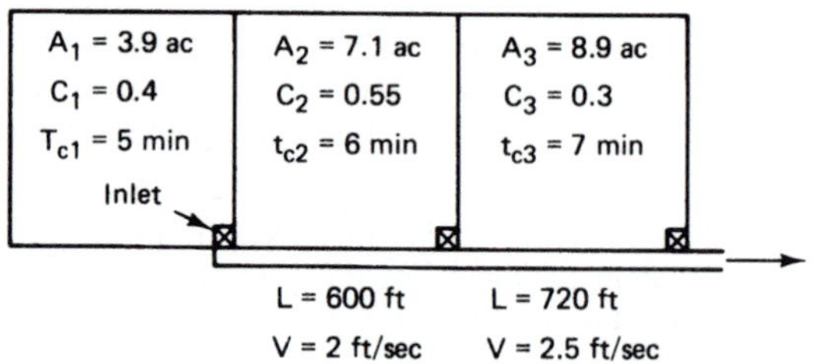

7–32. For the watershed information of Problem 7–31, compute the inlet discharge rates and discharge capacities of the pipes using the Rational Method if local drainage policies allow the use of a minimum time of concentration of 15 min.

7–33. For the watershed schematic shown below, find the peak discharge (using the Rational Method) into each inlet and the required discharge capacity for each pipe segment. Use a 50-yr return period and the *IDF* curve of Figure 4–4.

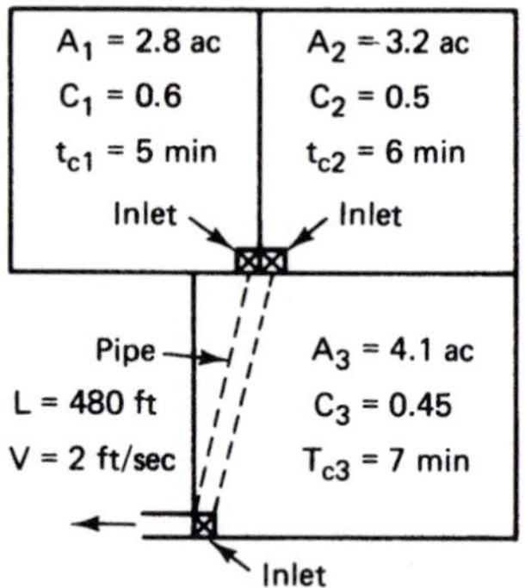

7–34. For the watershed information of Problem 7–33, compute the inlet discharge rates and discharge capacities of the pipes using the Rational Method if local drainage policies allow the use of a minimum time of concentration of 20 min.

7–35. For the watershed of Figure 7–2, compute the inlet flow rates and pipe capacities for Polonsky Alley from 3rd Street to Main Street using the Rational Method if the runoff coefficient is 0.6 below 3rd Street and 0.31 above 3rd Street and the inlet times of concentration are 6 min (rather than the 15 min used in Table 7–14). Use a 10-yr exceedence frequency.

7–36. Use the Rational Method to compute the 2-yr peak discharge for a 75-acre watershed, and compute and plot the flood frequency curve using the index-flood approach with the index ratios at the beginning of Section 7.4. Assume a C of 0.45, a time of concentration of 0.75 hr, and the *IDF* curve of Figure 4–4. Estimate the 40- and 75-yr peak discharges.

7–37. Use the Rational Method to compute the 2-yr peak discharge for a 64-acre watershed, and compute and plot the flood frequency curve using the index-flood approach with the index ratios at the beginning of Section 7.4. Assume a $C = 0.6$, a time of concentration of 0.5 hr, and the *IDF* curve of Figure 4–4. Estimate the 20- and 80-yr peak discharges.

7–38. For the conditions of Problem 7–31, use the Rational Method to compute the 2-, 5-, 10-, 25-, 50-, and 100-yr flood peaks for each inlet. Estimate the 40-yr peak.

7–39. Show the range of computed peak discharges for a 145-acre, light-industrial drainage area for the range of C values given in Table 7–10. Assume a 50-yr return period, a t_c of 35 min, and the *IDF* curve of Figure 4–4. Assume the diameter of a pipe culvert can be estimated by $D = 0.6467\ q_p^{0.375}$. Compute diameters for each C value.

7–40. A 42-acre watershed is 60% apartments and 40% single-family residential. Assuming a t_c of 22 min, use the Rational Method to compute the flood frequency curve (use the *IDF* curve of Figure 4–4).

7–41. A 136-acre watershed of unimproved land use has a t_c of 80 min. Use the Rational Method to compute the frequency curve. Using the 2-yr event as the index flood, compute moment ratios for return periods for 5, 10, 25, 50, and 100 yr. Compare the values with those given in Section 7.4.

7–42. The runoff volume from a 3.5-in. rainfall on a 32-acre watershed was 4.8 ac-ft. If 0.3 in. of rainfall occurred before the start of runoff, estimate the potential maximum retention. For this case, how accurate is the empirical relationship of Equation 7–41? Based on the empirical relationship of Equation 7–43, what is the storm event curve number?

7–43. Compute the weighted runoff curve number for the following conditions and use it to compute the depth of runoff (in.) for a storm depth of 4.5 in.

Area Fraction	Land Use/Condition	Soil Group
0.30	Lawns, open space/fair	C
0.25	Residential/0.5-acre lots	D
0.45	Woods/good	C

7–44. Compute the weighted runoff curve number for the following conditions. Compute the depth of runoff (in.) for storm depths of 3 and 5 in. using both a weighted *CN* approach and a weighted *Q* approach.

Area Fraction	Land Use/Condition	Soil Group
0.10	Commercial/70% imperviousness	C
0.20	Residential/55% imperviousness	D
0.20	Residential/25% imperviousness	C
0.35	Woods/good	C
0.15	Woods/good	D

7–45. Compute the curve number for a 20-acre drainage area having the characteristics shown below. Compute the depth of runoff (in.) for a storm of 5 in. Compute the runoff using both a weighted *CN* approach and a weighted *Q* approach.

Area (acres)	Imperviousness %	Soil Group
4	24	B
6	18	C
7	60	B
3	75	C

7–46. Compute the curve number for an 11-acre drainage area in residential land use with 28% impervious cover and C soils. (a) Assume that all of the impervious area is directly connected to the storm drain system; (b) assume that 75% of the impervious area is unconnected. For both cases, compute the depth of runoff (in.) from storms of 3.2, 4.8, and 7.2 in. Also, compute the reduction in Q due to disconnecting a portion of the impervious surface.

7–47. Assume a runoff CN of 80. Compute and plot the value of Q using Equation 7–42 but ignoring the constraint of Equation 7–44 for values of P from 0 to 1 at an increment of 0.1 in. Discuss the implications of ignoring Equation 7–44.

7–48. For the curve numbers of 60, 75, and 90 and rainfall depths of 3 and 7 in., compute the I_a/P ratio. Discuss the hydrologic implications of the results.

7–49. Assuming a C soil and the watershed conditions of Problem 7–27, estimate the peak discharge with the SCS Graphical Method.

7–50. Using a B soil and a brush land cover (poor condition), estimate the peak discharge with the SCS Graphical Method using the other watershed conditions of Problem 7–28.

7–51. For the conditions of Problem 7–29, estimate the peak discharge with the SCS Graphical Method.

7–52. Assuming CNs of 87, 81, and 78 for subareas 1, 2, and 3, respectively, estimate the inlet peak discharges for the watershed conditions of Problem 7–33 using the SCS Graphical Method.

7–53. Estimate the 100-yr peak discharge using the SCS Graphical Method for a 67–acre watershed in a coastal area. Assume the IDF curve of Figure 4–4 applies. The watershed with an average slope of 0.25% has a principal drainage path of 100 ft of overland flow in a sparsely wooded cover condition (soil group A) and 2700 ft of flow in small channels with R_h = 0.8 ft and n = 0.085. The watershed is wooded in poor hydrologic condition, with approximately 5% of the watershed considered swamp land.

7–54. The following measurements were made during a flood at two sections of a channel:

Variable	Upstream	Downstream
Depth (ft)	5.2	5.3
Width (ft)	142	161
n	0.038	0.037

Both sections are approximately rectangular. The invert elevation of the downstream section is 4.3 ft lower than that of the upstream section, and the reach is 1850 ft in length. Estimate the discharge using the slope-area method.

7–55. A 5-mi^2 watershed has a time of concentration of 6 hr and land use/soil that yields a CN of 94. Compute the 2-, 5-, 10-, 25-, 50-, and 100-yr peak discharges using the SCS Graphical Method. Assume Figure 4–4 is applicable. Also estimate the extreme maximum discharge from the envelope curve of Equation 7–55, region 4. What are the implications of these results in relationship to assigning an exceedence probability to extreme flood peaks?

7–56. Given the following maximum discharges (q, cfs) on watersheds with the indicated area A (sq. mi), develop and graph a discharge-drainage area envelope curve for Maryland watersheds. Compare it with the envelope curve of Crippen and Bue (1977) for region 4.

A	q	A	q	A	q	A	q
0.1	110	1.3	825	5.31	3500	34.8	21,800
0.26	515	1.5	1370	6.30	3960	59.8	38,000
0.35	695	2.28	1900	9.03	7060	93.4	38,600
0.47	750	2.45	3750	9.73	10,100	148.	76,100
		4.90	5500			345.	47,000

REVIEW QUESTIONS

7–1. Which one of the following is not true? The peak discharge is (a) a criterion for classifying hydrologic models; (b) a primary design variable; (c) the total volume of flood runoff; (d) the maximum volume flow rate passing the watershed outlet during a storm event.

7–2. Which one of the following watershed characteristics forms the basis for most of the earlier peak discharge equations? (a) The drainage area; (b) the land use; (c) the slope; (d) the flow length.

7–3. The effect of the exceedence probability on peak discharges computed using the Rational Method is accounted for with which of the following factor(s)? (a) C only; (b) i only; (c) C and i; (d) C, i, and A; (e) none of the above.

7–4. For designs on subdivided watersheds, weighted values of both the runoff coefficient and time of concentration are used so that (a) the maximum discharge is provided at the outlet; (b) an equal level of flood protection is provided for each structural element in the design; (c) the risk of flooding is minimized; (d) an adequate factor of safety is provided; (e) all of the above.

7–5. In developing the SCS rainfall-runoff relationship, the total rainfall was separated into three components. Which one of the following was not one of the three components? (a) The initial abstraction; (b) the antecedent soil moisture level; (c) the actual retention; (d) the direct runoff volume.

7–6. In developing the SCS rainfall-runoff relationship, the potential maximum retention is a function of three types of storage. Which one of the following was not one of the three storages? (a) Subsurface; (b) interception; (c) depression; (d) pond and wetland.

7–7. Which one of the following factors is the runoff curve number CN not a function of? (a) The antecedent soil moisture; (b) the soil groups; (c) the land use; (d) the hydrologic condition of the cover complex; (e) all of the above are factors.

7–8. Which one of the following is not a factor in estimating CNs for urban land uses? (a) The soil type; (b) the percent of imperviousness; (c) the percent of unconnected imperviousness; (d) the percent of watershed sewered; (e) all of the above are factors.

7–9. For a CN of 50 and a 24-hr rainfall of 3 in., an I_a/P value of _______ should be used: (a) 0.1; (b) 0.5; (c) 0.67; (d) 2.0; (e) none of the above.

7–10. For a drainage area of 64 acres, 2 in. of direct runoff, and a unit peak discharge of 600 $ft^3/sec/mi^2/in.$, the peak discharge (ft^3/sec) is (a) 120; (b) 1200; (c) 38,400; (d) 76,800; (e) none of the above.

7–11. Which one of the following is not a factor in determining the pond and swamp adjustment factor? (a) The I_a/P ratio; (b) the return period; (c) the percent of ponds and/or swamps; (d) the location of the ponds in the watershed; (e) all of the factors above.

7–12. Which one of the following is not a limitation of the SCS Graphical Method? (a) 0.1 hr $\le t_c \le$ 10 hr; (b) $CN \ge 50$; (c) the watershed is homogeneous in CN; (d) the computed value of I_a/P is between 0.1 and 0.5; (e) all of the above are limitations.

7–13. Which one of the following is not known in the analysis of single-return-period prediction equations? (a) The return period of the peak discharges; (b) the values for a set of predictor variables; (c) the model coefficients; (d) the values of the peak discharge.

7–14. Which one of the following is not known when making a design using the index-flood method? (a) The model coefficients of the index-flood equation; (b) the return period; (c) the values of the predictor variables; (d) the index ratio; (e) all of the above are known.

7–15. Which one of the following is not an advantage of the index-flood method? (a) The method only requires calibration of one index equation; (b) the method is computationally simple; (c) the index ratios assure consistency across return periods; (d) the method provides greater accuracy than the single-return-period equations; (e) all of the above are advantages.

7–16. Which one of the following is a disadvantage of the moment estimation method? (a) The analysis process requires extensive data; (b) it is difficult to develop an accurate equation for predicting the skew coefficient; (c) the log-Pearson Type III distribution must be used; (d) the method is not consistent across return periods.

7–17. Which of the following is not a factor in estimating the peak discharge using the slope-area method? (a) The slope of the channel reach; (b) the amount of vegetation in the stream; (c) the hydraulic radius of the reach; (d) the wetted perimeter of the section; (e) all of the above are factors.

7–18. Peak discharge envelope curves provide discharge estimates for a return period of (a) 500 years; (b) 100 years; (c) the PMF; (d) the return period is not specified; (e) none of the above.

7–19. Which one of the following is not a reason that regionalization is necessary in developing hydrologic models? (a) Regression modeling would not be applicable for a data base as large as that for the entire country; (b) the effect of variation in rainfall is too great across the country; (c) regionalization enables local conditions to be considered in model development; (d) regionalization improves the accuracy of prediction; (e) all of the above are reasons for regionalization.

DISCUSSION QUESTION

The technical content of this chapter is important to the professional hydrologist, but practice is not confined to making technical decisions. The intent of this discussion question is to show that hydrologists must often address situations where value issues intermingle with the technical aspects of a project. In discussing the stated problem, at a minimum include responses to the following questions:

1. What value issues are involved, and how are they in conflict?
2. Are technical issues involved with the value issues? If so, how are they in conflict with the value issues?

3. If the hydrologist attempted to rationalize the situation, what rationalizations might he or she use? Provide arguments to suggest why the excuses represent rationalization.
4. What are the hydrologist's alternative courses of action? Identify all alternatives, regardless of the ethical implications.
5. How should the conflict be resolved?

You may want to review Sections 1.6 to 1.12 in Chapter 1 in responding to the problem statement.

Case. A hydrologist who works for a land development company performs all of the flood peak estimations required for the company's projects. He routinely makes peak discharge estimates using both the Rational Method and the SCS Graphical Method and uses the smaller value because it can reduce project costs associated with the smaller inlets, pipes, culverts, and storm water detention basins required. The local drainage policy does not specifically state which method should be used for making peak discharge estimates, so it does not prohibit the practice.

8

Hydrologic Design Methods

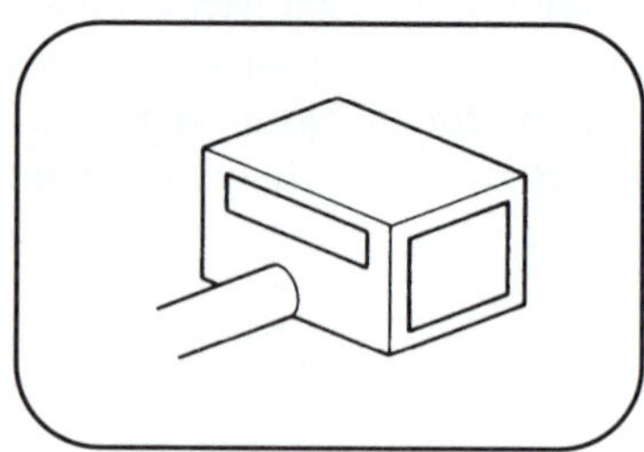

CHAPTER OBJECTIVES

1. Introduce both the conceptual basis and procedure for applying frequently used peak discharge design methods.
2. Demonstrate types of design problems where peak discharge estimates are the primary hydrologic input.
3. Introduce methods for sizing inlets, culverts, and roof drainage.
4. Provide methods for making planning estimates of detention storage volumes to control discharge rates.
5. Provide methods for sizing the outlet facilities for single-stage and two-stage structures.

8.0 NOTATION

A = area of cross section
A = drainage area (acres)
A = rooftop area
A = total area of inlet openings
A_o = area of orifice (ft^2)
C = runoff coefficient for Rational Method
C_d = orifice discharge coefficient
C_w = weir coefficient
C_* = conduit diameter coefficient
CN = runoff curve number
d = depth of flow
d = diameter of leader pipe
d_c = critical depth
D = conduit diameter (ft)
D = pipe diameter
E = elevation (ft)
E = specific energy
E_c = elevation of conduit centerline at the face of the riser (ft)
E_d = design efficiency
E_i = invert elevation of conduit at the face of the riser (ft)
E_o = elevation of orifice or low stage weir (ft)
E_1 = maximum water surface elevation in reservoir at routed low-stage hydrograph (ft)
E_2 = maximum water surface elevation at high-stage event (ft)
f = factor of safety
f_c = carryover reduction factor
$\mathbf{F}$ = Froude number
g = acceleration of gravity
h = head
h_i = depth of flow above the invert
h_i = hydrostatic pressure at section i
h_L = head loss
H_o = height of orifice opening
i = rainfall intensity
I_a = initial abstraction
K_p = head loss coefficient for circular pipe flowing full
L = flow length
L = length of curb opening
L = length of slotted drain
L = length of weir
L = watershed length
L_d = design length of drain
L_i = initial length of slotted drain
L_{w1} = low-stage weir length (ft)
L_{w2} = high-stage weir length (ft)
m = ratio of hydrograph recession duration to the time to peak
n = Manning's roughness coefficient
n = number of drains
n = number of openings in inlet
p_i = pressure head at section i
P = rainfall depth (or precipitation)
P_{24} = 24-hr rainfall depth (in.)
q = discharge rate
q_c = carryover discharge rate
q_d = design discharge
q_{o2} = flow through orifice or over low-stage weir during high-stage event (in.)
q_p = peak discharge
q_{pa} = post-development peak discharge
q_{pa1} = peak discharge (ft^3/sec), post-development, low-stage event
q_{pa2} = peak discharge (ft^3/sec), post-development, high-stage event
q_{pb} = pre-development peak discharge
q_{pb1} = peak discharge (ft^3/sec), pre-development, low-stage event (also, the allowable low-stage discharge from the basin post-development)
q_{pb2} = peak discharge (ft^3/sec), pre-development, low-stage event (also, the allowable high-stage discharge from the basin post-development)
q_u = unit discharge

and the slope of the horizontal drains located in the roof. The rainfall information, specifically the design return period and storm duration, is usually set by local building codes; for example, the code might state that the design rainfall is the 5-yr, 0.l-hr intensity. The design method assumes that all rainfall is converted to runoff; that is, the method assumes that there are no losses such as depression storage. The objective of a design is to size the vertical leader (see Figure 8–1), the vertical drains, and the horizontal drain pipes.

The design procedure is as follows:

1. Determine the number of drains.
2. Determine the design rainfall intensity based on the duration and frequency specified in the local policy.
3. Determine the flow capacity per drain.
4. Design the system by computing the size of the leader, vertical drains, and horizontal drains.

As a general rule, there should be one drain per 10,000 ft^2 of roof area, with a minimum of two drains per roof. Based on the policy statement or local building codes, the design rainfall intensity is obtained from the local *IDF* curve (for example, Figure 4–4). The intensity i in in./hr can be converted to a unit discharge rate Q_u in gal/min/sq. ft of roof area using

$$Q_u = 0.0104\, i \tag{8–1}$$

The total discharge rate Q_n for a rooftop of area A in sq. ft is

$$Q_n = Q_u A \tag{8–2}$$

The capacity per drain Q_d equals the total flow Q_n divided by the number of drains, n. Q_n and Q_d have units of volume per unit time and thus correspond to a peak discharge. The diameter d (in.) of both the leader and the vertical drains is given by

$$d = 0.6\, Q_d^{0.377} \tag{8–3}$$

in which Q_d is the capacity per drain in gal/min. The diameter of horizontal pipes in in. is a function of both Q_d and the slope of the horizontal drain S in in./ft

$$d = 0.53 S^{-0.188} Q_d^{0.377} \tag{8–4}$$

Diameters estimated with Equations 8–3 and 8–4 should be rounded up to the next larger commercially available size so that the pipe capacity will be sufficient to drain the design flow Q_d. For roof drains, the following pipe diameters (in.) are usually available commercially: 2, 4, 6, 8, 10, 12, 15, 18. Equation 8–3 should be accurate for flow rates less than 6000 gal/min, and Equation 8–4 should be valid for slopes from 0.125 to 0.5 in./ft and diameters less than 18 in.

Example 8–1

The design method will be illustrated for a commercial building with a flat roof of 80,000 ft^2 (200 × 400 ft). Thus according to the rule of thumb, a total of eight drains will be required. The layout of the drains is shown in Figure 8–2. If the local drainage policy requires a 10-yr, 0.l-hr

Final designs are based on hydrograph (see Chapter 9) or hydraulic analyses. The small storm water detention basin is one example where planning methods are based on peak discharge estimates while their design involves storage routing.

Design and planning problems that depend on peak discharge estimates are the subject of this chapter. The methods presented here are just examples of the methodologies that are available. For example, highway departments have inch-thick manuals for sizing culverts and for the design of inlets. The material presented in this chapter is only meant to be introductory material for the multitude of cases that exist.

8.2 DESIGN OF ROOF DRAINAGE

It should be obvious that rooftop drainage is an important element of building design. Improper drainage can lead to building damage and minor public health problems. A number of methods are available for designing roof drains; an approximation will be provided here.

Figure 8–1 shows a schematic of a roof drainage system for a large building. Rain falls on the roof, which has a very shallow slope, and collects at the roof drain inlet. After passing through the inlet the runoff falls through a leader pipe to a horizontal pipe that transports the water to a vertical drain pipe that is either attached to an exterior wall or within the wall. The runoff is discharged from the vertical drain into a horizontal storm sewer. Both the horizontal roof drain pipe and the storm sewer are laid at shallow slopes. In addition to the pipe system, roofs should be designed with scuppers, which are pipes in the parapet of the roof; scuppers serve as emergency drains that permit overflow in cases where the roof drain becomes clogged by debris or ice.

To use the design method provided here, the following variables are required input: the roof area and layout, the rainfall intensity-duration-frequency relationship for the location,

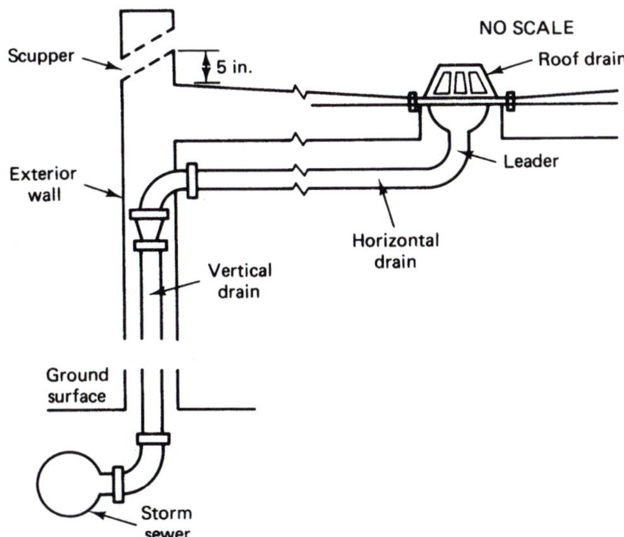

FIGURE 8–1 Schematic diagram of a roof drainage system.

and the slope of the horizontal drains located in the roof. The rainfall information, specifically the design return period and storm duration, is usually set by local building codes; for example, the code might state that the design rainfall is the 5-yr, 0.l-hr intensity. The design method assumes that all rainfall is converted to runoff; that is, the method assumes that there are no losses such as depression storage. The objective of a design is to size the vertical leader (see Figure 8–1), the vertical drains, and the horizontal drain pipes.

The design procedure is as follows:

1. Determine the number of drains.
2. Determine the design rainfall intensity based on the duration and frequency specified in the local policy.
3. Determine the flow capacity per drain.
4. Design the system by computing the size of the leader, vertical drains, and horizontal drains.

As a general rule, there should be one drain per 10,000 ft^2 of roof area, with a minimum of two drains per roof. Based on the policy statement or local building codes, the design rainfall intensity is obtained from the local *IDF* curve (for example, Figure 4–4). The intensity i in in./hr can be converted to a unit discharge rate Q_u in gal/min/sq. ft of roof area using

$$Q_u = 0.0104\, i \tag{8–1}$$

The total discharge rate Q_n for a rooftop of area A in sq. ft is

$$Q_n = Q_u A \tag{8–2}$$

The capacity per drain Q_d equals the total flow Q_n divided by the number of drains, n. Q_n and Q_d have units of volume per unit time and thus correspond to a peak discharge. The diameter d (in.) of both the leader and the vertical drains is given by

$$d = 0.6\, Q_d^{0.377} \tag{8–3}$$

in which Q_d is the capacity per drain in gal/min. The diameter of horizontal pipes in in. is a function of both Q_d and the slope of the horizontal drain S in in./ft

$$d = 0.53 S^{-0.188} Q_d^{0.377} \tag{8–4}$$

Diameters estimated with Equations 8–3 and 8–4 should be rounded up to the next larger commercially available size so that the pipe capacity will be sufficient to drain the design flow Q_d. For roof drains, the following pipe diameters (in.) are usually available commercially: 2, 4, 6, 8, 10, 12, 15, 18. Equation 8–3 should be accurate for flow rates less than 6000 gal/min, and Equation 8–4 should be valid for slopes from 0.125 to 0.5 in./ft and diameters less than 18 in.

Example 8–1

The design method will be illustrated for a commercial building with a flat roof of 80,000 ft^2 (200 × 400 ft). Thus according to the rule of thumb, a total of eight drains will be required. The layout of the drains is shown in Figure 8–2. If the local drainage policy requires a 10-yr, 0.l-hr

8.0 NOTATION

A = area of cross section
A = drainage area (acres)
A = rooftop area
A = total area of inlet openings
A_o = area of orifice (ft^2)
C = runoff coefficient for Rational Method
C_d = orifice discharge coefficient
C_w = weir coefficient
C_* = conduit diameter coefficient
CN = runoff curve number
d = depth of flow
d = diameter of leader pipe
d_c = critical depth
D = conduit diameter (ft)
D = pipe diameter
E = elevation (ft)
E = specific energy
E_c = elevation of conduit centerline at the face of the riser (ft)
E_d = design efficiency
E_i = invert elevation of conduit at the face of the riser (ft)
E_o = elevation of orifice or low stage weir (ft)
E_1 = maximum water surface elevation in reservoir at routed low-stage hydrograph (ft)
E_2 = maximum water surface elevation at high-stage event (ft)
f = factor of safety
f_c = carryover reduction factor
$\mathbf{F}$ = Froude number
g = acceleration of gravity
h = head
h_i = depth of flow above the invert
h_i = hydrostatic pressure at section i
h_L = head loss
H_o = height of orifice opening
i = rainfall intensity
I_a = initial abstraction
K_p = head loss coefficient for circular pipe flowing full
L = flow length
L = length of curb opening
L = length of slotted drain
L = length of weir
L = watershed length
L_d = design length of drain
L_i = initial length of slotted drain
L_{w1} = low-stage weir length (ft)
L_{w2} = high-stage weir length (ft)
m = ratio of hydrograph recession duration to the time to peak
n = Manning's roughness coefficient
n = number of drains
n = number of openings in inlet
p_i = pressure head at section i
P = rainfall depth (or precipitation)
P_{24} = 24-hr rainfall depth (in.)
q = discharge rate
q_c = carryover discharge rate
q_d = design discharge
q_{o2} = flow through orifice or over low-stage weir during high-stage event (in.)
q_p = peak discharge
q_{pa} = post-development peak discharge
q_{pa1} = peak discharge (ft^3/sec), post-development, low-stage event
q_{pa2} = peak discharge (ft^3/sec), post-development, high-stage event
q_{pb} = pre-development peak discharge
q_{pb1} = peak discharge (ft^3/sec), pre-development, low-stage event (also, the allowable low-stage discharge from the basin post-development)
q_{pb2} = peak discharge (ft^3/sec), pre-development, low-stage event (also, the allowable high-stage discharge from the basin post-development)
q_u = unit discharge

Q = volume of direct runoff
Q_a = after development runoff depth (in.) (or V_r)
Q_{a1} = runoff depth, post-development, low-stage event (in.)
Q_{a2} = runoff depth, post-development, high-stage event (in.)
Q_b = runoff depth, pre-development (in.) (or V_r)
Q_{b1} = runoff depth, pre-development, low-stage event (in.)
Q_{b2} = runoff depth, pre-development, high-stage event (in.)
Q_d = capacity per roof drain
Q_n = total discharge rate
Q_u = unit discharge rate
R_h = hydraulic radius
R_{q1} = discharge ratio, low-stage event
R_{q2} = discharge ratio, high-stage event
R_s = ratio of the depth of detention storage to the depth of post-development runoff
R_{s1} = storage volume ratio, low-stage event
R_{s2} = storage volume ratio, high-stage event
S = gutter slope
S = longitudinal slope
S = slope of horizontal drain
S_z = cross slope of roadway
t = time
t_c = time of concentration
t_{ca} = post-development time of concentration
t_p = time to peak of hydrograph
T = spread of water surface
T_b = time base of hydrograph
V = flow velocity
V_c = critical velocity
V_d = volume of dead storage (acre-ft)
V_i = flow velocity at section i
V_o = volume discharged through orifice
V_r = post-development runoff depth (Q_a)
V_s = volume of active (flood) storage
V_{s1} = net (detention) storage for low-stage event (in.)
V_{s2} = net (detention) storage for high-stage event (in.)
V_{st} = volume of detention storage
V_t = total volume of storage (flood + dead)
V_{t1} = total storage, low-stage event (acre-ft)
V_{t2} = total storage, high-stage event (acre-ft)
W_o = width of orifice aperture (ft)
z = cross slope
Z = difference in elevation heads
z_i = elevation head at section i
α = ratio of pre- to post-development peak discharges
Δh = change in head
γ = ratio of the before- to after-development time to peaks
γ = specific weight of water

8.1 INTRODUCTION

Numerous engineering design problems require peak discharge estimates as input. Storm drain inlets, culverts, and small storm drain pipe systems are examples of peak discharge design problems in urban and suburban development. Peak discharge estimates are also required for some large watershed design problems, such as in setting the height of levees and cofferdams. In general, peak discharge estimates are inadequate for designs where either storage or the duration of flooding is important; however, in these cases peak discharge estimates are commonly required for making planning estimates of such hydraulic structures.

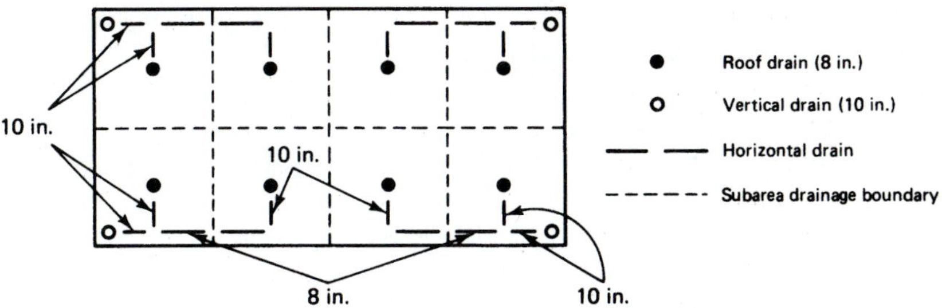

FIGURE 8–2 Roof drainage layout.

rainfall intensity, then the intensity, assuming that Figure 4–4 applies, would be 7.4 in./hr. Thus the total flow for the eight drains would be

$$Q_n = 0.0104\ (7.4)\ (80{,}000) = 6156.8 \text{ gal/min} \tag{8–5}$$

or 770 gal/min per drain. The diameters of the eight leaders are computed with Equation 8–3:

$$d = 0.6\ (770)^{0.377} = 7.35 \text{ in.} \tag{8–6}$$

which translates into an 8–in. leader. If the horizontal pipes leading from the roof drains to the wall are set at a slope of 0.125 in./ft, then Equation 8–4 yields a required diameter of

$$d = 0.53\ (0.125)^{-0.188}\ (770)^{0.377} = 9.6 \text{ in.} \tag{8–7}$$

Thus, a 10-in. diameter pipe would be used. If the horizontal pipe draining the four interior leaders in the interior walls are laid at a slope of 0.5 in./ft, these would require the following diameter:

$$d = 0.53\ (0.5)^{-0.188}\ (770)^{0.377} = 7.4 \text{ in.} \tag{8–8}$$

which translates into an 8–in. pipe. When the horizontal pipes from the interior leaders connect with the pipes from the exterior leaders, the total flow draining the 20,000 ft^2 will be 1540 gal/min. The pipe diameter for these sections, assuming a slope of 0.5 in./ft, is

$$d = 0.53\ (0.5)^{-0.188}\ (1540)^{0.377} = 9.6 \text{ in.} \tag{8–9}$$

which translates into a 10-in. pipe. Equation 8–3 would be used to compute the diameters of the four vertical drains:

$$d = 0.6\ (1540)^{0.377} = 9.6 \text{ in.} \tag{8–10}$$

which translates into a commercial pipe diameter of 10 in. The pipe sizes are also shown on Figure 8–2.

8.3 DESIGN OF SHALLOW CHANNELS

The concepts of Section 3.5.7 can be used to derive equations for computing the discharge in shallow channels such as roadway gutters, flow along curbed sections of parking areas, or shallow open channels. Figure 8–3 provides equations for computing the discharge rates from which flow velocities can be obtained with the continuity equation. The equation re-

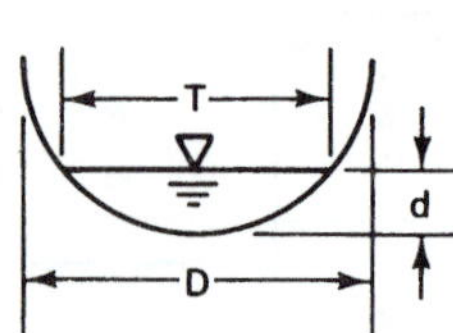

$$q = \frac{1.25}{n} d^{2.125} S^{0.5} D^{0.54}$$

$$\frac{d}{D} = \left(\frac{0.8qn}{D^{8/3} S^{1/2}}\right)^{0.47}$$

$$T = 2\left[\left(\frac{D}{2}\right)^2 - \left(\left(\frac{D}{2}\right) - d\right)^2\right]^{0.5}$$

(a) Circular cross section

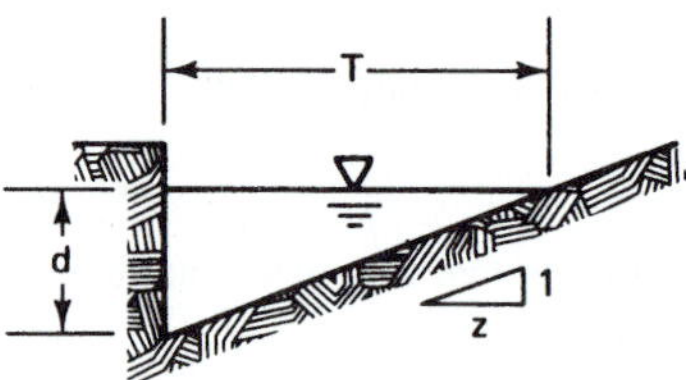

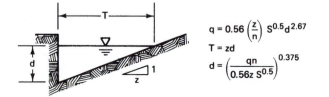

$$q = 0.56\left(\frac{z}{n}\right) S^{0.5} d^{2.67}$$

$$T = zd$$

$$d = \left(\frac{qn}{0.56z\, S^{0.5}}\right)^{0.375}$$

(b) Triangular cross section

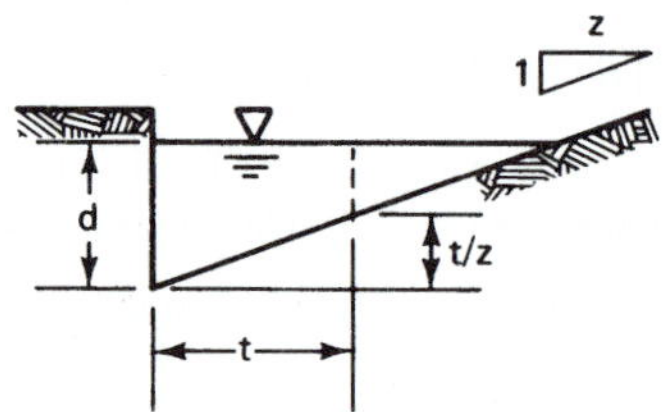

q_t = discharge in portion of channel having width t
= total discharge from part (b) minus discharge for triangular section with depth $d - (t/z)$

(c) Trapezoidal cross section

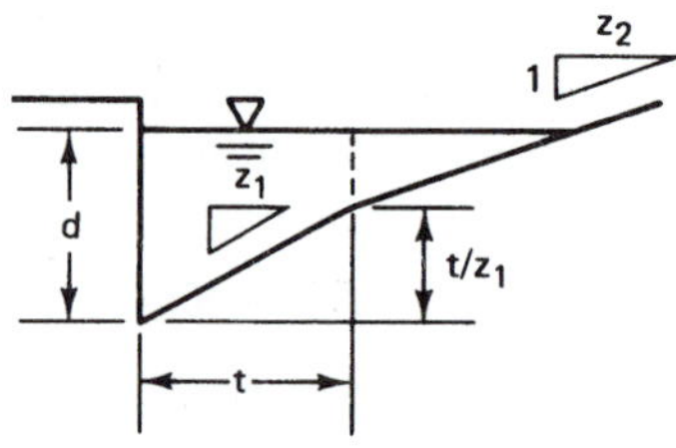

q_c = discharge from composite section
= discharge for section 1 from part (c) plus discharge for section 2 with depth $d - (t/z_1)$

(d) Composite cross section

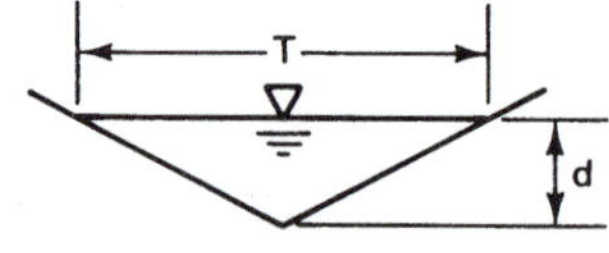

q_v = discharge for V-shaped section: use part (b) with $z = T/d$

(e) Shallow V-shaped section

FIGURE 8–3 Flow in shallow channels.

quires the roughness coefficient n, the gutter slope S, the cross slope z or section diameter D, and the depth of flow d.

The equations are useful for several aspects of design. If the discharge from a highway and adjacent land is computed with an equation such as the Rational Method, the depth of flow at the gutter and the spread of the water surface (T) over the highway surface can be estimated. In the design of a highway section, estimating the curb height necessary to contain the flow within the gutter is a part of the design. The allowable spread is set by policy and is an important design variable. The spread is a function of the type of highway, the traffic volume, and the importance of the highway in maintaining traffic patterns. For streets near hospitals or schools and major arteries, the policy may require the spread be limited to the adjacent right-of-way and not extend onto the roadway. For streets subject to low traffic volumes, a policy may allow one-half of the roadway to be covered with water. The equations of Figure 8–3 can be useful in solving such problems.

Example 8–2

To illustrate the use of the equations of Figure 8–3 consider a 5.8-acre watershed that drains a section of highway and an adjacent grass-covered area. Assuming that the weighted runoff coefficient for the Rational Method is 0.3 and the rainfall intensity is 5.5 in./hr, the Rational Formula yields a discharge of

$$q_p = CiA = 0.3\,(5.5)\,(5.8) = 9.6\ \text{ft}^3/\text{sec} \tag{8–11}$$

Assuming that the roadway gutter will have a longitudinal slope of 0.015 ft/ft, a cross slope of 0.5 in./ft ($z = 24$) and a roughness of 0.02, the equation from Figure 8–3 yields a maximum depth of flow for a triangular cross section of

$$d = \left[\frac{qn}{0.56zS^{0.5}}\right]^{0.375} = \left[\frac{9.6\,(0.02)}{0.56\,(24)\,(0.015)^{0.5}}\right]^{0.375} = 0.447\ \text{ft} \tag{8–12}$$

Thus the spread T equals dz, or 10.7 ft. Assuming that the policy prohibits ponded water on the roadway, a minimum shoulder with a width of 10.7 ft would be required.

8.4 WEIR AND ORIFICE EQUATIONS

Weirs and orifices are engineered devices that can be used to control and measure flow rates. While these devices can occur naturally, for the context of engineering design, the discussion will center on the equations used in the design of hydrologic/hydraulic facilities.

8.4.1 Orifice Equation

When an inlet is submerged, as a grate inlet would be during a heavy rainfall, it acts as an orifice, which is an opening to an area with a relatively large volume, such as a tank or manhole. Figure 8–4 shows a schematic of a tank with a hole of area A_2 in its bottom. If we assume all losses can be neglected, Bernoulli's equation can be written between a point on the surface of the pool (point 1) and a point in the cross section of the orifice (point 2):

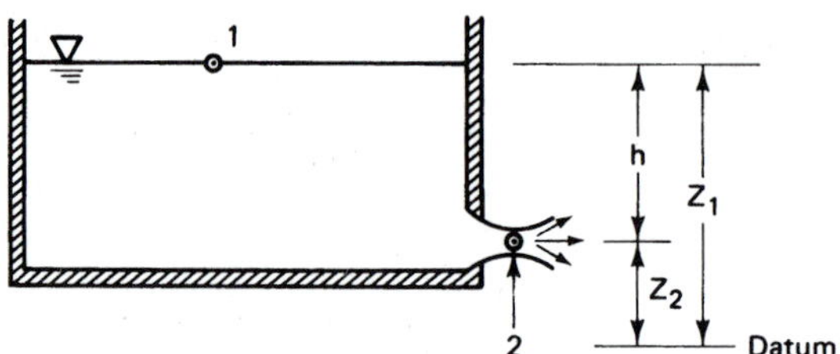

FIGURE 8–4 Schematic diagram of flow through an orifice.

$$\frac{p_1}{\gamma} + \frac{V_1^2}{2g} + z_1 = \frac{p_2}{\gamma} + \frac{V_2^2}{2g} + z_2 \tag{8–13}$$

This can be simplified by making the following assumptions: (1) The pressure at both points is atmospheric, therefore $p_1 = p_2$; (2) The surface area of the pool A_1 is very large relative to the area of the orifice A_2, so from the continuity equation, V_1 is essentially 0; and (3) $z_1 - z_2 = h$. Thus, Equation 8–13 becomes

$$h = V_2^2/2g \tag{8–14}$$

Solving for V_2 and substituting it into the continuity equation yields

$$q = AV = A\sqrt{2gh} \tag{8–15}$$

Equation 8–15 depends on two assumptions that are not always true: there are no losses and the pressure is atmospheric across the opening of the orifice; it is actually atmosphere at a point below the orifice, called the vena contracta, where the cross-sectional area is a minimum. Because of these violations, the discharge will be less than that given by Equation 8–15. The actual discharge through the orifice is estimated by applying a discharge coefficient C_d to Equation 8–15:

$$q = C_d A\sqrt{2gh} \tag{8–16}$$

in which C_d is called the discharge coefficient. Equation 8–16 is referred to as the orifice equation. For some design problems, the q of Equation 8–16 is multiplied by an efficiency factor f to reflect other types of losses that limit the discharge rate. Values of C_d range from 0.5 to 1.0, with a value of 0.6 often used. C_d is dimensionless.

Example 8–3

A forklift is accidentally driven into the side of an above-ground storage tank and it punctures an 11-in.^{2} hole 6 in. above the ground. In order to complete the environmental report, the engineer needs to estimate the amount of petroleum that leaked out of the tank during the 25 min between the time of the puncture and the time that the hole was plugged. The surface elevation of the petroleum above the ground was 8.46 ft after the leak was stopped. The tank has a diameter of 40 ft. The engineer estimates the discharge coefficient is 0.55, which is less than the 0.6 value because of the additional losses created by the jagged edges of the hole. Thus, the estimated discharge is

$$\begin{aligned} q &= C_d A\sqrt{2gh} = 0.55\left(\frac{11}{144}\right)[2\,(32.2)\,(8.46 - 0.5)]^{0.5} \\ &= 0.951 \text{ ft}^3/\text{sec} \end{aligned}$$

Thus, the volume (V_o) discharged was approximately

$$V_o = q\Delta t = 0.951\,\frac{\text{ft}^3}{\text{sec}} \times 25\text{ min} \times \frac{60\text{ sec}}{\text{min}} = 1427\text{ ft}^3$$

The volume used the final fluid height of 8.46 ft to estimate the discharge rate. At the time of puncture, the height and the discharge rate were higher. To evaluate the possible effect on the accuracy of the estimated volume, the change in height (Δh) from the time of puncture to the time of plugging is computed from geometry:

$$\Delta h = \frac{V_o}{A_o} = \frac{1427\text{ ft}^3}{\pi(40)^2/4} = 1.136\text{ ft}$$

If the initial height was 8.46 + 1.14 = 9.60 ft, then the initial discharge would have been

$$q = 0.55\left(\frac{11}{144}\right)[2\,(32.2)\,(9.60 - 0.5)]^{0.5} = 1.017\text{ ft}^3/\text{sec}$$

This discharge yields an estimate of the volume of

$$V_o = q\Delta t = 1.017\,\frac{\text{ft}^3}{\text{sec}} \times 25\text{ min} \times \frac{60\text{ sec}}{\text{min}} = 1526\text{ ft}^3$$

Thus, the actual volume discharged from the tank was less than 1526 ft^3 but more than 1427 ft^3.

Example 8–4

Because of corrosion in the sheet piling of a cofferdam, a 5-in.^{2} hole is opened, which allows water from the river that surrounds the cofferdam to enter the construction site. The river stage is 11.6 ft above the level of the hole. Assuming a discharge coefficient of 0.6, the discharge rate is

$$q = C_d A\sqrt{2gh} = 0.6\left(\frac{5}{144}\right)[2\,(32.2)\,(11.6)]^{0.5} = 0.57\text{ ft}^3/\text{sec}$$

8.4.2 Weir Equation

Consider the flow cross section shown in Figure 8–5. Point 1 is located at a point upstream of the obstruction at a distance where the obstruction cannot influence the flow characteristics at the obstruction. Point 2 is at the obstruction. The following analysis assumes (1) ideal flow, (2) frictionless flow, (3) critical flow conditions at the obstruction, and (4) the obstruction has a unit width perpendicular to the direction of flow. For the critical flow conditions, the following equations describe hydraulic conditions at the obstruction:

$$\mathbf{F} = 1 = \frac{V_c}{(gd_c)^{0.5}} \tag{8–17}$$

$$d_c = \left(\frac{q_u^2}{g}\right)^{1/3} \tag{8–18}$$

$$E = d_c + \frac{V_c^2}{2g} = \frac{3}{2}d_c \tag{8–19}$$

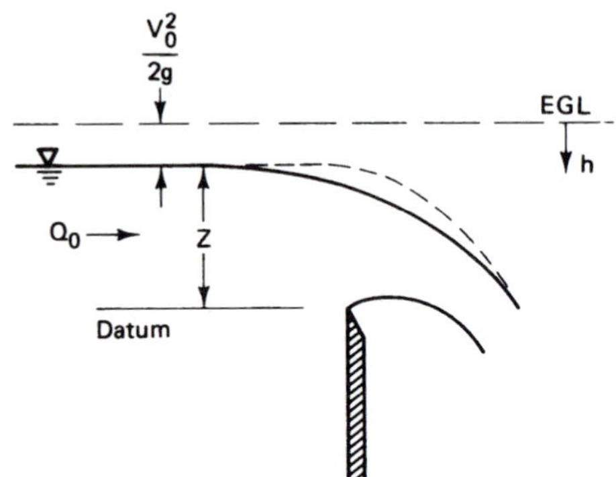

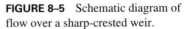

FIGURE 8–5 Schematic diagram of flow over a sharp-crested weir.

where **F** is the Froude number, V_c is the critical velocity, d_c is the critical depth, q_u is the discharge rate per unit width, and E is the specific energy. If hydrostatic pressure is assumed at sections 1 and 2, then $p_i/\gamma = h_i$. Thus, Bernoulli's equation is

$$h_1 + \frac{V_1^2}{2g} + z_1 = h_2 + \frac{V_2^2}{2g} + z_2 \tag{8–20}$$

Letting $Z = z_2 - z_1$ and assuming that the velocity head at section 1 is much smaller than the velocity head at section 2, Equation 8–20 reduces to

$$h_1 = \frac{V_c^2}{2g} + d_c + Z \tag{8–21}$$

Using Equations 8–17 and 8–19, the velocity head is $V_c^2/2g = d_c/2$. Letting $h = h_1 - z$, then $h = 0.5\,d_c + d_c = 1.5\,d_c$ or $d_c = 2h/3$. Solving Equation 8–18 for q_u, it then follows that

$$q_u = (gd_c^3)^{0.5} = \left[g\left(\frac{2}{3}h\right)^3\right]^{0.5} = \left(\frac{8g}{27}\right)^{0.5} h^{3/2} \tag{8–22}$$

For q_u[=] ft^3/sec, h [=] ft, and g [=] ft/sec^2, Equation 8–22 yields

$$q_u = 3.088\,h^{1.5} \tag{8–23}$$

Letting $q = q_u L$ and replacing the constant 3.088 with the weir coefficient C_w yields the general weir equation

$$q = C_w\,Lh^{1.5} \tag{8–24}$$

in which L is the length of the weir. Values of C_w can range from 2.3 to 3.3, depending on the losses that occur at the weir, but values from 2.6 to 3.1 are common for English units. An efficiency factor f can also be applied to reduce the discharge q for other losses.

Example 8–5

An engineering research unit is installing weirs on small experimental watersheds. Prior to installation they conduct laboratory tests of the weirs to determine the weir coefficients. For one test, a volume of 273 ft^3 passes over the rectangular weir in 30 sec. This yields a discharge of

273 ft^3/30 sec = 9.1 ft^3/sec. The water surface is 0.5 ft above the 8-ft weir. Thus, Equation 8–24 can be used to estimate the weir coefficient:

$$C_w = \frac{q}{Lh^{1.5}} = \frac{9.1}{8\,(0.5)^{1.5}} = 3.22$$

Example 8–6

In making a water balance of a wetland constructed adjacent to a development project, the engineer needs to estimate the discharge from the wetland during the design storm. The outlet from the wetland is a 400-ft grassed berm. The engineer estimates that during the design storm the water would be 0.6 ft above the top of the berm. The loss coefficient for the berm is estimated to be 2.9. Thus, the expected discharge is

$$q = C_w\,L\,h^{1.5} = 2.9\,(400)\,(0.6)^{1.5} = 64.1\ \text{ft}^3/\text{sec}$$

8.5 DESIGN OF PAVEMENT DRAINAGE INLETS

The design of drainage inlets is another example where peak discharge estimates are required. A complete discussion of their design is beyond the scope of the presentation in this section because of the large number of both the types of inlets and the factors that affect their sizing. Inlets for drainage from streets and highways, parking areas, and airport facilities can be divided into three major groups: curb-opening, gutter, and combination inlets; these are shown in Figure 8–6. Slotted drain inlets are an alternative for some drainage problems (see Section 8.6).

Factors that are considered in the design of inlets include the following:

1. The physical characteristics of the grate (length, width, and for inlets with curb openings, the height of the opening
2. The type of grate (parallel bar, curved vane, tilt bar, and reticuline), which are shown in Figure 8–7
3. The physical characteristics of the surface generating the flow to the grate (longitudinal and cross-section slopes, length, roughness)
4. The allowable spread, which is the proportion of a roadway cross section or traffic lane that the drainage policy allows to be covered by water
5. The return period of the design, which is also set by policy

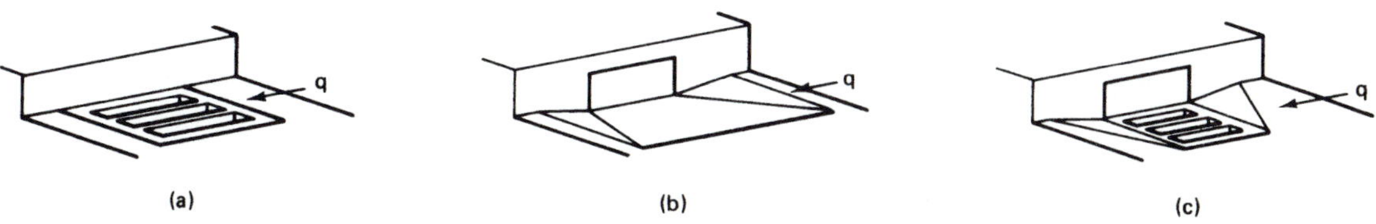

FIGURE 8–6 Views of pavement drainage inlets: (a) grate inlet; (b) curb-opening inlet; and (c) combination inlet.

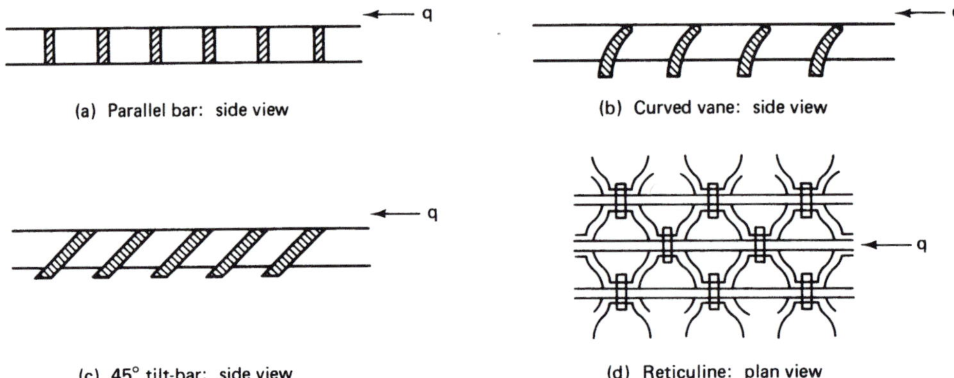

FIGURE 8–7 Grate configurations.

6. The required efficiency of the grate, which is the proportion of the total flow intercepted by the inlet for a specific set of conditions (the portion of the flow that is not intercepted is termed the carryover flow)

When using the weir equation for inlet evaluation, the perimeter of the grate inlet or the length of the curb opening can be used for L in Equation 8–24. When using Equation 8–16 for inlet evaluation the open area of the grate can be used as A; for a submerged curb-opening inlet, the area of the opening can be used as A, as long as the flow through the inlet facility does not have a control section elsewhere that places a greater limitation on the intercepted flow rate. Equations 8–16 and 8–24 represent upper limits on the flow that can be intercepted by an inlet. Safety factors are used to reduce the flow computed with these equations to reflect more accurately the amount of flow intercepted. The factors of safety and efficiency typically reduce the intercepted flow rate from 50 to 80% of the rates computed with Equations 8–16 and 8–24.

Example 8–7

Consider the case of a curb inlet that has a length of 14 ft. The relationship between the flow rate and depth is given by

$$q = fC_w\, Lh^{1.5} = 3.1(14)\, fh^{1.5} = 43.4\, fh^{1.5} \tag{8–25}$$

in which f is a factor of safety. The relationship of Equation 8–25 is shown in Figure 8–8 for factors of safety of 0.5, 0.67, and 0.8, as well as the theoretical discharge (that is, $f = 1$). For example, the capacity for the inlet, assuming that Equation 8–24 applies with no factor of safety, would be 7.1 ft^3/sec for a depth of 0.3 ft. If a factor of safety of 0.8 applied, the actual inlet flow rate would be 5.6 ft^3/sec. Only 3.6 ft^3/sec would enter if the factor of safety was 0.5.

Example 8–8

In Example 7–11, the initial subarea generated a peak discharge of 5.7 ft^3/sec. Assuming that this is divided equally into two inlets, Equation 8–24, with a factor of safety applied, can be used to size the inlets. Both inlets will be sized to handle a discharge of 2.85 ft^3/sec and limit the spread

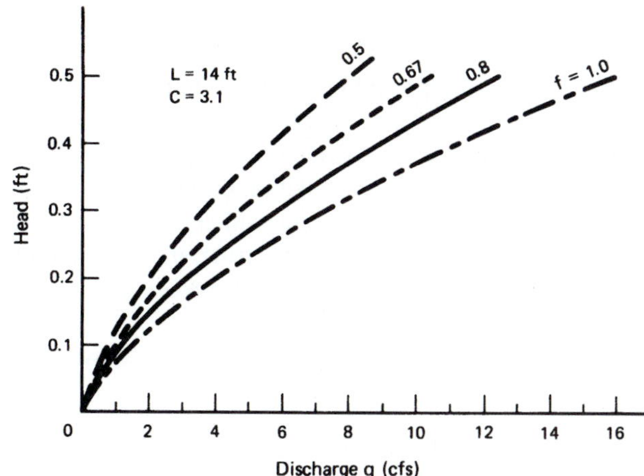

FIGURE 8–8 Depth-discharge relationships for grate inlet and selected factors of safety (f).

(T) to 5 ft. Assuming a cross slope of 0.5 in./ft, the relationship $T = zd$, where z is the reciprocal of the cross slope, can be used to solve for the allowable depth: $d = T/z = 5/24 = 0.208$ ft. Assuming a loss coefficient C_w of 3.1 and a factor of safety of 0.75, the required length of the inlet is

$$L = \frac{q}{fC_w d^{1.5}} = \frac{2.85}{0.75\,(3.1)\,(0.208)^{1.5}} = 12.9 \text{ ft} \tag{8–26}$$

Example 8–9

An on-grade grate inlet (see Figure 8–6a) has a grate with 18 openings, each with a length of 2.3 ft and a width of 0.12 ft. The design depth of submergence is 6 in., and a discharge coefficient of 0.6 is assumed. If the design is intended to intercept 75% of the flow, the discharge that passes into the grate is

$$\begin{aligned} q &= fC_d A \sqrt{2gh} \\ &= 0.75\,(0.6)\,[18\,(2.3)\,(0.12)]\,[2\,(32.2)\,(6/12)]^{0.5} \\ &= 12.7 \text{ ft}^3/\text{sec} \end{aligned}$$

Example 8–10

The Rational Formula was used to compute a design discharge of 4.3 ft^3/sec for an inlet. If a grate inlet is to be used with a design depth of 4 in., the grate opening area is

$$A = \frac{q}{C_d \sqrt{2gh}} = \frac{4.3}{0.6[2\,(32.2)\,(4/12)]^{0.5}} = 1.55 \text{ ft}^2$$

If each opening is 1.8 ft in length and 0.12 ft wide, the grate will need n openings:

$$n = \frac{1.55}{1.8\,(0.12)} = 7.2$$

Thus, eight openings will be needed to accommodate the design discharge.

8.6 SLOTTED DRAIN INLET DESIGN

In addition to the inlet configurations discussed in the preceding section, drainage from impervious surfaces such as highways, parking lots, and large sports complexes can be controlled with slotted drains. Figure 8–9 shows a schematic of a slotted drain inlet designed for a highway section. Two design situations will be discussed, installation on grade and installation in sag. Data required to compute the length of the slotted drain, which is one of the design variables, are the longitudinal slope S (ft/ft), the cross slope S_z (ft/ft), the information necessary to compute discharge rates, the design return period, the surface roughness n, and the allowable carryover efficiency for installation on grade.

The procedure for designing a slotted drain to be installed on grade is as follows:

1. Compute the discharge rate for the section extending to the previous drain using a method such as the Rational Formula or with one of the equations given in Figure 8–3.
2. Compute the design discharge q_d (ft^3/sec) as the sum of the discharge computed in Step 1 and any carryover flow q_c not intercepted by an up-gradient inlet.
3. Compute the initial length L_i (ft) by

$$L_i = 0.1233 q_d^{0.427} S^{0.305} z^{0.766} n^{-0.87} \tag{8–27}$$

in which S is the longitudinal slope (ft/ft), z is the reciprocal of the cross slope (ft/ft), and n is the roughness coefficient; use of Equation 8–27 beyond the following limits represents extrapolation:

$$0.8 \leq q_d \leq 6$$
$$10 \leq L_i \leq 45$$
$$0.001 \leq S \leq 0.09$$
$$16 \leq z \leq 48$$

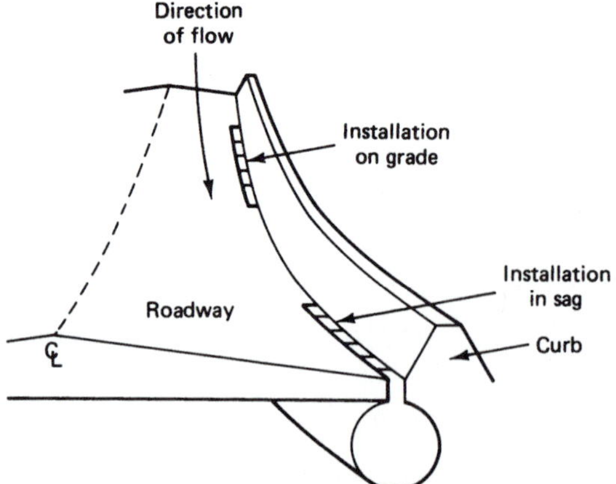

FIGURE 8–9 Schematic diagram of slotted drain installation.

4. Using the design efficiency E_d, which is the fraction of the flow that the policy requires to be intercepted, estimate the carryover reduction factor f_c by

$$f_c = \begin{cases} 20.7 - 44.9E_d + 25.2E_d^2 & \text{for } 0.9 \leq E_d \leq 1.0 \quad (8\text{–}28a) \\ 0.02 + 0.22E_d + 0.6E_d^2 & \text{for } 0.6 \leq E_d \leq 0.9 \quad (8\text{–}28b) \end{cases}$$

5. Compute the design length L_d by

$$L_d = f_c L_i \tag{8–29}$$

6. Compute the depth flow and required curb height using the equations of Figure 8–3.

For slotted drain inlets installed in sag locations, carryover and ponding are not acceptable for the design return period. For a flow depth of 0.2 ft or less, the inlet operates according to the weir equation. For depths greater than 0.4 ft, the inlet operates as an orifice. In the transition stage, between 0.2 and 0.4 ft, it is reasonable to assume that the inlet operates as an orifice. For orificial flow the discharge rate into the slotted drain is

$$q = C_d LW(2gd)^{0.5} \tag{8–30}$$

in which q is the discharge rate (ft^3/sec), C_d is the loss coefficient, L is the length of the slotted drain (ft), W is the width of the slotted drain (ft), and d is the depth of the inflow (ft) at the design discharge. The value of the loss coefficient C_d depends on the depth of the water and the length and width of the slot; typical values range from 0.6 to 1.0. A value of 0.8 will be used herein. The required length of the slotted drain inlet can be determined by transforming Equation 8–30:

$$L = \frac{q}{C_d W(2\,gd)^{0.5}} \tag{8–31}$$

A width of 1.75 in. is common, and for C_d equal to 0.8, Equation 8–31 reduces to

$$L = 1.07\, q\, d^{-0.5} \tag{8–32}$$

Equation 8–32 is valid for $C_d = 0.8$; however, for $C_d = 0.6$, which is also commonly used, the coefficient of Equation 8–32 would be 1.4 rather than 1.07. For computing the required length of the slotted drain installed in sag, follow these steps:

1. Compute the discharge, including any carryover from up-gradient inlets, that the inlet is required to control.
2. Determine the depth of flow for this discharge using the formulas of Figure 8–3.
3. Compute the required length using either Equation 8–31 or 8–32.

Example 8–11

A section of highway is shown in Figure 8–10 for which drainage will be provided using slotted drain. Discharge rates will be computed using the Rational Method with $C = 0.8$ and $i = 9$ in./hr. For a 300-ft length of roadway and a width for the roadway and shoulder of 28 ft, the drainage area is 0.193 acre, which yields a discharge of 1.39 ft^3/sec. Equation 8–27 yields an initial length of

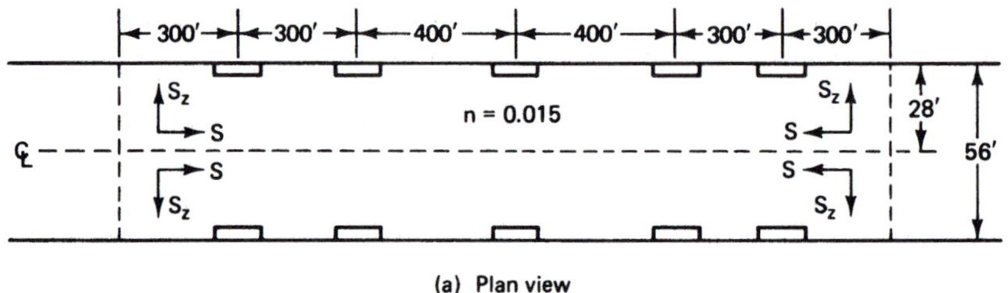

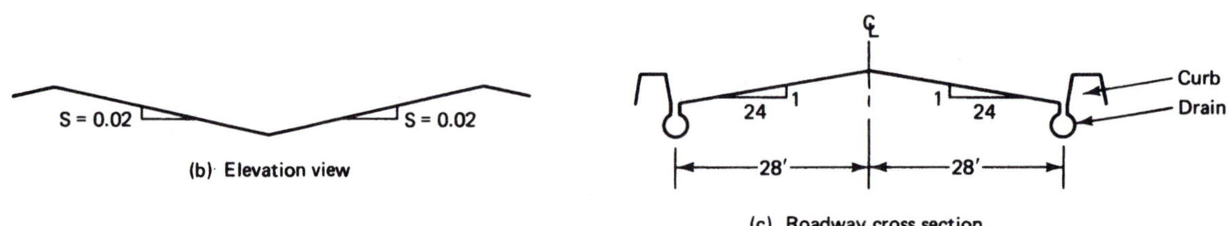

FIGURE 8–10 Example: Highway section with slotted drain inlets.

$$L_i = 0.1233\,(1.39)^{0.427}\,(0.02)^{0.305}\,(24)^{0.766}\,(0.015)^{-0.87} = 19.0 \text{ ft} \tag{8–33}$$

If a design efficiency of 80% is permitted, the carryover reduction factor can be computed using Equation 8–28b:

$$f_c = 0.02 + 0.22\,(0.8) + 0.6\,(0.8)^2 = 0.58 \tag{8–34}$$

Thus, the design length (Equation 8–29) is

$$L_d = f_c L_i = 0.58\,(19) = 11.0 \text{ ft} \tag{8–35}$$

Assuming a triangular-shaped gutter, the depth of flow and the spread can be determined from Figure 8–3:

$$d = \left[\frac{1.39\,(0.015)}{0.56\,(24)\,(0.02)^{0.5}}\right]^{0.375} = 0.184 \text{ ft} \tag{8–36}$$

$$T = zd = 24\,(0.184) = 4.42 \text{ ft} \tag{8–37}$$

The runoff will have a depth of slightly more than 2 in. and the spread will cover less than one-half of the width of the shoulder, which is 10 ft wide. If the inlet intercepts 80% of the flow, the carryover discharge is 0.28 ft^3/sec.

The second inlet must drain 300 ft of roadway and shoulder as well as the carryover flow of 0.28 ft^3/sec. Thus, the total discharge is 1.67 ft^3/sec. Using Equation 8–27 the initial drain length is

$$L_i = 0.1233\,(1.67)^{0.427}\,(0.02)^{0.305}\,(24)^{0.766}\,(0.015)^{-0.87} = 20.5 \text{ ft} \tag{8–38}$$

Equation 8–28b can be used to compute the carryover reduction factor, which is 0.56 for E_d = 80%. The design length is 11.5 ft. The depth of flow (d) and spread (T) are

$$d = \left[\frac{1.67\,(0.015)}{0.56\,(24)\,(0.02)^{0.5}}\right]^{0.375} = 0.20 \text{ ft} \tag{8–39}$$

$$T = zd = 24\,(0.20) = 4.8 \text{ ft} \tag{8–40}$$

If the inlet intercepts 80% of the flow, the carryover discharge is 0.33 ft^3/sec.

For the remaining 400-ft length of highway, the Rational Method gives a discharge rate of 1.85 ft^3/sec. Adding a carryover flow of 0.33 ft^3/sec yields a flow of 2.18 ft^3/sec. Since the highway has two sections entering the sag, each with a flow of 2.18 ft^3/sec, the drain at the sag must intercept 4.36 ft^3/sec. Using Figure 8–3, the depth of flow is

$$d = \left[\frac{4.36\,(0.015)}{0.56\,(24)\,(0.02)^{0.5}}\right]^{0.375} = 0.283 \text{ ft} \tag{8–41}$$

The required length of the slotted drain in the sag can be estimated using Equation 8–32:

$$L = 1.07\,(4.36)\,(0.283)^{-0.5} = 8.7 \text{ ft} \tag{8–42}$$

Given that both lanes of the highway cross section are the same, the same lengths of slotted inlets will be used on each side of the highway.

8.7 CULVERT DESIGN

Culverts are a common hydraulic structure. As such, there are highway design manuals that are devoted entirely to the wide range of possible culverts. A few basic concepts will be presented here, with the details of design, as well as graphical design aids, left to the design manuals.

A culvert is a pipe or box that is located under a roadway, embankment, or service area to allow the passage of storm runoff. While culverts are often circular pipes made of either concrete or corrugated metal, other shapes including elliptical and arch pipes and rectangular box culverts are widely used. A culvert consists of an inlet structure, the pipe or box, and an outlet structure that includes a device to dissipate the energy of the water before it is discharged downstream. In some cases, the inlet structure may include a trash rack to prevent debris from entering the pipe barrel and/or an anti-vortex device to ensure maximum hydraulic efficiency.

In addition to the shape, culverts are also classified according to the flow conditions at the inlet and outlet of the structure. Design procedures for four categories (see Figure 8–11) are given :

1. Unsubmerged inlet and outlet
2. Submerged inlet, unsubmerged outlet, partially full pipe
3. Submerged inlet, unsubmerged outlet, full-pipe flow
4. Submerged inlet and outlet

Procedures for design of three of the four are given here.

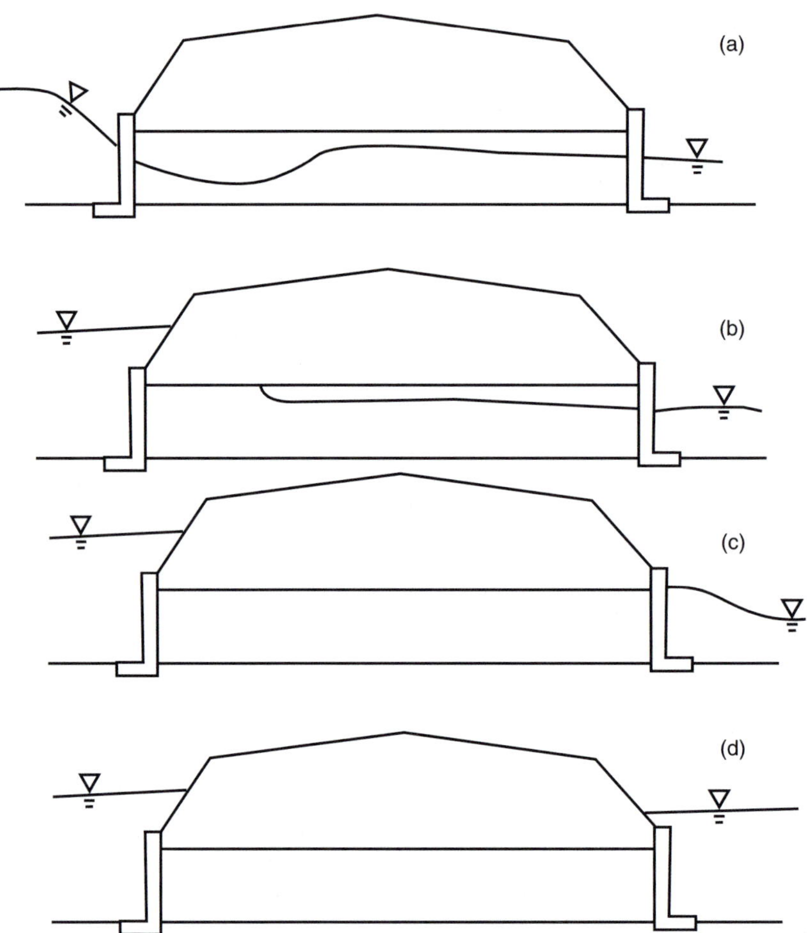

FIGURE 8–11 Classification of culverts: (a) unsubmerged inlet and outlet; (b) submerged inlet, unsubmerged outlet, partially full pipe; (c) submerged inlet, unsubmerged outlet, full-pipe flow; (d) submerged inlet and outlet.

8.7.1 Unsubmerged Inlet and Outlet

In this case, the runoff has a free surface, and a solution can be obtained using a hydraulic routing method. Because of its simplicity, Manning's Equation is commonly used for estimating the diameter of a circular pipe culvert:

$$q_p = \frac{1.49}{n} A R_h^{2/3} S^{1/2} = \frac{1.49}{n} \left(\frac{\pi D^2}{4}\right) \left(\frac{D}{4}\right)^{2/3} S^{1/2} \tag{8–43}$$

Solving for the diameter in feet yields

$$D = 1.333(q_p n)^{3/8} S^{-3/16} \tag{8–44}$$

Example 8–12

In Example 7–11 the discharges for the pipe between inlets 1 and 2 was 5.7 ft^3/sec. Assuming that the pipe will have a roughness coefficient of 0.014, will be laid at a slope of 0.36% (0.0036 ft/ft), and will flow almost full, Manning's Equation can be solved for the required diameter:

$$D = \left[\frac{2.15\,qn}{S^{0.5}}\right]^{0.375} = \left[\frac{2.15\,(5.7)\,(0.014)}{(0.0036)^{0.5}}\right]^{0.375} = 1.48 \text{ ft} \tag{8–45}$$

Therefore, an 18–in. pipe will be needed for the section of pipe from the inlet of subarea 1. The velocity can be computed using the continuity equation:

$$V = \frac{q}{A} = \frac{5.7 \text{ ft}^3/\text{sec}}{\pi(1.5)^2/4 \text{ ft}^2} = 3.2 \text{ ft/sec}$$

Flow in the pipe between subareas 2 and 3 must carry 20.9 ft^3/sec at a slope of 0.1% (0.001 ft/ft). For a roughness of 0.014, the required diameter is

$$D = \left[\frac{2.15\,(20.9)\,(0.014)}{(0.001)^{0.5}}\right]^{0.375} = 3.07 \text{ ft} = 39 \text{ in.} \tag{8–46}$$

The velocity is

$$V = \frac{q}{A} = \frac{20.9}{\pi(3.25)^2/4} = 2.52 \text{ ft/sec}$$

Example 8–13

Equation 8–44 can be applied to the layout of Figure 7–2 to determine the diameters for the pipes that make up the pipe network. The discharges were computed in Table 7–14. Assuming a pipe roughness of 0.012 and the slopes given in Table 7–14, the pipe diameters were computed with Equation 8–44 and are given in Table 8–1. A 30-in. pipe is required to pass the water under Main Street, which is the watershed outlet.

8.7.2 Submerged Inlet, Unsubmerged Outlet, Partially Full Pipe

In this case, the system can be treated as an orifice. Thus, the discharge is under entrance or inlet control and can be computed by

$$q = C_d A \sqrt{2gh} \tag{8–47}$$

where h is the hydrostatic head above the center of the pipe at the inlet. The depth of the ponded water above the invert of the inlet (h_i) is

$$h + \frac{D}{2} = h_i \tag{8–48}$$

Substituting Equation 8–48 into Equation 8–47 yields

$$q = C_d\left(\frac{\pi D^2}{4}\right)\sqrt{2g(h_i - 0.5D)} \tag{8–49}$$

TABLE 8–1 Determination of Pipe Diameters for Layout of Figure 7–2

Manhole at				Computed Pipe	Pipe Diameter
Upper End	Lower End	Discharge (cfs)	Slope (ft/ft)	Diameter (ft)	Installed (in.)
M_{54}	M_{44}	2.8	.024	0.75	10
M_{44}	M_{34}	6.9	.020	1.09	15
M_{34}	M_{24}	10.0	.028	1.18	15
M_{24}	M_{14}	13.6	.018	1.43	18
M_{14}	M_{13}	17.5	.004	2.09	27
M_{53}	M_{43}	5.8	.024	0.99	12
M_{43}	M_{33}	9.3	.020	1.22	15
M_{33}	M_{23}	11.8	.028	1.25	15
M_{23}	M_{13}	14.9	.018	1.48	18
M_{13}	M_{12}	34.8	.004	2.71	33
M_{52}	M_{42}	5.8	.024	0.99	12
M_{42}	M_{32}	9.3	.020	1.22	15
M_{32}	M_{22}	11.8	.028	1.25	15
M_{22}	M_{12}	14.9	.018	1.48	18
M_{12}	M_{11}	49.9	.004	3.10	39
M_{51}	M_{41}	2.8	.024	0.75	10
M_{41}	M_{31}	4.5	.020	0.93	12
M_{31}	M_{21}	5.7	.028	0.95	12
M_{21}	M_{11}	7.2	.018	1.13	15
M_{11}	outfall	56.5	.020	2.40	30

Solving for D yields the following polynomial:

$$D^5 - 2h_i D^4 + \frac{16\, q^2}{C_d^2\, g\, \pi^2} = 0 \tag{8–50}$$

A solution can be found numerically. Since Equation 8–50 is a fifth-order polynomial, it will have five roots. The appropriate diameter will need to be selected from the five roots.

Example 8–14

A 55 ft culvert will pass under a roadway. At flood stage, the water ponded behind the roadway will be 6 ft above the invert. The design discharge is 40 ft^3/sec, and a discharge coefficient of 0.6 is assumed. Thus, Equation 8–50 becomes

$$D^5 - 12\, D^4 + 223.76 = 0 \tag{8–51}$$

The fifth-order polynomial is easily solved using a numerical scheme such as Newton's method. There are two real, positive roots: 2.185 and 11.989 ft. Since the latter is not realistic, a diameter of 2.185 ft is used. A pipe smaller than this will restrict the flow, thus causing a head greater than the design value of 6 ft. Thus, the diameter is set as the next larger commercially available size, which would be 27 in. This will allow a slightly greater discharge than 40 ft^3/sec. The actual discharge of 42.3 cfs can be computed with Equation 8–49.

8.7.3 Submerged Inlet and Outlet

The third case is treated as a special case of the fourth case, specifically where the depth of flow at the outlet equals the pipe diameter. Thus, the development of the case for a submerged inlet and outlet will be presented briefly.

The energy equation between points on the headwater and tailwater surfaces is

$$\frac{p_1}{\gamma} + \frac{V_1^2}{2g} + z_1 - h_L = \frac{p_2}{\gamma} + \frac{V_2^2}{2g} + z_2 \tag{8–52}$$

in which p_i is the pressure at point i, V_i is the flow velocity at point i, z_i is the water surface elevation at point i, and h_L is the total head loss. If the surface areas of the headwater and tailwater ponds are large, then the velocities V_1 and V_2 can be considered negligible. The pressures are assumed to be hydrostatic: $h_i = p_i/\gamma$ where h_i is the elevation above the pipe invert. The invert elevation at the downstream end of the pipe is assumed to be the datum, thus $z_2 = 0$. For a pipe of length L laid at a slope of S_o, the invert at the inlet will be $z_1 = S_oL$. Thus, Equation 8–52 becomes

$$h_1 + 0 + S_oL - h_L = h_2 + 0 + 0 \tag{8–53}$$

which transforms to the following expression for h_L:

$$h_L = h_1 - h_2 + S_oL \tag{8–54}$$

The head losses include entrance and exit losses and friction losses along the interior of the pipe barrel. The entrance and exit losses are computed as minor losses, which are a function of the velocity head $V^2/2g$ where V is the velocity of the flow in the pipe. Friction losses are computed assuming that the friction slope can be approximated using Manning's Equation. Thus, the head losses are

$$h_L = K_{\text{ent}}\frac{V^2}{2g} + \left(\frac{n^2\,V^2}{2.22\,R_h^{4/3}}\right) L + K_{\text{exit}}\frac{V^2}{2g} \tag{8–55}$$

in which V is the velocity of the flow through the barrel and K_{ent} and K_{exit} are the loss coefficients at the entrance and exit, respectively. Using the continuity equation, $Q = VA$, and assuming a circular pipe with $A = \pi D^2/4$ and $R_h = D/4$, Equation 8–55 becomes

$$h_L = (K_{\text{ent}} + K_{\text{exit}})\left(\frac{0.0252\,Q^2}{D^4}\right) + \frac{4.637\,n^2\,Q^2\,L}{D^{16/3}} \tag{8–56}$$

Equating Equations 8–54 and 8–56 yields

$$h_1 - h_2 + S_oL = (K_{\text{ent}} + K_{\text{exit}})\left(\frac{0.0252\,Q^2}{D^4}\right) + \frac{4.637\,n^2\,Q^2\,L}{D^{16/3}} \tag{8–57}$$

Values of 0.5 and 1.0 are typically used for K_{ent} and K_{exit}, respectively. For given values of h_1, h_2, S_o, L, Q, and n, a value of D can be found numerically.

Example 8–15

A new highway section is being designed. At design conditions, ponding cannot exceed 8 ft above the pipe invert at the inlet. At the outlet, a maximum ponding depth of 5 ft is permitted at the inlet. The pipe ($n = 0.013$) will have a length of about 110 ft and will be laid at a slope of 0.02 ft/ft. The design discharge is 82 ft^3/sec. Thus, Equation 8–57 is

$$8 - 5 + 0.02\,(110) = (0.5 + 1.0)\left(\frac{0.0252\,(82)^2}{D^4}\right) + \frac{4.637\,(0.013)^2\,(82)^2\,(110)}{D^{16/3}}$$

$$5.2 = \frac{254.17}{D^4} + \frac{579.62}{D^{16/3}}$$

Using the half-interval method for finding the root of the equation yields a diameter of 2.945 ft. Thus, a 3-ft pipe will be needed.

8.8 ESTIMATING DETENTION BASIN VOLUMES

It is widely recognized that land development, especially in urban areas, is responsible for significant changes in runoff characteristics. Within the context of the hydrologic cycle of Chapter 1, especially Figures 1–1 and 1–2, land development decreases the natural storage of a watershed. The removal of trees and vegetation reduces the volume of interception storage. Grading of the site reduces the volume of depression storage and often decreases the permeability of the surface soil layer, which reduces infiltration rates and the potential for storage of rainfall in the soil matrix. In urban areas, increased impervious cover also reduces the potential for infiltration and soil storage of rainwater. This reduction of natural storage (that is, interception, depression, and soil storage) causes changes in runoff characteristics. Specifically, both the total volume and the peak of the surface (or direct) storm runoff increase. The loss of natural storage also causes changes in the timing of runoff, specifically a decrease in both the time to peak and the time of concentration. Runoff velocities are increased, which can increase surface rill and gully erosion rates (see Chapter 15). Higher stream velocities may also increase rates of bed-load movement. Land development is often accompanied by changes to drainage patterns and channel characteristics. For example, channels may be cleared of vegetation and straightened, with some also being lined with concrete or riprap. Modifications to the channel may result in decreases in channel storage and roughness, both of which can increase flow velocities and the potential for flooding at locations downstream from the developing area.

Recognizing the potential effects of these changes in runoff characteristics on the inhabitants of the local community, various measures have been proposed to offset these reductions in natural storage. The intent of storm water management (SWM) is to mitigate the hydrologic impacts of this lost natural storage, usually using manmade storage. Although a variety of SWM alternatives have been proposed, the storm water management basin remains the most popular. The SWM basin is frequently referred to as a detention or retention basin, depending on its effects on the inflow hydrograph. For our purpose, the terms will be used interchangeably because the fundamental concepts for estimating volumes to control discharge rates are the same.

8.8.1 SWM Policy Considerations

To mitigate the detrimental effects of land development, SWM policies have been adopted with the intent of limiting peak flow rates from developed areas to that which occurred prior to development. In addition to specifying the conditions under which SWM methods must be used, these policies indicate the intent of SWM. Specifically, the intent of many SWM policies is to limit runoff characteristics after development to those that existed prior to development. This intent can be interpreted to mean that the flood frequency curve for the after-development conditions coincides with the curve for the before-development conditions. However, policy statements usually specify one or two exceedence frequencies (that is, return periods) at which the after-development peak rate must not exceed the before-development peak rate for the same exceedence frequency. Such policies often use return periods of 2, 10, or 100 years as the target points on the frequency curve. Where channel erosion is of primary concern, a smaller return period such as the 6-month event may serve as the target event. Policies should also specify a specific design method to be used in the design of a SWM control method. Although data do not exist to show that any one method is best, the specification of a specific method as part of a SWM policy will ensure design consistency.

8.8.2 Elements of SWM Structures

Figure 8–12 shows a schematic of the cross section of a detention basin with a single-stage riser. A pool is formed behind the detention structure. The flood runoff enters the pool at the upper end of the detention basin. Water can be discharged from the pool through a pipe that passes through or around the detention structure. The invert elevation of the riser can serve to limit the outflow rate, thus forming a permanent pool, with the permanent pool elevation changing only through evaporation and infiltration losses. The use of a permanent pool has a number of advantages, including water quality control, aesthetic considerations, and wildlife habitat improvement. Of course, a permanent pool also increases the total storage volume, which requires both a larger retaining structure and a larger commitment of land, both of which increase the cost of the project. Storage below the riser invert elevation is sometimes referred to as dead storage. Storage allocated for flood runoff is referred to as active storage.

Figure 8–12 does not show several other elements of detention basin design. The riser inlet should be fitted with both an antivortex device and a trash rack. The anti-vortex device prevents the formation of a vortex, thus maintaining the hydraulic efficiency of the outlet structure. The trash rack prevents trash (and people) from being sucked into the riser by high-velocity flows. At least one antiseep collar is fitted to the outside of the discharge pipe to prevent erosion about the pipe within the retaining structure. All detention basins should have an emergency spillway to pass runoff from very large flood events, so the detention structure is not overtopped and washed out. The elevation of the bottom of the emergency spillway, which will pass high flows around the retaining structure, is above the elevation of the riser outlet but below the top of the retaining structure.

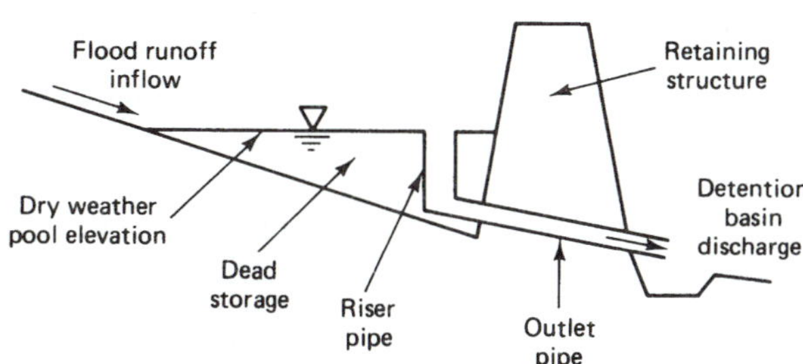

FIGURE 8–12 Schematic cross section of a detention basin with a single-stage riser.

8.8.3 Analysis Versus Synthesis

The problem of analysis versus synthesis is best evaluated in terms of systems theory (see Figure 1–3). The problem is viewed in terms of the input (inflow runoff hydrograph), output (outflow hydrograph), and the transfer function (stage storage-discharge relationship). In the analysis phase, the two hydrographs would be measured for an existing storm water management facility, and it would be necessary to calibrate the stage-storage-discharge relationship. While the stage-storage relationship is determined from topography, the stage-discharge relationship would have to be analyzed (calibrated). For a given storage facility, the physical characteristics of the outlet facility would be known. Therefore, the analysis would involve determining the best values of the weir and/or orifice coefficients for the outlet. Given the cost involved in data collection, analyses are rarely undertaken; therefore, only the synthesis case will be discussed in this chapter.

In the synthesis case, the objective is to make estimates of either the outflow hydrograph or the necessary characteristics of the proposed riser. For watershed studies where detention basins exist, it may be necessary to synthesize flood hydrographs for the detention basin outflow; in this case, the outflow hydrograph is estimated from a design storm. The standard procedure is to assume a design storm and a unit hydrograph and convolve the rainfall excess with the unit hydrograph. The resulting runoff hydrograph is used as the input (inflow runoff hydrograph). Weir coefficients are assumed along with the linear storage equation to compute the outflow hydrograph.

The second case of synthesis, which will be referred to as the problem of design, has the objective of estimating the characteristics of the riser/outlet facility in order to meet some design objective. In this case, the output of the design problem is the area of the orifice or weir length, riser and conduit diameters, and outlet facility elevation characteristics. Unlike the analysis case, the weir and/or orifice coefficients are assumed, as is the design criteria, unlike the watershed evaluation case outlined in the previous paragraph.

8.8.4 Planning versus Design

A number of detention basin planning methods for estimating the required volume of detention storage are introduced in this chapter. Methods will be provided for estimating the size of the outlet structure for both single-stage and two-stage risers. The methods of this chapter will be classed as planning methods, although they have been used for design. Design techniques differ in two ways from the SWM planning methods. First, the planning methods only require peak discharge estimates, as opposed to requiring entire flood hydrographs. Thus, routing hydrographs through the detention basin is not necessary when using these planning methods. Second, since routing is not required, a stage-storage-discharge relationship is not required; instead, the storage-discharge relationship is inherent in the planning methods. The design method introduced in Chapter 11 will use flood hydrographs, routing, and a site-specific stage-storage-discharge relationship. For this reason the design method is more accurate than the planning methods. The terms planning and design are used to distinguish between approaches to SWM problem solving that reflect differences in expected accuracy, as well as the cost and effort involved. The following summarizes the differences between methods:

Classification: Detention Volume Methods

Level	Basis	Timing	Storage	Routing	Relative Accuracy	Use
I	Volume	Not considered	Not considered	No	Low	Planning
II	Peak discharge	Constant time of conc.	Generalized stage-storage	No	Medium	Planning
III	Entire hydro-graph	Variable: time vs. storage	Site-specific stage-storage	Yes	High	Design

The problem of planning the detention facility is separated into two parts: estimating the volume of storage and sizing the characteristics of the outlet facility. In planning they are treated separately, whereas in design their determination is made simultaneously.

A number of methods have been proposed and are being used for estimating detention volumes. Recognizing that these methods often yield widely different estimates, a brief comparison of some of the more widely used methods is in order. A relationship between the ratio of the storage volume to the runoff depth and the ratio of the "pre-development" and "post-development" peak discharges is the basis for many of these methods. For SWM policies that require the peak discharge released from the SWM basin to be no greater than the pre-development peak discharge, the pre-to-post ratio is often referred to as the ratio of the outflow to inflow since the peak of the outflow from the detention basin equals the pre-development peak discharge, and the inflow to the detention basin equals the post-development peak discharge.

An estimate of the required volume of detention storage is necessary in both planning and design. Planning methods are usually based on one or more of the following parameters:

1. α: the ratio of the peak discharges before development (that is, the peak discharge from the detention facility) and after development
2. γ: the ratio of the times to peak for the outflow from and the inflow to the detention facility
3. Q_a: the depth (in.) of runoff after development

8.8.5 A Generalized Planning Model

Before considering the assumptions that are made about the parameters for each method, it is instructive to present a generalized model for relating these parameters. For planning purposes it is not unreasonable to use the assumptions that underlie the Rational Method. Specifically, a triangular hydrograph with a time to peak equal to t_c and a time base of $2t_c$ is assumed. The peak discharges for the pre- and post-development conditions are denoted as q_{pb} and q_{pa}, respectively; however, the peak discharges do not have to be computed using the Rational Method. They could, for example, be computed using the SCS Graphical Peak Discharge Method. The parameters α and γ are defined as

$$\alpha = \frac{q_{pb}}{q_{pa}} \tag{8–58}$$

$$\gamma = \frac{t_{pb}}{t_{pa}} \tag{8–59}$$

In most cases, α will be less than 1 because development causes increases in peak discharges, and γ will be greater than 1 because development causes decreases in the timing of runoff. For the assumption that the time to peak equals t_c or is a constant proportion of t_c, Equation 8–59 becomes

$$\gamma = \frac{t_{cb}}{t_{ca}} \tag{8–60}$$

A generalized planning model can be developed using the assumptions above about a triangular hydrograph; also, we will assume that the required volume of storage V_s is the volume under the post-development (inflow) hydrograph above the pre-development hydrograph when the discharge for the inflow hydrograph is greater than the discharge for the pre-development hydrograph. These assumptions yield the following general model for estimating the ratio of the volume of storage V_s to the post-development runoff depth Q_a, with both measured in inches:

$$\frac{V_s}{Q_a} = \begin{cases} \dfrac{\gamma + \alpha + \alpha\gamma(\gamma + \alpha - 4)}{\gamma - \alpha} & \text{for } \alpha < 2 - \gamma \quad (8\text{–}61a) \\[2ex] \dfrac{\gamma - \alpha}{\gamma + \alpha} & \text{for } \alpha \geq 2 - \gamma \quad (8\text{–}61b) \end{cases}$$

For Equation 8–61a the peak of the post-development hydrograph occurs prior to the time when the pre-development and post-development discharge rates are equal. For the case where, $\alpha \geq 2 - \gamma$, the time of the peak discharge for the pre-development hydrograph occurs after the time where the two hydrographs intersect; this is the case shown in Figure 8–13. For the special case where $\alpha = 2 - \gamma$, the time to peak of the pre-development hydrograph occurs at the time of intersection of the two hydrographs; in this case, Equation 8–61b reduces to

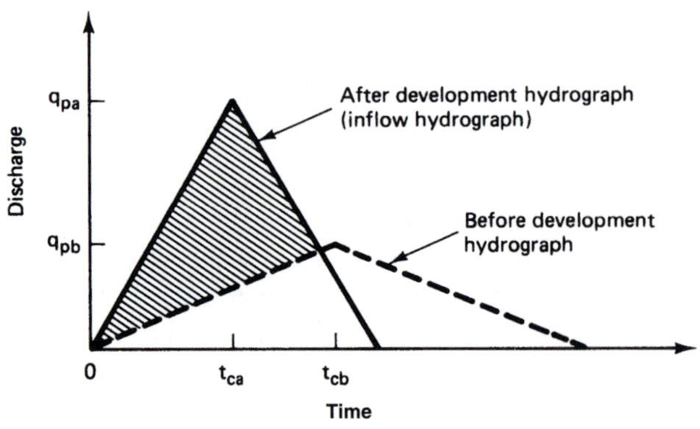

FIGURE 8–13 Generalized model for estimating the volume of required storage (cross-hatched area).

$$\frac{V_s}{Q_a} = \gamma - 1 \tag{8–62a}$$

$$= 1 - \alpha \tag{8–62b}$$

The computations necessary to estimate the volume of storage V_s in in. can be made on Worksheet 8–1. To convert V_s from in. to acre-ft of storage, the following relationship can be used:

$$V_{st} = \frac{V_s A}{12} \tag{8–63}$$

in which V_{st} is the volume of storage in acre-ft and A is the drainage area in acres.

8.8.6 The Loss-of-Natural-Storage Method

When storm water detention was first proposed as a method for controlling flood runoff, methods for estimating detention volumes were based on the idea that the volume of man-made storage should equal the volume of lost natural storage:

$$V_s = Q_a - Q_b \tag{8–64}$$

in which Q_a and Q_b are the depths (in.) of runoff for the post- and pre-development watershed conditions. To place Equation 8–64 in a general form, both sides can be divided by Q_a and rearranged:

$$\frac{V_s}{Q_a} = \frac{Q_a - Q_b}{Q_a} = 1 - \frac{Q_b}{Q_a} \tag{8–65}$$

The runoff depths Q_a and Q_b can be computed using any one of a number of methods. For the SCS method, the SCS runoff equation (Equation 7–42) can be used with the post- and pre-development *CN*s. For the Rational Method, runoff depths (in.) Q can be estimated using the peak discharge q_p (ft^3/sec), the time of concentration t_c (min), and the drainage area A (acres):

$$Q = \frac{2}{121}\left(\frac{q_p t_c}{A}\right) \tag{8–66}$$

in which the factor 2/121 is necessary to convert Q to inches. Equation 8–66 can be used for both the pre- and post-development conditions using the appropriate values of q_p and t_c. The computations can be made on Worksheet 8–1.

Example 8–16

A community is planning a 4.5-acre recreation center that will include basketball courts, tennis courts, a small building, a softball field, and a small parking area. The existing land is open space in fair condition on C soil. This yields a pre-development *CN* of 79. The proposed site will include 1.6 acres of impervious cover, 1.8 acres of open space in poor condition, and 1.1 acres of open space in good condition. Thus, the post-development *CN* is

$$CN_a = [1.6\,(98) + 1.8\,(86) + 1.1\,(74)]/(1.6 + 1.8 + 1.1) = 87.3 \text{ (use 87)}$$

WORKSHEET 8–1 Planning Estimates of Detention Storage

PARAMETER	SYMBOL	UNITS	VALUE	REQUIRED FOR METHOD 1	2	3	4	5	6	7
Drainage area	A	acres	———	✓	✓	✓	✓	✓	✓	✓
Runoff depth—before	Q_b	inches	———		✓					
Runoff depth—after	Q_a or V_r	inches	———	✓	✓	✓	✓	✓	✓	✓
Peak discharge—before	q_{pb}	ft³/sec	———	✓		✓	✓	✓	✓	✓
Peak discharge—after	q_{pa}	ft³/sec	———	✓		✓	✓	✓	✓	✓
Discharge ratio	α①	—	———	✓		✓	✓	✓	✓	✓
Time of concentration—before	t_{cb}	hours	———			✓				
Time of concentration—after	t_{ca}	hours	———					✓		
Time to peak—before	t_{pb}	hours	———	✓						
Time to peak—after	t_{pa}	hours	———	✓					✓	
Hydrograph time base—after	T_b	hours	———						✓	
Time ratio	γ②	—	———	✓						
Recession ratio	m	—	———					✓		

METHOD	COMPUTATIONAL FORM	R_v	V_s③ (in.)	V_{st}④ (ac-ft)
1. Generalized Model	$R_v = \begin{cases} \dfrac{\gamma + \alpha + \alpha\gamma(\gamma + \alpha - 4)}{\gamma - \alpha} & \text{for } \alpha < 2 - \gamma \\ (\gamma - \alpha)/(\gamma + \alpha) & \text{for } \alpha \geqslant 2 - \gamma \end{cases}$			
2. Natural Storage Lost	$V_s = Q_a - Q_b$			
3. Rational Hydrograph	$V_{st} = 0.08264 t_{ca} (q_{pa} - q_{pb})$			
4. Baker Method	$R_v = 1 - \alpha$			
5. Abt and Grigg	$V_{st} = 0.08264 \left(\dfrac{1 + m}{2}\right) q_{pa} t_{ca} (1 - \alpha)^2$			
6. Wycoff and Singh	$V_s = 1.291 Q_a (1 - \alpha)^{0.753} (T_b/t_{pa})^{-0.411}$			
7. SCS TR-55	$R_v = \begin{cases} 0.682 - 1.43\alpha + 1.64\alpha^2 - 0.804\alpha^3 & \text{for Type II and III} \\ 0.660 - 1.76\alpha + 1.96\alpha^2 - 0.730\alpha^3 & \text{for Type I and IA} \end{cases}$			

①$\alpha = q_{pb}/q_{pa}$
②$\gamma = t_{pb}/t_{pa}$ or t_{cb}/t_{ca}
③$V_s = R_v Q_a$[=] inches
④$V_{st} = V_s A/12$[=] acre-ft

The county design manual requires use of a 5-yr control. For the locality of the design, the *IDF* curve gives a 24-hr rainfall of 4.2 in. Thus, the pre- and post-development runoff depths (Equation 7–42) are 2.13 and 2.82 in., respectively. The volume of storage required to control the discharge out of the basin is

$$V_{st} = (2.82 - 2.13) \text{ in.} \times 4.5 \text{ ac}/12 \text{ in./ft} = 0.259 \text{ ac-ft}$$

For an average depth of 3 ft, the surface area of the pond would be 0.086 ac, or 3757 ft^2 (60 ft × 63 ft).

8.8.7 The Rational Formula Hydrograph Method

Given the popularity of the Rational Method, a number of detention volume estimation methods have been based on the Rational Method Hydrograph (see Chapter 9). Figure 8–14a shows the volume V_s that would need to be stored to control the peak discharge to the pre-development peak discharge. One method uses the difference between the post- and pre-development peak discharges in ft^3/sec and the post-development time of concentration t_{ca} in hours:

$$V_s = (q_{pa} - q_{pb})\, t_{ca}/A \tag{8–67}$$

in which t_{ca} is in hr, A is in acres, V_s is in in., and q_{pa} and q_{pb} are in ft^3/sec; this is shown schematically in Figure 8–14b. The computations can be made using Worksheet 8–1.

8.8.8 The Baker Method

The Baker Method is based on the premise that the volume of storage, V_s, equals the area between the post- and pre-development hydrographs (see Figure 8–15). Baker (1979) used the Rational Formula to develop the hydrographs. In such a case, it can be shown using the geometrics that the ratio of V_s to Q_a is given by

$$\frac{V_s}{Q_a} = 1 - \alpha \tag{8–68}$$

It should be evident that this is the special case (Equation 8–62b) of the generalized model.

8.8.9 The Abt and Grigg Method

Abt and Grigg (1978) provided an approximate method for sizing detention reservoirs. The volume of storage equals the volume between the inflow and outflow hydrographs (see Figure 8–16). The rising limbs of the inflow and outflow hydrographs are assumed to coincide until the limiting discharge rate (the pre-development peak discharge) is reached. The outflow and inflow hydrographs are assumed to be trapezoidal and triangular, respectively. For a single detention basin the volume of storage, V_{st}, in acre-ft, is given by

$$V_{st} = 0.08264 \left(\frac{1 + m}{2}\right) q_{pa}\, t_{ca}\, (1 - \alpha)^2 \tag{8–69}$$

in which m is the ratio of the duration of the hydrograph recession to the time to peak, t_{ca} is the post-development time of concentration in hours, q_{pa} is the post-development peak dis-

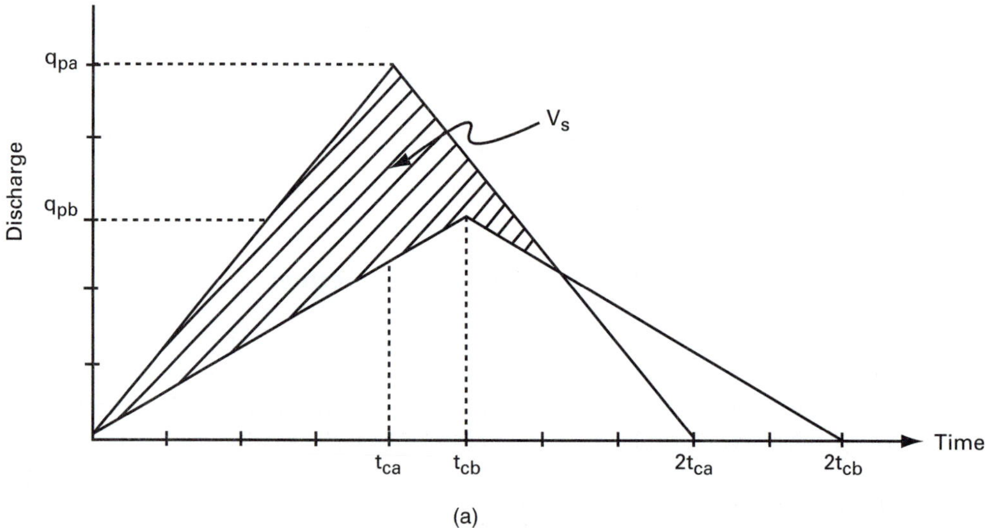

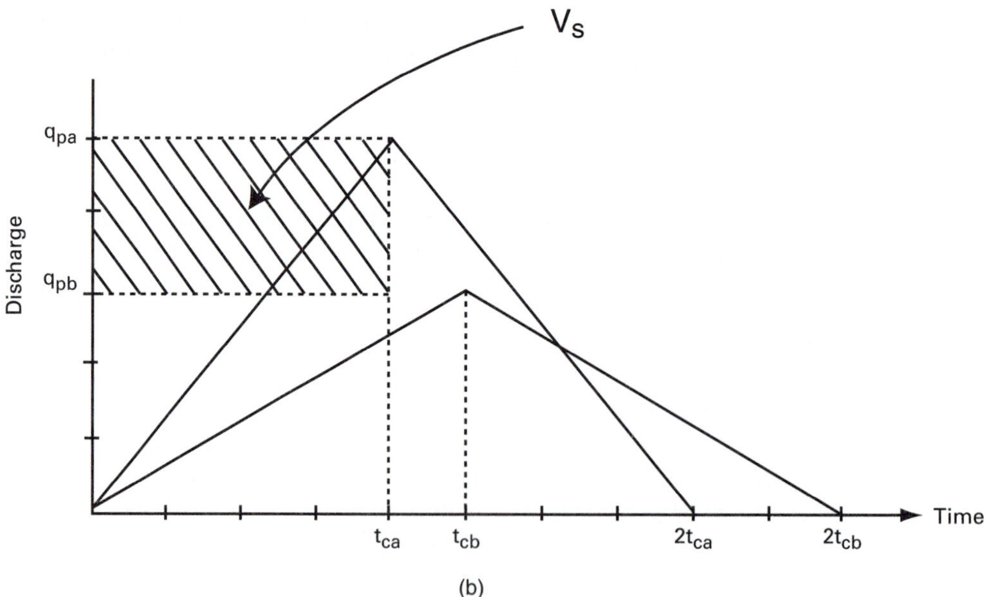

FIGURE 8–14 Schematic of the Rational Formula Hydrograph Method: (a) conceptual representation of the storage volume; (b) computational representation.

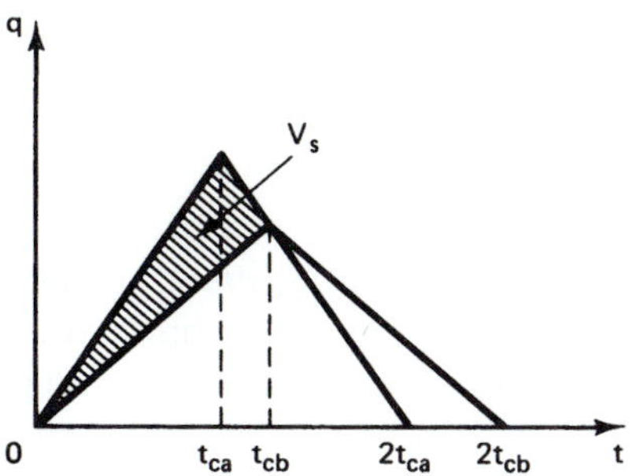

FIGURE 8–15 Schematic diagram of storage volume (V_s) determination using the Baker Method.

charge in ft³/sec, and α is the ratio of the peak discharges in ft³/sec for the outflow and inflow (see Equation 8–58). Worksheet 8–1 can be used in making estimates of V_{st} with Equation 8–69. The method assumes that the post-development time to peak equals the time of concentration.

Methods for computing the peak discharge, volumes of runoff, and times of concentration were not recommended by Abt and Grigg. This method is more realistic than some of the other methods in that it uses the inflow and outflow hydrographs rather than the pre- and post-development hydrographs. The peak outflow rate may be set equal to the pre-development peak discharge.

8.8.10 The Wycoff and Singh Method

Wycoff and Singh (1976) provide a method for making preliminary estimates of the volume of small flood detention reservoirs. The following relationship for computing the volume of storage was evaluated using a regression of data generated from a more detailed hydrologic model:

$$\frac{V_s}{Q_a} = \frac{1.291\,(1 - \alpha)^{0.753}}{(T_b/t_p)^{0.411}} \tag{8–70}$$

in which q_{pb} and q_{pa} are the peak discharge rates of the pre-development and post-development, respectively, and T_b and t_p are the time base and the time to peak of the post-development hydrograph, respectively. The time base is measured from the start of runoff to the time when the discharge on the recession equals 5% of the peak discharge rate. For the

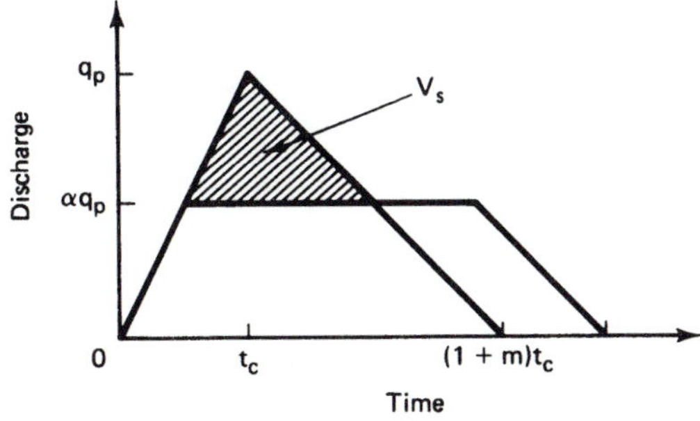

FIGURE 8–16 Estimate of volume of storage (V_s) using the Abt and Grigg Method.

calibration of Equation 8–70, values for α varied from 0.152 to 0.891 and values of the ratio T_b/t_p varied from 2.55 to 4.88.

8.8.11 The SCS TR-55 Method

Chapter 6 of Technical Release 55, or TR-55 (SCS, 1986), provides a method for quickly analyzing the effects of a storage reservoir on peak discharges. It is based on average storage and routing effects for many structures that were evaluated using the computerized TR-20 method (SCS, 1984). The ratio of the volume-of-storage to the volume-of-runoff (V_s/Q_a) is given as a function of the ratio of the peak rate of outflow to the peak rate of inflow, α. The relationship between V_s/Q_a and α is shown in Figure 8–17. Mathematically, the relationship is given by

$$R_s = \frac{V_s}{Q_a} = C_o + C_1\alpha + C_2\alpha^2 + C_3\alpha^3 \tag{8–71}$$

in which C_o, C_1, C_2, and C_3 are coefficients (see Table 8–2) that are a function of the SCS rainfall distribution. The computations can be made on Worksheet 8–2. The volume of storage (in.) is computed by

$$V_s = R_s \times Q_a \tag{8–72}$$

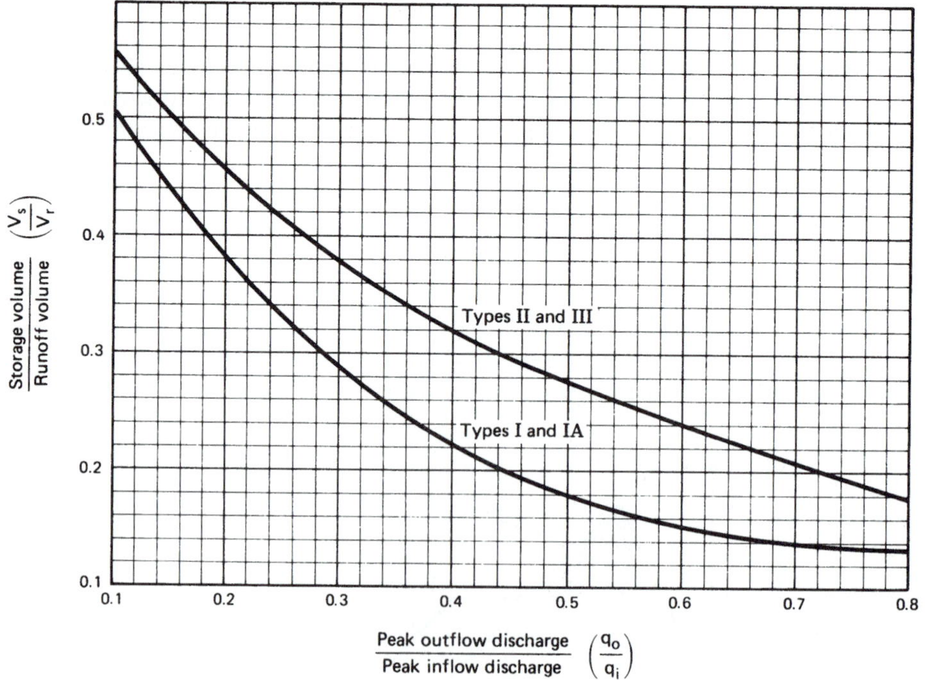

FIGURE 8–17 Approximate detention basin routing for rainfall types I, IA, II, and III.

WORKSHEET 8–2 Detention Volume Estimation: SCS TR-55 Method

Variable	Pre-development	Post-development
Drainage area, A (ac)		
Drainage area, A_m (mi^2)		
Time of Conc., t_c (hr)		
Curve number, CN		

Step	Variable	Low stage	High stage
1	Rainfall, P (in.)		
2	I_a/P (pre-dev.) I_a/P (post-dev.)		
3	Runoff depth, Q_b (pre-dev.) (in.) Runoff depth, Q_a (post-dev.) (in.)		
4	Unit peak, q_{ub} (pre-dev.) (ft^3/s/mi^2/in.) q_{ua} (post-dev.) (ft^3/s/mi^2/in.)		
5	Peak discharge, q_{pb} (pre-dev.) (ft^3/s) q_{pa} (post-dev.) (ft^3/s)		
6	q_p ratio = $q_{pb}/q_{pa} = R_q$		
7	Storage ratio, R_V		
8	Depth of storage, d_S (in.)		
9	Volume of storage, V_S (ac-ft)		
10	Elevation (ft)		

(2) $I_a = 0.2*S$ where $S = (1000/CN) - 10$

(3) $Q = (P\text{-}0.2*S)^2/(P+0.8*S)$

(4) Obtain unit peak discharge rates from Figure 7–9

(5) $q_p = q_u A_m Q$

(7) $R_V = C_o + C_1 R_q + C_2 R_q^2 + C_3 R_q^3$

Rainfall Type	C_o	C_1	C_2	C_3
I, IA	0.660	–1.76	1.96	–0.730
II, III	0.682	–1.43	1.64	–0.804

(8) $d_S = R_V Q_q$

(9) $V_S = d_S A/12$

(10) Use stage-storage relationship

TABLE 8–2 Coefficients for the SCS Detention Volume Method

Rainfall Distribution	C_o	C_1	C_2	C_3
I or IA	0.660	–1.76	1.96	–0.730
II or III	0.682	–1.43	1.64	–0.804

and the volume in acre-ft is

$$V_{st} = V_s A/12 \tag{8–73}$$

Example 8–17

Consider the conditions for the 4.5-acre development of Example 8–16. Assume the pre- and post-development times of concentration are 12 and 6 min., respectively. The I_d/P ratios are 0.13 and 0.1, respectively. The unit peak discharges computed using the SCS Graphical Method (Equation 7–45) are 784 and 1010 ft^3/sec/mi^2/in., respectively. Thus, the peak discharges are

$$q_{pb} = q_{ub}AQ_b = 784\ ((4.5/640)\ (2.13\text{ in.}) = 11.7\text{ cfs}$$

$$q_{pa} = q_{ua}AQ_a = 1010\ (4.5/640)\ (2.82\text{ in.}) = 20.0\text{ cfs}$$

Thus, the ratio $q_{pb}/q_{pa} = 0.585$, which is α of Equation 8–71. The storage volume ratio R_s is

$$R_s = 0.682 - 1.43\ (0.585) + 1.64(0.585)^2 - 0.804\ (0.585)^3$$
$$= 0.246$$

Using the post-development runoff depth of 2.82 in., the required volume of storage is

$$V_{st} = R_s QA/12 = 0.246\ (2.82)\ (4.5)/12$$
$$= 0.260\text{ ac-ft}$$

This is in unusually good agreement with the estimate from the loss-of-natural-storage method (Example 8–16).

8.8.12 Comparison of Planning Methods

It is of interest to compare the generalized model with other planning methods. The methods have slightly different input requirements, as is evident from the input summary shown on Worksheet 8–1. The input requirements are usually the same for many of the methods for estimating peak discharge, as discussed in Chapter 7. The relationship between V_s/Q_a and α provides another means of comparison. To compare the methods, an assumption has to be made about the relationship between the runoff depth and the peak discharge. For purposes of comparison, we can assume that the volume of runoff equals the product of the peak discharge and the time of concentration:

$$Q = q_p\, t_c \tag{8–74}$$

Substituting Equation 8–74 into the equation for the loss-of-natural-storage method of Equation 8–61 yields

$$\frac{V_s}{Q_a} = 1 - \frac{q_{pb}\,t_{cb}}{q_{pa}\,t_{ca}} = 1 - \frac{t_{cb}\,q_{pb}}{t_{ca}\,q_{pa}} = 1 - k\alpha \tag{8–75}$$

in which k is the ratio of the pre- to post-development times of concentration. Assuming that the ratio k equals the value of γ yields

$$\frac{V_s}{Q_a} = 1 - \alpha\gamma \tag{8–76}$$

Since land development usually decreases the time of concentration, γ is usually greater than 1.0. The relationship between V_s/Q_a and α is shown in Figure 8–18 for the cases where γ equals 1.0 and 1.25.

To compare the Rational Formula Hydrograph Method of Equation 8–67 with the other methods, we can divide both sides of Equation 8–67 by Q_a, and making use of Equation 8–74 yields

$$\frac{V_s}{Q_a} = \frac{(q_{pa} - q_{pb})\,t_{cb}}{Q_a} = \frac{(q_{pa} - q_{pb})\,t_{cb}}{q_{pa}t_{ca}} = k\left(1 - \frac{q_{pb}}{q_{pa}}\right) = \gamma(1 - \alpha) \tag{8–77}$$

Equation 8–77 is shown in Figure 8–18 for the cases where k equals 1.0 and 1.25.

If we assume that the inflow hydrograph for the Abt and Grigg Method is triangular with a time to peak equal to t_c and a time base equal to $2t_c$, Equation 8–69 can be used to give the ratio of the volume of storage to the depth of runoff:

$$\frac{V_s}{Q_a} = (1 - \alpha)^2 \tag{8–78}$$

This relationship is also shown in Figure 8–18.

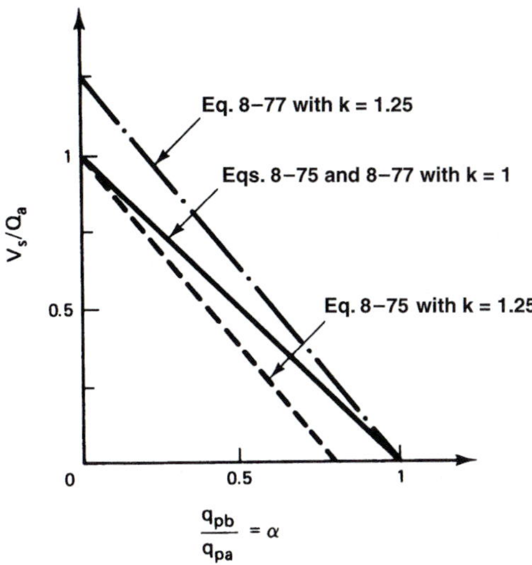

FIGURE 8–18 Relationship between the ratio of the volume of storage (V_s) to the after-development runoff depth (Q_a) and the ratio of the pre-development (q_{pb}) and post-development (q_{pa}) peak discharges.

Based on the assumption that $T_b = 2t_p$, the Wycoff and Singh Method of Equation 8–70 can be used to derive an expression for V_s/Q_a:

$$\frac{V_s}{Q_a} = 0.97\left(1 - \frac{q_{pb}}{q_{pa}}\right)^{0.753} = 0.97\,(1 - \alpha)^{0.753} \tag{8–79}$$

Example 8–18

Figure 8–19 shows the plot layout for a 250,000-ft^2 proposed multi-family housing development. The site is currently forested (fair condition) on a C soil. In the post-development condition, the pervious portions of the layout will be grass covered (good condition). The impervious portions are as follows: buildings (24,300 ft^2), pool area (11,250 ft^2), tennis courts (7500 ft^2), maintenance shed (500 ft^2), parking lots (48,150 ft^2), driveways (4500 ft^2), and sidewalks (4630 ft^2). The total imperviousness is 100,830 ft^2, which is 40.33% of the total area of the de-

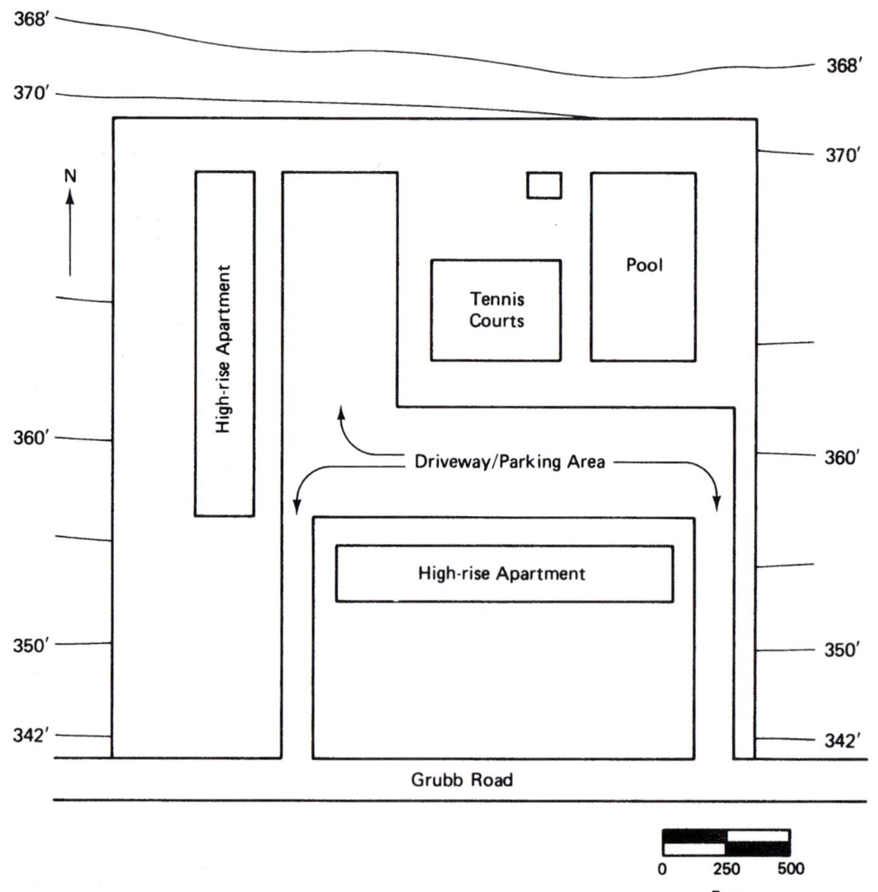

FIGURE 8–19 Layout of multi-family housing development for after-development condition.

velopment. The development is adjacent to an existing collector street (Grubb Road) and will be attached to the existing storm drainage system. The local storm drainage policy requires that the 25-yr peak discharge after development not exceed the 25-yr peak discharge for the pre-development conditions.

A planning estimate of the volume of detention storage that is required to meet the storm drainage policy can be obtained using any of the previous methods. The SCS method will be used to obtain estimates of the runoff depths and peak discharges. The watershed slope is 5.6%, with an elevation drop of 28 ft over a length of 500 ft.

The pre-development peak discharge can be computed using the SCS Graphical Method. For a slope of 5.6% on a forested watershed, the overland flow velocity is 0.6 ft/sec (see Figure 3–19). For a flow length of 500 ft, the velocity method (Equation 3–43) yields a time of concentration of 0.231 hr. A *CN* of 73 is obtained from Table 3–18 for a forested watershed in fair condition. For a 25-yr return period a 24-hr rainfall intensity of 0.245 in./hr is obtained from Figure 4–4; therefore, the 24-hr rainfall volume is 5.9 in. For a *CN* of 73, Equation 7–42 yields a runoff depth of 3.01 in. For a *CN* of 73 and a rainfall of 5.9 in., the I_a/P ratio is 0.12 (Table 7–14). Using the time of concentration and I_a/P as input to Figure 7–9, the unit peak discharge is 750 $ft^3/sec/mi^2/in.$ of runoff. Using Equation 7–45, the pre-development peak discharge is

$$\begin{aligned} q_{pb} &= (750 \text{ ft}^3/\text{sec/mi}^2/\text{in.})\ (0.008968 \text{ mi}^2)\ (3.01 \text{ in.}) \\ &= 20 \text{ ft}^3/\text{sec} \end{aligned}$$

A similar evaluation can be made for the post-development conditions. For a slope of 5.6% on a grassed overland flow area, the runoff velocity is 1.7 ft/sec. For a flow length of 500 ft, the velocity method yields a t_c of 0.082 hr; it is common practice to use a minimum value, such as 0.1 hr, when the computed t_c is less than the minimum. From Table 3–18, *CN*s of 98 and 74 are obtained for the impervious surfaces and grassed areas, respectively; therefore, the weighted *CN* is

$$CN_a = 98\ (0.4033) + 74\ (0.5967) = 84$$

An I_a/P ratio of 0.1 is obtained from Table 7–14. A runoff depth (Q_a) of 4.10 in. is obtained using Equation 7–42. For an I_a/P of 0.1 and a t_c of 0.1 hr, a unit peak discharge of 1048 $ft^3/sec/mi^2/in.$ is obtained from Figure 7–9. Using Equation 7–45, the after-development peak discharge is

$$\begin{aligned} q_{pa} &= (1048 \text{ ft}^3/\text{sec/mi}^2/\text{in.})\ (0.008968 \text{ mi}^2)\ (4.10 \text{ in.}) \\ &= 39 \text{ ft}^3/\text{sec} \end{aligned}$$

The post-development peak discharge represents a 95% increase in the pre-development peak discharge; therefore, R_q equals 0.513.

Having determined the pre- and post-development discharges and the after-development runoff depth (Q_a = 4.10 in.), the value of V_s/Q_a can be computed. Planning estimates of the required storage volumes are given in Table 8–3. The values of V_s/Q_a range from 0.020 to 0.637, with the largest estimate being 32 times greater than the smallest estimate. The depth of storage V_s in in. is found by multiplying the V_s/Q_a ratio by the post-development runoff depth of 4.10 in. For a drainage area of 250,000 ft^2, the required volume of storage would range from 0.038 acre-ft to 1.248 acre-ft. The site drains to the southwest corner of the lot, where there is an area of about 10,000 ft^2 that could be used for an on-site detention pond. If the site was excavated to a shallow bottom slope, storage could be provided on-site. Figure 8–20 provides a cross section of the site. For a basin with a bottom area of about 10,000 ft^2 the required storage depth would range from 0.164 to 5.44 ft. Four of the methods indicate a depth of less than 2.4 ft, which would be feasible. Two of the methods indicate a required depth from 4 to 5.5 ft; although this is physi-

TABLE 8–3 Alternative Estimates of the Required Detention Storage for the Multifamily Housing Development

Estimation Method	$\frac{V_s}{Q_a}$	V_s (in.)	V_s (ac-ft)
Equation 8–61	0.637	2.61	1.248
Equation 8–64	0.272	1.09	0.521
Equation 8–67	0.189	0.76	0.363
Equation 8–68	0.487	1.95	0.935
Equation 8–69	0.020	0.08	0.038
Equation 8–70	0.388	1.56	0.744
Equation 8–71	0.272	1.09	0.521

cally possible, it may not be practical from economic, safety, and aesthetic standpoints. Where the storage facility required is not feasible, other forms or sites for detention would be required. As an example, the buildings have a total rooftop area of 24,300 ft^2. At a depth of 5 in., rooftop storage could provide 10,125 ft^3 of detention storage. This represents a saving of 1 ft of depth in the 10,000 ft^2 on-site detention pond.

Example 8–19

Development within an 18-acre watershed is planned near a local roadway. A planning estimate of the storage required to detain runoff from a 4-in. storm is needed. The curve number for existing conditions is 70, and development within the watershed will increase the *CN* to 80. The pre- and post-development times of concentration are 0.55 and 0.37 hr, respectively. The pre- and post-development runoff depths can be obtained from Equation 7–42, with values of 1.33 and 2.04 in., respectively. From Table 7–14 the I_a/P ratios are 0.21 and 0.13. From Figure 7–9, the unit peak discharges are 450 csm/in. and 600 csm/in. Thus, the pre-development peak discharge is

$$q_{pb} = 450\ (18/640)\ (1.33) = 16.8 \text{ cfs} \tag{8–80}$$

The post-development peak discharge is

$$q_{pa} = 600\ (18/640)\ (2.04) = 34.4 \text{ cfs} \tag{8–81}$$

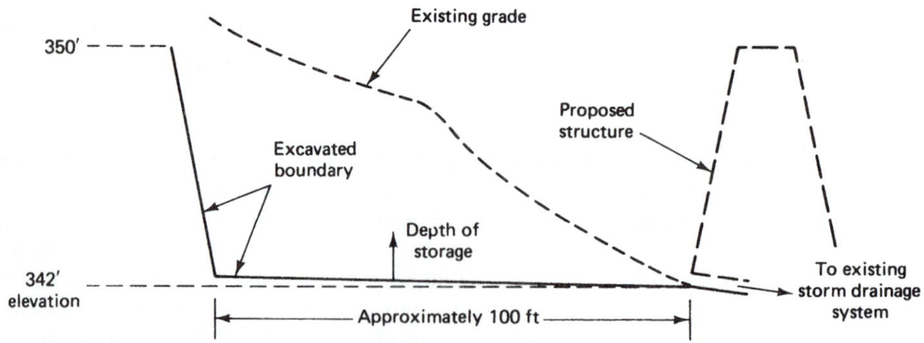

FIGURE 8–20 North-south cross section of southwest corner of the site for proposed detention basin.

These values yield a discharge ratio R_q of 0.489, which is used as input to Figure 8–17 to obtain the volume ratio R_s of 0.28. Thus, the volume of storage is computed using Equation 8–72:

$$\begin{aligned} V_s &= R_s\, Q_a = 0.28\,(2.04) = 0.571 \text{ in.} \\ V_{st} &= V_s\, A/12 = 0.571\,(18)/12 = 0.857 \text{ ac-ft} \end{aligned} \tag{8–82}$$

Assuming the culvert conditions for Equation 8–44 apply and the culvert has a slope of 0.02 ft/ft, the required diameter is

$$D = 1.33\,(16.8 \times 0.024)^{0.375}\,(0.02)^{-0.1875} = 1.97 \text{ ft}$$

Thus, a 2-ft pipe will limit the discharge to about 17 cfs.

8.9 SIZING OF DETENTION BASIN OUTLET STRUCTURES

The methods described in Section 8.8 can be used only to estimate the volume of detention storage. The second necessary step in sizing a detention basin is the determination of the physical characteristics of the outlet structure. The outlet may be based on a weir or an orifice, or both. Figure 8–12 shows a schematic of a basin with a pipe outlet. In addition to determining the diameter of the pipe barrel for a pipe outlet facility, it is also necessary to establish elevations of the pipe inlet and outlet. For those policies that require a permanent pool (wet pond), both the volume of dead storage and the corresponding elevation of the permanent pool must be set. Storage volumes computed with the methods of Section 8.8 are active storage estimates, which is added to the dead storage to estimate the total storage. Both the size and effectiveness of a detention basin are largely dependent on the exceedence frequency (return period). Studies have shown that a basin designed to control the frequent events (that is, 2- or 5-yr events) will tend to overcontrol the less frequent events (that is, 50- or 100-yr events). Conversely, a basin designed to control the less frequent events will tend to undercontrol the more frequent events. An outlet facility sized to pass the 2-yr event will not allow the 100-yr event to pass with the same speed that a pipe outlet sized for a 100-yr event will pass through; thus, overcontrol results. In the past, most SWM policies have required a single-stage riser. Recently, more enlightened SWM policies have been developed that require two-stage control because of the problems of undercontrol and overcontrol associated with single-stage risers. The sizing of both single-stage and multi-stage risers will be discussed here.

In the sizing of risers, it is necessary to determine both the required volume of storage and the physical characteristics of the riser. The physical characteristics include the outlet pipe diameter, the riser diameter, the length of the weir or the area of the orifice, and the elevation characteristics of the riser. Single-stage risers with weir flow and orifice flow are shown in Figures 8–21a and b, respectively. For weir flow control, Equation 8–24 defines the relationship between the discharge (q) and (1) the depth in feet (h) above the weir, (2) the discharge or weir coefficient C_w, and (3) the length of the weir L_w. Values of C_w depend on the characteristics of both the weir and the discharge rate, with values ranging from 2.6 to 3.3; however, a constant value of about 3.0 is often used. The general formula for flow through an orifice is given by Equation 8–16.

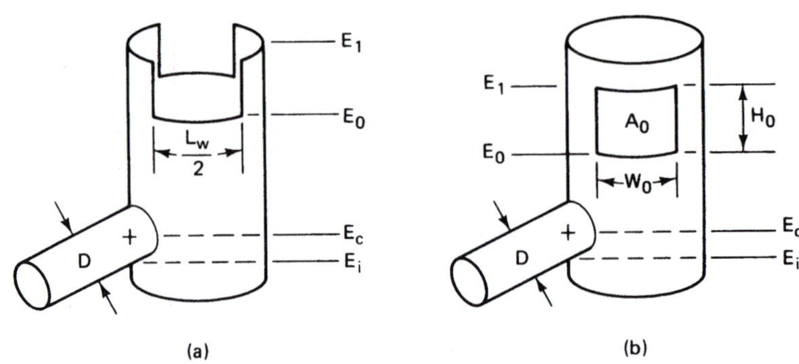

FIGURE 8–21 Single-stage riser characteristics for (a) weir flow and (b) orifice (or port) flow.

8.9.1 Procedure for Sizing Single-Stage Risers

To estimate the characteristics of a riser requires the following inputs:

1. Watershed characteristics, including area, pre- and post-development times of concentration, and pre- and post-development curve numbers
2. Rainfall depth(s) for the design storm(s)
3. Characteristics of the riser structure, including pipe roughness (n), length, and an initial estimate of the diameter
4. Elevation information, including the stage versus storage relationship, the wet pond elevation, and the elevation of the centerline of the pipe
5. Hydrologic and hydraulic models, including a model for estimating peak discharges and runoff depths, a model for estimating the volume of storage as a function of pre- and post-development peak discharges, and a model for estimating weir and orifice coefficients, as necessary

The output from the analysis yields the following: (1) the length of weir(s) and/or area of the orifice; (2) the depth and volume of storage; (3) elevations of riser characteristics; and (4) the diameter of the outlet pipe.

The following steps can be used to size a single-stage riser, with Worksheet 8–3:

1. a. Using the 24-hour rainfall depth and the pre-development CN, find the runoff depth, Q_b.
   b. Using the 24-hour rainfall depth and the post-development CN, find the runoff depth, Q_a.
2. a. Determine the pre-development peak discharge q_{pb}.
   b. Determine the post-development peak discharge, q_{pa}.

WORKSHEET 8–3 Single-Stage or Two-Stage Riser Computations

Variable	Pre-development	Post-development
Drainage area, A (ac)		
Drainage area, A_m (mi^2)		
Time of Conc., t_c (hr)		
Curve number, CN		

OUTLET CHARACTERISTICS			
Var	Units	Value	Comment
n	—		
L	ft		
D	ft		Initial estimate
K_p	—		Table 8–5
$K_p{*}L$	ft		
C_*			Eq. 8–86
E_o	ft		Wet pond
E_c	ft		Centerline
E_t	ft		Tailwater elev.
C_w	—		Weir coeff.

Step	Variable	Low stage	High stage
1	Rainfall, P (in.)		
2	I_a/P (pre-dev.) I_a/P (post-dev.)		
3	Runoff depth, Q_b (pre-dev.) (in.) Runoff depth, Q_a (post-dev.) (in.)		
4	Unit peak, q_{ub} (pre-dev.) ($ft^3/s/mi^2/in.$) q_{ua} (post-dev.) ($ft^3/s/mi^2/in.$)		
5	Peak discharge, q_{pb} (pre-dev.) (ft^3/s) q_{pa} (post-dev.) (ft^3/s)		
6	q_p ratio $= q_{pb}/q_{pa} = R_q$		
7	Storage ratio, R_V		
8	Depth of storage, d_S (in.)		
9	Volume of storage, V_S (ac-ft)		
10	Dead storage, V_d (ac-ft)		
11	Total storage, V_t (ac-ft)		
12	Elevation, E_j (ft)		
13	Diameter, D (ft)		
14	Orifice width, W_o Orifice area, A_o (ft^2) Orifice height, H_o (ft) Orifice discharge, q_{o2} (ft^3/s)		
15	Low-stage weir length, L_{w1} (ft) Weir discharge, q_{o2} (ft^3/s)		
16	High-stage weir length, L_{w2} (ft)		
17	Invert elevation, E_i (ft)		

(10) from stage-storage, with E_o

(11) $V_t = V_S + V_d$

(12) from stage-storage, with V_t

(13) $D = C_* q_{pb2}^{0.5}/(E_1 - E_c)^{0.25}$

(14) $W_o \sim 0.75{*}D$

$A_o = 0.2283 q_{pb1}/(E_1 - E_o)^{0.5}$

$H_o = A_o/W_o$

$q_{o2} = 4.82 A_o (E_2 - E_1)^{0.5}$

(15) $L_{w1} = q_{pb1}/[C_w(E_1 - E_o)^{1.5}]$

$q_{o2} = C_w L_{w1}(E_2 - E_o)^{1.5}$

(16) $L_{w2} = (q_{pb2} - q_{o2})/[C_w(E_2 - E_1)^{1.5}]$

(17) $E_i = E_c - 0.5D$

3. Compute the discharge ratio:

$$R_q = \frac{q_{pb}}{q_{pa}} \tag{8–83}$$

4. Enter the R_s-versus-R_q (Figure 8–17) curve with R_q to find the storage volume ratio R_s.
5. a. Compute the volume of storage in inches:

$$V_s = Q_a R_s \tag{8–84}$$

   b. Convert V_s to acre-feet by multiplying by $A/12$, where A is the watershed area in acres.
6. Using the elevation E_o of either the weir or the bottom of the orifice, obtain the volume of dead storage V_d from the stage-storage curve.
7. Compute the total storage in acre-feet:

$$V_t = V_d + V_s \tag{8–85}$$

8. Enter the stage-storage curve with V_t to obtain the maximum water surface elevation, E_1.
9. a. Obtain the friction head-loss coefficient K_p from Table 8–4.
   b. Using the product LK_p, which is denoted as X, compute C_*:

$$C_* = 0.456 + 0.047\,X - 0.0024\,X^2 + 0.00006X^3 \tag{8–86}$$

   c. Compute the conduit diameter:

$$D = C_* q_{pb}^{0.5}\, h^{-0.25} \quad \text{where } h = E_1 - E_c \tag{8–87}$$

   (assuming that E_c is greater than the tailwater elevation).
   d. Adjust D to the nearest larger commercial pipe size.

TABLE 8–4 K_p Values for Reinforced Concrete ($n = 0.013$) and Corrugated Metal Pipe ($n = 0.024$)

Pipe Diameter (in.)	RCP	CMP
24	0.0124	0.0423
27	0.01061	0.0362
30	0.00922	0.0314
36	0.00723	0.0246
42	0.00589	0.0201
48	0.00493	0.0168
54	0.00421	0.0144
60	0.00366	0.0125

For other values of n and d, K_p can be computed by $K_p = 5087n^2d^{-4/3}$, where d [=] inches.

10. If the outlet is an orifice, determine characteristics of the orifice:

a. Set the orifice width, W_o; as a rule of thumb, try $0.75D$.

b. Compute the area of the orifice:

$$A_o = \frac{0.2283 q_{pb}}{(E_1 - E_0)^{0.5}} \tag{8–88a}$$

c. Compute the height of the orifice:

$$H_o = \frac{A_o}{W_o} \tag{8–88b}$$

11. If the outlet is a weir, determine the characteristics of the weir length:

$$L_w = \frac{q_{pb}}{C_w (E_1 - E_0)^{1.5}} \tag{8–89}$$

where a value of C_w must be assumed.

12. Compute the conduit invert elevation (ft) at the face of the riser:

$$E_i = E_c - \frac{D}{2} \tag{8–90}$$

Example 8–20

The 23-acre watershed of Figure 7–10 can be used to illustrate the sizing of a single-stage riser. Given the small drainage area, a corrugated metal pipe will be used. A rectangular orifice will be cut into the riser.

Computations for the post-development peak discharge were given in Example 7–19. For the pre-development conditions, a *CN* of 61 will be assumed, and a time of concentration of 1.25 hr was computed using a flow velocity of 0.2 ft/sec over a length of 900 ft. For a *CN* of 61 and rainfall of 4.8 in., the runoff depth is 1.25 in. and the I_a/P equals 0.266. The unit peak discharge from Figure 7–9 is 290 csm/in., so the pre-development peak discharge is

$$q_{pb} = 290 \left(\frac{23}{640}\right)(1.25) = 13 \text{ cfs} \tag{8–91}$$

Thus, the pre- to post-development peak discharge ratio is 0.619. Using this as input to Figure 8–17 yields a storage volume ratio of 0.235. Thus, the volume of active flood storage is

$$V_s = R_s Q_a = 0.235\,(1.59) = 0.374 \text{ in.} \tag{8–92a}$$

$$V_{st} = V_s A/12 = 0.374\,(23)/12 = 0.716 \text{ ac-ft} \tag{8–92b}$$

The stage-storage relationship at the site of the detention structure is

$$V_s = 0.0444\, h^{2.17} \tag{8–93}$$

in which h is the stage (ft) measured above the datum and V_s is the storage (ac-ft).

Based on the invert elevation of the orifice, which is $E_o = 2$ ft, the dead storage can be computed from Equation 8–93 as 0.2 ac-ft. The total storage, flood plus dead, is 0.916 ac-ft (that is,

$V_t = V_{st} + V_d = 0.716 + 0.2$). Using V_t with Equation 8–93, the depth at flood stage can be computed by solving for h:

$$h = \left(\frac{V_t}{0.0444}\right)^{1/2.17} = \left(\frac{0.916}{0.0444}\right)^{0.461} = 4.03 \text{ ft} \tag{8–94}$$

The diameter of the outlet pipe is given by Equation 8–87

$$D = 0.63\,(13)^{0.5}\,(4.03 - 1.00)^{-0.25} = 1.7 \text{ ft} \tag{8–95}$$

As a rule of thumb, the width (W_o) of the orifice is taken as 75% of the conduit diameter; therefore, W_o equals 1.25 ft. The area of the orifice is

$$A_o = 0.2283\,(13)/(4.03 - 2.0)^{0.5} = 2.083 \text{ ft}$$

Thus, the height of the rectangular orifice is 2.08/1.25 = 1.666 ft. The invert of the outlet conduit (E_i) is

$$E_i = E_c - 0.5\,D = 1.0 - 0.5\,(1.75) = 0.14 \text{ ft}$$

The diameter of the riser barrel is usually 2 to 3 times the diameter of the outlet conduit, so a 4.5-ft (54-in.) corrugated metal pipe can be used for the riser. The computations are summarized on Worksheet 8–4.

8.9.2 Sizing of Two-Stage Risers

Where stormwater or drainage policies require control of flow rates of two exceedence frequencies, the two-stage riser is an alternative for control. The structure of a two-stage riser is similar to the single-stage riser except that it includes either two weirs or a weir and an orifice (see Figure 8–22). For the weir/orifice structure, the orifice is used to control the more frequent event, and the larger event is controlled using the weir. The runoff from the smaller and larger events are also referred to as the low-stage and high-stage events, respectively. Values for variables at low and high stages may be followed by a subscript 1 or 2, respectively; for example, q_{pb2} will indicate the pre-development peak discharge for the high-stage event. Recognizing that the two events will not occur simultaneously, both the low-stage weir or orifice and the high-stage weir are used to control the high-stage event.

The sizing of a two-stage riser is only slightly more complicated than the sizing of a single-stage riser. The procedure follows the same general format as for the single-stage riser, but both the high-stage weir and low-stage outlet characteristics must be determined. The input for sizing a two-stage riser is the same as that for a single-stage riser, but many of the values must be computed for both the low-stage and high-stage events. The input consists of watershed characteristics, rainfall depths, site characteristics, outlet characteristics, and the stage-storage relationship for the location. The input requirements are summarized in Worksheet 8–3. The following steps can be used to size a two-stage riser for the cases where the low-stage outlet is either a weir or an orifice:

1. a. Using the 24-hr rainfalls and the pre-development *CN*, find the runoff depth for both the low- and high-stage events, Q_{b1} and Q_{b2}

b. Using the 24-hr rainfalls and the post-development *CN*, find the runoff depth for both the low- and high-stage events, Q_{a1} and Q_{a2}

WORKSHEET 8–4 Sizing of Single-Stage Riser for Example 8–20

Variable	Pre-development	Post-development
Drainage area, A (*ac*)	23	23
Drainage area, A_m (mi^2)	0.03594	0.03594
Time of Conc., t_c (hr)	1.25	0.82
Curve number, CN	61	66

OUTLET CHARACTERISTICS			
Var	Units	Value	Comment
n	—	0.024	
L	ft	90	
D	ft	2	Initial estimate
K_p	—	0.05	Table 8–5
K_p*L	ft	4.5	
C_*		0.63	Eq. 8–86
E_o	ft	2	Wet pond
E_c	ft	1	Centerline
E_t	ft	—	Tailwater elev.
C_w	—	3.1	Weir coeff.

Step	Variable	Low stage	High stage
1	Rainfall, P (in.)	4.8	
2	I_a/P (pre-dev.) I_a/P (post-dev.)	0.266 0.210	
3	Runoff depth, Q_b (pre-dev.) (in.) Runoff depth, Q_a (post-dev.) (in.)	1.25 1.59	
4	Unit peak, q_{ub} (pre-dev.) (ft^3/s/mi^2/in.) q_{ua} (post-dev.) (ft^3/s/mi^2/in.)	290 370	
5	Peak discharge, q_{pb} (pre-dev.) (ft^3/s) q_{pa} (post-dev.) (ft^3/s)	13 21	
6	q_p ratio $= q_{pb}/q_{pa} = R_q$	0.619	
7	Storage ratio, R_V	0.235	
8	Depth of storage, d_S (in.)	0.374	
9	Volume of storage, V_S (ac-ft)	0.716	
10	Dead storage, V_d (ac-ft)	0.20	
11	Total storage, V_t (ac-ft)	0.916	
12	Elevation, E_j (ft)	4.03	
13	Diameter, D (ft)	1.72 (use 1.75)	
14	Orifice width, W_o Orifice area, A_o (ft^2) Orifice height, H_o (ft) Orifice discharge, q_{o2} (ft^3/s)	1.25 2.083 1.666 —	
15	Low-stage weir length, L_{w1} (ft) Weir discharge, q_{o2} (ft^3/s)		
16	High-stage weir length, L_{w2} (ft)		
17	Invert elevation, E_i (ft)	0.14	

(10) from stage-storage, with E_o

(11) $V_t = V_S + V_d$

(12) from stage-storage, with V_t

(13) $D = C_* q_{pb2}^{0.5}/(E_1 - E_c)^{0.25}$

(14) $W_o \sim 0.75*D$

$A_o = 0.2283 q_{pb1}/(E_1 - E_o)^{0.5}$

$H_o = A_o/W_o$

$q_{o2} = 4.82 A_o (E_2 - E_1)^{0.5}$

(15) $L_{w1} = q_{pb1}/[C_w(E_1 - E_o)^{1.5}]$

$q_{o2} = C_w L_{w1}(E_2 - E_o)^{1.5}$

(16) $L_{w2} = (q_{pb2} - q_{o2})/[C_w(E_2 - E_1)^{1.5}]$

(17) $E_i = E_c - 0.5D$

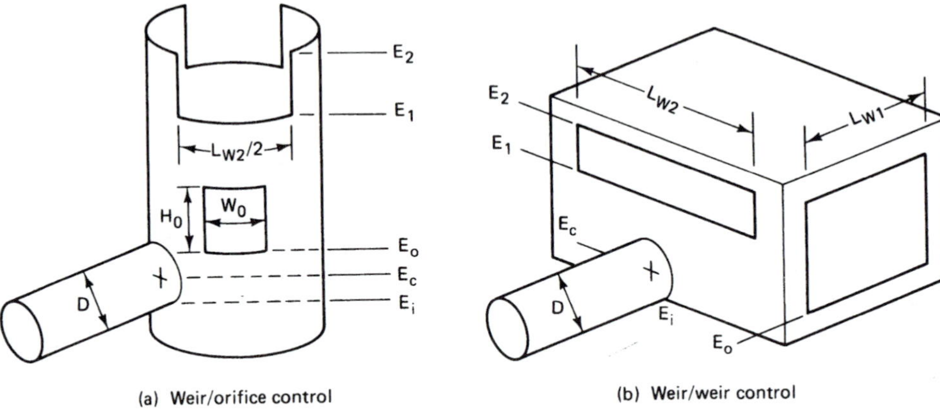

FIGURE 8–22 Two-stage outlet facility.

2. a. Determine the pre-development peak discharges for both the low- and high-stage events, q_{pb1} and q_{pb2}.
   b. Determine the post-development peak discharges for both the low- and high-stage events, q_{pa1} and q_{pa2}.
3. Compute the discharge ratios for both the low- and high-stage events:

$$R_{q1} = \frac{q_{pb1}}{q_{pa1}} \tag{8–96a}$$

$$R_{q2} = \frac{q_{pb2}}{q_{pa2}} \tag{8–96b}$$

4. Enter the R_s-versus-R_q curve with R_{q1} and R_{q2} to find the storage volume ratios R_{s1} and R_{s2}.
5. a. Compute the volume of active storage in inches for both the low- and high-stage events:

$$V_{s1} = Q_{a1}R_{s1} \tag{8–97a}$$

$$V_{s2} = Q_{a2}R_{s2} \tag{8–97b}$$

   b. Convert V_{s1} and V_{s2} to acre-feet by multiplying by $A/12$, where A is the drainage area in acres.
6. Using the elevation E_o obtain the volume of dead storage V_d from the elevation-storage curve.
7. Compute the total storage (acre-feet) for both the low- and high-stage events:

$$V_{t1} = V_d + V_{s1} \tag{8–98a}$$

$$V_{t2} = V_d + V_{s2} \tag{8–98b}$$

8. Enter the stage-storage curve with V_{t1} and V_{t2} to obtain the low- and high-stage water surface elevations, E_1 and E_2.
9. a. Obtain the friction head-loss coefficient K_p from Table 8–4.
   b. Using the product LK_p obtain C_* from Equation 8–86.
   c. Compute the conduit diameter:

$$D = C_* q_{pb2}^{0.5} h^{-0.25} \quad \text{where } h = E_1 - E_c \tag{8–99}$$

   (assuming that E_c is greater than the tailwater elevation).
   d. Adjust D to the nearest larger commercial pipe size.
10. If the low-stage control is an orifice, determine characteristics of the orifice:
   a. Set the orifice width, W_o; as a rule of thumb, try $0.75*D$.
   b. Compute the area of the orifice:

$$A_o = \frac{0.2283 q_{pb1}}{(E_1 - E_0)^{0.5}} \tag{8–100}$$

   c. Compute the height of the orifice:

$$H_o = \frac{A_o}{W_o} \tag{8–101}$$

   d. Estimate the flow through the low-stage orifice during the high-stage event:

$$q_{o2} = 4.82 A_o (E_2 - E_1)^{0.5} \tag{8–102}$$

11. If the low-stage control is a weir, determine the characteristics of the weir:
   a. Compute the weir length:

$$L_{w1} = \frac{q_{pb1}}{C_w (E_1 - E_0)^{1.5}} \tag{8–103}$$

   b. Compute the flow over the low-stage weir during the high-stage event:

$$q_{o2} = C_w L_{w1} (E_2 - E_o)^{1.5} \tag{8–104}$$

12. Compute the high-stage weir length:

$$L_{w2} = \frac{q_{pb2} \ 2 \ q_{o2}}{C_w (E_2 \ 2 \ E_1)^{1.5}} \tag{8–105}$$

13. Compute the conduit invert elevation (ft) at the face of the riser:

$$E_i = E_c - \frac{D}{2} \tag{8–106}$$

Example 8–21

Consider the forested (fair condition) watershed shown in Figure 8–23. A 52.8-acre tract within the watershed (dashed lines) is to be developed as a commercial/business center, with the total watershed having an area of 128.5 acres. To control the runoff rates from the developed site, an off-site detention structure is planned for the total watershed outlet. The local drainage policy re-

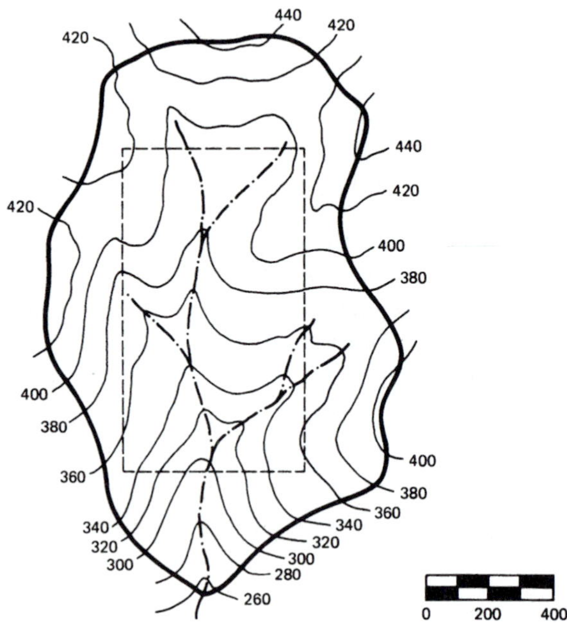

FIGURE 8–23 Watershed layout for Example 8–21.

quires control of both the 2- and 10-yr peak discharges. The SCS two-stage riser method can be used to develop a planning estimate of the storage volume and outlet facility characteristics. The input and calculations are given in Worksheet 8–5.

For the watershed of Figure 8–23, the pre-development *CN*, assuming a C soil group, is 73 (see Table 3–18). Assume that the 2-yr, 24-hr and 10-yr, 24-hr rainfall depths are 3.0 and 5.0 in., respectively; therefore, the runoff depths (Equation 7–42) are 0.86 and 2.28 in., respectively. From Table 7–14, the I_a/P ratios are 0.25 and 0.15, respectively, for the 2- and 10-yr events. The watershed has a total length of 3900 ft, with 50 ft of sheet flow in the forest, 250 ft of overland flow in the forest, 500 ft of upland gully flow, and 3100 ft of channel flow. Equation 3–47 can be used to compute the travel time for sheet flow. The flow velocities for the overland and gully flows can be obtained from Figure 3–19, with the velocity method used to compute the travel time. Channel flow velocities should be computed using Manning's Equation. The following tabular summary shows the calculation of the pre-development time of concentration:

Flow Path	n	L (ft)	S (ft/ft)	i (in./hr)	R_h (ft)	V (fps)	T_t (sec)
Sheet flow	0.2	50	.050	5	—	—	289
Forest (heavy litter)	—	250	.050	—	—	0.56	447
Concentrated (gully)	—	500	.050	—	—	5.25	95
Channel	0.065	1200	.044	—	1.2	5.43	221
Channel	0.060	1900	.040	—	1.3	5.92	321
		3900					1373

WORKSHEET 8–5 Two-Stage Riser Computations for Example 8–21

Variable	Pre-development	Post-development
Drainage area, A (ac)	128.5	128.5
Drainage area, A_m (mi^2)	0.2008	0.2008
Time of Conc., t_c (hr)	0.381	0.28
Curve number, CN	73	82

OUTLET CHARACTERISTICS			
Var	Units	Value	Comment
n	—	0.013	
L	ft	225	
D	ft	5	Initial estimate
K_p	—	0.00366	Table 8–5
K_p*L	ft	0.824	
C_*		0.493	Eq. 8–86
E_o	ft	2	Wet pond
E_c	ft	–6	Centerline
E_t	ft		Tailwater elev.
C_w	—	3.1	Weir coeff.

Step	Variable	Low stage	High stage
1	Rainfall, P (in.)	3.0	5.0
2	I_a/P (pre-dev.) I_a/P (post-dev.)	0.25 0.15	0.15 0.10
3	Runoff depth, Q_b (pre-dev.) (in.) Runoff depth, Q_a (post-dev.) (in.)	0.86 1.38	2.28 3.08
4	Unit peak, q_{ub} (pre-dev.) (ft^3/s/mi^2/in.) q_{ua} (post-dev.) (ft^3/s/mi^2/in.)	540 660	580 685
5	Peak discharge, q_{pb} (pre-dev.) (ft^3/s) q_{pa} (post-dev.) (ft^3/s)	93.2 182.9	265.5 423.6
6	q_p ratio $= q_{pb}/q_{pa} = R_q$	0.510	0.627
7	Storage ratio, R_V	0.272	0.230
8	Depth of storage, d_S (in.)	0.375	0.708
9	Volume of storage, V_S (ac-ft)	4.02	7.59
10	Dead storage, V_d (ac-ft)	1.41	—
11	Total storage, V_t (ac-ft)	5.43	9.00
12	Elevation, E_j (ft)	4.90	6.87
13	Diameter, D (ft)	4.42 (use 4.5)	—
14	Orifice width, W_o Orifice area, A_o (ft^2) Orifice height, H_o (ft) Orifice discharge, q_{o2} (ft^3/s)	8.33 12.5 1.5 84.6	
15	Low-stage weir length, L_{w1} (ft) Weir discharge, q_{o2} (ft^3/s)		
16	High-stage weir length, L_{w2} (ft)		21.1
17	Invert elevation, E_i (ft)	–8.25	

(10) from stage-storage, with E_o

(11) $V_t = V_S + V_d$

(12) from stage-storage, with V_t

(13) $D = C_* q_{pb2}^{0.5}/(E_1 - E_c)^{0.25}$

(14) $W_o \sim 0.75*D$

$A_o = 0.2283 q_{pb1}/(E_1 - E_o)^{0.5}$

$H_o = A_o/W_o$

$q_{o2} = 4.82 A_o (E_2 - E_1)^{0.5}$

(15) $L_{w1} = q_{pb1}/[C_w (E_1 - E_o)^{1.5}]$

$q_{o2} = C_w L_{w1} (E_2 - E_o)^{1.5}$

(16) $L_{w2} = (q_{pb2} - q_{o2})/[C_w (E_2 - E_1)^{1.5}]$

(17) $E_i = E_c - 0.5D$

This yields a time of concentration of 0.381 hr. Based on a t_c of 0.381 hr and the previously given values of I_a/P unit peak discharge estimates of 540 and 580 ft³/sec/mi²/in. are obtained from Figure 7–9 for the 2- and 10-yr events, respectively. Thus the 2- and 10-yr peak discharges are

$$q_{pb1} = (540\text{ ft}^3/\text{sec/mi}^2/\text{in.})\left(\frac{128.5}{640}\text{ mi}^2\right)(0.86\text{ in.}) = 93.2\text{ ft}^3/\text{sec} \tag{8–107}$$

$$q_{pb2} = (580\text{ ft}^3/\text{sec/mi}^2/\text{in.})\left(\frac{128.5}{640}\text{ mi}^2\right)(2.28\text{ in.}) = 266\text{ ft}^3/\text{sec} \tag{8–108}$$

For the post-development condition, 52.8 acres will be developed for commercial/business. Initial site plans indicate 88% of the developed area will be impervious cover, with the remainder in lawn cover (good condition). Thus the weighted *CN* for the 128.5 acre watershed is

$$CN = \frac{52.8}{128.5}[0.88\,(98) + 0.12\,(74)] + \frac{75.7}{128.5}(73) = 82 \tag{8–109}$$

For rainfall depths of 3.0 and 5.0 in., the runoff depths are 1.38 and 3.08 in., respectively. The post-development I_a/P ratios are 0.15 and 0.1, respectively. Using the assumption that the t_c decreased to 0.28 hr, the unit peak discharges are 660 and 685 ft³/sec/mi²/in., and the peak discharges are

$$q_{pa1} = (660\text{ ft}^3/\text{sec/mi}^2/\text{in.})\left(\frac{128.5}{640}\text{ mi}^2\right)(1.38\text{ in.}) = 183\text{ ft}^3/\text{sec} \tag{8–110}$$

$$q_{pa2} = (685\text{ ft}^3/\text{sec/mi}^2/\text{in.})\left(\frac{128.5}{640}\text{ mi}^2\right)(3.08\text{ in.}) = 424\text{ ft}^3/\text{sec} \tag{8–111}$$

Worksheet 8–5 provides a summary of the sizing of the detention structure. Using the ratios of the pre-development to post-development peak discharges with the SCS detention relationship (Figure 8–17), the depth of active storage for low-stage and high-stage control are 0.375 and 0.708 in., respectively, which translate to storage volumes of 4.02 and 7.59 acre-ft.

At the site of the detention structure, the stage-storage relationship is $V = 0.5\, h^{1.5}$ where V [=] ac-ft and L [=] ft. For a permanent pond depth of 2 ft, the dead storage is 1.41 acre-ft. Thus the total volume of storage at the low-stage and high-stage flood conditions would be 5.43 and 9.00 acre-ft, respectively; from the stage-storage curve these volumes correspond to depths of 4.90 and 6.87 ft.

In addition to the storage volumes, the riser must also be sized. Using Equation 8–99, the diameter of the pipe outlet is

$$D = 0.493\left[\frac{266}{(4.9 - (-6))^{0.5}}\right]^{0.5} = 4.42\text{ ft} \quad (\text{use 54 in.}) \tag{8–112}$$

Since the difference $E_2 - E_1$ is only 2 ft, the height of the orifice opening H_o will be set at 1.5 ft. The area of the orifice that would be required to limit the discharge through the orifice to the before-development peak discharge of 93.2 ft³/sec can be computed with Equation 8–100:

$$A_o = \frac{93.2}{4.38\ (4.9 - 2)^{0.5}} = 12.5\ \text{ft}^2 \tag{8–113}$$

Therefore, the width of the rectangular orifice is 8.33 ft, which is computed using Equation 8–101. To compute the length of the weir with Equation 8–105, the discharge (q_r) through the orifice when the high-stage event occurs must be estimated with Equation 8–102:

$$q_r = 4.82\ (12.5)\ (6.87 - 4.90)^{0.5} = 84.6\ \text{ft}^3/\text{sec} \tag{8–114}$$

Thus, with Equation 8–105, the weir length is

$$L_w = \frac{265.5 - 84.6}{3.1\ (6.87 - 4.9)^{1.5}} = 21.1\ \text{ft} \tag{8–115}$$

8.9.3 Sizing of Multiple-Stage Riser Facilities

A two-stage riser uses an orifice to control the high-frequency event (that is, short return period) and a weir to control the low-frequency event (that is, long return period). A two-stage configuration usually provides adequate control because it is not necessary to use the orifice to pass large flow rates. The orifice will not pass large flow rates at low heads. Therefore, for sites with mild slopes, such as coastal areas, the orifice is not an efficient design because a large opening is required to pass a small flow rate. Large openings may not be practical because of the structural integrity of the riser.

For a given head the weir is a more efficient means of passing storm water through the outlet facility. A simple example will be used to show this. Several assumptions are required, and while the specific figures are sensitive to these assumptions, the general conclusion is insensitive to the assumptions. Equation 8–16 can be used to predict flow through an orifice. For purposes of demonstration, it will be assumed that (l) the width of the orifice equals one-third of the circumference and (2) the height of the orifice is 2 ft less than the depth of flow; these assumptions are intended to reflect limits placed on the riser design because of structural considerations. These assumptions lead to the following equation for flow through an orifice:

$$q_o = 1.4\pi d\ (h - 0.5)\ h^{0.5} \tag{8–116}$$

where d is the diameter (ft) of the riser and h is the depth (ft) of flow. If we assume that the length of a weir equals the circumference of the riser, then discharge over a weir is given by

$$q_w = C_w \pi d h^{1.5} \tag{8–117}$$

The values of Table 8–5 show q_w and q_o for selected values of d and h with $C_w = 3.1$. These results are intended only to show that for low heads, a weir will be more effective in passing high discharge rates.

Recognizing that sites with mild slopes will not be capable of high heads, weir flow is desirable for controlling both the low- and high-stage events. Thus an alternative to the two-stage riser with an orifice and a weir is an outlet facility with two risers, with each riser serving as a weir. One riser can be used to control the low-stage event, while the other riser controls the high-stage event. A single-outlet conduit can be used to discharge flow from both risers.

TABLE 8–5 Comparison of Flow Rates (ft³/sec) for Orifice (q_o) and Weir (q_w) Flow

h (ft)	$d = 3$ ft		$d = 5$ ft		$d = 7$ ft	
	q_o	q_w	q_o	q_w	q_o	q_w
1	7	29	11	49	16	68
2	29	83	49	138	68	193
3	60	152	99	253	139	354
4	96	234	160	390	225	545
5	138	327	231	544	323	762

Example 8–22

Example 8–18 provided a design involving an orifice for control of the low-stage event. Using the same data, a low-stage weir can be designed using Step 11 of Worksheet 8–5. The low-stage weir length is

$$L_{wi} = \frac{93}{3.1\,(4.9 - 2.0)^{1.5}} = 6.07 \text{ ft} \tag{8–118}$$

The low-stage weir is 6.07 ft in length. During the high-stage event the flow over the low-stage weir is

$$q_{0.2} = 3.1\,(6.07)\,(6.87 - 4.90)^{1.5} = 52 \text{ ft}^3/\text{sec} \tag{8–119}$$

The length of the high-stage weir is

$$L_{w2} = \frac{266 - 52}{3.1\,(6.87 - 2.0)^{1.5}} = 6.42 \text{ ft} \tag{8–120}$$

PROBLEMS

8–1. Design a roof drainage system for a rectangular building 80 ft by 100 ft with a rooftop area of 8000 ft². Local drainage policy requires control of the 50-yr, 5-min rainfall (use the *IDF* curve of Figure 4–4). Use a slope of 0.25 in./ft for the horizontal drains.

8–2. Design a roof drainage system for a square building with a roof area of 90,000 ft². Local drainage policy requires control of the 25-yr, 10-min rainfall (use the *IDF* curve of Figure 4–4). Use a slope of 0.125 in./ft for horizontal drains located in the interior of the roof and 0.5 in./ft for those along the perimeter of the roof. Assume two vertical drains from the roof to ground level.

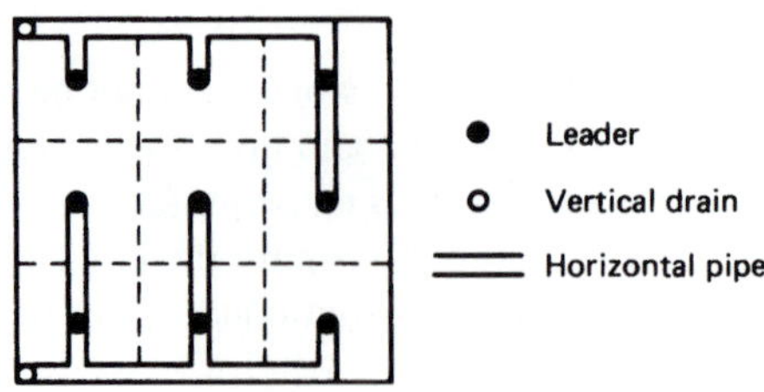

8–3. Design a roof drainage system for the commercial building shown. Local drainage policy requires control of the 50-yr, 0.1-hr storm event (use the *IDF* curve of Figure 4–4). Use a slope of 0.25 in./ft for all horizontal drains.

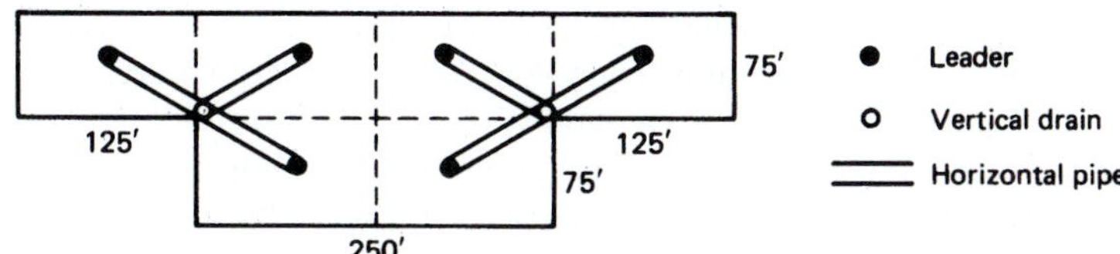

8–4. The schematic of the building shown is to be served by a single vertical pipe. The two-floor section has a roof area of 6000 ft^2 and the three-floor section has a roof area of 8000 ft^2. Horizontal drains should have slopes of 0.125 in./ft. Local drainage policy requires control of the 100-yr, 15-min rainfall (use the *IDF* curve of Figure 4–4).

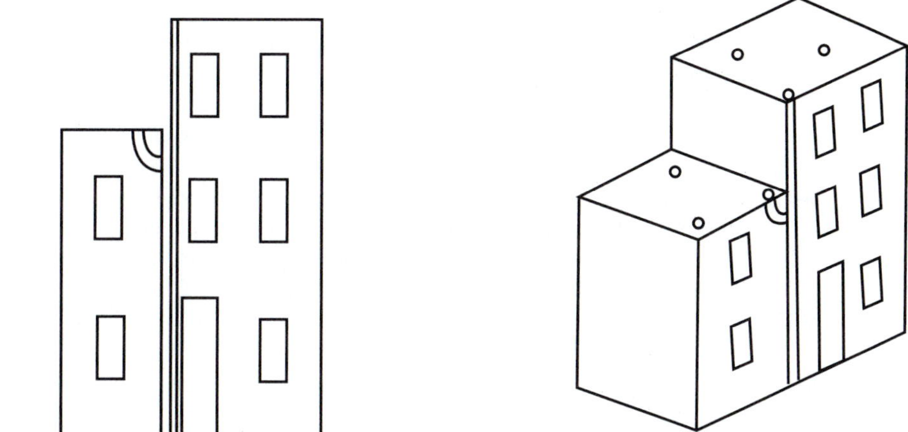

8–5. Estimate the discharge rate and the velocity of the flow in a triangular gutter having a longitudinal slope of 1%, a depth of 3 in., $n = 0.015$, and a cross slope of 0.0333 ft/ft. Compute the spread.

8–6. Because of right-of-way restrictions and policies on the allowable spread, the spread on a particular section of a proposed highway will be limited to 8 ft. If the 10-yr design discharge is 5.1 ft^3/sec, what cross slope z and curb height d will be necessary? Assume a triangular gutter section, with $n = 0.015$ and longitudinal slope S of 1.7%.

8–7. Runoff from a commercial area drains into a triangular gutter ($n = 0.013$). The curb is 5 in. high and drainage policy limits the spread to 10 ft. The longitudinal slope is 0.008 ft/ft and the cross slope z is 30. What discharge is permissible? If the design runoff coefficient and rainfall intensity has a product of 4.5, what is the maximum drainage area that could be served by an inlet?

8–8. Determine the depth of flow and top width in a 6-ft-diameter circular swale providing drainage from a highway section when the discharge rate is 8 ft^3/sec, the n is 0.019, and the longitudinal slope is 2.2%.

8–9. Design a semicircular swale made of corregated metal pipe to carry a discharge of 12.6 ft^3/sec at a slope of 0.9%.

8–10. A short-grass, *V*-shaped swale will be used to pass storm water past a commercial building. The swale must pass a flow of 12 ft^3/sec and limit the spread to 15 ft. The longitudinal slope is 2.6%. What side slopes will be necessary?

8–11. Surface runoff from a 3-acre section of a parking lot ($C = 0.9$) drains into a triangular gutter along one end of the lot. Assume a t_c of 10 min. Determine the curb height necessary to contain the 10-yr peak flow (use Figure 4–4). Assume $n = 0.015$, a cross slope of 0.04 and a longitudinal slope of 1.5%. Estimate the spread.

8–12. A study is undertaken to estimate the roughness coefficient for a new mix of asphalt. A triangular gutter is designed with a longitudinal slope of 1.8% and a cross slope z of 18. During the test, the depth of flow is held constant at 4 in. at the curb. During a period of 10 sec, 50 ft^3 of runoff is collected. Find the roughness coefficient.

8–13. A detention-basin riser has ten 1-in. diameter orifices at the base of the riser. Compute the total discharge through the orifices for depths of 0.5, 1, 2, 3, and 4 ft. Plot the depth-discharge relationship. Assume an orifice coefficient of 0.6.

8–14. A detention-basin riser has a rectangular orifice 6 in. wide and 4 in. high. When the water level is 2.5 ft above the orifice, a discharge rate of 1.22 ft^3/sec is measured. Estimate the discharge coefficient.

8–15. A detention-basin riser will be designed to pass discharge of 28.4 ft^3/sec. Assuming a discharge coefficient of 0.6 for the orifice, compute the area of the orifice if the design depth is 5 ft above the orifice.

8–16. A detention-basin riser uses a 2.75 ft weir ($C_w = 3.15$). Compute and plot the stage-discharge relationship for depths up to 5 ft.

8–17. Water can discharge from a controlled-outflow lake through two separate weirs; one with a length of 3 ft has an elevation of 127 ft and one with a length of 7 ft has an elevation of 129.5 ft. Both have a discharge coefficient of 3.0. Compute and plot the stage-discharge relationship for stages up to 6 ft above the low-elevation weir.

8–18. The design discharge for a detention-basin outlet is 139 ft^3/sec. Assuming a weir coefficient of 3, estimate the required weir length at a design depth of 4.5 ft.

8–19. Discharge measurements through a 5-ft weir were collected during four storm events, with the following values:

Storm	1	2	3	4
h (ft)	2.4	3.7	4.1	4.2
q (ft^3/sec)	55	102	126	127

Estimate the weir coefficient.

8–20. The design discharge for a detention basin outlet is 240 ft^3/sec. Assuming a weir coefficient of 3.1, estimate the required weir length at a design depth of 5.2 ft.

8–21. For the conditions of Problem 8–11, estimate the size of a grate inlet necessary to pass the flow into a subsurface pipe. Assume 80% efficiency and a loss coefficient of 0.6. Assume each opening has a length of 2.25 ft and a width of 0.15 ft.

8–22. For the conditions of Problem 8–6, estimate the size of a grate inlet necessary to pass the flow into a subsurface pipe. Assume a loss coefficient of 0.6 and 65% efficiency. Assume each opening has a length of 2 ft and a width of 0.12 ft.

8–23. For the conditions of Problem 8–5, estimate the size of a grate inlet necessary to pass the flow into a subsurface pipe. Assume a loss coefficient of 0.6 and 75% efficiency. Assume the grate has rectangular openings 1 ft by 0.1 ft.

8–24. A grate inlet has 15 rectangular openings, each 2.25 ft long and 0.15 ft wide. Assume the inlet has a C_d of 0.6 and an efficiency of 70%. If the curb is 5 in. high, what discharge will pass through the grate? If eight of the openings were clogged with litter, what discharge would pass through the grate? Assuming that the discharge rate is constant for 5 min, what volume of water would pond in the street when the grate inlet is partially clogged?

8–25. Design a curb-opening inlet to pass a discharge of 6.7 ft^3/sec at a maximum depth of 5 in. Assume a discharge coefficient of 3.2.

8–26. An existing 6-ft long curb-opening inlet with an opening height of 6 in. has a discharge coefficient of 3.2. Compute the discharge that can pass into the inlet.

8–27. A combination inlet has a 5-ft long curb opening with a height of 6 in. and a grate-opening area of 3 ft^2. Use loss coefficients of 3 and 0.55 for the curb opening and grate, respectively. Compute the stage-discharge relationship to a depth of 6 in.

8–28. At a particular intersection a combination inlet is needed to handle a discharge of 14.5 ft^3/sec. The curb opening will be 5 in. high. The standard grate inlet used by this highway department has 12 openings, each 2 ft by 0.12 ft. Design the combination inlet, assuming weir and orifice loss coefficients of 3 and 0.6, respectively.

8–29. For the highway section of Figure 8–10, design slotted drain inlets if the roadway is 80 ft wide (rather than 56 ft) and has a slope of 1.5% (rather than 2%).

8–30. For the highway section of Figure 8–10, design the first slotted drain inlet if the roadway is 44 ft wide (rather than 56 ft) and the policy requires control of the 6-min, 25-yr storm.

8–31. Assuming that the inlet is at a sag, design a slotted drain inlet to control runoff from the parking lot of Problem 8–11.

8–32. Assuming that the inlet is at a sag, design a slotted drain inlet to control the 25-yr runoff from the parking lot of Problem 8–11.

8–33. Design a slotted drain inlet to control the 10-yr runoff for 9-acre area of a sports complex that has a *CN* of 82. Assume that policy allows the use of a minimum time of concentration of 20 min. Use the *IDF* curve of Figure 4–4. Assume the longitudinal slope is 0.8%, the cross slope is 5%, and the maximum allowable depth is 6 in. Assume most surfaces are smooth concrete.

8–34. Design a pipe culvert (n = 0.013) that would carry the discharge for the following conditions: 25-yr return period, 17-acre watershed, short-grass land cover, slope of 3.5%, and a length of 1150 ft. Use Figure 4–4. Assume an unsubmerged inlet and outlet.

8–35. Design two equal pipe culverts (n = 0.018; S = 2%) that would carry the discharge for the following conditions: 10-yr return period, area = 45 acres, residential development with 0.5-acre lots on C soils. The principal flow path is 200 ft of paved-area sheet flow (3% slope), 350 feet of flow in a nonvegetated gully (2.5% slope), and 1500 ft of flow in a shallow channel (n = 0.035, S = 2%, R_h = 0.6 ft).

8–36. For the conditions of Problem 8–35, except using a 50-yr return period and all slopes of 4%, design the pipes (n = 0.013) required to carry the flow.

8–37. An existing 30-in. pipe passes a discharge of 15 ft^3/sec at full flow. If the pipe is laid at a slope of 0.9 ft per 100 ft, estimate the roughness coefficient of the pipe.

8–38. A 36-in. pipe (n = 0.012) has a discharge coefficient of 0.6. When the water is ponded 5 ft above the invert, what discharge will pass through the culvert?

8–39. A 24-in. pipe (n = 0.013) has a discharge coefficient of 0.6. When the water is ponded 5.5 ft above the invert, what discharge will pass through the culvert?

8–40. During a storm, a discharge of 120 ft^3/sec is passing through a 48–in. pipe (n = 0.013). At that time, the volume of runoff stored at the inlet has a depth of 5.5 ft above the invert. Estimate the discharge coefficient.

8–41. During a storm, a discharge of 92 ft^3/sec passes through a 42-in. pipe (n = 0.012). At that time, the storage at the inlet has a depth of 6 ft above the invert. Estimate the discharge coefficient.

8–42. Determine the pipe diameter that would be necessary to pass a discharge of 124 ft^3/sec when the water is ponded to a depth of 7.5 ft above the invert of the pipe. Assume $C_d = 0.6$.

8–43. Water is ponded behind a highway embankment to a depth of 5.6 ft above the invert of a culvert. Determine the pipe diameter needed to pass a flow of 74 ft^3/sec. Assume $C_d = 0.6$.

8–44. Determine the pipe diameter necessary to carry a discharge of 58 ft^3/sec under a roadway where water will be ponded to a depth of 7.2 ft above the invert of the inlet and 4.6 ft above the invert of the outlet. Assume the pipe will be laid at a slope of 1.1%, has a length of 82 ft, and a roughness coefficient of 0.024.

8–45. Determine the pipe size necessary to pass a discharge of 83 ft^3/sec under a roadway where water will be ponded to a depth of 6.7 ft above the invert of the inlet and 3.8 ft above the invert of the outlet. Assume the pipe will be laid at a slope of 2.4%, has a length of 75 ft, and a roughness coefficient of 0.023.

8–46. A 23-acre watershed undergoes development. The runoff coefficients (C) for pre- and post-development conditions are 0.2 and 0.4, respectively. The pre- and post-development times of concentration are 0.25 and 0.15 hr, respectively. Estimate the depth (in.) and volume (acre-ft) of detention storage required using the loss-of-natural-storage method. Assume that the *IDF* curve of Figure 4–4 applies and the design policy requires control for a 10-yr event.

8–47. The runoff coefficients (C) for pre- and post-development conditions of a 46-acre watershed are 0.3 and 0.55, respectively. The pre- and post-development times of concentration are 45 and 30 min, respectively. Assuming a 25-yr return period and the *IDF* curve of Figure 4–4, estimate the depth (in.) and volume (acre-ft) of detention storage required using the loss-of-natural-storage method.

8–48. The SCS curve numbers for pre- and post-development conditions of a 32-acre watershed are 66 and 78, respectively. The pre- and post-development times of concentration are 30 and 20 min, respectively. Assuming a 25-yr return period and the *IDF* curve of Figure 4–4, estimate the depth (in.) and volume (acre-ft) of detention storage using the loss-of-natural-storage method. Use the SCS rainfall-runoff equation to estimate depths of runoff.

8–49. The SCS curve numbers for pre- and post-development conditions of a 54-acre watershed are 72 and 82, respectively. The pre- and post-development times of concentration are 50 and 35 min, respectively. Assuming a 50-yr return period and the *IDF* curve of Figure 4–4, estimate the depth (in.) and volume (acre-ft) of detention storage using the loss-of-natural-storage method. Use the SCS rainfall-runoff equation to estimate depths of runoff.

For problems 8–50 to 8–53, estimate the volume of detention storage in both in. and acre-ft using the generalized planning method.

8–50. Use data and characteristics from Problem 8–46.

8–51. Use data and characteristics from Problem 8–47.

8–52. Use data and characteristics from Problem 8–48.

8–53. Use data and characteristics from Problem 8–49.

8–54. Use data and characteristics from Problem 8–46, and estimate the volume of detention storage required in both in. and acre-ft with the Rational Formula Hydrograph Method.

8–55. Use data and characteristics from Problem 8–47, and estimate the volume of detention storage required in both in. and acre-ft with the Rational Formula Hydrograph Method.

8–56. Use data and characteristics from Problem 8–46, and estimate the volume of detention storage required in both in. and acre-ft with the Baker Method.

8–57. Use data and characteristics from Problem 8–47, and estimate the volume of detention storage required in both in. and acre-ft with the Baker Method.

For problems 8–47 to 8–50, estimate the volume of detention storage required in both in. and acre-ft using the Abt and Grigg Model. Assume $m = 5/3$.

8–58. Use data and characteristics from Problem 8–46.

8–59. Use data and characteristics from Problem 8–47.

8–60. Use data and characteristics from Problem 8–48.

8–61. Use data and characteristics from Problem 8–49.

For Problems 8–62 to 8–65, estimate the volume of detention storage required in both in. and acre-ft using the Wycoff and Singh Model. Assume $T_b/t_p = 2.67$.

8–62. Use data and characteristics from Problem 8–46.

8–63. Use data and characteristics from Problem 8–47.

8–64. Use data and characteristics from Problem 8–48.

8–65. Use data and characteristics from Problem 8–49.

For Problems 8–66 and 8–67, estimate the volume of detention storage required in both in. and acre-ft using the SCS TR-55 Model.

8–66. Use data and characteristics from Problem 8–48.

8–67. Use data and characteristics from Problem 8–49.

For Problems 8–68 to 8–71, evaluate the characteristics of a single-stage riser with a weir for the data, characteristics, and planning method specified by the problem. Assume that $n = 0.013$, $L = 150$ *ft.*

8–68. Use the method and data of Problem 8–66. Derive the stage-storage relationship by assuming a detention basin with a rectangular bottom with an area of 0.9 acre and vertical sides. Use $E_o = 3$ ft; $E_c = 1.5$; $C_w = 3$.

8–69. Use the method and data of Problem 8–67. Derive the stage-storage relationship by assuming a detention basin with a rectangular bottom with a length of 300 ft, a width of 200 ft, and side slopes of 3:1 (*h*:*v*). Use $E_o = 2.5$ ft; $E_c = 1.5$ ft; $C_w = 3.1$.

8–70. Use the method and data of Problem 8–50. Derive the stage-storage relationship by assuming a detention basin with a rectangular bottom (length = 125 ft; width = 75 ft) and vertical sides. Use $E_o = 2$ ft; $E_c = 1$ ft; $C_w = 3$.

8–71. Use the method and data of Problem 8–51. Derive the stage-storage relationship by assuming a detention basin with a rectangular bottom that has a length of 300 ft, a width of 150 ft, and side slopes of 4.5:1 (*h*:*v*). Use $E_o = 2$ ft; $E_c = 1.5$ ft; $C_w = 3.1$.

For Problems 8–72 to 8–75, evaluate the characteristics of a single-stage riser using an orifice for controlling the outflow. Use the data, characteristics, and method specified by the problem.

8–72. Use the information provided with Problem 8–68.

8–73. Use the information provided with Problem 8–69.

8–74. Use the information provided with Problem 8–70.

8–75. Use the information provided with Problem 8–71.

For Problems 8–76 and 8–77, evaluate the characteristics of a single-stage riser using a weir to control the outflow for the larger event and an orifice for the smaller event.

8–76. Use the information provided with Problem 8–68 and a 5-yr event for the smaller event.

8–77. Use the information provided with Problem 8–69 and a 10-yr event for the smaller event.

REVIEW QUESTIONS

8–1. Detention storage is required because land development reduces natural watershed storage. Which one of the following would be least affected by land development? (a) Interception storage; (b) depression storage; (c) soil storage; (d) swale storage.

8–2. Which one of the following hydrologic runoff characteristics is not affected by land development? (a) The timing of runoff; (b) the peak runoff rate; (c) the velocity of the runoff; (d) the volume of runoff; (e) all of the above are affected.

8–3. Which one of the following is not an advantage of a policy that requires a permanent pool behind a detention structure (that is, a wet pond policy)? (a) Water quality enhancement; (b) reduction in land and construction costs; (c) aesthetical improvement of the site; (d) wildlife habitat improvement.

8–4. Which one of the following is not a component of a detention facility? (a) An antiseep collar; (b) a trash rack; (c) an emergency spillway; (d) an antivortex device; (e) all of the above are components.

8–5. Which one of the following is the reason for determining simultaneously the size of the riser outlet and the volume of detention storage? (a) Maximize design accuracy; (b) minimize cost; (c) maximize safety: (d) minimize required storage.

8–6. In the analysis phase of storm water management, which one of the following is the unknown? (a) The stage-storage relationship; (b) the stage-discharge relationship; (c) the upstream hydrograph; (d) the downstream hydrograph.

8–7. In storm water management synthesis, which one of the items in Review Question 8–6 is the unknown?

8–8. Which one of the following is not a difference between storm water management planning and design? (a) Design requires a stage-discharge relationship; (b) design requires a computerized solution; (c) design requires a hydrograph analysis; (d) design requires storage routing; (e) all of the above are differences.

8–9. If the pre- and post-development runoff depths are 2.4 and 3.2 in., respectively, the required volume of storage for a 0.25-mi^2 watershed, when estimated using the loss-of-natural-storage method, would be (a) 0.2 in.; (b) 3.2 in.; (c) 10.7 acre-ft; (d) 160 acre-in.; (e) none of the above.

8–10. If the value of the volume of detention storage/volume of runoff ratio is 0.4, the pre-and post-development runoff depths are 0.8 and 2.1 in., respectively, and the drainage area is 50 acres, the required storage volume (acre-ft) is (a) 3.5; (b) 5.42; (c) 16; (d) 42.

8–11. Which one of the following is not an input for sizing a single-stage riser? (a) The return period of the design storm; (b) the stage-storage-discharge relationship; (c) the length of the weir; (d) the post-development depth of runoff; (e) all are inputs.

8–12. Which one of the following energy loss factors is not needed for sizing a riser? (a) Manning's roughness coefficient n for the pipe; (b) the frictional head loss coefficient K_p; (c) the weir coefficient C_d; (d) the coefficient C_*, which reflects frictional losses in the conduit; (e) all are needed.

DISCUSSION QUESTION

The technical content of this chapter is important to the professional hydrologist, but practice is not confined to making technical decisions. The intent of this discussion question is to show that hydrologists must often address situations where value issues intermingle with the technical aspects of a project. In discussing the stated problem, at a minimum include responses to the following questions:

1. What value issues are involved, and how are they in conflict?
2. Are technical issues involved with the value issues? If so, how are they in conflict with the value issues?
3. If the hydrologist attempted to rationalize the situation, what rationalizations might he or she use? Provide arguments to suggest why the excuses represent rationalization.
4. What are the hydrologist's alternative courses of action? Identify all alternatives, regardless of the ethical implications.
5. How should the conflict be resolved?

You may want to review Sections 1.6 to 1.12 in Chapter 1 in responding to the problem statement.

Case. An engineer involved in land development frequently designs detention basins to control the additional storm water that runs off of the urbanized land. The local drainage policy requires use of the Rational Formula Hydrograph Method (Equation 8–67) in sizing the storage volume. The Rational Method is used to estimate the peak discharges. In computing the peak discharges with the Rational Method, the engineer uses the upper limit on the runoff coefficient C for the pre-development condition (for example, 0.3 for the unimproved category in Table 7–12) and the lower limit on the C value for the post-development condition (for example, 0.4 for detached, multiunit residential category in Table 7–10). This practice results in the smallest possible storage volume, which reduces both the cost of the detention facility and the amount of land that must be devoted to flood control. The company's project costs are lower and the profits maximized.

9

Hydrograph Analysis and Synthesis

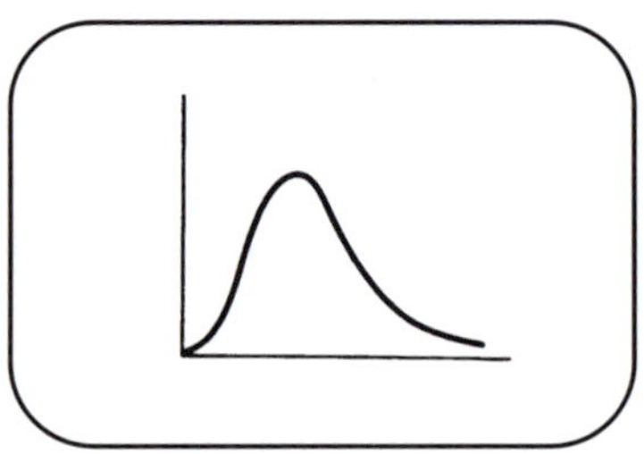

CHAPTER OBJECTIVES

1. Introduce a conceptual framework for representing and analyzing measured rainfall and runoff data.
2. Demonstrate hydrograph methods frequently used in engineering design.
3. Provide a discussion of empirical methods for representing infiltration capacity.
4. Discuss assumptions that underlie commonly used hydrograph methods.
5. Provide a summary of input to the computer model TR-20.

9.0 NOTATION

a = coefficient of gamma distribution
A = drainage area
b = coefficient of gamma distribution
C = runoff coefficient for Rational Method
C = constant of integration
C_p = coefficient of Snyder's Hydrograph Method
C_t = watershed storage coefficient
d_p = depth of ponding
D = duration of rainfall
D = duration of unit hydrograph
D_1 = a time duration
f = fraction of rainfall contributing to losses
f = infiltration capacity
f = ratio of time to time to peak
f_a = depth variable
f_c = ultimate infiltration capacity
f_o = initial infiltration capacity
f_r = infiltration capacity at start of recovery
F = total depth of water infiltrated
F_p = volume of water ponded
h = head
i = rainfall intensity
I_a = initial abstraction
K = rate constant for master depletion curve
K = Rational Method conversion factor
K = rate constant for infiltration
K_r = rate constant for recovery of infiltration capacity
K_s = hydraulic conductivity
L = flow length
L = watershed length
$L(t)$ = time-dependent loss function
L_{ca} = length to the center of area of a watershed
L_w = depth to wetting front
n_p = number of rainfall excess ordinates
n_u = number of unit hydrograph ordinates
n_{pe} = number of rainfall excess ordinates
n_{ro} = number of direct runoff ordinates
p_i = ith ordinate of rainfall excess distribution
P = rainfall depth (or precipitation)
$\mathbf{P}$ = vector of rainfall values
$P(t)$ = time distribution of rainfall
$PE(t)$ = distribution of rainfall excess
P_e = depth of rainfall excess
q = discharge rate
q_a = peak discharge per square mile for Snyder's Unit Hydrograph
q_b = baseflow discharge
q_d = direct runoff discharge
q_d = distribution of direct runoff
q_0 = initial discharge rate at time $t = 0$
q_p = peak discharge
$\bar{q}_p$ = average peak discharge
q_s = discharge at minimum flow rate
q_t = discharge rate at time t
q_u = unit discharge
Q = depth of direct runoff
Q_d = depth of direct runoff
$\mathbf{R}$ = vector of direct runoff distribution
S_i = ith ordinate of an S-hydrograph
S_w = soil water suction
t = time
t' = time as measured from start of recovery
t_b = time base of hydrograph
t_c = time of concentration
t_d = duration of rainfall excess
t_e = equivalent time to infiltrate a specified volume
t_L = lag time
t_{La} = adjusted lag time
t_p = time to peak of hydrograph
t_p = time to beginning of ponding
$\bar{t}_p$ = average time to peak
t_r = time at end of recession of a hydrograph
t_r = time at inflection point
t_s = time at start of direct runoff

T_b = time base of hydrograph
T_t = reach travel time
U = vector of unit hydrograph ordinates
U_j = j^{th} ordinate of a unit hydrograph
V = flow velocity
V_d = depth of direct runoff
V_L = total depth of losses
V_p = depth of rainfall
W_{50} = width (time) of Snyder's Unit Hydrograph at 50% of q_p
W_{75} = width (time) of Snyder's Unit Hydrograph at 75% of q_p
$x(t)$ = input at time t
$y(t)$ = output at time t
Δt = time increment
$\Delta\phi$ = change in phi-index value
θ_i = initial volumetric water content
θ_s = saturated volumetric water content
τ = time lag
ϕ = phi-index loss rate

9.1 INTRODUCTION

While peak discharge rates are adequate for many engineering design problems, they are inadequate for design problems where watershed or channel storage is significant. This can be storage in manmade structures, storage at a natural constriction in the flow path, or channel storage. Where storage is significant, design work is usually based on hydrographs rather than peak discharge rates. Hydrographs are also used for designs on nonhomogeneous watersheds, such as where significant variation in land use, soil types, or topography exists within the watershed. This is especially necessary where parts of watersheds undergo land use change such as urbanization or deforestation.

9.1.1 Hydrographs and the Systems Process

In discussing the concept of hydrograph analysis and synthesis, it is instructive to discuss the issue in terms of the concept of systems theory. This concept is represented in Figure 1–3. The system consists of three functions: the input function, the transfer function, and the output function. The rainfall hyetograph is the input function and the total runoff hydrograph is the output function. Although a number of concepts have been introduced to represent the transfer function, only the concept of a unit hydrograph, which is defined below, will be discussed herein.

One purpose of hydrograph analysis is to analyze measured rainfall and runoff data to obtain an estimate of the transfer function. Once the transfer function has been developed (identified), it can be used with design storms and measured rainfall hyetographs to compute (synthesize) the runoff that would be expected. The resulting runoff hydrograph could then be used for design purposes.

9.1.2 Hydrograph Analysis

While a number of conceptual frameworks are available for hydrograph analysis, the one presented herein will involve the following: (1) the separation of the rainfall hyetograph into three parts, (2) the separation of the runoff hydrograph into two parts, and (3) the identification of the unit hydrograph as the transfer function. The rainfall hyetograph is separated into three time-dependent functions: the initial abstraction, the loss function, and the rainfall excess; these are shown in Figure 9–1. The *initial abstraction* is that part of the rainfall that occurs prior to the start of direct runoff. The *rainfall excess* is that part of the rainfall that appears as direct runoff. The *loss function* is that part of the rainfall that occurs after the start of direct runoff but does not appear as direct runoff. The process is sometimes conceptualized as a two-part separation of the rainfall, with the initial abstraction being included as part of the loss function. The three components are used here for clarity and to emphasize the differences between the important processes of the hydrologic cycle.

The runoff hydrograph is conceptually separated into two parts: direct runoff and baseflow (Figure 9–1). The *direct runoff* is the storm runoff that results from rainfall excess; the volumes of rainfall excess and direct runoff must be equal. The *transfer function,* or unit hydrograph, is the function that transforms the rainfall excess into the direct runoff. For the purpose of our conceptual framework, *baseflow* is the runoff that has resulted from an accumulation of water in the watershed from past storm events and would appear as streamflow even if the rain for the current storm event had not occurred.

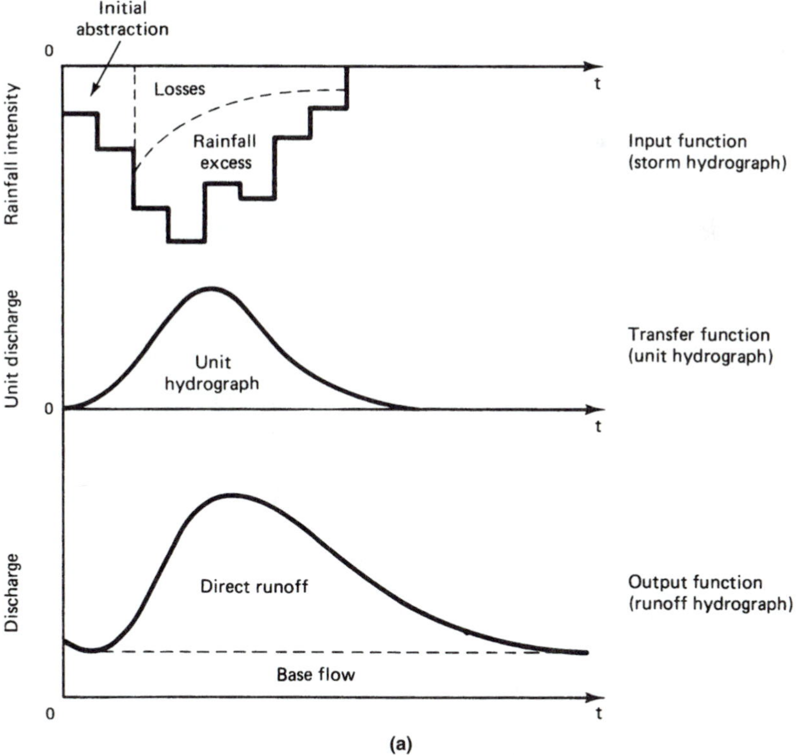

FIGURE 9–1 Rainfall-Runoff Process: (a) Conceptual Model; (b) Application for the Event of February 5–6, 1964, on Brush Creek Watershed W–1, Blacksburg, VA.

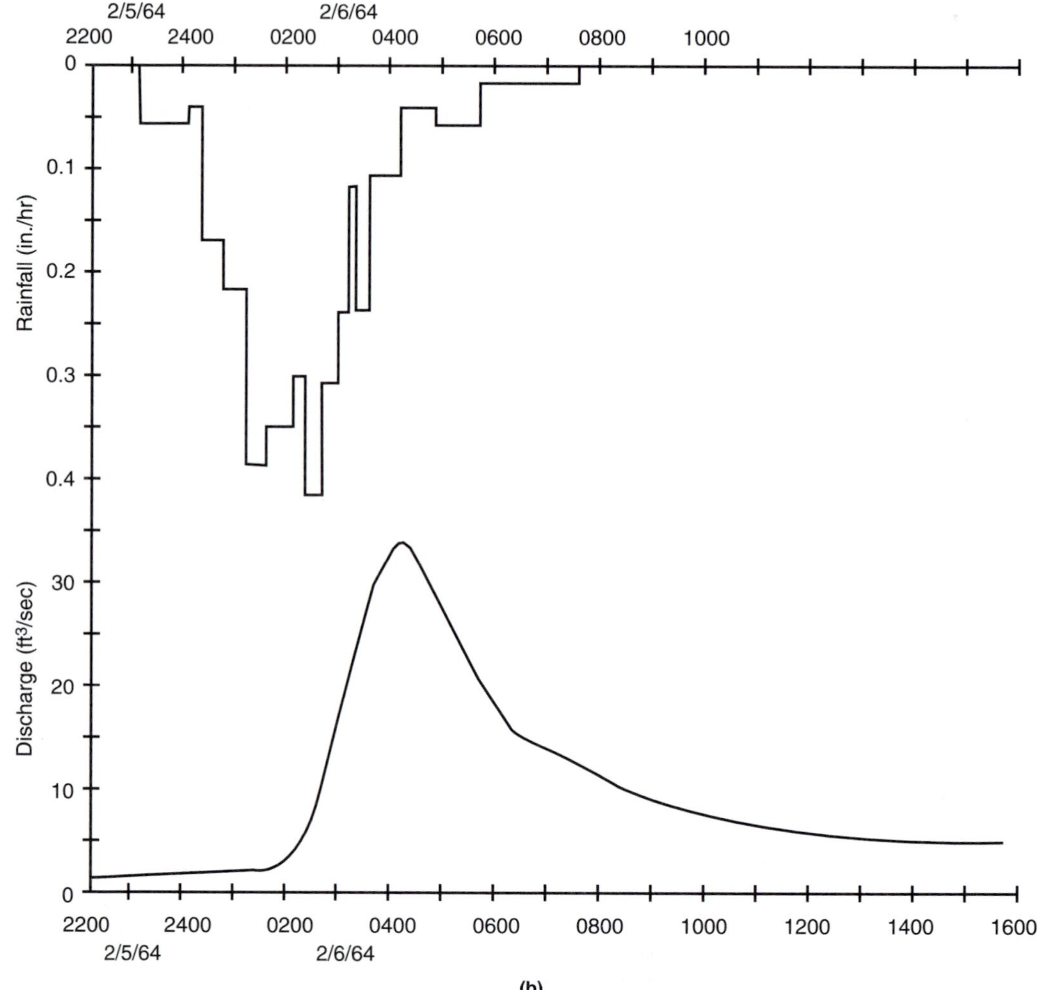

(b)

FIGURE 9–1 *Continued*

In performing a hydrograph analysis, it is common to begin by separating the baseflow from the total runoff hydrograph, which produces the direct runoff hydrograph. This is usually the first step because baseflow is usually a smooth function and it can be estimated more accurately than some of the other separation steps. Having estimated the baseflow and direct runoff hydrographs, the volume of direct runoff can be computed.

Having analyzed the runoff hydrograph, attention can be directed to the rainfall hyetograph. The initial abstraction is delineated if it is to be handled separately from the other losses. The losses are then separated from the total rainfall hyetograph such that the volume of rainfall excess equals the volume of direct runoff. The individual topics are introduced in this chapter in this order.

9.1.3 Hydrograph Synthesis

Having completed the analysis phase, the results of the analysis can be used with design storms to synthesize hydrographs either at ungaged locations, that is, at locations where data to conduct analyses are not available, or at gaged locations. In the synthesis phase, a rainfall hyetograph and a synthetic transfer function are used to compute a runoff hydrograph. The process of transforming the rainfall excess into direct runoff using the transfer function is called *convolution.* The rainfall hyetograph can be a synthetic design storm or a measured storm event. In summary, in the analysis phase, the hyetograph and hydrograph are known and the transfer function is estimated. In the synthesis phase, a hyetograph is used with the synthetic transfer function to compute a runoff hydrograph.

9.2 BASEFLOW SEPARATION

As indicated previously, the total runoff hydrograph can be viewed as consisting of two parts, direct runoff and baseflow. The baseflow is the water discharged from an extensive ground-water aquifer. The term hydrograph separation refers to the process of separating the time distribution of baseflow from the total runoff hydrograph. The direct runoff hydrograph is the difference between the total runoff hydrograph and the baseflow function. As with any problem in hydrologic analysis, a number of methods have been proposed for separating the direct runoff and baseflow. The methods reflect the different views on the hydrologic processes involved. The method selected for any one watershed analysis will depend on the type and amount of measured data available, the necessary accuracy for the design problem, and the effort that the modeler wishes to expend.

Baseflow rates for large watersheds are evident from mean daily discharge records. For example, in Chapter 5 (Table 5–1) the daily record for a 72.8-sq. mi watershed was given. The periods September 7–12 and September 17–24 indicate the baseflow is approximately 10 ft^3/sec. During the winter the baseflow appears to be higher. For example, in December and January, low flows of approximately 50 ft^3/sec are common.

9.2.1 Constant-Discharge Baseflow Separation

By far the easiest method to use is the constant-discharge method. Simply stated, the line separating baseflow and direct runoff begins at the point of the lowest discharge rate at the start of flood runoff and extends at a constant discharge rate until it intersects the recession limb of the hydrograph; this is shown in Figure 9–2. The baseflow is set by the lowest discharge rate, q_s, just prior to the start of the rising limb of the hydrograph. Mathematically, the baseflow function is

$$q_b = \begin{cases} q & \text{for} \quad t < t_s & \text{(9–1a)} \\ q_s & \text{for} \quad t_s \le t \le t_e & \text{(9–1b)} \\ q & \text{for} \quad t_e < t & \text{(9–1c)} \end{cases}$$

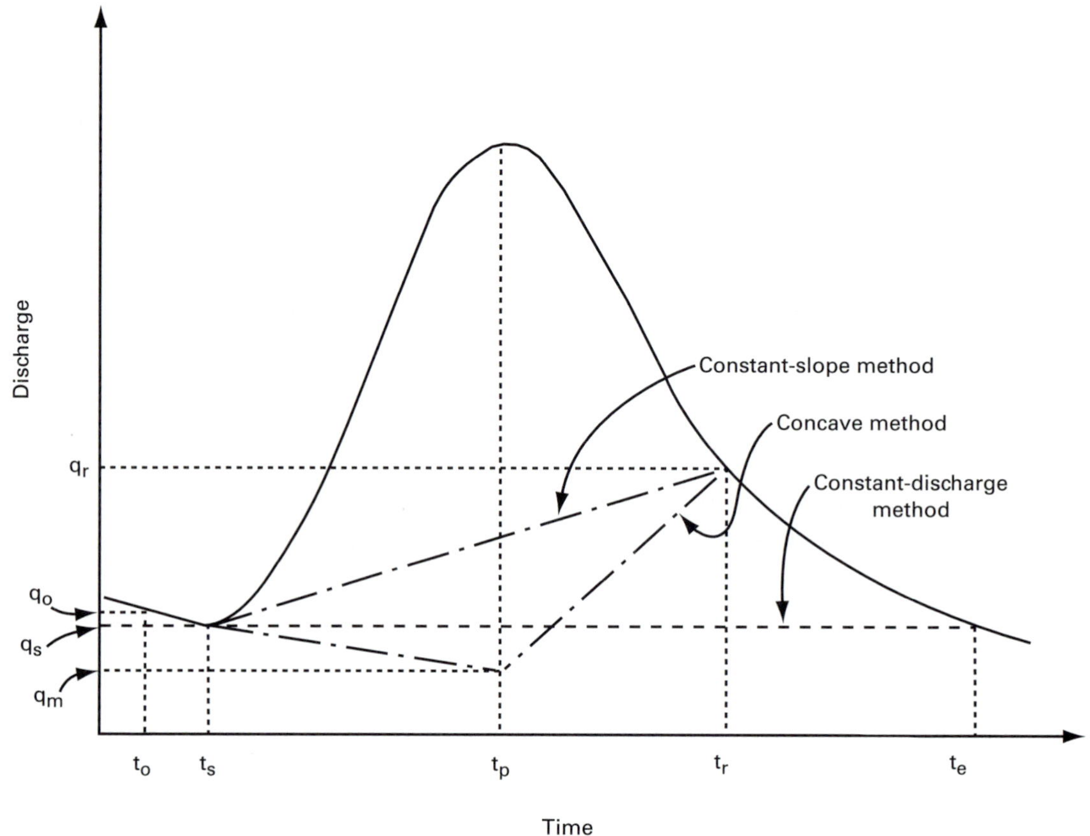

FIGURE 9–2 Baseflow separation methods.

where the variables are defined in Figure 9–2. The distribution of direct runoff, q_d, equals the difference between the total discharge, q, and the baseflow, q_b.

Although this method is easy to apply, it is easy to argue with the conceptual basis of the method. It can be argued that the baseflow will continue to decrease beyond the start of the flood runoff, possibly to the time when the flood runoff is maximum. It could also be argued that flow from ground water storage begins before the time when the total runoff equals the baseflow rate prior to the start of flood runoff. Such arguments have led to proposals of other methods.

9.2.2 Constant-Slope Baseflow Separation

If the runoff process is conceptualized in a way that flow from ground water aquifers begins on the hydrograph recession prior to the point used for the constant-discharge method, then it is only necessary to select the point on the recession curve where direct runoff ends. Several methods have been proposed for identifying this point. The most common conceptualization uses the inflection point on the hydrograph recession; the inflection point is the point where the hydrograph goes from being concave to convex (that is, the slope being greater than 1 to the slope being less than 1). A second method uses an empirical formula, such as the following equation that has been proposed for very large watersheds:

$$N = A^{0.2} \tag{9–2}$$

in which N is the number of days from the time to peak of the measured runoff hydrograph to the end of direct runoff and A is the drainage area in square miles. A third method is called the master depletion curve method, which is discussed below. While these three options are quite objective, the point is often selected subjectively using a graph of the runoff hydrograph.

Once the time at which direct runoff ends has been identified, direct runoff and baseflow are separated by connecting a straight line extending from the point of the lowest discharge rate at the start of the flood runoff to the point on the recession. The constant-slope method is also shown in Figure 9–2. The baseflow depends on the time (t_s) and discharge (q_s) at the low point of the discharge function, q, and the time (t_r) and discharge (q_r) at the inflection point. Mathematically, the baseflow function is

$$q_b = \begin{cases} q & \text{for} \quad t < t_s & \text{(9–3a)} \\ q_s + (t - t_s)\left[\dfrac{q_r - q_s}{t_r - t_s}\right] & \text{for } t_s \le t \le t_r & \text{(9–3b)} \\ q & \text{for } t_r < t & \text{(9–3c)} \end{cases}$$

where the variables are defined in Figure 9–2. The distribution of direct runoff, q_d, equals the difference between the total discharge, q, and the baseflow, q_b.

9.2.3 Concave Baseflow Separation

For the concave method, the starting and ending points for the line separating baseflow and direct runoff are the same as for the constant-slope method. However, for the concave method, baseflow continues to decrease until the time of the peak discharge of the storm hydrograph. At that time, the separation line is straight between that point and the inflection point on the recession. This is shown in Figure 9–2. While the concave method may require a little more effort to define than the other two methods, it is probably a more realistic representation of the actual separation of flow as determined by the physical processes that control flow during storm events. Mathematically, the baseflow function is

$$q_b = \begin{cases} q & \text{for } t < t_s & \text{(9–4a)} \\ q_s + (t - t_s)\left[\dfrac{q_s - q_o}{t_s - t_o}\right] & \text{for } t_s \le t \le t_p & \text{(9–4b)} \\ q_m + (t - t_p)\left[\dfrac{q_r - q_m}{t_r - t_p}\right] & \text{for } t_p \le t \le t_r & \text{(9–4c)} \\ q & \text{for } t_r < t & \text{(9–4d)} \end{cases}$$

where the variables are defined in Figure 9–2. The distribution of direct runoff, q_d, equals the difference between the total discharge, q, and the baseflow, q_b.

Example 9–1

The ordinates of a hydrograph measured on a 30-min interval are given in column 2 of Table 9–1. The watershed has a drainage area of 74 acres. The three baseflow separation methods were used to compute the direct-runoff distributions, and the trapezoidal rule was used to compute the volume of direct runoff.

For the constant-discharge method, the lowest discharge is 40 cfs, so the baseflow equals 40 cfs between storm times 1030 and 1530 hours; otherwise, the baseflow equals the total discharge. The volume of direct runoff is computed using the sum of the ordinates, the area, and the time increment:

$$490\,\frac{\text{ft}^3}{\text{sec}} \times 30\text{ min} \times \frac{60\text{ sec}}{\text{min}} \times \frac{1}{74\text{ ac}} \times \frac{1\text{ ac}}{43560\text{ ft}^2} \times \frac{12\text{ in.}}{\text{ft}} = 3.28\text{ in.}$$

The data of columns 1 and 2 were plotted; the resulting graph is characterized by a sharp break in the slope of the recession at a time of 1330 hours. This time is used to define the inflection point.

For the constant-slope method, the following equation defines baseflow for the period from storm time 1030 to the recession inflection point at 1330 hours:

$$q_b = 40 + (t - 10{:}30)\left[\frac{82 - 40}{13{:}30 - 10{:}30}\right] = 40 + (t - 10{:}30)\left(\frac{42\text{ cfs}}{180\text{ min}}\right)$$

The resulting baseflow is given in column 5 of Table 9–1. The computed volume of direct runoff is 1.84 in.

TABLE 9–1 Example 9–1: Computation of Baseflow (q_b) and Direct Runoff (q_d) Distributions

Time of Day (hr)	Flow Rate (ft³/sec)	Constant-Discharge Method q_b (ft³/sec)	Constant-Discharge Method q_d (ft³/sec)	Constant-Slope Method q_b (ft³/sec)	Constant-Slope Method q_d (ft³/sec)	Concave Method q_b (ft³/sec)	Concave Method q_d (ft³/sec)
0930	61	61	0	61	0	61	0
1000	49	49	0	49	0	49	0
1030	40	40	0	40	0	40	0
1100	54	40	14	47	7	31	23
1130	112	40	72	54	58	22	90
1200	153	40	113	61	92	13	140
1230	148	40	108	68	80	36	112
1300	113	40	73	75	38	59	54
1330	82	40	42	82	0	82	0
1400	71	40	31	71	0	71	0
1430	61	40	21	61	0	61	0
1500	52	40	12	52	0	52	0
1530	44	40	4	44	0	44	0
1600	36	36	0	36	0	36	0
1630	29	29	0	29	0	29	0
1700	22	22	0	22	0		
			490		275		419

For the concave method, the baseflow function is

$$q_b = \begin{cases} q & \text{for} \quad t \leq 10{:}30 \\ 40 + (t - 10{:}30)\left[\dfrac{49-40}{10{:}00-10{:}30}\right] & \text{for } 10{:}30 < t \leq 12{:}00 \\ 13 + (t - 12{:}00)\left[\dfrac{82-13}{13{:}30-12{:}00}\right] & \text{for } 12{:}00 < t \leq 13{:}30 \\ q & \text{for } 13{:}30 < t \end{cases}$$

The computed baseflows are given in column 7 of Table 9–1. The computed volume of direct runoff is 2.81 in.

In comparing the three distributions of direct runoff (columns 4, 6, and 8 of Table 9–1), the concave method yields the largest peak discharge even though the constant-discharge method yields the larger volume of direct runoff.

9.2.4 Master-Depletion-Curve Method

The master-depletion-curve method is used to provide a model of flow from ground water storage. Based on this, it can be used to identify a point on the recession where direct runoff ends and baseflow begins; however, it also provides a model of the recession limb. The procedure requires measured storm hydrographs for a good number of storm events covering a wide range of volumes and for different seasons of the year. The procedure is as follows:

1. Using semi-log paper, log q versus time, plot the recession curves for each storm event on separate pieces of tracing paper.
2. On a master sheet having a log q versus time-axis system (semi-log paper), plot the recession for the storm event having the smallest values of log q.
3. Using the recession curve with the next smallest values of log q values, position the tracing paper such that the curve appears to extend along a line coincident with the recession of the first event plotted.
4. Continue this process using successively larger magnitude log q recessions until all storm events are plotted.
5. Construct a master depletion curve that extends through the recessions of the observed storm events. Then fit a mathematical model to the master depletion curve; the following functional form often provides a reasonable fit to the data:

$$q_t = q_o e^{-Kt} \tag{9–5}$$

in which q_t is the discharge at time t, q_o is the discharge at time $t = 0$, and K is a fitting coefficient. The value of K can be determined using any two points of the master depletion curve. Letting one point be q_o, then making a natural-logarithm transformation of Equation 9–5 and solving for K yields

$$K = \frac{\ln_e q_o - \ln_e q_t}{t} \tag{9–6}$$

where t is the time at which discharge q_t is recorded. If sufficient scatter is evident in the recession line, then least squares can be used to estimate K. If the value of q_o is set, then the least squares estimator of K is

$$K = \frac{\sum_{i=1}^{n} t_i (\ln_e q_o - \ln_e q_i)}{\sum_{i=1}^{n} t_i^2} \tag{9–7}$$

in which n is the number of pairs of (q_i, t_i) points on the recession.

Example 9–2

The fitting and use of the master-depletion-curve method can be demonstrated using the data of Table 9–2, which gives parts of storm hydrographs for six storm events that were measured at 6-hr time intervals. Using five of the six storm events, the procedure above was used to obtain the master depletion curve for the watershed. Five storm events were plotted on tracing paper (Figure 9–3) to develop the master depletion curve, which is shown in Figure 9–4. The curve can be represented by the form of Equation 9–5 using any two points. Using the point where the master depletion curve has a discharge of 17.5 ft^3/sec at time $t = 0$, then $q_o = 17.5$ ft^3/sec. Taking the natural logs of Equation 9–5 yields:

$$\ln_e q_t = \ln_e q_o - Kt \tag{9–8}$$

TABLE 9–2 Discharges (ft^3/sec) for Fitting the Master Depletion Curve

Time Lag	Storm Event of 4/11/41	3/23/37	4/11/26	4/1/46	2/29/40	Test Event
1	44	32	84	19	9.6	7.3
2	33	27	23	9.3	5.7	12
3	28	21	12	6.9	4.3	23
4	24	18	10	5.8	4.1	38
5	21	16	8.2	5.1	3.6	44
6	18	14	6.2	4.6	3.0	47
7	17	12	6.0	4.1	2.6	46
8	15	11	5.9	3.5	2.4	40
9	14	9.5	5.8	3.1	2.3	31
10	13	8.0	5.8	3.0	2.2	23
11	12	7.7	5.7	2.9	2.1	18
12	12	7.5	2.9	2.1		14
13	11	7.2	2.9			11
						10
						9.6
						9.1
						8.6
						8.1
						7.9
						7.7

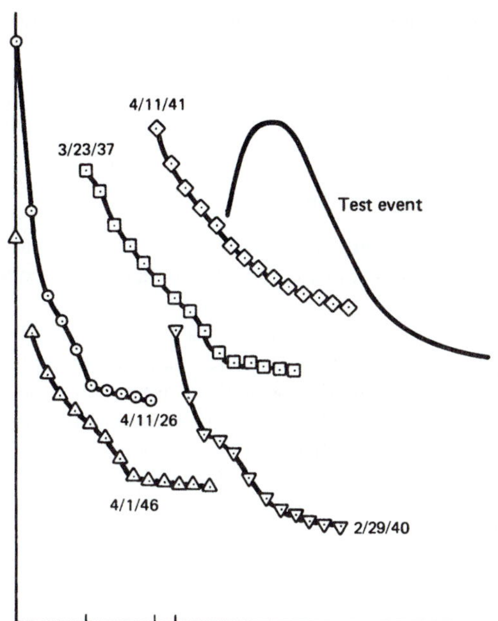

FIGURE 9–3 Plot of five storm events on tracing paper.

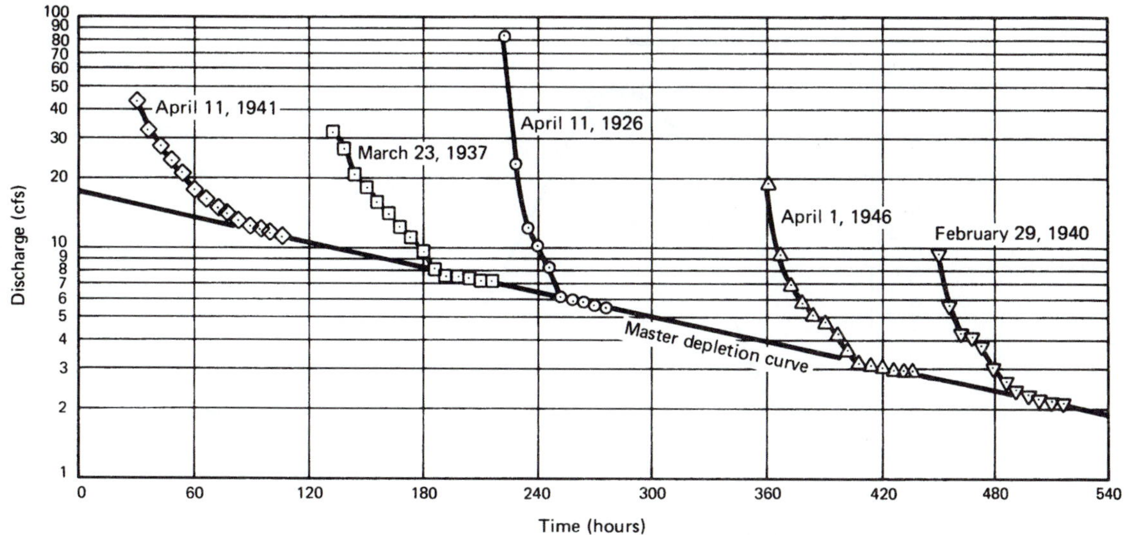

FIGURE 9–4 Fitting of master depletion curve using recessions of five storm events.

Using the discharge $q = 1.9$ ft^3/sec at $t = 540$ hr, Equation 9–8 yields the following:

$$\ln_e 1.9 = \ln_e 17.5 - K(540)$$

Solving for K yields $K = 0.00411$. Thus we get the following model for the master depletion curve:

$$q_t = 17.5e^{-0.00411t}$$

This equation can be used by defining any point on the master depletion curve at $t = 0$ and setting q_o equal to the discharge value at that point. Then the discharge at any other time can be determined from Equation 9–8.

The data for the test event of Table 9–2 can be used to show how the master depletion curve of Figure 9–4 can be used to define the end of direct runoff. Plotting the logarithms of the data on tracing paper (Figure 9–3) and superimposing the resulting recession onto the master depletion curve of Figure 9–4 indicates that the direct runoff ends at a discharge of 8 ft^3/sec. Thus the baseflow can be separated from the direct runoff by connecting a line from the discharge at time $t = 0$ to the discharge of 8 ft^3/sec; this is shown in Figure 9–5.

9.2.5 Baseflow and Direct Runoff

Regardless of the method used to define the time variation of baseflow, the curve that separates baseflow from the direct runoff varies smoothly. Since baseflow is water discharged from ground water storage, the physical processes provide considerable smoothing, regardless of the variability in the rainfall. The smoothness of the baseflow function makes it easy to compute the direct runoff, which is the difference between the measured runoff hydrograph and the baseflow. The volume of direct runoff can then be computed for use in separating the rainfall losses from the total rainfall.

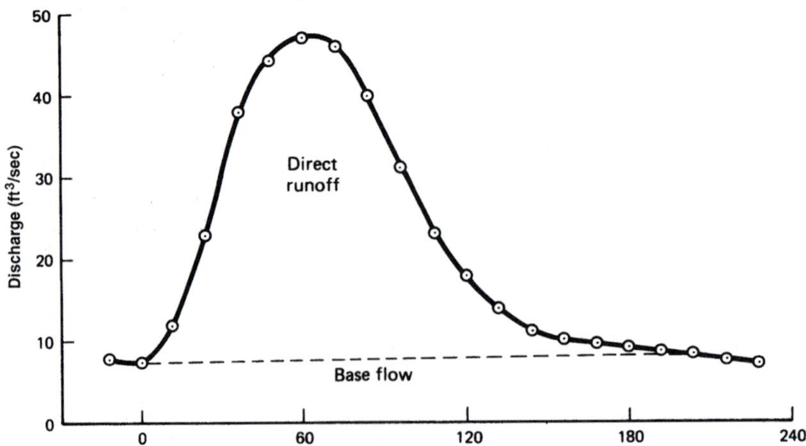

FIGURE 9–5 Baseflow separation for test storm event using a master depletion curve.

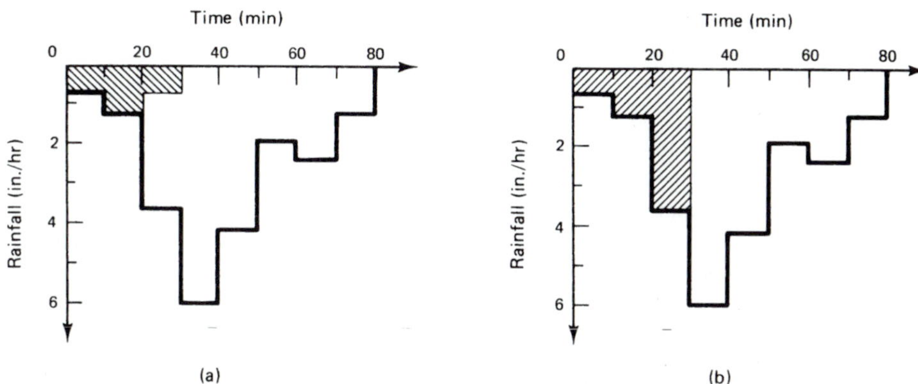

FIGURE 9–6 Separation of initial abstractions: (a) specific volume of 0.4 in.; (b) specific time at start of direct runoff.

9.3 ESTIMATION OF INITIAL ABSTRACTION

One definition of initial abstraction was previously stated. Specifically, one can assume that the initial abstraction consists of all rainfall that occurs prior to the start of direct storm runoff. As an alternative, the initial abstraction may be assumed to be a constant volume (depth). Viessman (1968) found that for small urban watersheds a value of 0.1 in. was reasonable. For rural and forested watersheds, a larger volume would be expected. The SCS method (which is discussed in Section 7.8) assumes that the initial abstraction is a fixed fraction of the maximum retention but varies with soil and land use since the retention is a function of the soil group and land use.

A number of methods have been proposed for estimating the initial abstraction. Two methods of identifying the initial abstraction are illustrated in Figure 9–6. Figure 9–6a shows a rainfall hyetograph with 0.4 in. of an initial abstraction. In this case, the depth of initial abstraction is constant regardless of the rainfall distribution. As an alternative, Figure 9–6b shows the initial abstraction when the direct runoff begins at a storm time of 30 min; in this case, the initial abstraction would have a depth of 0.9 in. The specific-volume method of Figure 9–6a requires specification of the depth of the initial abstraction; since only a few studies of this element of the runoff process have been made, it is difficult to justify a specific value. The specific-time method of Figure 9–6b can result in depths of initial abstraction that may be unusually high. Thus both the specific-volume and specific-time methods have drawbacks. For this reason, initial abstraction is sometimes considered to be part of the losses and not a separate calculation.

9.4 SEPARATION OF LOSSES USING INDEX METHODS

Losses reflect the ability of the watershed to retain water. At the beginning of a storm event that has been preceded by a period with no rainfall, the vegetation is dry, depressions are empty, and the upper layers of the soil structure have relatively low soil moisture. These watershed conditions represent the greatest potential for immediate storage of rainwater. Losses are usually thought to consist of water intercepted by vegetation (interception storage), water stored in small surface depressions (depression storage), and water that infiltrates into the soil (soil storage).

It is easy to become aware of interception storage. The interception of rain is evident from the dry area under the crown of a tree for the first few moments of a storm. The amount of interception storage could be measured using two rain gages, with one located under the crown of the tree and another placed outside the range of influence of the tree. This has been done at hydrologic research stations, and empirical formulas are available for estimating the interception storage as a function of the type and characteristics of the vegetation.

Depression storage, which is storage on the surface of the watershed in small depressions, depends on topography and land cover. Both depression and interception storage are depleted during the early part of a storm and thus are part of the initial abstraction if it is handled separately from other losses.

The primary component of losses is infiltration of rainwater into subsurface storage. In general, the available volume of subsurface storage is greatest at the start of rainfall and decreases over the duration of the storm. After a certain time, the infiltration capacity approaches a constant rate. Between storm events percolation and evapotranspiration serve to decrease water stored in the soil structure; this process is referred to as ground water storage recovery. A number of methods have been proposed for estimating losses.

9.4.1 Phi-Index Method

To this point in the analysis process, the baseflow has been estimated and subtracted from the total runoff to determine the distribution of direct runoff; the volume of direct runoff is then computed. Also, the initial abstraction has been identified and separated from the measured rainfall hyetograph. From a conceptual standpoint, the remaining part of the rainfall hyetograph must be separated into losses and rainfall excess, such that the volume of rainfall excess equals the volume of direct runoff.

By definition the phi index (ϕ) equals the average rainfall intensity (or depth) above which the volume of rainfall excess equals the volume of direct runoff. Thus the value of ϕ is adjusted such that the volumes of rainfall excess and direct runoff are equal. The procedure for computing the phi index is listed here:

1. Compute the depths of rainfall (V_p) and direct runoff (V_d).
2. Make an initial estimate of the phi index:

$$\phi = \frac{V_p - V_d}{D} \tag{9–9}$$

in which D is the duration of rainfall (excluding that part separated as initial abstraction) and ϕ is an intensity, with dimension of length per unit time such as inches per hour.

3. (a) Compute the loss function, $L(t)$:

$$L(t) = \begin{cases} \phi & \text{if } \phi \leq P(t) \quad (9\text{–}10a) \\ P(t) & \text{if } \phi > P(t) \quad (9\text{–}10b) \end{cases}$$

where $P(t)$ is the ordinate of the rainfall intensity hyetograph at time t.

(b) Compute the total depth of losses, V_L:

$$V_L = \Sigma\, L(t) * \Delta t \tag{9–11}$$

where V_L is a depth and the summation is over all ordinates where losses occur.

4. Compute $PE(t) = P(t) - L(t)$ for all ordinates in the rainfall hyetograph (excluding initial abstraction).

5. Compare V_L and $V_p - V_d$.

(a) If $V_L = V_p - V_d$, go to Step 6

(b) If $V_L < V_p - V_d$, compute the phi-index correction, $\Delta\phi$:

$$\Delta\phi = \frac{V_p - V_d - V_L}{D_1} \tag{9–12}$$

in which D_1 is the time duration over which $PE(t)$ is greater than zero.

(c) Adjust the phi index: $\phi_{\text{new}} = \phi_{\text{old}} + \Delta\phi$

(d) Return to Step 3.

6. Use the latest value of ϕ to define losses.

Example 9–3

Consider the rainfall hyetograph of Figure 9–7a. Assuming that the depth of direct runoff is 0.4 in., the value of ϕ must be set such that the depth of rainfall excess equals 0.4 in.; therefore, rainfall losses would equal 0.3 in. The value of ϕ is computed with Equation 9–9:

$$\phi = \frac{(0.7 - 0.4) \text{ in.}}{4 \text{ hr}} = \frac{0.3 \text{ in.}}{4 \text{ hr}} = 0.075 \text{ in./hr}$$

Subtracting a loss of 0.075 in./hr from each ordinate of the hyetograph yields a time distribution of rainfall excess of (0.025, 0.175, 0.075, 0.125) in./hr. Converting these to depths and summing yields a total depth of 0.4 in., which equals the volume of direct runoff. Thus, the computed value of ϕ is appropriate for this storm event.

Example 9–4

A rainfall hyetograph for a 70-min storm with a total depth of 3.5 in. is shown in Figure 9–7b. The depth of direct runoff is 1.90 in.; therefore, the depth of rainfall excess must be 1.9 in. and the depth of losses 1.6 in. One difference between this example and Example 9–3 is that the rainfall ordinates are not on a constant time interval; the phi-index method can still be applied.

The depth of rainfall is

$$V_p = \frac{1}{60}[1\,(25) + 8\,(10) + 5(15) + 1.5(20)] = 3.5 \text{ in.}$$

Using Equation 9–9, the initial estimate of ϕ is

$$\phi = \frac{V_p - V_d}{D} = \frac{3.5 - 1.9}{(70/60)} = 1.37 \text{ in./hr}$$

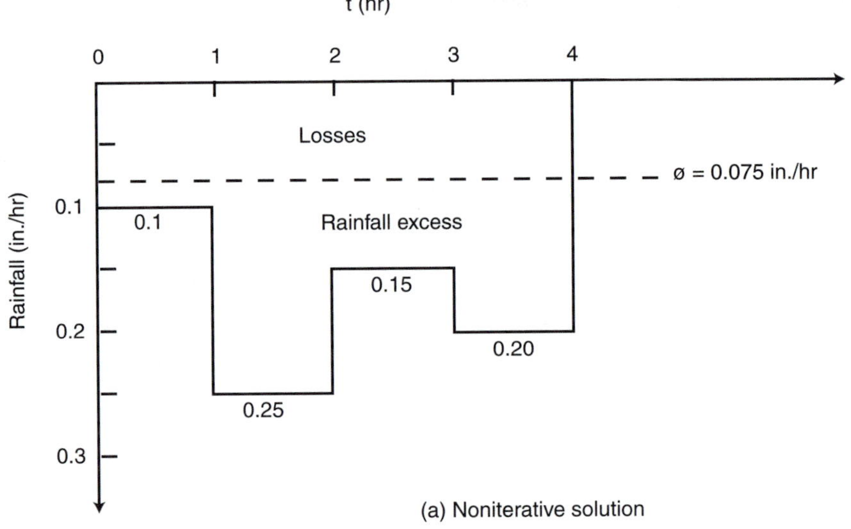

FIGURE 9–7 Separation of rainfall excess and losses using the phi-index method.

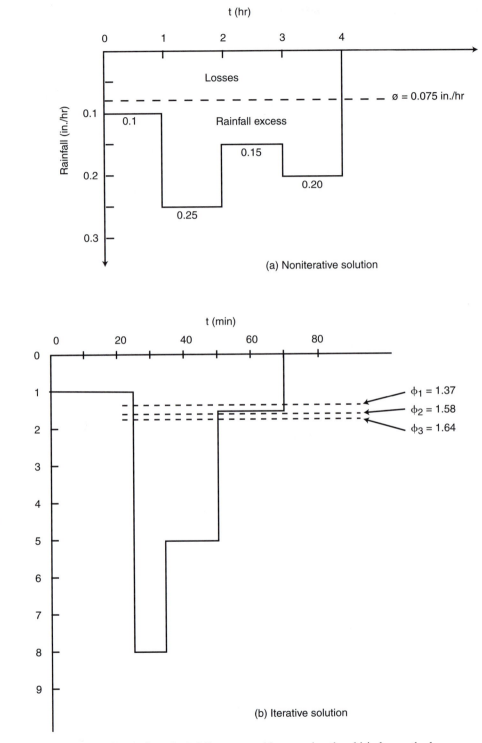

TABLE 9–3 Example 9–4: Calculation of Losses Using the Phi-Index Method

Time Period (min)	$i(t)$ (in./hr)	$L(t)$ (in./hr)	$V_L(t)$ (in.)	$L(t)$ (in./hr)	$V_L(t)$ (in.)	$L(t)$ (in./hr)	$V_L(t)$ (in.)	$i_e(t)$ (in./hr)	V_{pe} (in.)
0–25	1.0	1.00	0.417	1.00	0.417	1.00	0.417	0	0
25–35	8.0	1.37	0.228	1.58	0.263	1.64	0.273	6.36	1.06
35–50	5.0	1.37	0.342	1.58	0.395	1.64	0.410	3.36	0.84
50–70	1.5	1.37	0.457	1.50	0.500	1.50	0.500	0	0
		1.444			1.575		1.600		1.90

Using Equation 9–10, the loss function is given in column 3 of Table 9–3, and the depths of losses are given in column 4. The total depth of losses is 1.44 in. Since this is less than the difference between V_p and V_d, the value of ϕ must be corrected using Equation 9–12:

$$\Delta\phi = \frac{V_p - V_d - V_L}{D_1} = \frac{3.5 - 1.9 - 1.44}{45/60} = 0.21 \text{ in./hr.}$$

A value for D_I of 45 min is used because the initial value of ϕ is greater than $P(t)$ for the first 25 min. Thus, the adjusted ϕ is 1.37 + 0.21 = 1.58 in./hr, which is used to compute a revised loss function (column 5 of Table 9–3). The total loss is now 1.575 in., which is still less than the required depth of 1.6 in. Thus, Equation 9–12 is used to compute a new adjustment:

$$\Delta\phi = \frac{3.5 - 1.9 - 1.575}{25/60} = 0.06 \text{ in./hr.}$$

A duration D_1 of 25 min is used because now only the time period from 25 to 50 min has rainfall excess. The revised value of ϕ is 1.58 + 0.06 = 1.64 in./hr. The loss function for the revised ϕ value is given in column 7 of Table 9–3. Since the total loss is 1.6 in. (column 8), then the loss function of column 7 is the final value. The rainfall-excess intensity, $i_e(t)$, is computed (column 9) as the difference between the rainfall intensity (column 2) and the loss function of column 7. The depths of excess are given in column 10 with a total of 1.9 in., which equals the depth of direct runoff.

9.4.2 Constant Percentage Method

The constant percentage method assumes that losses are proportional to the rainfall intensity. Letting f be the fraction of rainfall that contributes to losses, then, assuming that depth of rainfall excess P_e equals the depth of direct runoff (Q_d), the value of f is given by

$$f = 1 - \frac{Q_d}{P} \tag{9–13}$$

in which P is the total depth of rainfall.

Example 9–5

For the hyetograph of Figure 9–7a, f would equal

$$f = 1 - \frac{0.4}{0.7} = 0.4286$$

Multiplying each ordinate of the rainfall hyetograph by f yields the loss function. Thus the loss function for the rainfall of Figure 9–7a would be (0.043, 0.107, 0.064, 0.086) in./hr, and the distribution of rainfall excess would be (0.057, 0.143, 0.086, 0.114) in./hour. This method does not require an iterative procedure.

9.5 SEPARATION OF LOSSES USING INFILTRATION CAPACITY CURVES

A dry soil has a certain capacity for infiltrating water. The capacity can be expressed as a depth of water that can be infiltrated per unit time, such as inches per hour. If rainfall supplies water at a rate that is greater than the infiltration capacity, water will infiltrate at the capacity rate, with the excess either ponding, draining as surface runoff, or evaporating. If rainfall supplies water at a rate that is less than the infiltration capacity, all of the water supply will infiltrate. In either case, as water infiltrates into the soil, the capacity to infiltrate more water decreases and approaches a minimum capacity. When the supply rate is equal to or greater than the capacity to infiltrate, the minimum capacity will be approached more quickly than when the supply rate is much less than the infiltration capacity. This conceptual description of infiltration can be used to develop a model for separating rainfall excess from losses.

9.5.1 Horton's Infiltration Capacity Curve

Horton (1937) suggested that the infiltration capacity (f) could be represented by the function

$$f = f_c + (f_o - f_c)\, e^{-Kt} \tag{9–14}$$

in which f_c is the infiltration capacity after a long period of continued wetting (that is, the ultimate capacity); f_o is the initial capacity, (that is, the capacity at the beginning of a storm event that occurs after a dry period); t is the time from beginning of rainfall; and K is a constant that describes the decay of the capacity with time and reflects the soil cover complex. Given estimates of f_o, f_c, and K, the infiltration capacity curve can be computed for the duration of a storm event and used to separate losses and rainfall excess.

All of the parameters of Equation 9–14 are a function of the surface texture and cover type. The parameter f_o is high for sandy soils; it increases for a turfed or vegetated surface and shows some seasonal variation. The ultimate infiltration capacity f_c is also relatively high for a sandy soil and increases for vegetated surfaces. Studies have suggested that it also varies with the slope of the watershed. Typical values of the parameters are difficult to present. Values of f_c can vary from 0.01 to 2 in./hr, with f_o 3 to 5 times greater than f_c. Values of K can range from less than 1 hr^{-1} to more than 20 hr^{-1}.

Example 9–6

Assume the rainfall hyetograph of Figure 9–8 and assume that field studies suggest the following values for the infiltration characteristics of a small watershed: $f_o = 0.6$ in./hr, $f_c = 0.2$ in./hr, and $K = 0.6\ hr^{-1}$. Therefore, the Horton Infiltration Capacity Curve is

$$f = 0.2 + 0.4e^{-0.6t} \tag{9–15}$$

The infiltration capacity f was computed for the period of storm times from 0 to 4 hr (see Table 9–4). Using the trapezoidal rule, the depth of infiltration for each increment of 0.25 hr is also given in Table 9–4. The total depth of infiltration for the 4 hr was 1.407 in., which yields a

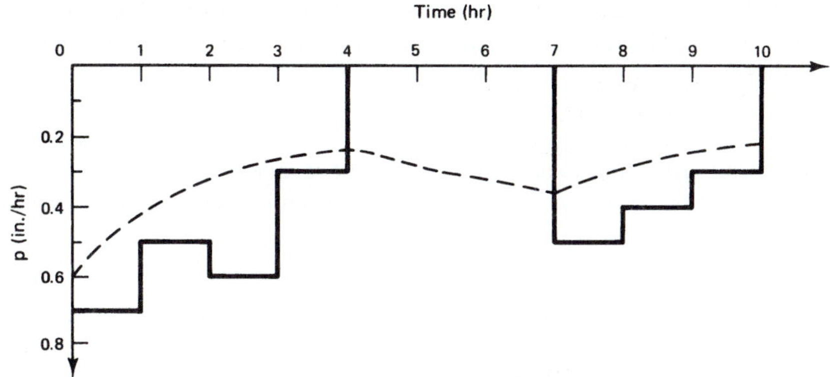

FIGURE 9–8 Illustration of Horton's equation in separation rainfall excess and losses.

rainfall-excess depth of 0.693 in. Integration of Equation 9–15 yields a total depth of infiltration of 1.406 in., which indicates that the trapezoidal rule provided an estimate within 0.001 in.

9.5.2 Recovery of Infiltration Capacity

Equation 9–14 assumes that the rainfall intensity is greater than the infiltration capacity. During periods when it is not raining, the infiltration capacity will recover. Although no method for estimating the recovery is widely accepted, it is not unreasonable to use an expo-

TABLE 9–4 Computation of Infiltration Capacity (f), Total Depth Infiltrated (F), and Rainfall Excess (P_e) for Storm Rainfall (P)

Time (hr)	P (in./hr)	f (in./hr.)	ΔF (in.)	P_e (in./hr.)	F (in.)
0	0.7	0.600	0	0.100	0
0.25	0.7	0.544	0.143	0.156	0.143
0.50	0.7	0.496	0.130	0.204	0.273
0.75	0.7	0.455	0.119	0.245	0.392
1.00	0.5	0.420	0.109	0.080	0.501
1.25	0.5	0.389	0.101	0.111	0.602
1.50	0.5	0.363	0.094	0.137	0.696
1.75	0.5	0.340	0.088	0.160	0.784
2.00	0.6	0.320	0.083	0.280	0.867
2.25	0.6	0.304	0.078	0.296	0.945
2.50	0.6	0.289	0.074	0.311	1.019
2.75	0.6	0.277	0.071	0.323	1.090
3.00	0.3	0.266	0.068	0.034	1.158
3.25	0.3	0.257	0.065	0.043	1.223
3.50	0.3	0.249	0.063	0.051	1.286
3.75	0.3	0.242	0.061	0.058	1.347
4.00	0.0	0.236	0.060		1.407
			1.409		

nential function that is similar in concept to the exponential function of Equation 9–14. Specifically, the following function can be used to approximate the recovery of the infiltration capacity during a dry period:

$$f = f_r + (f_o - f_r)(1 - e^{-K_r t'}) \tag{9–16}$$

in which t' is the time measured from the end of rainfall, K_r is the recovery coefficient, and f_r is the value of f at time t'. The recovery coefficient K_r is usually much smaller than the value of K.

Example 9–7

To illustrate the use of Equation 9–16, consider the example of Figure 9–8. Table 9–4 shows that f equals 0.236 in./hr at a storm time of 4 hr; thus f_r of Equation 9–12 will equal 0.236 in./hr. For purposes of illustration K_r will be assumed to equal 0.15 hr^{-1}. To use Equation 9–16, the time scale has to be shifted so that $t' = 0$ at the start of the recovery period. The following equation can be used to define the recovery:

$$f = 0.236 + 0.364(1 - e^{-0.15t'}) \tag{9–17}$$

The recovery of the infiltration capacity is computed with Equation 9–17, as shown in Figure 9–8 and Table 9–5. At the end of the recovery period (that is, storm time 7.0 hr or $t' = 3$ hr), the infiltration capacity recovered to a rate of 0.368 in./hr. The infiltration capacity recovers more slowly than it decays under the influence of rainfall.

9.5.3 Infiltration Capacity Estimation After Incomplete Recovery

If rain occurs prior to the time when the infiltration capacity has completely recovered (that is, before f recovers to f_o), Equation 9–14 cannot be directly applied to estimate the infiltration capacity. To use Equation 9–14, it is necessary to modify the value of t. The modification is simply a transformation of the infiltration capacity curve of Equation 9–14 so that the f curve intersects the recovery curve at the value of f at the time when the rainfall continues. This time transformation is more easily illustrated graphically. In Figure 9–9 the first burst of rain ended at time t_1; Equation 9–14 can be used to define f for the interval $0 \le t \le t_1$. Recovery begins at time t_1 and continues until time t_2, which is the time when the rain begins to fall again; the infiltration capacity can be computed using Equation 9–17 for the interval $t_1 \le t \le t_2$. At time t_2 the infiltration capacity will be denoted as f_2. Since $f_2 < f_o$, recovery has

TABLE 9–5 Computation of the Recovery of the Infiltration Capacity (f)

Time (hr)	t'	f (in./hr)	Time (hr)	f (in./hr)
4.0	0	0.236	7.0	0.368
4.5	0.5	0.262	7.5	0.324
5.0	1.0	0.287	8.0	0.292
5.5	1.5	0.309	8.5	0.268
6.0	2.0	0.330	9.0	0.251
6.5	2.5	0.350	9.5	0.237
7.0	3.0	0.368	10.0	0.228

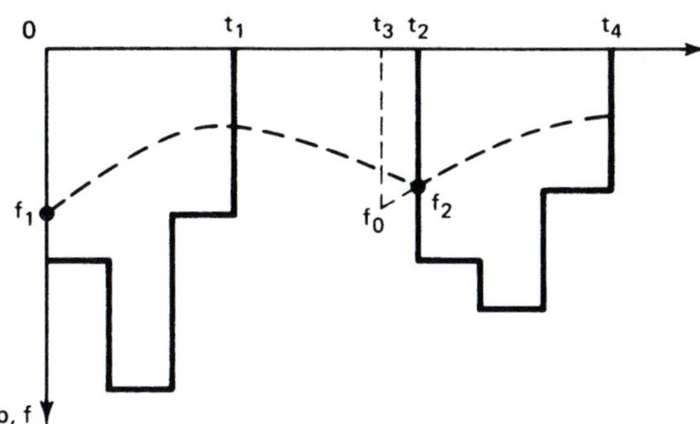

FIGURE 9–9 Time translation for infiltration capacity estimation after an incomplete recovery.

not been complete; thus we cannot apply Equation 9–14 directly using t_2 as the zero point. Instead, the infiltration capacity curve of Equation 9–15 is translated until the value f_2 intersects the f curve. This is shown in Figure 9–9 as the small dashed line. Thus it is necessary to find the time t_3 that would provide an infiltration capacity curve where $f = f_o$ at $t = t_3$ and $f = f_2$ at $t = t_2$. While the f curves over the intervals $0 \leq t \leq t_1$ and $t_3 \leq t \leq t_4$ are identical, infiltration capacity in the interval $t_3 \leq t \leq t_2$ is actually computed using Equation 9–17. The time t_3 is only necessary for computing the infiltration capacity curve for $t > t_2$. The value of t_3 can be determined by setting $f = f_2$ in Equation 9–14 and solving for t, which will be denoted as Δt:

$$\Delta t = \frac{-1}{K} \ln_e \frac{f_2 - f_c}{f_o - f_c} \tag{9–18}$$

Thus the infiltration capacity curve for the interval $t_2 \leq t \leq t_4$ is given by

$$f = f_c + (f_o - f_c)e^{-K(t - t_2 + \Delta t)} \tag{9–19}$$

where t_4 is the time at which the intensity of the second burst of rainfall is less than the infiltration capacity.

In applying the conceptual basis of Equations 9–18 and 9–19, it may be easier just to modify f_o of Equation 9–14 to the value of f_2 and measure time from t_2, not the original zero point. Thus, the following equation could replace Equation 9–19:

$$f = f_c + (f_2 - f_c)e^{-Kt}$$

where $t = 0$ at time t_2.

Example 9–8

The use of Equations 9–18 and 9–19 can be illustrated using the data of Figure 9–8. Assuming that the second burst of rain begins at storm time $t = 7$ hr, the infiltration capacity at $t = 7$ hr is 0.368 in./hr (see Table 9–5). Using Equation 9–18, Δt is computed as

$$\Delta t = \frac{-1}{0.6} \ln_e \left(\frac{0.37 - 0.20}{0.60 - 0.20} \right) = 1.446 \text{ hr}$$

Based on a value for Δt of 1.446 hr, we could superimpose an infiltration curve defined by Equation 9–15 onto Figure 9–8 with a starting time of $t_3 = t_2 - \Delta t = 7 - 1.446 = 5.554$ hr; such a curve would pass through the value $f = 0.368$ at $t = 7$ hr. The following equation can be used to compute the infiltration capacity curve for $t > 7$ hr:

$$f = 0.2 + 0.4e^{-0.6(t-5.554)}$$

The calculations are given in Table 9–5 and the resulting curve is shown in Figure 9–8.

9.5.4 Mass Infiltration Method

In the example of Figure 9–8, the rainfall intensity was always greater than the infiltration capacity; when such a condition exists, the available water is sufficient such that the rate of infiltration equals the capacity of the soil structure to store and transmit the water. In cases where the rainfall intensity is less than the infiltration capacity, all of the available soil storage is not used. Thus the capacity does not decrease at the same rate as it would if the rainfall intensity (that is, the available supply rate) were equal to or greater than the capacity. If one assumes that the capacity decreases at the rate given by Equation 9–14 when it would actually be less than the capacity, the volume of losses will be underestimated and the volume of rainfall excess will be overestimated. The mass infiltration method can be used to estimate the infiltration capacity when the rainfall intensity is less than the capacity.

The total mass of infiltrated water (F) at time t, where $t = 0$ at the start of rainfall, is given by:

$$F = \int_0^t f\,dt = \int_0^t [f_c + (f_0 - f_c)e^{-Kt}]\,dt \tag{9–20a}$$

$$= f_c t + \frac{f_o - f_c}{K}(1 - e^{-Kt}) \tag{9–20b}$$

A plot of F given by Equation 9–20b versus f is called the *mass infiltration curve*. Given f_c, f_o, and K, the values of F and f can be computed with Equations 9–20b and 9–14, respectively. Using the mass infiltration curve, the infiltration capacity can be computed for any value of F. When the rainfall intensity is less than the infiltration capacity, the volume of infiltration F, which is assumed to be equal to the total volume of rainfall up to the specific time t, can be computed and the infiltration capacity determined.

Example 9–9

The infiltration function is defined by Equation 9–15. However, for the rainfall hyetograph of Figure 9–10 the infiltration capacity curve will not be the same as for Figure 9–8 because the rainfall intensity is less than the infiltration capacity for $t < 2$ hours. Thus the mass infiltration curve must be computed. Equation 9–20b becomes

$$F = 0.2t + \frac{0.4}{0.6}(1 - e^{-0.6t}) \tag{9–21}$$

Using Equations 9–15 and 9–21 to compute f and F, respectively, the mass infiltration curve can be computed (see Table 9–4 and Figure 9–11). The infiltration capacity curve for the rainfall hyetograph of Figure 9–10 can be estimated using the mass infiltration curve. After 1 hr the total depth of rainfall is 0.1 in.; entering Figure 9–11 with $F = 0.1$ in., the value of f is 0.55 in./hr, which is shown in Figure 9–10. The infiltration capacity of 0.55 in./hr can be compared with the

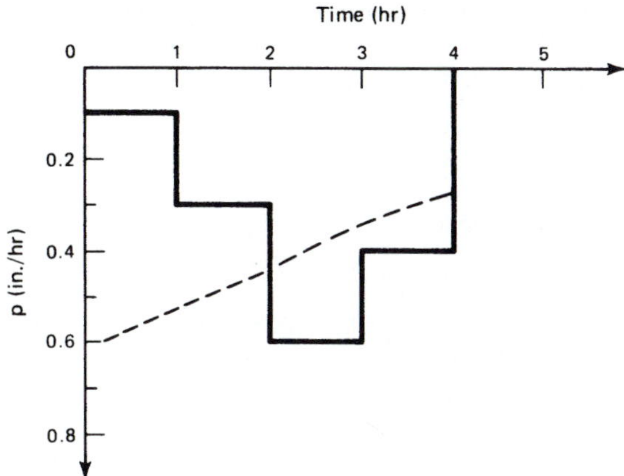

FIGURE 9–10 Estimating the infiltration capacity using the mass infiltration method.

case where the rainfall intensity exceeded the infiltration capacity. Figure 9–8 shows that the infiltration capacity was 0.42 when the rainfall rate exceeded the infiltration capacity during the first hour. Thus the low-intensity rainfall of Figure 9–10 during the first hour caused the infiltration capacity to decrease by an amount that was less than that when the rainfall intensity had been greater than the infiltration capacity, such as in Figure 9–8. After the second hour of the storm in Figure 9–10, a total rainfall depth of 0.4 in. had occurred. Entering Figure 9–11 with $F = 0.4$ in. yields an infiltration capacity f of 0.45 in./hr, rather than the 0.32 in./hr indicated in Figure 9–8 when the rainfall intensity exceeded the capacity.

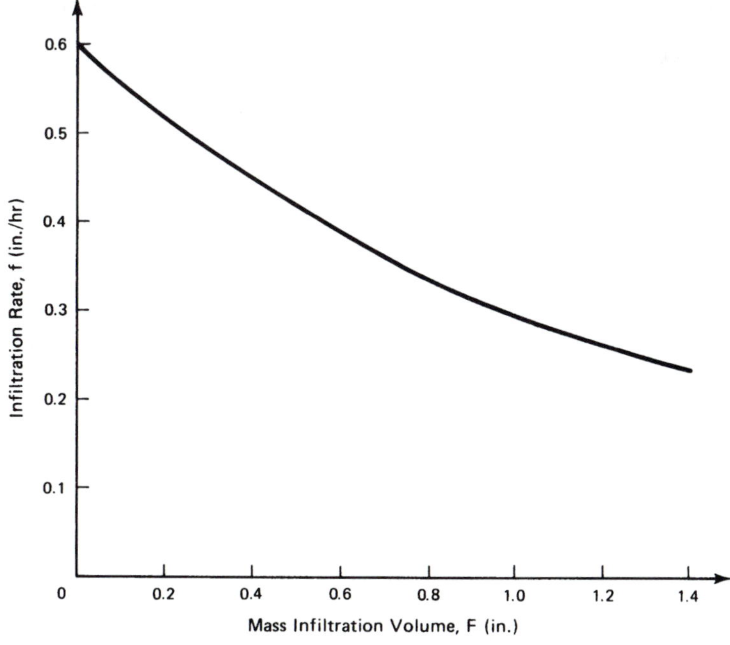

FIGURE 9–11 Mass infiltration curve.

Beginning at time $t = 2$ hrs, the rainfall intensity exceeds the capacity. Having determined that the capacity is 0.45 in./hr at time $t = 2$ hr, it is necessary to use the concepts behind Equations 9–18 and 9–19 to compute the infiltration capacity for $t > 2$. Equation 9–18 yields

$$\Delta t = \frac{-1}{0.6} \ln_e \left(\frac{0.45 - 0.20}{0.60 - 0.20} \right) = 0.783 \text{ hr}$$

Thus the capacity for $t > 2$ hr can be computed by

$$f = 0.2 + 0.4e^{-0.6(t-2+0.783)} \tag{9–22}$$

For times of $t = 3$ and $t = 4$ hr, Equation 9–22 yields infiltration capacities of 0.337 and 0.275 in./hr, respectively (see Figure 9–10). Using the trapezoidal rule, the volume of infiltration is 1.1 in., which is less than the total infiltration of 1.4 in. for the case where the rainfall intensity was always greater than the infiltration capacity (Figure 9–8).

9.5.5 Analysis and Synthesis

In the examples used to illustrate the infiltration capacity approach to compute losses, the parameters (f_o, f_c, K, and f_r) were assumed to be known. In practice, they are not known, so values must be estimated through data analysis. It is usually assumed that the initial abstraction is part of the loss function when using the infiltration capacity approach, but even with this simplifying assumption the analysis approach is not easily solved without a computer.

An approximate infiltration capacity curve can be drawn if values are available for two of the three coefficients of Equation 9–14. The ultimate capacity f_c and the decay coefficient K are probably easier to relate to soil and cover characteristics, so values for these variables may be available. If the depth of losses is known, F can be used to represent the depth of losses and Equation 9–20b can be solved for f_o:

$$f_o = f_c + K \left(\frac{F - f_c t}{1 - e^{-Kt}} \right) \tag{9–23}$$

In Equation 9–23, t is the duration of the storm event and f_o is the initial infiltration capacity.

Example 9–10

Figure 9–12a shows a 4-hr rainfall hyetograph. If we assume that analysis of the runoff hydrograph resulted in a depth of direct runoff of 1.1 in., there would be 1.1 in. of rainfall excess. Since the total depth of rainfall is 1.8 in., the infiltration capacity curve must separate 0.7 in. of losses. If soil studies suggest that $K = 1.2 \text{ hr}^{-1}$ and $f_c = 0.1$ in./hr, Equation 9–23 yields a value for f_o of

$$f_o = 0.1 + 1.2 \left[\frac{0.7 - 0.1(4)}{1 - e^{-1.2(4)}} \right] = 0.457 \text{ in./hr}$$

Based on this value of the initial capacity, the infiltration capacity curve of Figure 9–12a was drawn.

Example 9–11

For the case where the initial rainfall depth is less than the capacity, the depth for the initial part of the storm can be subtracted from the losses before using Equation 9–23; this assumes that all of the initial part of the storm is part of the loss function. Figure 9–12b shows a 4-hr rainfall

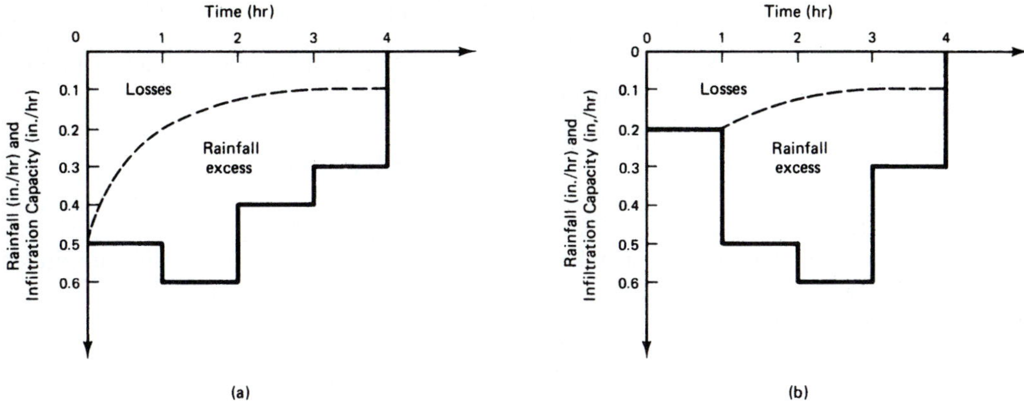

FIGURE 9–12 Analysis of infiltration capacity curves.

hyetograph with a total depth of 1.6 in. Assuming the depth of direct runoff was analyzed to be 1.0 in., there would be 0.6 in. of losses. Assuming that the initial infiltration capacity will be greater than 0.2 in./hr, the 0.2 in. of rainfall in the first hour of the storm will be part of the loss function. Thus 0.4 in. of losses will have to be taken from the remaining 3 hr of the storm. Solving Equation 9–23 for f_o yields

$$f_o = 0.1 + 1.2\left[\frac{0.4 - 0.1(3)}{1 - e^{-1.2(3)}}\right] = 0.223 \text{ in./hr}$$

This is the value of f at $t = 1$ hr. To use this initial infiltration capacity to separate the losses, this must be used for the initial value at $t = 1$ hr and the time (t of Equation 9–14) must be zero for $t = 1$ hr. The resulting infiltration capacity curve is shown in Figure 9–12b.

9.5.6 The Green-Ampt Model

The Green-Ampt Model is a conceptual representation of the infiltration process. Developed in 1911 by Green and Ampt, the model can be developed from Darcy's Law (Equation 6–1):

$$V = K_s \frac{dh}{dL} \tag{9–24}$$

in which V = the velocity of flow through the unsaturated soil, h is the head, and L is the length of the soil column. The hydraulic gradient in Darcy's Law equals the ratio of the sum of the depth of ponding d_p, the depth of the wetting front from the ground surface L_w, and the soil water suction S_w at the wetting front to the depth L_w:

$$\frac{dh}{dL} = \frac{d_p + L_w + S_w}{L_w} \tag{9–25}$$

The total depth of infiltration F to time t equals the product of the depth to the wetting front and the difference in the saturated θ_s and initial θ_i volumetric water contents:

$$F = (\theta_s - \theta_i) L_w \tag{9–26}$$

The method assumes that t_S and t_i are constant with time as the wetting front advances. Therefore, the change in F with time becomes

$$\frac{dF}{dt} = \frac{d[(\theta_s - \theta_i)L_w]}{dt} = (\theta_s - \theta_i)\frac{dL_w}{dt} \tag{9–27}$$

The rate dF/dt of Equation 9–27 equals Darcy's velocity of Equation 9–24. Using the assumption that the depth of ponding is very small, Equation 9–25 can be substituted into Equation 9–24, with the result equated to Equation 9–27:

$$(\theta_s - \theta_i)\frac{dL_w}{dt} = K_s\left(\frac{L_w + S_w}{L_w}\right) \tag{9–28}$$

Rearranging Equation 9–28 yields

$$\left(\frac{K_s}{\theta_s - \theta_i}\right) dt = \left(\frac{L_w}{L_w + S_w}\right) dL_w \tag{9–29}$$

Integrating Equation 9–29 yields

$$\frac{K_s t}{\theta_s - \theta_i} = S_w + L_W - S_W \ln(S_w + L_w) + C \tag{9–30}$$

in which C is the constant of integration. At time $t = 0$, $F = 0$; since $(\theta_s - \theta_i)$ in Equation 9–26 is a non-zero constant, then $L_w = 0$ when $F = 0$. Thus at $t = 0$, the constant of integration is

$$C = S_w \ln S_w - S_w \tag{9–31}$$

which when substituted into Equation 9–30 yields

$$\frac{K_s t}{(\theta_s - \theta_i)} = L_w - S_w \ln\left(1 + \frac{L_w}{S_w}\right) \tag{9–32}$$

or

$$K_s t = (\theta_s - \theta_i) L_w - S_w (\theta_s - \theta_i) \ln\left(1 + \frac{L_w}{S_w}\right) \tag{9–33}$$

Using Equation 9–26 yields a relationship between F and t:

$$K_s t = F - S_w(\theta_s - \theta_i) \ln\left[1 + \frac{F}{S_w(\theta_s - \theta_i)}\right] \tag{9–34}$$

The infiltration capacity f at any time t can be computed by taking the derivative of Equation 9–34:

$$f = K_s + \frac{K_s S_w (\theta_s - \theta_i)}{F} \tag{9–35}$$

In applying Equations 9–34 and 9–35, the underlying assumptions should be recognized:

1. As rain continues to fall and water infiltrates, the wetting front advances at the same rate throughout the ground water system, which produces a well-defined wetting front.
2. The volumetric water contents, θ_s and θ_i, remain constant above and below the wetting front as it advances.
3. The soil-water suction immediately below the wetting front remains constant with both time and location as the wetting front advances.

If the rainfall has constant intensity i and exceeds the infiltration capacity f, the infiltration rate will equal the intensity i until ponding occurs. At that time, a cumulative depth F_p will have infiltrated, which can be estimated from Equation 9–35:

$$F = \frac{K_s S_w (\theta_s - \theta_i)}{i - K_s} \tag{9–36}$$

A modification of Equation 9–34 was proposed by Mein and Larson (1973) to compute the relationship between F and t:

$$K_s(t - t_p + t_e) = F - S_w(\theta_s - \theta_i) \ln\left[1 + \frac{F}{S_w(\theta_s - \theta_i)}\right] \tag{9–37}$$

in which t_e is the equivalent time to infiltrate a depth F_p when subject to initially ponded surface conditions.

Example 9–12

In the synthesis mode, input values are required for K_s, S_w, θ_i, and θ_s, as well as the rainfall intensity i. The following values are assumed as input: $K_s = 0.02$ in./hr; $S_w = 10$ in.; $\theta_i = 0.15$; $\theta_s = 0.45$; $i = 0.4$ in./hour. The soil-water deficit at the beginning of rainfall is given by $(\theta_s - \theta_i)$ and equals 0.30 for the given values in this example. Solving Equation 9–36 for the depth of infiltration at the time of ponding, F_p, yields

$$F_p = \frac{S_w(\theta_s - \theta_i)}{(i/K_s) - 1} = \frac{10(0.3)}{(0.4/0.02) - 1} = 0.158 \text{ in.} \tag{9–38}$$

At the time of ponding, the infiltration capacity equals the rainfall intensity. For a constant rainfall intensity from the beginning of rainfall to the time at which ponding begins, the depth of infiltration is $F_p = it_p$; therefore, $t_p = F_p/i$. Based on the depth from Equation 9–38, the time to the beginning of ponding is

$$t_p = \frac{F_p}{i} = \frac{0.158 \text{ in.}}{0.4 \text{ in./hr}} = 0.395 \text{ hr} \tag{9–39}$$

The cumulative depth of infiltration F and the infiltration rate f can be determined from Equations 9–37 and 9–35, respectively. Equation 9–37 requires the equivalent time t_e that would have been required to infiltrate the depth F_p; this can be estimated by rearranging Equation 9–37 to compute t, which is t_e:

$$t_e = \frac{F_p}{K_s} - \frac{S_w(\theta_s - \theta_i)}{K_s} \ln_e\left[1 + \frac{F_p}{S_w(\theta_s - \theta_i)}\right] \tag{9–40}$$

TABLE 9–6 Computation of Infiltration Rate (f) and Times (t) at which the Cumulative Infiltration Volume Equals F using the Green-Ampt Method

F (in.)	t (hr)	f (in./hr)
0	0	0.40
0.158	0.395	0.40
0.200	0.514	0.32
0.250	0.689	0.26
0.500	2.072	0.14
0.750	4.223	0.10
1.000	7.043	0.08
1.500	14.375	0.06
2.000	23.571	0.05

For this problem, t_e is

$$t_e = \frac{0.158}{0.02} - \frac{10(0.45 - 0.15)}{0.02} \ln_e\left[1 + \frac{0.158}{10(0.45 - 0.15)}\right] \tag{9–41}$$

$$= 0.20 \text{ hr.}$$

Now, Equation 9–37 can be used to define the cumulative depth of infiltration. Since F appears nonlinearly in the equation, an iterative solution procedure can be used to estimate the value of F for any t; alternatively, F can be specified and the time at which the cumulative depth would occur computed. A plot of F versus t can be used to estimate either F for any value of t or t for any value of F. Solving Equation 9–37 for t gives

$$t = t_p - t_e + \frac{F}{K_s} - \frac{S_w(\theta_s - \theta_i)}{K_s} \ln_e\left[1 + \frac{F}{S_w(\theta_s - \theta_i)}\right] \tag{9–42a}$$

$$= 0.395 - 0.200 + \frac{F}{0.02} - \frac{10(0.45 - 0.15)}{0.02} \ln_e\left[1 + \frac{F}{10(0.45 - 0.15)}\right] \tag{9–42b}$$

The time corresponding to selected values of F were computed with Equation 9–42b and are given in Table 9–6. Equation 9–35 was then used to compute the infiltration capacity, with the values also given in Table 9–6 and plotted in Figure 9–13.

9.6 UNIT HYDROGRAPH CONCEPTS

9.6.1 Definitions

In Section 9.1 a conceptual framework was presented where a *unit hydrograph* (UH) was used as the system transfer function, that is, the function used to transform rainfall excess into direct runoff. By way of definition, a *unit hydrograph* is the hydrograph that results from l in. of rainfall excess generated uniformly over the watershed at a uniform rate during a specified period of time. There are four aspects of this definition that should be given special notice: (l) 1 in. of rainfall excess, (2) uniform spatial distribution of rainfall over the wa-

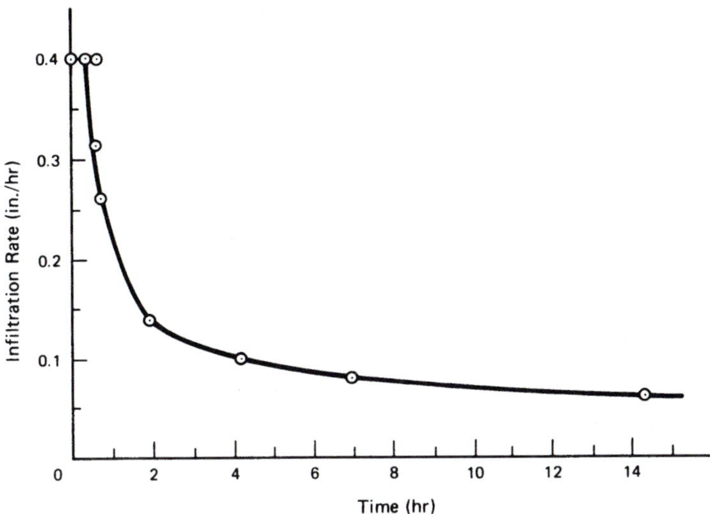

FIGURE 9–13 Infiltration rate (f) computed using the Green-Ampt Method.

tershed, (3) a rainfall excess rate that is constant with time, and (4) specific duration of rainfall excess. When developing a unit hydrograph, it is important to ensure that the sum of the ordinates is equivalent to 1 area-in. of direct runoff. A lack of spatial uniformity of rainfall can result in a unit hydrograph that does not reflect the temporal characteristics of runoff. The third part of the definition can be constraining because it is difficult to find storm events of significant volume where the excess rate is constant; it is usually necessary to accept some departure from the assumption of a constant rate. The peak and time to peak of a UH are sensitive to the duration of the rainfall excess, so it must be specified when developing a UH and considered when using a UH for design.

Several types of unit hydrographs can be defined. A *T-hour (or T-minute) unit hydrograph* is the hydrograph that results from a storm with a constant depth of rainfall excess of 1 in. over a duration of T hours (or T minutes). An *instantaneous unit hydrograph* (IUH) is a special case of the T-hour UH with the duration of rainfall excess being infinitesimally small; for such a UH to have a depth of 1 in., the intensity of the instantaneous rainfall excess is obviously not finite. The use of the concept of an IUH will be discussed later. The *dimensionless unit hydrograph,* which is a third form, is a direct runoff hydrograph whose ordinates are given as ratios of the peak discharge and whose time axis is defined as the ratio of the time to the time to peak; that is, a dimensionless UH has an axis system of q/q_p versus t/t_p, where q_p is the discharge rate at the time to peak t_p. Before a dimensionless UH can be used it must be converted to a T-hour UH.

9.6.2 Convolution

To this point, the elements of unit hydrograph analysis have been treated separately. Except for having the depths (or volumes) of rainfall excess and direct runoff equal, the separation of baseflow from direct runoff was independent of the separation of losses from rainfall excess. The purpose of the hydrograph separation analysis of Figure 9–1 is to identify the transfer function so that the analyzed transfer function can be used with a synthetic design storm to compute the design runoff hydrograph. The process by which the design storm is combined with the transfer function to produce the direct runoff hydrograph is called *convolution.*

Analytically speaking, convolution is referred to as the theory of linear superpositioning. Conceptually, it is a process of multiplication, translation with time, and addition. That is, the first burst of rainfall excess of duration D is multiplied by the ordinates of the unit hydrograph (UH). The UH is then translated for a time length of D, and the next burst of rainfall excess is multiplied by the UH. After the UH has been translated for all bursts of rainfall excess of duration D, the results of the multiplications are summed for each time interval. This process of multiplication, translation, and addition is the means of deriving a design runoff hydrograph from the rainfall excess and the UH.

The convolution process is best introduced using some very simple examples that illustrate the multiplication-translation-addition operation. First, consider a burst of rainfall excess of 1 in. that occurs over a period D. Assuming that the UH consists of two ordinates, 0.4 and 0.6, the direct runoff is computed by multiplying the rainfall excess burst by the UH; this is best presented graphically as in Figure 9–14a. It is important to note that the depth of direct runoff equals the depth of rainfall excess, which in this case is 1 in.

If 2 in. of rainfall excess occurs over a period of D, the depth of direct runoff must be 2 in. Using the same UH as the previous example, the resulting runoff hydrograph is shown in Figure 9–14b. In both this example and the previous example, computation of the runoff hydrograph consisted solely of multiplication; the translation and addition parts of the convolution process were not necessary because the rainfall excess occurred over a single time interval of D.

To illustrate the multiplication-translation-addition operation, consider 2 in. of rainfall excess that occurs uniformly over a period $2D$. This gives an intensity of 1 in. per time interval. In this case, the direct runoff will have a depth of 2 in., but the time distribution of direct runoff will differ from that of the previous example because the time distribution of rainfall excess is different. Figure 9–14c shows the multiplication-translation-addition operation. In this case, the time base of the runoff hydrograph is 3 time units ($3D$). In general, the time base of the runoff (t_{bRO}) is given by

$$t_{bRO} = t_{bPE} + t_{bUH} - 1 \tag{9–43}$$

in which t_{bPE} and t_{bUH} are the time bases of the rainfall excess and unit hydrograph, respectively. For the example above, both t_{bPE} and t_{bUH} equal 2, and therefore, according to Equation 9–43 t_{bRO} equals $3D$ time units.

One more simple example should illustrate the convolution process. In Figure 9–14d the depth of rainfall excess equals 3 in., with 2 in. occurring in the first time unit. The computation of the runoff hydrograph is shown in Figure 9–14d. In this case, the second ordinate of the runoff hydrograph is the sum of 2 in. times the second ordinate of the UH and 1 in. times the first ordinate of the translated UH:

$$2(0.6) + 1(0.4) = 1.6 \tag{9–44}$$

The convolution process can be used for processes with either a discrete or continuous distribution function. For a continuous process the multiplication-translation-addition operation is made using the convolution integral:

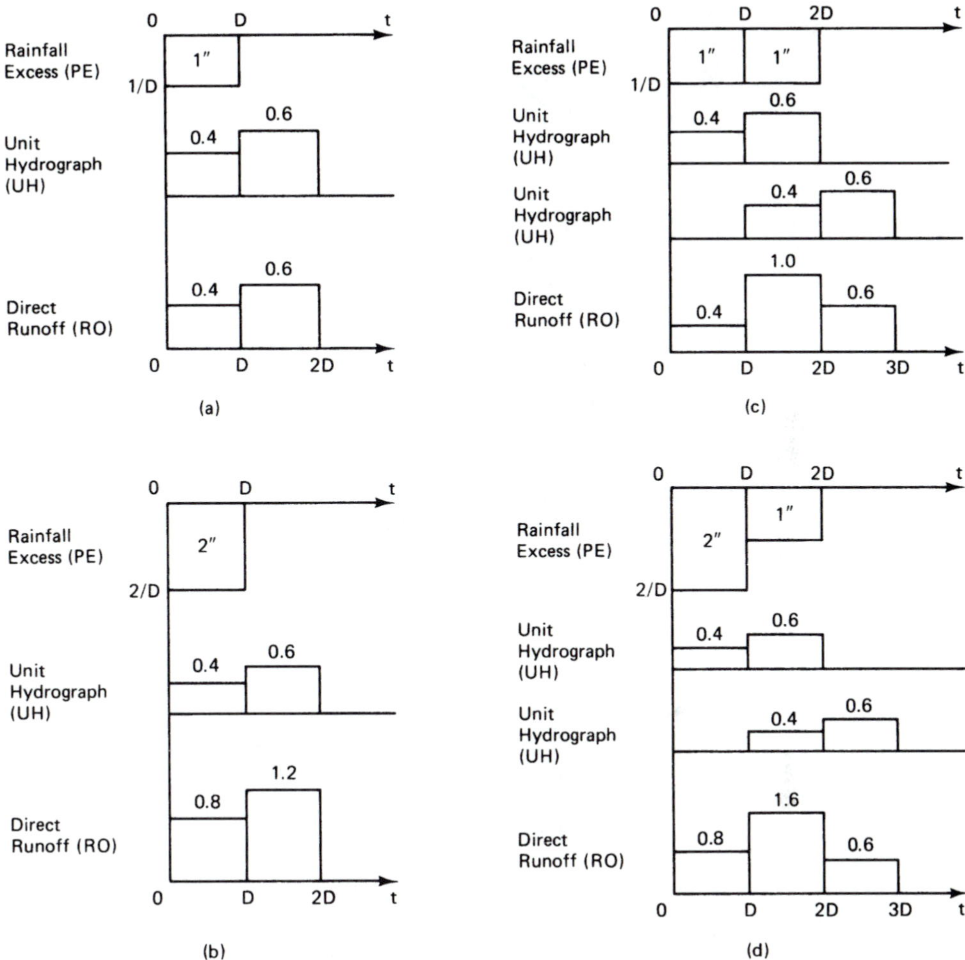

FIGURE 9–14 Convolution: a process of multiplication-translation-addition.

$$y(t) = \int_{o}^{t} x(\tau)\, U(t - \tau)\, d\tau \tag{9–45}$$

in which $U(t)$ is the time distributed UH, $y(t)$ is the time distribution of direct runoff, $x(\tau)$ is the computed time distribution of rainfall excess, and τ is the time lag between the beginning times of rainfall excess and the unit hydrograph. These elements are shown in Figure 9–15.

The convolution integral of Equation 9–45 can be placed in discrete form, which is the form used in hydrology with the digital computer. The discrete form relates the time distributions of rainfall excess $x(\tau)$, direct runoff $y(t)$, and the unit hydrograph $U(t - \tau)$:

$$y(t) = \sum_{r=0}^{t} x(\tau)\, U(t - \tau) \tag{9–46}$$

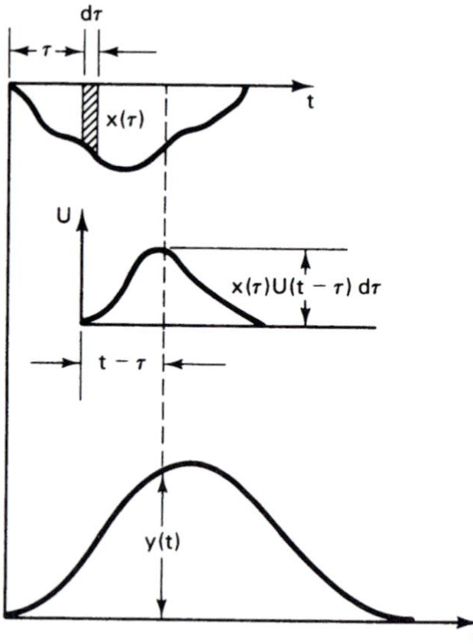

FIGURE 9–15 Schematic diagram of the convolution integral.

The following form can be used in place of Equation 9–46 for digital computations:

$$y(t) = \begin{cases} \sum_{j=1}^{t} x(j)\, U(t-j+1) \text{ if } i \leq n_p & (9\text{–}47\text{a}) \\ \sum_{j=1}^{n_p} x(j)\, U(t-j+1) \text{ if } n_p < i \leq n_u & (9\text{–}47\text{b}) \end{cases}$$

in which n_p is the number of ordinates in the discrete representation of the rainfall excess distribution $x(j)$, n_u is the number of ordinates in the discrete representation of the unit hydrograph $U(t - j + 1)$, and the following constraints apply:

$$U(i) = 0 \quad \text{for } i < 1 \text{ and } i > n_u \tag{9–48}$$

For the case where $n_u = 5$ and $n_p = 3$, Equations 9–47 yield the following equations for computing the direct runoff $y(i)$:

$$\begin{aligned} y(1) &= x(1)U(1) \\ y(2) &= x(1)U(2) + x(2)U(1) \\ y(3) &= x(1)U(3) + x(2)U(2) + x(3)U(1) \\ y(4) &= x(1)U(4) + x(2)U(3) + x(3)U(2) \\ y(5) &= x(1)U(5) + x(2)U(4) + x(3)U(3) \\ y(6) &= \qquad\qquad\quad x(2)U(5) + x(3)U(4) \\ y(7) &= \qquad\qquad\qquad\qquad\quad x(3)U(5) \end{aligned}$$

This representation of the convolution process should clearly indicate the multiplication, translation, and addition operations. In each case, the ith rainfall excess ordinate $x(i)$ is dis-

TABLE 9–7 Unit Hydrograph for Example 9–13

t (min)	0	6	12	18	24	30	36	42	48
UH (ft^3/sec)	0	1	18	140	85	38	9	2	0

tributed in time with the ordinates of the unit hydrograph. The number of ordinates in the direct runoff distribution is computed with Equation 9–43:

$$n_{ro} = n_{pe} + n_{uh} - 1 = 3 + 5 - 1 = 7$$

Example 9–13

A 0.1-hr unit hydrograph (Table 9–7) can be used to illustrate convolution in the synthesis phase. The rainfall excess and the unit hydrograph are known and the design runoff hydrograph must be computed. To illustrate this, a design storm with five 6-min increments will be used. The design storm will be derived from values of the intensity-duration-frequency relationship for Coshocton, Ohio. The 5-, 10-, 15-, and 30-min rainfall intensities for a 10-yr exceedence frequency are 6.1, 5.4, 4.5, and 3.2 in./hr, respectively. Using regression analysis, the following relationship between the rainfall intensity (i) in inches per hour and the duration in hours (D) was computed:

$$i = \frac{2.75}{0.357 + D} \tag{9–49}$$

This was used to compute the intensity, and then the depth, for durations of 6, 12, 18, 24, and 30 min. The estimated depths are given in Table 9–8. Computations for the development of the design storm are given in Table 9–8. From the estimated depths, the incremental depths were computed for each 6-min period. Based on these values the design storm pattern was developed using the method outlined in Section 4.6. The cumulative rainfall was computed, and the cumulative rainfall excess was computed using the SCS runoff depth equation (Equation 7–42), with a runoff CN of 84 assumed for estimating the runoff depths. The rainfall excess pattern (column 8 of Table 9–8) was computed by subtracting adjacent values of the cumulative rainfall excess distribution of column 7. The rainfall excess was then convolved with the 6-min unit hydrograph of Table 9–7. The convolution computations are shown in Table 9–9.

It is important to notice the multiplication, translation, and addition operations. The unit hydrograph is translated with each time increment, and each burst of rainfall excess is multiplied by the ordinates of the unit hydrograph. Summation yields the computed design hydrograph

TABLE 9–8 Computation of a 30-minute, 10-year Design Storm

Storm Time (hr)	Estimated Intensity (in./hr)	Estimated Depth (in.)	Incremental Depth (in.)	Design Storm (in.)	Cumulative Depth (in.)	Cumulative Rainfall Excess (in.)	Rainfall Excess (in.)
0.1	6.02	0.60	0.60	0.15	0.15	0.00	0.00
0.2	4.94	0.99	0.39	0.27	0.42	0.00	0.00
0.3	4.19	1.26	0.27	0.60	1.02	0.16	0.16
0.4	3.63	1.45	0.19	0.39	1.41	0.36	0.20
0.5	3.21	1.60	0.15	0.19	1.60	0.48	0.12

TABLE 9–9 Convolution of Rainfall Excess and the 6-minute UH

Storm Time (hr)	Convolution					Direct Runoff (ft³/sec)
0.1	0(1)					0
0.2	0(8)	+ 0(1)				0
0.3	0(140)	+ 0(18)	+ 0.16(1)			0
0.4	0(85)	+ 0(140)	+ 0.16(18)	+ 0.20(1)		3
0.5	0(38)	+ 0(85)	+ 0.16(140)	+ 0.20(18)	+ 0.12(1)	26
0.6	0(9)	+ 0(38)	+ 0.16(85)	+ 0.20(140)	+ 0.12(18)	44
0.7	0(2)	+ 0(9)	+ 0.16(38)	+ 0.20(85)	+ 0.12(140)	40
0.8		0(2)	+ 0.16(9)	+ 0.20(38)	+ 0.12(85)	19
0.9			0.16(2)	+ 0.20(9)	+ 0.12(38)	7
1.0				0.20(2)	+ 0.12(9)	1
1.1					0.12(2)	0
						140

given in Table 9–9. The sum of the hydrograph ordinates is 140 ft^3/sec. This translates to 0.479 in. for a 29–acre watershed:

$$140\,\frac{\text{ft}^3}{\text{sec}} * 6\text{ min} * \frac{60\text{ sec}}{\text{min}} * \frac{12\text{ in.}}{\text{ft}} * \frac{1}{29\text{ ac}} * \frac{1\text{ ac}}{43560\text{ ft}^2} = 0.479\text{ in.}$$

Thus the depth of the direct runoff hydrograph equals 0.48 in., which is the depth of rainfall excess (column 7 of Table 9–8).

9.7 UNIT HYDROGRAPH ANALYSIS

How is a unit hydrograph developed? In analysis, a unit hydrograph is computed from the time distributions of rainfall excess and direct runoff. If the two time distributions are simple (single-peaked with the primary ordinates in the rainfall excess having nearly equal intensities), then it may be acceptable to use the rainfall-excess reciprocal method, which is the traditional approach to unit hydrograph derivation. If the rainfall excess distribution is more complex, then a more sophisticated method of analysis, such as least squares, will need to be used. In either case, it will probably be necessary to smooth the ordinates of the resulting unit hydrograph and then adjust the smoothed curve to ensure that the ordinates are equivalent to 1 area-inch of direct runoff.

9.7.1 Rainfall-Excess Reciprocal Method

Before the availability of computers, it was difficult to derive estimates of unit hydrograph ordinates from complex storms. To minimize the necessary computations, only storms having simple patterns of rainfall excess were used. When storms having a rainfall excess duration of *T*-hours and a fairly uniform intensity could be found, the unit hydrograph was easily derived with the rainfall-excess reciprocal method. This was popular because it is computationally simple. The T-hour unit hydrograph is computed by multiplying each ordinate of the direct runoff hydrograph by the reciprocal of the depth of rainfall excess, which equals the depth of direct runoff. Since the unit hydrograph must have a depth of 1 in. and the direct runoff hydrograph has a volume equivalent to the depth of rainfall excess, the reciprocal of the depth of rainfall excess can be used as a proportionality constant to convert the direct runoff hydrograph to a unit hydrograph.

To illustrate the computations involved in unit hydrograph development, assume that a direct runoff hydrograph has the following ordinates: $DRO(t) = \{40, 70, 50, 20\}$ ft^3/sec. Assuming that the ordinates are on a 30-min time interval (the duration of direct runoff is 2 hr) and the watershed area is 35 acres, then the volume of direct runoff is

$$(40 + 70 + 50 + 20)\,\frac{\text{ft}^3}{\text{sec}}\,(30\text{ min})\left(\frac{60\text{ sec}}{\text{min}}\right)\left(\frac{1}{35\text{ ac}}\right)\left(\frac{1\text{ ac}}{43560\text{ ft}^2}\right)\left(\frac{12\text{ in.}}{\text{ft}}\right) = 2.55\text{ in.}$$

The unit hydrograph is obtained by dividing each ordinate of the direct runoff hydrograph by 2.55, which yields the unit hydrograph: $U(t) = \{15.7, 27.5, 19.6, 7.8\}$ ft^3/sec/in. To check the depth of runoff in the unit hydrograph, follow the same approach used to compute the depth of direct runoff:

$$(15.7 + 27.5 + 19.6 + 7.8)\,\frac{\text{ft}^3}{\text{sec}}\,(30\text{ min})\left(\frac{60\text{ sec}}{\text{min}}\right)\left(\frac{1}{35\text{ ac}}\right)\left(\frac{1\text{ ac}}{43560\text{ ft}^2}\right)\left(\frac{12\text{ in.}}{\text{ft}}\right) = 1.0\text{ in.}$$

Thus, $U(t)$ is a unit hydrograph.

Example 9–14

The rainfall (in./hr) and runoff (ft^3/sec) for the May 13, 1964, storm event on watershed 169, Coshocton, Ohio, are given in columns 8 and 2, respectively, in Table 9–10; the data were recorded at a nonconstant time interval. The watershed has an area of 29 acres, with land use as follows: 6% hardwood, 6% reforested, 48% grassland, 34% cultivated, and 6% miscellaneous, contour strip cropped. In comparing the distributions of rainfall and runoff, it is evident that significant runoff did not occur before hour 15:35, which is the time at which the peak intensity of rainfall began. Thus rainfall before hour 15:35 was considered to be lost to initial abstraction (column 9 of Table 9–10). The tail of the hydrograph is elongated by the low-intensity rainfall that occurred starting at hour 15:49.

Direct runoff was assumed to end with the measured runoff at hour 16:20. To account for the discharge that occurred after hour 16:20, a baseflow was assumed to be present. Use of a baseflow will eliminate sharp discontinuities at the beginning and end of the unit hydrograph. A constant-slope baseflow separation was used; subtracting the baseflow from the total measured runoff yielded the distribution of direct runoff (column 4 of Table 9–10). The accumulated runoff in inches was determined using the trapezoidal rule for the nonconstant time intervals (column 5 of Table 9–10), with a total runoff of 0.118 in. Since a unit hydrograph is the discharge for 1 in. of rainfall excess, the unit hydrograph can be determined by dividing the ordinates of the direct runoff hydrograph by the total runoff of 0.118 in. The unit hydrograph is given as column 6 of Table 9–10.

To determine the effective duration of the unit hydrograph, it is necessary to separate the measured rainfall into losses and rainfall excess. As indicated previously, rainfall prior to hour 15:35 was considered to be the initial abstraction. Rainfall after hour 16:18 was also eliminated as part of the losses. These two losses are given in column 9 of Table 9–10, with the difference

TABLE 9–10 Derivation of Rainfall Excess and a 6-minute Unit Hydrograph for the Storm of 5/13/64 on Watershed 169, Coshocton, Ohio

Time of Day (hr)	Flow Rate (ft^3/sec)	Base-Flow (ft^3/sec)	Direct Runoff (ft^3/sec)	Accumulation (in.)	6-Min UH (ft^3/sec)	Time of Day (hr)	Rainfall Intensity (in./hr)	I_a (in./hr)	$i - I_a$ (in./hr)	Accumulation (in.)	Losses (in./hr)	Rainfall Excess
1530	0.01	0.01	0.00	0.000		1523	0.00	0.00	0.00	0.00	0.00	0.00
1532	0.06	0.06	0.00	0.000		1528	0.96	0.96	0.00	0.00	0.00	0.00
1535	0.20	0.20	0.00	0.000	0	1535	3.43	3.43	0.00	0.00	0.00	0.00
1536	0.25	0.22	0.03	0.000	0.25	1541	6.00	0.00	6.00	0.60	4.82	1.18
1538	0.51	0.26	0.25	0.000	2.12	1549	1.12	0.00	1.12	0.75	1.12	0.00
1539	0.72	0.28	0.44	0.000	3.73	1618	0.17	0.00	0.17	0.83	0.17	0.00
1540	1.31	0.30	1.01	0.001	8.56	1703	0.03	0.03	0.00	0.83	0.00	0.00
1543	1.69	0.36	1.33	0.003	11.27	1833	0.01	0.01	0.00	0.83	0.00	0.00
1545	3.20	0.40	2.80	0.005	23.73							
1546	4.07	0.42	3.65	0.007	30.93							
1547	8.11	0.44	7.67	0.010	65.00							
1548	17.50	0.46	17.04	0.017	144.41							
1551	16.00	0.52	15.48	0.045	131.19							
1553	14.00	0.56	13.44	0.062	113.90							
1555	11.70	0.60	11.10	0.076	94.07							
1557	9.11	0.64	8.47	0.087	71.78							
1559	7.53	0.68	6.85	0.095	58.05							
1601	6.05	0.72	5.33	0.102	45.17							
1603	4.38	0.76	3.62	0.107	30.68							
1605	3.20	0.80	2.40	0.111	20.34							
1608	2.39	0.86	1.53	0.114	12.97							
1610	2.03	0.90	1.13	0.116	9.58							
1612	1.63	0.94	0.69	0.117	5.85							
1616	1.24	1.02	0.22	0.118	1.86							
1620	1.10	1.10	0.00	0.118	0.00							
1632	0.60	0.60	0.00	0.118								
1640	0.46	0.46	0.00	0.118								

given in column 10. The accumulated rainfall minus initial losses is given in column 11. The total accumulated difference $(i - I_a)$ equals 0.83 in. Since the depth of direct runoff equaled 0.118 in., the depth of rainfall excess must equal 0.118 in., so additional losses of 0.712 in. must be subtracted from the values in column 10. The phi-index method can be used to determine the remaining losses. A first estimate of the phi index is

$$\phi = \frac{0.712 \text{ in.}}{(16{:}18 - 15{:}35) \text{ min}/60} = 0.993 \text{ in./hr}$$

Since this value exceeds the rate of $(i - I_a)$ from storm time 15:49 to 16:18, all of the excess during this period, which is 0.082 in., will be losses. Thus losses of (0.712 – 0.083) in. = 0.629 in. must be subtracted from the rainfall from storm time 15:35 to 15:49. Thus the revised phi index is

$$\phi = \frac{0.629 \text{ in.}}{(15{:}49 - 15{:}35) \text{ min}/60} = 2.70 \text{ in./hr}$$

Since this rate exceeds the $(i - I_a)$ rate from storm time 15:41 to 15:49, all rainfall excess during this 9–min period will be losses, which represents a loss of 0.149 in. Thus losses for the period between storm times 15:35 and 15:41 is (0.629 – 0.149) = 0.48 in. Thus the revised phi index is

$$\phi = \frac{0.48 \text{ in.}}{(15{:}41 - 15{:}35) \text{ min}/60} = 4.8 \text{ in./hr}$$

During the period between storm time 15:35 and 15:41, the loss rate is 4.8 in./hr and the rainfall excess is 1.2 in./hr, which yields an excess volume of 0.12 in.; within roundoff accuracy this agrees with the volume of direct runoff of 0.118 in. The rainfall excess has a duration of 6 min, so the unit hydrograph of column 6 in Table 9–10 is considered to be a 0.l-hr unit hydrograph.

9.7.2 Least Squares Analysis of Unit Hydrographs

Least squares regression is an alternative to the method of Section 9.7.1. The rainfall excess and direct runoff become the predictor and dependent variable, respectively. The least squares procedure was discussed in Chapter 2. While the least squares method has some limitations, which will be discussed, it is a convenient method for deriving unit hydrograph ordinates from rainfall excess and direct runoff ordinates. It is easiest to introduce the least squares method using an example. The example will assume that the distribution of rainfall excess has two ordinates, P_1 and P_2, and the unit hydrograph has three ordinates, U_1, U_2, and U_3. Expressing the relationship between rainfall excess, the unit hydrograph, and the ordinates of the direct runoff hydrograph (r_i, $i = 1, 2, \ldots, n$) using the discrete form of the convolution integral provides the following:

$$\begin{aligned}
P_1U_1 \qquad\qquad &= P_1U_1 &&= R_1 \\
P_1U_2 + P_2U_1 &= P_2U_1 + P_1U_2 &&= R_2 \\
P_1U_3 + P_2U_2 &= \qquad\quad P_2U_2 + P_1U_3 &&= R_3 \\
+ P_2U_3 &= \qquad\qquad\quad P_2U_3 &&= R_4
\end{aligned} \tag{9–50}$$

Equations 9–50 represent four equations and three unknowns, which are the unit hydrograph ordinates. Additionally, the depth of direct runoff must equal 1 area-in. These equations can be expressed in matrix form as follows:

$$\begin{bmatrix} P_1 & 0 & 0 & 0 \\ P_2 & P_1 & 0 & 0 \\ 0 & P_2 & P_1 & 0 \\ 0 & 0 & P_2 & 0 \end{bmatrix} \begin{bmatrix} U_1 \\ U_2 \\ U_3 \\ 0 \end{bmatrix} = \begin{bmatrix} R_1 \\ R_2 \\ R_3 \\ R_4 \end{bmatrix} \tag{9–51}$$

which can be simplified to

$$\mathbf{PU} = \mathbf{R} \tag{9–52}$$

in which **P** is a 4×4 matrix, and both **U** and **R** are column vectors with four elements. Since the elements of **P** and **R** are known, while the elements of **U** are unknown, Equation 9–52 can be transformed so that the knowns are on one side of the equation and the unknowns on the other side. Specifically, concepts in matrix algebra provide the following:

$$\begin{aligned} \mathbf{PU} &= \mathbf{R} \\ (\mathbf{P}^T\mathbf{P})\mathbf{U} &= \mathbf{P}^T\mathbf{R} \\ (\mathbf{P}^T\mathbf{P})^{-1}(\mathbf{P}^T\mathbf{P})\,\mathbf{U} &= (\mathbf{P}^T\mathbf{P})^{-1}\mathbf{P}^T\mathbf{R} \\ \mathbf{IU} &= (\mathbf{P}^T\mathbf{P})^{-1}\mathbf{P}^T\mathbf{R} \\ \mathbf{U} &= (\mathbf{P}^T\mathbf{P})^{-1}\mathbf{P}^T\mathbf{R} \end{aligned} \tag{9–53}$$

in which $\mathbf{P}^T$ is the transpose of **P**, $\mathbf{P}^{-1}$ is the matrix inverse of **P**, and **I** is the identity matrix. The unit hydrograph obtained from Equation 9–53 will very likely need adjusting because of irregularities in the shape. Specifically, either negative ordinates or undulations in the ordinates of the unit hydrograph recession will probably occur. These irregularities, which are a by-product of the regression method, can be eliminated through either manual smoothing or by fitting a synthetic functional form to the resulting moments. It is also important to ensure that the ordinates have a sum of 1 area-in. One possible method of smoothing the ordinates of a UH with a synthetic form is given in Section 9.8.2. The solution of Equation 9–53 must be obtained using a computer because of the length and complexity of the calculations.

Example 9–15

To illustrate the use of Equations 9–53 for deriving a unit hydrograph from the time distributions of rainfall excess (*PE*) and direct runoff (*DRO*), the simplest possible problem will be used. The example will use data where the true solution is known. Consider the case where the rainfall excess consists of two ordinates (0.4, 0.8); and the unit hydrograph, which is unknown in the analysis case, has the ordinates (0.6, 0.4). The goal will be to derive these values of the unit hydrograph. From convolution, the direct runoff ordinates would be (0.24, 0.64. 0.32). For this case, the rainfall excess and direct runoff are assumed to be known, and the objective is to estimate the unit hydrograph.

Placing the rainfall excess and direct runoff in matrix form, Equation 9–52 is

$$\begin{bmatrix} 0.4 & 0 & 0 \\ 0.8 & 0.4 & 0 \\ 0 & 0.8 & 0 \end{bmatrix} \begin{bmatrix} U_1 \\ U_2 \\ 0 \end{bmatrix} = \begin{bmatrix} 0.24 \\ 0.64 \\ 0.32 \end{bmatrix} \tag{9–54}$$

The inverse matrix $(\mathbf{P}^T\mathbf{P})^{-1}$ of Equation 9–53 is

$$(\mathbf{P}^T\mathbf{P})^{-1} = \left(\begin{bmatrix} 0.4 & 0.8 & 0 \\ 0 & 0.4 & 0.8 \\ 0 & 0 & 0 \end{bmatrix} \begin{bmatrix} 0.4 & 0 & 0 \\ 0.8 & 0.4 & 0 \\ 0.0 & 0.8 & 0 \end{bmatrix} \right)^{-1} \tag{9–55}$$

$$= \begin{bmatrix} \frac{2.5}{1.68} & \frac{-1}{1.68} & 0 \\ \frac{-1}{1.68} & \frac{2.5}{1.68} & 0 \\ 0 & 0 & 0 \end{bmatrix}$$

Postmultiplying this inverse by $\mathbf{P}^T$ yields

$$(\mathbf{P}^T\mathbf{P})^{-1}\mathbf{P}^T = \begin{bmatrix} \frac{2.5}{1.68} & \frac{-1}{1.68} & 0 \\ \frac{-1}{1.68} & \frac{2.5}{1.68} & 0 \\ 0 & 0 & 0 \end{bmatrix} \begin{bmatrix} 0.4 & 0.8 & 0 \\ 0 & 0.4 & 0.8 \\ 0 & 0 & 0 \end{bmatrix} \tag{9–56}$$

$$= \frac{1}{1.68} \begin{bmatrix} 1 & 1.6 & -0.8 \\ -0.4 & 0.2 & 2.0 \\ 0 & 0 & 0 \end{bmatrix}$$

Completing the matrix computations of Equation 9–53 yields

$$(\mathbf{P}^T\mathbf{P})^{-1}\mathbf{P}^T\mathbf{R} = \frac{1}{1.68} \begin{bmatrix} 1 & 1.6 & -0.8 \\ -0.4 & 0.2 & 2.0 \\ 0 & 0 & 0.0 \end{bmatrix} \begin{bmatrix} 0.24 \\ 0.64 \\ 0.32 \end{bmatrix} = \begin{bmatrix} 0.6 \\ 0.4 \\ 0 \end{bmatrix} = \begin{bmatrix} U_1 \\ U_2 \\ 0 \end{bmatrix} \tag{9–57}$$

The matrix computations yields the unit hydrograph that was used originally to derive the distribution of direct runoff by convolution with the rainfall excess.

9.8 UNIT HYDROGRAPH ADJUSTMENTS

In concept, a unit hydrograph reflects the effects that both watershed characteristics and watershed conditions have on rainfall excess. An implicit assumption that the characteristics and conditions are constant for each storm is made. Additionally, it is assumed that rainfall characteristics are not a factor in the shape of the unit hydrograph. In practice, watershed conditions show considerable storm-to-storm variation and the UH analysis procedure does not eliminate the effects of rainfall characteristics. Thus, unit hydrographs analyzed from different storms on the same watershed show considerable variation. Storms that dumped a larger portion of the rainfall near the watershed outlet will typically have unit hydrographs that peak earlier than average. A unit storm on a relatively dry watershed will have a delayed peak that is lower than average. The longer time to peak occurs because a larger time period is needed to fulfill initial abstractions and a lower-than-average soil moisture deficit. These differences in storm characteristics and watershed conditions can produce widely different unit hydrographs.

Figure 9–16 shows the unit hydrographs for five storm events on the White Oak Bayou, Texas. While three of the five have the shape typical of unit hydrographs, one of the five (the 1959 event) has a flat portion on the rising limb that was probably caused by a period of low rainfall. The 1955 UH has a low peak and a slower recession than the other

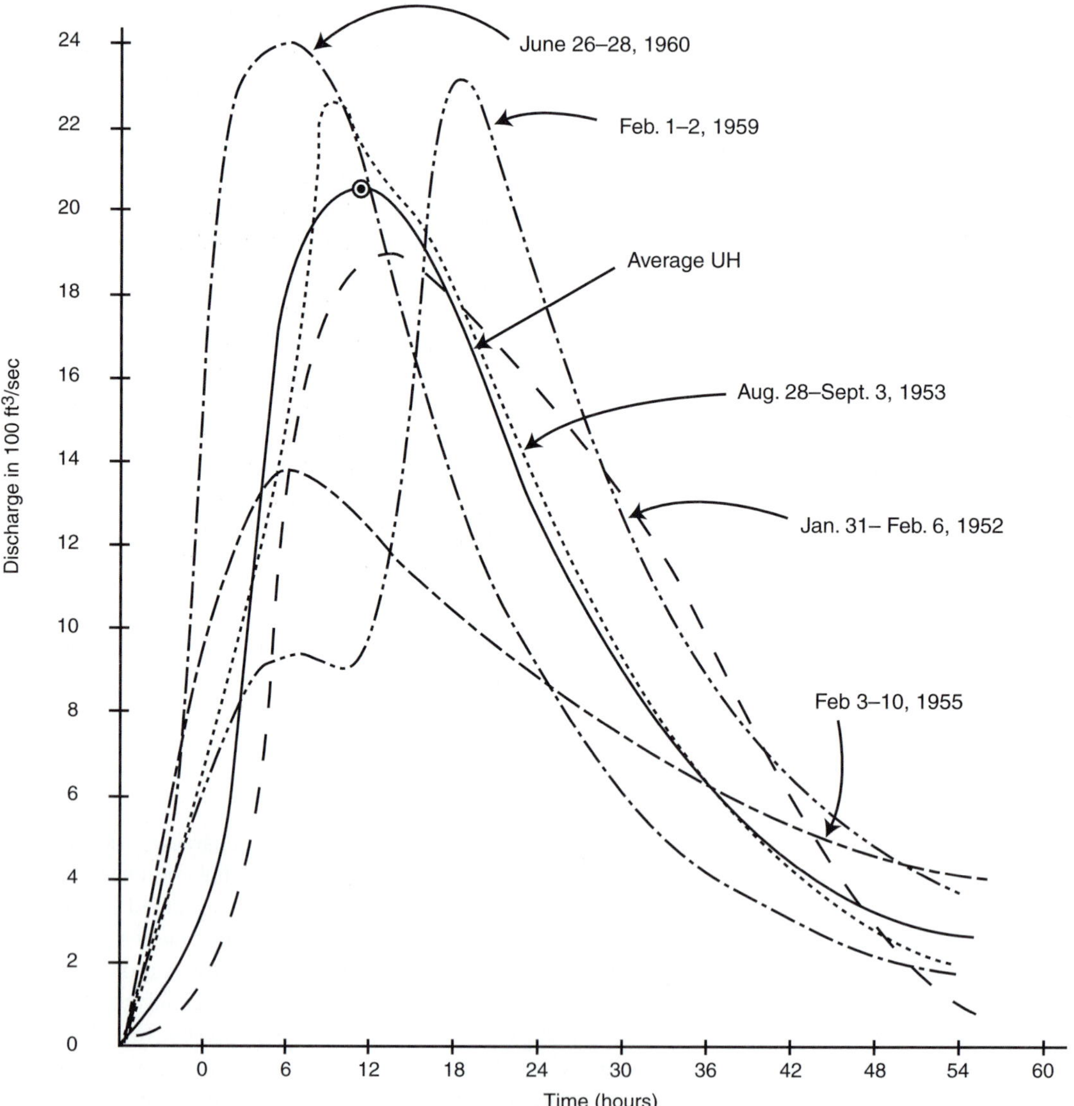

FIGURE 9–16 Fitted unit hydrographs from five storm events on White Oak Bayou, TX. (Hare, 1970)

events; these characteristics could have been caused by rainfall characteristics, nonuniformity of rainfall over the storm duration, or unsaturated watershed conditions. This is one factor that makes adjustments of a UH necessary.

Four adjustments are occasionally used in unit hydrograph analyses. First, it may be necessary to adjust storm event UHs to a common duration of rainfall excess. The S-hydrograph method is used for changing the duration of a unit hydrograph. Second, a computed hydrograph that has an irregular shape can be smoothed using the gamma UH form. Third, if several unit hydrographs are available from different storm events on the same watershed, a watershed-averaged UH can be computed. Fourth, if unit hydrographs from several watersheds in a region are available, it is necessary to put them in dimensionless form to derive a regionalized UH. Each of these adjustments are discussed in this section.

9.8.1 S-Hydrograph Method

In deriving a unit hydrograph from measured storm data, the characteristics of the unit hydrograph depend on the effective duration of the rainfall excess. The duration of the excess is important because it must be coordinated with the watershed time of concentration when using the unit hydrograph for design. If it is necessary to transform the unit hydrograph so that it represents a different effective duration of the rainfall excess, the S-hydrograph method can be used. The S-hydrograph method is based on the idea that, whereas an infinite number of effective durations are possible, only one S-hydrograph exists for a watershed. The S-hydrograph is sometimes referred to as either an S-graph or a summation hydrograph.

Given a unit hydrograph for effective duration *T*, the S-curve can be computed by adding an infinite series of *T*-hour unit hydrographs, with each *T*-hour unit hydrograph lagged by *T* hours. This is shown in Figure 9–17. A series of triangular unit hydrographs are lagged by a time increment of *T* hours, which is the effective duration of the rainfall excess. The rainfall excess is shown at the top of Figure 9–17, with each burst of rainfall excess hav-

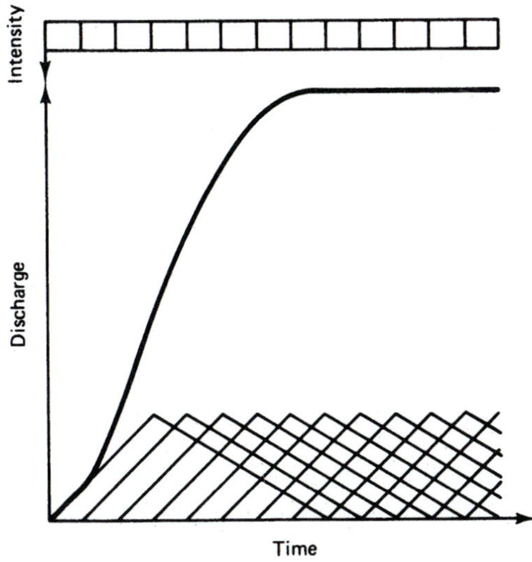

FIGURE 9–17 Derivation of an S-hydrograph.

ing a depth of 1 in. in a period of T hours. Summing the triangular unit hydrographs yields the S-graph shown. The S-graph is unique for each watershed. Mathematically, the ith ordinate of the S-graph, which is denoted as S_i, is the sum of all unit hydrograph ordinates, U_j, from the first ordinate to the ith ordinate:

$$S_i = \sum_{j=1}^{i} U_j \tag{9–58}$$

This computation applies when the time interval of the unit hydrograph equals the unit duration of the unit hydrograph.

When using an S-curve to derive a unit hydrograph for a duration other than the duration of the unit hydrograph that was used to develop the S-curve, it is necessary for the ordinates to be recorded on the time interval of the shorter of the two durations. For example, if an S-curve is derived from a 10-min unit hydrograph and a 20-min unit hydrograph is needed, the ordinates must be recorded on a 10-min interval. However, if a 5-min unit hydrograph is needed, the ordinates must be recorded on a 5-min interval. In this case, the intermediate points must be interpolated from the S-curve. When values are interpolated from the S-graph, it is important to ensure that the sum of the ordinates is equivalent to 1 area-in. This will occur if linear interpolation between points on the S-graph is used; however, linear interpolation from the S-graph should be avoided because it will always result in a peak discharge for the shorter-duration unit hydrograph that is smaller than the peak of the longer-duration unit hydrograph. This is not rational since the peak discharge of the unit hydrograph should increase as the duration of the rainfall excess decreases. Thus, selecting intermediate points from the S-graph should take into account the nonlinear nature of the S-graph, but the interpolation must be done in a way that ensures a volume of 1.0 area-in. of direct runoff.

Example 9–16

The 6-min unit hydrograph of Table 9–10 can be used to illustrate the derivation of an S-graph. However, to use the unit hydrograph it must first be restructured so that it is measured on an equal-interval time scale, with the scale equal to the effective duration of 6 min; at the same time, it is necessary to maintain a volume of 1 area-in. An objective means for converting a unit hydrograph recorded on an unequal interval to a unit hydrograph for a constant time interval is not available. However, an approximation is shown in Figure 9–18. The ordinates are given in column 2 of Table 9–11a. The ordinates were adjusted to ensure that the volume equaled 1 area-in. and that both the shape and magnitude were similar to the unit hydrograph derived for the unequal time increment.

Based on the 6-min unit hydrograph of equal time intervals, the S-graph was derived with Equation 9–58. The S-graph is given as column 3 in Table 9–11a. The S-graph was then offset by 12 min (column 4 of Table 9–11a) and the differences between the two S-graphs computed (column 5 of Table 9–11a). Since the differences have a volume of 2 area-in., each ordinate was halved, which provides a 12-min unit hydrograph (column 6 of Table 9–11a). The 12-min unit hydrograph is shown in Figure 9–18. In general, the longer duration unit hydrograph will be less peaked and may have a delayed time to peak; this is evident from the 12-min unit hydrograph in Figure 9–18.

To compute a unit hydrograph for effective durations shorter than that used to drive the S-graph, it is necessary to interpolate interior points from the S-graph and use a time interval equal to the effective duration for which the unit hydrograph is needed. To derive a 3-min unit

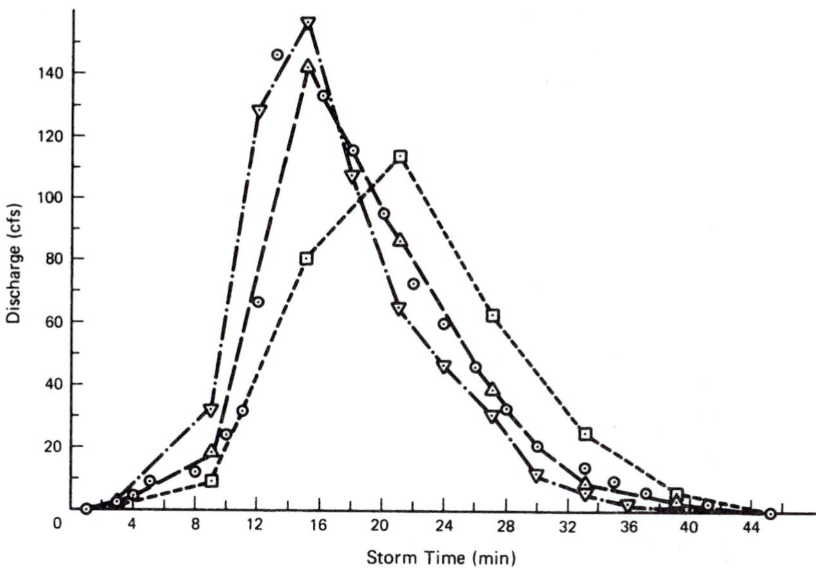

FIGURE 9–18 Unit hydrographs for 3 minutes (▽—•▽), 6 minutes (△— —△), and 12 minutes (⊡– – –⊡), with 6-minute unit hydrograph from measured data (⊙⊙).

hydrograph from the 6-min unit hydrograph, values were interpolated at a 3-min increment from the S-graph of Figure 9–19. The values are given in column 7 of Table 9–11a. The S-graph was then offset by 3 min (column 8) and the differences (column 9) computed. For the 3-min unit hydrograph to have a unit volume, it is necessary to double each ordinate of column 9, which yields the unit hydrograph of column 10 in Table 9–11a. The 3-min unit hydrograph is shown in Figure 9–18. The peak is slightly greater than that of the 6-min unit hydrograph and the rising limb occurs sooner.

Example 9–17

A unit hydrograph recorded on a time interval of 20 min is given in column 2 of Table 9–11b. The S-hydrograph is given in column 3. A 10-min unit hydrograph can be derived from the 20-min unit hydrograph by interpolating between the ordinates of the 20-min S-hydrograph. Since linear interpolation might seem like the obvious choice, an S-graph for a 10-min interval was derived using linear interpolation (see column 4). The S-graph was offset by 10 min (see column 5), and the differences computed (column 6). The 10-min unit hydrograph is given in column 7. Two points are especially noteworthy. First, the peak discharge did not change significantly, increasing slightly because of rounding errors from 1981 ft^3/sec to 1982 ft^3/sec. Second, adjacent ordinates of the unit hydrograph are essentially the same, which is not rational. This example shows that linear interpolation should not be used when forming unit hydrographs of shorter durations.

An S-graph was formed using nonlinear interpolation (see column 8 of Table 9–11b). The S-curve was offset and the differences computed. The 10-min unit hydrograph was computed (column 11). Both problems identified for linear interpolation were avoided by using nonlinear interpolation. The peak discharge for the 10-min unit hydrograph, 2012 ft^3/sec, represents an increase of 1.6% over the peak of the 20-min unit hydrograph.

TABLE 9–11 S-graph Analysis Results: (a) Derivation of 3-minute and 12-minute Unit Hydrographs from 6-minute Unit Hydrographs using S-graph Offset Method

Storm Time (min)	6-min UH (ft^3/sec)	S-graph (ft^3/sec)	Offset S-graph (ft^3/sec)	ΔS (ft^3/sec)	12-min UH (ft^3/sec)	S-graph (ft^3/sec)	Offset S-graph (ft^3/sec)	ΔS (ft^3/sec)	3-min UH (ft^3/sec)
0	0	0	0	0	0	0	0	0	0
3	1	1	0	1	0.5	1	0	1	2
6						3	1	2	4
9	18	19	0	19	9.5	19	3	16	32
12						82	19	63	126
15	140	159	1	158	79.0	159	82	77	154
18						212	159	53	106
21	85	244	19	225	112.5	244	212	32	64
24						267	244	23	46
27	38	282	159	123	61.5	282	267	15	30
30						288	282	6	12
33	9	291	244	47	23.5	291	288	3	6
36						292	291	1	2
39	2	293	282	11	5.5	293	292	1	2
42						293	293	0	0
45	0	293	291	2	1.0	293	293	0	0
					293.0				586

(b) Comparison of the Effect of Linear and Nonlinear Interpolation of S-graphs on Computed Unit Hydrograph

Time (min)	20-min UH (ft^3/sec)	S-graph (ft^3/sec)	Linear S-graph (ft^3/sec)	Offset S-graph (ft^3/sec)	ΔS (ft^3/sec)	10-min UH (ft^3/sec)	Nonlinear S-graph (ft^3/sec)	Offset S-graph (ft^3/sec)	ΔS (ft^3/sec)	Nonlinear 10-min UH (ft^3/sec)
0	0	0	0	0	0	0	0	0	0	0
10			60	0	60	120	30	0	30	60
20	121	121	121	60	61	121	121	30	91	182
30			485	121	364	728	402	121	281	562
40	728	849	849	485	364	728	849	402	447	894
50			1,528	849	679	1,358	1,454	849	605	1,210
60	1,359	2,208	2,208	1,528	680	1,360	2,208	1,454	754	1,508
70			3,100	2,208	892	1,784	3,055	2,208	847	1,694
80	1,785	3,993	3,993	3,100	893	1,786	3,993	3,055	938	1,876
90			4,982	3,993	989	1,978	4,965	3,993	972	1,944
100	1,978	5,971	5,971	4,982	989	1,978	5,971	4,965	1,006	2,012
110			6,962	5,971	991	1,982	6,967	5,971	996	1,992
120	1,981	7,952	7,952	6,962	990	1,980	7,952	6,967	985	1,970
130			8,879	7,952	927	1,854	8,899	7,952	947	1,894
140	1,854	9,806	9,806	8,879	927	1,854	9,806	8,899	907	1,814
150			10,634	9,806	828	1,656	10,660	9,806	854	1,708
160	1,655	11,461	11,461	10,634	827	1,654	11,461	10,660	801	1,602
170			12,174	11,461	713	1,426	12,203	11,461	742	1,484
180	1,425	12,886	12,886	12,174	712	1,424	12,886	12,203	683	1,366

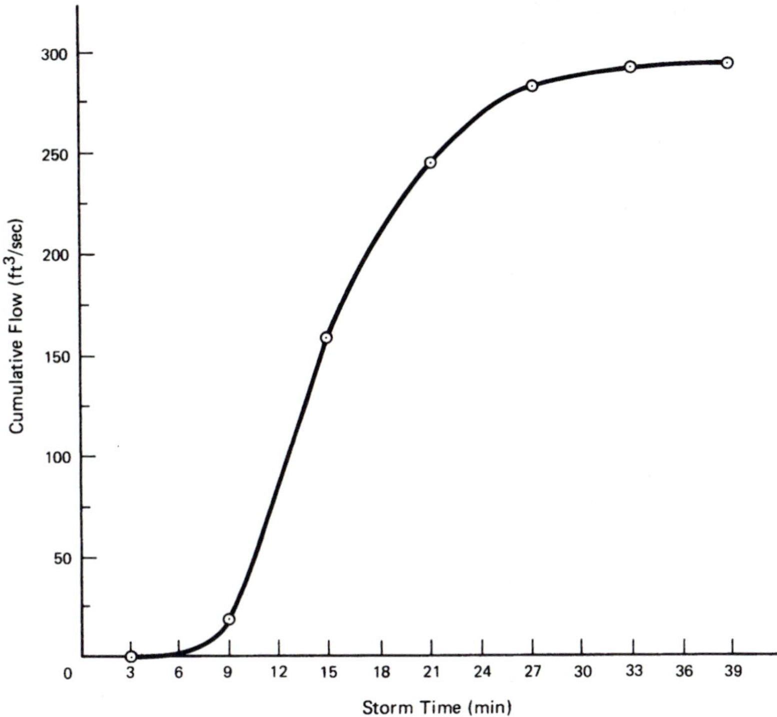

FIGURE 9–19 S-hydrograph for storm event of 5/13/64 on Watershed 169, Coshocton, Ohio.

This example clearly illustrates a major problem with the S-graph method. It is difficult to accurately and systematically produce a unit hydrograph for a shorter duration. Attempts to fit a cumulative gamma distribution to the S-curve and interpolate from the resulting curve were not very fruitful.

9.8.2 Gamma Distribution Unit Hydrograph

When unit hydrographs are derived from measured data, the ordinates must be smoothed to provide a reasonable shape. To provide for a systematically varying shape, it may be preferable to use a probability function and use the moments of the measured-data unit hydrograph to fit the parameters of the probability function. Aron and White (1982) provided a method of estimating the parameters of a gamma probability distribution using the peak discharge and time to peak of a computed unit hydrograph. The gamma distribution has the form

$$f(t;\ a,\ b) = \frac{t^a\, e^{-t/b}}{b^{a+1}\,\Gamma\,(a+1)} \tag{9–59}$$

in which a and b are called the shape and scale parameters, respectively; and $\Gamma(a)$ is the gamma function with argument a. The parameters a and b are obtained from the computed peak discharge q_p and the time to peak t_p of the unit hydrograph using the following steps:

1. Compute the depth variable (f_a):

$$f_a = \frac{q_p t_p}{A} \tag{9–60}$$

in which q_p has units of ft^3/sec, t_p has units of hours, and A is the drainage area in acres.

2. Find the value of a from the following:

$$a = 0.045 + 0.5f_a + 5.6f_a^2 + 0.3f_a^3 \tag{9–61}$$

3. Compute the parameter b from

$$b = \frac{t_p}{a} \tag{9–62}$$

Given the values of the parameters a and b, the ordinates of the gamma distribution unit hydrograph at time t can be computed from the following:

$$q(f) = q_p f^a e^{a(1-f)} \tag{9–63}$$

in which f is the ratio of time t to the time to peak, ($f = t/t_p$). Variable $q(f)$ can be computed for any value of t. Typically, f is varied from 0 to 5.

Example 9–18

Unit hydrographs computed from measured storm events on a single watershed can take on a variety of shapes, with some shapes not being indicative of general runoff characteristics of the watershed. The peak discharge and time to peak of such unit hydrographs can be used to fit a synthetic unit hydrograph, such as the gamma function of Equation 9–59. Figure 9–20 shows the unit hydrograph for the October 19, 1960, storm event on Buffalo Bayou, which is a 58-mi^2 watershed in Houston, Texas. The computed unit hydrograph has a peak of 1456 ft^3/sec and a time to peak of 16 hr. From Eqs. 9–60, 9–61, and 9–62, f_a, a, and b are 0.62759, 2.6386, and 6.0638, respectively. Thus the synthetic unit hydrograph of Equation 9–59 is

$$q(t) = \frac{t^{2.6386} e^{-t/6.0638}}{6.0638^{2.6386+1}\,\Gamma(3.6386)} \tag{9–64}$$

Ordinates of the unit hydrograph can be computed using Equation 9–63, which yields

$$q(f) = q_p f^{2.6386} e^{\{2.6386(1-f)\}} \tag{9–65}$$

The ordinates are given in Table 9–12 for values of f from 0 to 5, and the synthetic unit hydrograph is shown in Figure 9–20. The measured and synthetic unit hydrographs are noticeably different. The synthetic UH is a reasonable replacement for the measured UH if one suspects that the shape of the measured UH has been distorted by other factors such as rainfall characteristics or nonuniformity of rainfall over the watershed. This was the case for the UH on Buffalo Bayou shown in Figure 9–20.

9.8.3 Averaging Storm Event Unit Hydrographs

Unit hydrographs analyzed from different storm events on the same watershed will have widely different shapes even if the duration of rainfall excesses are similar. These differences are illustrated in Figure 9–16 and can be due to differences in storm patterns, storm volumes, storm-cell movement, and antecedent watershed conditions.

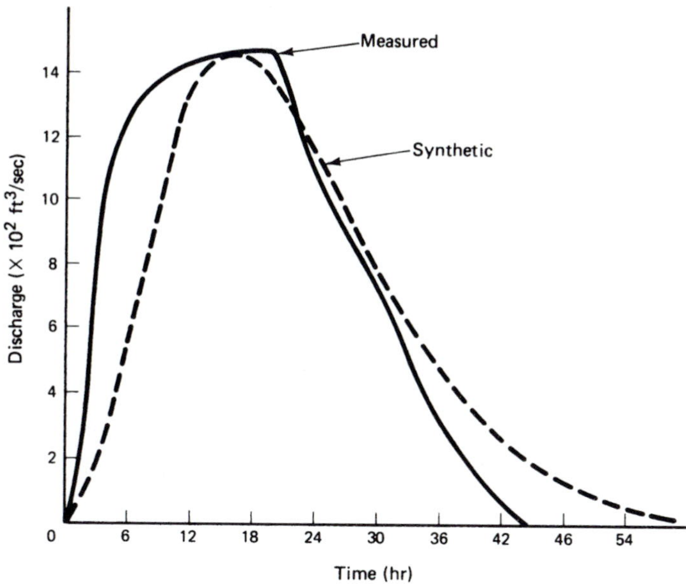

FIGURE 9–20 Comparison of measured and synthetic unit hydrographs for the 10/19/60 storm event on the Buffalo Bayou Water shed.

Design work usually involves a single representative unit hydrograph. Where multiple unit hydrographs are available, such as on the White Oak Bayou of Figure 9–16, it is necessary to "average" the unit hydrographs from the individual storm events.

The following steps are used to average two or more unit hydrographs that were computed from different storm events on the same watershed:

1. Compute the average peak discharge.
2. Compute the average time to peak.
3. Plot each of the storm event UHs on a single graph.
4. Locate the point defined by the average peak discharge and average time to peak from Steps 1 and 2.
5. Sketch a unit hydrograph that represents an average of the shapes of the storm event UHs and passes through the point defined in Step 4.
6. Read off the ordinates of the average unit hydrograph sketched in Step 5 and compute the volume of the average UH.
7. Adjust the ordinates of the sketched UH so that it has a volume of 1 area-in.; the adjustments are usually made in the recession of the UH.

This averaging method assumes that all of the storm event UHs have approximately the same unit duration. If they do not, they should be adjusted using the S-hydrograph method prior to averaging.

TABLE 9–12 Ordinates of the Measured (q) and Synthetic ($\hat{q}$) Unit Hydrographs at Buffalo Bayou for the Storm Event of October 19, 1960

Time (hr)	q (ft³/sec)	f	$\hat{q}$ (ft³/sec)
0	0	0	0
4	1050	0.25	272
8	1340	0.50	875
12	1400	0.75	1318
16	1456	1.00	1456
20	1400	1.25	1356
24	1060	1.50	1135
28	800	1.75	881
32	540	2.00	648
40	110	2.50	312
48	0	3.00	135
56	0	3.50	54
64	0	4.00	21
72	0	4.50	8
80	0	5.00	3

It is important to emphasize that it is incorrect to compute an average of the ordinates for each time. If this were done, the watershed-average UH would have a low peak discharge and the shape would not be representative of the true unit hydrograph.

Example 9–19

Figure 9–16 shows unit hydrographs for five storm events on White Oak Bayou, Texas; the data were adapted from graphs provided by Hare (1970). The ordinates are given in Table 9–13. White Oak Bayou has a drainage area of 92 sq. mi. The peak discharge and time to peak are given in Table 9–13 for each storm UH; the averages are also given. The point defined by $\bar{q}_p$ and $\bar{t}_p$ is located on Figure 9–16 with the five storm event UHs. A smooth distribution was sketched through the point, with consideration given to the shapes of the five storm event unit hydrographs. The ordinates at 2-hr intervals were taken from the initial sketch of the average UH and the volume under the curve computed using the trapezoidal rule. The ordinates were adjusted because the volume of the initially sketched UH was greater than 1 area-in. The sum of the ordinates for the final UH are shown in Table 9–13. The volume is

$$V_{UH} = 28{,}870\,\frac{\text{ft}^3}{\text{sec}} * 2\text{ hr} * \frac{3600\text{ sec}}{\text{hr}} * \frac{12\text{ in.}}{1\text{ ft}} * \frac{1}{92\text{ mi}^2} * \left(\frac{1\text{ mi}}{5280\text{ ft}}\right)^2$$
$$= 0.973\text{ in.}$$

The difference of 0.027 in. is assumed to be the volume in the recession of the UH beyond a storm time of 60 hr.

9.8.4 Dimensionless Unit Hydrographs

Where UHs are developed from different watersheds in a relatively homogeneous region, a regionalized UH cannot be developed directly using the averaging procedure given in Section 9.8.3. To develop a regionwide UH, it is necessary to convert the individual watershed

TABLE 9–13 Unit Hydrograph Characteristics: White Oak Bayou, Texas

Storm dates	Peak Discharge (ft^3/sec)	Time to Peak (hours)
Jan. 31–Feb. 6, 1952	1893	19.0
Aug. 28–Sept. 3, 1953	2231	14.8
Feb. 3–10, 1955	1388	11.4
Feb. 1–2, 1959	2278	23.4
June 26–28, 1960	2407	10.9
	mean = 2039	15.9

Time (hours)	Discharge (ft^3/sec) for UH of 1952	1953	1955	1959	1960	Average UH (ft^3/sec)	Dimensionless UH q/q_p	t/t_p
0	0	0	0	0	0	0	0	0
2	60	190	210	100	280	170	0.083	0.125
4	130	520	680	240	800	380	0.186	0.250
6	225	710	1060	450	1770	630	0.309	0.375
8	400	950	1220	820	2300	1000	0.490	0.500
10	970	1210	1320	880	2400	1650	0.809	0.625
12	1370	1520	1370	870	2400	1910	0.936	0.750
14	1660	2230	1310	830	2330	2000	0.980	0.875
16	1830	2220	1230	900	2200	2040	1.000	1.000
18	1890	2170	1190	1010	2010	2000	0.980	1.125
20	1890	2010	1140	1470	1700	1910	0.936	1.250
22	1825	1830	1080	2250	1400	1800	0.882	1.375
24	1750	1680	1030	2250	1270	1640	0.804	1.500
26	1670	1510	980	2100	1110	1490	0.730	1.625
28	1570	1320	920	1810	980	1300	0.637	1.750
30	1500	1190	880	1610	880	1170	0.574	1.875
32	1420	1070	830	1420	750	1050	0.515	2.000
34	1290	920	780	1210	650	900	0.441	2.125
36	1210	820	740	1110	600	800	0.392	2.250
38	1090	720	700	980	510	710	0.348	2.375
40	960	630	650	850	460	630	0.309	2.500
42	860	560	630	770	410	560	0.275	2.625
44	740	500	600	680	370	510	0.250	2.750
46	600	440	570	600	330	460	0.225	2.875
48	490	410	540	540	300	430	0.211	3.000
50	390	370	510	480	270	390	0.191	3.125
52	290	320	480	420	240	340	0.167	3.250
54	250	280	460	390	220	300	0.147	3.375
56	180	250	440	350	200	260	0.127	3.500
58	140	220	420	320	180	230	0.113	3.625
60	100	200	410	300	160	210	0.103	3.750
Sum	28750	28970	24380	28010	29480	28870		
UH depth	0.969	0.976	0.821	0.944	0.993	0.973		

UHs to a dimensionless form by dividing the ordinates by the peak discharge and the time values by the time to peak. The resulting dimensionless UHs can then be plotted and an average of the individual dimensionless UHs drawn. In order to use the average dimensionless UH on watersheds in the same region, it is also necessary to have relationships for computing the peak discharge and time to peak. An example that uses a dimensionless UH will be given in Section 9.9. The dimensionless UH for the watershed-averaged UH for White Oak Bayou is given in Table 9–13.

9.9 SYNTHETIC UNIT HYDROGRAPHS

To this point the discussion has centered about the analysis problem. The objective has been to derive the unit hydrograph from the measured storm data. The example of Section 9.6.2 (Example 9–13) assumed that the design storm was applied to the same watershed from which the unit hydrograph was derived. But at almost all locations where a design is necessary, rainfall and runoff data are not available to derive both a unit hydrograph and an S-curve for the watershed. Therefore, a need to have unit hydrographs that can be used at such ungaged locations exists. These are often referred to as regionalized or synthetic unit hydrographs. It is preferable to use a unit hydrograph only for designs on watersheds that are similar to the watersheds used in developing the unit hydrograph. Thus the term *regional unit hydrograph* would suggest that the unit hydrograph was developed from watersheds in a region that have similar watershed characteristics and meteorological conditions. The regional or synthetic unit hydrographs are developed from the analysis of measured storm data and then used on ungaged locations. To use the regionalized unit hydrographs on other watersheds, they are usually made dimensionless so that they can be applied to watersheds that have a wide range of watershed characteristics.

A number of synthetic unit hydrographs have been proposed and are used. Time-area diagrams, which are based on the cumulative proportion of a watershed that supposedly contributes runoff to the watershed outlet as the storm time increases, have been used in a number of models as the basis for representing the unit hydrograph. Dimensionless unit hydrographs are also available that can be used on ungaged watersheds. Each of these will be discussed.

9.9.1 Time-Area Unit Hydrographs

A time-area curve is an S-graph with the abscissa as the travel time and the ordinate as the cumulative proportion of the total watershed that contributes runoff for a specified travel time. The ordinate can alternatively be shown as the cumulative area contributing. The derivative of the S-graph would be a unit hydrograph or, if expressed as a proportion, a distribution graph. The time base of the unit hydrograph equals the time of concentration, so it must be a special form of a unit hydrograph called the instantaneous UH (IUH). The time-area UH is the result of a burst of rainfall that has 1-in. of excess and occurs over an infinitesimally small duration.

Consider the watershed shown in Figure 9–21a. The watershed is divided into 59 subareas of equal area. The travel time from the subarea to the outlet is shown in the section, with all subareas contributing to the outlet within 14 time units. The distribution of subareas

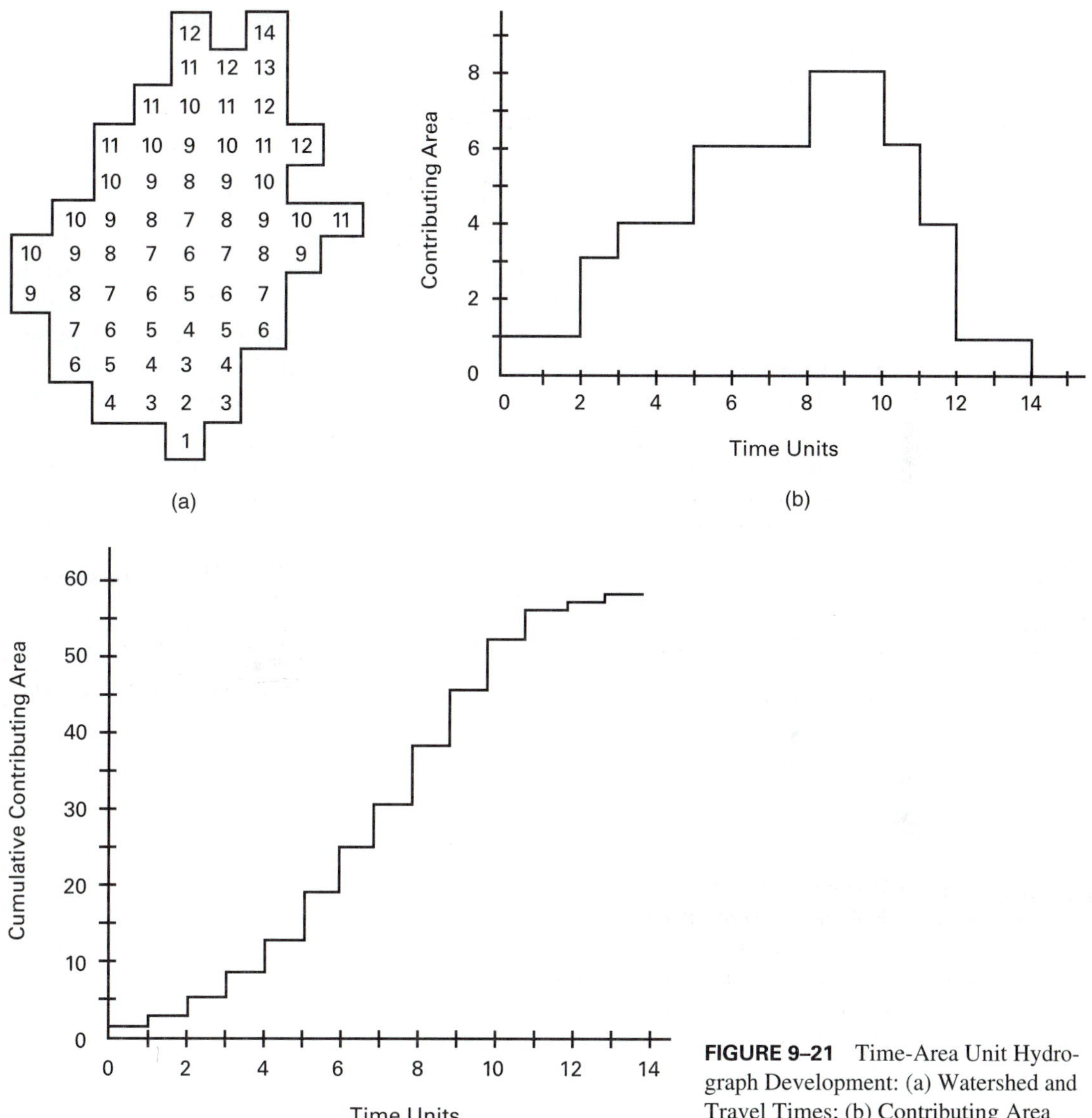

FIGURE 9–21 Time-Area Unit Hydrograph Development: (a) Watershed and Travel Times; (b) Contributing Area versus Time; (c) S-Hydrograph.

contributing for each time is shown in Figure 9–21b. The cumulative distribution of Figure 9–21b is shown in Figure 9–21c; this is the time-area diagram. The time base of the time-area diagram equals the time of concentration, which is 14 time units. The instantaneous unit hydrograph of Figure 9–21b indicates that runoff ceases at the time of concentration, which occurs because rainfall occurred over an infinitesimally small time interval.

The time-area diagram and the instantaneous unit hydrograph reflect runoff due to translation only, with no storage effects. Since this is not realistic to use for design, it is common to route the instantaneous UH through a single linear reservoir (SLR) to represent the effects of storage. The single linear reservoir is given by

$$h(t) = \frac{e^{-t/K}}{K} \tag{9–66}$$

in which K is the routing coefficient. It is common practice to assume that the routing coefficient equals the time lag, which is usually assumed to equal 60% of the time of concentration. Routing the time-area UH through a single linear reservoir will lower the peak discharge, delay the time to peak, and lengthen the time base; these changes are assumed to reflect the effects of watershed storage. When applying the SLR of Equation 9–66, it is necessary to have the sum of the ordinates equal 1. The usual practice for applying the SLR is to present the SLR as a series of discrete ordinates and convolve the SLR with the time-area UH.

The advantage of the time-area curve is that it has a realistic conceptual basis. The problem with its use is the fact that, in spite of the number of studies of the time-area SLR approach, its accuracy has not been convincingly verified. In some cases, other models, most notably the gamma distribution or Nash Model, have shown to provide results that are more accurate than those from Equation 9–66. However, accurate methods of computing K of the SLR model or the parameters of the gamma distribution have proven to be illusive. Thus, the approach is not widely used on ungaged watersheds.

An alternative to the routing through the SLR is to use a storm having a duration other than the infinitesimally small duration of the IUH. For example, a constant-intensity storm that has a duration equal to the time of concentration is widely used, with the belief that such a storm would produce the maximum peak discharge. The following computations illustrate the process.

Assume that the watershed consists of six equal subareas, as shown in Figure 9–22a, and that the rainfall hyetograph is as shown in Figure 9–22b. The rainfall duration equals the time of concentration for the watershed. The storm has a constant intensity i. The distribution of rainfall excess is also shown in Figure 9–22b, with a magnitude of Ci, where C is a runoff coefficient of the Rational Method; thus the loss function is constant with a magnitude of $i * (1\text{-}C)$. The initial abstraction is assumed to be part of the losses. Based on the assumption that the rainfall is uniformly distributed over the watershed, the depth of rainfall excess Ci is assumed to fall on each subarea of the watershed. At the end of time $t_c/3$ runoff from only subarea 1 will be arriving at the watershed outlet. Assuming that runoff from subareas 2, 3, and 4 have equal travel times to the watershed outlet, which equal $2t_c/3$, then at the end of time $2t_c/3$ subareas 1 to 4 will be contributing runoff at the outlet. This is shown in the hydrograph of Figure 9–22c. At time $2t_c/3$ rain that fell on subareas 5 and 6 would not contribute to direct runoff at the outlet. At time t_c all six subareas are contributing to runoff at the outlet, which is reflected in the ordinate of the hydrograph. At time t_c the storm has reached its duration. During the time interval from t_c to $4t_c/3$ rain that fell during time interval $t_c/3$ to $2t_c/3$ is still contributing runoff at the outlet from subareas 5 and 6, and rain that fell during the time interval $2t_c/3$ to t_c is contributing runoff from subareas 2, 3, and 4. Thus the runoff ordinate equals 5 units (see Figure 9–22c). Rain that fell during the time interval from $2t_c/3$ to t_c on subareas 5 and 6 appears as runoff at the outlet in the time interval $4\,t_c/3$ to $5t_c/3$ (see Table 9–14). It is also important to observe that the depth of rainfall excess, Cit_c, equals the depth of direct runoff; the depths can be converted to volumes by multiplying by the total area, A.

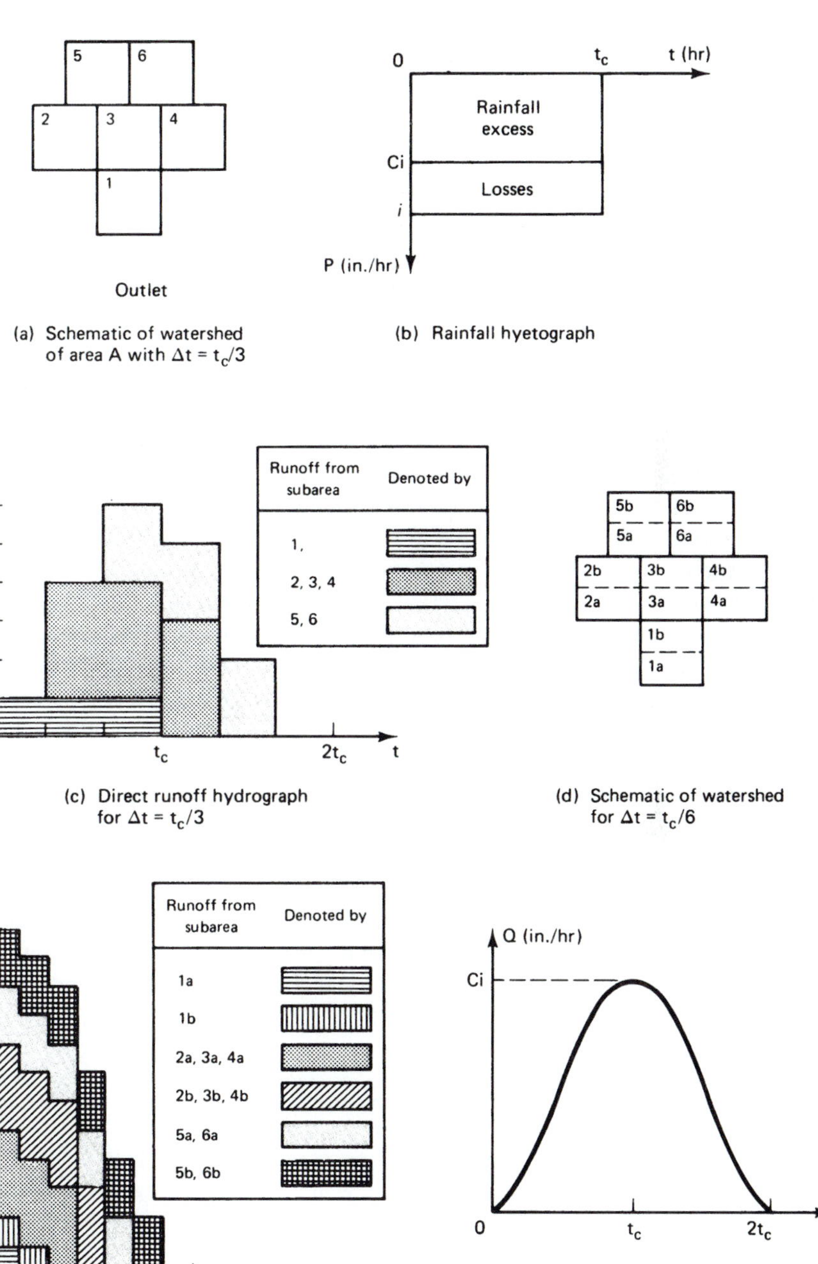

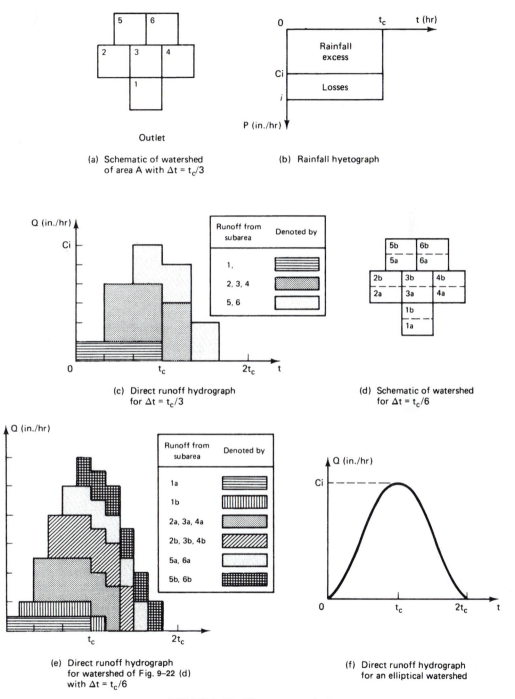

(a) Schematic of watershed of area A with $\Delta t = t_c/3$

(b) Rainfall hyetograph

(c) Direct runoff hydrograph for $\Delta t = t_c/3$

(d) Schematic of watershed for $\Delta t = t_c/6$

(e) Direct runoff hydrograph for watershed of Fig. 9–22 (d) with $\Delta t = t_c/6$

(f) Direct runoff hydrograph for an elliptical watershed

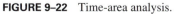

FIGURE 9–22 Time-area analysis.

TABLE 9–14 Time-Area Analysis of Hypothetical Watershed

Runoff Measured at Outlet During Time Interval	Subarea Contributing Runoff to Outlet			Rainfall Occurring During Time Interval
	1	2, 3, 4	5, 6	
0 to $t_c/3$	✓			
$t_c/3$ to $2t_c/3$	✓	✓		
$2t_c/3$ to t_c	✓	✓	✓	0 to $t_c/3$
t_c to $4t_c/3$		✓	✓	$t_c/3$ to $2t_c/3$
$4t_c/3$ to $5t_c/3$			✓	$2t_c/3$ to t_c

The above analysis is somewhat misleading because the rainfall excess and direct runoff are used with a relatively large, discrete time interval, $t_c/3$. The last particle of rainfall excess falling at the upper end of subarea 5 or 6 at time t_c will require a travel time to the outlet of t_c, which means that it will appear as runoff at the outlet at time $2t_c$. The runoff hydrograph of Figure 9–22c suggests that this particle of rainfall reaches the outlet at $5t_c/3$. The difference is due to the discretizing of the calculations. If the time interval, Δt, goes from $t_c/3$ to an infinitesimally small time period, the analysis will yield a hydrograph with a shape similar to that of Figure 9–22e but with a time base equal to $2t_c$. The peak still equals Ci and occurs at time t_c. This can be shown empirically by using successively smaller time increments. For a time increment of $t_c/6$, which is one-half of the time increment used above, we separate the watershed as shown in Figure 9–22d. This produces the direct runoff hydrograph shown in Figure 9–22e. In this case, the time base of the direct runoff hydrograph is $11t_c/6$; this supports the statement that the time base will approach $2t_c$ as the time increment Δt goes to 0.

For a rectangular watershed of length L and width W the direct runoff hydrograph will have the shape of an isosceles triangle, with a peak Ci and a time base of $2t_c$. Actual watersheds do not have square edges like the schematic Figure 9–22a, and they are not rectangular. Instead, they appear more elliptical. In such a case, the hydrograph will have a shape such as that shown in Figure 9–22f.

Example 9–20

Figure 9–23a shows a topographic map of watershed W1 in Treynor, Iowa. The watershed has a drainage area of 74.5 acres, with 39, 13, 36, and 12% of the watershed in the slope ranges of 2–4, 5–8, 9–13, and 14–19%, respectively; a slope map is shown as Figure 9–23b. The principal waterway has a length of 3500 ft, with 5% of the drainage area as gully and grassed waterways. The remaining 95% of the land is in corn and contoured with a high level of fertility and good farming practices.

To compute the time-area curve, the relationship between slope and velocity for the grassed waterways is $V = 1.5S^{0.5}$ (Figure 3–19) and for contoured corn, the relationship is $V = 0.5S^{0.5}$. Using the slope and topographic maps (Figure 9–23b and a, respectively), iso-time lines were constructed using a time increment of 2 min (see Figure 9–23c). The travel distance for the 2-min time increments was determined using Equation 3–42, with the velocity computed as a function of slope and flow regime. The iso-time map of Figure 9–23c was used to compute the time-area curve shown in Figure 9–24. This is a valid S-graph.

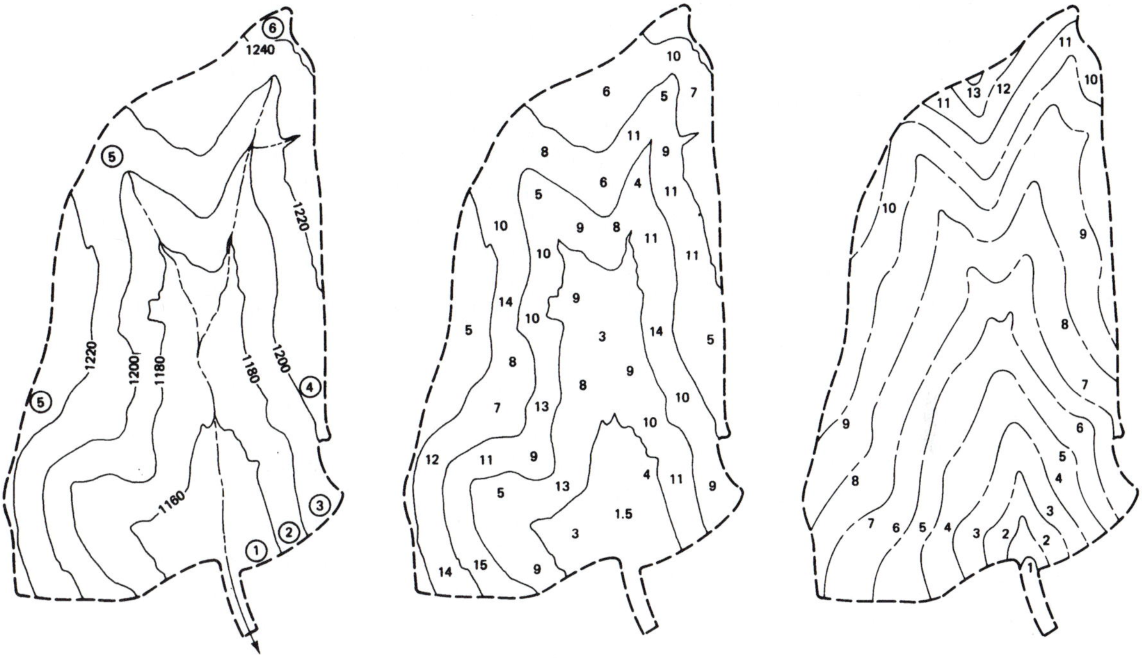

(a) Topographic map (b) Slope (c) Iso-time lines

FIGURE 9–23 Watershed W1, Treynor, Iowa (A = 74.5 acres).

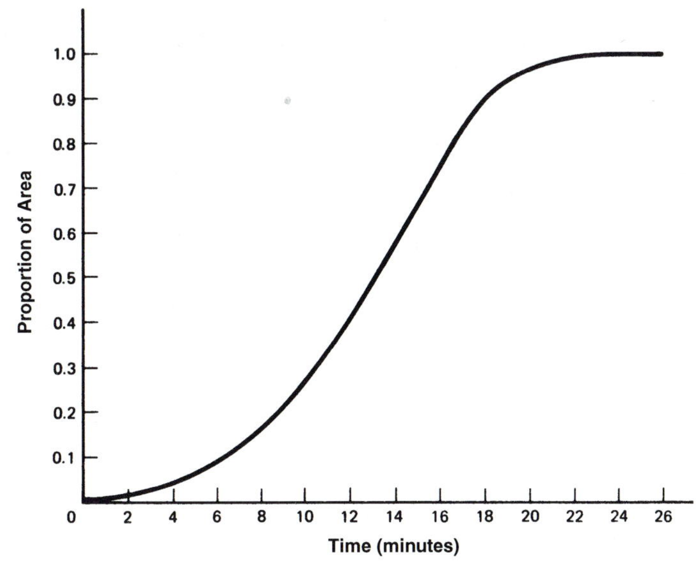

FIGURE 9–24 Time-area curve for Watershed W1, Treynor, Iowa.

The time-area curve was numerically differentiated to obtain the IUH. The ordinates of both the time-area curve and the IUH are given in columns 2 and 4 of Table 9–15, respectively. The time of concentration for the watershed is 25.5 min, which yields a time lag of 15.3 min; this is used as the routing coefficient of Equation 9–66, which was used to compute the SLR function given in column 5 of Table 9–15. The sum of the ordinates computed with Equation 9–66 was used to adjust the ordinates of the SLR to a unit volume. The SLR was convolved with the IUH to produce the routed unit hydrograph (column 6 of Table 9–15). The IUH has a peak of 410 ft^3/sec and a time to peak of 14 min. The routed UH has a peak of 220 ft^3/sec and a time to peak of 18 min. A time to peak of 18 min seems reasonable since the relationship $t_p = 2t_c/3$ is commonly used. The changes in q_p and t_p reflect the smoothing effect of the routing. The time base of the routed unit hydrograph doubled because the time base of the SLR equaled the time base of the IUH.

TABLE 9–15 Development of Unit Hydrograph from Time-Area Curve: Watershed W-1, Treynor, Iowa

Time (min)	Dimensionless S-curve	Ordinates	IUH (ft^3/s)	SLR Ordinates	Routed UH (ft^3/sec)
0	0	0	0	0.14595	0
2	0.0150	0.0150	34	0.12806	5
4	0.0425	0.0275	62	0.11237	13
6	0.0863	0.0438	99	0.09860	26
8	0.1618	0.0755	170	0.08652	48
10	0.2645	0.1027	231	0.07592	76
12	0.4051	0.1406	317	0.06661	113
14	0.5871	0.1820	410	0.05845	159
16	0.7557	0.1686	380	0.05129	195
18	0.9050	0.1493	337	0.04500	220
20	0.9676	0.0626	141	0.03949	214
22	0.9906	0.0230	52	0.03465	195
24	0.9982	0.0076	17	0.03041	174
26	1.0000	0.0018	4	0.02668	153
28		1.0000	2254		134
30					117
32					101
34					86
36					72
38					58
40					43
42					28
44					16
46					6
48					2
50					1
52					0
					2254

9.9.2 Hydrograph Assumptions of the Rational Method

The Rational Method was previously introduced as a method for estimating peak discharges. The development of the Rational Method made several assumptions:

1. The rainfall intensity, i, is constant over the storm duration.
2. The rainfall is uniformly distributed over the watershed.
3. The maximum rate of runoff will occur when runoff is being contributed to the outlet from the entire watershed.
4. The peak rate of runoff equals some fraction of the rainfall intensity.
5. The watershed system is linear.

These assumptions can be used to develop a hydrograph.

One of several possible assumptions can be made to formulate the hydrograph. The easiest solution would be as follows:

1. Estimate the peak discharge of the runoff hydrograph from Equation 7–20.
2. Assume that the runoff hydrograph is an isosceles triangle with a time to peak equal to t_c and a time base of $2t_c$.

This method would produce a hydrograph with 50% of the volume under the rising limb of the hydrograph and a total volume of $CiAt_c$. The assumption of an isosceles triangle would probably be reasonable for most design problems on small urban watersheds. To obtain a more realistic description of the runoff hydrograph, the shape of the hydrograph can be determined from the time-area curve, as suggested earlier.

If assumption 1 above is used, namely $q_p = CiA$, the time base of the hydrograph must equal $2t_c$ if a triangular shape is assumed. A time base longer than $2t_c$ will result in more runoff than rainfall. To demonstrate this, assume the rainfall hyetograph has a constant intensity i over a duration t_c. The volume of rainfall is iAt_c, and the volume of rainfall excess is $CiAt_c$. If the peak discharge of the direct runoff hydrograph is CiA then the volume of direct runoff, assuming a triangular shape, is 1/2 $CiAT_b$ where T_b is the time base. In order for the volume of direct runoff to equal the volume of rainfall excess, T_b must equal $2t_c$. If it were assumed to equal $4t_c$, then the volume of direct runoff would be twice the volume of rainfall excess.

Regardless of the assumption about the shape of the hydrograph, the use of the Rational Method has advantages and disadvantages. Certainly, it is subject to the limitations of the Rational Equation (Equation 7–4). However, it is easy to develop, and the accuracy should be sufficient for designs on small, highly urbanized watersheds.

A unit hydrograph is inherent in the Rational Method. Since the volume of the unit hydrograph must equal 1 area-in., the ordinates of the UH can be determined by multiplying each ordinate of the direct runoff hydrograph of the Rational Method by the conversion factor K:

$$K = \frac{1}{Cit_c} \tag{9–67}$$

in which i and t_c are in./hr and hr, respectively. Thus the peak discharge of the unit hydrograph will be Kq_p.

Example 9–21

For $C = 0.4$, $i = 3$ in./hr, $A = 14$ acres, and $t_c = 0.25$ hr, the direct runoff would have a peak discharge of 16.8 ft^3/sec, the conversion factor K would equal 3.33, and the peak discharge of the unit hydrograph would be 56 ft^3/sec. For a 14-acre watershed that has a UH with a time base of 0.5 hr and a peak discharge of 56 ft^3/sec, the volume is 1 in.

Example 9–22

The case of Example 7–11 and Figure 7–1 can be used to illustrate the application of the Rational Formula with runoff hydrographs. The assumption is made that the passage of the hydrograph from a subarea through a pipe occurs with translation and no attenuation. This may be a reasonable assumption for short pipe lengths. This method also assumes that the pipe has the capacity to pass the peak discharge and that the inlet does not constrict the flow. These are reasonable assumptions for design. The isosceles-triangle hydrographs have a time base of $2t_c$, a peak computed with the Rational Formula, and a beginning time equal to the travel time from the subarea outlet through the pipe to the design point.

The discharge hydrograph from subarea 1 is translated to the outlet of subarea 2, where it is added to the discharge hydrograph from subarea 2 (see Table 9–16). Notice that the subarea-2 hydrograph begins at time $t = 0$ because translation is not necessary; however, the subarea-1 hydrograph has a 3-min travel time through the pipe to the junction at subarea 2. Therefore, it begins at a time of 3-min. At subarea 2, the peak of the total discharge hydrograph occurs at $t = 9$ min and has a magnitude of 18.2 cfs, which is 2.6 cfs greater than the peak of the subarea-2 hydrograph.

At the junction at subarea-3 inlet, the total hydrograph has a peak of 28.1 cfs. Most of this is due to subarea-3 runoff, although runoff from all three subareas contribute to the discharge at that time ($t_p = 7$ min). The subarea-1 hydrograph begins at $t = 6$, with a 3-min translation between inlet 1 and junction 2 and another 3-min translation between junctions 2 and 3.

In Example 7–11, the method provided an estimate of 37.4 ft^3/sec for the entire 18.9 acres. The analysis in this example provides an estimate of 28.1 ft^3/sec, which is 23% lower. Which is correct? Unfortunately, we cannot say that one estimate is correct and the other is incorrect. Both are based on assumptions for which valid arguments can be made. The method of Example 7–11 lumps the watershed characteristics together by computing mean values of the runoff coefficient; thus, it does not take into consideration the spatial variation in land use. The method of this section assumes perfect translation of the pipeflow. This may not be realistic. The method of Example 7–11 ignores the difference between pipeflow and surface runoff. Procedures in addition to the two described here are used. In practice, the method used depends on local policies and practices.

9.9.3 The SCS Dimensionless Unit Hydrographs

The SCS hydrograph program TR-20 uses a dimensionless unit hydrograph that is based on an extensive analysis of measured data. Unit hydrographs were evaluated for a large number of actual watersheds and then made dimensionless by dividing all discharge ordinates by the peak discharge and all time ordinates by the time to peak. An average of these dimensionless unit hydrographs was computed. The time base of the dimensionless UH was approximately 5 times the time to peak, and approximately 3/8 of the total volume occurred before the time

TABLE 9–16 Rational Method Hydrographs for Example 9–22

	Discharge (ft^3/sec) Hydrographs						
Time (min)	Sub–1 at Pipe B	Sub-2 at Inlet	Total at Pipe B	Sub–1 at Pipe C	Sub-2 at Pipe C	Sub-3 Inlet	Total at Pipe C
0	.0	.0	.0	.0	.0	.0	.0
1	.0	1.7	1.7	.0	.0	3.0	3.0
2	.0	3.5	3.5	.0	.0	5.9	5.9
3	.0	5.2	5.2	.0	.0	8.9	8.9
4	.4	6.9	7.4	.0	1.7	11.8	13.6
5	.9	8.7	9.5	.0	3.5	14.8	18.3
6	1.3	10.4	11.7	.0	5.2	17.7	22.9
7	1.8	12.1	13.9	.4	6.9	20.7	28.1
8	2.2	13.9	16.1	.9	8.7	17.7	27.3
9	2.6	15.6	18.2	1.3	10.4	14.8	26.5
10	3.1	13.9	16.9	1.8	12.1	11.8	25.7
11	3.5	12.1	15.6	2.2	13.9	8.9	24.9
12	3.9	10.4	14.3	2.6	15.6	5.9	24.1
13	4.4	8.7	13.1	3.1	13.9	3.0	19.9
14	4.8	6.9	11.8	3.5	12.1	.0	15.6
15	5.3	5.2	10.5	3.9	10.4	.0	14.3
16	5.7	3.5	9.2	4.4	8.7	.0	13.1
17	5.3	1.7	7.0	4.8	6.9	.0	11.8
18	4.8	.0	4.8	5.3	5.2	.0	10.5
19	4.4	.0	4.4	5.7	3.5	.0	9.2
20	3.9	.0	3.9	5.3	1.7	.0	7.0
21	3.5	.0	3.5	4.8	.0	.0	4.8
22	3.1	.0	3.1	4.4	.0	.0	4.4
23	2.6	.0	2.6	3.9	.0	.0	3.9
24	2.2	.0	2.2	3.5	.0	.0	3.5
25	1.8	.0	1.8	3.1	.0	.0	3.1
26	1.3	.0	1.3	2.6	.0	.0	2.6
27	.9	.0	.9	2.2	.0	.0	2.2
28	.4	.0	.4	1.8	.0	.0	1.8
29	.0	.0	.0	1.3	.0	.0	1.3
30	.0	.0	.0	.9	.0	.0	.9
31	.0	.0	.0	.4	.0	.0	.4
32	.0	.0	.0	.0	.0	.0	.0

to peak. The inflection point on the recession limb occurred at approximately 1.7 times the time to peak, and the UH had a curvilinear shape. The average dimensionless UH is shown in Figure 9–25, and the discharge ratios for selected values of the time ratio are given in Table 9–17.

The curvilinear unit hydrograph can be approximated by a triangular UH that has similar characteristics. Figure 9–26 shows a comparison of the two dimensionless unit hydrographs. While the time base of the triangular UH is only 8/3 of the time to peak (compared to 5 for the curvilinear UH), the areas under the rising limbs of the two UHs are the same (37.5%).

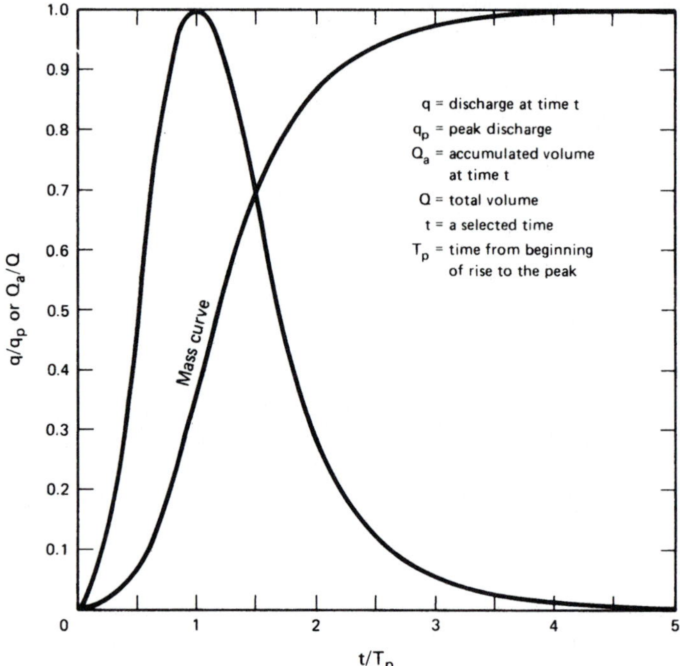

FIGURE 9–25 Dimensionless unit hydrograph and mass curve.

The area under a unit hydrograph represents the depth of direct runoff Q, which is 1 in. for a unit hydrograph; based on geometry the runoff depth is related to the characteristics of the triangular unit hydrograph by

$$Q = \frac{1}{2} q_p t_b = \frac{1}{2} q_p (t_p + t_r) \tag{9–68}$$

in which t_p and t_r are the time to peak and the recession time, respectively; and q_p is the peak discharge. Solving Equation 9–68 for q_p and rearranging yields

$$q_p = \frac{Q}{t_p}\left[\frac{2}{1 + t_r/t_p}\right] \tag{9–69}$$

Letting K replace the contents within the brackets yields

$$q_p = \frac{KQ}{t_p} \tag{9–70}$$

To have q_p in ft^3/sec, t_p in hr, and Q in in., it is necessary to divide q_p by the area A in mi^2 and to multiply the right-hand side of Equation 9–70 by the constant 645.3; also because $t_r = 1.67\ t_p$, Equation 9–70 becomes

TABLE 9–17 Ratios for SCS Dimensionless Unit Hydrograph and Mass Curve

Time Ratios t/T_p	Discharge Ratios q/q_p	Mass Curve Ratios Q_a/Q
0	0.000	0.000
0.1	0.030	0.001
0.2	0.100	0.006
0.3	0.190	0.012
0.4	0.310	0.035
0.5	0.470	0.065
0.6	0.660	0.107
0.7	0.820	0.163
0.8	0.930	0.228
0.9	0.990	0.300
1.0	1.000	0.375
1.1	0.990	0.450
1.2	0.930	0.522
1.3	0.860	0.589
1.4	0.780	0.650
1.5	0.680	0.700
1.6	0.560	0.751
1.7	0.460	0.790
1.8	0.390	0.822
1.9	0.330	0.849
2.0	0.280	0.871
2.2	0.207	0.908
2.4	0.147	0.934
2.6	0.107	0.967
2.8	0.077	0.953
3.0	0.055	0.977
3.2	0.040	0.984
3.4	0.029	0.989
3.6	0.021	0.993
3.8	0.015	0.995
4.0	0.011	0.997
4.5	0.005	0.999
5.0	0.000	1.000

$$q_p = \frac{484AQ}{t_p} \tag{9–71}$$

The constant 484 reflects a unit hydrograph that has 3/8 of its area under the rising limb. For mountainous watersheds the fraction could be expected to be greater than 3/8, and therefore, the constant of Equation 9–71 may be near 600. For flat, swampy areas the constant may be less than 300.

The time to peak of Equation 9–71 can be expressed in terms of the duration of unit rainfall excess and the time of concentration. Figure 9–26 provides the following two relationships:

$$t_c + D = 1.7\, t_p \tag{9–72}$$

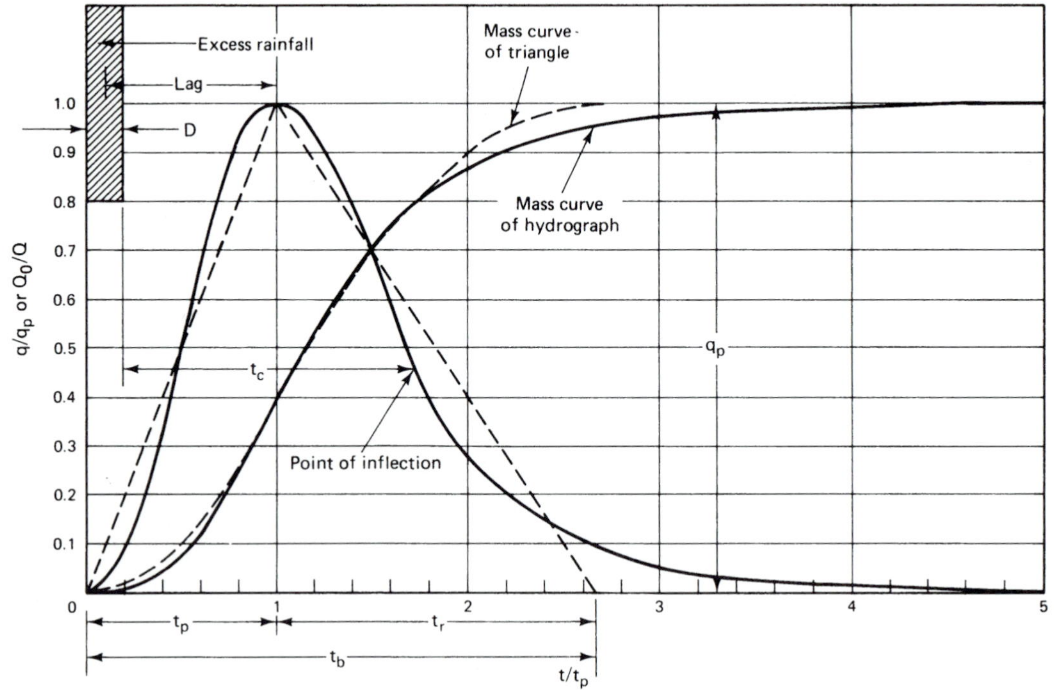

FIGURE 9–26 Dimensionless curvilinear unit hydrograph and equivalent triangular hydrograph.

and

$$\frac{D}{2} + 0.6t_c = t_p \tag{9–73}$$

Solving Equations 9–72 and 9–73 for D yields

$$D = 0.133t_c \tag{9–74}$$

and

$$t_p = \frac{D}{2} + 0.6t_c = \frac{2}{3}t_c \tag{9–75}$$

Expressing Equation 9–71 in terms of t_c rather than t_p yields

$$q_p = \frac{726AQ}{t_c} \tag{9–76}$$

For a unit hydrograph, the depth of runoff Q would equal 1 in.

Example 9–23

The objective is to determine the triangular UH for a 300-acre watershed that has been commercially developed. The flow length is 6500 ft, the slope is 1.3%, and the soil is of group B.

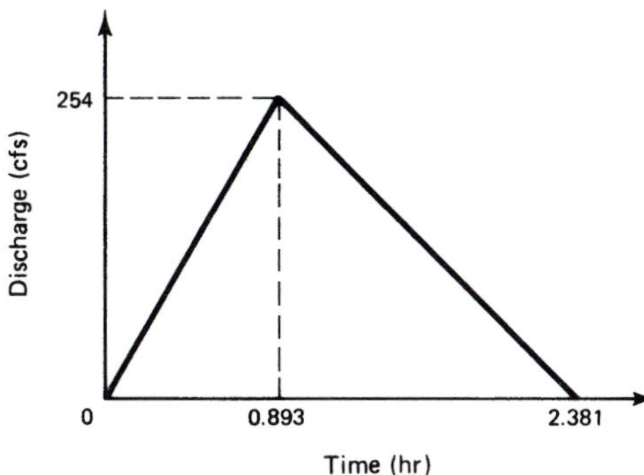

FIGURE 9–27 The triangular unit hydrograph for Example 9–23.

For commercial land use and soil group *B*, the watershed *CN* is 92 (see Table 3–18). The watershed t_c from Equation 3–56 is 1.34 hr. For 1 in. of rainfall excess, Equation 9–76 provides a peak discharge of

$$q_p = \frac{726\ (300\ \text{acres})\ (1\ \text{in.})}{(640\ \text{acres/mi}^2)\ (1.34\ \text{hr})} = 254\ \frac{\text{ft}^3}{\text{sec}} \tag{9–77}$$

The time to peak is

$$t_p = \frac{2}{3} t_c = 0.893\ \text{hr} \tag{9–78}$$

and the time base of the triangular UH is

$$t_b = \frac{8}{3} t_p = 2.381\ \text{hr} \tag{9–79}$$

The resulting triangular UH is shown in Figure 9–27.

Example 9–24

For the hypothetical watershed defined above, the SCS curvilinear UH can be determined using the values of Table 9–17. The curvilinear UH will be approximated for selected values of t/t_p; the TR-20 computer program uses all of the values shown in Table 9–17. For selected values of t/t_p, the curvilinear UH is computed in Table 9–18 and shown in Figure 9–28. Although the triangular and curvilinear UHs have the same q_p and t_p, they have different time bases and the por-

TABLE 9–18 Calculation of SCS Curvilinear Unit Hydrograph

t/t_p	q/q_p	t (hr)	q (ft³/sec)
0.0	0.000	0.	0
0.4	0.310	0.357	79
0.7	0.820	0.625	208
1.0	1.000	0.893	254
1.5	0.680	1.340	173
2.0	0.280	1.786	71
3.0	0.055	2.679	14
4.0	0.011	3.572	3
5.0	0.000	4.465	0

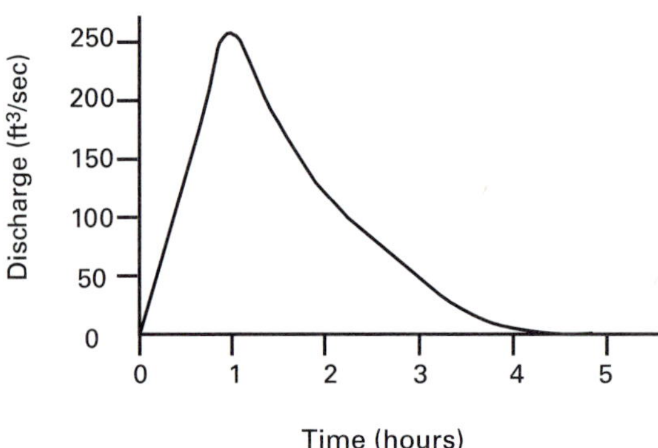

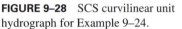

FIGURE 9–28 SCS curvilinear unit hydrograph for Example 9–24.

tion of the curvilinear UH near the peak is higher than that of the triangular UH. Both the triangular and curvilinear UHs can be considered D-hour UHs, with D computed by Equation 9–74 as

$$D = 0.133t_c = 0.178 \text{ hr}$$

Thus the UH should be reported on an interval of about 0.178 hr, such as 0.2 hr., and all computations should be performed at the interval.

It is important to clearly distinguish between the triangular and curvilinear hydrographs presented in SCS documents. The triangular UH is useful for demonstrating the source of the assumptions and the underlying equations. However, it is the curvilinear UH that is actually used in the SCS methods. The triangular UH is a useful educational tool, but design work should be based on the curvilinear UH. If the triangular UH is convolved with a rainfall-excess hyetograph, the resulting peak discharge will be less than the peak discharge that results when the curvilinear UH is convolved with the same rainfall-excess hyetograph.

9.9.4 Snyder's Synthetic Unit Hydrograph

F. F. Snyder (1938) provided a method of developing a synthetic unit hydrograph that is widely used. The method is most appropriate for large watersheds, but calibration of the coefficients is recommended. Formation of the unit hydrograph depends on a number of elements, including the time to peak, the time base, the duration of the rainfall excess, the peak discharge, and the width of the unit hydrograph at both 50% and 75% of the peak discharge.

The time of the peak of the unit hydrograph depends on two time elements, the duration of the rainfall excess (t_d) and the time lag (t_L), which is defined for this model as the time between the center of mass of rainfall excess and the time of the peak discharge. Snyder related the time lag (t_L, hours) to (1) the length of the main channel (L, miles) measured from the watershed outlet to the basin divide; (2) a watershed shape parameter (L_{ca} in miles), which is the length measured along the main channel from the watershed outlet to a point on the main channel that is perpendicular to the center of the area of the watershed (see Section 3.5); and (3) a watershed storage coefficient C_t, which should be calibrated. The following relationship is used to compute the lag time:

$$t_L = C_t (LL_{ca})^{0.3} \tag{9–80}$$

Values of C_t are typically believed to range from 1.8 to 2.2. The duration of rainfall excess (t_d, in hours) can be computed by the empirical equation

$$t_d = t_L/5.5 \tag{9–81}$$

Equation 9–81 indicates that the relationship between lag time and the duration of rainfall excess is constant. The following equation can be used to provide adjusted values of the lag time for other values of the duration of rainfall excess:

$$t_{La} = t_L + 0.25\,(t_{da} - t_d) \tag{9–82}$$

in which t_{da} is the alternative unit hydrograph duration (hr) and t_{La} is the adjusted time lag (hr). The time to peak is given by

$$t_p = t_L + 0.5\,t_d \tag{9–83}$$

The adjusted lag of Equation 9–82 can be used in place of t_L in Equation 9–80.

The time base of the unit hydrograph (t_b) is a function of the time lag:

$$t_b = 3 + \frac{t_L}{8} \tag{9–84}$$

in which t_b has units of days. Equation 9–84 is intended for use on large watersheds so it would give excessively large values for small watersheds. For small and moderately sized watersheds, the time base can be set as a ratio of the time to peak, with three to five times being reasonable values according to the studies behind the SCS methods, as well as others.

The unit peak discharge (q_p, ft^3/sec/in.) of the unit hydrograph is given by

$$q_p = \frac{640\,AC_p}{t_L} \tag{9–85}$$

in which A is the drainage area (mi^2) and C_p is the second empirical coefficient of the Snyder Method. Both C_p of Equation 9–85 and C_t of Equation 9–80 should be calibrated. However, values of C_p are believed to range from 0.5 to 0.7.

The shape of Snyder's Unit Hydrograph is largely controlled by two time parameters, W_{50} and W_{75}. These are the time widths of the unit hydrograph at discharges of 50% and 75% of the peak discharge computed with Equation 9–85. Values for W_{50} and W_{75} are given by

$$W_{50} = 756q_a^{-1.081} \tag{9–86a}$$

and

$$W_{75} = 450q_a^{-1.081} \tag{9–86b}$$

in which q_a is the peak discharge per square mile (q_p/A, ft^3/sec/mi^2). As a rule of practice, the location of the end points of W_{50} and W_{75} are placed such that one-third of both W_{50} and W_{75} occur prior to the time to peak and the remaining two-thirds after the time to peak. Thus seven points are used to define critical points of the unit hydrograph (see Figure 9–29). A smooth curvilinear shape should be drawn through these seven points and the volume of the UH computed. Adjustments to the ordinates can then be made so that the shape of the unit hydrograph is reasonable and the volume equals 1 area-in. of direct runoff. The adjustments are usually made in the recession of the UH.

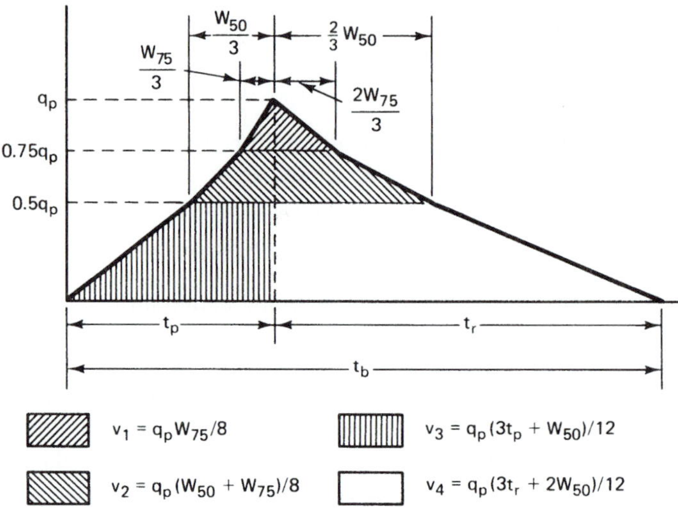

FIGURE 9–29 Configuration of Snyder's Synthetic Unit Hydrograph.

Example 9–25

To develop a unit hydrograph based on Snyder's Method, five inputs are required: the drainage area, A; the watershed length, L; the length to the center of area, L_{ca}; and the fitting coefficients, C_p and C_t. To illustrate the development of a unit hydrograph, the following values will be assumed: $A = 2375$ mi^2; $L = 86$ mi; $L_{ca} = 41$ mi; $C_t = 2.0$; and $C_p = 0.5$. The time lag is computed from Equation 9–80:

$$t_L = C_t\,(LL_{ca})^{0.3} = 2.0\,[86\,(41)]^{0.3} = 23.2 \text{ hr} \tag{9–87}$$

The duration of rainfall excess and the duration of the unit hydrograph is computed from Equation 9–81:

$$t_d = \frac{t_L}{5.5} = \frac{23.2}{5.5} = 4.2 \text{ hr} \tag{9–88}$$

Equation 9–82 can be used to compute the adjusted time lag for a 4-hr storm event:

$$t_{La} = t_L + 0.25\,(t_{da} - t_d) = 23.2 + 0.25\,(4 - 4.2) = 23.15 \text{ hr}$$

The time to peak is computed using Equation 9–83:

$$t_p = 23.15 + 0.5\,(4.0) = 25.15$$

where both the adjusted lag and adjusted duration are used. Equation 9–84 yields a time base of

$$t_b = 3 + \frac{T_L}{8} = 3 + \frac{23.15}{8} = 5.9 \text{ days}$$

The peak discharge is computed with Equation 9–85:

$$q_p = \frac{640\,AC_p}{t_L} = \frac{640\,(2375)\,(0.5)}{23.15} = 32,800 \text{ ft}^3/\text{sec/in.}$$

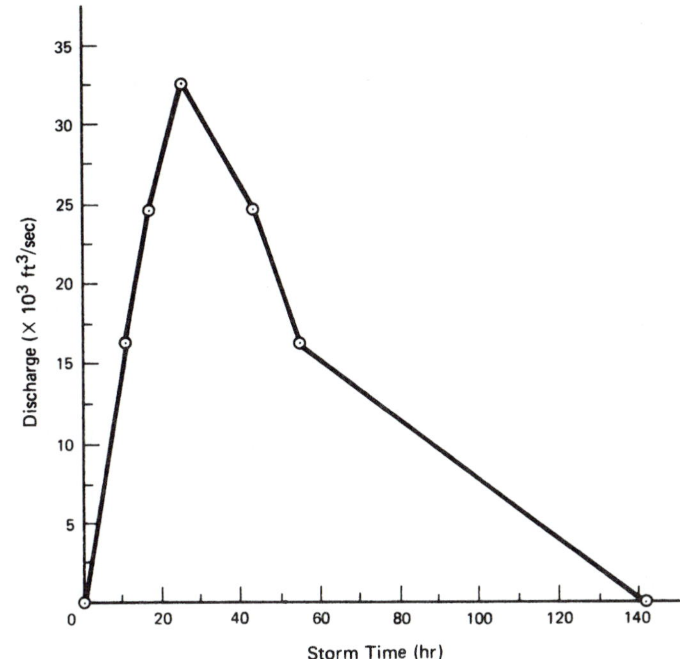

FIGURE 9–30 Computed Synder's Unit Hydrograph for Example 9–25.

which is 13.82 ft^3/sec/mi^2. The widths W_{50} and W_{75} are computed with Equations 9–86:

$$W_{50} = 756q_a^{-1.081} = 756\,(13.82)^{-1.081} = 44.2 \text{ hr}$$
$$W_{75} = 450q_a^{-1.081} = 450\,(13.82)^{-1.081} = 26.3 \text{ hr}$$

These values provide seven points, denoted (discharge, time) (0, 0), (16,400, 10.42), (24,600, 16.38), (32,800, 25.15), (24,600, 42.68), (16,400, 54.62), (0, 141.6); the 4-hr unit hydrograph defined with these points, connected by straight lines, is shown in Figure 9–30. The volume under the unit hydrograph computed with the equations of Figure 9–29 is 6.628×10^9 ft^3. For 1 in. of direct runoff, the volume would be 5.518×10^9 ft^3. Thus the straight-line segments would have to be adjusted to reduce the volume to 5.518×10^9 ft^3.

In practice, curved lines should be used to connect the points, although the straight lines can be used to provide a first approximation. The area near the peak is especially important, and the curve should be flatter at the peak than that shown with the straight lines.

9.10 DESIGNING WITH HYDROGRAPHS

Peak discharge methods are widely used in hydrologic design, especially for small watersheds and for highway drainage. Urban drainage design also makes use of peak discharge methods. These methods have the advantage of being computationally simple to apply and they provide a reasonable level of accuracy. Historically, peak discharge methods were used almost exclusively because of the complexity of hydrograph design. The availability of computers has eliminated this constraint.

Computer models based on hydrographs are believed to be more accurate than peak discharge methods, at least when the assumptions commonly specified for peak discharge methods are obviously violated. Watershed size is considered to be a very important criterion for selection of a design method, with peak discharge equations appropriate for small watersheds and hydrograph methods preferred for moderate and large watersheds. Unfortunately, it is difficult to decide on a line of demarcation between small and not-so-small watersheds. It is difficult to specify a delimiting area since peak discharge equations may be appropriate for an area A in a steeply sloped region but inappropriate for a watershed of the same size in a coastal area, where low slopes and sandy soils dominate.

Hydrograph methods are believed to be appropriate where one or more of the following conditions exist: (1) significant amounts of natural or developed storage, such as naturally swampy areas or small urban detention reservoirs; (2) significant variations in watershed or hydrometeorological conditions exist, including differences in land use, soil, hydrologic condition, topography or rainfall characteristics; (3) projects where watershed subdivision will be necessary for coordinating the design of different flood and drainage projects; and (4) where there are constraints in the principal flow paths, such as those caused by bridge abutments or encroachments in the flood plain. Where such conditions exist, hydrograph designs are more appropriate than peak discharge methods.

Where hydrograph design is warranted, the following procedure should be followed:

1. Select a design storm and design return period; this is discussed in Chapter 4.
2. For larger areas, make appropriate depth-area adjustments.
3. Compute initial abstractions and losses to separate the rainfall excess from the total rainfall.
4. Select a unit hydrograph model and obtain the inputs necessary to put it in dimensional form; make sure that the resulting UH has a volume of 1 acre-in.
5. Convolve the rainfall excess hyetograph with the unit hydrograph to compute the direct runoff hydrograph.
6. Add baseflow to the direct runoff hydrograph to compute the total runoff hydrograph.
7. Where the watershed has been subdivided, the total runoff hydrograph will have to be routed through the channel system (see Chapter 10).

This procedure forms the conceptual basis of computer programs such as the SCS TR-20 and Corps of Engineers HEC-1.

9.11 SCS WATERSHED MODEL: TR-20

In comparison with many other watershed models, the SCS TR-20 model is relatively simple, both in terms of the input and the conceptual framework. It is a single-event model that is usually used with a design storm as input. The program was formulated to develop runoff hydrographs, route hydrographs through channel reaches and reservoirs, and combine or separate hydrographs at confluences. Runoff hydrographs are developed using the SCS runoff equation and the SCS curvilinear unit hydrograph. Hydrographs are routed through channel reaches using the modified Att-Kin Method (see Chapter 10), and the storage-indication method is used for reservoir routing. The program is designed to make multiple analyses in a single run so that various alternatives can be evaluated in one pass through the program; this leads to more efficient use of computer time. Additionally, a single option can be used; for example, the storage routing algorithm can be used in the design of a detention basin.

9.11.1 Watershed Modeling

The TR-20 program, as well as other watershed models, are especially useful for developing design hydrographs on nonhomogeneous watersheds. Consider a watershed where four distinct subwatersheds with different land covers have been delineated. The two upper subwatersheds, both of which drain to the headwater of a channel, differ in land use, with one being developed in single-family residential units and the other predominantly forested. The subwatershed immediately downstream of the two upper subwatersheds is multi-family residential and has different soil types. The subwatershed at the watershed outlet is primarily light commercial. If a design hydrograph were needed at the watershed outlet, it would be incorrect to treat the watershed as a single watershed. To properly evaluate the design problem, it is necessary to determine the hydrologic runoff for each subarea and then combine them appropriately, including routing the runoff throughout the streams. Many of the elements discussed in Section 9.10 are involved. The following elements are necessary to model this simple case:

- For each of the four subwatersheds the rainfall excess distribution is computed using the SCS rainfall distribution for the region and the subarea curve number. Equation 7–42 is used to compute the rainfall excess.
- The SCS dimensionless unit hydrograph is dimensionalized using the peak discharge (Equation 9–71) and time to peak ($t_p = 2/3\ t_c$) for each subwatershed.
- The rainfall excess is convolved with the unit hydrograph to obtain the direct runoff hydrograph for each subwatershed.
- The runoff hydrographs for the two subwatersheds are added together and then routed through the channel reach of watershed 3.
- The reach-routed hydrograph is then added to the runoff hydrograph of subwatershed 3 and routed through the channel reach of subwatershed 4.
- The reach routed hydrograph is then added to the runoff hydrograph of subwatershed 4.

Each operation is executed by entering the appropriate command and necessary input data. The process is quite simple once these commands are understood.

9.11.2 Summary of Input Requirements

Even though the computer is used to solve problems, the input data requirements are surprisingly minimal, with the amount of data depending on the complexity of the problem to be solved. If actual rainfall events are not going to be used, the depth of precipitation is the only meteorological input. For each subarea, the drainage area, runoff curve number, and time of concentration are required; the antecedent soil moisture condition (I, II, or III) must also be specified. For each channel reach, the length is required along with the channel cross-section description, which is the elevation, discharge, and end-area data; while optional, a routing coefficient may also be input. If the channel routing coefficient is not given as input, it will be computed using Equation 10–15 and the cross-section data. For each structure it is necessary to describe the outflow characteristics with the elevation-discharge-storage relationship. The time increment for all computations must be specified, and any baseflow in a channel reach must be specified.

The discussion provided herein will be that for a generic application of the program. This will use the column numbers originally assigned to the input. Since numerous versions of TR-20 are available in microsoftware form, the input structure will probably vary with the software package. Most of these packages permit free-field input that does not require use of specific columns.

9.11.3 Basic Structure of Input Data

Input requirements for a single program execution can be separated into three parts: tabular data, standard control, and executive control. The tabular data, which is input first, consists of the following:

1. The routing coefficient table, which is a tabular statement of Equation 10–15
2. The dimensionless hydrograph table, which is a tabular representation of the curvilinear unit hydrograph of Table 9–17
3. One or more cumulative rainfall tables, which provide the depth-time relationship of rainfall input
4. Stream cross-section data, which describes the elevation-discharge-end area relationship
5. Structure data, which defines the elevation-discharge-storage relationship.

The tabular data serves as support data for the problem description.

Following the tabular data, the user must specify the standard control, which consists of records that establish the sequence in which runoff hydrographs are determined, routed through reaches and/or structures, and combined at tributary junctions. That is, standard control is used to describe the configuration of the watershed.

Executive control follows the standard control. Executive control controls the computational process. That is, it initiates the execution of the standard control sequence. The data arrangement is shown schematically in Figure 9–31.

In addition to tabular data, standard control, and executive control, operations called "modify standard control" are available for solving problems. The modify standard control operations are used when more than one computational sequence is to be tested for a watershed in a single execution. These operations will be discussed later. However, the modify standard control usually follows the executive control records for the previous problem execution. Additionally, the modify standard control records are followed by another set of ex-

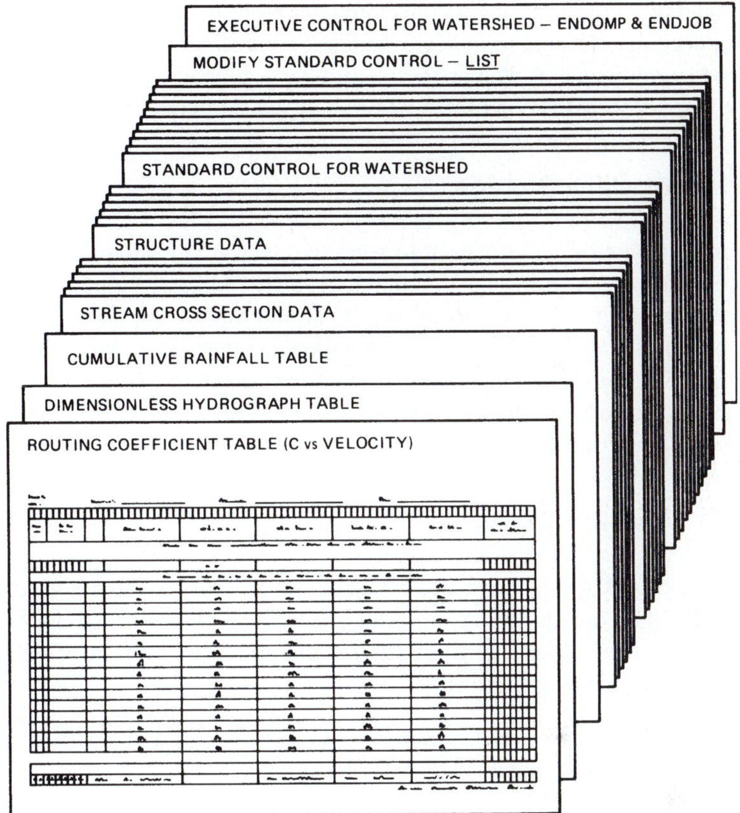

FIGURE 9–31 Schematic diagram of TR-20 card sequence.

ecutive control records that serve to initiate execution of the problem as described by the computational sequence as altered by the modified standard control.

9.11.4 Standard Control Operations

Standard control includes six subroutine operations:

1. RUNOFF: an instruction to develop an inflow hydrograph
2. RESVOR: an instruction to route a hydrograph through a structure
3. REACH: an instruction to route a hydrograph through a channel reach
4. ADDHYD: an instruction to combine two hydrographs
5. SAVMOV: an instruction to move a hydrograph from one computer memory storage location to another
6. DIVERT: an instruction for a hydrograph to be separated into two parts

The operations are indicated on the computer records in columns 2, 4–9, and 11. The number 6 is placed in column 2 to indicate standard control. The name of the subroutine operation is placed in columns 4–9. The operation number, which is shown preceding the operation name (that is, 1 for RUNOFF, 5 for SAVMOV), is placed in column 11.

9.11.5 Machine Storage

Internal machine storage is used to store the hydrograph ordinates. Machine storage is indicated by a number from 1 to 7. Columns 19, 21, and 23 are used on standard control records to indicate machine storage locations. It is important to emphasize that only one hydrograph can occupy any one storage element at a time and that there must be a hydrograph in the storage element from which a subroutine references a storage location. The actual use of the machine storage numbers (1 to 7) and the columns (19, 21, and 23) will be discussed separately for each subroutine operation.

9.11.6 Output Options

Columns 61, 63, 65, 67, 69, and 71 of standard control records are used to specify the output. The individual output options are selected by placing a 1 in the appropriate column. If the column is left blank or a zero is inserted, the option will not be selected. The following list summarizes the output options:

Output Option	A "1" in Column	Produces the Following Printout
PEAK	61	Peak discharge and corresponding time of peak and elevation (maximum stage for a cross section and maximum storage elevation for a structure)
HYD	63	Discharge hydrograph ordinates
ELEV	65	Stage hydrograph ordinates (reach elevations for a cross section and water surface elevation for structures)
VOL	67	Volume of water under the hydrograph in in.-depth, acre-ft, and ft^3/sec-hours
SUM	71	Requests the results of the subroutine be inverted in the summary tables at the end of the job

9.11.7 Content of Standard Control Records

In addition to the reserved spaces discussed previously, three data fields are available for input on each standard control record. The input for each subroutine will be discussed separately.

RUNOFF. The RUNOFF subroutine is intended to develop runoff hydrographs for subareas of a watershed. It uses Equation 7–42 in computing the rainfall excess for a given *CN* and the cumulative rainfall table. The curvilinear dimensionless unit hydrograph is made

TABLE 9–19 The RUNOFF Record

Columns	Content
2	Data code—use 6
4–9	Indicator—use RUNOFF
11	Subroutine number—use 1
13–17	Cross-section (13–15) or structure (16–17) number identifying the location in the watershed; do not insert both a cross section and a structure number
23	Hydrograph number: insert number of machine storage element (1–7) where hydrograph is stored
25–36[a]	Drainage area (mi^2)
37–48[a]	Runoff curve number[b]
49–60[a]	Time of concentration (hr)
61	Output option PEAK
63	Output option HYD
65	Output option ELEV
67	Output option VOL
71	Output option SUM
73–80	Record number/identification: available for user

[a]Include decimal point but do *not* include commas to indicate thousands.

[b]Insert runoff curve number for antecedent moisture condition II; for either condition I or III the machine will make the adjustment in the curve number as instructed on the COMPUT record (column 69).

dimensional and used in convolving the rainfall excess to develop a runoff hydrograph. The input for the RUNOFF subroutine is given in Table 9–19.

RESVOR. The RESVOR subroutine uses the concepts presented in the section on reservoir routing (Chapter 11) to route a hydrograph through a structure. The structure number is placed in columns 16 and 17 and the starting surface elevation in columns 25–36. The machine storage numbers of the inflow and outflow hydrographs are placed in columns 19 and 23, respectively. Table 9–20 shows the content of a RESVOR record.

TABLE 9–20 The RESVOR Record

Columns	Content
2	Data code—use 6
4–9	Indicator—use RESVOR
11	Subroutine number—use 2
16–17	Number identifying the structure
19	Hydrograph number: insert number of machine storage element that contains the inflow hydrograph
23	Hydrograph number: insert number of machine storage into which the outflow hydrograph is stored
25–36[a]	Surface elevation (ft) at time zero
61	Output option PEAK
63	Output option HYD
65	Output option ELEV
67	Output option VOL
71	Output option SUM
73–80	Record number/identification: available for user

[a]Include decimal point but do *not* include commas to indicate thousands.

REACH. The REACH subroutine is used to route a hydrograph through a channel reach using the modified Att-Kin Method. The machine storage element that contains the upstream hydrograph is inserted into column 19; the routed hydrograph is identified by the machine storage element indicted in column 23. The length of the reach, in feet, is inserted in columns 25–36. The routing coefficient of Equation 10–15 can be inserted in columns 37–48; if these columns are left blank, the routing coefficient will be computed using the cross-section data and the routing coefficient table, both of which are described in the section on tabular data. The number of routings, which is also optional, is placed in columns 49–60. The input for the REACH subroutine is given in Table 9–21.

ADDHYD. The ADDHYD subroutine is used to add the two hydrographs that are located in the machine storage elements in columns 19 and 21; the resulting hydrograph is stored in the machine storage element indicated in column 23. The cross-section number where the two hydrographs are added is indicated in columns 13–15. The input for an ADDHYD record is outlined in Table 9–22.

SAVMOV. The SAVMOV subroutine is used to move a hydrograph from one machine storage element to another; the respective machine storage elements are indicated in columns 19 and 23. The cross section or structure number is indicated in columns 13–17. Table 9–23 summarizes the input requirements for the SAVMOV subroutine.

DIVERT. The DIVERT subroutine is used to separate an upstream hydrograph into hydrographs. The record input is given in Table 9–24. Note that for option 2, columns 25–36 will be left blank.

TABLE 9–21 The REACH Record

Columns	Content	
2	Data code—use 6	
4–8	Indicator—use REACH	
11	Subroutine number—use 3	
13–15	Cross-section number identification	
19	Hydrograph number: insert number of machine storage element that contains the inflow hydrograph	
23	Hydrograph number: insert number of machine storage element into which the outflow hydrograph is stored	
25–36[a]	Length (ft) of stream reach	
37–48	Coefficient, x in $Q = xA^m$ must be positive number	} if both are blank actual cross section will be used
49–60	Exponent, m, in $Q = xA^m$	
61	Output option PEAK	
63	Output option HYD	
65	Output option ELEV	
67	Output option VOL	
71	Output option SUM	
73–80	Record number/identification: available for user	

[a]Include decimal point but do *not* include commas to indicate thousands.

TABLE 9–22 The ADDHYD Record

Columns	Content
2	Data code—use 6
4–9	Indicator—use ADDHYD
11	Subroutine number—use 4
13–15	Cross-section number indicating where the two hydrographs are added
19	Hydrograph number: insert number of machine storage element that contains the inflow hydrograph to be added
21	Hydrograph number: insert number of machine storage element that contains the other inflow hydrograph to be added
23	Hydrograph number: insert number of machine storage element into which the sum of the two inflow hydrographs is stored
61	Output option PEAK
63	Output option HYD
65	Output option ELEV
67	Output option VOL
71	Output option SUM
73–80	Record number/identification: available for user

Example 9–26

A 1.5-mi^2 watershed is subdivided, with the characteristics as follows

Subarea	Area (mi^2)	CN	tc (hr)
1	0.8	81	0.5
2	0.7	67	0.75

A detention basin is located at the outlet of the upper subarea, with a surface elevation of 57.7 ft at the start of precipitation. The outflow from the detention basin passes through a 4000-ft reach, which flows through subarea 2.

TABLE 9–23 The SAVMOV Record

Columns	Content
2	Data code—use 6
4–9	Indicator—Use SAVMOV
11	Subroutine number—use 5
13–17	Cross-section number (13–15) or structure number (16–17) indicating the reach or structure hydrograph
19	Hydrograph number: insert number of machine storage element that contains the hydrograph to be moved
23	Hydrograph number: insert number of machine storage element into which the hydrograph is to be stored after moving it
73–80	Record number/identification: available for user

TABLE 9–24 The DIVERT Record

Column	Value[a]	Description
2	6	Data code signifying standard control statement
4–9	DIVERT	Operation name
11	6	Operation number
13–15	var	Cross-section numbers 1–200, right justified } associated with output hydrograph 1
16–17	var	Structure number, 1–99, right justified } associated with output hydrograph 1
19	var	Input hydrograph storage location number, 1–7
21	var	Output hydrograph 1 storage location number, 1–7
23	var	Output hydrograph 2 storage location number, 1–7
25–36	b or var[b]	Specified discharge (ft^3/sec); invokes procedure 1, blank invokes procedure 2
37–48	var[b]	Decimal fraction of drainage area to be associated with output hydrograph 1 (use only to split the drainage area)
49–60	var[b]	Cross-section number associated with output hydrograph 2 (1–200); required for procedure 2
61	b or 1	Output option, print peak discharge[c]
63	b or 1	Output option, print discharge hydrograph[a]
65	b or 1	Output option, print elevations associated with hydrograph[a]
67	b or 1	Output option, print runoff volume[a]
69	b or 1	Output option, generate discharge hydrograph file[a]
71	b or 1	Output option, save results for summary tables[a]
73–80	b or 1	Record/statement identification, optional

[a]var, variable data to be entered by user; b, blank

[b]Decimal required in these fields except when blank

[c]Enter 1 to turn the option on; otherwise, leave it blank

A schematic diagram of the watershed is shown in Figure 9–32a. The standard control record images that are necessary to compute the flood hydrograph for the 1.5-mi^2 watershed are given in Table 9–25.

Example 9–27

If the channel characteristics of the first 2200 ft in Example 9–26 are different from those in the lower 1800 ft, the problem would have to include a cross section at the point where the channel

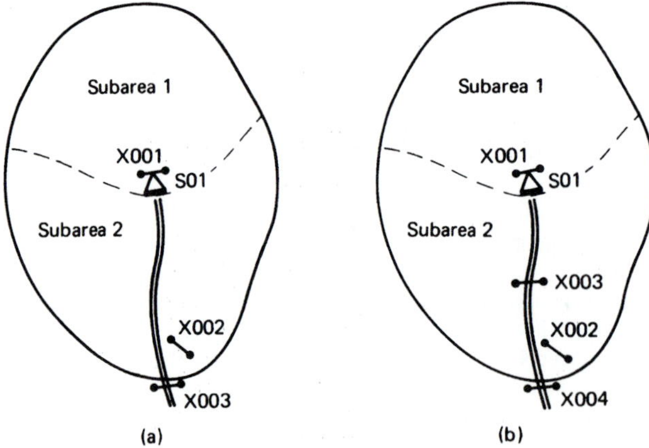

FIGURE 9–32 Schematic diagrams of (a) watershed for Example 9–26 and (b) watershed for Example 9–27.

TABLE 9–25 Standard Control for Watershed

Watershed ______ Hydrologist ______ Date ______ SHEET ______ OF ______

Card columns: 1-10 | 11-20 | 21-30 | 31-40 | 41-50 | 51-60 | 61-70 | 71-80

DATA CODE	OPERATION NAME	NO.	XSECT. I.D.	STRUCT. I.D.	IN-PUT #1	IN-PUT #2	OUT-PUT	DATA FIELD #1	DATA FIELD #2	DATA FIELD #3	PEAK	HYD	ELEV	VOL	FILE	SUM	RECORD IDENT.
6	RUNØFF	1					6	AREA, SQ. MI.	RUNOFF CURVE NO.	T_c, HRS.							EXAMPLES DO NOT ENTER
6	RESVØR	2			6		7	SURF EL. AT T=0, FT.									
6	REACH	3			7		5	LENGTH, FT.	(OPTIONAL) END AREA COEFF (X)	(OPTIONAL) EXPONENT (M)							
6	ADDHYD	4			5	6	7										
6	SAVMØV	5			FROM		TO										
6	DIVERT	6			IN	OUT1	OUT2	OUT1 DISCHARGE, CFS	OUT 1, Decimal % D.A.	OUT2, I.D.							
	ENDATA							This record is to be used only at end of all standard control records.									
			(Right justify these fields)					IMPORTANT: Data Fields No. 1, 2, and 3 require decimal points.									
6	RUNOFF	1	001				5	0.8	81.0	0.5	1						1
6	RESVOR	2		01	5		6	57.7			1						2
6	RUNOFF	1	002				5	0.7	67.0	0.75	1						3
6	REACH	3	003		6		7	4000.0			1						4
6	ADDHYD	4	003		7	5	6				1	1	1	1			5
	ENDATA																

characteristics show significant change. The schematic diagram is given in Figure 9–32b, and the standard control record images are shown in Table 9–26.

Example 9–28

A schematic diagram of a 60.8-mi^2 watershed is shown in Figure 9–33. The characteristics of the subareas are given in Table 9–27 and the channel reaches in Table 9–28. The reservoir located at the confluence of subareas 2 and 4 has an initial water surface elevation of 20.5 ft. The standard control record images are given in Table 9–29.

TABLE 9–26 Standard Control for Watershed (continuation sheet)

Watershed ______ Hydrologist ______ Date ______ SHEET ______ OF ______

Card columns: 1-10 | 11-20 | 21-30 | 31-40 | 41-50 | 51-60 | 61-70 | 71-80

CODE	OPERATION		XSEC	STRUC	HYD. NO.			DATA FIELD NO. 1	DATA FIELD NO. 2	DATA FIELD NO. 3	OUTPUT OPTIONS						RECORD ID.
			(Right justify these fields)					IMPORTANT: Data Fields No 1, 2, and 3 require decimal points.			P	H	E	V	F	S	
6	RUNOFF	1	001				5	0.8	81.0	0.5	1						1
6	RESVOR	2		01	5		6	57.7			1						2
6	RUNOFF	1	002				5	0.7	67.0	0.75	1						3
6	REACH	3	003		6		7	2200.0			1						4
6	REACH	3	004		7		6	1800.0			1						5
6	ADDHYD	4	004		6	5	7				1	1	1	1			6
	ENDATA																7

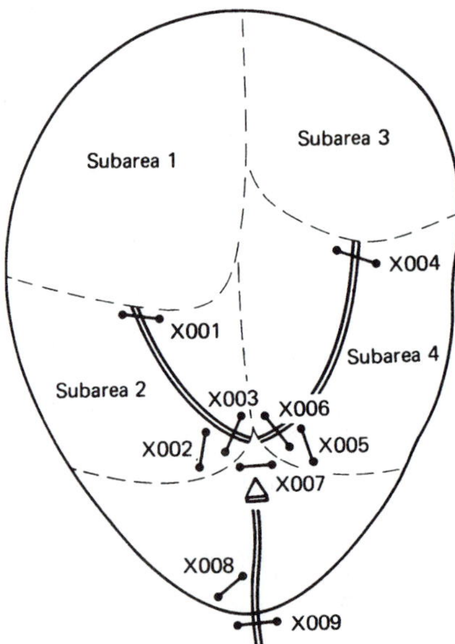

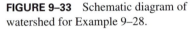

FIGURE 9–33 Schematic diagram of watershed for Example 9–28.

9.11.8 Tabular Data

Tabular data, which precede both the standard and executive control, includes six tables:

1. *Routing coefficient table:* the relationship between the streamflow routing coefficient (C of Equation 10–15) and velocity
2. *Dimensionless hydrograph table*: the dimensionless curvilinear unit hydrograph ordinates as a function of dimensionless time
3. *Cumulative rainfall tables*
   (a) *One-day watershed evaluation storm:* the 24-hr dimensionless, cumulative precipitation distribution
   (b) *Emergency spillway or freeboard hydrograph:* a cumulative precipitation table in which both the precipitation and time axes are dimensionless
   (c) *Actual:* a cumulative precipitation table in which both the precipitation and time axes are in actual units

TABLE 9–27 Watershed Characteristics for Example 9–28

Subarea	Area (mi^2)	Runoff *CN*	t_c (hr)
1	10.0	75	2.0
2	12.0	68	2.7
3	7.8	80	1.5
4	11.0	71	3.0
5	20.0	72	6.1

TABLE 9–28 Channel Characteristics for Example 9–28

Channel Reach	Length (ft)	Routing Coefficient
003	2000	0.4
006	1800	0.45
009	2700	0.3

4. *Stream cross-section data tables*: a tabular summary of the water surface elevation-discharge-cross-sectional end-area relationship
5. *Structure data table*: a tabular summary of the water surface elevation-discharge-reservoir storage relationship
6. *Read-discharge hydrograph data table*: a table containing an actual hydrograph

These tables serve as support data for the standard and executive control record images. Not all problems will include all of the tables.

9.11.9 The Routing Coefficient Table

The routing coefficient table gives the tabular relationship between the routing coefficient (C) of Equation 10–15 and the channel velocity (V). This table is required when a routing coefficient is not inserted on a REACH record.

TABLE 9–29 Standard Control for Watershed (continuation sheet)

Watershed ________ Hydrologist ________ Date ________ SHEET ____ OF ____

Columns: 1-10 | 11-20 | 21-30 | 31-40 | 41-50 | 51-60 | 61-70 | 71-80

IMPORTANT: Data Fields No 1, 2, and 3 require decimal points. (XSEC/STRUC: Right justify these fields)

CODE	OPERATION		XSEC	STRUC	HYD. NO.			DATA FIELD NO. 1	DATA FIELD NO. 2	DATA FIELD NO. 3	OUTPUT OPTIONS P	H	E	V	F	S	RECORD ID.
6	RUNOFF	1	001				5	10.0	75.0	2.0	1						1
6	RUNOFF	1	002				6	12.0	68.0	2.7	1						2
6	REACH	3	003		5		7	2000.0	0.4		1						3
6	ADDHYD	4	003		7	6	5				1						4
6	SAVMOV	5	003		5		1				1						5
6	RUNOFF	1	004				5	7.8	80.0	1.5	1						6
6	RUNOFF	1	005				6	11.0	71.0	3.0	1						7
6	REACH	3	006		5		7	1800.0	45.0		1						8
6	ADDHYD	4	006		7	6	5				1						9
6	SAVMOV	5	003		1		6										10
6	ADDHYD	4	007		5	6	7				1						11
6	RESVOR	2		01	7		5	20.5			1						12
6	RUNOFF	1	008				6	20.0	72.0	6.1	1						13
6	REACH	3	009		5		7	2700.0	0.3		1						14
6	ADDHYD	4	009		6	7	5				1	1	1	1			15
	ENDATA																16

Table 9–30 provides a summary of the input that is required to use the routing coefficient table. The tabular data must be preceded by a CTABLE record and followed by and ENDTBL record. The velocity increment used in the table is included in columns 25–36 of the CTABLE record. The C value given by Equation 10–15 is given for values of the velocity starting at 0 ft/sec and incremented by the constant given on the CTABLE record. Table 9–31 shows the routing coefficient table for an increment of 0.2 ft/sec.

9.11.10 The Dimensionless Hydrograph Table

The dimensionless hydrograph table gives the ordinates of the curvilinear unit hydrograph for a constant time increment. Both the discharge and time axes are dimensionless. The discharge axis is given as a ratio of the discharge to the peak discharge; therefore, the peak ordinate equals 1. The time axis is given as a ratio of the time to the length of the time base; this differs from the previous discussion in that the time axis had been given as a ratio of time to the time to peak. Because the time to peak is one-fifth the time base, the peak discharge of the unit hydrograph occurs at dimensionless time of 0.2.

Table 9–32 summarizes the input required for the dimensionless hydrograph table. The tabular data are preceded by a DIMHYD record and followed by a ENDTBL record. The dimensionless time increment must be specified in columns 25–36 of the DIMHYD record and followed by a ENDTBL record. The tabular data are given in constant increments of dimensionless time, with the first ordinate for a time of 0. Table 9–33 gives an example of the dimensionless hydrograph table.

TABLE 9–30 Routing Coefficient Table[a]

Columns	Content
	CTABLE Record
2	Data code—use 1
4–9	Table name—use CTABLE
25–36[a]	Velocity increment (ft/sec)
73–80	Record number/identification: available for user
	Routing Coefficient Records
2	Data code—use 8
13–24[b]	Data field: enter C coefficients
25–36	Data field: enter C coefficients
37–48	Data field: enter C coefficients
49–60	Data field: enter C coefficients
61–72	Data field: enter C coefficients
73–80	Record number/identification
	ENDTBL Record
2	Data code—use 9
4–9	Table name—use ENDTBL
73–80	Record number/identification: available for user

[a]Must be included in every run.

[b]Include decimal point but do *not* include commas to indicate thousands.

TABLE 9–31 Routing Coefficient Table C Versus Velocity

Watershed ________ Hydrologist ________ Date ________

DATA CODE	TABLE NAME		DATA FIELD NO. 1	DATA FIELD NO. 2	DATA FIELD NO. 3	DATA FIELD NO. 4	DATA FIELD NO. 5	CARD NO./ IDENTIFICATION
				VELOCITY INCREMENT, FT/SEC				
1	CTABLE			0.20				
			Enter successive entries left to right with initial entry for velocity = 0. Fill last row of data with last entry of table					
8			0.00	0.08	0.18	0.25	0.32	
8			0.37	0.41	0.45	0.49	0.51	
8			0.54	0.57	0.59	0.61	0.63	
8			0.65	0.66	0.67	0.69	0.70	
8			0.71	0.72	0.73	0.74	0.75	
8			0.76	0.77	0.77	0.78	0.79	
8			0.79	0.80	0.81	0.81	0.82	
8			0.82	0.83	0.83	0.84	0.84	
8			0.84	0.85	0.85	0.86	0.86	
8			0.86	0.86	0.87	0.87	0.87	
8			0.88	0.88	0.88	0.89	0.89	
8			0.89	0.89	0.89	0.89	0.90	
8			0.90	0.90	0.90	0.90	0.91	
8			0.91	0.91	0.91	0.91	0.91	
8			0.92	0.92	0.92	0.92	0.92	
8			0.92	0.92	0.92	0.93	0.93	
9	ENDTBL							

TABLE 9–32 Dimensionless Hydrograph Table[a]

Columns	Content
2	Data code—use 4
4–9	Table name—use DIMHYD
25–36[b]	Dimensionless time increment (last entry must be for dimensionless time = 1.0)
73–80	Record number/identification: available for user
	DISCHARGE RATIO Records[c]
2	Data code—use 8
13–24[b] 25–36 37–48 49–60 61–72	Data field—enter discharge ratios
73–80	Record number/identification: available for user
	ENDTBL Record
2	Data code—use 9
4–9	Table name—use ENDTBL
73–80	Record number/identification: available for user

[a]Must be included in every run.

[b]Include decimal point but do *not* include commas to indicate thousands.

[c]Number of entries must not exceed 75.

TABLE 9–33 Dimensionless Hydrograph Table: Discharge Versus Time

Watershed ________ Hydrologist ________ Date ________ SHEET ____ OF ____

DATA CODE	TABLE NAME		DATA FIELD NO. 1	DATA FIELD NO. 2	DATA FIELD NO. 3	DATA FIELD NO. 4	DATA FIELD NO. 5	RECORD IDENT.
1-10			11-20	21-30	31-40	41-50	51-60	61-70 / 71-80
			IMPORTANT: Line out unused lines. Data fields require decimal points.					
			DIMENSIONLESS TIME INCREMENT (Last entry must be made for dimensionless time 1.0)					
4	DIMHYD		0.02					
			NOTE: A maximum of 55 entries (11 records) are allowed.					
			Enter successive entries left to right. First and last entries must be = 0. Fill last row of data with zeros.					
8			0.0	0.030	0.100	0.190	0.310	
8			0.470	0.660	0.820	0.930	0.990	
8			1.000	0.990	0.930	0.860	0.780	
8			0.680	0.560	0.460	0.390	0.330	
8			0.280	0.207	0.147	0.107	0.007	
8			0.055	0.040	0.029	0.021	0.015	
8			0.011	0.005	0.000			
8								
8								
8								
8								
9	ENDTBL		NOTE: This record must be the last record of this table					

9.11.11 Cumulative Rainfall Tables

Three types of rainfall tables are available, which are labeled in TR-20 as follows: (1) the 1-day watershed evaluation storm, (2) emergency spillway or freeboard hydrograph; and (3) actual. These names reflect their intended use. The three types may be easier understood if they are classified on the basis of the scale used for the precipitation and time axes. Table 9–34 summarizes the scales of the rainfall tables and indicates the input that is required to reference the tables. It is especially important to emphasize that each table must be expressed in cumulative form.

The one-day watershed evaluation storm. Table 9–34 indicates that the precipitation scale for this cumulative rainfall table is dimensionless and the time scale ranges from 0 to 24 hr. This table is, therefore, of use when the standard type I, Ia, II, or III is used. It is assigned a table number of 1, which is indicated in column 11 of the RAINFL record. Since the precipitation scale is dimensionless, an actual rainfall volume in inches must be specified on the COMPUT record. Because an actual time scale is used, a duration of 1.0 is specified on the COMPUT record. The actual rainfall volume and the duration of 1.0 serve as multipliers to convert the rainfall table to a dimensional form. This rainfall table is suitable for the evaluation of watersheds in which the travel time through the watershed is approximately 2 days or less.

TABLE 9–34 Summary of Precipitation Tables

Type	Rainfall Table No.	Precipitation Scale	Time Scale	Record Input	
				Rainfall Volume	Rainfall Duration
Cumulative rainfall table: one-day storm	1	Dimensionless (0 to 1)	Actual (24 hr)	Actual (in.)	1.0
Cumulative rainfall table: emergency spillway	2	Dimensionless (0 to 1)	Dimensionless (0 to 1)	Actual (in.)	Actual (hr)
Cumulative rainfall table: actual	3–9	Actual (in.)	Actual (hr)	1.0	1.0

Table 9–35 summarizes the record images that are necessary to create the 1-day watershed evaluation rainfall table. The rainfall ratio records contain the dimensionless ordinates of the rainfall scale, and they are given at the time increment specified on the RAINFL record. The rainfall ratio records are preceded by the RAINFL record and followed by an ENDTBL record. An example of these records is shown in Table 9–36 which gives the ordinates of the SCS type I storm.

Emergency spillway or freeboard hydrograph table. Table 9–34 indicates that both the precipitation and time scales for this table are dimensionless. Therefore, the actual rainfall volume and rainfall duration must be specified on the COMPUT record; these

TABLE 9–35 Cumulative Rainfall Table: One-Day Storm

Columns	Content
	RAINFL Record
2	Data code—use 5
4–9	Table name—use RAINFL
11	Table number—use 1
25–36	Velocity increment (hr)
73–80	Record number/identification: available for user
	RAINFALL RATIO Records
2	Data code—use 8
13–24[a]	Data field: enter rainfall ratios
25–36	Data field: enter rainfall ratios
37–48	Data field: enter rainfall ratios
49–60	Data field: enter rainfall ratios
61–72	Data field: enter rainfall ratios
73–80	Record number/identification: available for user
	ENDTBL Record
2	Data code—use 9
4–9	Table name—use ENDTBL
73–80	Record number/identification: available for user

[a]Include decimal point but do *not* include commas to indicate thousands.

TABLE 9–36 Cumulative Rainfall Table, for One-Day SCS Type-I Precipitation Distribution

Watershed ______________ Hydrologist ______________ Date ______________

DATA CODE	TABLE ID NAME	TABLE ID NO.	DATA FIELD NO. 1	DATA FIELD NO. 2	DATA FIELD NO. 3	DATA FIELD NO. 4	DATA FIELD NO. 5	CARD NO./ IDENTIFICATION
		TABLE NO.		TIME INCREMENT*				
5	RAINFL	1		0.5				
			Enter successive entries left to right with first entry for time = 0. Fill last row of data with last entry of table.					
8			0.000	0.008	0.017	0.026	0.035	
8			0.045	0.055	0.065	0.076	0.087	
8			0.099	0.112	0.125	0.140	0.156	
8			0.174	0.194	0.219	0.254	0.303	
8			0.515	0.583	0.624	0.654	0.682	
8			0.705	0.727	0.748	0.767	0.784	
8			0.800	0.816	0.830	0.844	0.857	
8			0.870	0.882	0.893	0.905	0.916	
8			0.926	0.936	0.946	0.955	0.965	
8			0.974	0.983	0.992	1.000	1.000	
9	ENDTBL							

*Time increment is 0.5 hour. On "Executive Control for Watershed" (SCS-274) form set DATA FIELD NO. 2 to actual rainfall depth and rainfall duration, DATA FIELD NO. 3, to 1.0.

values are used as multipliers to form a dimensional table. This table is assigned a value of 2 for the rainfall table number. Tables 9–37 and 9–38 show the input and an example, respectively.

Actual storm. Table 9–34 indicates that both axes of this cumulative rainfall table are given in actual amounts, with the precipitation volume in inches and the storm duration in hours. Thus both scaling factors on the COMPUT record are set at 1.0.

Table 9–39 summarizes the record images that are necessary to formulate this rainfall table. The number of rainfall depth records can not exceed 20 (100 field widths), and all

TABLE 9–37 Cumulative Rainfall Table, for Emergency Spillway or Freeboard Hydrograph Storms

Columns	Content
	RAINFL Record
2	Data code—use 5
4–9	Table name—use RAINFL
11	Table number—use 2
25–36[a]	Time increment (hr) for rainfall ratio records
73–80	Record number/identification: available for user
	RAINFALL RATIO Records (same as for one-day storm table)
	ENDTBL Record (same as for the one-day storm table)

[a]Include decimal point but do *not* include commas to indicate thousands.

TABLE 9–38 Cumulative Rainfall Table: Storm Type or Date

Watershed ______ Hydrologist ______ Date ______ SHEET ______ OF ______

DATA CODE	TABLE I.D. NAME	NO.	DATA FIELD NO. 1	DATA FIELD NO. 2	DATA FIELD NO. 3	DATA FIELD NO. 4	DATA FIELD NO. 5	RECORD IDENT.
		(1-9)	RUNOFF OPTION	TIME INCREMENT	IMPORTANT: Line out unused lines. Data fields require decimal points.			
5	RAINFL	2		0.02				
			NOTE: A maximum of 100 entries (20 records) are allowed. Enter successive entries left to right with first entry for time = 0. Fill last row of data with last entry of table.					
8			0.00	0.01	0.02	0.02	0.03	
8			0.04	0.05	0.06	0.07	0.08	
8			0.10	0.11	0.13	0.14	0.17	
8			0.19	0.22	0.27	0.35	0.44	
8			0.52	0.60	0.63	0.66	0.68	
8			0.70	0.72	0.74	0.76	0.77	
8			0.79	0.80	0.82	0.83	0.84	
8			0.85	0.87	0.88	0.89	0.90	
8			0.91	0.92	0.93	0.94	0.95	
8			0.9567	0.9633	0.97	0.98	0.99	
8			1.00	1.00	1.00	1.00	1.00	
9	ENDTBL		NOTE: This record must be the last record of this table.					

five field widths on each record, including the last, must be filled in. Table 9–40 shows a cumulative rainfall table: actual for an actual 24-hr storm having a total depth of 4 in., the cumulative depths are given in increments of 2 hr, which is indicated on the RAINFL record. The first entry in the rainfall depth records is for a time of 0 hours (the start of precipitation).

TABLE 9–39 Cumulative Rainfall Table for Actual Storms

Columns	Content
	RAINFL Record
2	Data code—use 5
4–9	Table name—use RAINFL
11	Table number—use an integer from 3 to 9
25–36	Time increment (hr) for rainfall depth records
73–80	Record number/identification: available for user
	RAINFALL DEPTH Records
2	Data code—use 8
13–24[a]	Rainfall depths (in.)
25–36	Rainfall depths (in.)
37–48	Rainfall depths (in.)
49–60	Rainfall depths (in.)
61–72	Rainfall depths (in.)
73–80	Record number/identification: available for user
	ENDTBL Record (same as for the one-day storm table)

[a]Include decimal point but do *not* include commas to indicate thousands.

TABLE 9–40 Cumulative Rainfall Table: Storm Type or Date

Watershed ________ Hydrologist ________ Date ________ SHEET ______ OF ______

DATA CODE	TABLE I.D. NAME	NO.	DATA FIELD NO. 1	DATA FIELD NO. 2	DATA FIELD NO. 3	DATA FIELD NO. 4	DATA FIELD NO. 5	RECORD IDENT.
		(1-9)	RUNOFF OPTION	TIME INCREMENT	IMPORTANT: Line out unused lines.	Data fields require decimal points.		
5	RAINFL	3		2.0				
			NOTE: A maximum of 100 entries (20 records) are allowed. Enter successive entries left to right with first entry for time = 0. Fill last row of data with last entry of table.					
8			0.0	0.1	0.7	1.4	1.8	
8			2.0	2.0	2.1	2.3	2.7	
8			3.4	3.9	4.0	4.0	4.0	
9	ENDTBL		NOTE: This record must be the last record of this table.					

9.11.12 Stream Cross-Section Data Table

A stream cross-section data table is required for each channel reach that is numbered in columns 13–15 of the standard control. The data table summarizes the water surface elevation-discharge-cross-sectional end-area relationship. It consists of a series of cross-section data records that are preceded by an XSECTN record and followed by an ENDTBL record. There can not be more than 20 cross-section data records for any one cross section. Values between the points that are used to define the cross section are estimated using straight-line interpolation. For extrapolation beyond the highest elevation, a straight-line extension is made using the last two values of data shown. If the discharge is given in $ft^3/sec/mi^2$ (CSM), the total drainage area above the cross section must be shown on the XSECTN record. The computer multiplies the value shown in this field by the discharge in $ft^3/sec/mi^2$ in order to convert to ft^3/sec. If the discharge is given in ft^3/sec, a value of 1.0 must be inserted for the drainage area on the XSECTN record (Table 9–41).

Table 9–42 shows the records that comprise the cross-section data table. The drainage area above cross section 001 is 5.4 mi^2, as indicated on the XSECTN record. Because a drainage area is given on the XSECTN record, the discharge values on the cross-section data records are given in $ft^3/sec/mi^2$.

9.11.13 Structure Data Table

A structure data table is required for each structure number identified in columns 16 and 17 of the standard control. The table summarizes the water surface elevation-discharge-reservoir storage relationship. It consists of a series of elevation-discharge-storage records preceded by a STRUCT record and followed by an ENDTBL record. The number of coordinates that are used to describe a structure cannot exceed 20 elevations. The 0 discharge on the first record must be oriented to the crest elevation of the low-stage outlet in the principal spillway (the elevation on the RESVOR record of standard control at time 0). A maximum of

TABLE 9–41 Stream Cross-Section Data Table

Columns	Content
	XSECTN Record
2	Data code—use 2
4–9	Table name—use XSECTN
13–15	Cross-section number (001–120)
25–36[a]	Drainage area[b] (square miles)
73–80	Record number/identification: available for user
	CROSS-SECTION Data Record
2	Data code—use 8
25–36[a]	Elevation (ft)
37–48[a]	Discharge[b] ($ft^3/sec/mi^2$ or ft^3/sec)
49–60[a]	End area (ft^2)
73–80	Record number/identification: available for user
	ENDTBL Record
2	Data code—use 9
4–9	Table name—use ENDTBL
73–80	Record number/identification: available for user

[a]Include decimal point but do *not* include commas to indicate thousands.

[b]If discharge is given in cubic feet per square mile the *total* drainage area above the cross section must be shown on the XSECTN Record. The computer multiplies the value shown in this space by the discharge in $ft^3/sec/mi^2$ in order to convert to ft^3/sec. If the discharge is given in ft^3/sec, a value of 1.0 must be inserted for the drainage area.

TABLE 9–42 Stream Cross-Section Data: Cross Section Number

Watershed ______ Hydrologist ______ Date ______ SHEET ____ OF ____

DATA CODE (1-10)	TABLE NAME		X SECTION ID. (001-200)	DATA FIELD NO. 1 (21-30, 31-40)	DATA FIELD NO. 2 (41-50)	DATA FIELD NO. 3 (51-60)	(61-70)	RECORD IDENT. (71-80)
				IMPORTANT: Line out unused lines. Data fields require decimal points.				
				Drainage Area, Sq. Mi.	Bankfull Elevation, Ft.			
2	XSECTN		001	5.4				
				(Use 1.0 unless discharge in CSM.)				
				NOTE: Maximum of 20 data records allowed.				
				Elevation, Ft.	Discharge, CFS	End Area, Sq. Ft.		
8				742.	0.0	0.0		
8				743.	8.0	20.		
8				744.	20.0	80.		
8				746.	75.0	190.		
8				748.	200.0	350.		
8				750.	450.	650.		
8				752.	800.	1350.		
8				754.	1400.	2450.		
8				755.	1800.	3150.		
9	ENDTBL			NOTE: This record must be the last record of this table.				

99 structures can be used. The structure number placed in columns 16 and 17 of the STRUCT record will correspond to the structure number located in columns 16 and 17 of the RESVOR record in standard control. Table 9–43 summarizes the records that are used to form a structure data table. Table 9–44 provides an illustration.

Null structure. Many problems involve an analysis in which one objective is to determine the effect of a structure. In such cases, a structure data table that includes just one elevation-discharge-storage record can be created. This record includes the elevation of 0 discharge, a value of 0 for the discharge, and the corresponding storage. The problem would be analyzed by executing the standard control using the null structure table and then executing the modify standard control with the RESVOR record altered so that the elevation-discharge-storage relationship for the proposed structure is used.

9.11.14 Discharge Hydrograph Table

The drainage hydrograph table provides the means of inputting a hydrograph. Thus it serves as an alternative to using the RUNOFF subroutine to develop a hydrograph. The discharge hydrograph table consists of two READHD records followed by a series of discharge records and ENDTBL record. The first READHD record indicates the machine storage element

TABLE 9–43 Structure Data Table

Columns	Content
	STRUCT Record
2	Data code—use 3
4–9	Table name—use STRUCT
16–17	Structure number (01 to 60)
73–80	Record number/identification: available for user
	Elevation-Discharge-Storage Records[a]
2	Data code—use 8
25–36[a]	Elevation[c] (ft)
37–48[b]	Discharge (ft^3/sec)
49–60[b]	Storage (acre-ft)
73–80	Record number/identification: available for user
	ENDTBL Record
2	Data code—use 9
4–9	Table name—use ENDTBL
73–80	Record number/identification: available to user

[a]The number of coordinates that are used to describe a structure cannot exceed the data field spaces on a single input form (20 elevations).

[b]Include decimal point but do *not* include commas to indicate thousands.

[c]The zero discharge on the first line must be oriented to the crest elevation of the low-stage outlet in the principal spillway (surface elevation at $T = 0$).

TABLE 9–44 Structure Data: Structure Number

Watershed ______ Hydrologist ______ Date ______ SHEET ______ OF ______

DATA CODE	TABLE NAME			DATA FIELD NO. 1	DATA FIELD NO. 2	DATA FIELD NO. 3		RECORD IDENT.
		STRUCTURE ID. (01-99)	IMPORTANT: Line out unused lines. Data fields require decimal points.					
3	STRUCT	05						
			NOTE: Maximum of 20 data records allowed.					
				Elevation, Ft.	Discharge, CFS	Storage, Acre, Ft.		
8				663.	0.0	200.		
8				664.	58.	250.		
8				668.	256.	375.		
8				672.	300.	575.		
8				676.	352.	860.		
8				680.	371.	1225.		
8				684.	396.	1650.		
8				688.	418.	2200.		
8				690.4	440.	2575.		
8				691.4	1286.	2740.		
8				692.4	3440.	2900.		
8				693.4	6802.	3075.		
8				694.4	10950.	3250.		
8				695.4	15677.	3425.		
8				696.4	21034.	3600.		
9	ENDTBL			NOTE: This record must be the last record of this table.				

number (1 to 7) into which the hydrograph will be placed. This READHD record has an 8 in column 11, while the second READHD record has a 9 in column 11. The content of the discharge hydrograph table is given in Table 9–45 and an example is shown in Table 9–46. The hydrograph is limited to 300 ordinates.

9.11.15 Executive Control

The executive control serves both to execute the standard control and to provide data that are necessary for execution. Executive control consists of three subroutine operations: the INCREM record, the COMPUT record, and the ENDCMP record. The executive control records are placed after the tabular data and the standard control.

INCREM record. The sole purpose of the INCREM record is to specify the main time increment in hours (columns 25–36). All hydrographs generated by the program will be determined at this time increment. It is important that the main time increment be made short enough to describe adequately the hydrographs for smaller subwatersheds and large enough that when multiplied by the number of coordinates, it will extend through the peak period. An INCREM record must precede the first COMPUT record, and it remains in force until su-

TABLE 9–45 Read Discharge Hydrograph Data Table

Columns	Content
	READHD Record No. 1
2	Data code—use 7
4–9	Table name—use READHD
11	Indicator—use 8
17	Computer storage element number (1–7)
73–80	Record number/identification: available to user
	READHD Record No.2
2	Data code—use 7
4–9	Table name—use READHD
11	Indicator—use 9
13–24[a]	Starting time (hr)
25–36[a]	Time increment (hr)
37–48[a]	Drainage area (mi^2)
49–60[a]	Base flow (ft^3/sec)
73–80	Record number/identification: available for user
	Discharge Records
2	Data code—use 8
13–24[a] 25–36[a] 37–48[a] 49–60[a] 61–72[a]	Discharge (ft^3/sec), with first value in table relating to starting time shown on READHD Record No. 2
73–80	Record number/identification: available to user

[a]Include decimal point but do *not* include commas to indicate thousands.

perseded by the insertion of a new INCREM record. The content of the INCREM record is given in Table 9–47.

COMPUT record. The COMPUT record specifies (1) the cross sections or structure locations where routings are to begin and end; (2) the rainfall starting time, depth, and duration; (3) the rain table number; and (4) the antecedent soil moisture condition (I, II, or III). The contents of a COMPUT record are summarized in Table 9–48. The rainfall depth and duration are placed in columns 37–48 and 49–60, respectively. If actual rainfall depths and times are given on the cumulative rainfall table for either the depth or the duration, a 1.0 is inserted into the appropriate column. If either the rainfall depth or duration is dimensionless on the cumulative rainfall table, the actual depth or duration is given on the COMPUT record. The rain table number is placed in column 65 of the COMPUT record.

ENDCMP and ENDJOB records. After the executive control records are completed, an ENDCMP record should follow. The content of the ENDCMP record is given in Table 9–49. If the standard control is modified, appropriate executive control records must follow. After all computations are complete, the program must be ended with an ENDJOB record, the contents of which are specified in Table 9–50.

TABLE 9–46 Read Discharge Hydrograph

Watershed ______ Hydrologist ______ Date ______ SHEET ______ OF ______

1-10			11-20	21-30	31-40	41-50	51-60	61-70	71-80
DATA CODE	TABLE NAME		DATA FIELD NO. 1	DATA FIELD NO. 2	DATA FIELD NO. 3	DATA FIELD NO. 4	DATA FIELD NO. 5		RECORD IDENT.
			IMPORTANT: Line out unused lines. Data fields require decimal points.						
			HYDROGRAPH NO.						
7	READHD	8	6						
			STARTING TIME, HRS.	TIME INCREMENT, HRS.	DRAINAGE AREA, SQ. MI.	BASE FLOW, CFS			
7	READHD	9	0.0	2.0	26.84				
			Enter successive entries left to right with first entry for time = Baseflow, if any. Fill last row of data with last entry of table ≥ Baseflow.						
8			0.	100.	300.	550.	1350.		
8			1900.	1800.	1200.	950.	700.		
8			500.	300.	225.	250.	700.		
8			1450.	1350.	1100.	925.	550.		
8			625.	575.	525.	500.	600.		
8			1000.	775.	600.	400.	400.		
8			750.	500.	325.	300.	300.		
8			300.	300.	275.	225.	175.		
8			125.	90.	80.	50.	40.		
8			30.	25.	20.	15.	10.		
8			5.	0.	0.	0.	0.		
8									
8									
8									
8									
9	ENDTBL		NOTE: This record must be the last record of this table.						

9.11.16 Modify Standard Control

The modify standard control subroutines are used to insert new routines into the standard control sequence, to alter the input data for existing operations, and to delete any operations in the sequence. Many hydrologic analyses are intended to examine various alternatives, such as changes in land use or the insertion of a control structure into the watershed system. The modify standard control operations enable the user to modify the original standard con-

TABLE 9–47 The INCREM Record

Columns	Content
2	Data code—use 7
4–9	Indicator—use INCREM
11	Operation number—use 6
25–36[a]	Time increment (hr)
73–80	Card number/identification: available for user

[a]Include decimal point but do *not* include commas to indicate thousands.

TABLE 9–48 The COMPUT Record[a]

Columns	Content
2	Data code—use 7
4–9	Indicator—use COMPUT
11	Operation number—use 7
13–17	Cross-section number (13–15) or structure number (16–17) where computation begins
19–23	Cross-section number (19–21) or structure number (22–23) where computation ends (Note: computation includes this cross-section or structure)
25–36[b]	Time at which computation starts (hours)
37–48[b]	Rainfall depth (in.)
49–60[b]	Rainfall duration
65	Rain table no. (1–9)
69	Soil (1: dry; 2: normal; 3: wet)
73–80	Record number/identification: available for user

[a]More than one COMPUT record can be used when different precipitation or soil moisture conditions exist within a watershed.

[b]Include decimal point but do *not* include commas to indicate thousands.

trol sequence and analyze an alternative. A separate executive control sequence is placed after each modify standard control sequence.

There are six modify standard control subroutine operations:

1. INSERT: an operation used to insert new standard control subroutines into the standard control sequence
2. ALTER: an operation that is used to change any input on a standard control record image, including machine storage numbers, data field input, and output options
3. DELETE: an operation that causes one or more standard control operations to be deleted from the standard control sequence

TABLE 9–49 The ENDCMP Record

Columns	Content
4–9	Indicator—use ENDCMP
11	Operator number—use 1
73–80	Record number/identification: available for user

TABLE 9–50 The ENDJOB Record

Columns	Content
4–9	Operation—insert ENDJOB
11	Operator number—use 2
73–80	Record number/identification: available for user

TABLE 9–51 The INSERT Record

Columns	Content
2	Data code—use 7
4–9	Operation name—use INSERT
11	Operation number—use 2
13–15[a]	Cross-section numbers: the cards following the INSERT record are inserted in standard control following the first record that uses this *X*-section number
16–17[a]	Structure number: the records following the INSERT record are inserted in standard control following the first record that uses the structure number.
73–80	Record number/identification: available for user

[a]Insert either a cross-section number or a structure number, but not both.

4. LIST: an operation that causes the tabular and standard control data to be printed out
5. UPDATE: an operation that is used when the library tape for a watershed is to be retained for subsequent processing, and it is desired that all new tabular data and modifications to standard control operations should replace the original counterparts to become permanent record
6. BASFLO: an operation that allows a uniform rate of baseflow (ft^3/sec) to be introduced into reach routings at any location

The INSERT record. TR-20 is quite frequently used to evaluate the effect of structures in a developing watershed. This use represents an example where an INSERT record would be of value. A standard sequence of tabular data, standard control, and executive control would be used to analyze the watershed without the structure. Then the modify standard control and necessary executive control records could be used to analyze the watershed with the structure.

The contents of an INSERT record is summarized in Table 9–51. The INSERT record is followed by one or more standard control records. These records are inserted into the standard control sequence immediately following the first reference to a cross-section number or a structure that has the same value as that located in columns 13–15 or 16 and 17, respectively, of that value on the INSERT record.

TABLE 9–52 The ALTER Record[a]

Columns	Content
2	Data code—use 7
4–8	Operation number—use ALTER
11	Operation number—use 3
73–80	Record number/identification: available for user

[a]Data given in columns 1–18 of the records following the ALTER record must be identical with the standard control record to be altered.

TABLE 9–53 The DELETE Record[a]

Columns	Content
2	Data code—use 7
4–9	Operation number—use DELETE
11	Operation number—use 4
73–80	Record number/identification: available for user

[a]Data given in columns 1–18 of the records following the DELETE record must be identical with the standard control record to be deleted.

The ALTER record. The ALTER record indicates that one or more records in the standard control sequence are to be altered. Following the ALTER record is a sequence of standard control records. These records will replace the records in the standard control sequence that are identical to the values in columns 1–18. The input for the ALTER record is shown in Table 9–52.

The DELETE record. The DELETE record is used to eliminate records from the standard control sequence. The DELETE record is followed by one or more standard control records; the records in the standard control sequence that are identical in columns 1–18 to these records are eliminated. Table 9–53 provides a summary of the contents of the DELETE record.

The BASFLO record. The BASFLO record is used to add a constant baseflow (ft^3/sec) to each hydrograph. If a second BASFLO record is inserted, the difference between the old baseflow and the new baseflow is added to the inflow hydrograph when the next REACH subroutine is referenced. Table 9–54 summarizes the input for the BASFLO record.

TABLE 9–54 The BASFLO Record

Column	Value[a]	Description
2	7	Data code signifying executive control statement
4–9	BASFLO	Operation name
11	5	Operation number
		Procedure 1: Baseflow is constant
25–36	var.[b]	Baseflow (ft^3/sec) to be added to next reach inflow hydrograph
37–48	b or 0	
49–60	b or 0	
		Procedure 2: Baseflow is triangular hydrograph
25–36	var[b]	Volume of baseflow in watershed inches
37–48	var[b]	Time (hr) that baseflow hydrograph peaks (measured from beginning of storm)
49–60	var[b]	Time (hr) of triangular hydrograph
73–80	var[b]	Record/statement identification: optional

[a]var., variable data to be entered by the user; b, blank field.

[b]Decimal required in these fields, omitted in all others.

9.11.17 Execution Sequence

In general, a TR-20 program will be structured with the tabular data first, followed by the standard control and executive control operations. If additional analyses are to be executed using modify standard control operations, the modifications follow the ENDCMP record. An executive control sequence will then follow the modify standard control operations. When additional sequences of modify standard control-executive control are not included, the program run is terminated with an ENDJOB record.

PROBLEMS

9–1. For the mean daily discharge record of Table 5–1, estimate the baseflow for each month and plot the values.

9–2. Runoff processes of the hydrologic cycle were discussed in Chapter 1. Compare the importance of the processes that control baseflow from a 100-acre watershed versus those for a 100 mi^2 watershed.

9–3. The following points define a hydrograph for a 115 mi^2 watershed, with the discharge varying linearly with time between the given values. Compute the baseflow function, the distribution of direct runoff, and the volume of direct runoff (ac-ft) for each of the following: (a) The constant-discharge method; (b) the constant-slope method if direct runoff ends at storm hour 10; (c) the constant-slope method if the direct runoff ends at storm hour 20; (d) the concave method if direct runoff ends at storm hour 10.

t (hr)	0	2	5	10	20	30
q (ft^3/sec)	25	20	140	60	30	20

9–4. The following points define a hydrograph, with the discharge varying linearly with time between the given values. Compute the baseflow function, the distribution of direct runoff, and the volume of direct runoff for each of the following: (a) The constant-discharge method; (b) the constant-slope method if direct runoff ends at storm hour 15; (c) the constant-slope method if the direct runoff ends at storm hour 21; (d) the concave method if direct runoff ends at storm hour 15.

t (hr)	0	3	8	11	15	21	31
q (ft^3/sec)	40	35	110	170	120	60	35

9–5. The following data were measured at stream gage station F318-R, Eaton Wash at Loftus Drive, Los Angeles, for the storm of March 1, 1983. Compute the direct runoff and baseflow distributions and the volume of direct runoff for each of the following: (a) The constant-discharge method assuming direct runoff begins at 0404 standard time; (b) the constant-slope method assuming direct runoff begins at 0404 and ends at 1604.

Time	q (ft³/sec)	Time	q (ft³/sec)	Time	q (ft³/sec)
0204	155	0704	1020	1204	2940
0304	231	0804	1590	1304	1120
0404	191	0904	4300	1404	1370
0504	612	1004	1320	1504	915
0604	1310	1104	1240	1604	765

9–6. The following is the daily runoff hydrograph (ft³/sec) for a 327-mi² watershed: 126, 124, 123, 230, 640, 1280, 1190, 1070, 850, 580, 420, 310, 265, 195, 180. Use Equation 9–2 to find the end of direct runoff. Find the distributions of baseflow and direct runoff using both the constant-slope and concave methods. Compute the volume of direct runoff (both ac-ft and in.) for each method.

9–7. (a) The following data are daily discharge rates from recession curves for six storms events. Use the master-depletion-curve method to characterize the recession and estimate the coefficients for Equation 9–5.

Storm 1	Storm 2	Storm 3	Storm 4	Storm 5	Storm 6
2100	440	1310	250	1040	900
1890	410	1210	220	920	780
1710	375	1120	196	810	700
1570	355	1040	181	760	630
1460	330	975	170	680	580
1400	312	925	159	640	550
1330	298	880	150	610	525
1280	286	840	144	580	505
1210	273	800	138	560	485
1160		770	130	535	460
		740	126		440
					425

(b) For the following storm event, use the recession derived in part (a) to compute the time when direct runoff ends.

t (days)	0	1	2	3	4	5	6	7	8	9
q (ft³/sec)	350	315	290	275	255	240	228	219	209	199

9–8. (a) The following data are daily discharge rates from recession curves for five storms events. Use the master-depletion-curve method to characterize the recession and estimate the coefficients for Equation 9–5.

Storm 1	Storm 2	Storm 3	Storm 4	Storm 5
1650	160	680	290	260
1420	146	590	265	225
1300	131	530	245	205
1190	120	480	225	185
1090	110	445	208	170
1010	104	415	197	160
965	98	390	186	151
910	92	370	176	143
850	87	350	165	134
805	81	330	155	
			147	

(b) For the following storm event, use the recession derived in part (a) to compute the time when direct runoff ends.

t (days)	0	1	2	3	4	5	6	7	8	9
q (ft^3/sec)	1440	1200	1030	940	890	815	765	720	680	640

9–9. Show the development of Equations 9–6 and 9–7.

For Problems 9–10 to 9–14, compute the value of the phi index for the given hyetograph and the depth of direct runoff specified. Show the distributions of losses and rainfall excess.

9–10. Depth of direct runoff = 1.8 in.

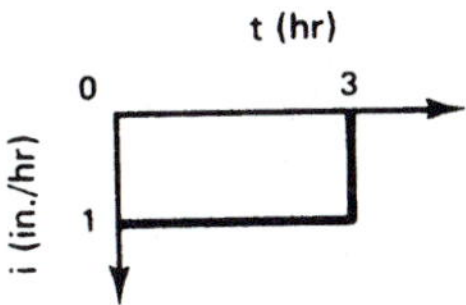

9–11. Depth of direct runoff = 3.2 in.

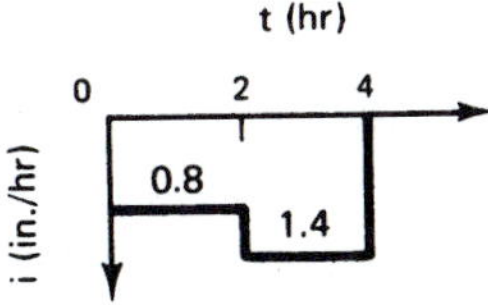

9–12. Depth of direct runoff = 4.5 in.

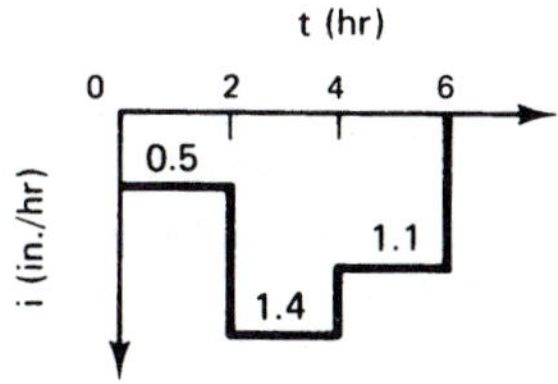

9–13. Depth of direct runoff = 1.2 in, with the hyetograph of Problem 9–12.

9–14. Depth of direct runoff = 0.6 in., with the hyetograph of Problem 9–11.

9–15. Determine the value of ϕ for the rainfall hyetograph shown. Assume the depth of direct runoff is 2.5 in.

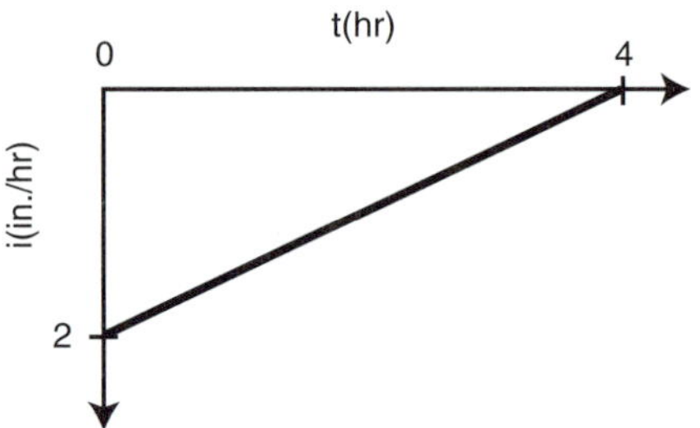

9–16. The following data are from the 2.15-acre watershed W–15, Cherokee, Oklahoma, for the storm of November 16, 1964. Assume that baseflow equals 0 and all rainfall prior to the start of runoff is initial abstraction. Determine the distribution of rainfall excess using the phi-index method. [Note: Rainfall values give the average intensity over the interval; runoff values are ordinates.]

Hyetograph		Hydrograph			
Time of Day	Rate (in./hr)	Time of Day	Rate (in./hr)	Time of Day	Rate (in./hr)
0530	0.00	0732	0.0000	0826	0.543
0630	0.01	0736	0.0014	0828	0.418
0730	0.00	0738	0.0049	0830	0.322
0745	0.44	0754	0.0049	0834	0.218
0800	0.12	0756	0.0096	0838	0.152
0805	1.92	0758	0.0228	0842	0.104
0810	1.92	0800	0.0441	0900	0.0904
0815	1.20	0802	0.0597	0910	0.0654
0820	0.72	0804	0.0972	0920	0.0490
0825	0.36	0806	0.127	0930	0.0393
0830	0.24	0808	0.218	0940	0.0266
0845	0.12	0809	0.335	0950	0.0155
0900	0.16	0810	0.494	1000	0.0124
0945	0.00	0812	0.687	1030	0.0124
0950	0.24	0814	0.807	1056	0.0096
1000	0.00	0816	0.937	1110	0.0049
1015	0.16	0817	1.05	1130	0.0049
1035	0.03	0820	0.871	1220	0.0014
		0824	0.668	1350	0.0000

9–17. The data are from the 1.68-acre watershed W–10, Cherokee, Oklahoma, for the storm of October 25, 1964. Assume that baseflow equals 0 and all rainfall prior to the start of runoff is initial abstraction. Determine the distribution of rainfall excess using the phi-index method. [Note: Rainfall values give the average intensity over the interval; runoff values are ordinates.]

Hyetograph		Hydrograph	
Time of Day	Intensity (in./hr)	Time of Day	Rate (in./hr)
1605	0.12	1912	0.0000
1705	0.01	1914	0.0204
1805	0.01	1916	0.0665
1905	0.00	1918	0.122
1920	2.24	1920	0.285
2020	0.02	1922	0.524
2030	0.12	1923	1.51
2045	0.04	1924	2.41
		1925	1.39
		1926	0.804
		1927	0.636
		1928	0.490
		1930	0.365
		1934	0.237
		1936	0.174
		1938	0.130
		1944	0.0990
		1947	0.0852
		1948	0.0501
		1952	0.0360
		2000	0.0240
		2012	0.0138
		2030	0.0084
		2042	0.0043
		2106	0.0012
		2150	0.0000

For Problems 9–18 to 9–21, use the constant percentage method to determine the distribution of rainfall excess for the data indicated.

9–18. Use the data of Problem 9–11.

9–19. Use the data of Problem 9–13.

9–20. Use the data of Problem 9–14.

9–21. Use the data of Problem 9–17.

9–22. Using the constant-percentage method determine both the loss function and the rainfall excess distribution for the conditions of Problem 9–15.

9–23. If $f_o = 0.7$ in./hr, $f_c = 0.2$ in./hr, and $K = 0.5$ hr^{-1}, determine the distribution of losses and the depth of rainfall excess for the hyetograph given in Problem 9–10.

9–24. If $f_o = 0.6$ in./hr, $f_c = 0.15$ in./hr, and $K = 0.3$ hr^{-1}, determine the distribution of losses and the depth of rainfall excess for the hyetograph given in Problem 9–11.

9–25. If $f_o = 0.5$ in./hr, $f_c = 0.15$ in./hr, and $K = 0.4$ hr^{-1}, determine the distribution of losses and the depth of rainfall excess for the hyetograph given in Problem 9–12.

9–26. If $f_o = 0.65$ in./hr, $f_c = 0.25$ in./hr, and $K = 0.35$ hr^{-1}, determine the distribution of losses and the depth of rainfall excess for the following hyetograph:

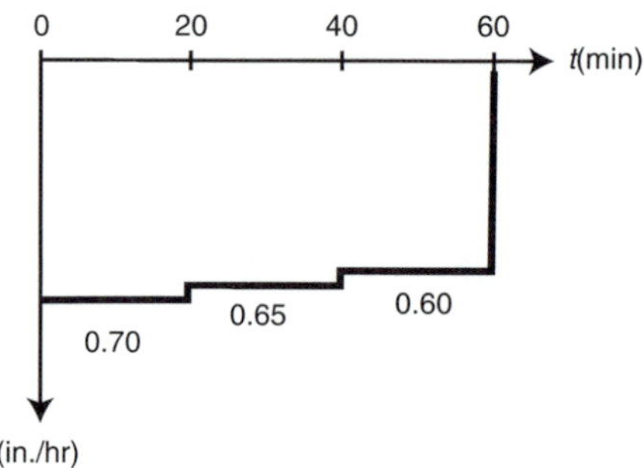

9–27. If $K_r = 0.1\ \text{hr}^{-1}$ and $f_o = 0.4$ in./hr, compute the infiltration capacity curve during 6 hr of recovery, if $f = 0.25$ in./hr at the end of the first burst of rainfall.

9–28. If $K_r = 0.06\ \text{hr}^{-1}$, $f_o = 0.35$ in./hr, compute the infiltration capacity curve during 4 hr of recovery, if $f = 0.12$ in./hr at the end of the first burst of rainfall.

9–29. Using the data from Problem 9–27, compute the infiltration capacity for 2 hr following the end of the 6 hr recovery period. Assume the rainfall intensity exceeds the infiltration capacity, $f_c = 0.1$ in./hr, and $K = 0.45\ \text{hr}^{-1}$. Use a time increment of 0.5 hr.

9–30. Using the data from Problem 9–28, compute the infiltration capacity for 4 hr following the end of the 4 hr recovery period. Assume the rainfall intensity exceeds the infiltration capacity, $f_c = 0.10$ and $K = 0.25\ \text{hr}^{-1}$. Use a time increment of 1 hr.

9–31. Given $f_o = 0.85$ in./hr, $f_c = 0.3$ in./hr, $K = 0.45\ \text{hr}^{-1}$, $K_r = 0.15\ \text{hr}^{-1}$, and the following hyetograph, compute the infiltration capacity curve and the depth of rainfall excess.

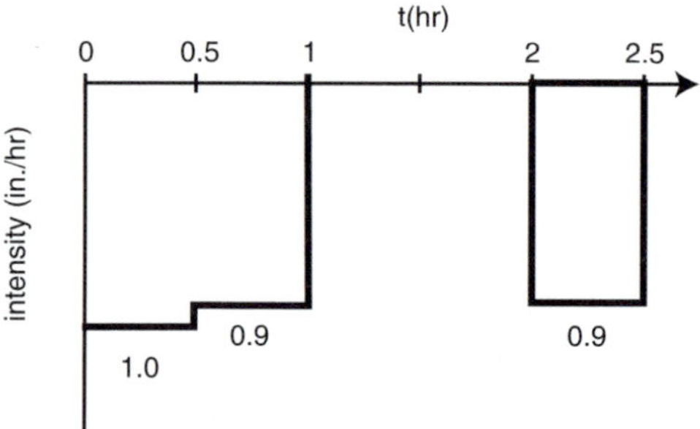

9–32. Given $f_o = 0.6$ in./hr, $f_c = 0.15$ in./hr, $K = 0.7\ \text{hr}^{-1}$, $K_r = 0.1\ \text{hr}^{-1}$, and the following hyetograph, compute the infiltration capacity curve and the depth of rainfall excess.

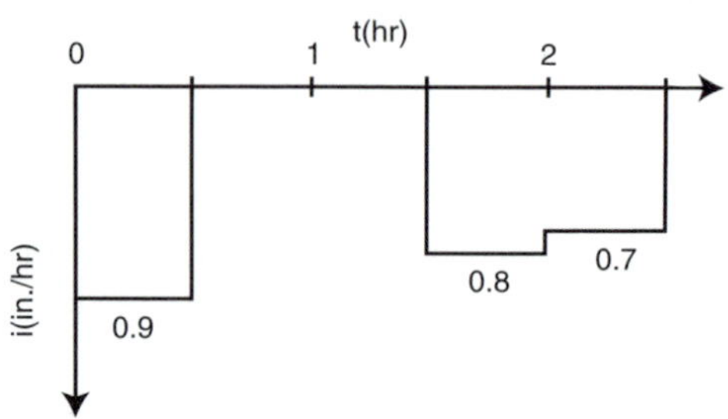

9–33. Given $f_o = 0.8$ in./hr, $f_c = 0.2$ in./hr, and $K = 0.4$ hr^{-1}, compute the loss function for the storm hyetograph of Figure P9–33 below. Compute the depths of losses and rainfall excess.

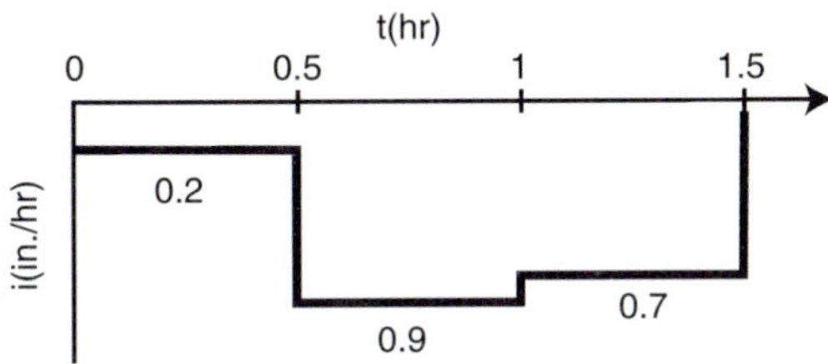

9–34. Given $f_o = 0.75$ in./hr, $f_c = 0.25$ in./hr, and $K = 0.3$ hr^{-1}, compute the loss function for the storm hyetograph of Figure P9–34. Compute the depths of losses and rainfall excess.

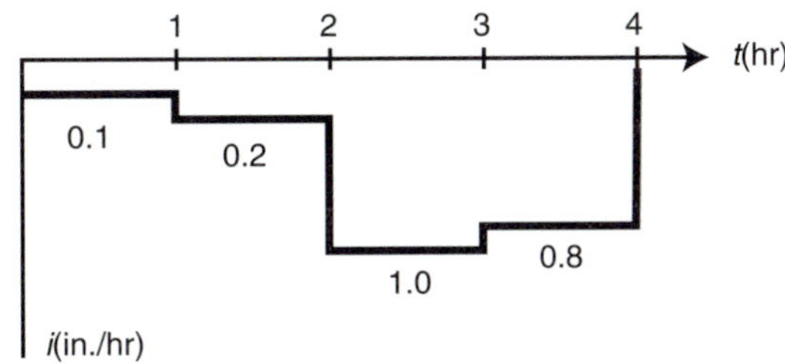

9–35. If $f_o = 0.5$ in./hr, $f_c = 0.1$ in./hr, and the depth of direct runoff is 2.25 in., determine the value of K for Equation 9–14 and the hyetograph of Problem 9–10.

9–36. If $K = 0.35$ hr^{-1}, $f_c = 0.25$ in./hr, and the depth of direct runoff equal 4.0 in., determine the value of f_o for Equation 9–14 and the hyetograph of Problem 9–12.

9–37. If $K = 0.5$ hr^{-1}, $f_o/f_c = 3$, and the depth of direct runoff equals 2.9 in., find the values of f_o and f_c for Equation 9–14 and the hyetograph of Problem 9–11.

9–38. Using the Green-Ampt Model, compute the time distribution of infiltration capacity for a site with the following parameter values: $i = 0.5$ in./hr, $S_w = 8$ in., $\theta_s = 0.55$, $\theta_i = 0.20$, $K_s = 0.025$ in./hr.

9–39. What is the drainage area (acres) of the watershed for the unit hydrograph of Problem 9–45?

9–40. What is the drainage area (acres) of the watershed for the unit hydrograph of Problem 9–46?

9–41. Unit hydrographs are commonly shown with units of ft^3/sec. What are the actual units of a unit hydrograph and why is there a difference?

9–42. Develop a triangular unit hydrograph with a time base of 6 hr for a 525-acre watershed.

9–43. Is the following hydrograph a legitimate unit hydrograph for a 297-acre watershed?

t (hr)	0	0.4	1.0	1.3	1.7	2.5	5.0	8.0
q (cfs)	0	35	105	135	90	50	15	0

9–44. If the rainfall excess consists of two bursts of 3.3 in./hr. and 1.8 in./hr., each for 20 min, and the unit hydrograph has ordinates of 15, 40, 25, 10 ft^3/sec, determine the distribution of direct runoff.

9–45. Compute the direct runoff hydrograph for the following rainfall excess and unit hydrograph.

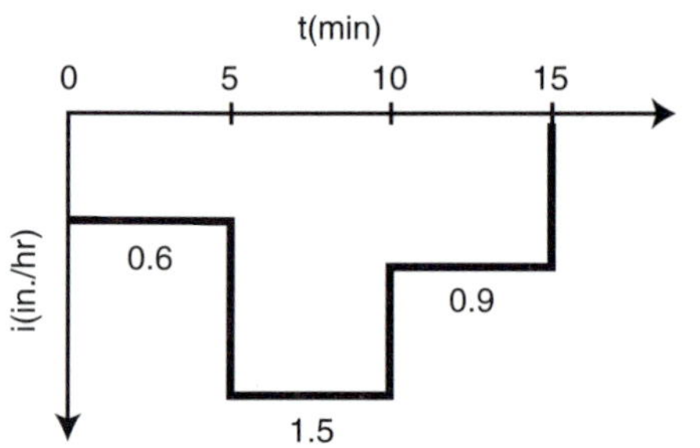

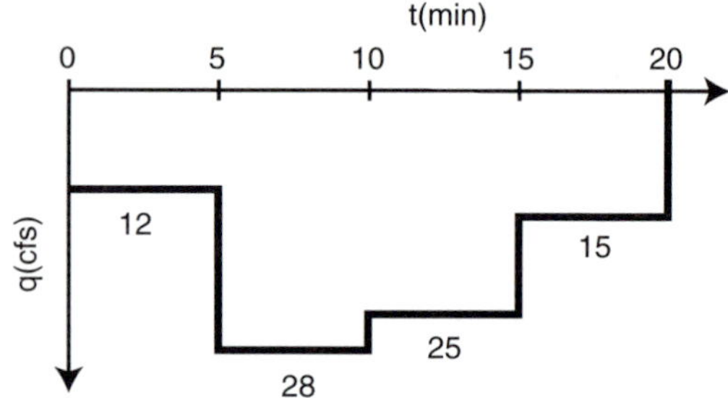

9–46. Compute the direct runoff hydrograph for the following rainfall excess and unit hydrograph.

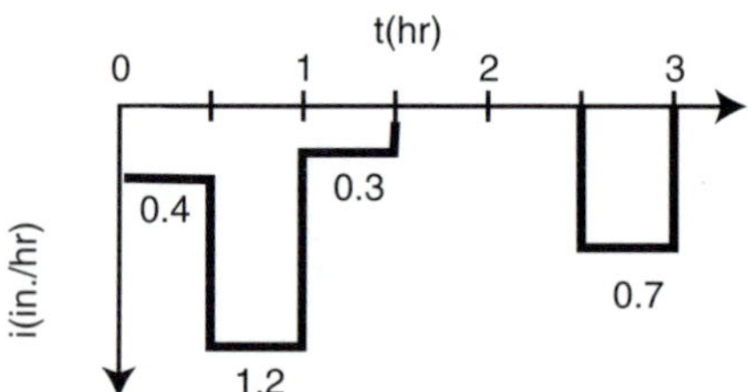

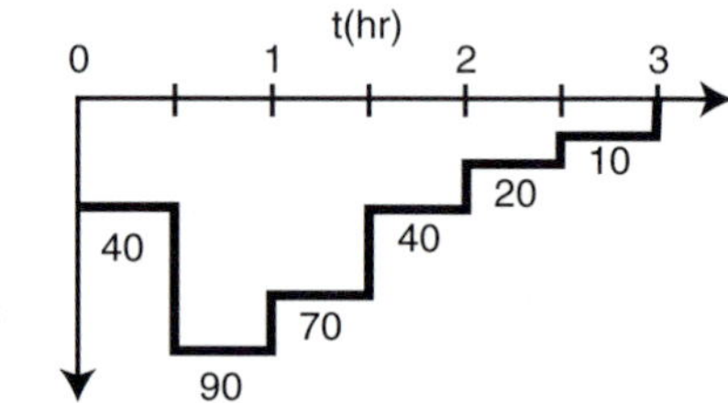

9–47. A 1.5-hr storm occurs on a 290-acre watershed, which has an average phi-index value of 1.7 in./hr. The hyetograph ordinates are as follows:

time (hr)	0		0.25		0.5		0.75		1.25		1.5
i(in./hr.)		1.5		3.7		4.7		2.9		0.5	

The UH ordinates on a 0.25-hr interval are {0, 30, 100, 180, 210, 200, 170, 130, 80, 40, 20, 10, 0} ft^3/sec. Assume a seasonal baseflow of 5 ft^3/sec. Compute the total storm hydrograph.

9–48. The following storm event data are for a 7-acre watershed. Assuming that baseflow equals 0 and that all rainfall prior to the start of runoff is initial abstraction, find the unit hydrograph and the distribution of rainfall excess. Use the phi-index method to separate losses.

T (min)	0	5	10	15	20	25	30
i (in./hr)	0	2	14	0	0	0	0
q (ft^3/sec)	0	0	9	21	17	13	0

9–49. The following storm event data are for a 25-acre watershed. Assuming that baseflow equals 0 and that rainfall prior to the start of runoff is initial abstraction, find the unit hydrograph and the distribution of rainfall excess. Use the phi-index method to separate losses.

t (min)	0	10	20	30	40	50	60	70	80	90
i(in./hr)	0	0.3	5.5	5.3	0	0	0	0	0	0
q (ft^3/sec)	0	0	10	25	60	50	35	15	5	0

9–50. The storm event data (June 17, 1944) given is for a 300-acre watershed (watershed W–1 in Fennimore, Wisconsin). Assume that rainfall excess and direct runoff begin at 0910. Also, assume that direct runoff ends at 1024. Use the phi-index method to separate losses and rainfall excess.

Use the constant-slope method to separate baseflow from direct runoff. Rainfall prior to 0910 is considered to be initial abstraction. Compute the unit hydrograph and determine the duration of the unit hydrograph.

Hyetograph		Hydrograph	
Time of Day	Intensity (in./hr)	Time of Day	Rate (in./hr)
0857	0.00	0902	0.0010
0903	2.90	0912	0.0108
0906	4.00	0914	0.0414
0910	0.75	0915	0.0432
0916	0.20	0916	0.0363
0920	4.80	0918	0.0484
0925	1.92	0922	0.0714
0930	0.84	0926	0.1062
0935	0.12	0928	0.1306
		0929	0.1512
		0930	0.1737
		0931	0.2030
		0936	0.3090
		0937	0.3270
		0940	0.3540
		0942	0.3720
		0944	0.4070
		0945	0.4380
		0946	0.4450
		0947	0.4410
		0948	0.4410
		0949	0.4380
		0950	0.4200
		0954	0.3750
		1000	0.2890
		1006	0.2215
		1012	0.1710
		1018	0.1306
		1024	0.1014
		1030	0.0790
		1036	0.0645
		1048	0.0439
		1100	0.0322

9–51. Given the rainfall excess distribution (0.6, 0.3) and the direct runoff distribution (0.12, 0.36, 0.34, 0.15), find the unit hydrograph using the least squares method.

9–52. Given the rainfall excess distribution (0.6, 0.2) and the direct runoff distribution (0.03, 0.35, 0.28, 0.12), find the unit hydrograph using the least squares method.

9–53. The rainfall hyetograph P and runoff hydrograph R were measured on a 0.5-hr time interval. A phi-index of 0.5 in./hr is appropriate for the watershed. Use the least squares method to derive the unit hydrograph.

$$P = \{0.8, 2.4, 1.6, 0.0, 1.0, 1.8\} \text{ in./hr}$$

$$R = \{10, 90, 250, 410, 450, 440, 450, 300, 210, 120, 50, 30, 10\} \text{ ft}^3\text{/sec}$$

9–54. Construct an S-hydrograph for the unit hydrograph computed in Problem 9–48. Use the S-hydrograph to derive a 10-min UH.

9–55. Construct an S-hydrograph for the unit hydrograph computed in Problem 9–49. Use the S-hydrograph to derive a 40-min UH.

9–56. Given the following 30-min UH for a 167.6-acre watershed, use the S-hydrograph method to derive a 60-min and a 15-min unit hydrographs:

$$UH_{30} = \{8, 32, 54, 58, 54, 40, 28, 22, 16, 12, 8, 4, 2\}\ \text{ft}^3/\text{sec}$$

9–57. For the 30-min unit hydrograph computed in Problem 9–56, compute a synthetic unit hydrograph with a gamma distribution.

9–58. Using the least-squares UH developed from Problem 9–53, fit a gamma UH to the least-squares UH. Area = 629.2 ac. Time increment = 30 min. Compare the least-squares and gamma synthetic UH's.

9–59. The following are 10-min unit hydrographs (ft^3/sec) computed from four storm events on a 70.25-acre watershed. Develop an average 10-min unit hydrograph.

Storm 1: {3, 7, 13, 18, 25, 30, 35, 37, 38, 37, 34, 31, 27, 24, 19, 16, 12, 9, 6, 3, 1}
Storm 2: {4, 12, 21, 26, 30, 31, 32, 31, 30, 28, 26, 25, 23, 21, 19, 16, 15, 13, 10, 7, 4, 1}
Storm 3: {9, 26, 37, 38, 37, 36, 35, 32, 29, 26, 23, 20, 17, 14, 12, 10, 8, 7, 5, 3, 1}
Storm 4: {6, 19, 36, 42, 44, 42, 38, 36, 34, 31, 26, 20, 17, 14, 10, 7, 3}

9–60. Compare the dimensionless UH of White Oak Bayou (Table 9–13) with a gamma UH that has a peak discharge of 1 and a time to peak of 1.

9–61. For the watershed of Figure 3–3, compute a time-area curve. Assume a forested watershed condition for overland flow and a velocity of 2.5 ft/sec for flow in the drainage network. Develop an instantaneous UH and route it with Equation 9–66.

9–62. Construct a unit hydrograph with the Rational Formula assuming a commercial land use (C = 0.7), an area of 17 acres, a time of concentration of 25 min, and a 50-yr design frequency.

9–63. Show that the volumes of direct runoff and rainfall excess are not equal when a hydrograph time base other than $2t_c$ is used with the Rational Formula hydrograph.

9–64. Equation 9–71 can be rearranged to the form $q_p t_p = 484\ AQ$. Provide an interpretation of each side of this equation.

9–65. Develop an expression relating the proportion of the hydrograph under the rising limb and the constant in the relationship of Equation 9–71.

9–66. Develop a SCS triangular unit hydrograph for a 900-acre watershed with a time of concentration of 4.5 hr.

9–67. Develop a SCS triangular unit hydrograph for a 740-acre watershed that has an average runoff curve number of 67 and an average slope of 1.3%

9–68. Develop a SCS triangular unit hydrograph for a 360-acre watershed that has an average slope of 2% and is wooded (fair condition) on a C soil.

9–69. Develop a SCS curvilinear unit hydrograph for the conditions of Problem 9–67. Use a time interval of 0.536 hr.

9–70. Develop a SCS curvilinear unit hydrograph for the conditions of Problem 9–68. Use a time interval of 0.25 hr.

9–71. Develop a unit hydrograph using Snyder's Method for a 150 mi^2 watershed, with L_{ca} = 6.5, L = 12.2 mi, C_t = 2.15, C_p = 0.62.

REVIEW QUESTIONS

9–1. Which of the following is not in the hydrograph separation process? (a) Direct runoff; (b) initial abstraction; (c) the time of concentration; (d) rainfall excess; (e) baseflow.

9–2. Convolution is the process of transforming (a) rainfall excess to a peak discharge; (b) rainfall excess to direct runoff; (c) direct runoff to a unit hydrograph; (d) an upstream hydrograph to a downstream hydrograph; (e) rainfall volume to a runoff volume.

9–3. Which of the following is not a method of baseflow separation? (a) Constant-discharge method; (b) master-depletion-curve method; (c) concave method; (d) constant-slope method; (e) phi-index method.

9–4. The master-depletion-curve method (a) assumes that runoff from ground water storage follows a systematic pattern; (b) is not valid for small watersheds; (c) provides a description of the rainfall loss function; (d) requires a straight-line decay function on log paper.

9–5. The initial abstraction (a) is dependent on the separation of baseflow; (b) can be a constant volume; (c) would be larger for an urban watershed than for a rural watershed; (d) is not a function of land use.

9–6. Interception storage does not depend on (a) the season of the year; (b) the amount of crown cover of trees; (c) the period of time from the last storm event; (d) the level of soil moisture.

9–7. The phi-index method (a) requires that the initial abstraction equal zero: (b) provides an index of rainfall excess intensities; (c) assumes that the loss rate is constant with time; (d) is a direct function of the total runoff volume; (e) all of the above.

9–8. If the volume of direct runoff equals 2.7 in., the total rainfall volume is 4.2 in., and the rainfall duration is 10 hr, the phi index is (a) 0.42 in./hr; (b) 0.27 in./hr; (c) 1.5 in.; (d) 0.0025 in./min; (e) none of the above.

9–9. The constant percentage method of estimating losses (a) assumes that the intensities of the rainfall excess are a constant proportion of the intensities of the total rainfall hyetograph; (b) requires the volume of rainfall excess to be less than the volume of direct runoff; (c) is only valid for design storm methods; (d) requires an initial abstraction model to be used.

9–10. For the data of Review Question 9–8, the constant percentage loss function would be (a) 0.643; (b) 0.357; (c) 0.556; (d) 0.0556; (e) 0.42.

9–11. The parameters of Horton's Method are a function of (a) rainfall excess intensities; (b) land cover type; (c) slope of the watershed; (d) antecedent soil moisture; (e) all of the above.

9–12. The recovery coefficient of an infiltration capacity model is primarily a function of (a) watershed slope; (b) volume of rainfall excess; (c) soil texture; (d) location of the water table; (e) none of the above.

9–13. For the mass infiltration method, the infiltration capacity model is a direct function of: (a) total volume of water infiltrated during the storm; (b) rainfall excess rate; (c) time from the start of the storm; (d) volume of rainfall excess; (e) none of the above.

9–14. A unit hydrograph (a) results from 1 in. of rainfall; (b) is independent of the duration of rainfall excess; (c) is not a function of the land use or cover; (d) can only be derived from storms of constant rainfall intensity; (e) none of the above.

9–15. A T-hour unit hydrograph: (a) has a time base of T hours; (b) has a depth of 1 in. per T hours; (c) has time to peak of T hours; (d) is based on a rainfall excess duration of T hours; (e) none of the above.

9–16. The units of a unit hydrograph are (a) acre-ft; (b) ft^3/sec/in.; (c) in./hr; (d) ft^3/sec; (e) none of the above.

9–17. An S-graph (a) is independent of the unit hydrograph duration; (b) is computed from a series of T-hour unit hydrographs; (c) does not require the ordinates to be at an interval of T hours; (d) may be different for adjacent watersheds; (e) all of the above.

9–18. A time-area curve has a time base equal to (a) time of concentration, t_c; (b) time to peak for the unit hydrograph; (c) $2t_c$; (d) duration less than t_c.

9–19. The shape of a time-area curve reflects (a) drainage density of channels in the watershed; (b) distribution of slopes in the watershed; (c) proportion of the area contributing runoff for times less than t_c; (d) nonhomogeneity of land covers in the watershed; (e) all of the above.

9–20. The SCS curvilinear unit hydrograph (a) has a time base of 8/3 the time to peak; (b) has a time to peak of 3/2 of the time of the concentration; (c) assumes a duration of about 2/15 of the time of concentration; (d) has a recession five times the time to peak; (e) none of the above.

DISCUSSION QUESTION

The technical content of this chapter is important to the professional hydrologist, but practice is not confined to making technical decisions. The intent of this discussion question is to show that hydrologists must often address situations where value issues intermingle with the technical aspects of a project. In discussing the stated problem, at a minimum include responses to the following questions:

1. What value issues are involved, and how are they in conflict?
2. Are technical issues involved with the value issues? If so, how are they in conflict with the value issues?
3. If the hydrologist attempted to rationalize the situation, what rationalizations might he or she use? Provide arguments to suggest why the excuses represent rationalization.
4. What are the hydrologist's alternative courses of action? Identify all alternatives, regardless of the ethical implications.
5. How should the conflict be resolved?

You may want to review Sections 1.6 to 1.12 in Chapter 1 in responding to the problem statement.

Case. A professional engineer submits a proposal to a local governmental agency in response to an open request for highway drainage work. The proposal carries a price of $40,000 for the completed work. The local government official in charge of awarding the contract meets with the professional engineer after receiving the engineer's proposal. The official indicates that the contract will be awarded to the engineer's company if the engineer will revise the cost figure in the proposal to $48,000 and then make $8,000 in payments for unspecified work to a company that is essentially the public official. The engineer recognizes this as a kickback.

10

Channel Routing

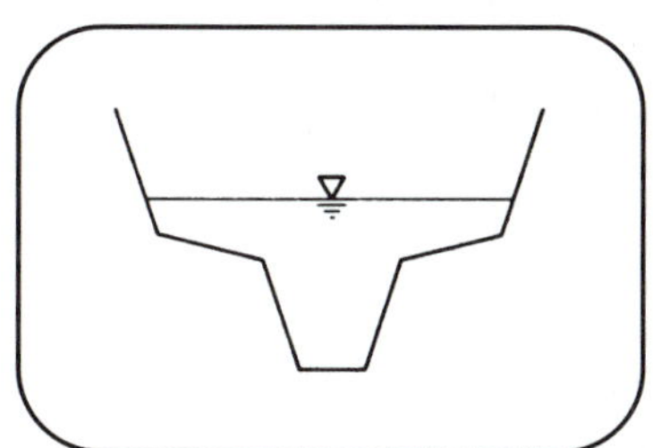

CHAPTER OBJECTIVES

1. Introduce and compare hydrologic methods for routing flood hydrographs through channel reaches.
2. Discuss the development of a rating table.
3. Identify requirements for the analysis and synthesis of the coefficients of routing methods.

10.0 NOTATION

A = area of cross section
c = kinematic wave celerity
C = routing coefficient (Convex Method)
C_m = routing coefficient (modified Att-Kin Method)
$C_0\ C_1, C_2$ = Muskingum Routing Coefficients
D = average detention storage over routing interval
D_e = surface flow detention storage
F = Froude number
h = depth of flow
i = rainfall intensity
$i_e{}^*$ = excess inflow rate
I = inflow or hydrograph ordinate at upstream section
I_a = average inflow
k = routing coefficient for modified Att-Kin Method
K = routing coefficient
L = length of channel reach
m = time coefficient for estimating travel time
n = Manning's roughness coefficient
O = outflow or hydrograph ordinate at downstream section
p = power-model exponent of discharge cross-sectional area curve
P = wetted perimeter
q = discharge
q_I = peak discharge of upstream hydrograph
q_o = discharge per unit width
q_{po} = peak discharge of the downstream hydrograph
Q = discharge
R_h = hydraulic radius
S = channel slope
S = slope of overland flow surface
S = storage in channel reach
S_f = friction slope
S_I = storage at upstream section
S_o = slope of channel
S_O = storage at downstream section
S_{po} = a measure of the reach storage at the peak outflow
t = time
t_e = equilibrium time
t_{pi} = time to peak of upstream hydrograph
t_{po} = time to peak of downstream hydrograph
T_t = reach travel time
V = velocity of flow
V_0 = average velocity
w = channel width
W_t = top width of water surface
x = distance along slope from top end
x = routing coefficient
y = flow depth
y_e = flow depth at equilibrium time
y_O = full flow depth
z = depth of flow
β = proportion of inflow appearing as outflow
Δt = time interval for routing
Δt_p = kinematic travel time
Δt_{pS} = time-shift parameter equal to the difference in the time to peak between the downstream and upstream hydrographs

10.1 INTRODUCTION

For designs on small upland watersheds, there may be none or insignificant amounts of flow in defined stream channels. In terms of the hydrologic cycle, the channel processes for such drainage areas are unimportant with respect to design. As the size of the drainage area increases or when a design is required at a location with a large contributing area upstream of the design location, channel processes must be considered to maintain a reasonable level of design accuracy. As the lengths of the stream channels and the size of the contributing areas increase, the storage in the stream channels usually increases and the channel processes become more important. In these cases, designs require accounting for the effects of channel storage. Channel routing is the term applied to methods of accounting for the effects of channel storage on the runoff hydrograph as the flood runoff moves through the channel reach.

10.1.1 Analysis Versus Synthesis

Channel routing can be introduced using the systems-theory model (see Figure 10–1). The input and output functions consist of the runoff hydrographs at the upstream and downstream sections of a channel, respectively; the transfer function is the channel routing procedure that is used to translate and attenuate the upstream runoff hydrograph into a downstream hydrograph. The transfer function actually consists of two parts, the routing method and the physical characteristics of the stream reach. In practice, we usually assume that the channel characteristics are known, so the routing method becomes the transfer function.

The systems-theory representation of channel routing can also be used to distinguish between the concepts of analysis and design. In the case of analysis, the hydrographs at the upstream and downstream sections of the channel reach are known and the coefficients of the transfer function (that is, routing method or channel characteristics) are evaluated. In the design or synthesis case, both the hydrograph at the upstream section and the transfer function are known, and the flood hydrograph at the downstream section must be computed.

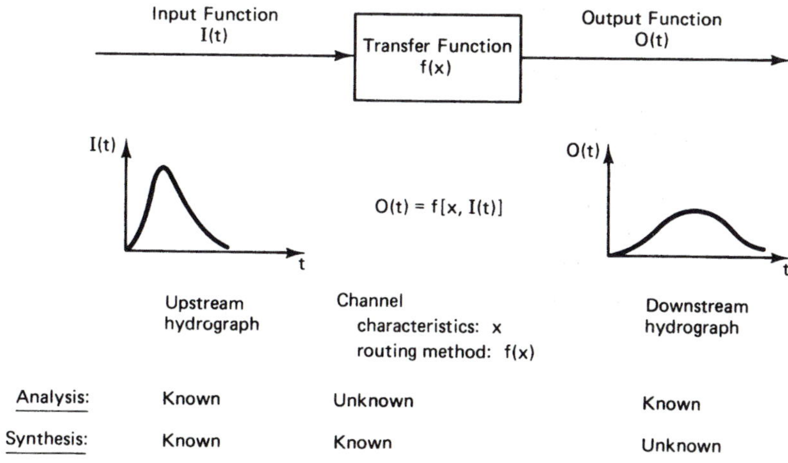

FIGURE 10–1 Channel routing: analysis versus synthesis.

In engineering design work, which represents synthesis, a routing method is usually assumed or dictated by the drainage or flood control policy, with values for the coefficients established using relationships between channel characteristics and the coefficients. Ideally, we would like to assume that channel properties could be assessed directly, which would mean that the problem of analysis would only involve fitting the coefficients of the routing method. However, while channel characteristics such as the length can be measured with a reasonable level of accuracy, the estimates of variables such as Manning's roughness coefficient are only approximations. Therefore, it may be necessary to check the value of channel parameters for which accurate estimates may not be possible.

For a given upstream hydrograph and channel system, a number of routing methods are available to route the hydrograph through the stream reach. Without passing judgment on the methods, a number of routing methods will be introduced and compared on the basis of their underlying assumptions. The methods to be introduced here are examples of the simplified routing methods that are often referred to as hydrologic routing methods. More complex methods referred to as hydraulic routing methods are available, but they are beyond the scope of this introduction.

10.2 DEVELOPMENT OF THE ROUTING EQUATION

For the purpose of developing an equation for routing through either stream channels or reservoirs, the continuity of mass can be expressed as

$$I - O = \frac{dS}{dt} \sim \frac{\Delta S}{\Delta t} \qquad (10\text{–}1)$$

in which I and O are the inflow and outflow, respectively, during the incremental time dt (or Δt), and S is the storage. For stream flow routing, I and O would be the upstream and downstream discharge hydrographs. For the hydrographs shown in Figure 10–2, the continuity equation can be expressed in terms of the inflow (upstream) and outflow (downstream) at

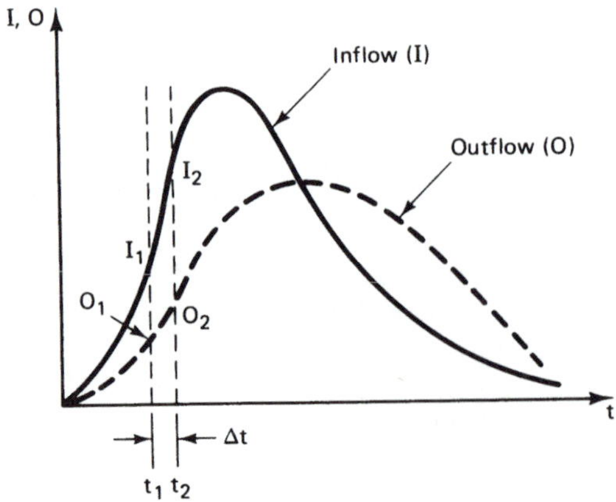

FIGURE 10–2 Schematic diagram of upstream (I) and downstream (O) flood hydrographs.

two times t_1 and t_2, which are separated by the incremental time $\Delta t = t_2 - t_1$. Expressing Equation 10–1 in terms of the average inflow and average outflow at times t_1 and t_2 gives

$$\frac{1}{2}(I_1 + I_2) - \frac{1}{2}(O_1 + O_2) = \frac{S_2 - S_1}{\Delta t} \tag{10–2}$$

Equation 10–2 is the numerical form of the routing equation. Assuming that the inflow hydrograph is known for all t and that the initial outflow and storage, O_1 and S_1, are known at t_1, then Equation 10–2 contains two unknowns, O_2 and S_2. Thus, to use the routing equation, a second relationship is needed.

For steady, uniform channel flow, the inflow, outflow, and storage are a function of the depth of flow, h. If we assume that the stage-discharge relationship (that is, the rating curve) is a straight line on log-log paper, the discharges I and O can be represented by the following:

$$I = ah^d \tag{10–3a}$$

and

$$O = ah^d \tag{10–3b}$$

Equations 10–3 have the same coefficients a and d, which implies that the properties of the stream reach are relatively constant. The storage characteristics at the upstream and downstream sections are also assumed constant, with storage and depth also assumed to be related through a log-log relationship:

$$S_1 = bh^m \tag{10–4a}$$

and

$$S_O = bh^m \tag{10–4b}$$

in which b and m are the constants that define the log-log relationship between storage and depth. Solving Equations 10–3a and 10–3b for h and substituting these into Equations 10–4a and 10–4b, respectively, yields

$$S_I = b\left[\left(\frac{I}{a}\right)^{I/d}\right]^m = b\left(\frac{I}{a}\right)^{m/d} = \left[b\left(\frac{1}{a}\right)^{m/d}\right] I^{m/d} = KI^{m/d} \tag{10–5a}$$

and

$$S_O = b\left[\left(\frac{O}{a}\right)^{I/d}\right]^m = b\left(\frac{O}{a}\right)^{m/d} = \left[b\left(\frac{1}{a}\right)^{m/d}\right] O^{m/d} = KO^{m/d} \tag{10–5b}$$

If we assume that the weighted-average storage S is a weighted function of the storage at the upstream and downstream cross sections, we get

$$S = xS_I + (1 - x)\, S_O \tag{10–6a}$$

$$= xKI^{m/d} + (1 - x)\; KO^{m/d} \tag{10–6b}$$

in which x is a coefficient that reflects the importance of storage at the two cross sections.

For uniform flow in a prismatic channel, it can be shown that $d = 5/3$ and $m = 1$; thus the storage function is

$$S = xKI^{0.6} + (1 - x)\,KO^{0.6} \tag{10–7}$$

The value of x depends on the characteristics of the channel reach. For natural streams, x is often in the range from 0.4 to 0.5. For a reservoir, x is usually assumed to equal zero since storage then only depends on the outflow. For a stream where storage is more dependent on outflow, such as in a low-sloped channel with wide floodplains, the value of x would be smaller than the typical range from 0.4 to 0.5.

The use of the routing equation to derive a storage-discharge curve can be illustrated using a rectangular channel of width w. For a depth h, the cross-sectional area equals wh and the wetted perimeter equals $(w + 2h)$. For the case where w is much greater than $2h$, the hydraulic radius is approximately equal to the depth of flow. Using Manning's Equation to compute the velocity of flow, the continuity equation gives

$$q = AV = wh\left(\frac{1.49}{n}R_h^{2/3}S_o^{1/2}\right) = \frac{1.49wS_0^{1/2}}{n}h^{5/3} \tag{10–8}$$

Comparing Equation 10–8 with Equations 10–3 implies that $d = 5/3$ and $a = 1.49wS_o^{1/2}n^{-1}$. Thus I or O in Equations 10–3 are a function of the 5/3 power of the flow depth.

If we assume that the change in depth over the channel reach of length L is not sufficiently large and that the reach length is less than the wavelength of the unsteady flow flood wave, the storage in the reach can be represented by the volume of a parallelepiped of length L, width w, and mean depth h:

$$S = Lwh \tag{10–9a}$$

Comparing Equation 10–9a to Equation 10–4 implies that $m = 1$ and $b = Lw$. Based on these values of a and b, the value of K in Equations 10–5 can be computed by solving Equation 10–8 for h and substituting it into Equation 10–9a:

$$S = Lw\left[\frac{n}{1.49\,w\,S_o^{0.5}}\right]^{0.6} q^{0.6} = (Lw)^{0.4}\left(\frac{nL}{\sqrt{S_o}}\right)^{0.6} q^{0.6} \tag{10–9b}$$

Comparing Equation 10–9b with Equation 10–5 yields

$$K = 0.7872\,(Lw)^{0.4}\left(\frac{nL}{S_o^{0.5}}\right)^{0.6} \tag{10–9c}$$

In addition to the stage-discharge relationship of Equation 10–3 and stage-storage relationship of Equation 10–4, the area-discharge relationship is used in channel routing:

$$q = fA^p \tag{10–10}$$

in which A is the cross-sectional area, and f and p are constants.

Example 10–1

The annual maximum flood series is given in Table 10–1 for the 42-yr record for a watershed located in Annandale, VA. The measured stage is also given. The data were fitted using least squares analysis following a logarithmic transformation. The computed rating curve for predicting discharge for a given state is

$$Q = 11.48\,h^{2.4477}$$

TABLE 10–1 Rating Table for USGS Gage No. 01654000

Water Year	Gage Height	Discharge	Water Year	Gage Height	Discharge
1947	9.90	3950	1968	9.46	3500
1948	5.20	780	1969	11.85	7870
1949	5.80	940	1970	7.10	1300
1950	5.74	910	1971	9.20	2380
1951	6.73	1300	1972	15.96	12000
1952	9.10	2560	1973	10.69	3270
1953	8.41	2120	1974	8.79	1890
1954	4.20	584	1975	12.90	6420
1955	9.05	2500	1976	9.14	2180
1956	7.05	1270	1977	8.91	2020
1957	6.12	990	1978	9.98	2900
1958	5.69	865	1979	11.54	4480
1959	5.99	965	1980	10.23	3100
1960	6.62	1140	1981	10.27	3170
1961	7.77	1600	1982	7.41	1190
1962	5.81	915	1983	10.22	3120
1963	8.82	2240	1984	11.31	4250
1964	7.86	1650	1985	9.03	2100
1965	7.58	1500	1986	6.49	777
1966	9.74	3400	1987	9.81	2710
1967	11.84	7870	1988	8.08	1520

in which Q is the measured discharge in ft^3/sec and h is the stage in ft. Figure 10–3 shows the measured data, except for the 1972 water year, and the fitted rating curve. The measured data fit the computed curve reasonably well.

10.3 THE CONVEX ROUTING METHOD

The Convex Routing Method is a simplified procedure for routing hydrographs through stream reaches. It is based on Equations 10–1 and 10–5b. The method assumes that $x = 0$ in Equation 10–6, and $m/d = 1$ in Equation 10–5b; Equation 10–6b then reduces to

$$S = KO \tag{10–11}$$

Assuming that I and O of Equation 10–1 can be represented by I_1 and O_1, respectively, then substituting Equation 10–11 into Equation 10–1 for both time 1 and 2 yields

$$I_1 - O_1 = \frac{K(O_2 - O_1)}{\Delta t} \tag{10–12}$$

Rearranging and solving for O_2 (that is, the unknown) yields

$$O_2 = I_1\left(\frac{\Delta t}{K}\right) + O_1\left(\frac{\Delta t}{k}\right) \tag{10–13}$$

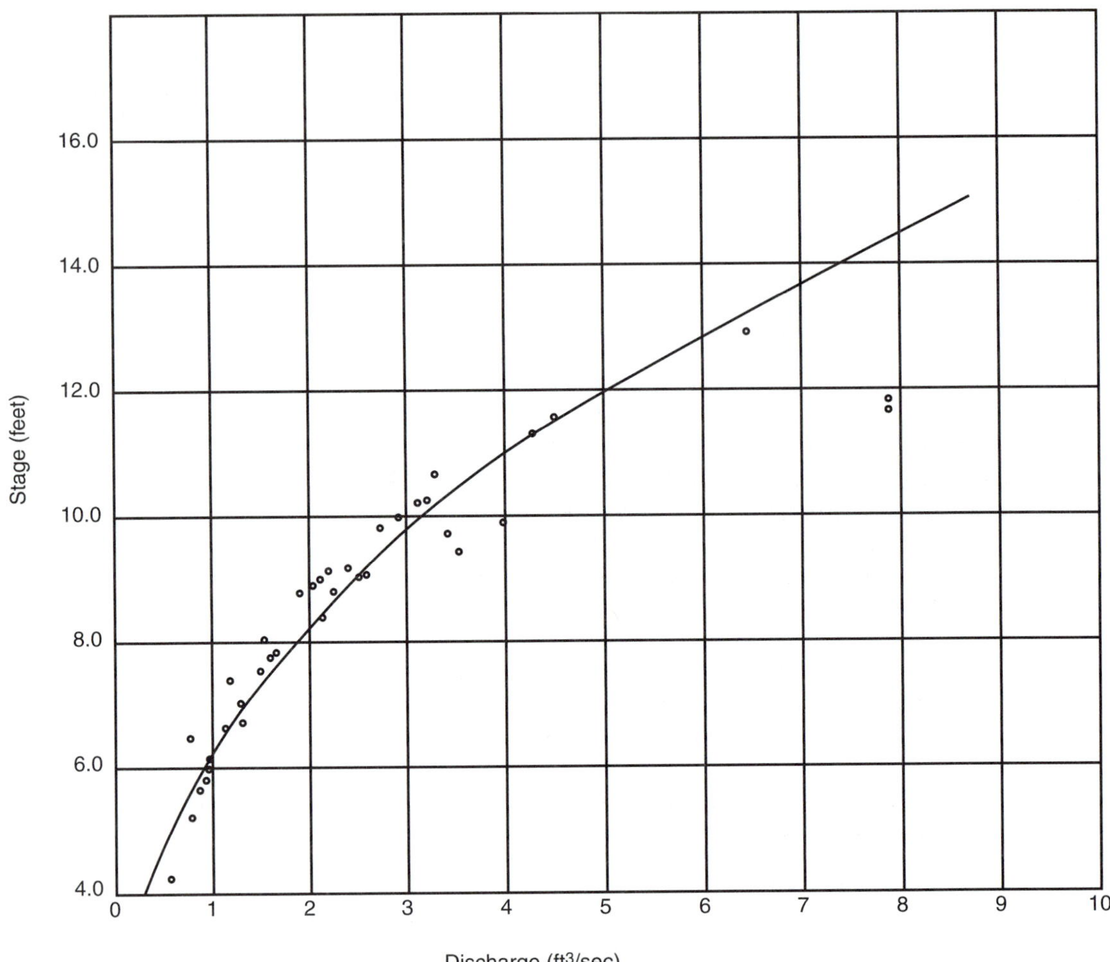

FIGURE 10–3 Rating Curve for Example 10–1.

Letting $C = \Delta t/K$ yields the routing equation for the Convex Method:

$$O_2 = CI_1 + (I - C)\, O_1 \tag{10–14}$$

10.3.1 Application of the Convex Method

In terms of analysis and synthesis, the upstream hydrograph is known in both cases. For analysis, the downstream hydrograph is also known, but the routing coefficient C is not known. An estimate of C is the end product of the analysis process.

When applying the Convex Routing Method in a design problem, the task is to find the outflow, or downstream, hydrograph. The known information is the upstream hydrograph (or inflow), the routing coefficient, and the outflow at time $t = 0$. The routing interval must also be known; as a rule of thumb, the routing interval should be less than one-fifth of the time to peak of the inflow hydrograph. Given this information, the downstream hydrograph can be computed using Equation 10–14.

To illustrate the use of Equation 10–14 to compute a downstream hydrograph, assume the routing coefficient C equals 0.3. Thus, the routing equation is

$$O_{t+\Delta t} = 0.3\,I_t + 0.7\,O_t$$

Assume current values of I_t and O_t are 25 and 13, respectively, and the next three values of I_t are 28, 33, and 41. The values of $O_{t+\Delta t}$ are computed as follows:

I	O
25	13
28	**17** = 0.3(25) + 0.7(13)
33	**20** = 0.3(28) + 0.7(17)
41	**24** = 0.3(33) + 0.7(20)

Note that the computed values of O are used to compute subsequent values of O.

10.3.2 Estimation of the Routing Coefficient (C)

A number of methods have been suggested for computing the routing coefficient. First, the value of C could be analyzed using measured flood waves through the stream reach. The analysis is performed by finding the value of C that provides the best fit between the measured and estimated hydrograph ordinates at the downstream section, where "best fit" is measured using an objective function such as the sum of the squares of the errors (that is, differences between the ordinates of the measured and predicted hydrographs); this approach will be described below and illustrated in an example that follows. Second, the optimum value of C could be found using the mean of C values computed for each time interval. Third, SCS presents the following relationship as one option:

$$C = \frac{V}{V + 1.7}$$

in which V is the steady-flow velocity (ft/sec). A fourth alternative is that C is approximately $2x$, where x is a coefficient in the Muskingum Routing Method (see Equation 10–51). The fifth alternative is to let C equal the ratio of the time interval Δt to a constant K:

$$C = \frac{\Delta t}{K} \qquad (10\text{–}15)$$

where K equals the ratio of the reach length to the flow velocity, where the velocity could be estimated using Manning's Equation. The value of Δt should be much smaller than the time to peak of the upstream hydrograph; this ensures that the routing will provide accurate definition of the curvature in the hydrograph at the downstream cross section.

10.3.3 Regression Analysis of the Routing Coefficient

Where measured storm hydrographs are available, the routing coefficient of the Convex Method can be analyzed using the principle of least squares. Using the Convex Routing Equation (Equation 10–14), with a more general notation, algebraic manipulation yields the following form:

$$(O_{t+1} - O_t) = C(I_t - O_t) \tag{10–16}$$

Equation 10–16 has the form of a zero-intercept regression model, with $y = O_{t+1} - O_t$, $x = I_t - O_t$, and $b = C$. The regression equation $y = bx$ has the following solution for the slope coefficient b: $b = \Sigma xy/\Sigma x^2$. Therefore, the least squares estimator of the routing coefficient C can be obtained from

$$\hat{C} = \frac{\sum_{\text{all } t} [(O_{t+1} - O_t)(I_t - O_t)]}{\sum_{\text{all } t} [(I_t - O_t)^2]} \tag{10–17}$$

The solution requires the two summations based on the ordinates of the measured upstream and downstream hydrographs.

Example 10–2

The data of Table 10–2 can be used to illustrate the calculations. The ordinates of the measured upstream (I) and downstream (O) hydrographs are given in Table 10–2; the table also includes the computations of the two summations of Equation 10–17. Thus, the best estimate of the routing coefficient is

$$C = \frac{25{,}878}{73{,}370} = 0.353 \tag{10–18}$$

The routing equation is

$$O_{t+1} = 0.353\, I_t + 0.647\, O_t \tag{10–19}$$

The predicted values are given in Table 10–2. The largest error is 4 ft^3/sec, with many errors being 0. Thus for this storm the statistical method provided an estimate of the routing coefficient that provided a good fit to the measured data.

Example 10–3

The inflow hydrograph of Figure 10–4 will be used to illustrate the solution procedure. For a velocity of 0.75 ft/sec, Equation 10–15 would yield a value of 0.3 for the routing coefficient. By applying Equation 10–14 with a time increment of 0.5 hr, the outflow hydrograph can be computed. The computations are given in Table 10–3, and the resulting outflow (downstream) hydrograph is shown in Figure 10–4. The actual peak discharge of the outflow hydrograph does not necessarily occur at 4.0 hr and may be greater than the computed value of 1417 ft^3/sec. The actual peak may be obtained either by routing at a smaller routing increment Δt or through curvilinear interpolation of points around the peak value. In any case, Δt should be made small enough so that the peak is accurately estimated and large enough to avoid unnecessarily lengthy calculations. It is recommended that the time increment be no larger than one-fifth of the time to peak.

TABLE 10–2 Example 10–2: Convex Method: Analysis

Upstream (I) (ft³/sec)	Downstream (O) (ft³/sec)	$O_{t+1} - O_t$	$I_t - O_t$	$(O_{t-1} - O_t)(I_t - O_t)$	$(I_t - O_t)^2$	$\hat{Q}$ (ft³/sec)
0	0	0	0	0	0	0
10	0	3	10	30	100	0
35	3	11	32	352	1,024	4
75	14	22	61	1,342	3,721	15
145	36	38	109	4,142	11,881	36
210	74	48	136	6,528	18,496	74
225	122	40	103	4,120	10,609	122
165	162	3	3	9	9	159
160	165	–5	–5	25	25	161
175	160	4	15	60	225	161
185	164	8	21	168	441	166
130	172	–14	–42	588	1,764	172
90	158	–25	–68	1,700	4,624	157
60	133	–25	–73	1,825	5,329	134
40	108	–22	–68	1,496	4,624	108
20	86	–24	–66	1,584	4,356	84
10	62	–18	–52	936	2,704	61
0	44	–14	–44	616	1,936	43
0	30	–10	–30	300	900	28
0	20	–8	–20	160	400	18
0	12	–5	–12	60	144	12
0	7	–4	–7	28	49	8
0	3	–3	–3	9	9	5
0	0	0				3
				25,878	73,370	

10.4 MODIFIED ATT-KIN METHOD

The modified Att-Kin Method transforms the continuity-of-mass relationship of Equation 10–1 to the following:

$$I_1 - \frac{O_2 + O_1}{2} = \frac{S_2 - S_1}{\Delta t} \tag{10–20}$$

which is slightly different from the general form of Equation 10–2. Substituting Equation 10–11 into Equation 10–20 and solving for O_2 yields the following:

$$O_2 = \left(\frac{2\Delta t}{2K + \Delta t}\right) I_1 + \left(1 - \frac{2\Delta t}{2K + \Delta t}\right) O_1 \tag{10–21a}$$

$$= C_m I_1 + (1 - C_m)\, O_1 \tag{10–21b}$$

in which C_m is the routing coefficient for the modified Att-Kin Method. The value of K is assumed to be given by

$$K = \frac{L}{mV} \tag{10–22}$$

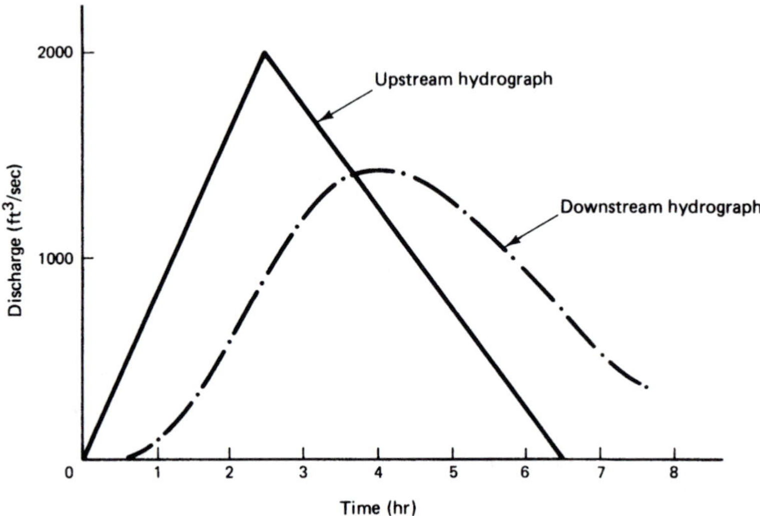

FIGURE 10–4 Upstream and convex-routed downstream hydrographs for Example 10–3.

in which L is the reach length and V is the velocity, defined by the continuity equation:

$$V = \frac{q}{A} \tag{10–23}$$

in which A is related to q by the rating curve equation:

$$q = xA^m \tag{10–24}$$

TABLE 10–3 Computing of Convex-Routed Downstream Hydrograph for Example 10–3

Time (hr)	Inflow (ft^3/sec)	Computation $(1 - C)O_t + CI_t$	Outflow (ft^3/sec)
0	0	0	0
0.5	400	0.7 (0) + 0.3 (0)	0
1.0	800	0.7 (0) + 0.3 (400)	120
1.5	1200	0.7 (120) + 0.3 (800)	324
2.0	1600	0.7 (324) + 0.3 (1200)	587
2.5	2000	0.7 (587) + 0.3 (1600)	891
3.0	1750	0.7 (891) + 0.3 (2000)	1224
3.5	1500	0.7 (1224) + 0.3 (1750)	1381
4.0	1250	0.7 (1381) + 0.3 (1500)	1417
4.5	1000	0.7 (1417) + 0.3 (1250)	1367
5.0	750	0.7 (1367) + 0.3 (1000)	1257
5.5	500	0.7 (1257) + 0.3 (750)	1105
6.0	250	0.7 (1105) + 0.3 (500)	923
6.5	0	0.7 (923) + 0.3 (250)	721
7.0	0	0.7 (721) + 0.3 (0)	504
7.5	0	0.7 (504) + 0.3 (0)	353

Equation 10–24 corresponds directly to Equation 10–3 since A is a function of h. If the discharge is derived using Manning's Equation, then

$$q = \frac{1.49}{n} R_h^{2/3} A S^{1/2} = \frac{1.49}{n} S^{1/2} A\left(\frac{A}{P}\right)^{2/3} = \frac{1.49 S^{1/2}}{n P^{2/3}} A^{5/3} \tag{10–25}$$

Comparing Equations 10–24 and 10–25 indicates that

$$m = \frac{5}{3} \tag{10–26a}$$

and

$$x = \frac{1.49 S^{1/2}}{n P^{2/3}} \tag{10–26b}$$

Therefore, m is a function of the velocity-versus-area relationship, and x is a function of the characteristics of the cross section.

10.4.1 Calculation of the Rating Table Coefficients

The rating table of Equation 10–24 assumes that the flow (q) and cross-sectional area (A) data measured from numerous storm events will lie about a straight line when plotted on log-log paper. That is, taking the logarithms of Equation 10–24 yields the straight line

$$\log q = \log x + m \log A \tag{10–27}$$

Thus the intercept is $\log x$, and m is the slope of the line. For the form of Equation 10–27, the intercept $\log x$ equals the logarithm of the discharge at an area of 1.0.

The coefficients of Equation 10–27 can be fit with any one of several methods. Visually, a line could be drawn through the points and the slope computed; the value of x would equal the discharge for the line when A equals 1.0.

The coefficients of Equation 10–27 could also be fitted using linear regression analysis after making the logarithmic transform of the data. This is identical to the analysis of the rating-table fitting in which stage is the predictor variable. The statistical fit may be more rational and consistent than the visual fit, especially when the scatter of the data is significant and the visual fit is subject to a lack of consistency.

As indicated before, Manning's Equation can be used where rating table data (q versus A) are not available. Manning's Equation can be applied for a series of depths and the rating table constructed. Of course, the rating table values and, thus, the coefficients are dependent on the assumptions underlying Manning's Equation.

In many cases, the graph of $\log q$ versus $\log A$ will exhibit a nonlinear trend, which indicates that the model of Equation 10–24 is not correct. The accuracy in using Equation 10–24 to represent the rating table will depend on the degree of nonlinearity in the plot. The SCS TR-20 manual provides a means of deriving a weighted value of m, which is as follows. The slope between each pair of points on the rating curve is estimated numerically:

$$S_i = \frac{\log q_i - \log q_{i-1}}{\log A_i - \log A_{i-1}} \tag{10–28}$$

in which S_i is the slope between points i and $(i–1)$. The weighted value of m, which is denoted as $\overline{m}$, is

$$\overline{m} = \frac{q_3 S_3 + \sum_{j=4}^{k} (q_j - q_{j-1}) S_j}{q_i} \tag{10–29}$$

in which k is the number of pairs of points on the rating table. The weighting of Equation 10–29 provides greater weight to the slope between points for which the range of $(q_j - q_{j-1})$ is larger.

10.4.2 Modified Att-Kin Procedure

The modified Att-Kin Method uses Equation 10–21b to perform the routings necessary to derive the downstream hydrograph. Whereas the Convex Routing Method provides only attenuation, the modified Att-Kin Method provides for both attenuation and translation. To apply the modified Att-Kin Method, the values of m and x in Equation 10–24 are evaluated using either cross-section data or a rating table developed from measured runoff events. The value of K is then computed from Equation 10–22 and used to compute C_m with Equation 10–21. The routing equation (Equation 10–21b) can then be used to route the upstream hydrograph. After deriving the first estimate of the downstream hydrograph, it is necessary to check whether or not hydrograph translation is necessary. The hydrograph computed with Equation 10–21b is translated when the kinematic travel time (Δt_p) is greater than Δt_{ps} where Δt_{ps} is the difference in the times to peak between the upstream hydrograph I and the hydrograph computed with Equation 10–21:

$$\Delta t_{ps} = t_{po} - t_{pi} \tag{10–30}$$

in which t_{po} and t_{pi} are the times to peak of the downstream (outflow) and upstream (inflow) hydrographs, respectively. The kinematic travel time is given by

$$\Delta t_p = \frac{S_{po}}{q_{po}} \left[\frac{(q_I/q_{po})^{1/m} - 1}{(q_I/q_{po}) - 1} \right] / \, 3600 \tag{10–31}$$

in which q_{po} is the peak discharge of the downstream hydrograph, q_I is the peak discharge of the upstream hydrograph, and S_{po} is given by

$$S_{po} = \left(\frac{q_{po}}{k} \right)^{1/m} \tag{10–32}$$

and

$$k = \frac{x}{L^m} \tag{10–33}$$

If $\Delta t_p > \Delta t_{ps}$, the storage-routed hydrograph from Equation 10–21b is translated by an amount $(\Delta t_p - \Delta t_{ps})$; this is shown in Figure 10–5.

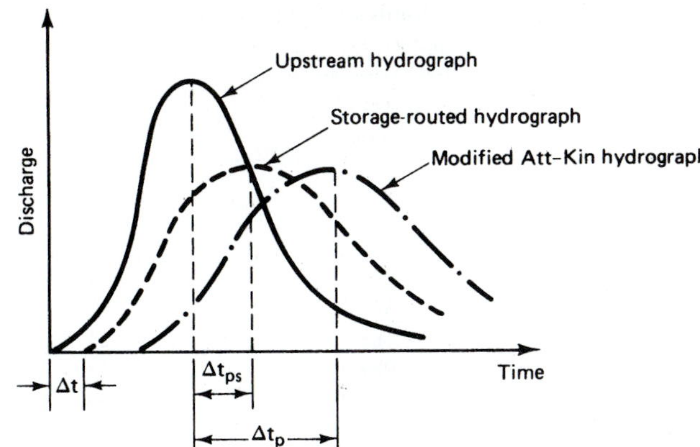

FIGURE 10–5 Translation of routed hydrograph using the modified Att-Kin Method.

This procedure can be summarized by the following steps:

1. From cross-section information, evaluate the rating table coefficients m and x.
2. Compute K and then C_m. (Note: It is necessary for $C_m < 1$ and preferable that $C_m < 0.67$.)
3. Use the routing equation (Equation 10–21b) to route the upstream hydrograph.
4. Compute Δt_{ps} from Equation 10–30.
5. Compute the kinematic travel time Δt_p (Equation 10–31).
6. If $\Delta t_p > \Delta t_{ps}$, translate the computed downstream hydrograph (see Figure 10–5).

Example 10–4

To apply the modified Att-Kin Method, the upstream hydrograph for an appropriate time increment Δt must be known; additionally, channel characteristics must be known, including the reach length and rating curve information. For this example, the upstream hydrograph for $\Delta t = 0.25$ hr is given in Table 10–4. Assume that the rating curve coefficients are $x = 0.1$ and $m = 1.6$. The problem will be to compute the downstream hydrograph for a reach length (L) of 10,000 ft. Using Equation 10–33, k is

$$k = \frac{x}{L^m} = \frac{0.1}{10{,}000^{1.6}} = 3.98 \times 10^{-8} \tag{10–34}$$

and the cross-sectional area at the maximum discharge computed from Equation 10–24 is

$$A = \left(\frac{q}{x}\right)^{1/m} = \left(\frac{2000}{0.1}\right)^{1/1.6} = 487.7 \text{ ft}^2 \tag{10–35}$$

From the continuity equation, Equation 10–23, the velocity is

$$V = \frac{q}{A} = \frac{2000}{487.7} = 4.1 \text{ ft/sec} \tag{10–36}$$

From Equation 10–22 the routing coefficient K is

$$K = \frac{L}{mV} = \frac{10{,}000}{1.6\,(4.1)\,(3600 \text{ sec/hr})} = 0.423 \text{ hr} \tag{10–37}$$

TABLE 10–4 Example 10–4: Modified Att-Kin Method: Synthesis of Downstream Hydrographs O_1 for $L = 10{,}000$ ft and O_2 for $L = 20{,}000$ ft

t (hr)	I (ft^3/sec)	O_1 (ft^3/sec)	O_2 (ft^3/sec)
0	0	0	0
0.25	250	0	0
0.50	500	114[a]	64[b]
0.75	800	290[c]	75[d]
1.00	1100	523[e]	333[f]
1.25	1450	786	524
1.50	1800	1089	753
1.75	1900	1413	1009
2.00	2000	1635	1220
2.25	1900	1801	1398
2.50	1800	1846	1502
2.75	1600	1825	1552
3.00	1400	1723	1536
3.25	1250	1575	1474
3.50	1100	1427	1416
3.75	1000	1278	1335
4.00	900	1151	1249
4.25	650	1036	1159
4.50	400	860	1028
4.75	200	650	867
5.00	0	445	695
5.25	0	242	516
5.50	0	132	384

[a]0.456 (250) + 0.544 (0) = 114
[b]0.2572(250) + 0.7428 (0) = 64
[c]0.456 (500) + 0.544 (114) = 290
[d]0.2572(500) + 0.7428 (64) = 175
[e]0.456 (800) + 0.544 (290) = 523
[f]0.2572 (800) + 0.7428 (175) = 333

and the routing coefficient from Equation 10–21 is

$$C_m = \frac{2\Delta t}{2K + \Delta t} = \frac{2\,(0.25)}{2\,(0.423) + 0.25} = 0.456 \tag{10–38}$$

Therefore, the routing equation (Equation 10–21b) is

$$O_{t+\Delta t} = 0.456 I_t + 0.544\, O_t \tag{10–39}$$

Using the routing equation, the downstream hydrograph is given in Table 10–4 and shown in Figure 10–6. The time of peak for the upstream and downstream hydrographs are 2.0 and 2.5 hr, respectively; therefore, from Equation 10–30, Δt_{ps} is 0.5 hr. Using Equation 10–21, S_{po} is

$$S_{po} = \left(\frac{q_{po}}{k}\right)^{1/m} = \left(\frac{1846}{3.98 \times 10^{-8}}\right)^{1/1.6} = 4.64 \times 10^6 \tag{10–40}$$

and from Equation 10–31,

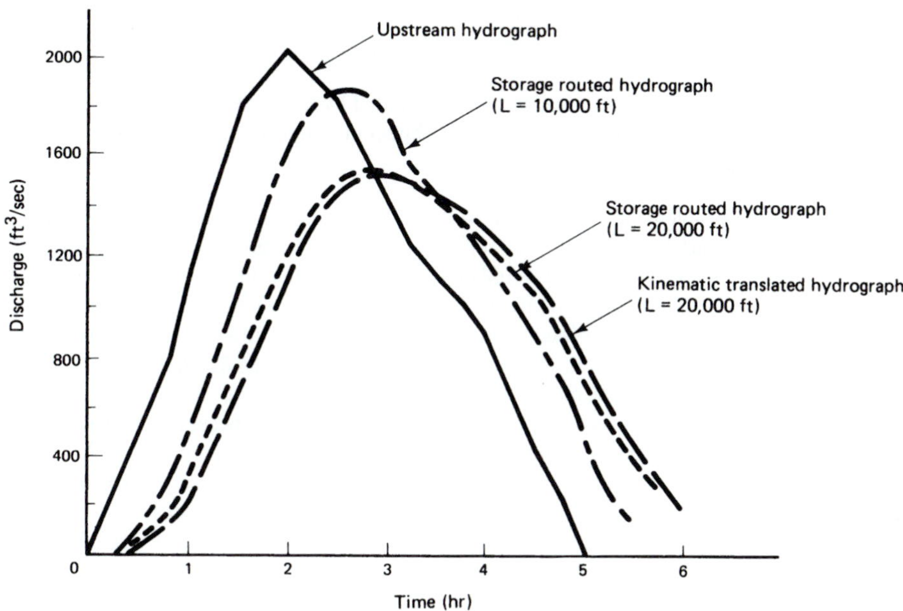

FIGURE 10–6 Hydrographs for Example 10–4: modified Att-Kin Method.

$$\Delta t_p = \left(\frac{4.64 \times 10^6}{1846}\right)\left[\frac{(2000/1846)^{1/1.6} - 1}{(2000/1846) - 1}\right]/3600 = 0.46 \tag{10–41}$$

Since $\Delta t_{ps} > \Delta t_p$, translation is not required.

If instead of a reach length of 10,000 ft, the upstream hydrograph is routed through a reach of 20,000 ft, the following calculations provide the routing equation:

$$k = \frac{0.1}{20{,}000^{1.6}} = 1.313 \times 10^{-8} \tag{10–42}$$

$$A = \left(\frac{2000}{0.1}\right)^{1/1.6} = 487.7 \text{ ft}^2 \tag{10–43}$$

$$V = \frac{2000}{487.7} = 4.1 \text{ ft/sec} \tag{10–44}$$

$$K = \frac{20{,}000}{1.6\,(4.1)\,(3600)} = 0.847 \tag{10–45}$$

$$C_m = \frac{2\,(0.25)}{2\,(0.847) + 0.25} = 0.257 \tag{10–46}$$

$$O_{t+\Delta t} = 0.257\, I_t + 0.743\, O_t \tag{10–47}$$

Using Equation 10–47, the downstream storage-routed hydrograph is given in Table 10–4 and shown in Figure 10–6. The peak of the downstream hydrograph occurs at 2.75 hr; therefore, from Equation 10–30, Δt_{ps} is 0.75 hr. Using Equation 10–32, S_{po} equals 8.3234×10^6 and from Equation 10–31, Δt_p is

$$\Delta t_p = \left(\frac{8.323 \times 10^6}{1552}\right)\left[\frac{(2000/1552)^{1/1.6} - 1}{(2000/1552) - 1}\right]/3600 = 0.886 \text{ hr} \quad (10\text{–}48)$$

Therefore, since $\Delta t_{ps} < \Delta t_p$ the storage-routed hydrograph must be translated by 0.886 – 0.75 = 0.136 hr. The resulting hydrograph is shown in Figure 10–6.

Example 10–5

In the problem of analysis, the following information is known: the measured upstream and downstream hydrographs, the appropriate routing increment, and some physical characteristics such as the reach length. The unknowns center on the routing coefficient and the translation time difference ($\Delta t_p - \Delta t_{ps}$). While these two unknowns are sufficient for defining the routing procedure, it may also be of interest to know values for other unknowns, such as the rating curve parameters x and m, as well as the kinematic time parameter K. These unknowns can be estimated after values for the primary unknowns have been estimated.

In the analysis of the Convex Method, it was only necessary to estimate the routing coefficient C. The same statistical procedure could be used for the modified Att-Kin Method, except that the time difference ($\Delta t_p - \Delta t_{ps}$) must also be estimated. While an effort-intensive numerical method could be used, it may be sufficient to estimate the time difference, back-translate the hydrograph in time, and then use the same statistical procedure that was outlined for the Convex Method. Where it is difficult to estimate the time difference, several trial estimates may be made and selection based on the best overall fit.

The approximate fitting method will be illustrated using the data of Table 10–5. The measured upstream (I) and downstream (O) hydrographs are given in Table 10–5 for a 0.l-hr time increment; the hydrographs are also shown in Figure 10–7. There appears to be a time translation of 0.1 hr at the start of the downstream hydrograph, so a value of 0.1 hr is used for the time difference ($\Delta t_p - \Delta t_{ps}$). The downstream hydrograph is back-translated by 0.1 hr and is indicated by O' in Table 10–5. The two hydrographs I and O' are then subjected to the statistical analysis of Equation 10–17. The calculations are shown in Table 10–5. From Equation 10–17 we get the following estimate of the routing coefficient:

$$C_m = \frac{10{,}950}{35{,}000} = 0.313 \quad (10\text{–}49)$$

Therefore, the routing equation is

$$O'_{t+1} = 0.313\, I_t + 0.687\, O'_t \quad (10\text{–}50)$$

The flood hydrograph O' is then translated by 0.1 hr, which is the estimate of the time difference ($\Delta t_p - \Delta t_{ps}$), to get the predicted downstream hydrograph O_{t+1} which is given in Table 10–5 and shown in Figure 10–7. The errors are computed in Table 10–5, with the largest absolute error being 6 ft^3/sec. The fitting procedure provided a reasonably good fit to the measured data.

10.5 MUSKINGUM METHOD

The Muskingum Method is a third alternative for routing hydrographs through stream reaches. The method is based on the routing equation (Equation 10–2) and the storage function of Equation 10–6b. Letting $m/d = 1$ and $K = b/a$, Equation 10–6b reduces to

$$S = K[xI + (1 - x)\,O] \quad (10\text{–}51)$$

TABLE 10–5 Example 10–5: Modified Att-Kin Method Analysis

t (hr)	I (ft^3/sec)	O (ft^3/sec)	O' (ft^3/sec)	$O'_{t+1} - O'_t$ (ft^3/sec)	$I_t - O'_t$ (ft^3/sec)	$[(O'_{t+1} - O'_t)(I_t - O_t)]$	$(I_t - O'_t)$	$\hat{Q}_t$ (ft^3/sec)	$\hat{Q}_{t+\Delta t}$ (ft^3/sec)	Error = $\hat{Q}_{t+\Delta t} - O$
0	0	0	0	0	0	0	0	0	0	0
0.1	25	0	0	10	25	250	625	0	0	0
0.2	60	0	10	10	50	500	2,500	8	0	0
0.3	100	10	20	25	80	2,000	6,400	24	8	–2
0.4	130	20	45	35	85	2,975	7,225	48	24	4
0.5	145	45	80	20	65	1,300	4,225	74	48	3
0.6	140	80	100	5	40	200	1,600	96	74	–6
0.7	125	100	105	5	20	100	400	110	96	–4
0.8	105	105	110	0	–5	0	25	115	110	5
0.9	85	110	110	–5	–25	125	625	112	115	5
1.0	65	110	105	–10	–40	400	1,600	103	112	2
1.1	50	105	95	–15	–45	675	2,025	91	103	–2
1.2	35	95	80	–15	–45	675	2,025	78	91	–4
1.3	25	80	65	–15	–40	600	1,600	65	78	–2
1.4	15	65	50	–10	–35	350	1,225	52	65	0
1.5	10	50	40	–10	–30	300	900	41	52	2
1.6	5	40	30	–5	–25	125	625	31	41	1
1.7	0	30	25	–5	–25	125	625	23	31	1
1.8	0	25	20	–5	–20	100	400	16	23	–2
1.9	0	20	15	–5	–15	75	225	11	16	–4
2.0	0	15	10	–5	–10	50	100	7	11	–4
2.1	0	10	5	–5	–5	25	25	5	7	–3
2.2	0	5	0	0	0	0	0	4	5	0
2.3	0	0						2	4	4
						10,950	35,000			–6

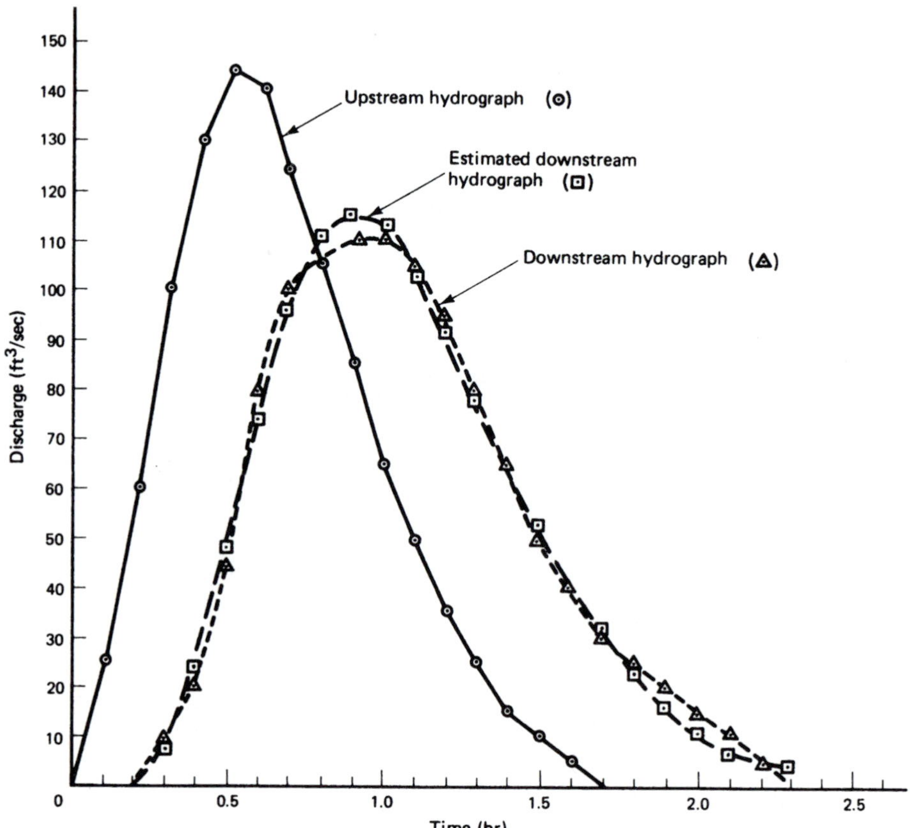

FIGURE 10–7 Example 10–5: modified Att-Kin Method analysis.

Substituting Equation 10–51 into Equation 10–2 gives

$$\frac{I_1 + I_2}{2} - \frac{O_1 + O_2}{2} = \frac{K[xI_2 + (1 - x)\,O_2] - K[xI_1 + (1 - x)O_1]}{\Delta t} \tag{10–52}$$

Solving Equation 10–52 by collecting and rearranging terms yields the routing equation

$$O_2 = C_o I_2 + C_1 I_1 + C_2 O_1 \tag{10–53}$$

where

$$C_o = -\frac{Kx - 0.5\Delta t}{K - Kx + 0.5\Delta t} \tag{10–54a}$$

$$C_1 = \frac{Kx + 0.5\Delta t}{K - Kx + 0.5\Delta t} \tag{10–54b}$$

$$C_2 = \frac{K - Kx - 0.5\Delta t}{K - Kx + 0.5\Delta t} \tag{10–54c}$$

Summing the three coefficients shows

$$C_0 + C_1 + C_2 = 1 \tag{10–55}$$

The upstream hydrograph is routed through the reach using Equation 10–53. The initial estimate, O_1, must be specified, as with the Convex and modified Att-Kin Methods. It is also necessary to specify the routing interval Δt and provide estimates of the routing coefficients K and x.

10.5.1 Estimation of the Muskingum Routing Coefficients

Again, a distinction must be made between the two cases of analysis and synthesis. In the case of analysis, measured hydrographs are available for both the upstream and downstream cross sections, and the objective is to analyze the data to find the best estimates of x and K. In the synthesis case, the values of x and K can be computed using site characteristics and known relationships between the site characteristics and x and K. The objective is to compute the hydrograph at the downstream cross section for a given hydrograph at the upstream cross section. This discussion, so far, has assumed that a single site is of interest; that is, the values of x and K estimated from the analysis of hydrographs are to be used for synthesizing flood hydrographs for that same stream reach.

A second type of estimation problem exists. Where measured hydrographs are not available (the site is ungaged), estimated values of x and K must be obtained from other sites. In this "ungaged" problem, analysis would refer to the process of developing prediction equations relating physical site characteristics (that is, Manning's n, channel slope, hydraulic radius) to many sets of values of x and K obtained from analysis of measured hydrograph data. Synthesis would refer to the use of the prediction equations with physical site characteristics to estimate the values of x and K. In this "ungaged" case, no hydrographs exist on the stream reach of interest. The accuracy of the estimated values of x and K will depend on the similarity between the ungaged stream reach and the characteristics of the stream reaches used in developing the prediction equations.

While the Muskingum Method has been widely used for decades, values of the coefficients K and x are not available from the analysis of measured hydrographs. A general rule of thumb is that K can be estimated by the travel time through the reach and a value of 0.2 can be used for x. A value of 0.5 for x is usually considered to be the upper limit of rationality; this value suggests equal weighting of inflow I and outflow O, and it appears primarily as translation of the hydrograph with little attenuation (see Figure 10–8). If $x = 0.5$ and $\Delta t << K$, then $C_0 = -1$, $C_1 = 1$, and $C_2 = 1$. If Δt is small, then I_1 is approximately equal to I_2, which means that they will cancel and the hydrograph will be translated by one time increment. A value of 0 for x is the practical lower limit and suggests that the inflow has little effect; this would reflect reservoir-storage type effects in which attenuation would be dominant. Some have reported values for x in the range from 0.4 to 0.5 for natural streams, while others suggest values from 0.1 to 0.3 for natural streams, with $x = 0.2$ commonly assumed. The value of x would probably be smaller for stream reaches with relatively wide flood plains. Unfortunately, these guidelines are imprecise and somewhat conflicting.

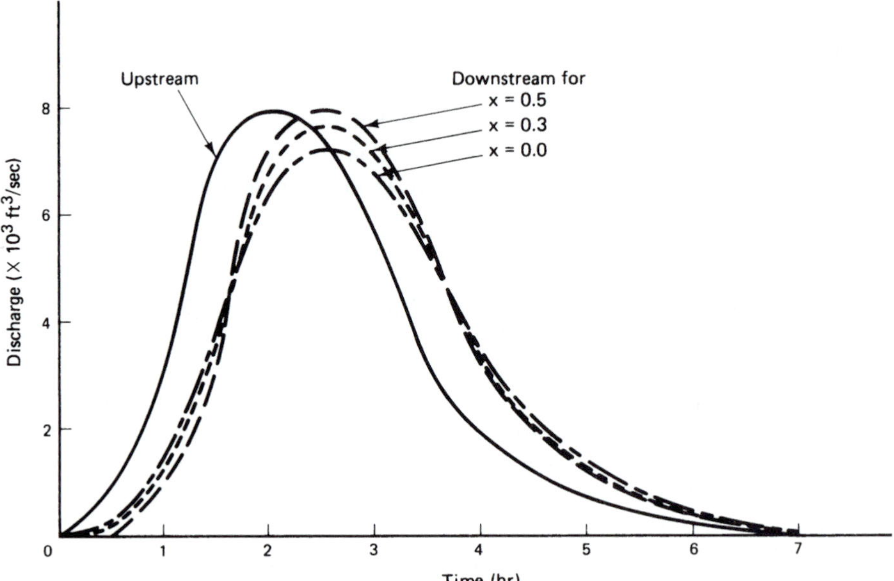

FIGURE 10–8 Effect of Muskingum parameter x in a routed hydrograph.

10.5.2 Analysis of the Muskingum Coefficients

The problem of estimating values for the routing coefficients can be solved where measured hydrographs are available. Values of K and x can be estimated with the measured data using either a graphical or statistical approach. Given the greater effort required by the graphical solution and the greater precision with the statistical approach, the statistical approach is far superior to the graphical approach. This is more true now that simple regression algorithms are available on hand-held calculators. The graphical and statistical approaches that are presented here are based on the same formulation.

The graphical approach uses Equations 10–2 and 10–51, with the following procedure used to obtain estimates of K and x:

1. For each point in time, compute the storage S_{t+1} by rearranging Equation 10–2:

$$S_{t+1} = S_t + \Delta t\left(\frac{I_t + I_{t+1}}{2} - \frac{O_t + O_{t+1}}{2}\right) \tag{10–56}$$

   Note that S_t is usually assumed to be 0 for the initial condition.

2. Using a trial value of x, compute $[xI + (1 - x)O]$ for each point in time.
3. Plot the computed storage S from Step 1 as the ordinate versus $[xI + (1 - x)O]$ from Step 2 as the abscissa for each point in time.

4. Revise the value of x and repeat Steps 1 to 3 until the plot shows a minimum amount of deviation from a straight line.
5. Use the slope of the line as the best estimate of K and the value of x that produced the minimum deviation from the line in Step 4 as the estimate of x.

The graphical approach is somewhat subjective in that for a given set of data, exactly equal estimates will not be obtained when two or more people independently analyze the set.

Estimates of x and K can be obtained using least squares regression. Equation 10–53 can be inserted into Equation 2–52 using O_2 as the value of y. Differentiating F with respect to the coefficients C_0, C_1, and C_2 yields the following three simultaneous equations:

$$C_0 \Sigma I_2^2 + C_1 \Sigma I_1 I_2 + C_2 \Sigma O_1 I_2 = \Sigma O_2 I_2 \tag{10–57}$$

$$C_0 \Sigma I_1 I_2 + C_1 \Sigma I_1^2 + C_2 \Sigma O_1 I_1 = \Sigma O_2 I_1 \tag{10–58}$$

$$C_0 \Sigma O_1 I_2 + C_1 \Sigma O_1 I_1 + C_2 \Sigma O_1^2 = \Sigma O_1 O_2 \tag{10–59}$$

where each of the summations is over all elements of the upstream and downstream hydrographs. Equation 10–55 must also be satisfied, so it becomes the fourth equation. The four simultaneous equations can be solved analytically or numerically.

Equations 10–54a and 10–54b represent two equations with two unknowns, x and K. Using the values of C_0 and C_1 obtained by regression analysis, the values of x and K are

$$x = \frac{C_1 - C_0}{2(1 - C_0)} \tag{10–60a}$$

$$K = \frac{\Delta t(1 - C_0)}{C_0 + C_1} \tag{10–60b}$$

The accuracy of the regression estimates of x and K can be evaluated by regenerating the values of O_2 using Equation 10–53 and assessing the errors (that is, $\hat{O}_2 - O_2$). A standard error can be computed.

10.5.3 Synthesis of the Muskingum Coefficients

It should be evident that the complexity of hydrologic processes is such that the analysis of hydrographs for different storm events will produce different estimates of the routing coefficients. Therefore, one should expect that the analysis of hydrographs for a number of different storm events on the same watershed will improve the accuracy of future estimates of routing. Where more than one set of flood hydrographs are available, values of K and x can be computed for each pair of upstream and downstream hydrographs; the vectors of values of K and x can then be used to derive "best" estimates, such as mean values.

The values of x and K can also be estimated using Manning's Equation. The value of x can be taken from Equation 10–26b, while K is set equal to b/a, where the evaluation of b and a was discussed with respect to Equation 10–7.

The selection of the routing parameters is important to ensure rationality. Improper selection can result in negative ordinates of the computed downstream hydrograph. Hjelmfelt (1985), as well as others, has pointed out that negative flows can be avoided if K and x are chosen such that

$$x \leqslant \frac{0.5\Delta t}{K} \leqslant 1 - x \quad \text{and} \quad x \leqslant 0.5 \tag{10–61}$$

One rule of thumb used in practice is that the ratio $\Delta t/K$ should be approximately 1 and x should be in the range 0 to 0.5. When this is applied, the restriction of Equation 10–61 will be met. Negative ordinates may also be the result of the use of a reach length that is too long. In such cases, it may be preferable to separate the channel into two reaches and route through two reaches using the outflow from reach 1 as the inflow to reach 2.

Explicit methods for selecting values of the coefficients are not readily available for locations where estimates are not made from the analysis of measured hydrographs. In most cases, empirical guidelines are used. The parameter K can be approximated by the travel time through the reach. Rules for estimating x are less readily available. As indicated previously, standard practice is to assume a value of 0.2 for x, with a smaller value for channel systems with large floodplains and larger values, near 0.4, for natural channels. Dooge et al. (1982) provided the following relationships for estimating K and x:

$$K = \frac{0.6L}{V_o} \tag{10–62}$$

and

$$x = 0.5 - 0.3\left(1 - \frac{4\mathbf{F}^2}{9}\right)\frac{y_o}{S_o L} \tag{10–63}$$

in which L is the reach length, V_o is the average velocity, y_o is the full flow depth, S_o is the slope of the channel bottom, and $\mathbf{F}$ is the Froude Number.

Example 10–6

The hydrograph data of Table 10–2 can also be used to demonstrate the analysis of data for estimating the Muskingum coefficients. Values for K and x are first estimated using the graphical method. The computed storages (Equation 10–56) and the values of $[xI + (1 - x)O]$ for values of x of 0.2, 0.3, and 0.4 are given in Table 10–6. Plots of storage S versus $[xI + (1 - x)O]$ for the three cases are shown in Figure 10–9. It is evident that there is little deviation from the line for x equal to 0.2; however, loops are apparent for x equal to both 0.3 and 0.4. Therefore, a value of 0.2 can be used for x. An estimate of the coefficient K is obtained from the slope of the line; using the largest storage value and the origin, the estimate of K is

$$K = \frac{569{,}400 - 0}{168 - 0} = 3389 \tag{10–64}$$

For $K = 3389$, $x = 0.2$, and $\Delta t = 1200$, the routing coefficients from Equations 10–54 are $C_o = -0.023$, $C_1 = 0.386$, and $C_2 = 0.637$; therefore, the routing equation is

$$O_2 = -0.023\, I_2 + 0.386\, I_1 + 0.637\, O_1 \tag{10–65}$$

The accuracy of Equation 10–65 could be assessed by regenerating the downstream hydrograph and comparing the measured and predicted hydrographs.

TABLE 10–6 Graphical and Statistical Evaluation of Muskingum Coefficient for Example 10–6

I (ft³/sec)	O (ft³/sec)	S	$0.2I + 0.8O$	$0.3I + 0.7O$	$0.4I + 0.6O$	$\hat{Q}$ (ft³/sec)
0	0	0	0	0	0	0
10	0	6,000	2	3	4	0(−0.2)
35	3	31,200	9	13	16	3
75	14	87,000	26	32	38	14
145	36	189,000	58	69	80	34
210	74	336,000	101	115	128	72
225	122	479,400	143	153	163	122
165	162	543,000	163	163	163	160
160	165	541,800	164	164	163	162
175	160	547,800	163	164	166	161
185	164	569,400	168	170	172	166
130	172	556,800	164	159	155	174
90	158	490,800	144	138	131	159
60	133	406,200	118	111	104	135
40	108	321,600	94	88	81	108
20	86	241,200	73	66	60	84
10	62	170,400	52	46	41	61
0	44	112,800	35	31	26	43
0	30	68,400	24	21	18	28
0	20	38,400	16	14	12	18
0	12	19,200	10	8	7	11
0	7	7,800	6	5	4	7
0	3	1,800	2	2	2	5
0	0	0	0	0	0	

The data were also analyzed using regression analysis. The data of Table 10–6 yields the following simultaneous equations:

$$271{,}075C_0 + 260{,}000C_1 + 185{,}610C_2 = 212{,}055$$

$$260{,}000C_0 + 271{,}075C_1 + 212{,}055C_2 = 233{,}475$$

$$185{,}610C_0 + 212{,}055C_1 + 226{,}396C_2 = 221{,}747$$

Using Equation 10–60 yields $K = 3410$ and $x = 0.199$, which are close to the values that resulted from the graphical analysis. Using Equation 10–54, the routing coefficients are $C_0 = -0.024$, $C_1 = 0.384$, and $C_2 = 0.640$; thus the routing equation is

$$\hat{O}_2 = -0.024\, I_2 + 0.384\, I_1 + 0.640\, O_1 \tag{10–66}$$

Equation 10–66 was used with the upstream hydrograph to predict the downstream hydrograph; the resulting ordinates $\hat{Q}$ are given in Table 10–6. The largest error is 3 ft³/sec, and the computed hydrograph shows good agreement with the measured hydrograph.

Applying Equation 10–61, the value of $0.5\Delta t/K$ equals 0.179; this value is less than the value of x, so negative ordinates occur. In this case, the second ordinate equaled −0.2, with the remaining ordinates being positive. Thus the effect was not significant.

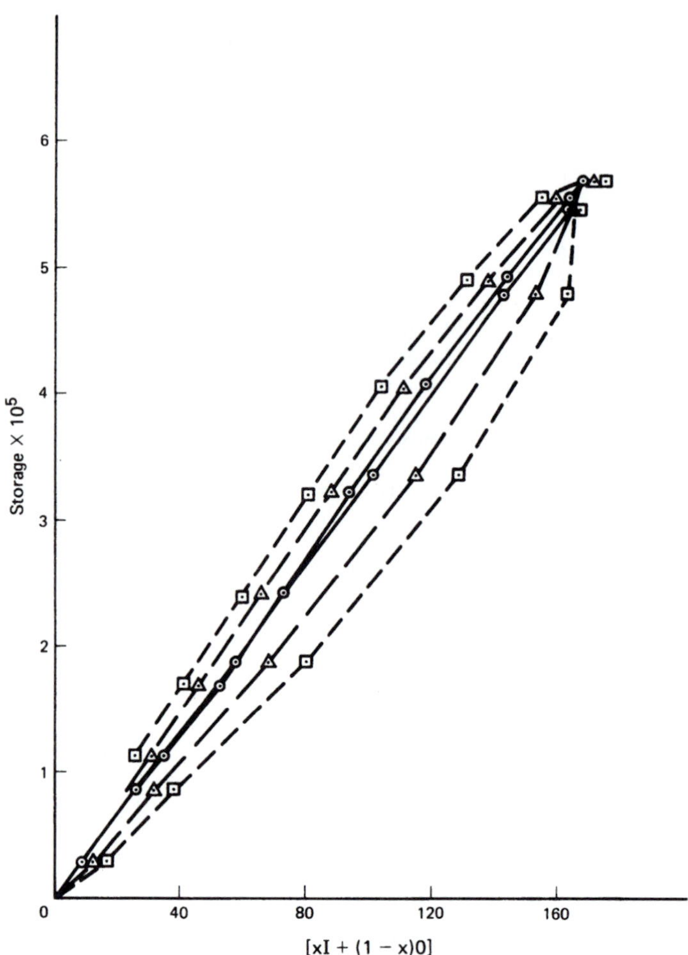

FIGURE 10–9 Graphical estimation of the Muskingum Coefficients K and x for $x = 0.2$ (⊙——⊙), $x = 0.3$ (△ — — △), $x = 0.4$ (⊡ - - - ⊡).

10.6 COMPARISON OF ROUTING METHODS

The three routing methods (Convex, modified Att-Kin (MAK), and Muskingum) are based on the same numerical approximation of the continuity equation (Equation 10–1). Considering only the routing portion of the modified Att-Kin Method, both the MAK and the Convex Methods are based on the linear storage function of Equation 10–11; the Muskingum Method assumes the more general storage function of Equation 10–51. If $x = 0$, Equations 10–11 and 10–51 are identical. Given that values of x for natural channels are often in the range 0.1 to 0.3, the assumption that $x = 0$ may not be a realistic assumption. The translation correction of the modified Att-Kin Method may partially compensate for the assumption that $x = 0$.

In applying the Convex and MAK Methods, the same routing coefficient (either C or C_m) is used to weight the inflow hydrograph ordinates. This can be shown by expressing Equation 10–12 in terms of I_1 and I_2 rather than I_a:

$$O_2 = C\,I_1 + (1 - C)\,O_1 \tag{10–67}$$

Whereas the Convex and MAK Methods apply weightings of C and 0 to I_1 and I_2, the Muskingum Method uses different weights; specifically, C_0 and C_1 of Equations 10–54a and 10–54b are applied to I_2 and I_1, respectively. The use of different weights for I_1 and I_2 provides somewhat greater flexibility in fitting the Muskingum Routing Method with measured storm hydrographs, at least in comparison to the routing equations of Equations 10–14 and 10–21b.

The weights applied to the downstream hydrograph ordinate O_1 can also be compared. For the Convex and MAK Methods, using Equation 10–21a gives

$$1 - C = 1 - \frac{2\Delta t}{2K + \Delta t} = \frac{K - 0.5\Delta t}{K + 0.5\Delta t} \tag{10–68}$$

The form of Equation 10–68 is similar to the form of Equation 10–54c; the two weights are equal when $x = 0$, and the two weights will be nearly equal when Kx is small compared with $(K - 0.5\Delta t)$, which is when x is near 0.

In terms of analysis, the Muskingum Method requires fitting values for the two coefficients x and K. The Convex and MAK Methods only require fitting the parameter C or C_m, respectively. Given the ease with which parameters can be fit by computer, the analysis of two parameters is just as easy as fitting the single routing coefficient. However, it is recognized that when fitting multicoefficient models in which the coefficients are not independent, the less sensitive coefficients may not approach their true value. When this occurs, the methods may not provide accurate predictions.

10.6.1 Application of the Routing Methods

In addition to comparing the methods on the basis of their conceptual framework, it is of interest to compare the routing methods using data. For purposes of comparison, a hypothetical channel, 5400 ft in length, will be used. A slope of 0.0015 ft/ft is assumed. The rating curve for the channel is

$$q = 0.4\,A^{4/3} \tag{10–69}$$

where q is the discharge in ft^3/sec and A is the cross-sectional area in ft^2. A roughness coefficient (n) of 0.06 is assumed. Computations will be based on a routing increment of 0.1 hr ($\Delta t = 0.1$ hr). The hydrograph at the upstream reach is given in Table 10–7 and shown in Figure 10–10.

To use the Convex Routing Method, it is necessary to make an estimate of the parameter C of Equation 10–14. The rating table of Equation 10–69 was used to find the cross-sectional area for the peak discharge of the hydrograph at the upstream cross section:

$$A = \left(\frac{q}{x}\right)^{1/m} = \left(\frac{924}{0.4}\right)^{0.75} = 333\text{ ft}^2 \tag{10–70}$$

TABLE 10–7 Computed Hydrographs at Downstream Cross Section Using the Convex, Modified Att-Kin, and Muskingum Methods

Time (hr)	Upstream Hydrograph (ft^3/sec)	Discharge (ft^3/sec) by method		
		Convex	Modified Att-Kin	Muskingum
0.0	7	0	0	0
0.1	14	7	7	7
0.2	17	8	9	8
0.3	20	10	11	10
0.4	24	12	13	12
0.5	29	15	16	14
0.6	34	18	19	17
0.7	42	22	23	20
0.8	56	27	29	25
0.9	85	35	38	31
1.0	196	54	60	45
1.1	379	98	109	80
1.2	618	172	194	142
1.3	821	273	307	235
1.4	917	383	429	343
1.5	924	483	535	445
1.6	882	560	615	528
1.7	821	614	666	588
1.8	754	646	692	627
1.9	688	660	698	647
2.0	627	660	689	651
2.1	570	648	670	645
2.2	518	629	643	629
2.3	471	604	611	608
2.4	429	576	576	582
2.5	391	545	540	553
2.6	358	513	504	523
2.7	328	482	469	493
2.8	301	451	436	462
2.9	277	421	404	433
3.0	258	392	374	404
3.1	236	365	347	377
3.2	219	340	321	352
3.3	204	316	297	327
3.4	191	294	276	305
3.5	179	274	256	284
3.6	168	255	238	265
3.7	158	238	222	248
3.8	148	223	207	231
3.9	140	208	193	216
4.0	132	195	181	202

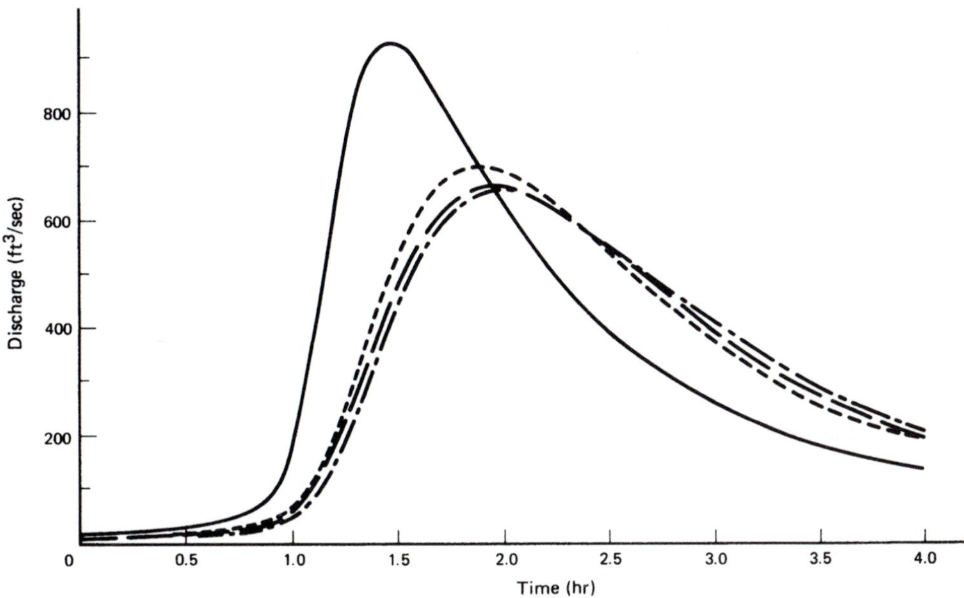

FIGURE 10–10 Hydrographs at upstream section (———) and the downstream section using the Convex (— —), modified Att-Kin (– – – –), and Muskingum (— - — - —) Methods.

Using the continuity equation, the average velocity is

$$V = \frac{Q}{A} = \frac{924\ \text{ft}^3/\text{sec}}{333\ \text{ft}^2} = 2.77\ \text{ft/sec} \tag{10–71}$$

Using the ratio of the reach length to the flow velocity as an estimate of K, the Convex routing coefficient can be computed by Equation 10–15:

$$C = \frac{\Delta t}{K} = \frac{\Delta t}{L/V} = \frac{0.1\ (3600)}{5400/2.77} = 0.185 \tag{10–72}$$

Thus Equation 10–14 becomes

$$O_{t+\Delta t} = CI_a + (1 - C)\ O_t = 0.185\ I_a + 0.815\ O_t \tag{10–73}$$

Equation 10–73 was used to derive the hydrograph at the downstream cross section, which is given in Table 10–7 and shown in Figure 10–10. Flow through the reach, with local inflow assumed to be 0, attenuated the peak flow rate by about 29%.

The value of the routing coefficient C_m for the modified Att-Kin Method can be estimated using Equations 10–21a and 10–22. The value of K is

$$K = \frac{L}{mV} = \frac{5400}{(4/3)\ (2.77)} = 1462\ \text{sec} \tag{10–74}$$

TABLE 10–8 Summary of Coefficients of Routing Equations

	Coefficient			
Method	$I_{t+\Delta t}$	I_t	O_t	Equation
Convex	0	0.185	0.815	10–73
Modified Att-Kin	0	0.22	0.78	10–75
Muskingum	0.040	0.137	0.823	10–79

and the routing coefficient is

$$C_m = \frac{2\,(360)}{2\,(1462) + 360} = 0.22 \tag{10–75}$$

Thus the routing equation is

$$O_{t+\Delta t} = 0.22\, I_a + 0.78\, O_t \tag{10–76}$$

Equation 10–82 was used to derive the hydrograph at the downstream end of the reach; the hydrograph is shown in Figure 10–10, with the ordinates given in Table 10–7.

To use the Muskingum Routing Equation (Equation 10–53), values must be provided for the routing coefficients, K and x. Equation 10–26b was used to estimate x:

$$x = \frac{1.49\, S^{0.5}}{nP^{2/3}} = \frac{1.49\,(0.0015)^{0.5}}{0.06[75 + 2(4.5)]^{2/3}} = 0.05 \tag{10–77}$$

The value of the coefficient K was set equal to the ratio of the reach length to the velocity:

$$K = \frac{L}{V} = \frac{5400}{2.77} = 1949 \text{ sec} = 0.5414 \text{ hr} \tag{10–78}$$

Based on these values of x and K, Equations 10–54 were used to derive values for C_0, C_1, and C_2, which results in the following routing equation:

$$Q_{t+\Delta t} = 0.04\, I_{t+\Delta t} + 0.137\, I_t + 0.823\, O_t \tag{10–79}$$

Equation 10–79 was used to compute the hydrograph at the downstream section, which is shown in Figure 10–10.

For this case, the three methods give reasonably similar computed hydrographs. The coefficients of the routing equations are summarized in Table 10–8.

10.7 ST. VENANT EQUATIONS

The routing methods discussed previously represent a class of methods that, while widely used, may not be as accurate or conceptually sound as we would like. Other more theoretically detailed routing methods exist. Assuming an open channel where lateral inflow can be ignored, the continuity equation for gradually varied, unsteady flow is

$$\frac{\partial A}{\partial t} + \frac{\partial Q}{\partial x} = 0 \tag{10–80}$$

in which A is the cross-sectional area of flow, Q is the discharge, and t and x are the time and the distance along the channel, respectively; both A and Q are a function of t and x. The conservation of mass indicated by Equation 10–1 can be derived by integrating Equation 10–80 between the two ends of the channel reach:

$$\int_1^2 \frac{\partial A}{\partial t}\, dx = -\int_1^2 \frac{\partial Q}{\partial x}\, dx \tag{10–81}$$

$$\frac{d}{dt}\int_1^2 A dx = -\int_1^2 \partial Q \tag{10–82}$$

$$\frac{dS}{dt} = Q_1 - Q_2 \tag{10–83}$$

in which Q_1 and Q_2 correspond to I and 0, respectively.

In addition to the conservation of mass, the conservation of momentum can be written after St. Venant:

$$\frac{1}{g}\frac{\partial V}{\partial t} + \frac{V}{g}\frac{\partial V}{\partial x} + \frac{\partial h}{\partial x} + (S_f - S_o) = 0 \tag{10–84}$$

in which V is the velocity, h is the depth of flow, and S_f and S_o are the friction and bed slopes, respectively. The terms of Equation 10–84 represent local acceleration, convective acceleration, hydrostatic pressure, and frictional and gravity forces, respectively; each of the terms are dimensionless. The first term ($\partial V/\partial t$) is important for unsteady, nonuniform flow. The second term ($\partial V/\partial x$) is important for steady nonuniform flow, and the third term ($\partial h/\partial x$) is significant for steady uniform flow. Thus, as the complexity of the flood routing problem increases, the additional terms must be evaluated in order to have an accurate solution.

Equations 10–80 and 10–84 represent a system of equations for gradually varied, unsteady flow in open channels in which lateral inflow is not significant. Since there is no known analytical solution, the equations must be solved numerically. The numerical solution requires an initial condition and boundary conditions. Finite-difference schemes are commonly used for the numerical solution; however, the numerical solution is beyond the intent here.

Equation 10–80 was shown to be the basis for the storage equation of Equation 10–1, which forms the basis for the simple routing methods described previously. Dooge et al. (1982) used the results of the solution of the hydrodynamic equation of motion as the basis of linking the parameters of the Muskingum Model with the hydraulic parameters of the open channel reach; one such solution was given in Equations 10–62 and 10–63.

10.8 KINEMATIC FLOOD ROUTING

Kinematic wave theory can be applied to shallow water flow, such as overland runoff from flat surfaces or flow in wide channels. For such cases, the friction slope is approximately equal to the bed slope ($S_o = S_f$ in Equation 10–84). This implies that the remaining terms of Equation 10–84 are not significant. The equality of the bed and friction slopes implies that the gravitational and shear forces are equal and that the discharge is proportional to the depth of flow:

$$q = ah^b \tag{10–85}$$

in which a and b are rating curve coefficients that reflect cross-section properties. Calculation of a and b was discussed in Section 10.4.1. The simultaneous solution of Equations 10–83 and 10–85 represents the kinematic wave solution.

10.8.1 Kinematic Wave Routing of Sheet Flow

Sheet flow can be described by the continuity equation for two-dimensional flow:

$$\frac{\partial q}{\partial x} = i_e^* - \frac{\partial y}{\partial t} \tag{10–86}$$

in which q is the discharge per unit width, i_e^* is the excess inflow rate in flow rate per unit area, y is the flow depth at any point at distance x from the top of the sloping surface, and t is time. Two conditions are important to the solution. First, before equilibrium occurs, $\partial q/\partial x = 0$, which yields the following constraint: $y = i_e^* t_e$. This implies that the flow velocity equals the excess inflow rate i_e^* which introduces increasingly greater error in the solution as the wave moves down the surface. The second condition is that $\partial y/\partial t = 0$ at equilibrium. At the time of peak, this condition should be realistic at the end of the slope.

The solution also uses the conservation of mass: total inflow equals the sum of the total outflow and the water stored on the surface, which in the literature is referred to as detention storage. At the time of equilibrium (t_e), the total outflow equals $i_e^* L t_e$. The outflow is assumed to be a proportion β of the inflow. The detention storage (D_e) at equilibrium is assumed to equal the volume of water still on the surface. Thus, the conservation of mass is

$$i_e^* t_e L = \beta i_e^* t_e L + D_e \tag{10–87}$$

Solving Equation 10–87 for the equilibrium time yields

$$t_e = \frac{D_e}{i_e^* L (1 - \beta)} \tag{10–88}$$

The depth-discharge relationship is assumed as

$$q = ay^b \tag{10–89}$$

A linear depth-time relationship for rainfall excess is assumed, which at equilibrium requires $i_e^* = y_e/t_e$. The resulting outflow hydrograph has no attenuation and leads to the conclusion that the constant β is only a function of the coefficient b of Equation 10–89. Consequently, the detention storage is given by

$$D_e = \frac{b\,(i_e^*)^{1/b}\, L^{1+1/b}}{a^{1/b}(1 + b)} \tag{10–90}$$

which can be substituted into Equation 10–88 to compute the equilibrium time. Using Manning's Equation to define the coefficients of the depth-discharge relationship of Equation 10–89 and adjusting for the units of the variables yields

$$D_e = \frac{0.000818\, i^{0.6} n^{0.6} L^{1.6}}{S^{0.3}} = 0.000818\, L i^{0.6} \left(\frac{nL}{\sqrt{S}}\right)^{0.6} \tag{10–91}$$

Equation 10–91 is intended to give the detention storage at equilibrium. The use of Equation 10–91 carries with it the assumption that there is no attenuation at the end of the slope, although this does not agree with the empirical evidence of Izzard (1946) and others. Crawford and Linsley (1966) indicated that Equation 10–91 provided good estimates of detention only for smooth surfaces. However, Crawford and Linsley (1966) also indicated that the ratio of detention storage for turbulent flow to laminar flow (D_{et}/D_{el}) on a paved surface equals 1 for iL of 260, which is below Izzard's limit of 500. Even for iL equal to 2000, they indicated that the ratio D_{et}/D_{el} is less than 1.75. Thus, detention is not overly sensitive to changes in iL, and the concepts may still provide reasonable approximations for turbulent flow. Chow (1964) stated that applications of these concepts indicated that Equation 10–91 was satisfactory for turbulent flow, which supports the findings of Crawford and Linsley.

As an alternative to Equation 10–91, Izzard (1946) developed the following relationship for detention storage:

$$D_e = 0.0285\,(0.0007\, i + C)\, L \left(\frac{Li}{S}\right)^{1/3} \tag{10–92}$$

in which C is a coefficient that varies with the roughness of the surface. Use of Equation 10–92 produces attenuation of the sheet-flow runoff hydrograph. Izzard indicated that Equation 10–92 is accurate where the product iL is less than about 500.

An algorithm can be formulated based on the above equations. Izzard's empirical equation (Equation 10–92) for detention is used in place of the theoretical model of Equation 10–91. The time of concentration can be used as the equilibrium time. Sheet-flow discharge rates can be determined from the following (Viessman et al., 1996):

$$q = \frac{1.486}{n} S^{1/2} \left(\frac{D}{L}\right)^{5/3} \left(1.0 + 0.6\left(\frac{D}{D_e}\right)^3\right)^{5/3} \tag{10–93}$$

in which D is the average detention during the routing interval. The continuity of mass can use i_e^* as the inflow, q as the outflow, and storage defined by Equation 10–92.

10.9 MUSKINGUM-CUNGE METHOD

Cunge developed a routing method that is more flexible than the hydrologic routing methods discussed in Sections 10.3, 10.4, and 10.5. It can be shown that for certain conditions, Cunge's development is equivalent to the Muskingum Method, thus it is referred to as the Muskingum-Cunge Method. Variations of the Muskingum-Cunge Method, most notably by Ponce and Yevjevich (1978), have been published. The Muskingum-Cunge Method uses physical characteristics to compute the coefficients, rather than the somewhat arbitrary, empirical approach used for estimating x for the Muskingum Method. Specifically, the Muskingum-Cunge Method uses the reach length L and slope S_o, the kinematic wave celerity c, and a characteristic unit discharge q_o. The kinematic wave celerity is defined as

$$c = \frac{dq/dh}{W_t} \tag{10–94}$$

in which W_t is the top width of the water surface. The numerator of Equation 10–94 is the slope of the stage-discharge relationship, which should be evaluated at the water surface elevation. As an alternative, the celerity is assumed to equal pV, where V is the velocity and p is the power-model exponent of the discharge cross-sectional area curve (see Equation 10–10). Manning's Equation provides a value of 5/3 for p. The characteristic unit discharge is the actual discharge per unit width at any time, although the mean or peak discharge can be used as an approximation.

Cunge provided the following expression for estimating x of Equations 10–51 and 10–54:

$$x = 0.5\left(1 - \frac{q_o}{c\,S_o\,L}\right) \tag{10–95}$$

where L is the reach length. The Muskingum-Cunge Method uses the following expression for estimating K:

$$K = \frac{L}{c} \tag{10–96}$$

where c is computed from Equation 10–94. The equation for routing is the same as that used for the Muskingum Method except it is most often expressed in finite-difference form:

$$Q_{i+1,\,t+1} = C_o\,Q_{i,\,t+1} + C_1\,Q_{i,\,t} + C_2\,Q_{i+1,\,t} \tag{10–97}$$

where i is the station indicator (i = upstream section, $i + 1$ = downstream section) and t is the time indicator ($t + 1$ = next time, t = current time); thus, in comparing Equation 10–53 with Equation 10–97, the following shows that the quantities differ only in notation:

$$Q_{i+1,t+1} = O_2$$
$$Q_{i,t+1} = I_2$$
$$Q_{i,t} = I_1$$
$$Q_{i+1,t} = O_1$$

The equations for computing C_0, C_1, and C_2 for the Muskingum-Cunge Method are identical to those for the Muskingum Method, although different notation is often used.

The difference in the Muskingum and Muskingum-Cunge Method lies in the way that the parameters are evaluated and the equations applied. If constant values of c and q_o are used for all ordinates of the hydrograph, then K and x are constant for all computations, and the computed downstream hydrograph will be identical to that computed using the Muskingum Method with the same parameters. However, if c is computed for each ordinate, using Equation 10–94 and q_o is the discharge for that ordinate, then x and K will vary with each time step and thus the computed hydrograph will differ from that of the one computed using the Muskingum Method.

If the coefficients are allowed to vary with the discharge, then dq/dh and W_t of Equation 10–94 and q_o of Equation 10–95 will vary with both time and cross section. Thus, there

TABLE 10–9 Example 10–7: Three-Point Muskingum-Cunge Routing

(1) Time	(2) $U(t)$	(3) h_1	(4) h_2	(5) w_1	(6) w_2	(7) C_1	(8) C_2	(9) q_{o1}	(10) q_{o2}	(11) x_1	(12) x_2	(13) K_1	(14) K_2	(15) x	(16) K	(17) C_0	(18) C_1	(19) C_2	(20) $D(t)$
0	1000.	3.96	3.96	212.	212.	2.98	2.98	4.72	4.72	.412	.412	3022.	3022.	.408	2830.	.186	.850	–.036	1000.
60	1400.	4.53	4.08	215.	213.	3.58	3.10	6.50	5.05	.399	.409	2512.	2905.	.397	2419.	.258	.847	–.104	1074.
120	2100.	5.33	4.80	220.	217.	4.48	3.88	9.55	7.44	.382	.393	2010.	2322.	.378	1929.	.357	.843	–.200	1614.
180	3200.	6.31	5.80	225.	222.	5.65	5.03	14.25	11.66	.360	.371	1594.	1791.	.360	1591.	.435	.842	–.277	2590.
240	3900.	6.83	6.67	227.	226.	6.29	6.09	17.20	16.25	.348	.352	1430.	1477.	.347	1413.	.481	.841	–.322	3674.
300	4400.	7.17	7.04	228.	228.	6.73	6.57	19.28	18.51	.341	.343	1338.	1370.	.343	1366.	.494	.841	–.335	4213.
360	4100.	6.97	7.11	227.	228.	6.47	6.65	18.03	18.93	.345	.342	1391.	1352.	.346	1403.	.484	.841	–.325	4314.
420	3700.	6.69	6.78	226.	227.	6.12	6.24	16.36	16.93	.351	.349	1472.	1443.	.354	1508.	.456	.841	–.298	3837.
480	3100.	6.23	6.45	224.	225.	5.55	5.82	3.83	15.03	.362	.357	1622.	1545.	.364	1656.	.420	.842	–.261	3385.
540	2500.	5.72	5.96	222.	223.	4.93	5.22	1.27	12.44	.373	.368	1826.	1724.	.374	1846.	.375	.843	–.218	2774.
600	2100.	5.33	5.52	220.	221.	4.48	4.70	9.55	10.37	.382	.377	2010.	1916.	.382	2013.	.339	.844	–.182	2290.
660	1900.	5.12	5.23	219.	219.	4.24	4.36	8.68	9.11	.386	.384	2123.	2066.	.386	2124.	.316	.844	–.160	1998.

(1) Time (minutes)
(2) Upstream hydrograph (cfs)
(3) Upstream depth (ft)
(4) Downstream depth (ft)
(5) Upstream top width (ft)
(6) Downstream top width (ft)
(7) Upstream celerity (ft/sec)
(8) Downstream celerity (ft/sec)
(9) Upstream unit discharge (cfs/ft)
(10) Downstream unit discharge (cfs/ft)
(11) Upstream coefficient x
(12) Downstream coefficient x
(13) Upstream coefficient K
(14) Downstream coefficient K
(15) Reach average coefficient x
(16) Reach average coefficient K
(17) Reach average coefficient C_0
(18) Reach average coefficient C_1
(19) Reach average coefficient C_2
(20) Downstream hydrograph (cfs)

would be four values, one corresponding to each of the discharges of Equation 10–97. When computing $Q_{i+1,t+1}$, values of $\partial q/\partial h$, W_t, and q_o can be computed for $Q_{i,t+1}$, $Q_{i,t}$, and $Q_{i+1,t}$; however, since $Q_{i+1,t+1}$, is not known initially, the values cannot be computed for that point. There are two options. First, values of dq/dh, W_t, and q_o could be computed for the three known conditions and averaged to compute c and x. Second, the values for the three sections can be averaged as in the first option, which are then used to compute an initial estimate of $Q_{i+1,t+1}$, which itself is used to compute values of c and x. The four values are then used to compute averages of c and x, which are then used to recompute $Q_{i+1,t+1}$. The two options will not usually yield significantly different computed hydrographs; thus, the second option is not usually worth the additional effort.

Example 10–7

A portion of a hydrograph is routed using the 3-point algorithm. The details are given in Table 10–9. The 9000-ft channel has a slope of 0.001 ft/ft. The stage-discharge relationship is

$$Q = 32\, h^{2.5}$$

where Q [=] ft^3/sec and h [=] ft. This yields the following:

$$\frac{dQ}{dh} = 80\, h^{1.5}$$

The top width-discharge relationship is

$$W = 150\, Q^{0.05}$$

in which W is the top width (ft). The unit discharge is computed by dividing the discharge by the top width. Equations 10–95 and 10–96 are used to compute values of x and K for each cross section and each depth. The average value of each time period is computed with the three individual

TABLE 10–10 Example 10–7: Computed Downstream Hydrographs for Muskingum, Three-point Muskingum-Cunge, and Four-point Muskingum-Cunge Routing

		Computed Downstream Hydrographs		
Time (min)	Upstream Hydrograph (cfs)	Muskingum (cfs)	3-Point M-C (cfs)	4-Point M-C (cfs)
0	1000	1000	1000	1000
60	1400	1159	1074	1073
120	2100	1736	1614	1619
180	3200	2625	2590	2602
240	3900	3617	3674	3681
300	4400	4167	4213	4213
360	4100	4336	4314	4314
420	3700	3884	3837	3838
480	3100	3417	3385	3387
540	2500	2785	2774	2777
600	2100	2272	2290	2292
660	1900	1979	1997	1998
720	1800	1841	1853	1853

values corresponding to $Q_{i,t+1}$, $Q_{i,t}$, and $Q_{i+1,t}$. The average values of x and K are used with Equations 10–54 to compute the routing coefficients. The routed hydrograph is given in column 20 of Table 10–9.

The same upstream hydrograph was routed using the Muskingum and the four-point Muskingum-Cunge algorithms. The mean discharge from the upstream section was used as the reference discharge for the Muskingum routing; then Equations 10–95 to 10–97 were used to compute the coefficients. The three computed hydrographs for the downstream section are given in Table 10–10. In general, there is little variation in the resulting hydrographs.

PROBLEMS

10–1. Which of the following are problems of analysis and which are problems of synthesis:
(a) Given measured upstream and downstream hydrographs; find the channel roughness coefficient.
(b) Given a design-storm upstream hydrograph and channel characteristics, compute the downstream hydrograph.
(c) Given the measured upstream hydrograph and the routing coefficients, compute the downstream hydrograph.
(d) Given an upstream design-storm hydrograph and the downstream hydrograph computed using the St. Venant equation, find the convex routing coefficient.

10–2. The following shows the upstream (I) and downstream (O) hydrographs for a storm event. Assuming a constant time increment, compute the change in storage and the total accumulated storage at each time interval. On the same graph, plot the two hydrographs and the incremental change in storage. Where does the storage reach a maximum?

I	0	2	6	10	16	20	22	20	18	16	12	8	6	4	2	0	0	0	0	0	0
O	0	0	1	3	6	10	14	16	18	18	16	14	12	10	8	6	5	3	1	1	0

10–3. The following stage and discharge measurements were made at a channel section. Compute the parameters of the rating curve $q = ah^d$.

h (ft)	1.4	1.8	1.9	2.3	2.7	2.7	3.1	3.6
q (ft^3/sec)	50	120	190	320	640	710	1230	2190

10–4. The following stage and discharge measurements were made at a channel section. Compute the parameters of the rating curve $q = ah^d$.

h (ft)	0.8	1.1	1.2	1.5	1.7	2.1	2.1	2.4	2.9
q (ft^3/sec)	8	19	34	66	84	157	181	277	475

10–5. A rectangular channel section has a width of 8 ft and is at a slope of 0.8%. The roughness coefficient at the section is 0.031. Compute and plot the rating table for depths up to 3 ft.

10–6. The following are characteristics of a rectangular channel section: width = 6 ft; slope = 1.2%; roughness coefficient = 0.046. Compute and plot the rating table for depths up to 3.0 ft.

10–7. A rectangular channel section has a width of 80 ft, a roughness coefficient of 0.04, and a slope of 0.005. Use Equation 10–8 to compute a rating table for depths up to 6 ft. Plot the rating table.

10–8. Develop a discharge cross-sectional area rating curve, $q = cA^b$, using Manning's Equation. Assume a rectangular channel.

10–9. Discuss the rationale of C from Equation 10–15 as a weight applied to ordinates of the upstream hydrograph I.

10–10. Assuming that $C = 0.25$, compute the downstream hydrograph using the Convex Method for the following upstream hydrograph: (0, 12, 27, 56, 53, 47, 38, 23, 15, 8, 4, 0) ft^3/sec. Assume that the initial discharge at the downstream section is 0 ft^3/sec.

10–11. Assuming that $C = 0.40$, compute the downstream hydrograph using the Convex Method for the following upstream hydrograph: (0, 7,18, 39, 46, 45, 38, 29, 17, 9, 3, 0) ft^3/sec. Assume that the initial discharge at the downstream section is 0 ft^3/sec.

10–12. The following hydrograph (ft^3/sec) measured at the upstream section of a 12,000-ft section is given for a 30-min time increment. The flow through the reach has an average velocity of 4 ft/sec. Using the Convex Routing Method, compute the hydrograph at the downstream end of the reach: (0, 210, 840, 1350, 1520, 1460, 1210, 790, 540, 370, 190, 80, 40, 15). Assume that the initial discharge at the downstream section is 0 ft^3/sec.

10–13. The following hydrograph (ft^3/sec) was measured at the upstream section of a river, with the data recorded on a 0.4-hr time increment. Route the hydrograph using the Convex Method through two reaches that have the following characteristics: reach 1 (length = 7200 ft, velocity = 2.5 ft/sec); reach 2 (length = 6480 ft, velocity = 1.8 ft/sec). The upstream hydrograph ordinates are: (0, 210, 530, 840, 920, 870, 610, 380, 190, 80, 40, 20). Assume that the initial discharge at the downstream sections is 0 ft^3/sec.

10–14. Using the principle of least squares, derive the equation for computing the routing coefficient C of the Convex Method (Equation 10–17).

10–15. The following hydrographs (ft^3/sec) are for the upstream (I) and downstream (O) sections of a river. Estimate the routing coefficient C of the Convex Method.

I (ft^3/sec)	0	210	530	840	920	870	610	380	190	80	40	20
O (ft^3/sec)	0	10	90	230	480	620	710	690	540	440	270	190

10–16. The following hydrographs (ft^3/sec) are for the upstream (I) and downstream (O) sections of a river. Estimate the routing coefficient C of the Convex Method.

I (ft^3/sec)	0	55	105	170	205	195	160	115	85	60	25	10	5
O (ft^3/sec)	0	5	10	30	70	100	125	130	115	110	100	75	65

10–17. A rectangular channel with $n = 0.035$ has a bed slope of 0.5%. The average hydraulic radius is 1 ft. The following flood hydrograph was measured during a storm at a downstream section, with a time increment Δt of 0.5 hr: (10, 10, 10.6, 12.2, 14, 15, 14.9, 14.6, 14.1). The flood had caused damage at a site 18,000 ft upstream. Estimate the flood hydrograph at the upstream section using the Convex Method.

10–18. A portion of a river is modeled with two reaches, A and B. Reach A extends from section 1 to section 2 and has a length of 15,000 ft, $n = 0.07$, $S = 1\%$, and $R_h = 4$ ft. Reach B extends from section 2 to section 3 and has a length of 17,000 ft, $n = 0.05$, $S = 0.5\%$, and $R_h = 5$ ft. The hydrograph at section 1 is: (200, 300, 700, 600, 500, 400, 350, 300, 250, 225, 210, 205, 200), with a time increment Δt of 10 min. Assume the initial discharge at downstream sections is

200. Using the Convex Routing Method: (a) route the hydrograph from section 1 to section 2 and then from section 2 to section 3; (b) route the hydrograph from section 1 to section 3 using weighted averages of the n, S, and R_h; (c) compare the two computed hydrographs at section 3 and provide an explanation for the differences.

10–19. Determine the downstream hydrograph for Example 10–4 using the modified Att-Kin Method and a reach length of 10,000 ft if the rating table coefficients are $x = 0.075$ and $m = 1.8$.

10–20. Using the upstream hydrograph of Problem 10–12 and the modified Att-Kin Method, route the hydrograph through a channel that has a length of 7500 ft. Use a time increment of 15 min and assume rating table coefficients of $x = 0.12$ and $m = 1.5$.

10–21. Using the upstream hydrograph of Problem 10–11 and the modified Att-Kin Method, route the hydrograph through a channel having a length of 2200 ft. Assume an average velocity of 5 ft/sec. Use a time increment of 0.1 hr and assume rating table coefficients of $x = 0.15$ and $m = 1.4$.

10–22. Route the hydrograph of Problem 10–18 through reach A using the modified Att-Kin Method. Assume rating table coefficients of $x = 0.11$ and $m = 1.6$.

10–23. Route the hydrograph of Problem 10–10 using the Muskingum Method with $x = 0.25$, $K = 0.75$, and $\Delta t = 0.25$. Assume that the initial discharge at the downstream section is 0 ft^3/sec.

10–24. Route the hydrograph of Problem 10–11 using the Muskingum Method with $C_0 = -0.1$, $C_1 = 0.6$, $C_2 = 0.5$. Assume that the initial discharge at the downstream section is 0 ft^3/sec. Use $\Delta t = 10$ min.

10–25. Assuming a routing time increment Δt of 15 min, determine the routing coefficient x and K for the Muskingum Method and the hydrographs of Problem 10–16.

10–26. Assuming a routing time increment Δt of 15 min, determine values of the routing coefficients x and K for the Muskingum Method and the hydrographs of Problem 10–15.

10–27. Solve Problem 10–18 using the Muskingum Method assuming $x = 0.2$. Use $\Delta t = 10$ min.

10–28. Route the following measured flows through a 2-mi reach that has a slope of 0.003 ft/ft, $n = 0.045$, a stage-discharge curve of $Q = 26h^{2.4}$, and a top width-discharge curve of $W = 135Q^{0.06}$.

$$Q = \{800, 925, 1350, 2120, 2080, 1870, 1710, 1630, 1560, 1490\}$$

Assume $\Delta t = 0.8$ hr and the initial downstream flow is 800 ft^3/sec. Compare the Muskingum and 3-point Muskingum-Cunge computed hydrographs.

10–29. Route the following hydrograph through a reach with length = 13,000 ft, slope = 0.005 ft/ft, $n = 0.05$, a stage-discharge curve of $Q = 30h^{2.6}$, a top width-discharge curve of $W = 130Q^{0.07}$, and routing interval of 0.75 hrs:

$$Q = \{1200, 1550, 2410, 3630, 4150, 3990, 3620, 3250, 2980, 2760\}$$

Assume the initial downstream flow is 1200 ft^3/sec. Compare the Muskingum and 3-point Muskingum-Cunge computed hydrographs.

REVIEW QUESTIONS

10–1. In channel routing analysis, which one of the following is the unknown? (a) The downstream hydrograph; (b) the routing equation; (c) the channel characteristics; (d) the routing coefficients; (e) all of the above are known.

10–2. Channel routing methods are often based on (a) the continuity of mass; (b) the continuity equation; (c) Darcy's law; (d) Bernoulli's equation.

10–3. A rating curve relates (a) discharge to velocity; (b) velocity to cross-sectional area; (c) discharge to depth of flow; (d) discharge to storage; (e) none of the above.

10–4. When applying channel routing methods, the routing interval should be (a) approximately equal to the reach travel time; (b) greater than the time between the times to peak of the upstream and downstream hydrographs; (c) about 20% of the time to peak of the upstream hydrograph; (d) greater than the time interval on which the hydrograph ordinates are recorded.

10–5. Which one of the following would not affect the value of the routing coefficient of the Convex Method? (a) The velocity of flow in the stream reach; (b) Manning's Roughness Coefficient of the stream reach; (c) the time increment used for routing; (d) the time to peak of the upstream hydrograph.

10–6. In using regression analysis to fit the routing coefficient of the Convex Method, which one of the following is needed? (a) The reach travel time; (b) the downstream hydrograph; (c) the hydraulic radius of the reach; (d) the length of the reach; (e) all of the above.

10–7. Which one of the following is provided by the modified Att-Kin Method but not the Convex Method? (a) Hydrograph attenuation; (b) conservation of momentum; (c) hydrograph translation; (d) continuity of mass.

10–8. Unlike the Convex and modified Att-Kin Methods, routing with the Muskingum Method makes use of (a) the travel time of the flow in the reach; (b) the ordinate of the upstream hydrograph for the current time; (c) the continuity equation; (d) the routing increment.

10–9. The St. Venant Equations (a) assume no lateral inflow; (b) are valid for steady flow only; (c) do not require continuity of momentum; (d) require use of the linear storage equation; (e) none of the above.

DISCUSSION QUESTION

The technical content of this chapter is important to the professional hydrologist, but practice is not confined to making technical decisions. The intent of this discussion question is to show that hydrologists must often address situations where value issues intermingle with the technical aspects of a project. In discussing the stated problem, at a minimum include responses to the following questions:

1. What value issues are involved, and how are they in conflict?
2. Are technical issues involved with the value issues? If so, how are they in conflict with the value issues?
3. If the hydrologist attempted to rationalize the situation, what rationalizations might he or she use? Provide arguments to suggest why the excuses represent rationalization.
4. What are the hydrologist's alternative courses of action? Identify all alternatives, regardless of the ethical implications.
5. How should the conflict be resolved?

You may want to review Sections 1.6 to 1.12 in Chapter 1 in responding to the problem statement.

Case. An engineer is in charge of the hydraulics and hydrology division of his company. It is his responsibility to meet with clients and establish the costs and technical details of all H&H projects. Since he has worked for the firm for over twenty years, he has considerable flexibility in choosing work for his company. Because of his reputation, he works with very little oversight by the company president. In recent months, the engineer has given prospective clients a very high price tag on projects. When the clients indicate that they will take their work elsewhere, the engineer gives the prospective clients his personal business card and asks them to call him at home that evening, with the remark that he is sure a more favorable price can be negotiated with him personally. The effect on the company's bottom line due to this practice is minimal, although the amount of work in the H&H division declines slightly.

11

Reservoir Routing

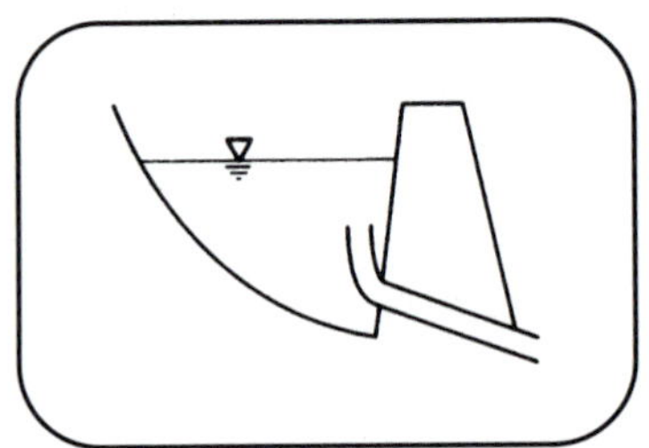

CHAPTER OBJECTIVES

1. Develop the routing equation for use in routing a flood hydrograph through a reservoir.
2. Discuss the derivation of a stage-storage-discharge relationship.
3. Introduce commonly used storage routing methods: the storage-indication and modified Puls Method.
4. Introduce the procedures for designing the outlet facility of a reservoir.

11.0 NOTATION

A = length of major axis of elliptical basin
A_b = bottom area of storage basin
A_i = surface area of facility at depth i
A_o = area of orifice
A_2 = minimum area of jet discharging from orifice
B = length of minor axis of elliptical basin
C_d = coefficient of discharge through orifice
D = diameter of outlet pipe
E_1 = maximum stage at low-stage control
E_2 = maximum stage at high-stage control
g = acceleration of gravity = 32.2 ft/sec^2
h = depth of water above bottom of storage facility
h_o = elevation of the bottom of the riser
h_1 = elevation of the weir
H_o = maximum height of the riser
H_1 = maximum elevation of water for low-stage design
H_2 = maximum elevation of water for high-stage design
I = inflow
L = bottom length of storage facility
L = weir length
L_w = length of weir
O = outflow
p_{atm} = atmospheric pressure
q_o = discharge per unit length of weir
q_o = target discharge in design of detention basin
Q = discharge
Q_r = orifice flow in two-stage riser
r = ratio of A to B
s_b = slope in direction of minor axis of elliptical basin
S = storage
t = time
V = flow velocity
V_o = velocity of flow through orifice
V_s = storage volume
W = bottom width of storage facility
Z = depth of flow over the weir
Z = elevation head
α = angle of side slopes of storage basin
γ = specific weight of water
ΔS = change in storage
Δt = time interval for routing

11.1 INTRODUCTION

Flood routing refers to the process of calculating the passage of a flood hydrograph through a system. If the system is a channel, the general term flood routing is replaced with the specific term channel or streamflow routing. If the system is a reservoir, the term storage routing or reservoir routing is applied. Channel routing was the subject of Chapter 10, and storage routing is the topic of this chapter.

Storage routing is similar in many respects to channel routing. The inflow hydrograph to the storage reservoir corresponds to the hydrograph at the upstream cross section. The hydrograph of the outflow from the reservoir corresponds to the hydrograph at the downstream cross section. Just as the storage in a channel reach affects the shape of the downstream hydrograph, the storage in a reservoir also affects the characteristics of the outflow hydrograph. As will be shown, the conservation-of-mass equation that was used to develop the routing equation for channel flow is also used in the development of the storage routing equation.

A number of design (or synthesis) problems depend on the routing of a flood hydrograph through a storage facility. In the past, the greatest use of storage routing methods had been for the design of large multipurpose dams. These dams are used for flood control, power generation, to provide for recreation, and to provide a water supply for irrigation. In such designs, the sizing of the spillway requires an understanding of the interrelationships between the storage characteristics at the site and both the physical and energy characteristics of the spillway.

Knowledge of storage routing has been most necessary in design problems involving the creation of detention facilities used to control runoff from development taking place on small urban and suburban watersheds. Similar to the design of a large dam, the design of a small detention facility for control of urban flooding involves the sizing of the outlet facility. Several hundred thousand small dams have been constructed to provide water supply for farms; the water is used for irrigation and livestock, as well as water supply. Intermediate-sized dams are located close to almost all large urban areas to provide a public water supply. These examples suggest that the design of storage facilities is a common hydrologic design problem.

Storage routing is also used extensively in conducting watershed studies for watersheds where one or more storage facilities exist. Specifically, for watersheds in which existing reservoirs are located, a storage routing method is necessary to evaluate alternative watershed development plans. These plans may be necessary in order to evaluate the effects of such watershed-wide alternatives as (1) zoning plans, (2) location of water supply structures, or (3) regional flood control measures, as well as planning for ultimate watershed development conditions.

11.1.1 Analysis Versus Synthesis

In comparison to other hydrologic problems, storage routing is relatively complex. The following variables are involved:

1. The input (upstream) hydrograph
2. The output (downstream) hydrograph
3. The stage-storage relationship
4. The energy loss (weir and/or orifice) coefficients
5. Physical characteristics of the outlet facility (weir length, diameter of the riser pipe, length of the discharge pipe)
6. The storage volume-versus-time relationship
7. The depth-discharge relationship
8. The target peak discharge allowed from the reservoir.

The problem is further complicated in that the inflow and outflow hydrographs can be from either storms that have occurred (actual measured events) or from design storm models.

In the analysis case, the hydrographs (inflow and outflow) are measured runoff events. The stage-storage relationship is known, as it can be determined from site characteristics.

Since the storage facility actually exists, the physical characteristics of the outlet facility are also known. The variables that are not known are the loss coefficients and the stage-discharge relationship, which depend on the loss coefficients. Thus the objective in analysis is to determine values of the loss coefficients.

In the synthesis or design case, the inflow hydrograph is usually derived from a design storm model, although the design could be based on a measured hydrograph for a severe storm. The outflow hydrograph is not known, although a target peak discharge, such as the predevelopment peak discharge, is known. The stage-storage relationship is known, and values for the loss coefficients are assumed. The design problem would be to determine (1) the physical characteristics of the outlet facility, which would determine the outflow hydrograph that meets the target discharge, (2) the storage volume-versus-time relationship, and (3) the depth-discharge relationship.

Given the importance of detention basin design to practicing engineers, it serves as a useful example that illustrates the difference between analysis and design (synthesis). For our purpose, a detention basin with a two-stage riser will be the problem. In systems analysis, the input and output functions are known and the transfer function is unknown. One problem in detention basin analysis is the evaluation of the discharge coefficients (weir and orifice loss coefficients) of a two-stage riser. While values for single-stage risers are available, at the present time, accurate estimates of the discharge coefficients for two-stage risers are not available. The accuracy of single-stage riser coefficients in the design of two-stage risers is questionable because experimental measurements have not shown the degree to which flow over the high-stage weir interferes with the flow through the low-stage orifice. Some design engineers assume that flow from the high-stage outlet drowns out all flow from the low-stage outlet. Other designers assume that the two outlets do not interfere with each other. Thus a valid analysis study would be to build a series of outlet facilities with varying designs of the high- and low-stage outlets and analyze inflow and outflow hydrographs to find accurate estimates of the weir/orifice coefficients of the multistage riser designs.

For the design or synthesis problem, the physical characteristics of the outlet facility would be evaluated for a particular site using the weir/orifice coefficients computed in the analysis phase. When using weir/orifice coefficients in design, the designer assumes that the values of the coefficients are accurate. Although this may not be the case, an analysis of the coefficients, sometimes called a calibration, would be prohibitive in cost, especially when one considers that this would involve constructing the facility and then making any necessary retrofitting. It is beyond the scope of this chapter to provide further discussion of the analysis phase; the remaining portion will deal with the design phase.

11.1.2 Watershed Planning and Legal Case Studies

In addition to the analysis and design problems, two other types of hydrologic studies occur frequently, the watershed planning study and studies required in legal conflicts. Watershed planning studies are often undertaken to assess the hydrologic effects of alternative patterns of land development. These are usually undertaken by county or local governments or by regional planning councils. Such studies often use the design-storm concept. Both existing and planned storage facilities may be involved. In both cases, the objective is to predict the outflow hydrographs for the ultimate conditions of watershed development.

Where an existing facility is involved, both the physical characteristics of the outlet and the loss coefficients are known. For a planned facility, a stage-storage-discharge relationship is assumed.

For legal cases that involve flooding during an actual storm event, measured rainfall and runoff data would have to be obtained. The data could be used with a watershed model that includes a component for reservoir routing. The effect of a reservoir on downstream flooding might be important in the resolution of the problem. The measured rainfall and the watershed model can be used for the hydrologic part and a hydraulic water surface profile model, such as the Corps of Engineers HEC-2 model, could be used to delineate the water surface profiles.

11.2 THE ROUTING EQUATION

The equation for storage routing is based on the conservation of mass. Specifically, the inflow (I), outflow (O), and storage (S) are related by

$$I - O = \frac{dS}{dt} \tag{11–1}$$

where dS is the change in storage during the time interval dt. Both I and O are time-varying functions, with I and O being the inflow and outflow hydrographs, respectively. Approximating dS/dt by $\Delta S/\Delta t$, Equation 11–1 can be rewritten as

$$I\Delta t - O\Delta t = \Delta S \tag{11–2}$$

If the subscripts 1 and 2 are used to indicate values at times t and $(t + \Delta t)$, respectively, then Equation 11–2 can be written as

$$\frac{1}{2}(I_1 + I_2)\,\Delta t - \frac{1}{2}(O_1 + O_2)\,\Delta t = S_2 - S_1 \tag{11–3}$$

While I_1, I_2, O_1, and S_1 are known at any time t, values for O_2 and S_2 are unknown. Equation 11–3 can, therefore, be rearranged such that the knowns are placed on the left side of the equal sign and the unknowns on the right side:

$$\frac{1}{2}(I_1 + I_2)\,\Delta t + \left(S_1 - \frac{1}{2}O_1\Delta t\right) = \left(S_2 + \frac{1}{2}O_2\Delta t\right) \tag{11–4}$$

Equation 11–4 represents one equation with two unknowns (S_2 and O_2). To find a solution to the problem, it is necessary to have a second equation. In storage routing the storage-discharge relationship is the second equation. The outflow is usually assumed only to be a function of the storage. Although this may not be exactly true for small reservoirs, it is the assumption that is usually made. In practice, the storage-discharge relationship is derived from the stage-discharge and stage-storage relationships.

In summary, Equation 11–4 represents the equation for routing a hydrograph through a storage facility. Given the stage-storage-discharge relationship for a site and the facility, the outflow hydrograph can be computed with Equation 11–4.

11.3 DERIVATION OF A STAGE-STORAGE-DISCHARGE RELATIONSHIP

Routing a hydrograph through a reservoir or detention structure requires the relationship between the stage, storage, and discharge to be known. The stage-storage-discharge (SSD) relationship is a function of both the topography at the site of the storage structure and the characteristics of the outlet facility. The topographic features of the site control the relationship between stage and storage, and the relationship between stage and discharge is primarily a function of the characteristics of the outlet facility. Although it may appear that the SSD relationship consists of two independent functions, in practice it is important to view the relationship between stage, storage, and discharge as a single function because changes can be made to the topography at the site in order to change the stage-discharge relationship. However, to introduce the concept, the two will be treated separately.

11.3.1 The Stage-Storage Relationship

The stage-storage relationship depends on the topography at the site of the storage structure. Consider the unrealistic case of a site where the topography permits a storage structure that has a horizontal rectangular bottom area with vertical sides. In this case, the storage is simply the bottom area (length times width) multiplied by the depth of storage. If the relationship is plotted on a Cartesian axis system with storage as the ordinate and stage as the abscissa, the stage-storage relationship will be a straight line with a slope equal to the surface area of the storage facility. For the hypothetical case where the bottom of the storage facility is a rectangle ($L \times W$), the longitudinal cross section is a trapezoid with base W and side slopes of angle α, and the ends are vertical, the stage-storage relationship is given by

$$S = \frac{L}{\tan \alpha} h^2 + (LW)\, h \tag{11–5}$$

in which h is the height above the bed of the storage facility. If Equation 11–5 is plotted on a graph, the stage-storage relationship has the shape of a second-order polynomial with a zero intercept and a shape that depends on the values of L, W, and α.

Unless the site undergoes considerable excavation, the simple forms described previously are not "real world." However, the concepts used to derive the stage-storage relationship for the simple forms are also used to derive the stage-storage relationship for an actual site. Instead of a continuous function such as Equation 11–5, the stage-storage relationship is derived as a discrete function (a set of points). The area within contour lines of the site can be planimetered, with the storage in any depth increment Δh equal to the product of the average area and the depth increment. Thus the storage increment ΔS is given by

$$\Delta S = \frac{1}{2} (A_i + A_{i+1})\, \Delta h \tag{11–6}$$

where A_i and A_{i+1} are the surface areas for the ith and $(I + 1)^{\text{th}}$ contours.

Example 11–1

Consider the storage facility of Equation 11–5. If the basin has a length of 625 feet, a width of 360 ft, and side slopes of $2H{:}1V$, then Equation 11–5 becomes

$$S = 1250\, h^2 + 225{,}000\, h \tag{11–6}$$

At a depth of 5 ft, the facility would have a storage volume of 1.156×10^6 ft^3. If the invert for the outlet was located at a depth of 1.5 ft, then the volume of dead storage would be 340,313 ft^3. The active storage would be the difference, 815,687 ft^3.

Example 11–2

To illustrate the use of Equation 11–6, consider the site shown in Figure 11–1. The area bounded by each contour line was estimated (Table 11–1) and the average area within adjacent contours computed. The topolines are drawn with a 2-ft contour interval. The stage-storage relationship was computed using Equation 11–6 and is given in Table 11–1 and Figure 11–2.

11.3.2 The Stage-Discharge Relationship

The discharge from a reservoir or detention facility depends on the depth of flow and the characteristics of the outlet facility. The outlet facility can include either a weir, an orifice, or both. The weir and orifice equations were introduced in Chapter 8. A brief summary is presented here.

Weir equation. Figure 11–3 shows a schematic of flow over a sharp-crested weir. Upstream of the weir, the flow rate is Q_o, with a velocity V_o, and a depth Z. Using the top of the sharp-crested weir as the datum and assuming atmospheric pressure over the flow cross

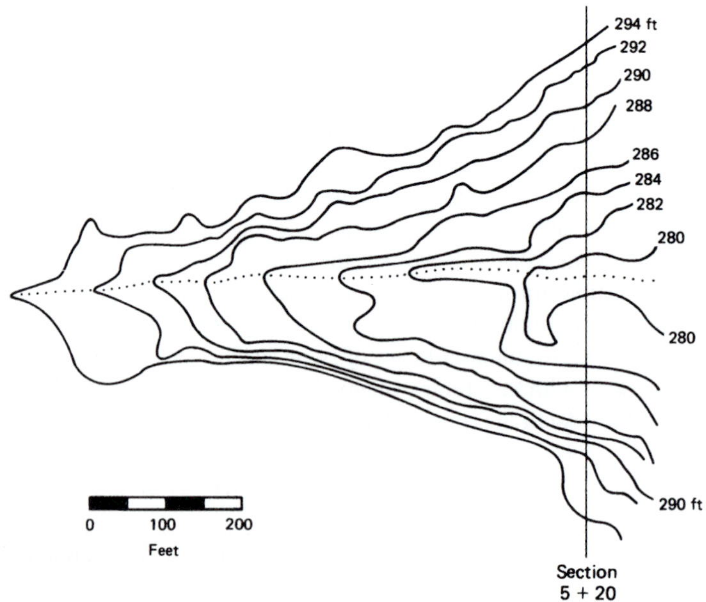

FIGURE 11–1 Topographic map deriving stage-storage relationship.

TABLE 11–1 Derivation of Stage-Storage Relationship for Watershed of Figure 11–1

Contour Elevation (ft)	Area (acres) within Contour Elevation	Average Area (acres)	Contour Interval (ft)	Depth *h* (ft)	Change in Storage (acre-ft)	Storage (acre-ft)
278	0			0		0
280	0.12	0.060	2	2	0.12	0.12
282	0.43	0.275	2	4	0.55	0.67
284	1.06	0.745	2	6	1.49	2.16
286	1.70	1.380	2	8	2.76	4.92
288	2.55	2.125	2	10	4.25	9.17
290	3.17	2.860	2	12	5.72	14.89
292	3.83	3.500	2	14	7.00	21.89
294	4.91	4.370	2	16	8.74	30.63

section, the energy grade line (EGL) equals the sum of the elevation head (Z) and the velocity head ($V^2/2g$). The discharge over the weir is defined by the continuity equation:

$$Q = \int V dA \tag{11–7}$$

If we assume that the water surface of the cross section above the weir does not contract (the dashed line of Figure 11–3), the flow velocity at any point in the cross section equals

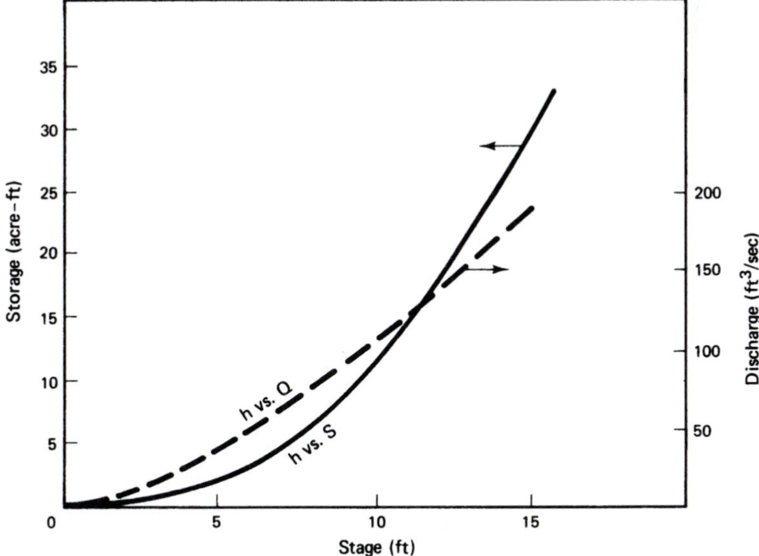

FIGURE 11–2 Stage-storage and stage-discharge relationships for watershed of Figure 11–1.

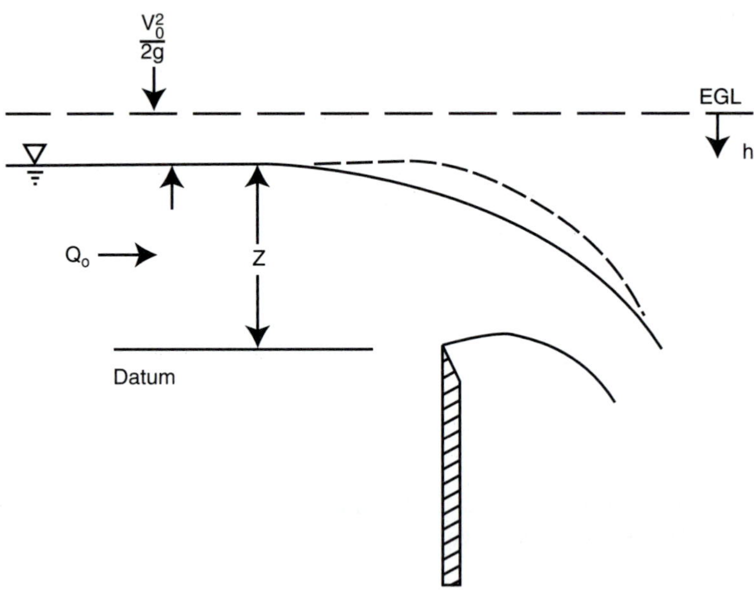

FIGURE 11–3 Schematic diagram of flow over a sharp-crested weir.

$(2gh)^{0.5}$, where h is measured from the EGL. The cross-sectional area equals the depth times the length, L (where L is measured into the page in Figure 11–3). Expressing the discharge as a rate per unit length (q_o), Equation 11–7 becomes

$$q_o = \int (2\,gh)^{0.5}\,dh \tag{11–8}$$

Equation 11–8 must be integrated from the water surface (at $h = V_0^2/2g$) to the top of the weir (at $h = Z + V_0^2/2g$):

$$\begin{aligned} q_o &= \int_{V_o^2/2g}^{Z+V_o^2/2g} (2gh)^{0.5}\,dh \\ &= \frac{2}{3}(2\,g)^{0.5}\left[\left(Z + \frac{V_0^2}{2\,g}\right)^{1.5} - \left(\frac{V_0^2}{2\,g}\right)^{1.5}\right] \end{aligned} \tag{11–9}$$

In truth, the water surface contracts at the cross section above the weir (solid line in Figure 11–3), and thus the discharge is less than that indicated by Equation 11–9. Also, the derivation of Equation 11–9 assumes that losses are negligible. A discharge coefficient C_d will be used to account for the losses and flow contraction; thus the flow per unit length q_o through the section is

$$q_o = \frac{2}{3}\,C_d\,(2\,g)^{0.5}Z^{1.5} \tag{11–10a}$$

$$= C_w\,Z^{1.5} \tag{11–10b}$$

where C_w is the weir coefficient. Experiments have indicated that C_d of Equation 11–10a varies from about 0.6 to 0.75, with a value of about 0.6 for the case where the depth of flow over the weir (Z in Figure 11–3) is small compared with the length of the weir. Expressing Equation 11–10b in terms of the total flow Q, where $Q = q_o L$, we get

$$Q = C_w \, L \, Z^{1.5} \tag{11–11}$$

The value of the weir coefficient of Equation 11–11 varies from 2.8 to 3.33, with a value of 3.1 commonly used. Equation 11–11 can be used to derive the stage-discharge (Z versus Q) relationship for a sharp-crested weir of length L. In Equation 11–11, Q has units of ft^3/sec and L and Z have units of ft; for other dimension systems the constant will be different from the value of 3.1 in Equation 11–11.

Example 11–3

Consider the storage facility of Example 11–1. Assume that a 4-ft weir exists at an elevation of 2 ft. Thus, water below the 2-ft datum will not flow over the weir so it is considered dead storage; active storage consists of all storage above the weir. At an elevation of 2 ft, the storage would be 455,000 ft^3. If a flood resulted in a depth of 5 ft, the total storage would be 1,156,250 ft^3, which means that there is 701,250 ft^3 of active storage. At a depth of 5 ft, the discharge over the weir would be

$$Q = C_w \, L Z^{1.5} = 3.1\,(4)\,(5 - 2)^{1.5} = 64.4 \text{ cfs}$$

The stage-discharge relationship could be graphed by computing values of the following equation for selected depths:

$$Q = 12.4\,(h - 2)^{1.5}$$

Example 11–4

Equation 11–11 can be illustrated for the topographic site in Figure 11–1. Assuming that a riser spillway with a sharp-crested weir is required, the stage-discharge relationship was computed for a unit length of 1 ft and is plotted with the stage-storage relationship in Figure 11–2.

Orifice equation. In a number of geographic areas, storm water management policies require alternative designs such as two-stage control or dry detention ponds. In either case, an orifice or port may be part of the detention basin outlet facility. The equation that can be used to estimate flow rates through orifices in a riser can be developed from Bernoulli's equation:

$$\frac{p_1}{\gamma} + \frac{V_1^2}{2g} + Z_1 = \frac{p_2}{\gamma} + \frac{V_2^2}{2g} + Z_2$$

in which p_i, V_i, and Z_i are the pressure, velocity, and elevation of section i, respectively; and γ is the specific weight of the fluid. For the system shown in Figure 11–4, point 1 is at the water surface of the reservoir with surface area A_1, and point 2 is at the center of the water jet discharging through the orifice at elevation Z_2. The jet of water has a minimum cross-sectional area A_2', with the orifice having a slightly larger area A_2. The surface area of the reservoir is considered to be much larger than the area of the orifice. The pressure at both the water surface (point 1) and across the jet are assumed to be atmospheric; therefore, $p_1 = p_2$.

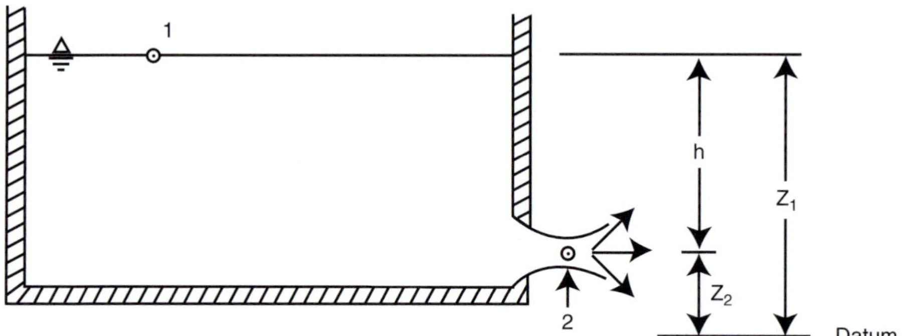

FIGURE 11–4 Schematic diagram of flow through an orifice.

The difference in elevation heads between points 1 and 2 ($Z_1 - Z_2$) equals the stage in the pond, h. Based on the continuity equation, $Q = VA$, the velocity of water at the surface of the reservoir (point 1) will be very small compared with V_2 because A_1 is so much larger than A_2; therefore, it is reasonable to assume that $V_1 = 0$. Thus Equation 11–12 becomes

$$\frac{p_{\text{atm}}}{\gamma} + 0 + h = \frac{p_{\text{atm}}}{\gamma} + \frac{V_2^2}{2g} + 0 \tag{11–13}$$

Solving for the velocity of the jet yields

$$V_2 = (2gh)^{0.5} \tag{11–14}$$

The discharge through the orifice is

$$Q = V_2 A_2 = A_2 (2gh)^{0.5} \tag{11–15}$$

in which Q is the discharge rate in ft^3/sec, A_2 is in ft^2, and h is in ft. Equation 11–15 assumes that there are no losses in the system and that the area where the pressure across the cross section of the jet equals atmospheric is A_2. In practice, there are significant losses, and the cross-sectional area where the pressure equals atmospheric pressure across the section is smaller than the A_2. Thus the actual discharge through the orifice is less than that given by Equation 11–15. The coefficient of discharge C_d is used to adjust for losses and area difference (A_2 versus A_2'):

$$Q = C_d A_2 (2gh)^{0.5} \tag{11–16}$$

Experimental studies indicate that C_d is approximately 0.6.

Example 11–5

The use of Equation 11–16 can be illustrated for a hypothetical watershed in which the topography at the detention basin riser can be represented by a half-ellipse with major and minor axes of a and b, respectively (see Figure 11–5). The area of a half-ellipse with semimajor axis A and semiminor axis B is $0.5\pi AB$. If the ratio of A to B, which is denoted as r, remains constant for all b, then the area can be written in terms of r and B:

$$A = 0.5\pi r B^2 \tag{11–17}$$

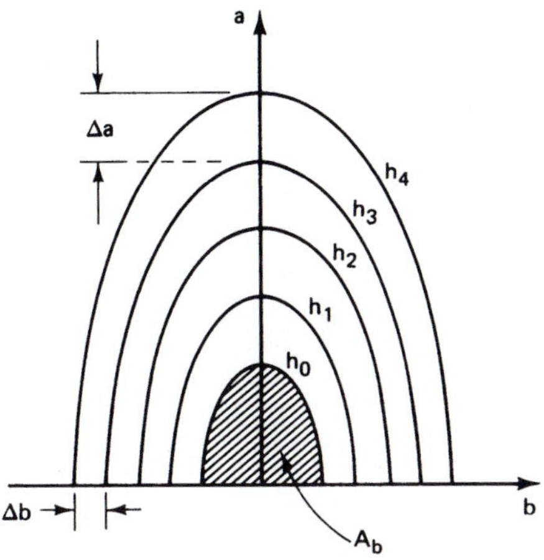

FIGURE 11–5 Topographic map of half-ellipse with bottom area A_b, minor-axis side slope s, and major to minor axis length ratio $r = a/b$ (h_i = stage for contour line i).

For the purpose of illustrating Equation 11–16, we will assume that the area within the first contour line h_o, denoted as A_b, is a horizontal plane. Beyond the contour line h_o, the semiminor axis has a slope s_b; thus, for the ratio r to remain constant, the slope of the contours along the semimajor axis is r^*s_b. Assuming that the site shown in Figure 11–5 has a scale of 1 in. = 360 ft and a contour interval of 4 ft, the bottom of the basin has an area (A_b) of 20,600 ft^2. The surface area for each contour interval is shown in Table 11–2; these are used with Equation 11–6 to compute the change in storage ΔS for each contour interval. The cumulative storage S is also shown in Table 11–2. The stage-discharge curve was computed for a riser with an orifice of 12 ft^2. The stage-storage-discharge relationship is shown in Figure 11–6.

The stage-discharge relationship can be computed for various values of h, once the physical characteristics of the weir or orifice are defined. For the case of a multistage outlet, it will be necessary to specify the characteristics of both the weir and orifice.

11.3.3 Stage-Storage-Discharge Relationships for Two-Stage Risers

There has been very little theoretical research or empirical studies of the hydraulics of two-stage risers; thus a number of procedures have been proposed for routing through such facilities. Some proposals have assumed that, when weir flow begins, flow through the orifice ceases. At the other end of the spectrum, other proposals have assumed that flow through the orifice is independent of flow over the weir. It would probably be more realistic to assume

TABLE 11–2 Stage-Storage-Discharge Relationship for the Site in Figure 11–5

h	b	a	Area (ft^2)	ΔS (acre-ft)	Storage (acre-ft)	Discharge (ft^3/sec)
0	81	162	20,600	—	0	0
4	119	238	44,500	2.99	2.99	116
8	156	312	76,500	5.55	8.54	163
12	194	388	118,200	8.94	17.48	200
16	231	462	167,200	13.13	30.61	231

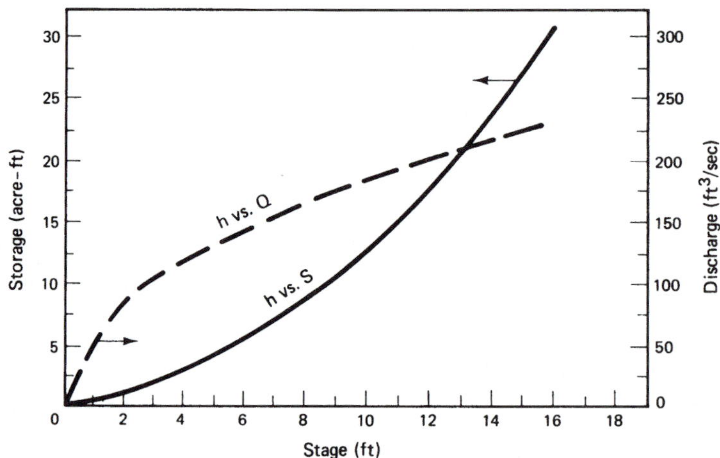

FIGURE 11–6 Stage-storage discharge realtionship for site of Figure 11–5 and a riser with an orifice of 12 ft^2.

that the two are not independent and that, as the depth of storage above the elevation of the weir increases, the interdependence between weir flow and orifice flow increases. This interdependence is recognized in the calculations of the orifice flow Q_r in the SCS two-stage riser method; specifically, Q_r is calculated by

$$Q_r = C_d A_o \,[2\,g\,(E_2 - E_1)]^{0.5} = 4.82 A_o\,(E_2 - E_1)^{0.5} \tag{11–18}$$

in which C_d is the discharge coefficient and is assumed equal to 0.6, and E_2 and E_1 are the maximum water surface elevations in the reservoir at routed high and low stages, respectively. Equation 11–18 implies that the effective head controlling flow through the orifice is the same head as that controlling flow over the weir, which is a reasonable assumption if the riser pipe is reasonably full. Under such an assumption the general form of the stage-discharge relationship is

$$Q = \begin{cases} 0 & h \leq h_o \\ C_w L\,(h - h_o)^{1.5} & h_o \leq h \leq h_o + H_o \\ C_d\,A_o[2\,g\,(h - h_o)]^{0.5} & h_o + H_o \leq h \leq h_1 \\ C_w\,L\,(h - h_1)^{1.5} + C_d\,A_o[2\,g\,(h - h_1)]^{0.5} & h_1 < h \end{cases} \tag{11–19}$$

in which h_o is the elevation of the bottom of the riser, H_o is the maximum height of the orifice, and h_1 is the elevation of the weir.

Example 11–6

To illustrate the use of Equation 11–16 in deriving a stage-discharge relationship, assume that the riser of a detention structure has the physical characteristics shown in Figure 11–7. The riser has a diameter of 5 ft, and the top of the riser acts as a sharp-crested weir of length $L = 5\pi$ ft. The orifice has an area of 8 ft^2. Assuming that the weir coefficients are 3.1 and the discharge coefficient for the orifice is 0.6, the stage-discharge relationship is given by

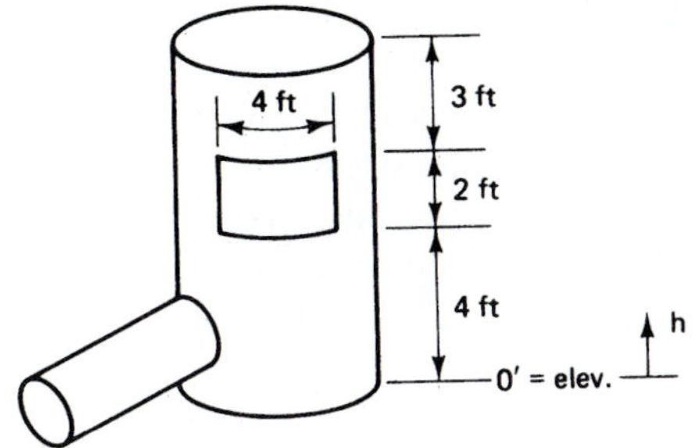

FIGURE 11–7 Two-stage riser configuration.

$$Q = \begin{cases} 0 & h \leqslant 4 \\ 12.4\,(h - 4)^{1.5} & 4 \leqslant h \leqslant 6 \\ 4.8\sqrt{2\,g\,(h - 4)} & 6 \leqslant h \leqslant 9 \\ 16.5\pi\,(h - 9)^{1.5} + 4.8\sqrt{2\,g\,(h - 4)} & 9 < h \end{cases} \tag{11–20}$$

In this relationship h is measured from the bottom of the riser. Since the bottom of the low-stage orifice is at a stage of 4 ft, there will be a wet pond to a depth of 4 ft before discharge begins. For the low-stage event, the orifice will act as a weir until the orifice is submerged. When the stage exceeds 6 ft, the orifice will act as an orifice rather than a weir. Above a stage of 9 ft, water passes through the orifice and over the crest of the riser.

The outlet pipe would have to be sized so that it does not limit the flow through the outlet structure. The relationship of Equation 11–20 also assumes that the flow through the orifice is not inhibited by the flow over the weir. For this purpose, the diameter of the riser should be at least three times the diameter of the outlet pipe (D).

The stage-discharge relationship can be combined with the stage-storage relationship to produce a stage-storage-discharge relationship. For example, if we assume a basin has a rectangular floor with a length of 1000 ft and a width of 400 ft and sloping sides with $\alpha = 30°$, the stage-storage relationship can be represented by Equation 11–5. If the riser defined by Figure 11–7 and Equation 11–20 is used and the floor of the basin is at an elevation of 0 ft, the stage-storage-discharge curve is given in Table 11–3.

TABLE 11–3 Stage-Storage-Discharge Relationship for Figure 11–7 and Equation 11–18

Elevation, h (ft)	Storage (ft^3) × 10^5	Discharge (ft^3/sec)
0	0	0
2	8.1	0
4	16.3	0
6	24.6	54
8	33.1	77
9	37.4	86
10	41.7	146
12	50.5	378
14	59.4	701

11.4 STORAGE-INDICATION ROUTING

As indicated previously, the routing equation of Equation 11–4 can be used to derive the downstream hydrograph (O_2) when the stage-storage-discharge relationship is known. The storage-discharge relationship is used to derive the storage-indication curve, which is a relationship between O and $(S + O\Delta t/2)$, where the latter is the right-hand side of Equation 11–4. Given the storage-discharge curve, O versus S, the following four-step procedure can be used to develop the storage-indication curve:

1. Select a value of O.
2. Determine the corresponding value of S from the storage-discharge curve.
3. Use the values of S and O to compute $(S + O\Delta t/2)$.
4. Plot a point on the storage-indication curve O versus $(S + O\Delta t/2)$.

These four steps are repeated for a sufficient number of values of O to cover the range of discharges expected in the outflow hydrograph. The set of values of O and $(S + O\Delta t/2)$ define the storage-indication curve, which can be graphed with O as the ordinate.

The objective of the storage-indication method is to derive the outflow hydrograph. There are five data requirements:

1. The storage-discharge relationship
2. The storage-indication curve
3. The inflow hydrograph
4. Initial values of the storage and outflow rate
5. The routing increment (Δt)

While the outflow hydrograph is the primary output for most design problems, the storage function (S versus t) is also a necessary output of the storage-indication routing procedure. The maximum value of S from the S-versus-t relationship is the required storage volume at maximum flood stage. The maximum storage occurs when the outflow rate first exceeds the inflow rate, which is the point on the recession curve where the inflow hydrograph becomes less than the outflow hydrograph.

The following five-step procedure can be used to derive the outflow hydrograph, with the storage-time relationship as a by-product:

1. Determine the average inflow: $0.5\Delta t(I_1 + I_2)$.
2. Compute: $(S_1 - {}^1\!/_2\, O_1\, \Delta t)$.
3. Using Equation 11–4 and the values from Steps 1 and 2, compute $S_2 + {}^1\!/_2\, O_2 \Delta t$.
4. Using the value computed in Step 3 as input, find O_2 from the storage-indication curve.
5. Use O_2 with the storage-discharge relationship to obtain S_2.

These five steps are repeated for the next time increment using I_2, O_2, and S_2 as the new values of I_1, O_1, and S_1, respectively. The process is solved iteratively until the outflow hydrograph is computed. Worksheet 11–1 can be used to organize the computations.

WORKSHEET 11–1 Storage-Indication Routing

t	Time ( )	Inflow I_i ( )	Average inflow rate, $\bar{I}_i$ $0.5\,(I_i + I_{i-1})$ ( )	Average inflow volume $\bar{I}_i\,\Delta t$ ( )	$S_i - 1/2\,O_i\,\Delta t$ ( )	$S_i + 1/2\,O_i\,\Delta t$ ( )	Outflow, O_i ( )	Storage, S_i ( )
0								
1								
2								
3								
4								
5								
6								
7								
8								
9								
10								
11								
12								

TABLE 11–4 Stage-Storage-Discharge Relationship for Example 11–7

Stage (ft)	Storage (ft^3)	Storage (ac-ft)	Discharge (ft^3/sec)	$S + 0\Delta t/2$ (ft^3)	$2s/\Delta t + 0$ (ft^3/sec)
0	0	0	0	0	0
1	6,350	0.1458	5	7,850	26.2
2	13,000	0.2984	10	16,000	53.3
4	26,650	0.6118	20	32,650	108.8
6	40,550	0.9309	30	49,550	165.2
8	54,600	1.2534	40	66,600	222.0

In using the routing equation of Equation 11–4, the inflow and outflow hydrographs are represented as series of discrete points, rather than as continuous functions. This can be a source of error if the time increment Δt is not chosen correctly. The averaging of the inflow [that is, $(I_1 + I_2)/2$] assumes linearity; therefore, Δt should be sufficiently short so that the assumption of linearity is not violated and that the peak of the outflow hydrograph is not missed, which can occur if Δt is too large.

Example 11–7

As a first example, we assume that the stage-storage-discharge relationship is as given in Table 11–4 and the time increment Δt equals 600 sec. The stage-discharge and storage-indication curves are shown in Figure 11–8. Assuming an initial storage and outflow of zero and the inflow

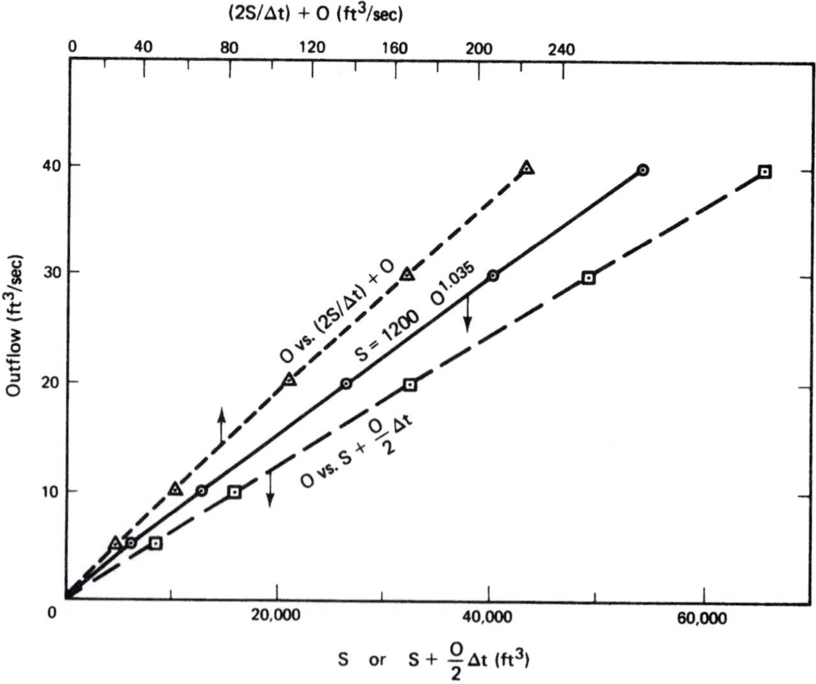

FIGURE 11–8 The storage-discharge and storage-indication curves for Example 11–7.

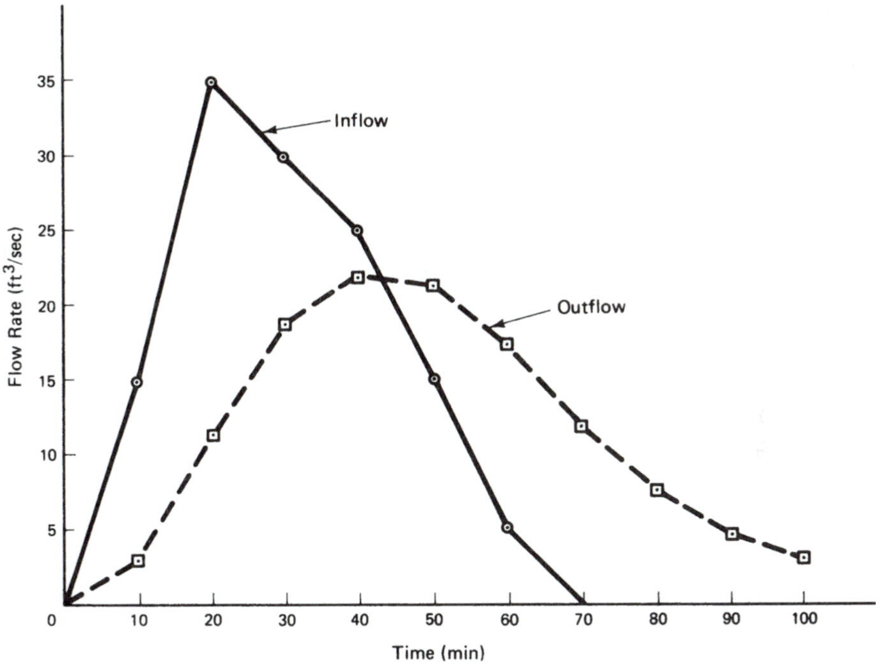

FIGURE 11–9 Inflow and outflow hydrographs for Example 11–7.

hydrograph as shown in Figure 11–9, the computations of the outflow hydrograph are given in Table 11–5. Both the inflow and outflow hydrograph are shown in Figure 11–9.

11.5 MODIFIED PULS ROUTING METHOD

An alternative formulation for routing can be derived by algebraic transformation of Equation 11–4. Multiplying Equation 11–4 by $2/\Delta t$ yields

$$(I_1 + I_2) + \left(\frac{2S_1}{\Delta t} - O_1\right) = \frac{2S_2}{\Delta t} + O_2 \tag{11–21}$$

Equation 11–21 is the routing equation known as the modified Puls Method. The routing can be accomplished graphically in the same way as the storage-indication method, except a graph of O versus $(2S/\Delta t) + O$ is developed instead of the graph of O versus $(S + O\,\Delta t/2)$. As with the storage-indication method, Δt must be chosen so that the hydrographs are approximately linear over the time increment.

To use the modified Puls Method the stage-storage-discharge relationship must be developed from topographic characteristics of the site and the hydraulic characteristics of the reservoir outlet. From this relationship the relationship of outflow O versus $[(2S/\Delta t) + \Delta O]$ is computed for various stages. The number of points used to represent the relationship should be sufficient such that the use of linear interpolation between adjacent points will not lead to

TABLE 11–5 Computation of Reservoir Outflow Hydrograph for Example 11–7 of Storage-Indication Method

Time (min)	Inflow (ft^3/sec)	Average Inflow (ft^3/sec)	$0.5(I_1 + I_2)\,\Delta t$	$S - ½\Delta t$	$S + ½O\Delta t$	Outflow, O (ft^3/sec)	Storage S (ft^3)
0	0					0.0	0
		7.5	4,500	0	4,500		
10	15					2.9	3,620
		25.0	15,000	2,749	17,748		
20	35					11.1	14,430
		32.5	19,500	11,113	30,613		
30	30					18.8	24,977
		27.5	16,500	19,342	35,842		
40	25					21.9	29,273
		20.0	12,000	22,704	34,704		
50	15					21.2	28,337
		10.0	6,000	21,971	27,971		
60	5					17.2	22,808
		2.5	1,500	17,647	19,147		
70	0					11.9	15,576
		0	0	12,005	12,005		
80	0					7.6	9,735
		0	0	7,468	7,468		
90	0					4.7	6,035
		0	0	4,606	4,606		
100	0					3.0	3,707
	125					120.3	

significant errors. Given the inflow hydrograph, a tabular computation sheet can be developed to iteratively solve Equation 11–21. Assuming that initial values of outflow O_1 and storage S_1 are known, the terms on the left side of Equation 11–21 are known. Unlike the storage-indication method, it is not necessary to compute the average inflow; the sum of inflow ordinates for adjacent time periods is used. Given I_1, I_2, S_1, O_1, and Δt, the value of the left side of Equation 11–21 is computed and used as input to the graph of O versus $[(2S/\Delta t) + O]$, from which the value for O_2 is obtained. The storage S_2, which corresponds to O_2, is then computed from the storage-outflow relationship. This completes an iteration, and the values of O and S for the next time period are computed using the most recent values as the input to Equation 11–21. The iteration procedure is continued until the entire outflow hydrograph is generated.

As a check, the volume of outflow should be computed and compared with the volume of inflow; the volumes equal the product of the time increment and the sum of the hydrograph ordinates. The volume of outflow may be slightly less than the volume of the inflow when the iteration procedure is discontinued before the outflow is essentially zero. The volumes of inflow and outflow are equal because the routing is based on the conservation of mass, which is represented by Equation 11–1. Small differences in the volumes of inflow and outflow can also result from errors in reading the O-versus-S and O-versus-$[(2S/\Delta t) + O]$ graphs.

TABLE 11–6 Computation of Reservoir Outflow Hydrograph for Example 11–8 of Modified Puls Method

Time (min)	Inflow (ft^3/sec)	$I_1 + I_2$	$\frac{2S}{\Delta t} - O$	$\frac{2S}{\Delta t} + O$	Outflow (ft^3/sec)	Storage (ft^3)
0	0	—	0	—	0	0
10	15	15	8.7	15.0	3.0	3,500
20	35	50	36.8	58.7	11.0	14,350
30	30	65	63.5	101.8	18.5	24,600
40	25	55	74.2	118.5	21.5	28,700
50	15	40	72.3	114.5	21.0	28,000
60	5	20	72.3	92.3	17.0	22,500
70	0	5	58.0	63.0	11.5	15,000
80	0	0	38.5	38.5	7.5	9,650
90	0	0	24.7	24.7	4.5	5,700
100	0	0	14.5	14.5	3.0	3,500
110	0	0	8.7	8.7	2.0	2,450
120	0	0	6.1	6.1	1.0	1,200
	125				121.5	

Example 11–8

The data of Tables 11–4 and 11–5 can be used to illustrate the modified Puls Method. The graph of O versus $(2S/\Delta t) + O$ is derived from the storage-discharge curve of Figure 11–8 and is also shown in Figure 11–8. The calculations of the outflow hydrograph and storage-versus-time relationship are given in Table 11–6. The outflow hydrograph is essentially identical to the outflow hydrograph computed by the storage-indication method because the two procedures are essentially identical. The area between the inflow and outflow hydrographs represents the storage. The maximum storage occurs where the two hydrographs intersect, which is at a storm time of approximately 43 min; after this point the storage is reduced since the outflow rate is greater than the inflow rate. From the conservation-of-mass equation (Equation 11–1), O is greater than I, so $\Delta S/\Delta t$ is negative.

The volume of inflow equals 125 ft^3/sec (10 min)(60 sec/min) or 75,000 ft^3. The volume of outflow equals 121.5 (ft^3/sec) (10 min)(60 sec/min) = 72,900 ft^3. The difference of 2100 ft^3 represents the errors from discontinuing computation of O at a time of 120 min and from reading the curve of Figure 11–8.

There are some slight differences between the outflow and storage function of Tables 11–5 and 11–6. These differences are only the result of errors from reading the graphs. The two methods should produce identical results.

11.6 DESIGN PROCEDURE

It is worthwhile to review the two general types of hydrologic synthesis problems for which storage routing methods are used. First, watershed studies are frequently conducted where structures currently exist. In this case, the objective is to route the design-flood hydrograph through the existing structure using the known stage-storage-discharge relationship. Very often watershed studies will be undertaken to evaluate the hydrologic effects of various land use conditions, such as a natural state, current conditions, and completely developed, or for different zoning practices. For such watershed studies the routing is not undertaken as part of the design of the structure.

The second case of hydrologic synthesis is the design of the outlet facility. Quite often, the design of a storage facility centers around a target discharge. For example, in the design of urban detention basins, storm water management policies may require the post-development peak discharge to be no greater than the peak discharge for the pre-development watershed conditions at a selected exceedence frequency (return period). In designing the riser to meet the target discharge, one must identify the riser characteristics (weir length and/or area of the orifice) that will limit the computed post-development peak discharge out of the detention basin so that it does not exceed the target discharge, which is usually the peak discharge that would occur if development had not taken place within the watershed. The stage-storage-discharge relationship is a function of both the target discharge and the riser characteristics. Because of this interdependence, the design procedure is iterative. That is, the design procedure begins with some assumption about the riser characteristics. Then the storage routing procedure is applied to determine whether or not the target discharge requirement has been met. The riser characteristics are continually modified until the target discharge is met. If the weir length is too long or the orifice too large, the target discharge will be exceeded. If the computed discharge is less than the target discharge (that is, the weir length is too short or the orifice is too small), the design will result in storage that is greater than that really required.

The design procedure is summarized in the flowchart of Figure 11–10. There are three initial steps: (1) establish the stage-storage relationship; (2) establish the stage-discharge relationship; and (3) compute the design inflow hydrograph and the target discharge (that is, the allowable discharge out of the basin). The stage-storage curve is computed from topographic information at the site of the storage facility. To compute the stage-discharge relationship requires assumptions about the riser outlet, including the number of stages and whether weirs, orifices, or both are used to control the outflow. The assumed weir lengths and/or orifice areas and the elevations are used to compute the stage-discharge relationship. The stage-storage and stage-discharge curves are then used to compute both the storage-indication and storage-discharge curves. The final step in the input is to obtain the initial conditions. This includes the routing interval, the target outflow discharge, initial values for the outflow and active storage (which are both usually zero), and the inflow hydrograph, which is usually the hydrograph for the post-development conditions computed with a design rainfall hyetograph and a unit hydrograph.

Based on these inputs and initial computations, the inflow hydrograph can be routed through the storage basin; this begins at node B in Figure 11–10. The routing procedure was detailed in Section 11.4. Once the outflow hydrograph is computed, its peak discharge is compared to the target discharge q_o. If it is greater than q_o, then the assumed outlet configuration has too large a capacity. Thus, the weir lengths or orifice areas should be decreased (return to node A in Figure 11–10). If the peak discharge of the outflow is less than the target discharge q_o, then the assumed outlet configuration does not provide for sufficient outflow. The weir lengths or orifice areas can be increased (return to node A in Figure 11–10). This will require recomputing the stage-discharge relationship, the storage-indication curve, and the storage-discharge curve. The routing process, which begins at node B of Figure 11–10, is then repeated with the new stage-storage-discharge information, and the maximum discharge

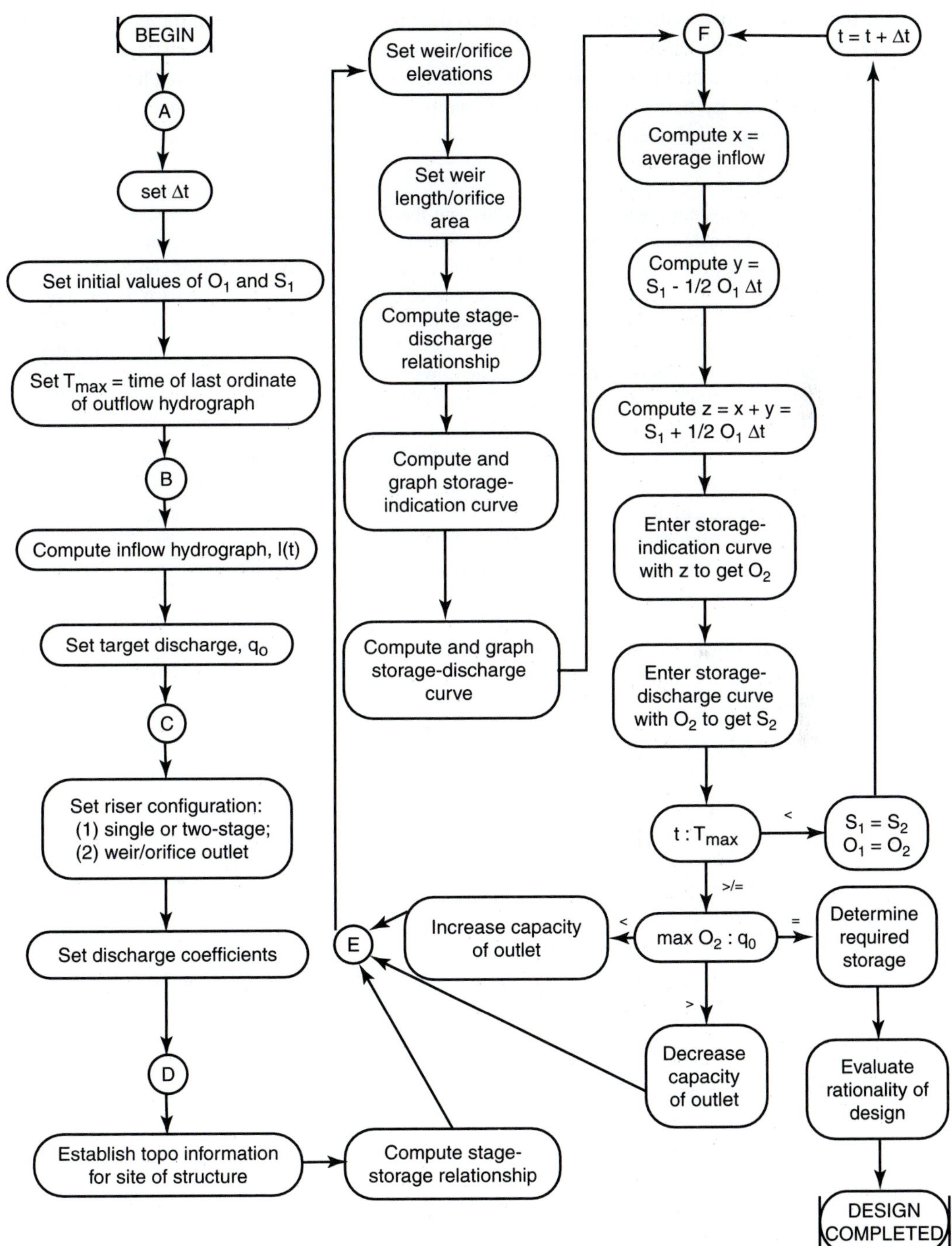

FIGURE 11–10 Reservoir Storage Design Process.

of the new outflow hydrograph compared to the target discharge. When the peak outflow approximately equals the target discharge, then the assumed outlet facility is a reasonable design. The required storage is estimated by the largest value of storage S_2. The depth of storage is estimated by using the maximum S_2 as input to the stage-storage curve. The computed design should be evaluated for safety and cost.

Example 11–9

A detailed example will be used to illustrate the design process. The watershed has a drainage area of 38.3 acre and is entirely in C soil. The design will use a 10-yr exceedence frequency. In the before-development condition, the watershed was 40% forest in good condition ($CN = 70$) and 60% in brush in good condition ($CN = 65$); therefore, the weighted CN is 67. For after-development conditions, the watershed has the following land cover: 10 acres of 1/8-acre lots ($CN = 90$); 14 acres of 1/4-acre lots ($CN = 83$); 5 acres of light commercial (75% impervious); and 9.3 acres of open space in good condition ($CN = 74$). The CN for the commercial area is $0.75(98) + 0.25(74) = 92$. Thus, the weighted CN for the watershed is

$$CN_a = \frac{1}{38.2}[10\,(90) + 14\,(83) + 5\,(92) + 9.3\,(74)] = 83.8 \text{ (use 84)}$$

Characteristics of the principal flow paths for both watershed conditions are given in Table 11–7. For a 10-yr exceedence frequency, Figure 4–4 gives a 24-hr rainfall depth of 4.8 in.

Using the before- and after-development CNs, the rainfall depths are computed from Equation 7–42 as 1.67 and 3.09 in., respectively. The initial abstractions are 0.895 and 0.381 in., which yields I_a/P ratios of 0.21 and 0.08 (use 0.1) for pre- and post-development, respectively. Using the times of concentration and the I_a/P ratios, unit peak discharges of 500 and 710 csm/in. are taken from Figure 7–9 for the before and after conditions, respectively. These yield peak discharges of

$$q_{pb} = q_{ub}\,Q_bA = 500\left(\frac{38.3}{640}\right)(1.67) = 50.0 \text{ cfs}$$

$$q_{pa} = q_{ua}\,Q_aA = 710\left(\frac{38.3}{640}\right)(3.09) = 131.3 \text{ cfs}$$

TABLE 11–7 Computation of Times of Concentration

Condition	Land Cover	Length (ft)	n	Slope (%)	Rainfall i (in./hr)	Hydr. Radius (ft)	Method	Travel Time (min)	t_c (min)
Before	Forest	150	0.3	2.0	7.0	—	Kinematic	16.3	
	Grassed swale	200	—	1.7	4.5	—	Figure 3–13	1.7	
	Gully	350	—	1.5	—	—	Figure 3–13	2.5	
	Open Channel	1200	0.05	1.2	—	0.8	Manning's Equation	7.1	27.6
After	Forest	50	0.3	2.0	7.0	—	Kinematic	7.1	
	Asphalt	100	0.012	2.0	7.8	—	Kinematic	1.5	
	Asphalt gutter	300	0.012	1.6	—	0.10	Manning's Equation	1.5	
	Circular pipe	600	0.024	1.4	—	0.25	Manning's Equation	3.4	
	Open channel	750	0.04	1.3	—	1.00	Manning's Equation	2.9	16.4

TABLE 11–8 Derivation of Stage-Storage Relationship

Depth, h (ft)	Surface Area (ft^2)	Average Area (ft^2)	Incremental Storage (ft^3)	Storage Volume, V_s (ft^3)
0	33,500			0
		33,650	16,825	
0.5	33,800			16,825
		34,050	17,025	
1.0	34,300			33,850
		34,450	17,225	
1.5	34,600			51,075
		34,950	17,475	
2.0	35,300			68,550
		35,550	17,775	
2.5	35,800			86,325
		36,000	18,000	
3.0	36,200			104,325
		36,500	18,250	
3.5	36,800			122,575
		37,150	18,575	
4.0	37,500			141,150
		37,750	18,875	
4.5	38,000			160,025
		38,400	19,200	
5.0	38,800			179,225

TABLE 11–9 Stage-Storage-Discharge and Storage-Indication Curves for a Weir Length of 1.50 Feet

Storage Stage (ft)	Storage Discharge (cfs)	Storage (cu ft)	Storage Indication (cu ft)
.00	.0	0.	.0
.50	1.6	16641.	16937.1
1.00	4.6	33911.	34748.0
1.50	8.5	51426.	52964.1
2.00	13.2	69103.	71470.6
2.50	18.4	86901.	90209.6
3.00	24.2	104796.	109145.0
3.50	30.4	122772.	128252.3
4.00	37.2	140817.	147513.4
4.50	44.4	158924.	166914.1
5.00	52.0	177085.	186443.3
5.50	60.0	195296.	206092.0
6.00	68.3	213551.	225852.5

TABLE 11–10 Storage-Indication Analysis for Weir Length of 1.5 Feet

Time (min)	Inflow (cfs)	Average Inflow (cfs)	Average Inflow (cu ft)	$S - OdT/2$ (cu ft)	$S + OdT/2$ (cu ft)	Outflow (cfs)	Storage (cu ft)
0	.0					.0	0.
		.5	180.	0.	180.		
6	1.0					.0	177.
		1.5	540.	174.	714.		
12	2.0					.1	701.
		2.5	900.	689.	1589.		
18	3.0					.2	1561.
		3.5	1260.	1533.	2793.		
24	4.0					.3	2744.
		4.5	1620.	2696.	4316.		
30	5.0					.4	4240.
		5.5	1980.	4165.	6145.		
36	6.0					.6	6037.
		6.5	2340.	5930.	8270.		
42	7.0					.8	8126.
		8.0	2880.	7981.	10861.		
48	9.0					1.1	10671.
		11.5	4140.	10482.	14622.		
54	14.0					1.4	14366.
		17.5	6300.	14111.	20411.		
60	21.0					2.2	20009.
		31.5	11340.	19608.	30948.		
66	42.0					4.0	30226.
		63.5	22860.	29505.	52365.		
72	85.0					8.4	50850.
		107.0	38520.	49335.	87855.		
78	129.0					17.7	84665.
		130.0	46800.	81475.	128275.		
84	131.0					30.5	122793.
		110.5	39780.	117311.	157091.		
90	90.0					40.7	149756.
		72.5	26100.	142421.	168521.		
96	55.0					45.0	160419.
		46.0	16560.	152316.	168876.		
102	37.0					45.2	160749.
		32.0	11520.	152621.	164141.		
108	27.0					43.4	156336.
		23.5	8460.	148531.	156991.		
114	20.0					40.7	149663.
		17.5	6300.	142335.	148635.		
120	15.0					37.6	141864.
		14.0	5040.	135093.	140133.		
126	13.0					34.6	133903.
		12.5	4500.	127673.	132173.		
132	12.0					31.8	126445.
		11.5	4140.	120717.	124857.		

Thus, development within the watershed increased the 10-yr peak discharge by 162%. The detention basin must be designed to reduce the peak discharge from 131 cfs to 50 cfs.

The stage-storage relationship is computed from topographic information. At the site where the detention structure will be located, surface areas were estimated from the topographic map at an increment of 0.5 ft (see Table 11–8). The product of the average area (column 3) and the incremental depth of 0.5 ft yields the incremental storage (column 4). The storage for any stage is the accumulated incremental storage. A graphical presentation of columns 1 and 5 would be the stage-storage relationship. The values were used to fit the following power model:

$$V_s = 33{,}911\, h^{1.027}$$

The design policy requires a one-stage riser with a weir. The inflow hydrograph will be routed using the storage-indication method. An initial estimate of the weir length is made using the weir equation, with an assumed depth of 5 ft:

$$L_w = \frac{q_o}{C_w\, h^{1.5}} = \frac{50}{3.1\,(5)^{1.5}} = 1.44 \text{ ft}$$

Thus, an initial length of 1.5 ft will be used. The stage-storage-discharge relationship and the storage-indication curve are given in Table 11–9.

For design, an input hydrograph is necessary. The TR-20 computer program was used to derive the input hydrograph. The ordinates on a 0.1-hr increment are given in Table 11–10. Only the values for storm times from 11.0 to 13.2 hr are given.

The inflow hydrograph is routed with the storage-indication method; the results are given in Table 11–10. The largest outflow occurred at 102 min, which is storm time 12.7 hr. The peak outflow is 45.2 cfs, which is less than the allowable peak of 50 cfs. The maximum storage is 161,000 ft^3.

Since the computed peak is less than the allowable peak, the required storage volume can be reduced by allowing a higher peak discharge out of the basin. Thus, the weir length was increased to 2.0 ft and the computations repeated. Since the weir length is changed, the stage-discharge and storage-indication curves must be recomputed (see Table 11–11). The inflow hy-

TABLE 11–11 Stage-Storage-Discharge and Storage-Indication Curves for a Weir Length of 2.0 Feet

Stage (ft)	Discharge (cfs)	Storage (cu ft)	Storage Indication (cu ft)
.00	.0	0.	.0
.50	2.2	16641.	17035.7
1.00	6.2	33911.	35027.0
1.50	11.4	51426.	53476.6
2.00	17.5	69103.	72259.8
2.50	24.5	86901.	91312.4
3.00	32.2	104796.	110594.8
3.50	40.6	122772.	130079.2
4.00	49.6	140817.	149745.4
4.50	59.2	158924.	169577.4
5.00	69.3	177085.	189562.7
5.50	80.0	195296.	209690.8
6.00	91.1	213551.	229952.9

TABLE 11–12 Storage-Indication Analysis for Weir Length of 2.0 Feet

Time (min)	Inflow (cfs)	Average Inflow (cfs)	Average Inflow (cu ft)	$S - OdT/2$ (cu ft)	$S + OdT/2$ (cu ft)	Outflow (cfs)	Storage (cu ft)
0	.0					0.0	0.
		0.5	180.	0.	180.		
6	1.0					0.0	176.
		1.5	540.	172.	712.		
12	2.0					0.1	695.
		2.5	900.	679.	1579.		
18	3.0					0.2	1542.
		3.5	1260.	1506.	2766.		
24	4.0					0.4	2702.
		4.5	1620.	2637.	4257.		
30	5.0					0.5	4159.
		5.5	1980.	4060.	6040.		
36	6.0					0.8	5900.
		6.5	2340.	5760.	8100.		
42	7.0					1.0	7913.
		8.0	2880.	7725.	10605.		
48	9.0					1.4	10360.
		11.5	4140.	10114.	14254.		
54	14.0					1.8	13924.
		17.5	6300.	13594.	19894.		
60	21.0					2.8	19385.
		31.5	11340.	18875.	30215.		
66	42.0					5.1	29292.
		63.5	22860.	28369.	51229.		
72	85.0					10.8	49293.
		107.0	38520.	47356.	85876.		
78	129.0					22.5	81823.
		130.0	46800.	77770.	124570.		
84	131.0					38.2	117689.
		110.5	39780.	110808.	150588.		
90						50.2	

TABLE 11–13 Stage-Storage-Discharge and Storage-Indication Curves for a Weir Length of 1.75 Feet

Stage (ft)	Discharge (cfs)	Storage (cu ft)	Storage Indication (cu ft)
.00	.0	0.	.0
.50	1.9	16641.	16986.4
1.00	5.4	33911.	34887.5
1.50	10.0	51426.	53220.4
2.00	15.3	69103.	71865.2
2.50	21.4	86901.	90761.0
3.00	28.2	104796.	109869.9
3.50	35.5	122772.	129165.8
4.00	43.4	140817.	148629.4
4.50	51.8	158924.	168245.7
5.00	60.7	177085.	188003.0
5.50	70.0	195296.	207891.4
6.00	79.7	213551.	227902.7

TABLE 11–14 Storage-Indication Analysis for Weir Length of 1.75 Feet

Time (min)	Inflow (cfs)	Average Inflow (cfs)	Average Inflow (cu ft)	*S-OdT*/2 (cu ft)	*S+OdT*/2 (cu ft)	Outflow (cfs)	Storage (cu ft)
0	0.0					0.0	0.
		0.5	180.	0.	180.		
6	1.0					0.0	176.
		1.5	540.	173.	713.		
12	2.0					0.1	698.
		2.5	900.	684.	1584.		
18	3.0					0.2	1552.
		3.5	1260.	1519.	2779.		
24	4.0					0.3	2723.
		4.5	1620.	2666.	4286.		
30	5.0					0.5	4199.
		5.5	1980.	4112.	6092.		
36	6.0					0.7	6968.
		6.5	2340.	5844.	8184.		
42	7.0					0.9	8018.
		8.0	2880.	7852.	10732.		
48	9.0					1.2	10514.
		11.5	4140.	10296.	14436.		
54	14.0					1.6	14142.
		17.5	6300.	13849.	20149.		
60	21.0					2.5	19692.
		31.5	11340.	19235.	30575.		
66	42.0					4.6	29751.
		63.5	22860.	28926.	51786.		
72	85.0					9.6	50056.
		107.0	38520.	48326.	86846.		
78	129.0					20.2	83214.
		130.0	46800.	79581.	126381.		
84	131.0					34.5	120178.
		110.5	39780.	113974.	153754.		
90	90.0					45.6	145548.
		72.5	26100.	137342.	163442.		
96	55.0					49.7	154490.
		46.0	16560.	145538.	162098.		
102	37.0					49.2	153249.
		32.0	11520.	144401.	155921.		
108	27.0					46.5	147548.
		23.5	8460.	139175.	147635.		
114	20.0					43.0	139895.
		17.5	6300.	132155.	138455.		
120	15.0					39.3	131385.
		14.0	5040.	124314.	129354.		
126	13.0					35.6	122946.
		12.5	4500.	116538.	121038.		
132	12.0					32.4	115200
		11.5	4140.	109362.	113502.		

drograph was routed (see Table 11–12) with a resulting peak discharge of 50.2 cfs. Since this exceeds the allowable, another trial was made for a weir length of 1.75 ft (see Tables 11–13 and 11–14). The peak equals 49.7 cfs and the required storage equals 154,490 ft^3. The larger weir length reduced the required storage by about 6,500 ft^3 and thus the maximum water surface elevation by about 0.1 ft.

PROBLEMS

11–1. An underground storage tank with rectangular sections (length = 100 ft, width = 70 ft, height = 10 ft) is used for temporary control of storm runoff. Determine the stage-storage-discharge relationship assuming outflow from the tank is controlled by an orifice at the base of the tank. The rectangular orifice has a length of 3 ft and a height of 0.5 ft. The loss coefficient is 0.6.

11–2. An underground storage tank with rectangular sections (length = 75 ft, width = 30 ft, height = 8 ft) is used for temporary control of storm runoff. Determine the stage-storage-discharge relationship assuming outflow from the tank is controlled by an orifice at the base of the tank. The circular orifice has a cross-sectional area of 2 ft^2. The loss coefficient is 0.65.

11–3. A surface storage facility has a prismatoid configuration with a rectangular bottom, 60 ft × 100 ft, and side slopes of 4:1(h:v). The facility is used to control runoff from a small commercial area. Determine the stage-storage-discharge relationship assuming that outflow from the basin is controlled by a weir with a 3-ft length and $C_w = 3.1$. The weir invert is the same as the elevation of the bottom of the storage basin and the basin has a maximum height of 6 ft.

11–4. If the tank of Problem 11–3 contains 3 ft of dead storage (that is, the invert of the weir is 3 ft above the bottom of the storage basin) determine the stage-storage-discharge relationship with the stage measured from the invert of the weir.

11–5. A surface storage facility has a rectangular bottom ($L = 175$ ft, $W = 75$ ft), with the ends being vertical (that is, the length remains constant) and the sides having slopes of 5:1(h:v). The facility is used to control runoff from a residential area. Determine the stage-storage-discharge relationship assuming that outflow from the basin is controlled by a weir with a 6-ft length and $C = 3.1$. The weir invert is the same as the elevation of the bottom of the storage basin.

11–6. If the reservoir of Problem 11–5 has 2.5 ft of dead storage (that is, the invert of the weir is 2.5 ft above the bottom of the storage basin) determine the stage-storage-discharge relationship with stage measured from the invert of the weir. Also, determine the volume of dead storage.

11–7. For the storage facility of Problem 11–5, determine the stage-storage-discharge relationship assuming the outflow from the basin is controlled by a pipe riser with a rectangular orifice outlet, which has length of 8 ft, a height of 0.5 ft, and a loss coefficient of 0.6. The base of the orifice is 1 ft above the bottom of the storage basin.

11–8. For the storage facility of Problem 11–5, determine the stage-storage-discharge relationship if the outlet is a two-stage riser with a low-stage orifice and a high-stage weir. The rectangular orifice has $C_d = 0.6$, a length of 7 ft, and a depth of 0.6 ft. The weir has a length of 3 ft, a discharge coefficient of 3, and an invert elevation of 2.5 ft. The facility does not have dead storage.

11–9. Determine the stage-storage discharge relationship for the basin of Problem 11–8 if the low-stage outlet is a 2.0-ft weir ($C_w = 3.1$) rather than the orifice described in Problem 11–8 and a high-stage weir as described in Problem 11–8.

11–10. Derive and plot the storage-indication curve for the system described in Problem 11–1. Plot the storage-discharge relationship on the same figure. Assume a Δt of 5 min.

11–11. Derive and plot the storage-indication curve for the system described in Problem 11–5. Plot the storage-discharge relationship on the same figure. Assume a Δt of 15 min.

11–12. Derive and plot the storage-indication curve for the system described in Problem 11–9. Plot the storage-discharge relationship on the same figure. Assume a Δt of 20 min.

11–13. Assuming that the initial storage and discharge are zero, route the following hydrograph through the structure of Problem 11–1 using the storage-indication method. Use the storage-indication curve derived in Problem 11–10.

t (min)	0	5	10	15	20	25	30	35	40	45	50
q (ft^3/sec)	0	10	30	50	60	55	45	30	15	5	0

11–14. Assuming the initial storage and discharge are zero, route the following hydrograph through the structure of Problem 11–5 using the storage-indication method. Use the storage-indication curve derived in Problem 11–11.

t (min)	0	15	30	45	60	75	90	105	120	135	150	165	180
q (ft^3/sec)	0	30	75	120	175	160	125	85	55	35	20	10	0

11–15. Assuming the initial storage and discharge are zero, route the following hydrograph through the structure of Problem 11–9 using the storage-indication method. Use the storage-indication curve derived in Problem 11–12.

t (min)	0	20	40	60	80	100	120	140	160	180	200	220	240
q (ft^3/sec)	0	30	80	140	205	210	190	160	115	85	55	25	0

11–16. Use the modified Puls Method to derive the outflow hydrograph of Problem 11–13.

11–17. Use the modified Puls Method to derive the outflow hydrograph of Problem 11–14.

11–18. Use the modified Puls Method to derive the outflow hydrograph of Problem 11–15.

11–19. A detention basin that has a rectangular bottom (200 ft x 100 ft) and slopes of 4:1(h:v) on the sides; the ends are vertical (that is, the width increases with height but the length remains constant). Design a single-stage outlet facility that has a weir ($C_w = 3.0$) and will limit the peak outflow rate to 35 ft^3/sec when the hydrograph of Problem 11–13 is used as the inflow. Determine the volume of storage and depth of storage for the design.

11–20. Solve Problem 11–19 for an outlet facility that consists of a rectangular orifice with a height of 0.5 ft and $C_d = 0.6$.

11–21. Design a detention basin that has a maximum depth of 4.5 ft for active storage, side slopes that are not greater than 4:1, and uses a single-stage weir ($C_w = 3.0$) to limit the peak discharge to 105 ft^3/sec. Use the hydrograph of Problem 11–14 as the inflow. Assume topography will require both the length and width to vary with depth.

11–22. For the conditions of Problem 11–21, design a single-stage outlet with an orifice that limits the discharge to 120 ft^3/sec.

11–23. Using the detention basin of Problem 11–21, design a two-stage riser with a low-stage orifice ($C_o = 0.6$) and a high-stage weir ($C_w = 3.1$) that will limit the low-stage and high-stage peak discharges to 40 and 125 ft^3/sec, respectively. The depth of the basin cannot exceed 4 ft. The low-stage q_1 and high-stage q_2 inflow hydrographs are as follows:

t (min)	0	15	30	45	60	75	90	105	120	135	150	165	180
q_1 (ft^3/sec)	0	5	25	45	60	50	40	25	15	5	0	0	0
q_2 (ft^3/sec)	0	30	80	110	150	170	160	135	100	70	40	15	0

11–24. For the conditions of Problem 11–23, design a two-stage outlet with weirs ($C_w = 3.1$) for both stages.

REVIEW QUESTIONS

11–1. In reservoir routing synthesis, which one of the following is the unknown? (a) The downstream hydrograph; (b) the spillway energy loss coefficient; (c) the stage-storage-discharge relationship; (d) the physical characteristics of the outlet facility; (e) none of the above.

11–2. Reservoir routing methods are based on which of the following pair? (a) Bernoulli's Equation and the continuity equation; (b) the conservation of both mass and energy; (c) the conservation of mass and the stage-storage-discharge relationship; (d) the continuity equation and the conservation of energy; (e) Bernoulli's Equation and the stage-storage-discharge relationship.

11–3. The stage-discharge relationship for flow through a weir is not a function of (a) the upstream (inflow) hydrograph; (b) the length of the weir; (c) the depth of flow above the weir; (d) the energy loss characteristics of the weir; (e) all of the above are factors.

11–4. The stage-discharge relationship for flow through an orifice is based on which pair given in Review Question 11–2?

11–5. The storage-indication curve is a relationship between (a) I-O versus $\Delta S/\Delta t$; (b) S versus $S + O\Delta t/2$; (c) S versus O; (d) $S + O\Delta t/2$ versus O.

11–6. The routing increment Δt should be (a) small enough so that the assumption of linearity is not violated; (b) about 10% of the time to peak of the inflow hydrograph; (c) independent of the volume of storage in the reservoir; (d) about 30% of the time to peak of the outflow hydrograph.

DISCUSSION QUESTION

The technical content of this chapter is important to the professional hydrologist, but practice is not confined to making technical decisions. The intent of this discussion question is to show that hydrologists must often address situations where value issues intermingle with the technical aspects of a project. In discussing the stated problem, at a minimum include responses to the following questions:

1. What value issues are involved, and how are they in conflict?
2. Are technical issues involved with the value issues? If so, how are they in conflict with the value issues?
3. If the hydrologist attempted to rationalize the situation, what rationalizations might he or she use? Provide arguments to suggest why the excuses represent rationalization.

4. What are the hydrologist's alternative courses of action? Identify all alternatives, regardless of the ethical implications.
5. How should the conflict be resolved?

You may want to review Sections 1.6 to 1.12 in Chapter 1 in responding to the problem statement.

Case An engineer has been involved in the design of many large reservoir spillways for flood control and water supply dams. The engineer attends a workshop on the subject and based on the material presented deduces that he has not been applying the design concepts correctly. After the workshop, he does some calculations and, while almost all of his past design projects are safe, two spillways may not pass the design discharge that was required. He believes that the dams are safe for the moment but could fail for some floods less than the design flood. He recognizes that if he makes it known that the two projects are underdesigned, his reputation will be tarnished and his company will be responsible for costly retrofitting of the two spillways.

12

Water Yield and Snowmelt Runoff

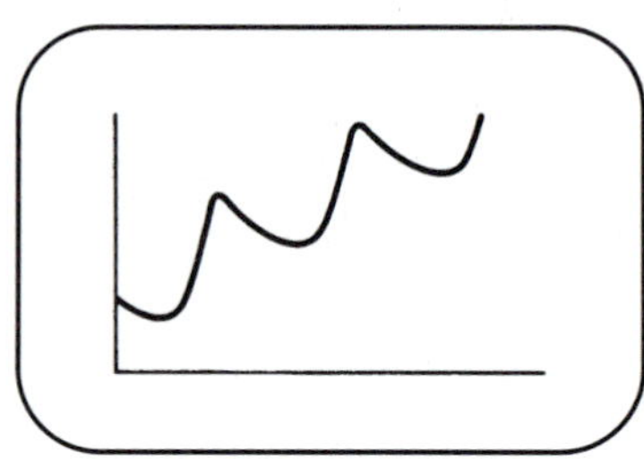

CHAPTER OBJECTIVES

1. Introduce methods for estimating snowmelt runoff that can be used for water management decisions.
2. Discuss methods used in evaluating the accuracy of snowmelt runoff models.
3. Provide an introduction to a range of models: conceptual versus empirical, short-term estimation versus long-term estimation, and simple structure versus complex structure.

12.0 NOTATION

a = degree-day factor
A = watershed area
C = characteristic that is an indicator of the volume of water available for runoff
C = runoff coefficient for Martinec Model
D = number of degree-days
K = daily runoff recession coefficient
K = proportionality constant between snowmelt runoff depth and the number of degree-days
$\overline{K}$ = mean value of K
k = recession coefficient
M = snowmelt runoff depth
m = number of gages
P = precipitation
P_w = winter precipitation
Q = snowmelt runoff volume or rate
Q_n = average daily discharge for time period n
Q_o = initial runoff rate
Q_s = seasonal snowmelt runoff
Q_t = recession runoff rate at time t
Q_w = winter runoff
S = fraction of snow
T = number of degree-days for Martinec Model
T = mean daily temperature
T_m = maximum daily temperature
t = time
$\hat{y}$ = estimated snowmelt runoff
$\overline{y}$ = regionalized estimate of snowmelt runoff

12.1 INTRODUCTION

Most hydrologic design problems involve either peak discharge estimation or flood hydrograph development. These methods are important because they are associated with the type of engineering analysis in which the greatest volume of design work is involved. Specifically, they are required for designs associated with urban and suburban development.

For other types of hydrologic problems, volumes of runoff over extended periods of time (up to 180 days) are important. Water supply for agriculture, power generation, urban water uses, recreation, and pollution abatement are just some of the needs where the volume of water is the design variable. Although some of the methods of analysis and synthesis used in peak discharge and flood hydrograph modeling are applicable to water yield estimation, the importance of the subject justifies a separate discussion of the application of these methods.

Snowmelt-runoff analysis is a special case of water-yield analysis. Snow is the major source of water supply and the major cause of flood-producing runoff in many parts of the western United States. Planning for agriculture, municipal water supply, energy generation, recreation, navigation, flood control, and pollution control depend on reliable forecasts of snowmelt runoff. Due to the increasing complexity and interaction of the various water uses, more accurate forecasting is essential for optimum water resources management. The continuing need for improved water supply forecasts is accentuated by the monetary benefit derived from a given percentage increase in forecast accuracy. For example, based only on improved management of power production and irrigation, Hamon (1972) estimated a \$6.2 million annual return for a 1% improvement in the January 1, 1972, Columbia River forecast at the Dallas, Oregon. Decades later, the benefits are most likely much greater.

Recognizing the importance of estimates of water yield and snowmelt runoff, methods of estimation are required. Ideally, theory would provide the basis for such estimation. Unfortunately, the inputs required to use theoretical developments in this area are difficult to collect or require excessive resources; furthermore, little historical data records of such measurements are available, which would make it difficult to test the theoretical models.

Fortunately, empirical models, which are an alternative to the theoretical models, have proven to provide reasonably accurate estimates of water yield and snowmelt runoff, especially for durations longer than 60 days. The data required to use these empirical models are readily available in many locations where such estimates are required.

Although the empirical formulas are used almost exclusively for actual water-yield estimation, it is worthwhile to take a cursory look at the theoretical approach to snowmelt-runoff estimation.

12.2 SNOWMELT-RUNOFF MODELING

Models for estimating seasonal water yield often relate factors, such as snow water equivalent observations, monthly precipitation, and observations on baseflow during the fall of the previous year, to seasonal water yield (SCS, 1970; Zuzel, 1975). Because of the lengthy time duration involved (3 to 5 months), much of the variation apparent over shorter time intervals is smoothed out. This data smoothing makes it possible to obtain fairly accurate results from models with simple structures and small input data requirements. The reliability of such models is often assessed using goodness-of-fit statistics, such as the correlation coefficient and the standard error of estimate. Although these criteria reflect the ability of a proposed model to fit the observed data, they may not be a good indication of the accuracy of future estimates, especially when the model was developed from a small sample of limited range. This is especially true in years of extreme, either large or small, snowfalls.

Besides the statistical criteria of goodness of fit, an investigator should (l) examine the empirical coefficients for rationality in sign and magnitude; (2) compare the relative importance of the predictor variables, as identified by the model and the investigator's knowledge of the system being modeled; and (3) examine the residuals (that is, the predicted value of the criterion variable minus the observed value) to ensure that the model provides reasonable estimates over the entire range of prediction.

To obtain a reliable model, an investigator should give consideration to the data analysis concepts of model formulation and model calibration. Besides the linear structure, other model structures (that is, a square-root or logarithmic transformation) should be investigated. A correlation analysis and transformations of the data are useful in selecting the best relationship between water yield and the predictor variables. Data transformations may significantly increase the goodness-of-fit statistics and decrease the correlation between predictor variables, thus providing a more reliable model in terms of model selection criteria other than the goodness-of-fit statistics.

Besides the model formulation stage, an investigator should give consideration to model calibration. A linear model that relates water yield to a set of predictor variables is often selected because experience has shown such a model yields high correlation coefficients; these models are most often calibrated using the principle of least squares. However, the frequently used multiple regression technique has some serious pitfalls; most notable are the irrational coefficients that result when two or more of the predictor variables are highly correlated among themselves. In such cases, an investigator should select a model calibration procedure that is not sensitive to such intercorrelation between predictors. Principal components regression analysis (McCuen, 1993), which is a feasible alternative to multiple regression, attempts to avoid irrational coefficients by discarding some of the troublesome intercorrelation through eigenvalue-eigenvector analysis. Stepwise regression (McCuen, 1993) avoids intercorrelation by removing those predictor variables that cause statistically significant intercorrelation. In summary, the use of a systematic and comprehensive strategy for model formulation and calibration should lead to a more reliable model.

12.3 EMPIRICAL FORMULAS

Probably the most widely used snowmelt forecast model is the linear regression equation, with both the bivariate and multivariate models being used. Hawley et al. (1980) showed that the accuracy of the regression model increased significantly as the length of the forecast period increased, with high accuracy for forecast periods of 60 days and longer. The volume of snowmelt runoff is used as the criterion (dependent) variable. The predictor (independent) variables may include snow water equivalent measurements, previous season runoff volumes (baseflow rates at the end of the previous year), and monthly or seasonal precipitation volumes. In many cases, several snow stations may be located in or near the watershed; in such cases, the individual measurements are used as separate predictor variables, while in some cases a weighted average of the individual measurements is used as a predictor variable. When numerous potential predictor variables are available, the model is usually fit using stepwise regression analysis. In this case, the computer analysis selects the subset of predictor variables that provides the maximum explained variance within the statistical constraint placed on the data. Other methods are available for fitting the coefficients of the model to data. For example, principal components regression analysis (PCRA) is useful when the predictor variables are highly correlated; PCRA overcomes the problem of irrational coefficients for highly correlated predictor variables. The use of statistical techniques and the interpretation of the results are best illustrated using an example.

Example 12–1

The area chosen for the analysis was the Upper Sevier River basin above Hatch, in south central Utah (Figure 12–1). The Sevier River above Hatch has a drainage area of 340 mi^2 (880.6 km^2). Elevations range from about 6500 ft (1981 m) to 11,000 ft (3353 m), with about 70% of the land area being above 9000 ft (2743 m). Native vegetation above 7000 ft (2133 m) consists primarily of aspen, Ponderosa pine, spruce, and Douglas fir with the most dense stands occurring on the plateaus and mountain slopes with a northern exposure. Vegetation below 7000 ft (2133 m) consists primarily of sagebrush, juniper, and scrub oaks. There are a few small diversions and a storage reservoir with a capacity of 10,700 acre-ft (1320 m-ha) (Carpenter et al., 1967).

The natural streamflow regime consists of low flows from August through March and high snowmelt flows between April and August. The average yearly discharge is about 94,000 acre-ft (11,560 m-ha) with about 75% occurring between April and August. The 1952–1976 baseflow data for the fall months and the flow for April to July and May to July are summarized

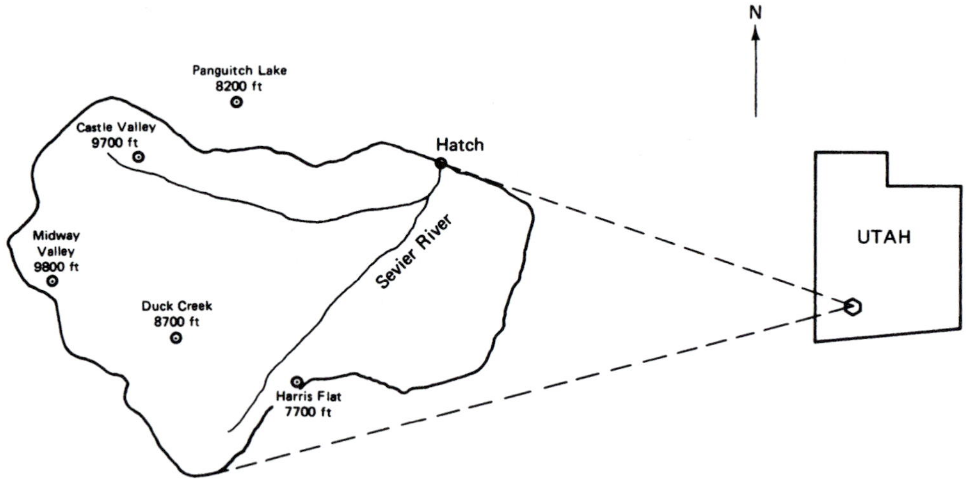

FIGURE 12–1 Location of Sevier River watershed and data stations.

in Table 12–1. The relatively low coefficients of variation for baseflow suggest that there is little yearly variation.

The basin has five operating snow courses and three rain gages (Figure 12–1). About 75% of the precipitation occurring during the winter months is in the form of snow. The mean monthly precipitation in April and May is small with a moderate degree of yearly variation (Table 12–1).

Snow water equivalent measurements (Table 12–1) vary considerably, with Midway Valley having the largest monthly averages and standard deviations. However, the monthly coefficients of variation are smaller than those for the other stations, which indicates that there is more yearly relative stability in the observed values. The monthly distribution of snow water equivalent at Midway Valley differs from the distributions at the other four sites since the mean for May is not very different from the mean for April, whereas there is a noticeable difference in the means for April and May for the other four snow stations. These characteristics suggest that the observed values at Midway Valley represent information that is different from the information at the other four sites.

Panguitch Lake has the smallest mean snow water equivalent values of the five stations and, while the standard deviations are also small, the relative variation is larger than for any of the other four sites. The monthly distribution of snow water equivalent is characterized by a significant decrease from April to May. This decrease is probably associated with its comparatively low elevation; thus, it appears to be somewhat different from the snow regime of Midway Valley.

Castle Valley and Duck Creek have similar monthly mean snow water equivalent values. The standard deviations and coefficients of variation for Duck Creek are somewhat larger than those for Castle Valley. This may be indicative of the elevation effect of the two sites. Castle Valley has an elevation quite similar to that of Midway Valley, which was also characterized by comparatively low relative variation.

Snow water equivalent measurements at Harris Flat are characterized by large relative variation and a significant decrease in the mean value from April to May. Again, this is probably due to the elevation effect.

TABLE 12–1 Summary Statistics for the Sevier River Watershed, 1952–1976

	Mean	Standard Deviation	Coefficient of Variation
	SNOW WATER EQUIVALENT (in.)[a]		
Castle Valley			
March 1	11.152	4.092	0.367
April 1	12.436	4.876	0.392
May 1	7.124	5.892	0.827
Midway Valley			
March 1	17.944	7.224	0.403
April 1	22.280	9.137	0.410
May 1	21.716	8.712	0.401
Harris Flat			
March 1	7.604	5.067	0.666
April 1	7.612	6.703	0.881
May 1	1.440	8.712	0.401
Duck Creek			
March 1	11.872	6.723	0.566
April 1	13.324	8.448	0.634
May 1	6.936	7.548	1.088
Panguitch Lake			
March 1	3.916	2.555	0.652
April 1	3.460	3.505	1.013
May 1	1.000	2.485	2.485
	PRECIPITATION (in.)[a]		
Castle Valley			
April	2.830	1.894	0.669
May	1.489	1.131	0.760
Duck Creek			
April	2.686	2.131	0.793
May	1.610	1.187	0.737
Hatch			
April	0.514	0.315	0.613
May	0.722	0.517	0.716
	MONTHLY BASEFLOW (10^3 acre-ft)[b]		
October	3.692	1.200	0.325
November	3.572	1.010	0.283
December	3.560	1.034	0.290
	FLOW (10^3 acre-ft)[b]		
April 1 to July 31	39.276	25.931	0.660
May 1 to July 31	35.504	23.404	0.699

[a]To convert inches to centimeters, multiply by 2.54.

[b]To convert acre-feet to ha-m, multiply by 0.123.

Examination of the correlations between the hydrologic variables indicated that the October, November, and December baseflow values are significantly correlated among themselves (0.959, 0.780, and 0.775). The correlations between baseflow and both snow water equivalent and precipitation are very small, less than 0.3. The intercorrelations between snow water equivalent for the March and April observations are very high (greater than 0.8), while the intercorrelations are much less (greater than 0.5) for the May observations; most of the correlations between the precipitation variables and the snow water equivalent variables are less than 0.3 in absolute value. Most of the precipitation variables are moderately correlated with each other. In summary, baseflow, snow water equivalent, and precipitation seem to be statistically independent, while the variables within each subgroup are highly correlated. In general, the correlations between the April-to-July or May-to-July flow and snow water equivalent are quite high (0.672 to 0.855), whereas the correlations between the April-to-July or May-to-July flow and precipitation and baseflow are small (–0.072 to 0.286).

Models for the April-to-July water-yield forecast were calibrated using each of the five calibration methods. The models were then compared with each other on the basis of the five criteria for model selection. The eight models calibrated are listed in Table 12–2. The correlation coefficient, percent explained variance (R^2), standard error of estimate, and the standardized partial regression coefficients (shown in parentheses below the partial regression coefficients) are also given in Table 12–2.

Multiple regression was used to calibrate models 1 and 2. Model 1 includes only one predictor variable, the April snow water equivalent at Midway Valley, which is the variable most highly correlated with the April-to-July streamflow. Model 2 includes the October baseflow and the April 1 snow water equivalents for all snow courses. Stepwise regression was used to calibrate model 3, which includes the March 1 snow water equivalents at Duck Creek and those for April 1 at Duck Creek and the December baseflow. Model 4 was derived using principal components regression analysis, where only two of the six principal components were used. Included as predictor variables were October baseflow and the April 1 snow water equivalents for each of the five snow courses. Model 5 has a polynomial form and was calibrated using principal components analysis. This model uses a single predictor variable (April 1 snow water equivalent at Midway Valley) and powers of it as the predictors instead of a set of predictor variables. Models 6 and 7 were calibrated using the numerical optimization technique known as pattern search. In model 6 the predictor variables include the October baseflow and the April 1 snow water equivalents at all snow courses. The model was formulated so that it included an intercept constant; both the intercept coefficient and coefficient for the April 1 snow water equivalent at Castle Valley were evaluated to be zero by the optimization process. Model 7 resembles model 6 except that baseflow was not included as a predictor variable and the intercept coefficient was allowed to take on negative values. This model is similar to the one presented by Zuzel and Ondrechen (1975) and Zuzel et al. (1975), except the April-to-July streamflow was not converted to area-inches. Model 8, which was calibrated using multiple regression, uses the December baseflow and a weighted sum of the April 1 snow water equivalents for Duck Creek, Harris Flat, and Midway Valley as predictor variables. Weighting of snow water equivalent measurements has the advantage of including measures of water content from different parts of the basin while keeping the statistical degrees of freedom at a maximum.

Correlation coefficient and standard error of estimate. The explained variance (R^2) for the eight models range from 0.778 to 0.921, a difference of 0.143 or 14.3%. The standard errors of estimate ranged from 7.62 to 13.74 × 10^3 acre-ft (937 to 1690 ha-m). Although these differences are significant, these statistical criteria cannot be used as the sole criteria in distinguishing between models, especially because of the small sample size (length of record) and the irrationality of the coefficients for some models.

TABLE 12–2 Models for Predicting April-July Streamflow, Sevier River Watershed

Model Number	Method[a]	Correlation Coefficient (R)	Explained Variation (R^2)	Standard Error of Estimate (10^3 acre-ft)	Model[b,c]
1	MR	0.887	0.787	12.22	$Q = -16.828 + 2.518x_5$ (0.887)
2	MR	0.906	0.820	12.71	$Q = -9.3086 + 2.321x_1 - 1.283x_2 - 1.226x_3 + 1.011x_4 - 1.499x_5 - 0.094x_6$ (0.107) (−0.241) (0.399) (0.261) (0.510) (−0.013)
3	SR	0.924	0.853	10.62	$Q = -22.14 + 6.092x_8 - 1.400x_7 + 4.013x_3$ (0.275) (−0.363) (1.307)
4	PCRA(2)	0.889	0.790	13.74	$Q = -12.97 + 2.964x_1 + 0.924x_2 + 0.569x_3 + 0.713x_4 + 0.535x_5 + 1.366x_6$ (0.136) (0.177) (0.185) (0.184) (0.188) (0.185)
5	PPC(1)	0.882	0.778	13.07	$Q = 6.265 + 0.838x_5 + 0.0153x_5{}^{**}2 + 0.0003231x_5{}^{**}3$ (0.295) (0.300) (0.296)
6	PS	0.897	0.804	13.35	$Q = 0.965x_1 + 1.618x_2 + 0.499x_4 + 0.424x_5 + 0.449x_6$ (0.045) (0.057) (0.129) (0.149) (0.061)
7	PS	0.895	0.801	13.01	$Q = -4.067 + 1.324x_3 + 0.290x_4 + 0.998x_5 + 0.381x_6$ (0.490) (0.086) (0.398) (0.058)
8	MR	0.960	0.921	7.62	$Q = -27.834 + 4.90747x_8 + 3.4925x_9$ (0.196) (1.012)

[a]MR = multiple regression
SR = stepwise regression
PCRA = principle components regression analysis with two principal components
PPC = polynomial fitting using PCRA.
PS = pattern search numerical optimization

[b]Q = April-July streamflow ($\times 10^3$ acre-feet)
x_1 = October baseflow ($\times 10^3$ acre-feet)
x_2 = April 1 snow water equilivant: Castle Valley (inches)

[b](*continued*)
x_3 = April 1 snow water equilivant: Duck Creek (inches)
x_4 = April 1 snow water equilivant: Harris Flat (inches)
x_5 = April 1 snow water equilivant: Midway Valley (inches)
x_6 = April 1 snow water equilivant: Panguitch Lake (inches)
x_7 = March 1 snow water equilivant: Duck Creek (inches)
x_8 = December baseflow ($\times 10^3$ acre-feet)
x_9 = $(x_3 + x_4 + x_5)/3$ (inches)

[c] Values in parentheses are the standardized partial coefficients.

Rationality of coefficients. Except for models 5 and 6 (model 6 was constrained not to have a negative intercept), all of the models had a negative intercept that ranged from –27.83 ac-ft for model 8 to –4.07 ac-ft for model 7. The negative intercepts are irrational because they suggest that when values for all the predictor variables are either zero or very small, the flow will be negative. A model with a zero intercept will produce a zero estimate when all the predictor variables equal zero. This may not be rational if factors other than the predictor variables that are included in the model affect the system; for example, precipitation and baseflow contribute to the total flow, and thus an intercept equal to zero may not be rational for a model that does not account for these variables. Based on the intercept coefficient the best models would be the pattern-search models (models 6 and 7) or the polynomial (model 5).

Besides the intercept coefficients, the partial regression slope coefficients must also be checked for rationality in both magnitude and sign. Because water yield should increase as the snow water equivalent measurements increase, the corresponding partial regression coefficients should be positive. Models 2, 3, and 8 include one or more irrational coefficients.

Relative importance. To assess the rationality of the magnitude of a regression coefficient, it is necessary to convert it to standardized form. Using the standardized partial regression coefficients as a measure of the relative importance of the corresponding predictor variable, the rationality of the magnitudes can be assessed. Models 6 and 8 have only one large standardized coefficient, which for model 6 corresponds to the variable having the largest predictor-criterion correlation (Midway Valley SWE for April 1), with the remaining standardized coefficients being significantly smaller. However, models 2, 4, 5, and 7 have more than one coefficient of similar magnitudes. Whether or not it is rational to have an equation dominated by a single predictor variable depends on the physical situation. Statistically, if only one predictor variable is dominant, it may be feasible to use a model that uses only that predictor variable. For example, the goodness-of-fit statistics suggest that model 1 is not significantly different from model 2, both of which have Midway Valley as the dominant variable. But one must also consider that an equation dominated by just one variable will give poor results if measurements on that one variable are in error or not representative of the state of the system. In summary, according to the magnitude and sign of the partial regression coefficients, models 4 and 7 are the best if a single dominant variable is not acceptable.

Residuals. Examination of the residuals represent the last criterion for model selection. Because accurate estimates of volumes during years with low streamflow are more critical than years with higher flows, low streamflows are considered more important in evaluating the residuals. The residuals for the eight models are given in Table 12–3, with the residuals given by year in order of increasing streamflow. All the models except model 7 tended to overpredict for the low values of streamflow and underpredict for the high values of streamflow. Models 1 and 5 had the most reasonable errors for low streamflow, while the other models had similar residuals at low flows. The residuals for model 8 are smaller than those of other models.

12.4 CONCORDANT FLOW METHOD

The regression method discussed previously is intended to be used only where gaged data exist to calibrate a prediction model for the watershed. Where watershed-specific data do not exist, it is necessary to transfer information from other watersheds. The concordant flow

TABLE 12–3 Residual Errors for Snowmelt Runoff Models, Sevier River Watershed

Year	Observed April to July Streamflow (10^3 acre ft)[a]	Residual Error[b] (percent of observed flow)							
		Model 1	Model 2	Model 3	Model 4	Model 5	Model 6	Model 7	Model 8
1959	13.8	24.0	97.8	−90.6	78.9	53.3	83.7	−49.9	−44.0
1974	16.6	26.1	48.9	−48.8	35.8	40.7	33.7	−16.4	−27.3
1953	16.7	29.9	41.1	−52.2	66.0	42.6	49.0	−31.5	0.7
1960	17.4	49.3	24.4	5.7	38.5	52.3	52.1	−51.1	25.3
1955	17.5	77.3	95.0	−63.4	105.5	70.9	106.6	−95.4	70.9
1961	18.2	58.0	24.2	−48.3	57.0	55.7	56.6	−56.8	−4.1
1963	18.9	−71.8	−78.5	58.2	−47.6	−20.4	−47.4	63.3	44.4
1972	20.9	15.9	−18.5	50.0	−35.7	21.4	−34.8	29.5	−15.3
1956	21.0	52.5	16.9	20.5	−7.7	45.8	25.0	−32.8	13.5
1970	22.9	−10.7	−3.9	−42.4	15.7	0.7	−11.5	26.1	−13.4
1971	24.3	10.0	5.8	21.8	18.1	11.1	17.1	−12.7	8.5
1976	25.2	44.1	46.6	−48.8	69.0	34.0	67.9	−61.9	−53.3
1964	29.1	−44.5	−37.7	13.1	−24.6	−29.2	−14.7	23.7	23.6
1954	35.4	54.5	48.4	−11.9	38.2	40.5	40.1	−44.6	−23.4
1975	37.4	11.2	−0.4	−5.3	4.9	1.4	9.1	−10.7	−18.8
1966	39.4	5.6	4.6	−12.2	−15.8	−3.7	−5.6	4.6	9.5
1957	40.2	−4.7	0.3	4.5	−21.9	−12.1	4.0	−3.6	−4.4
1965	51.9	−35.9	−33.2	−32.8	−36.6	−39.3	−27.1	28.7	14.9
1962	53.8	23.2	21.3	−27.3	19.9	15.4	24.0	−26.0	−0.8
1967	55.7	−41.1	−43.5	29.4	−54.3	−44.1	−47.7	47.9	26.1
1968	56.4	−17.8	−3.6	8.0	−4.7	−25.5	−6.4	11.3	7.4
1958	62.8	−5.3	−13.1	8.3	−0.4	−13.0	−10.9	9.2	−3.1
1952	88.0	13.7	8.4	−10.8	4.5	25.2	3.6	−9.1	−9.1
1973	91.3	−21.3	−18.5	18.7	19.0	−24.7	−22.9	22.1	16.4
1969	107.1	−16.9	−8.8	8.4	−8.5	−13.9	−11.2	11.9	1.0

[a]To convert acre-ft to ha-m, multiply by 0.123.

[b]Positive, underpredicted; negative, overpredicted.

method assumes that volumes of snowmelt runoff are available for similar watersheds in the region. If such data are available, it is possible to use indices of snowmelt similarities between the regional stations to make an estimate at the ungaged site. An estimate of the snowmelt runoff for the ungaged watershed ($\hat{y}$) is assumed to be a function of the regional snowmelt estimate ($\bar{y}$) and the ratio of a snowmelt runoff characteristic (C) for the ungaged and gaged watersheds:

$$\hat{y} = \bar{y}\left(\frac{C_u}{C_g}\right) \tag{12–1}$$

where the subscripts u and g on C indicate the ungaged and gaged watersheds, respectively. The characteristic C could be the percentage of the watershed covered with snow or the average snow depth for the watershed. If more than one gaged watershed with similar characteristics exists in the region, Equation 12–1 can be generalized to handle m gaged watersheds:

$$\hat{y} = \frac{C_u}{m} \sum_{i=1}^{m} \frac{y_i}{C_i} \tag{12–2}$$

in which y_i is the snowmelt estimate for gaged watershed i and C_i is the characteristic for that watershed. The concordant flow method is similar to the station-average method (see Section 4.5.1) used to determine a weighted estimate of rainfall. The concordant method uses the external factor C to weight the gaged value.

Example 12–2

To illustrate the use of Equation 12–1, assume that an estimate of snowmelt runoff of 0.80 in./day is available for a gaged watershed in which the average snow depth is 23.4 in. If the average snow depth for the ungaged watershed is 18.7 in., the estimated snowmelt runoff for the ungaged watershed is

$$\hat{y} = 0.8 \left(\frac{18.7}{23.4}\right) = 0.64 \text{ in./day} \tag{12–3}$$

If the ungaged watershed has an area of 54 mi^2, the estimated volume of runoff ($\hat{Q}$) is:

$$\begin{aligned} \hat{Q} &= 0.64 \frac{\text{in.}}{\text{day}} (54 \text{ mi}^2) \left(\frac{\text{ft}}{12 \text{ in.}}\right) \left(\frac{640 \text{ acres}}{\text{mi}^2}\right) \\ &= 1843.2 \text{ acre-ft/day} \end{aligned} \tag{12–4}$$

Example 12–3

If the fraction of snow-covered area for two gaged watersheds are 84 and 63% and the snowmelt runoff depths are 1.6 and 2.1 in./day, respectively, then Equation 12–2 can be used to estimate the snowmelt runoff from the ungaged watershed. If the snow-covered area for the ungaged watershed is 88%, Equation 12–2 yields:

$$\hat{y} = \frac{88}{2} \left(\frac{1.6}{84} + \frac{2.1}{63}\right) = 2.30 \text{ in./day} \tag{12–5}$$

The flow rate in acre-ft/day can be computed if the drainage area of the ungaged watershed is available. The weighting scheme of Equation 12–2 assumes that each of the gaged watersheds are given equal weight. Equation 12–2 could be modified to allow for unequal weights.

12.5 DEGREE-DAY METHOD

Temperature is an important determinant of snowmelt runoff. Given the general availability of temperature data, a method requiring only mean daily temperature would represent a most fundamental prediction method. The degree-day method, which requires only temperature as input, has the form

$$M = KD \tag{12–6}$$

in which K is a proportionality constant, D is the number of degree-days for a given day, and M is the snowmelt runoff in inches per day. The volume rate of melt (Q) can be computed by multiplying M by the drainage area or the snow-covered area:

$$Q = MA = KDA \tag{12–7}$$

The number of degree-days per day equals the difference between the mean temperature and 32°F. Thus Equation 12–6 becomes

$$M = \begin{cases} K(T - 32) & \text{for } T \geqslant 32°\text{F} \quad (12\text{–}8a) \\ 0 & \text{for } T < 32°\text{F} \quad (12\text{–}8b) \end{cases}$$

in which T is the mean temperature for the day.

It should be evident that the form of Equation 12–7 is similar to the form of the Rational Method used to compute peak discharges. The proportionality constant K of Equation 12–7 corresponds to the C of the Rational Equation. While the rainfall intensity of the Rational Equation indicates the available rainfall, the value of D in Equation 12–7 reflects the amount of heat available to cause melt. The similarity between the degree-day method and the Rational Formula is pointed out only to suggest the potential accuracy of the method.

The value of K in Equation 12–8 is a function of a number of factors that reflect the potential of a watershed to produce daily melt, including factors such as the depth of snow, the watershed orientation, the percent of forest cover, the slope, and the time of year. Given that it does not seem possible to develop a means of computing the daily variation of K, a constant value is usually assumed for a watershed. For watersheds where data to fit K are not available, a value can be taken from Table 12–4.

12.5.1 Analysis

The value of K in Equation 12–6 shows wide variation, both spatially and temporally. Therefore, if data for a watershed exist, an average value of K can be estimated for a watershed. For a record of n days, the average value of K could be computed using

$$\overline{K} = \frac{1}{n} \sum_{i=1}^{n} \frac{M_i}{D_i} \tag{12–9}$$

TABLE 12–4 Values of K for the Degree-Day Method

Condition	K
Extremely low runoff potential	0.02
Heavily forested areas; north-facing slopes of open country	0.04–0.06
Average runoff potential	0.06
South-facing slopes of forested watersheds; average open country	0.06–0.08
Extremely high runoff potential	0.30

where the subscript i is used to indicate the measurement of M or D for day i. Depending on the characteristics of the watershed, the ratio M_i/D_i may show wide scatter from day to day.

Example 12–4

The analysis for K can be demonstrated using data from the Conejos River in Colorado. The Conejos River basin has a drainage area of 282 mi^2 and a range in elevation of 4908 ft (8272 to 13,180 ft). The watershed can be separated into three elevation zones, which correspond roughly to the mixed conifer-aspen band (zone A), the spruce-fir zone (zone B), and the area above the timberline (zone C), with area fractions of 0.201, 0.459, and 0.340, respectively. Degree-day data for each zone are given in Table 12–5, along with the mean weighted according to the areal fraction for each zone. After converting the flow, which is given in Table 12–5 in ft^3/sec to in./day, the daily value of K was computed as the ratio M_i/D_i from Equation 12–9, with a mean value of 0.022 resulting. The daily values for this short record ranged from 0.016 to 0.028. For such a large and varied watershed, one would expect the daily values to range considerably.

12.5.2 Synthesis

Values of the parameter K can be obtained by analysis of measured data for the watershed of interest or from average values obtained by analyses of measured data for watersheds in the region (regional estimates). The accuracy of estimates obtained from data for the watershed of interest will depend on the representativeness of the data used for calibration. The accuracy of both regionalized estimates or generalized values obtained from Table 12–4 will depend on the similarity of the snowmelt characteristics on the watershed of interest and those used to derive the regionalized or generalized values of K. Most often, this is difficult or impossible to evaluate.

The potential accuracy of a method can be further assessed by split-sample testing. In its most basic form, split-sample testing involves fitting the model coefficients, such as K in Equation 12–9, with part of the available data and using the fitted coefficients to make pre-

TABLE 12–5 Analysis of Flow and Degree-Day data estimating K for the Conejos River Basin, Colorado

1979	Degree-Days for Zone:			Weighted	Flow	
May	A	B	C	Mean	(ft^3/sec)	K
20	16.7	13.0	9.3	12.5	1665	0.018
21	15.1	11.9	8.7	11.5	1630	0.019
22	18.3	14.1	10.0	13.6	1605	0.016
23	18.8	13.0	7.3	12.2	1680	0.018
24	16.7	11.8	7.0	11.2	1965	0.023
25	17.0	11.5	6.0	10.7	1925	0.024
26	18.1	12.1	6.2	11.3	2010	0.023
27	19.7	13.2	6.8	12.3	2200	0.024
28	18.7	12.2	5.8	11.3	2325	0.027
29	19.8	12.9	6.0	11.9	2485	0.028
					Sum	0.220
					Mean	0.022

TABLE 12–6 Synthesis of flow data for the Conejos River Basin, Colorado, for May 20–29, 1976.

1979 May	Degree-Days for Zone: A	B	C	Weighted Mean	Computed Flow (in./day)	Computed Flow (ft^3/sec)	Actual Flow (ft^3/sec)	Error (ft^3/sec)
20	14.6	6.5	0.0	5.9	0.13	984	1190	−206
21	15.8	9.5	3.8	8.8	0.19	1468	1240	228
22	11.5	7.9	4.6	7.5	0.16	1251	1190	61
23	12.6	7.3	2.5	6.7	0.15	1118	1090	28
24	14.2	8.4	3.2	7.8	0.17	1301	1160	141
25	16.7	12.1	7.8	11.6	0.26	1935	1150	785
26	14.7	10.1	5.6	9.5	0.21	1585	966	619
27	18.3	13.3	8.6	12.7	0.28	2119	1070	1049
28	20.7	14.9	9.7	14.3	0.31	2386	1300	1086
29	20.1	14.8	10.0	14.2	0.31	2369	1400	969
							Sum	4760
							Mean	476

dictions that can be compared with the remaining part of the measured data. The bias and accuracy for the synthesized data can then be computed and used as independent estimates of the validity of the model coefficients.

Example 12–5

Based on the K value computed with the data of Table 12–5, values of M (Equation 12–6) and Q (Equation 12–7) can be predicted. The value of K was tested using values of D for the period of May 20–29, 1976. The values of D for the three zones are shown in Table 12–6. The areal weighted value of D was also computed. The value of K of 0.022 and the weighted values of D were used to compute the values of M and Q, which are given in Table 12–6 along with the actual values of Q. The errors were computed. For the first 5 days of the test period, the model gave reasonable estimates; however, for the second period of 5 days, the computed values show a large positive bias. The bias for the 10-day test period was 476 ft^3/sec. These results indicate that a more complete analysis would be needed for this watershed.

12.5.3 Adjustment of Temperatures for Altitude

An indirect relationship between air temperature and altitude is commonly used. Measurements have shown that for each 1000-ft increase in altitude, the air temperature will decrease from 3°F to 5°F. As an approximation, a correction of 4°F per 1000-ft change is reasonable. The adjustment of temperatures for changes in altitude is often called a *lapse rate* adjustment.

12.6 TEMPERATURE INDICES OF SNOWMELT

Air temperature data are widely available, and studies have shown that reliable estimates of snowmelt runoff can be made using indices based on temperature data. Recognizing that snowmelt runoff rates vary with factors other than temperature, the accuracy of estimates

TABLE 12–7 Point Melt Rates: Degree-Day Factors (inches of melt per degree-day above 32°F) for Three Snow Laboratories

Laboratory	Maximum Melt	Minimum Melt	Mean Melt
	Based on mean daily temperatures		
CSSL	0.128	0.066	0.106
CSL	0.131	0.054	0.090
WBSL	0.108	0.026	0.060
Mean	0.122	0.049	0.085
	Based on maximum daily temperatures		
CSSL	0.054	0.029	0.045
UCSL	0.041	0.020	0.035
WBSL	0.060	0.015	0.034
Mean	0.052	0.021	0.038

Source: U.S. Army Corps of Engineers, *Runoff from Snowmelt*, EM-1110-2–1046, 75 pp., Jan. 5, 1960.

made with such indices will depend highly on the representativeness of the conditions inherent in the data used to calibrate the temperature index.

Based on data from three snow laboratories, the Corps of Engineers (1960) reported the point melt rates shown in Table 12–7. The values given in the table are the inches of melt per degree-day above 32°F. These values are valid only for the spring snowmelt season, when rain and wind are not factors affecting melt rates.

Daily snowmelt rates (M in inches) can also be estimated using equations based on either the mean temperature (T, in °F) or maximum temperature (T_m in °F) index. The following are examples of such equations:

Open sites:

$$M = 0.06\,(T - 24) \quad \text{for } 34^\circ \leq T \leq 66^\circ \qquad (12\text{–}10a)$$
$$M = 0.04\,(T_m - 27) \quad \text{for } 44^\circ \leq T_m \leq 76^\circ \qquad (12\text{–}10b)$$

Forested sites:

$$M = 0.05\,(T - 32) \quad \text{for } 34^\circ \leq T \leq 66^\circ \qquad (12\text{–}10c)$$
$$M = 0.04\,(T_m - 42) \quad \text{for } 55^\circ \leq T_m \leq 76^\circ \qquad (12\text{–}10d)$$

It is evident from Equations 12–10 that melt rates are greater for open sites than for forested sites. Equations 12–10 are valid only for springtime snowmelt estimation.

12.7 DAILY SNOWMELT ESTIMATION

To this point, the snowmelt models have been designed to make long-duration forecasts, with durations from 30 to 180 days. These models use input variables that reflect the amount of water available either in the snowpack at the start of the forecast period or the amount of water that existed at the end of the previous year. The coefficients of the models reflect the expected rate of melt, the expected losses, and the precipitation that can be expected during the forecast period. Such models are very useful for planning purposes.

A number of design and forecast problems require short-duration forecasts (forecasts for 30 days or less). Snowmelt runoff is responsible for flooding on many watersheds. Reservoir operation depends on short-duration forecasts as well as long-duration forecasts. Water allocation is also dependent on daily or weekly flows. Because of these needs, a number of models have been developed specifically for making short-duration forecasts.

Because of the greater amount of variation that one can expect over short durations, forecast models must use input variables that reflect the causes of the variation. Additionally, one would expect the accuracy of forecasts to be greater when the model is based on a rational conceptual analysis of the snowmelt runoff process. Therefore, the models developed for making short-duration forecasts differ from the long-duration forecast models in both the type of input and the model structure.

12.7.1 The Martinec Model

The Martinec Model (1975) is an example of a simple, short-duration forecast model. It has the following simple form:

$$Q_{n+1} = CA\,(aTS + P)\,(1 - k) + kQ_n \tag{12–11}$$

in which C is the runoff coefficient, A the watershed area, a the degree-day factor, T the number of degree-days for the forecast period, S the fraction of the snow-covered area, P the precipitation contributing to runoff, k the recession coefficient, and Q_{n+1} and Q_n the average daily discharges for forecast period $(n + 1)$ and n, respectively. Before discussing the input to the model, it is worthwhile examining the structure of the model. The second term in the model is a term that reflects the autocorrelation between flows in adjacent time periods. On this basis, k would be an autoregression coefficient. For large watersheds one would expect greater autocorrelation between flows in adjacent forecast periods, and in such cases, the second term would have greater significance. For small watersheds and short forecast durations, flows are more sensitive to the daily fluctuations in meteorological conditions, and therefore the autocorrelation term becomes less significant. As the autoregression coefficient k becomes smaller, the factor $(1 - k)$ increases and the first term on the right side of Equation 12–11 becomes more significant. In summary, the second term on the right side of Equation 12–11 represents the degree to which flows in adjacent forecast periods are similar, and the value of k reflects the importance of the autoregressive nature of the flows relative to the importance of the meteorological inputs during the forecast period.

Whereas the autoregressive term of Equation 12–11 reflects the effect of past input on the current flow, the first term on the right side of Equation 12–11 reflects the result of snowmelt and precipitation during the current forecast interval. The product aTS reflects the melt from the snowpack, and P reflects the precipitation input for the forecast interval. The available water "depth" $(aTS + P)$ is multiplied by the area to give a volume and by the runoff coefficient to reflect the proportion of the available supply that contributes directly to runoff during the forecast interval. As a side note, the similarity between the first term on the right side of Equation 12–11 and the Rational Equation for computing peak discharges should be evident. Also, the form of the product aTS is similar to the form of the degree-day method of Equation 12–7. The degree-day factor (a for Equation 12–11) is similar to the constant K of the degree-day method (Equation 12–6). In the term aTS, the value of S is used to adjust the melt aT for the proportion of the watershed that contributes melt.

12.7.2 Analysis of Coefficients

The conceptual framework of the Martinec Model is obviously rational. However, to apply the model it is necessary to provide values for both the input variables A, T, S, P, and Q_n and the empirical coefficients C, a, and k. Martinec (1975) provided fitted values of the degree-day factor, which reflects the amount of snow that melts for each degree-day. Values are typically in the range from 0.35 to 0.60 cm per degree(Celsius)-day. The variation reflects variation in factors that affect the rate of snowmelt, including the available energy, the state of the snowpack, and meteorological factors. Thus the value of a will vary on a day-to-day basis, with an obvious systematic variation over the course of a snowmelt season. The value will also vary with the elevation, and for a watershed that is divided into zones, different values may have to be fit for each elevation zone. One problem in fitting C, a, and k to data is the structural interdependence between the three coefficients.

Least-squares estimates of the coefficients k and C can be obtained from the following:

$$C = \frac{\Sigma\, Q_{n+1}V - k\Sigma\, Q_n V}{(1 - k)\,\Sigma V^2} \tag{12–12}$$

$$k = \frac{C^2\Sigma V^2 - C\,[\Sigma V Q_n + \Sigma V Q_{n+1}] + \Sigma Q_n Q_{n+1}}{C^2\Sigma V^2 - 2C\Sigma V Q_n + \Sigma Q_n^2} \tag{12–13}$$

in which $V = A(aTS + P)$. The values of C and k can be obtained iteratively by assuming a value of k, solving Equation 12–12 for C, substituting this value of C into Equation 12–13, and comparing the initial value of k with the value computed with Equation 12–13. If they are different, the new estimate must be used for a second iteration, with the iterating continued until the values of C and k converge. Once the process has converged, the flows Q_{n+1} can be estimated, residuals ($\hat{Q}_{n+1} - Q_{n+1}$) computed, and an estimate of the standard error made.

12.8 SNOWMELT RECESSION ESTIMATION

While runoff volumes for periods of 30 days or more are often the primary variables used in water resources management of snowmelt runoff supplies, many models use physically based models that involve moisture accounting on a daily basis. For such models an understanding and representation of the snowmelt process is important. While data bases used for these models may consist of daily totals or daily means, it is recognized that melt and runoff rates are not constant over a 24 hr period. The daily fluctuations in temperature are reflected in the snowmelt runoff hydrograph, with the runoff rates lagging the changes in temperature. Figure 12–2 shows a snowmelt runoff hydrograph for a 4-day period. The schematic shows that all of the water that melts on a given day does not appear as runoff on that day; in fact, the melt can appear over a period of many days, just as baseflow from runoff that results from rain can occur over an extended period of time. Figure 12–2 also suggests that the water in the snowmelt recession can represent a significant part of the total runoff.

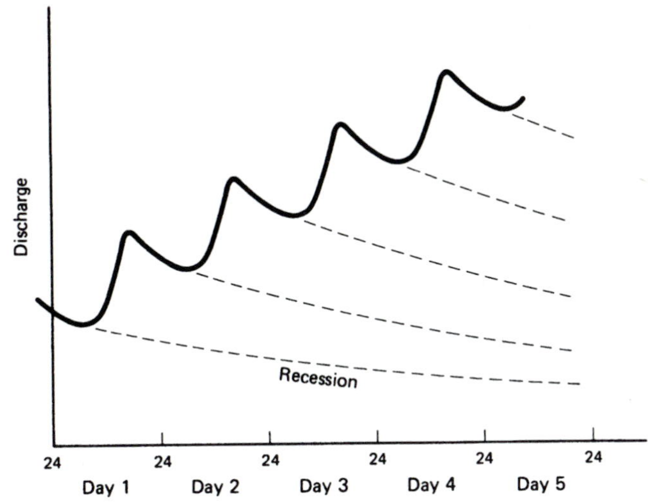

FIGURE 12–2 Continuous snowmelt hydrograph showing contribution from snowmelt recessions.

For short-term modeling of snowmelt runoff, the accuracy of estimated runoff rates can be enhanced by including a model component that represents the effect of recession runoff. A number of models have been proposed. Barnes (1939) proposed a decay function having the form

$$Q_t = Q_o K^t \tag{12–14}$$

in which Q_o is the initial runoff rate of the recession on day $t = 0$, t is the number of days since the start of the recession, Q_t is the recession runoff rate on day t, and K is the daily runoff recession coefficient. A least squares estimate of K can be obtained using Equation 2–100, with t equal to 1 and values of Q_o and Q_t selected from recessions of measured snowmelt streamflow. Equation 12–14 can also be stated as

$$Q_t = Q_o e^{-kt} \tag{12–15}$$

The parameters k and K are related by

$$k = -ln_e K \tag{12–16}$$

Other functional forms can also be used to represent the recession, such as the simple power model:

$$Q_{t+1} = aQ_t^b \tag{12–17}$$

Equation 12–17 has the advantage that the coefficients are easily computed using readily available computer software for least squares.

Example 12–6

The calibration of a recession model can be demonstrated using data from the Conejos River. Table 12–8 provides the daily mean streamflow for the period from April 1, 1976, to September 30, 1976. For this illustration, recessions were defined as periods of 4 or more days during which

TABLE 12–8 Measured Streamflow (ft^3/sec), Conejos River, near Magote, Colorado, for 1976

Day	April	May	June	July	August	September
1	101	460	1460	564	184	64
2	125	530	1780	512	204	68
3	148	633	1695	502	180	66
4	201	740	1705	476	180	53
5	212	730	1805	436	154	51
6	184	770	2140	344	131	48
7	161	640	1700	285	101	55
8	187	525	1870	292	88	53
9	236	460	1840	285	88	55
10	306	464	1790	282	96	59
11	368	559	1570	271	109	60
12	416	727	1240	250	106	64
13	376	690	1110	250	96	59
14	282	920	1055	229	98	59
15	229	1155	940	201	98	59
16	194	1300	885	222	98	60
17	174	1450	850	240	84	59
18	154	1785	685	198	72	50
19	154	1645	765	177	72	48
20	139	1350	820	181	77	48
21	174	1360	860	145	72	48
22	198	1425	895	167	64	57
23	236	1305	920	164	62	55
24	288	1315	725	136	66	51
25	344	1345	668	125	79	59
26	364	1235	610	128	84	84
27	372	1250	600	148	96	86
28	444	1665	516	208	84	122
29	520	2035	494	164	68	131
30	484	1885	476	128	62	123
31		1365		119	62	

the runoff rate did not increase. This provided 63 values of Q_t and Q_{t+1} from which the following model was calibrated:

$$\hat{Q}_{t+1} = 0.8973\, Q_t^{0.99945} \tag{12–18}$$

This model resulted in a correlation coefficient of 0.9968 and is shown in Figure 12–3.

12.9 WATER BALANCE MODELS

Water balance models are formulated with a simple structure to reflect causality of the physical factors that control the melt process and runoff. The simple structure is accompanied by a reduction in the data requirements and serves to simplify the calibration. The intent behind water balance models is to account for all water entering and leaving the basin. The volume

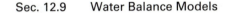

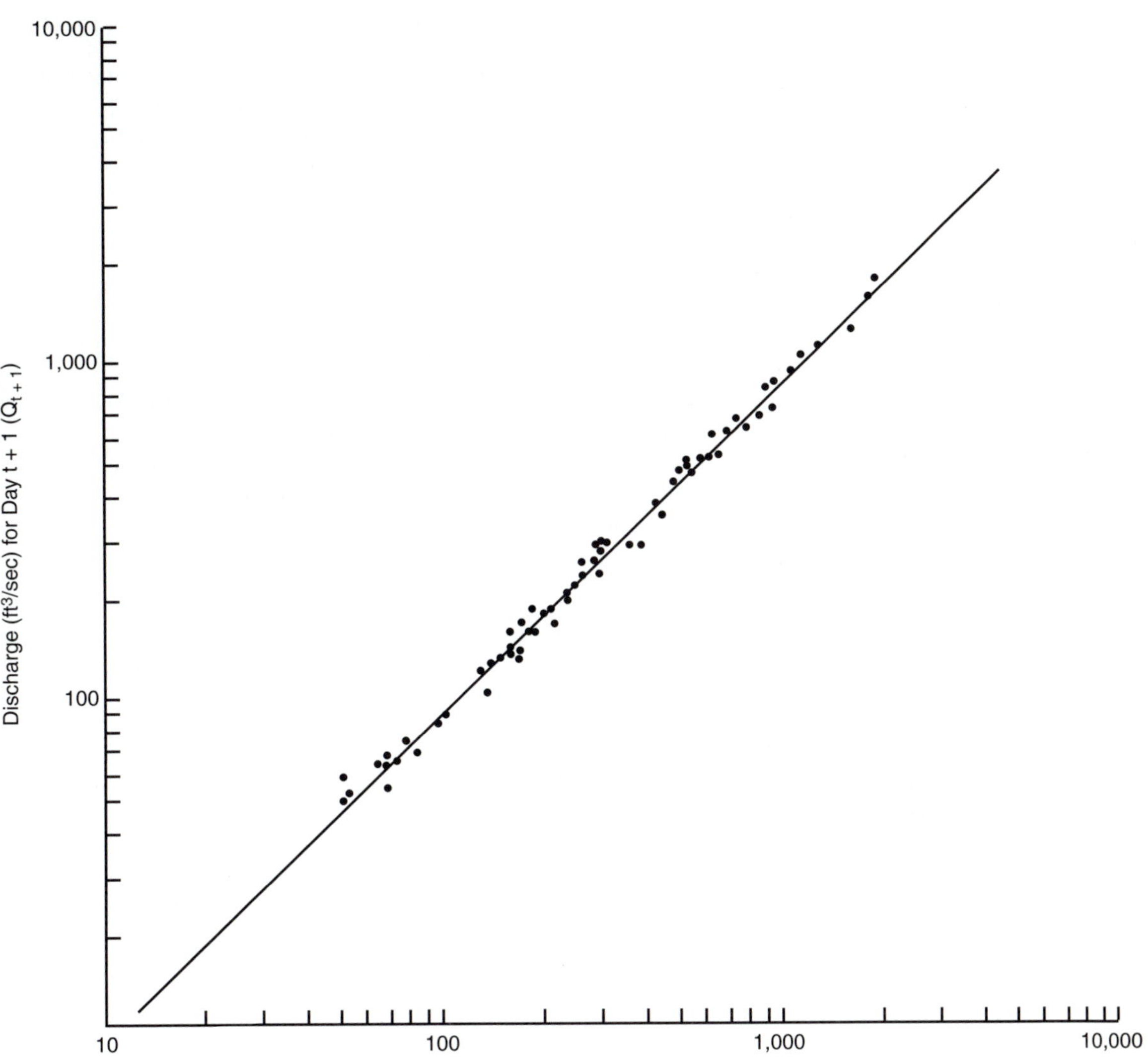

FIGURE 12–3 Recession model for the Conejos River watershed (April 1, 1976, to September 30, 1976).

of water stored in the snowpack is represented through input requirements such as total precipitation volumes for long time periods or snow water equivalent measurements. Some models may make allowances for losses due to evaporation, transpiration, and ground water storage. For some models, loss rates are used to reflect the losses, with no regard for the specific processes governing losses.

Tangborn (1978) formulated a model that can be considered as a water balance model. The model relates the seasonal snowmelt runoff Q_s to the measured winter precipitation P_w and winter runoff Q_w, by

$$Q_s = a + bP_w - Q_w \tag{12–19}$$

in which a and b are fitting coefficients. The coefficients a and b can be fit using least squares because Q_w is not associated directly with a coefficient; however, Equations 2–56 cannot be used directly. Instead, the model of Equation 12–19 must be inserted into Equation 2–52 and the equations for estimating a and b derived, which are as follows:

$$b = \frac{\Sigma Q_s P_w - (\Sigma Q_s \Sigma P_w/n) + \Sigma Q_w P_w - (\Sigma Q_w \Sigma P_w/n)}{\Sigma P_w^2 - [(\Sigma P_w)^2/n]} \tag{12–20}$$

and

$$a = \overline{Q_s} + \overline{Q_w} - b\overline{P_w} \tag{12–21}$$

in which $\overline{Q}_s$, $\overline{Q}_w$, and $\overline{P}_w$ are the means of Q_s, Q_w, and P_w, respectively. To calibrate Equation 12–19 it is necessary to have values of Q_s, Q_w, and P_w for n years of record. Equation 12–19 should not be calibrated by subtracting Q_w from Q_s and regressing the difference $(Q_s + Q_w)$ on P_w because then the values of a and b relate $(Q_s + Q_w)$ to P_w, and the computed correlation coefficient would reflect the accuracy in estimates of $(Q_s + Q_w)$ but not Q_s.

Example 12–7

The data of Table 12–9 shows the winter precipitation (P_w), winter runoff (Q_w), and seasonal snowmelt runoff Q_s (all in in.) for a record of 9 years. The summations of Equation 12–20 are given in Table 12–9 and were used to compute the model coefficients with Equations 12–20 and 12–21:

$$\hat{Q}_s = -16.63 + 1.030\,P_w - Q_w \tag{12–22}$$

Equation 12–22 was used to regenerate the values of Q_s for the 9 years of record and the errors (e_j) computed (see Table 12–9). The sum of the errors equals zero, which indicates that the model is unbiased. Equation 2–64 cannot be used to compute the correlation coefficient for Equation 12–22, so the accuracy of Equation 12–22 was assessed by computing the standard error of estimate with the sum of the squares of the errors; Equation 2–65 yields a S_e of 2.63 in., which can be compared with the standard error of the Q_s values, which was 4.38 in. Thus the expected error was reduced by 40%.

TABLE 12–9 Example 12–7: Water Balance Model

Year	P_w (in.)	Q_w (in.)	Q_s (in.)	$\hat{Q}_s$ (in.)	e (in.)
1971	42	8	17	18.6	1.63
1972	37	6	16	15.5	–0.52
1973	51	14	26	21.9	–4.10
1974	44	7	19	21.7	2.69
1975	33	4	12	13.4	1.36
1976	46	12	18	18.7	0.75
1977	38	5	16	17.5	1.51
1978	53	17	25	21.0	–4.04
1979	45	10	19	19.7	0.72
Mean	43.2	9.2	18.7		$\Sigma e = 0$
Std. dev.	6.5	4.4	4.4		$\Sigma e^2 = 48.48$

PROBLEMS

12–1. The following data consist of April 1 snow water equivalent measurements (x, in.) and streamflow (y, acre-ft) for the 4/1 to 7/31 period over a span of seven years. Derive and evaluate both linear and power least squares models for estimating the streamflow.

y (* 10^4)	8.0	10.7	9.6	11.8	8.7	7.4	9.8
x	16	26	23	28	19	15	22

12–2. The following eight-year record consists of April 1 snow water equivalent measurements (in.) at two stations (a high-elevation (x_1) and a low-elevation (x_2) station) and the streamflow (y, acre-ft) for the 4/1 to 7/31 period. Derive and evaluate both linear and power least squares models for estimating streamflow.

y (* 10^4)	7.0	8.9	10.3	9.4	6.1	10.1	8.6	10.8
x_1	22	26	31	29	14	32	28	33
x_2	1.4	2.2	6.7	6.4	1.1	7.8	3.9	8.6

12–3. The following data consist of March 1 snow water equivalent measurements (x_1, in.), April 1 snow-covered area measurements (x_2, %) for 9 years, and the streamflow for the 4/1 to 6/30 forecast period. Derive and evaluate both linear and power least squares models for estimating streamflow.

y (* 10^4)	3.2	4.2	4.0	4.9	6.2	5.6	6.8	5.0	5.5
x_1	10.9	11.5	13.1	10.3	11.2	13.2	14.8	15.0	16.4
x_2	58	55	71	41	60	74	76	84	95

12–4. A regression equation for watershed A yields a snowmelt runoff estimate of 8300 acre-ft; with a snow-covered area of 75%. The snow-covered area for watershed B, which is hydrologically similar to watershed A, is 64%. Estimate the snowmelt runoff for watershed B.

12–5. A regression equation for watershed C yields a snowmelt runoff estimate of 92,000 acre-ft, with an average snow water equivalent of 33 in. The average snow water equivalent for watershed D, which is hydrologically similar to watershed C, is 28 in. Estimate the snowmelt runoff for watershed D.

12–6. Watershed E, which is similar in size as well as other characteristics to both watersheds A and B of Problem 12–4, has a snow-covered area of 81% and a snowmelt runoff estimate of 9100 acre-ft. Estimate the snowmelt runoff for watershed B using the information from watersheds A and E.

12–7. The snow pack on watershed F has an average snow water equivalent of 24 in.; however, in comparison to watersheds C and D of Problem 12–5, watershed F has a few dissimilar characteristics. Therefore, weights of 0.3 and 0.7 are applied to watersheds F and C, respectively. If the snowmelt runoff estimate for watershed F is 65,000 acre-ft, estimate the snowmelt runoff for watershed D using the information for watersheds C and F.

12–8. The following data are the maximum and minimum air temperatures for Pagosa Springs, Colorado, for 10 days in April 1979. Determine the daily number of degree-days, the melt (in./day), and the volume rate of melt (acre-ft) from a 79-mi^2 watershed. Compare the melt assuming the watershed has (a) extremely low, (b) average, and (c) extremely high runoff potential.

Day	Maximum Temperature (°F)	Minimum Temperature (°F)
1	40	18
2	40	5
3	43	16
4	43	7
5	53	14
6	63	19
7	60	6
8	64	12
9	59	23
10	45	29

12–9. The following data are the maximum and minimum air temperatures for Wolf Creek, Colorado, for 10 days in April 1979. Determine the daily number of degree-days, the melt (in./day), and the volume rate of melt (acre-ft) from a 203-mi^2 watershed. Compare the melt assuming the watershed has (a) extremely low, (b) average, and (c) extremely high runoff potential.

Day	Maximum Temperature (°F)	Minimum Temperature (°F)
1	33	−5
2	24	2
3	29	10
4	28	11
5	34	19
6	44	31
7	48	30
8	40	18
9	50	21
10	29	4

12–10. (a) The following data are the average degree-day values (°F/day) and the streamflow (ft^3/sec) in the South Fork of the Rio Grande of the South Fork, Colorado, which drains an area of 216 mi^2. Compute the daily values of the proportionality constant K in Equation 12–6 and the average for the 11-day period. Compare the value with those of Table 12–4.

May 1975	Flow (ft^3/sec)	Degree-Days (°F)
8	205	1.6
9	260	3.1
10	431	5.6
11	610	5.7
12	765	4.2
13	864	5.2
14	978	6.5
15	1130	12.6
16	1240	13.9
17	1300	10.7
18	1320	10.1

(b) Using the average value of K determined in part (a) and the data below, compute the snowmelt runoff (ft^3/sec) using Equation 12–7 and compare the computed and actual runoff, which is also given.

April 1976	Flow (ft^3/sec)	Degree-Days (°F)
20	135	0.8
21	148	1.5
22	173	2.7
23	208	3.0
24	238	1.9
25	298	2.3
26	316	2.6
27	340	3.5
28	391	4.0
29	439	4.3
30	383	1.7

12–11. (a) The following data are the average degree-day values (°F/day) and the streamflow (ft^3/sec) in the Conejos River near Magote, Colorado, which drains an area of 282 mi^2. Compute the daily values of the proportionality constant K in Equation 12–6 and the average for the 11-day period. Compare the value with those of Table 12–4.

April 1979	Flow (ft^3/sec)	Degree-Days (°F)
21	378	3.37
22	418	4.05
23	486	9.71
24	506	5.83
25	534	5.51
26	530	3.61
27	488	4.10
28	583	3.89
29	619	2.78

(b) Using the average value of K determined in part (a) and the data below, compute the snowmelt runoff (ft^3/sec) using Equation 12–7 and compare the computed and actual runoff, which is also given.

April 1975	Flow (ft^3/sec)	Degree-Days (°F)
15	77	0.5
16	93	3.5
17	131	1.4
18	128	0.0
19	106	0.0
20	142	1.4
21	194	3.7
22	257	5.2
23	324	5.6
24	360	3.1
25	456	6.7

12–12. The following mean daily temperatures were recorded near a watershed that is 70% open and 30% forested: 37, 42, 29, 35, 41, 39, 44°F. Use Equations 12–10 to estimate the total melt for the week.

12–13. The following mean daily temperatures were recorded near a watershed that is 40% open and 60% forested: 43, 37, 44, 49, 52, 26, 31°F. Use Equations 12–10 to estimate the total melt for the week.

12–14. Graph the melt rates for 100% open and 100% forested sites for mean daily temperatures from 35°F to 65°F. Provide an explanation as to why the melt is higher in open areas and why the rate of change of melt with temperature is greater in open areas.

12–15. Using a general form of Equation 12–10:

$$M = M_o(T - T_o)$$

Calibrate values for M_o and T_o using the following record of melt rates (M, in.) and mean daily temperature (T, °F):

T	36	39	46	51	54	62	66
M	0.42	0.59	0.95	1.21	1.38	1.80	2.02

12–16. (a) Using the data of Problem 12–10(a) and the following values of the fraction of snow-covered area and precipitation (in.), evaluate the coefficients C and k of the Martinec Model (Equation 12–11). The flow on May 19, 1975, was 1330 ft^3/sec. Assume $a = 0.5$.

May 1975	S	P (in.)
8	0.867	0
9	0.853	0
10	0.837	0
11	0.820	0

(continued)

May 1975	S	P (in.)
12	0.801	0
13	0.780	0
14	0.755	0
15	0.722	0
16	0.691	0
17	0.652	0.02
18	0.614	0

(b) Using the values of C and k computed in part (a) and the following data, compute the snowmelt runoff (ft^3/sec) using Equation 12–11. Using the data below and in Problem 12–10(b), compare the computed and actual daily runoff. On April 19, 1976, the flow was 141 ft^3/sec and the degree-day was 0.9°F. Compute the bias and standard error of estimate.

April 1976	S	P (in.)
19	0.93	0.03
20	0.92	0.00
21	0.91	0.00
22	0.90	0.00
23	0.89	0.00
24	0.88	0.00
25	0.86	0.00
26	0.85	0.00
27	0.84	0.00
28	0.82	0.00
29	0.81	0.00

12–17. The following data are snowmelt runoff (ft^3/sec) for the South Fork (area = 216 mi^2) of the Rio Grande at South Fork, Colorado, for April 1, 1976, to September 30, 1976. If a recession is defined as a period of four or more days in which the runoff rate did not increase, derive a recession curve with the model of Equation 12–17 for the months of April and May.

Day	April	May	June	July	August	September
1	90	364	1240	464	111	42
2	120	431	1330	364	143	42
3	160	496	1300	316	120	67
4	192	532	1380	283	124	69
5	178	550	1470	260	99	88
6	158	610	1430	238	85	137
7	141	572	1370	222	79	141
8	158	509	1380	220	85	160
9	210	419	1400	208	101	158
10	265	391	1330	178	115	158
11	310	447	1180	165	148	158

(continued)

Day	April	May	June	July	August	September
12	325	545	990	158	137	156
13	278	554	870	160	120	148
14	230	695	800	165	113	148
15	202	942	715	141	84	148
16	180	1130	655	135	54	143
17	160	1260	605	143	58	143
18	148	1390	554	129	66	150
19	141	1330	550	122	74	145
20	135	1190	572	126	85	176
21	148	1240	605	128	78	220
22	173	1190	630	116	70	89
23	208	1090	625	111	69	53
24	238	1160	522	103	79	42
25	298	1150	443	99	78	64
26	316	966	395	115	75	158
27	340	1070	364	126	74	129
28	391	1300	344	110	79	96
29	439	1400	328	98	81	79
30	383	1370	334	93	79	70
31		1150		95	65	

12–18. Using the data for June and July from Problem 12–17, derive a recession curve of Equation 12–17, if a recession is defined as a period of four or more days in which the runoff rate did not increase.

12–19. Using the data for August and September from Problem 12–17, derive a recession curve of Equation 12–17, if a recession is defined as a period of four or more days in which the runoff rate did not increase.

12–20. Estimate the coefficients (Q_o and k) of Equation 12–15 using the following daily melt data for a 1-week period: 0.31, 0.27, 0.23, 0.21, 0.18, 0.16, 0.14 in.

12–21. Show the development of Equations 12–20 and 12–21.

12–22. Given the following values of Q_s, P_w, and Q_w, estimate the coefficients (a and b) of Equation 12–19:

P_w	62	71	67	85	78	57	74	82
Q_w	11	16	17	24	21	10	19	19
Q_s	29	35	34	39	38	26	38	42

Assess the prediction accuracy.

REVIEW QUESTIONS

12–1. Simple empirical models are sufficient for making long-term forecasts of two to five months because (a) inaccurate forecasts can be updated as new data are obtained; (b) highly accurate forecasts are not required; (c) variations in measured long-term water yields are relatively small, which contributes to high correlations with measured predictor variables; (d) theoretical models are difficult to formulate; (e) none of the above.

12–2. Which one of the following is not a criterion for evaluating an empirical snowmelt runoff model? (a) Confidence intervals; (b) the standard error of estimate; (c) the correlation coefficient; (d) the residuals; (e) all of the above are important criteria.

12–3. A non-zero intercept coefficient for a multiple linear regression snowmelt runoff model indicates (a) model irrationality; (b) poor correlation; (c) a nonlinear model should be used; (d) variables other than those included in the equation are important.

12–4. The concordant flow method is used to estimate snowmelt runoff (a) by transferring information from other nearby watersheds; (b) based solely on watershed characteristics, (c) without requiring any measured snowmelt runoff data; (d) by indexing runoff volumes with historical data; (e) none of the above.

12–5. Which one of the following is not a realistic predictor variable for the concordant flow method? (a) The snow-covered area; (b) the number of degree-days per day; (c) incoming solar radiation; (d) the snow water equivalent; (e) all of the above are valid predictor variables.

12–6. Which one of the following is not required to estimate snowmelt runoff volume with the degree-day method? (a) The drainage area; (b) the precipitation volume; (c) the temperature; (d) the amount of forest cover; (e) all of the above are required.

12–7. The lapse rate adjustment adjusts measured air temperature for variation in (a) watershed slope; (b) altitude; (c) atmospheric pressure; (d) incoming solar radiation.

12–8. Which one of the following is not an input for making daily snowmelt runoff estimates with the Martinec Model? (a) The precipitation depth; (b) the temperature; (c) the depth of the snowpack; (d) the snow-covered area.

12–9. The recession coefficient of the Martinec Model indicates (a) the correlation between flows on adjacent days; (b) the effect of temperature on the depletion of the snowpack; (c) the volume of water remaining in the snowpack; (d) the lag between the time when snow melts and when it appears as runoff.

12–10. A water balance model uses which one of the following to estimate snowmelt runoff? (a) The depth of the snowpack; (b) a degree-day index; (c) the snow-covered area; (d) the winter precipitation depth.

DISCUSSION QUESTION

The technical content of this chapter is important to the professional hydrologist, but practice is not confined to making technical decisions. The intent of this discussion question is to show that hydrologists must often address situations where value issues intermingle with the technical aspects of a project. In discussing the stated problem, at a minimum include responses to the following questions:

1. What value issues are involved, and how are they in conflict?
2. Are technical issues involved with the value issues? If so, how are they in conflict with the value issues?
3. If the hydrologist attempted to rationalize the situation, what rationalizations might he or she use? Provide arguments to suggest why the excuses represent rationalization.

4. What are the hydrologist's alternative courses of action? Identify all alternatives, regardless of the ethical implications.
5. How should the conflict be resolved?

You may want to review Sections 1.6 to 1.12 in Chapter 1 in responding to the problem statement.

Case. A hydrologist obtains a commercially marketed piece of software that advertisements indicate will provide snowmelt runoff predictions given measurements of snow water equivalent values and daily temperature values, which is the data that the hydrologist has available for a particular project. The hydrologist has experience in rainfall-runoff estimation, but does not have practical experience in snow hydrology. No one else in the hydrologist's company has any experience in snow hydrology. The software comes with a brief manual that describes how to use the program, but it lacks any discussion of the mathematical model that transforms the input into snowmelt runoff estimates. The hydrologist applies the software and uses the computed runoff in the design.

13

Water Quality Estimation

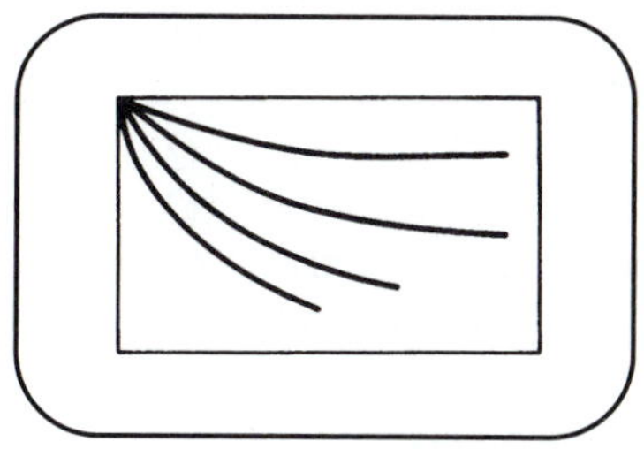

CHAPTER OBJECTIVES

1. Discuss concentration and load estimation at ungaged locations.
2. Present empirical methods for estimating concentrations of water quality constituents.
3. Introduce the basics of the Streeter-Phelps Oxygen Sag Curve.
4. Show how to route pollutant loads through channel sections.

13.0 NOTATION

A = area being developed
A = total contributing drainage area
C = concentration
C = flow-weighted mean concentration
C = observed dissolved oxygen
$\hat{C}$ = predicted concentration
C_0 = initial concentration
C_o = initial dissolved oxygen concentration
D_c = critical deficit
D_o = initial dissolved oxygen deficit
D_t = oxygen consumed by the wastes
I = inflow
I = watershed imperviousness
I_a = percentage of impervious land cover
I_c = indicator variable of commercial and industrial land cover
k_i = rate coefficient at temperature T_i
k_0 = rate constant at 20°C
k_1 = rate constant for deoxygenation
k_2 = rate constant for reoxygenation
k_{2o} = value of k_2 at temperature T_o
K = rate coefficient
K_u = conversion constant
L = load
L_c = distance downstream where critical condition occurs
L_o = initial BOD
L_o = ultimate demand
L_t = amount of carbonaceous waste available at time t
n = number of storm events
N_s = number of storm events in an average year
O = overflow
P = increase or loss of mass due to production or decay
P = rainfall depth
P_a = mean annual rainfall
P_j = correction factor for storms not producing runoff
Q = discharge rate
$\overline{Q}_a$ = long-term average volume of runoff per storm
Q_i = flow rate in
Q_i = volume of runoff for storm i
Q_o = flow rate out
r = mass rate of decay
R_v = fraction of rainfall converted to runoff
S = storage
t = time
t = travel time in reach
T_J = mean minimum January temperature
V = flow velocity
V = volume
W = load
W_i = load for storm i
$\overline{W}_a$ = mean annual load
$\overline{W}_e$ = event-mean load
x = distance along the channel reach
$\hat{Y}$ = predicted value of a constituent
θ = temperature coefficient

13.1 INTRODUCTION

Decisions often depend on water quality information. In some cases, policy makers need to project the effects of land development on regional water quality resources. Engineers may need to assess the effect of engineering projects on stream or lake quality. In resolving legal cases, engineers may need to identify the most likely source of a contaminant. The potential for improving water quality by requiring vegetative buffer strips along streams requires knowledge of the potential of the buffer to trap water quality constituents such as suspended solids. Water quality data are not readily available and rarely are they available at the specific site where decisions need to be made. A definite need for water quality estimates exists, but the data base necessary to support this need is very minimal.

The solutions to some engineering problems require models that use a time-dependent variable such as discharge as a predictor variable. The use of concentration-flow or load-flow relationships are examples. First, the time-dependent change in the dissolved oxygen concentration in a stream is a problem common in stream analysis. Second, a relationship between the pollutant concentration and time is called a pollutograph; a load-time relationship is similarly called a loadograph. These can be important in cases such as accidental spills of a toxic chemical into streams.

13.2 CONCENTRATION AND LOADS

13.2.1 Water Quality Measurements

Where water quality information is important in the decision-making process, the engineer may find it necessary to obtain data in the form of actual measurements of water quality levels or regional information based on measured data. Water quality measurements are not readily available. Local, state, and national agencies now collect water quality data, but the records are often short and the measurements taken irregularly. The first step in assembling a water quality data base is to contact agencies who regularly collect such data, and talk with local planning offices who may have used such data in the past. The United States Geological Survey (U.S.G.S.) collects water quality data and regularly publishes the results. Figure 13–1 shows an example of a water quality report published by the Geological Survey.

Different types of engineering projects require information about different constituents. Nitrogen and phosphorus data are important in areas where agricultural runoff is significant and the control of eutrophication is desirable; these constituents act as nutrients for unwanted biological and vegetative growth in streams. Suspended solids can be a significant pollutant in urban areas where construction activity is taking place. In addition to increasing the sediment flux, suspended solids also increase the turbidity of a stream, which can be detrimental to aquatic life, and they can increase the cost of purifying water for public water supplies. Table 13–1 lists the wide array of water quality constituents that are associated with highway runoff. Heavy metals, such as zinc, are a major public health consideration.

Where water quality measurements are not available, but projections of water quality are needed, it may be necessary to use regionalized values. These are often mean values based on measurements collected within the region. Mean values are usually given for different land use conditions. Where regional data are adequate, prediction equations may have been developed. Prediction equations based on nationwide data are also available for some constituents. On-site measurements would be expected to provide the most accurate estimates followed by regional and national data. Accuracy assessments should be made of all projections and reported with the projections.

PATAPSCO RIVER BASIN 123

01589000 PATAPSCO RIVER AT HOLLOFIELD, MD--Continued

WATER-QUALITY RECORDS

PERIOD OF RECORD.--Water years 1969-74, 1976 to current year.

WATER QUALITY DATA, WATER YEAR OCTOBER 1980 TO SEPTEMBER 1981

DATE	TIME	STREAM-FLOW, INSTAN-TANEOUS (CFS)	SPE-CIFIC CON-DUCT-ANCE (UMHOS)	PH (UNITS)	TEMPER-ATURE, AIR (DEG C)	TEMPER-ATURE (DEG C)	COLOR (PLAT-INUM-COBALT UNITS)	OXYGEN, DIS-SOLVED (MG/L)	HARD-NESS (MG/L AS CACO3)	HARD-NESS NONCAR-BONATE (MG/L AS CACO3)	CALCIUM DIS-SOLVED (MG/L AS CA)	MAGNE-SIUM, DIS-SOLVED (MG/L AS MG)
NOV 04...	0945	53	190	8.7	12.0	7.5	0	10.0	--	--	--	--
JAN 28...	0945	45	150	6.5	.0	.3	<5	13.5	52	17	14	4.2
JUN 30...	1530	72	175	8.5	32.0	27.0	2	11.0	62	12	18	4.1
SEP 18...	1015	53	202	7.9	--	19.0	5	--	67	9.0	18	5.3

DATE	SODIUM, DIS-SOLVED (MG/L AS NA)	PERCENT SODIUM	SODIUM AD-SORP-TION RATIO	POTAS-SIUM, DIS-SOLVED (MG/L AS K)	POTAS-SIUM 40 DIS-SOLVED (PCI/L AS K40)	ALKA-LINITY LAB (MG/L AS CACO3)	SULFATE DIS-SOLVED (MG/L AS SO4)	CHLO-RIDE, DIS-SOLVED (MG/L AS CL)	FLUO-RIDE, DIS-SOLVED (MG/L AS F)	SILICA, DIS-SOLVED (MG/L AS SIO2)	SOLIDS, RESIDUE AT 180 DEG. C DIS-SOLVED (MG/L)	SOLIDS, SUM OF CONSTI-TUENTS, DIS-SOLVED (MG/L)
NOV 04...	--	--	--	--	--	41	13	16	.2	--	111	--
JAN 28...	10	28	.6	2.0	1.5	35	10	15	.2	8.9	106	85
JUN 30...	7.2	19	.4	3.1	2.3	50	10	13	.1	9.1	116	95
SEP 18...	9.0	22	.5	2.4	--	58	12	14	.3	9.6	123	106

DATE	SOLIDS, DIS-SOLVED (TONS PER AC-FT)	SOLIDS, DIS-SOLVED (TONS PER DAY)	NITRO-GEN, NO2+NO3 TOTAL (MG/L AS N)	PHOS-PHORUS, TOTAL (MG/L AS P)	PHOS-PHORUS TOTAL (MG/L AS PO4)	IRON, TOTAL RECOV-ERABLE (UG/L AS FE)	IRON, SUS-PENDED RECOV-ERABLE (UG/L AS FE)	IRON, DIS-SOLVED (UG/L AS FE)	MANGA-NESE, TOTAL RECOV-ERABLE (UG/L AS MN)	MANGA-NESE, SUS-PENDED RECOV. (UG/L AS MN)	MANGA-NESE, DIS-SOLVED (UG/L AS MN)
NOV 04...	.15	15.8	1.9	.150	.46	640	--	--	70	--	--
JAN 28...	.14	12.9	2.3	.140	.43	310	280	30	60	10	50
JUN 30...	.16	22.6	1.5	.120	.37	840	800	40	100	40	60
SEP 18...	.17	17.6	1.6	.170	.52	890	850	40	110	70	40

FIGURE 13–1 Example of U.S.G.S. water quality records. Source: U.S. Geological Survey, *Water Resources Data for Maryland and Delaware*, Part 2, Water Quality Records, Towson, MD. 1981.

13.2.2 Definitions

It is important to distinguish between the concentration and the load or flux of a water quality constituent. Concentration (C) has dimensions of mass per unit volume. The load (W) of a constituent has dimensions of weight per unit of time. The concentration and load are related through the flow rate Q (volume per unit of time) by

$$W = K_u C Q \tag{13–1}$$

TABLE 13–1 Common Highway Runoff Constituents and their Primary Sources

Constituent	Primary Sources
Particulates	Pavement wear, vehicles, atmosphere, maintenance
Nitrogen	Atmosphere, roadside fertilizer application phosphorus
Lead	Leaded gasoline (auto exhaust), tire wear (lead oxide filler material), lubricating oil and grease, bearing wear
Zinc	Tire wear (filler material), motor oil (stabilizing additive), grease
Iron	Autobody rust, steel highway structures (guard rails, etc.), moving engine parts
Copper	Metal plating, bearing and bushing wear, moving engine parts, brake lining wear, fungicides and insecticides applied by maintenance operations
Cadmium	Tire wear (filler material), insecticide application
Chromium	Metal plating, moving engine parts, brake lining wear
Nickel	Diesel fuel and gasoline (exhaust) lubricating oil, metal plating, bushing wear, brake lining wear, asphalt paving
Manganese	Moving engine parts
Bromide	Exhaust
Cyanide	Anticake compound (ferric ferrocyanide, Prussian Blue or Sodium ferrocyanide, Yellow Prussiate of Sode) used to keep deicing salt granular
Sodium	Deicing salts, grease
Calcium	
Chloride	Deicing Salts
Sulphate	Roadway beds, fuel, deicing salts
Petroleum	Spills, leaks, or blow-up of motor lubricants, antifreeze and hydraulic fluids, asphalt surface leachate
Polychlorinated	Spraying of highway right-of-ways, background atmospheric deposition, byphenyis, PCB catalyst in synthetic tires, pesticides
Pathogenic	Soil litter, bird droppings and trucks hauling livestock and stockyard bacteria (indicators) waste
Rubber	Tire wear
Asbestos	Clutch and brake lining wear

Source: *Federal Highway Administration, 1990. FHWA-RD-88-006. *Pollutant Loadings and Impacts from Highway Stormwater Runoff*, Vol. 1: Design Procedure.

in which K_u is a conversion constant that depends on the units of W, C, and Q. If W [=] g/sec, Q [=] m^3/sec, and C [=] mg/L (or g/m^3), then $K_u = 1$. If W [=] lb/day, Q [=] ft^3/sec, and C [=] mg/L, then $K_u = 5.39$.

While the variables of Equation 13–1 are shown as single values, it is important to recognize that water quality constituents vary in both time and space. Many constituents show significant diurnal and seasonal variations. As material flows downstream, a portion of the load may settle and/or undergo biological or chemical transformation; thus, a significant spatial trend may be evident along the stream. Equation 13–1 is sometimes written as

$$W(x;t) = K_u C(x;t)\, Q(x;t) \tag{13–2}$$

This form indicates that the load, concentration, and discharge show temporal and spatial variation.

Water quality data are most often measured and reported as a concentration. For many decisions, values expressed as a concentration are the appropriate form. When contaminated water is withdrawn from a stream, the concentration of various constituents is measured and used to determine the appropriate level of clean-up. When considering the effect of a constituent on aquatic life, the concentration and duration are the appropriate form for presenting the data. Measurements of nonconservative water quality constituents are presented in concentration form, rather than as a load. Conservative pollutants that can settle out and accumulate in the environment are most often expressed as a load. The long-term mass of the pollutant would then be used to assess the importance of action to clean-up the material. For example, suspended solids that enter a reservoir or small retention pond can settle to the bottom and represent a long-term loss of storage. If other water quality constituents are attached to the solids, then they can also accumulate in the sediment. In fact, up to half of the nitrogen input entering lakes has been reported to be associated with the particles, and hence, settled in the sediment.

Example 13–1

Data from a 12.8 acre urban area were obtained for part of a storm event (October 18, 1981) in Baltimore, MD. Table 13–2 gives the measured instantaneous flow rates and total recoverable lead concentrations to the Hampden storm sewer. The loading rates (pounds/day) were computed using Equation 13–1 ($K_u = 5.39$), and the total load (lbs) for part of the storm was estimated using the trapezoidal rule. For the measured data, the total load is 106 lbs. At a specific weight of 708 lb/ft^3, the discharge included lead with a volume of 0.15 ft^3.

13.2.3 Concentration-Flow Relationships

Studies have shown that the concentration of a constituent in a stream is proportional to the flow rate Q, with the power-model form commonly used:

$$C = aQ^b \tag{13–3}$$

TABLE 13–2 Estimation of Lead Loadings for Hampden Storm Sewer, Baltimore, MD (Gage No. 01589460) for Part of the October 18, 1981, Event

Time	Flow (ft^3/sec)	Concentration of Lead (mg/L)	Load (lb/day)	Load (lb)
1807	2.3	780	9670	
				32.2
1814	2.3	290	3595	
				29.9
1834	0.95	140	717	
				8.9
1854	0.87	120	563	
				22.4
2002	0.80	90	388	
				6.7
2032	0.40	120	259	
				5.5
2132	0.01	80	4	
				0.1
2232	0.01	40	2	
				105.7

where a and b are fitting coefficients, which are commonly fitted using least squares following a logarithmic transform of C and Q. Equation 13–3 is valid for continuous flows. For intermittent flows, such as with washloads from storm events, an alternative approach is used. Once the concentration is determined from Equation 13–3, the load can then be estimated using Equation 13–1. Some empirical formulas are designed to predict the load directly, while others provide predictions of concentrations.

Examples 13–2

Table 13–3 gives measurements of the instantaneous streamflow and chloride concentration at the Conowingo, MD, station (01578310) on the Susquehanna River. The measurements were made over a period of four water-years. Streamflow rates varied from about 3,000 ft^3/sec to almost 450,000 ft^3/sec. The data are shown in Figure 13–2. The power model form of Equation 13–3 was fitted to the data, with the following result:

$$\hat{C} = 113.7Q^{-0.2071} \qquad (13\text{–}4)$$

TABLE 13–3 Instantaneous Streamflow Q and Dissolved Chloride C: Susquehanna River at Conowingo, MD (Gage No. 01578310)

Q (cfs)	C (mg/L)	Date
444000	7.5	2–17–84
265000	6.4	4–09–84
185000	18.0	2–13–81
173000	9.7	3–17–82
163000	6.1	2–27–81
81800	12.0	4–15–81
78600	9.5	1–12–82
76700	9.2	4–29–81
72800	9.5	3–12–81
68900	9.0	6–22–81
67800	7.7	11–9–81
67500	8.0	5–21–84
60200	27.0	11–04–80
60000	9.8	5–12–82
51000	16.0	12–10–80
41600	14.0	1–04–84
39900	15.0	9–09–81
34900	12.0	3–25–81
34300	7.5	5–27–81
32700	17.0	9–29–81
32400	24.0	10–23–80
31700	10.0	7–28–81
31400	22.0	11–24–80
31000	13.0	12–30–80
30800	13.0	9–19–84
22200	18.0	1–29–81
17000	16.0	9–22–82
5500	10.0	7–15–82
3370	25.0	11–07–83

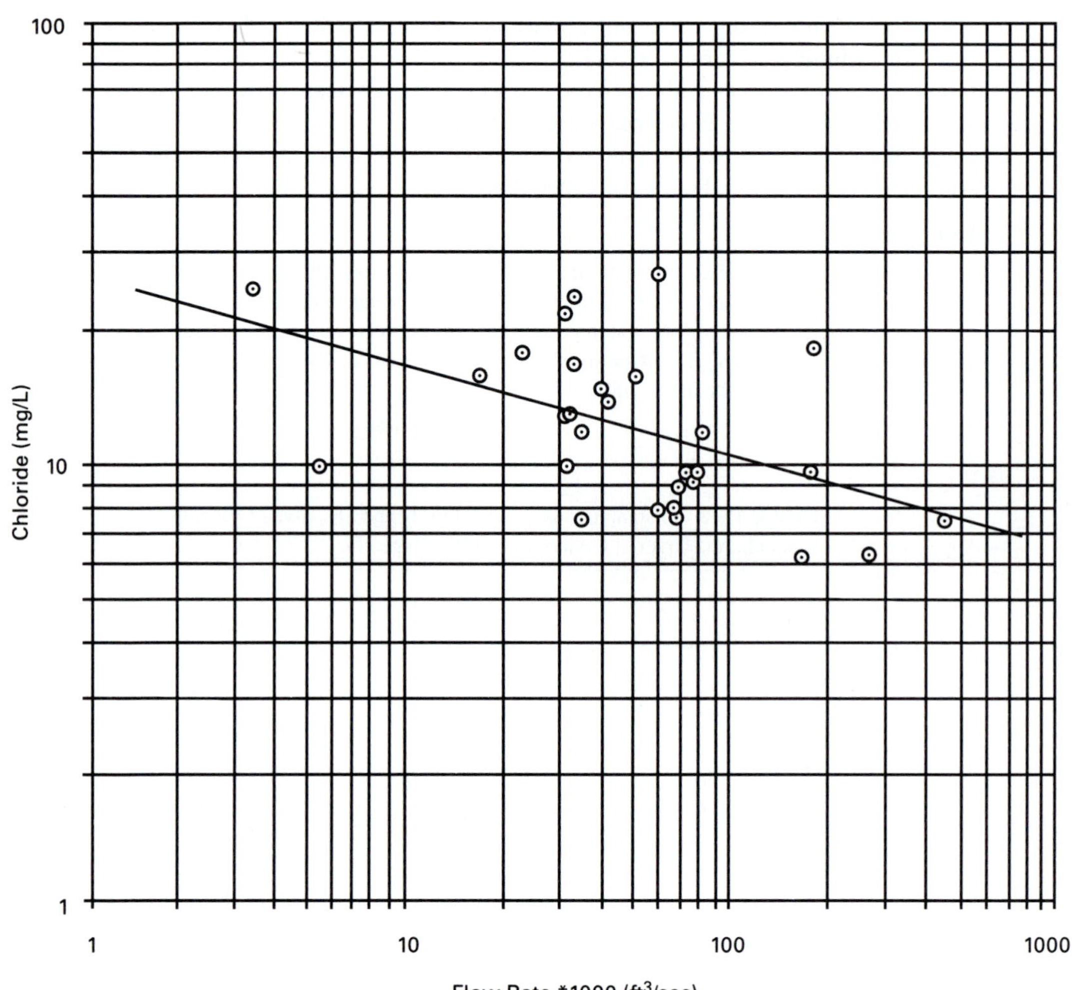

FIGURE 13–2 Concentration-flow relationship for chloride: Susquehanna River at Conowingo, MD (01578310)

in which $\hat{C}$ is the predicted value of the concentration C. The model had a bias (mean error) of –0.81 mg/L, a standard error of estimate of 5.29 mg/L ($S_e/S_y = 0.918$) and an adjusted correlation coefficient of 0.43 ($R^2 = 0.187$). Thus, the model does not provide a good fit to the data, which is evident from Figure 13–2. However, such scatter is typical of concentration-flow relationships.

Example 13–3

The author used data collected from two small urban watersheds (45 acres and 148 acres) to develop regression relationships relating several water quality constituents and discharge. The equations predict contaminant loads in pounds per second (see Table 13–4). These equations

TABLE 13–4 Regression Relationships for Selected Water Quality Parameters and Discharge Rate Q (ft^3/sec)

Parameter, Y	Units of Y	Equation
5-day BOD	10^{-3} lb/sec	$Y = 3.138\ Q^{0.4749}$
Zinc	10^{-6} lb/sec	$Y = 0.02945\ Q^{0.9959}$
Cadmium	10^{-6} lb/sec	$Y = 0.0010\ Q^{1.2344}$
Fecal Coliform	No./100mL	$Y = 13.95\ Q^{1.2417}$
Nitrates and Nitrites	10^{-3} lb/sec	$Y = 0.0202\ Q^{2.4478}$
Chromium	10^{-6} lb/sec	$Y = 0.00308\ Q^{1.1687}$
Orthophosphates	10^{-3} lb/sec	$Y = 0.00308\ Q^{1.0470}$
Total Phosphorus	10^{-3} lb/sec	$Y = 0.01479\ Q^{1.14896}$

were developed from data in an area just north of Washington, D.C. The application of such equations to other sites outside of the immediate vicinity is limited. Many of the exponents indicate near-linear relationships.

13.2.4 Frequency Analysis

Frequency analyses are widely used to analyze rainfall data in developing intensity-duration-frequency curves (Chapter 4), to characterize the random nature of flood data (Chapter 5), and to provide the discharges necessary to fit regional peak discharge equations (Chapter 7). While frequency analyses could provide useful information for water quality planning, the necessary input data are rarely available. If the interest was annual maximum concentrations, it is unlikely that the random sampling programs that exist would produce the annual maximum. Thus, using the measured maximum for each year would probably produce a frequency curve that would underestimate the annual maximum, with the underprediction possibly being very significant. Where accumulation of a constituent over a period of time is important for planning, annual loads become the random variable of interest. If probability estimates of annual loads were needed and both continuous records of streamflow and selected measurements of the constituent were available, the selected measurements could be used to fit a concentration-flow relationship, which would then be used with the streamflow to estimate annual loads. The procedure is as follows:

1. Using measured concentrations from a random sampling program, fit the concentration-flow relationship of Equation 13–3 using the concentrations and the corresponding discharge measurements.
2. Using the concentration-flow relationship and the continuous streamflow relationship, estimate the concentration for all discharges and convert these to loads using Equation 13–1.
3. Compute the annual load by integrating the computed loads.
4. Perform a frequency analysis on the annual loads.

Of course, the accuracy of annual loads estimated from the resulting frequency curve will depend, in part, on the accuracy of the concentration-flow relationship. As an alternative to Steps 2 and 3, the annual flow could be input to the concentration-flow relationship of Step 1 to estimate the annual loads, which are then used to make a frequency curve.

13.2.5 Mean Concentration and Load Estimation at Ungaged Locations

Ideally, historical water quality data would be available at sites where decisions must be made. Unfortunately, site-specific data are rarely available. In such cases, it is best to obtain data for the same region and, to the extent possible, the same land use/activity conditions. Such data are often difficult to locate, although regional planning agencies are often aware of water quality data that have been collected in the region. Table 13–5 provides mean runoff water quality concentrations for numerous land uses and activities in the area around Lake Tahoe. Table 13–6 provides concentrations for five land-use categories in the Washington, D.C. area. Table 13–7 gives the means in terms of loads for land use activities in Washington, D.C. In the absence of site-specific data, regional means may provide sufficient accu-

TABLE 13–5 Comparison of Mean Runoff Water Quality for Several Land Uses and Activities in the Tahoe Area

Land Use or Activity	Suspended Solids	Turbidity (Ftu)	NO_3-N	Total Nitrogen	Total Phosphate	Total Iron	Oil and Grease
Unpaved parking lots	16,600	1,000	—	9.2	3.4	3.4	76.0
Bare areas	989	319	0.3	4.0	1.7	1.9	8.0
Unsurfaced roads and driveways	7,780	5,060	0.9	2.6	1.2	3.2	38.1
Paved parking lots	320	107	0.6	3.8	1.6	1.0	42.6
Dirt roadside ditches	648	175	—	3.2	1.0	1.1	28.4
Unstable dirt channels	613	305	0.1	1.2	1.0	0.8	31.3
Paved streets	680	280	0.1	1.2	0.9	0.9	23.8
Snow storage	136	90	0.1	3.5	0.6	0.2	9.6
Rooftop drainage	30	7	0.02	0.8	0.5	4.7	7.1
Roadway slopes	443	304	0.2	1.0	0.7	0.5	6.7
Construction sites	8,630	764	0.1	4.0	0.5	2.3	0.1
Corporation yards	435	142	0.1	3.3	0.8	7.7	56.6
Mobile home parks	5,680	931	0.1	0.9	0.8	4.4	23.9
Service stations	281	112	0.2	0.8	0.9	1.3	11.7
Stable	71	27	0.02	1.8	2.2	6.2	9.1
Land Use Types							
Tourist commercial	4,020	1,084	0.4	1.3	0.8	4.2	67.7
General commercial	773	832	0.2	1.7	1.3	1.1	33.0
Public service	323	105	0.1	1.9	0.8	4.3	23.8
High-density residential	249	92	0.1	0.7	0.8	1.4	20.0
Medium-density residential	489	52	0.04	0.6	0.5	0.4	3.6
Low-density residential	613	169	0.1	1.2	0.7	0.3	0.8
Recreation	48	21	0.1	0.6	0.4	0.5	5.3
General forest	66	6	0.03	0.2	0.1	0.4	0.6
General urbanized area	482	242	0.1	1.1	0.8	1.3	34.4

* *Source:* Tahoe Regional Planning Agency, 1978. Lake Tahoe Basin Water Quality Management Plan, Approved Summary.
Note: All figures in mg/L except as noted.

TABLE 13–6 Pollution Concentrations in Urban Runoff for Several Land Uses in the Washington, D.C. Area

	Pollutant concentration, mg/L				
Land Use	COD	Total Nitrogen	Total Phosphorus	Lead	Zinc
Low-density residential	70–120	2–4	0.3–0.4	0.05–0.1	0.02–0.1
Medium-density residential	80–130	2–3	0.3–0.4	0.1–0.2	0.05–0.2
High-density residential	70–90	2–3	0.3–0.5	0.1–0.2	0.05–0.3
High-rise residential	50–100	1–2	0.2–0.3	0.1–0.2	0.10–0.2
Commercial	90–120	2–3	0.2–0.3	0.3–0.5	0.1–0.4

Source: Northern Virginia Planning District Commission and Virginia Polytechnic Institute and State University, 1977. *Planning for Nonpoint Pollution Management.* Prepared for EPA Conference on Watershed Management R&D, Athens, GA.

racy for decisions about expected loadings in storm runoff. For example, if suspended solids are a major problem, then the values in Table 13–5 indicate that efforts should be directed to control runoff from unpaved parking lots, construction sites, unsurfaced roads and driveways, and mobile home parks.

Example 13–4

Table 13–5 shows that unpaved and paved parking lots yield mean suspended solids concentrations of 16,600 mg/L and 320 mg/L, respectively. Therefore, if an ordinance that required parking lots larger than 0.25 acres to be paved were passed, a reduction of 16,280 mg/L of suspended solids could be expected. For a 1-acre parking lot with a discharge of 0.5 cfs, this would represent a reduction of

$$\begin{aligned} W &= 5.39\,(16{,}280\text{ mg/L})(0.5\text{ ft}^3/\text{sec}) \\ &= 43{,}875\text{ lb/day per acre} \end{aligned} \tag{13–5}$$

TABLE 13–7 Storm Pollutant Yields for Several Land Uses in the Washington, D.C. Area

		Pollutant Yield, lb/acre/day					
		Suspended Solids		Total Nitrogen		Total Phosphorus	
Land Use	Number of Storms	Mean	Range	Mean	Range	Mean	Range
Low-density residential	23	15	0–110	0.2	0–0.7	0.04	0–0.2
Medium-density residential	42	21	1–270	0.7	0.1–5.1	0.12	0.01–0.8
High-density residential	64	31	0–190	0.6	0–2.9	0.11	0–0.5
High-rise residential	21	18	0–120	0.8	0–3.7	0.10	0–0.4
Commercial	50	43	0–320	1.0	0–6.5	0.12	0–0.8
Rural	6	29	0–170	0.1	0–0.3	0.01	0–0.1
Agricultural	32	83	0–620	0.6	0–4.9	0.18	0–2.1

Source: Griffin, D. M., Jr., Grizzard, T. J., Randall, C. W., and Hartigan, J. P., 1978. *An Examination of Nonpoint Pollution Export from Various Land Use Types,* International Symposium on Urban Stormwater Management, Lexington, Ky.

13.2.6 Mean Weighted Load

Sampling programs often collect multiple samples per storm event from which estimates are made of the event mean load or event mean concentration. The mean annual load ($\overline{W}_a$) can then be estimated as the product of the event mean load ($\overline{W}_e$) and the number of events in an average year, N_s:

$$\overline{W}_a = N_s \overline{W}_e \tag{13–6}$$

Table 13–8 gives values of N_s for selected cities. The event mean is usually estimated as the average load of several storm events. Obviously, the larger the number of events used to estimate the event mean load, the greater the accuracy of the estimated annual load. Where the data record consists of only a few events, greater accuracy can be achieved by weighting the individual storm loads by the runoff volumes for the storm event:

$$\overline{W}_e = \frac{\overline{Q}_a}{n} \sum_{i=1}^{n} \frac{W_i}{Q_i} \tag{13–7}$$

in which $\overline{Q}_a$ is long-term average volume of runoff per storm, Q_i is the volume of runoff for storm i, W_i is the load for storm i, and n is the number of storm events in the sample. The ratio W_i/Q_i of Equation 13–7 represents the average concentration of storm i; thus, Equation

TABLE 13–8 Mean Number of Storms per year (N_s)

City	N_s
Austin, TX*	54
Bellevue, WA*	98
Boston, MA+	84
Chicago, IL+	87
Cleveland, OH+	100
Dallas, TX+	53
Denver, CO+	57
Fresno, CA*	41
Houston, TX+	70
Knoxville, TN*	92
Los Angeles, CA+	32
Miami, FL*	100
New Orleans, LA+	83
New York, NY+	83
Philadelphia, PA+	77
Phoenix, AZ+	31
Portland, OR*	96
St. Louis, MO+	74
San Francisco, CA+	46
Seattle, WA+	97
Tampa, FL*	79
Winston-Salem, NC*	77

* From Driver and Tasker (1988)

+Estimated

TABLE 13–9a Computation of Total Organic Nitrogen for October 1–2, 1981, Storm Event: Jones Falls at Lake Roland Dam (01589452)

Time (hr)	(min)	Discharge (cfs)	Volume (in.)	Total Volume (in.)	Concen-tration (mg/L)	Flux (lb/day)	Load (lb)	Total Load (lb)
21	15	7			1.3	49		
			0.00	0.001			17.55	18
22	15	16			9.2	793		
			0.00	0.002			27.44	45
23	15	36			2.7	524		
			0.00	0.003			16.42	61
0	7	55			1.3	385		
			0.01	0.009			68.93	130
2	6	70			3.4	1283		
			0.01	0.015			73.89	204
4	6	70			1.3	490		
			0.01	0.026			53.00	257
8	6	45			0.6	146		
			0.00	0.029			80.13	337
10	6	34			9.7	1778		
			0.00	0.032			82.41	420
12	7	31			1.1	184		
			0.00	0.034			16.44	436
14	7	23			1.7	211		

TABLE 13–9b Computation of Total Organic Nitrogen for October 18–19, 1981, Storm Event: Jones Falls at Lake Roland Dam (01589452)

Time (hr)	(min)	Discharge (cfs)	Volume (in.)	Total Volume (in.)	Concen-tration (mg/L)	Flux (lb/day)	Load (lb)	Total Load (lb)
19	23	26			7.0	981		
			0.00	0.003			122.59	123
21	25	39			9.1	1913		
			0.00	0.007			92.91	215
23	25	42			1.4	317		
			0.00	0.010			20.21	236
1	25	39			0.8	168		
			0.00	0.013			14.64	250
3	25	34			1.0	183		
			0.00	0.016			23.65	274
5	25	31			2.3	384		
			0.00	0.020			55.25	329
7	59	28			4.3	649		
			0.00	0.022			34.27	364
9	59	23			1.4	174		
			0.00	0.024			18.55	382
11	59	21			2.4	272		

13–7 becomes the product of the mean concentration for all storms and the average runoff volume. The load is then computed with Equation 13–1.

Example 13–5

Random samples of the total organic nitrogen concentrations were collected for six storm events at the station known as Jones Falls at Lake Roland Dam (01589452), which has a watershed area of 34.6 mi^2. The discharges and concentrations for the six storms are given in Table 13–9 a–f. The total load and total runoff were computed for each storm (see Table 13–9), which are summarized in Table 13–10. The station is located near Baltimore. Since Table 13–8 does not include a value for Baltimore, a total of 77 storms per year was assumed based on the value for Philadelphia, which is the city in the table closest to Baltimore. The long-term annual runoff for the station was not available, but the 16-yr average runoff for Jones Falls at Sorrento, MD (01589440), is 18.11 in. Since rainfall and runoff conditions at the two stations are similar, $\overline{Q}_a$ is assumed to equal

$$\overline{Q}_a = 18.11 \text{ in./yr}/77 \text{ storms/yr} = 0.235 \text{ in./storm} \tag{13–8}$$

TABLE 13–9c Computation of Total Organic Nitrogen for November 5–6, 1981, Storm Event: Jones Falls at Lake Roland Dam (01589452)

Time (hr)	(min)	Discharge (cfs)	Volume (in.)	Total Volume (in.)	Concentration (mg/L)	Flux (lb/day)	Load (lb)	Total Load (lb)
21	24	14			4.6	347		
			0.00	0.001			17.45	17
22	54	14			2.8	211		
			0.00	0.002			13.21	30
23	54	14			5.6	423		
			0.00	0.004			25.24	56
1	54	34			1.0	183		
			0.00	0.006			30.70	87
2	54	63			3.8	1290		
			0.00	0.009			105.02	192
3	54	98			7.1	3750		
			0.00	0.014			119.91	311
4	54	120			3.1	2005		
			0.01	0.020			90.28	402
5	54	120			3.6	2328		
			0.01	0.025			90.56	492
6	54	107			3.5	2019		
			0.01	0.033			124.67	617
8	54	86			2.1	973		
			0.01	0.044			88.22	705
12	6	59			1.1	350		
			0.00	0.049			103.51	809
14	6	45			8.8	2134		
			0.00	0.052			147.34	956
16	1	39			7.4	1556		

TABLE 13–9d Computation of Total Organic Nitrogen for December 1–3, 1981, Storm Event: Jones Falls at Lake Roland Dam (01589452)

Time (hr)	(min)	Discharge (cfs)	Volume (in.)	Total Volume (in.)	Concen-tration (mg/L)	Flux (lb/day)	Load (lb)	Total Load (lb)
9	57	18			0.6	58		
			0 .01	0.006			14.82	14
14	57	39			0.4	84		
			0.00	0.008			9.93	25
15	57	52			1.4	392		
			0.01	0.014			25.99	51
17	51	70			0.7	264		
			0.02	0.033			271.54	322
23	6	98			4.2	2219		
			0.01	0.040			145.50	468
0	21	125			5.0	3369		
			0.01	0.047			137.56	605
1	36	142			2.0	1913		
			0.01	0.062			169.84	775
4	6	125			2.0	1347		
			0.01	0.069			50.67	826
5	21	111			1.0	598		
			0.01	0.080			40.82	867
7	51	86			0.4	185		
			0.01	0.086			17.52	884
9	31	74			0.8	319		
			0.01	0.098			119.44	1004
13	50	52			3.6	1009		
			0.01	0.106			330.00	1334
18	4	39			13.0	2732		
			0.01	0.113			272.02	1606
22	4	34			2.9	531		
			0.01	0.119			74.92	1680
2	4	31			2.2	368		
			0.01	0.125			49.50	1730
6	4	42			1.0	226		

Thus, the weighted load from Equation 13–7 is

$$\overline{W}_e = \frac{0.235}{6}\,(208{,}141 \text{ lbs/storm-in.}) = 8{,}152\ \frac{\text{lbs}}{\text{storm}} \tag{13–9}$$

The annual total is computed from Equation 13–6:

$$\begin{aligned}\overline{W}_a &= N_s\overline{W}_e = 77 \text{ storms/yr } (8{,}152 \text{ lbs/storm}) = 627{,}704 \text{ lbs/yr} \\ &= 313.9 \text{ tons/yr}\end{aligned} \tag{13–10}$$

The arithmetic average would lead to an estimate of 182.8 tons/yr, which underestimates the mean by 42%.

TABLE 13–9e Computation of Total Organic Nitrogen for Dec. 14–18, 1981, Storm Event: Jones Falls at Lake Roland Dam (01589452)

Time (hr)	(min)	Discharge (cfs)	Volume (in.)	Total Volume (in.)	Concen-tration (mg/L)	Flux (lb/day)	Load (lb)	Total Load (lb)
15	46	18			3.5	340		
			0.00	0.004			57.61	58
18	46	45			2.4	582		
			0.01	0.016			145.53	203
21	46	135			2.4	1746		
			0.01	0.030			186.54	390
23	48	170			2.9	2657		
			0.01	0.042			222.51	612
1	18	184			4.5	4463		
			0.02	0.065			413.68	1026
4	18	160			2.5	2156		
			0.01	0.078			176.07	1202
6	18	120			3.2	2070		
			0.03	0.103			420.26	1622
12	15	70			3.5	1321		
			0.00	0.108			87.81	1710
13	46	66			4.1	1459		
			0.03	0.136			4315.89	6026
22	39	78			52.0	21862		
			0.03	0.166			7544.69	13571
10	32	34			47.0	8613		
			0.03	0.185			4481.79	18052
22	32	36			48.0	9314		
			0.02	0.201			3630.17	21683
10	32	23			42.0	5206		
			0.01	0.208			874.25	22556
17	56	21			4.1	464		
			0.01	0.216			165.01	22721
2	28	21			4.1	464		

TABLE 13–9f Computation of Total Organic Nitrogen for Feb. 9–10, 1982, Storm Event: Jones Falls at Lake Roland Dam (01589452)

Time (hr)	(min)	Discharge (cfs)	Volume (in.)	Total Volume (in.)	Concen-tration (mg/L)	Flux (lb/day)	Load (lb)	Total Load (lb)
4	39	36			9.9	1921		
			0.01	0.007			335.26	335
8	39	39			10.0	2102		
			0.01	0.014			352.51	688
12	39	42			9.4	2128		
			0.01	0.023			495.34	1183
16	39	59			12.0	3816		
			0.01	0.034			501.81	1685
20	39	66			6.2	2206		
			0.01	0.045			261.77	1947
0	39	56			3.1	936		
			0.01	0.054			320.53	2267
4	39	45			12.0	2911		

TABLE 13–10 Computation of Weighted Mean Annual Load

Event Date	Runoff Volume (in.)	Total Organic N (lbs/storm)	Load/ Runoff (lbs/storm-in.)
October 1–2, 1981	0.034	436	12,824
October 18–19, 1981	0.024	382	15,917
November 5–6, 1981	0.052	956	18,385
December 1–3, 1981	0.125	1,730	13,840
December 14–18, 1981	0.216	22,722	105,194
February 9–10, 1982	0.054	2,267	41,981
		28,493	208,141 = sum

13.3 EMPIRICAL MODELS

13.3.1 Nationwide Equations for Urban Watersheds

Where site-specific or local data are not available to estimate water quality concentrations, it may be necessary to use regional or nationwide equations. If the conditions for which estimates are required are similar to those for which the nationwide equations were developed, then they may yield reasonably accurate predictions. Nationwide equations often require more detailed input than just land use activity, such as the mean concentrations of Tables 13–5 and 13–6.

Urban areas are of special concern with respect to water quality. While the area of a city might be small compared with the area of agricultural lands, urban areas produce high concentrations of pollutants per unit area and they lack the natural buffering that often exists in agricultural areas. Urban planners, therefore, need estimates of loads from urban drainage basins.

Driver and Tasker (1988) provided nationwide equations that can be used to estimate the mean loads of a storm for ten constituents (see Table 13–11). The regression coefficients given in Table 13–11 are for the model

$$\hat{Y} = b_c 10 ** [b_0 + b_1 A ** 0.5 + b_2 I_a + b_3 P_a + b_4 T_j + b_5 I_c] \quad (13\text{–}11)$$

in which $\hat{Y}$ is the predicted value of the constituent; b_c is a bias correction factor; A is the total contributing drainage area (sq. mi); I_a is the impervious land cover as a percentage of the total area (%); P_a is the mean annual rainfall (in.); T_J is the mean minimum January temperature (°F); and I_c is an indicator variable of commercial and industrial land cover exceeding ($I_c = 1$) or not exceeding ($I_c = 0$) 75% of the drainage area. The annual load can be estimated by multiplying $\hat{Y}$ of Equation 13–11 by the mean number of runoff-producing storms in a year. Table 13–8 gives values for selected U.S. cities. Equation 13–11 was developed using measured data for drainage areas in the range from approximately 0.02 to 0.8 sq. mi.

Example 13–6

A small town (area = 0.74 sq. mi) discharges storm runoff into an adjacent river. The mean annual low flow in the river is 20 cfs and the suspended solids concentration upstream of the town

TABLE 13–11 Regression Models of Mean Loads of a Storm for Selected Constituents

Constituent*	b_c	b_o	b_1	b_2	b_3	b_4	b_5
COD	1.298	1.1174	2.0069	0.0051	0	0	0
CU	1.403	−1.4824	1.8281	0	0	−0.0141	0
DP	1.469	−1.3661	1.3955	0	0	0	0
DS	1.251	1.8449	2.5468	0	0	−0.0232	0
PB	1.365	−1.9679	1.9037	0	0	−0.0232	0
SS	1.521	1.5430	1.5906	0	0.0264	−0.0297	0
TKN	1.277	−0.7282	1.6123	0.0064	0.0226	−0.0210	−0.4345
TN	1.345	−0.2433	1.6383	0.0061	0	0	−0.4442
TP	1.314	−1.3884	2.0825	0	0.0234	−0.0213	0
ZN	1.322	−1.6302	2.0392	0.0072	0	0	0

*COD Chemical oxygen demand in mean seasonal or mean annual load, in pounds
CU Total recoverable copper in mean seasonal or mean annual load, in pounds
DP Dissolved phosphorus in mean seasonal or mean annual load, in pounds
DS Dissolved solids in mean seasonal or mean annual load, in pounds
PB Total recoverable lead in mean seasonal or mean annual load, in pounds
SS Suspended solids in mean seasonal or mean annual load, in pounds
TN Total nitrogen in mean seasonal or mean annual load, in pounds
TKN Total ammonia plus organic nitrogen as nitrogen in mean seasonal or mean annual load, in pounds
TP Total phosphorus in mean seasonal or mean annual load, in pounds
ZN Total recoverable zinc in mean seasonal or mean annual load, in pounds

is 100 mg/L. The town engineer needs to assess the effect of storm runoff on the suspended solids load. Using the coefficients of Table 13–11 with Equation 13–11, the estimated load in pounds is

$$\begin{aligned}\hat{W} &= 1.521 * 10 ** [1.543 + 1.5906(0.74) ** 0.5 \\ &\quad + 0.0264\,(37\text{ in.}) - 0.0297\,(25^\circ\text{F})] \qquad (13\text{–}12) \\ &= 2127 \text{ pounds per storm}\end{aligned}$$

For an average storm duration of 6 hr and a storm discharge of 1.2 ft^3/sec, a load of 2127 lbs translates to a concentration of 1316 mg/L. Since the recommended limit in this locality is 200 mg/L in urban runoff, the town must adopt a program for reducing suspended solids loads.

13.3.2 Estimation of Stream Nutrient Levels

Omernik (1976) used the regression-equation approach to develop relationships for predicting stream nutrient levels in the eastern U.S. Of particular interest was the relationship between land use and nutrient levels. Equations for predicting total and inorganic forms of phosphorus and nitrogen were developed using the combined percentage of agricultural and urban land use (X) as the predictor variable. The exponential model form was used:

$$\hat{C} = 10 ** [b_o + b_1 X] \qquad (13\text{–}13)$$

where $\hat{C}$ is the predicted concentration (mg/L). The coefficients are given in Table 13–12. These equations were developed using data from drainage areas from about 0.2 to 90 sq. mi.

TABLE 13–12 Regression Coefficients for Predicting Stream-Nutrient Concentrations

Nutrient	b_o	b_1
Mean total phosphorus	−1.831	0.0093
Mean total nitrogen	−0.278	0.0088
Mean orthophosphorus	−2.208	0.0089
Mean inorganic nitrogen	−0.873	0.0136

Source: Federal Highway Administration, 1990. FHWA-RD-88-006. *Pollutant Loadings and Impacts from Highway Stormwater Runoff*, Vol. I: Design Procedure.

Example 13–7

A land-use planner is assessing the potential effect of changes in zoning regulations that would allow urban development to replace existing forest and nonagricultural lands. The 7.2 sq. mi drainage area currently uses 4.5% of the area for agricultural and urban uses. The changes in zoning regulations would allow this to increase to 35%. The estimated mean total nitrogen for existing conditions is

$$\hat{N} = 10 ** [-0.278 + 0.0088\,(4.5)] = 0.578 \text{ mg/L} \tag{13–14}$$

For phosphorus, orthophosphorus, and inorganic nitrogen, the estimated concentrations are 0.01625 mg/L, 0.006793 mg/L, and 0.1542 mg/L, respectively. For projected land use conditions, the total nitrogen will be

$$\hat{N} = 10 ** [-0.0278 + 0.0088\,(35)] = 1.072 \text{ mg/L} \tag{13–15}$$

For these conditions, the total phosphorus, orthophosphorus, and inorganic nitrogen are 0.03122 mg/L, 0.0127 mg/L, and 0.4009 mg/L, respectively. These represent increases of 92, 86, 87, and 160% for total phosphorus, total nitrogen, orthophosphorus, and inorganic nitrogen, respectively.

13.3.3 The Simple Method

Schueler (1987) developed a method for estimating pollutant loads for urban areas. The method is based on data from the National Urban Runoff Program. It is intended to be used for time periods ranging from a single storm event to mean annual rates and for areas less than a square mile. The structure of the method known as the "Simple Method" is similar to that of the Rational Formula for estimating peak discharges; it combines the discharge computed in a manner similar to the Rational Method with the relationship of Equation 13–1. Specifically, the load L (lbs) is computed by

$$L = 0.2267\, P P_j R_v A C \tag{13–16}$$

in which P is the rainfall depth (in.) for the time interval for which L is needed, P_j is a factor that corrects P for storms that do not produce runoff, R_v is the fraction of rainfall that is converted to runoff, A is the area of the development site (acres), and C is the flow-weighted mean concentration (mg/L) of the pollutant. Equation 13–16 has the form of Equation 13–1 if the product PP_jR_vA corresponds to the discharge. This similarity should be evident in that R_v would correspond to the runoff coefficient of the Rational Method and PP_j to the intensity. Given that the Rational Formula is widely used and accepted, the Simple Method is appropriate for its intended use.

The value of P in Equation 13–16 can be the total depth for a storm event or a value as large as the annual rainfall. The duration of the value of P will determine the duration of the computed load. If the annual rainfall is used in Equation 13–16, then the computed load is in pounds per year. The correction factor P_j is the most difficult of the five inputs. For an individual storm, P_j equals one. When computing an annual load, the value of P_j may be less than 0.5 depending on the location. P_j is the proportion of storms during the duration specified for P that produce measurable runoff.

Schueler provides the following relationship for estimating R_v:

$$R_v = 0.05 + 0.009\,I \tag{13–17}$$

in which I is the watershed imperviousness (%).

The mean pollutant concentration can be obtained from either local data or averages from other studies. Mean values from different locations can be quite variable, and considerable judgment is necessary in selecting a value. Schueler provided the values shown in Table 13–13 for urban areas. Values from Tables 13–5 and 13–6 can also be used for C in Equation 13–16.

TABLE 13–13 Urban "C" Values For Use with the Simple Method (mg/L)

Pollutant	New Suburban NURP Sites (Wash., DC)	Older Urban Areas (Baltimore)	Central Business District (Wash., DC)	National NURP Study Average	Hardwood Forest (Northern Virginia)	National Urban Highway Runoff
PHOSPHORUS						
Total	0.26	1.08	—	0.46	0.15	—
Ortho	0.12	0.26	1.01	—	0.02	—
Soluble	0.16	—	—	0.16	0.04	0.59
Organic	0.10	0.82	—	0.13	0.11	—
NITROGEN						
Total	2.00	13.6	2.17	3.31	0.78	—
Nitrate	0.48	8.9	0.84	0.96	0.17	—
Ammonia	0.26	1.1	—	—	0.07	—
Organic	1.25	—	—	—	0.54	—
TKN	1.51	7.2	1.49	2.35	0.61	2.72
COD	35.6	163.0	—	90.8	>40.0	124.0
BOD (5/day)	5.1	—	36.0	11.9	—	—
METALS						
Zinc	0.037	0.397	0.250	0.176	—	0.380
Lead	0.018	0.389	0.370	0.180	—	0.550
Copper	—	0.105	—	0.047	—	—

13.4 MATERIALS BALANCE

In Chapter 1, the conservation of mass was introduced as the linear storage equation (see Figure 1–1):

$$I - O = \frac{dS}{dt} \tag{13–18}$$

It was also used in Chapter 10 (see Equation 10–1) as the basis for routing a flood hydrograph through a stream reach and in Chapter 11 (see Equation 11–1) for routing a hydrograph through a storage structure. While it will be necessary to generalize the mathematical description of the conservation of mass given by Equation 13–18, the principle can be used for describing the behavior of water quality parameters. So far in this text, the form of Equation 13–18 has been used solely for water, which does not change form. Some water quality parameters change form because of biological or chemical reactions. Specifically, some substances are referred to as nonconservative because the mass changes due to a reaction. In such a case, the conservation of mass is given by

$$I - O \pm P = \frac{dS}{dt} \tag{13–19}$$

in which P is the increase or loss of mass due to production or decay and I, O, and S are the same as previously defined except they refer to the inflow, outflow, and rate of accumulation of the water quality substance rather than water. A minus sign before P is used for decay, while a plus sign is used for production. It is not unreasonable to view the decay of Equation 13–19 as a part of the outflow O; in a sense, the decay is a loss of mass that must be subtracted from the inflow just as the outflow O from the system is subtracted.

13.4.1 Batch Reactors

For a water quality constituent, the conservation of mass of Equation 13–19 can be rewritten as

$$CQ_i - CQ_o \pm rV = \frac{dC}{dt}V \tag{13–20}$$

in which V is the volume and r is the mass rate of decay. For a batch reactor $Q_i = Q_o = 0$. Assuming decay, rather than production, Equation 13–20 reduces to

$$\frac{dC}{dt} = -r \tag{13–21}$$

Assuming the rate of decay follows the first-order reaction, or $r = KC$, Equation 13–21 yields

$$\frac{dC}{dt} = -KC \tag{13–22}$$

in which K is the reaction rate coefficient, with a unit of time^{-1} (that is, per day). Based on this assumed form, the concentration at any time t is

$$C = C_0 e^{-Kt} \tag{13–23}$$

in which C_0 is the initial concentration.

Equation 13–23 is commonly used for expressing the temporal decay of a substance. In practice, it is used to estimate the concentration C at time t, which requires a value of the rate coefficient K. Such use represents synthesis. The value of K is often obtained from a table of values. However, for a particular case, it may be necessary to have a site-specific value of K. Thus, it would be necessary to analyze measured data. Two measurements of the concentration would be needed. The initial concentration at an assumed time of 0 is denoted as C_0. A second measurement at time t yields C. Rearranging Equation 13–23 and making a logarithmic transform yields

$$K = -\ln (C/C_0)/t \tag{13–24}$$

which uses the concentration at time t and the value of C. C, C_0, and t are then used to compute K, which can then be used for synthesis at the location and for the conditions at which the measurements were made.

While two measured values and the time interval between measurements can be used, it is more common to collect a series of values and perform a regression analysis of concentration on time. For such an analysis, a logarithmic transformation of Equation 13–23 is made:

$$\ln C = \ln C_0 - Kt$$

$$y = a + bt \tag{13–25}$$

in which $y = \log C$ and $a = \ln C_0$. The coefficients a and b are fitted with measurements, with the resulting values used to quantify $K = b$ and $C_0 = e^a$.

Since measured values may contain a significant amount of error, the use of two points may yield inaccurate values of the coefficients C_0 and K. The regression approach with several measurements is preferable.

Example 13–8

A stream with a flow velocity of 5 fps has an initial coliform number of 100/100mL. The coliform level measured 20 mi downstream is 50/100 mL. Assuming that coliform follows the first-order decay, the rate can be determined from Equation 13–24:

$$K = -\ln (C/C_0)/(L/V)$$

$$= \frac{-\ln\left(\frac{50}{100}\right)}{\frac{20\,(5280)}{5\,(3600)}} = 0.118 \text{ hr}^{-1}$$

The time interval t of Equation 13–24 was computed as the ratio of the reach length to the flow velocity.

Example 13–9

Assuming the rate coefficient is known, Equation 13–24 can be used to estimate the time that the inflow must be retained so that the concentration in the outflow meets a stated water quality criterion. Solving Equation 13–24 for the time yields

$$t = -\ln(C/C_0)/K \tag{13–26}$$

Assuming that the coliform count in combined sewer overflow is 10^4/100mL and that the state water quality criteria require the coliform to be equal to or less than 200/100 mL, the outflow will have to be retained for a period of time t:

$$t = -\ln(200/10000)/0.12\ \text{hr}^{-1}$$
$$= 33\ \text{hr} \tag{13–27}$$

13.4.2 Effect of Temperature

The rates of a reaction are temperature dependent. Therefore, the reaction rate coefficients depend on the temperature at which the reaction takes place. It is common to assume that the rate coefficients at temperatures T_1 and T_2 are related by

$$k_1 = k_2\,\theta^{(T_1 - T_2)} \tag{13–8}$$

in which k_i is the rate coefficient at temperature T_i and θ is the temperature coefficient. T_2 is often taken as 20°C. A temperature coefficient ranging from 1.04 to 1.06 is frequently used for a variety of rate constants.

13.5 FLOW DILUTION

Equations 13–1 and 13–2 apply to a single flow path in a system. It is frequently necessary to combine or separate flows at junctions within the system. For example, storm runoff systems originating in two parts of a watershed may combine to form a main pipe, and the flow rates and loadings within the two-pipe segments may be quite different. A similar case arises when two streams join to form a larger stream. In both instances, it is necessary to compute the concentration of pollutants in the combined flow.

In solving such problems, these two points must be considered:

1. At a junction, the flow into the junction must equal the flow out of the junction.
2. The load into the junction must equal the load out of the junction.

These concepts assume that storage in the junction is not possible. These concepts are expressed mathematically by

$$\Sigma Q_{\text{in}} = \Sigma Q_{\text{out}} \tag{13–29}$$

$$\Sigma W_{\text{in}} = \Sigma W_{\text{out}} \tag{13–30}$$

As an example, consider the case where two streams join to form a third stream. Then Equation 13–29 becomes

$$Q_1 + Q_2 = Q_3 \tag{13–31}$$

and Equation 13–30 becomes

$$W_1 + W_2 = W_3 \tag{13–32}$$

To find the concentration in the main stream, Equation 13–1 is used with Equation 13–32:

$$C_1Q_1 + C_2Q_2 = C_3Q_3 \tag{13–33}$$

Solving for the unknown C_3 and using Equation 13–31 yields

$$C_3 = \frac{C_1Q_1 + C_2Q_2}{Q_1 + Q_2} \tag{13–34}$$

Equation 13–34 assumes that the flow from the two inputs completely mix. This may or may not happen. It depends on the nature of the flow and the characteristics of the water quality constituents. It is not uncommon to see aerial photographs of the confluence of two rivers, one reasonably free of sediment and the other sediment laden. The photo shows that the two streams have not mixed. Conversely, one would expect the flows from two 1-ft diameter pipes from different parts of a watershed to mix very quickly. Thus, the assumption of complete mixing, which underlies Equation 13–34, would be valid.

Example 13–10

A cannery has a daily production rate of 600 cases of green beans. Records indicate that this production requires 100 gallons of water per case, or 0.06 mgd (that is, million gallons per day). The effluent has a BOD concentration of 95 mg/L and a suspended solids concentrations of 290 mg/L. The untreated waste water is discharged into a local stream, where regulations limit BOD and suspended solids concentrations to 20 mg/L and 100 mg/L, respectively.

During a drought, flow in the stream decreases to 5 ft^3/sec. The local water agency is charged with determining whether or not the limits specified in the regulations are being exceeded in the cannery effluent. Measurements at a station upstream indicate BOD and SS concentrations of 2 and 15 mg/L, respectively.

The discharge of 0.06 mgd is equivalent to 0.09283 ft^3/sec. Following Equation 13–34, the concentration of BOD in the stream (C_{sb}) below the discharge point from the cannery is

$$C_{sb} = \frac{C_eQ_e + C_sQ_s}{Q_e + Q_s} = \frac{95\,(0.09283) + 2(5)}{0.09283 + 5} = 3.7 \text{ mg/L} \tag{13–35}$$

For the suspended solids, the downstream concentration of the total flow is

$$C_{ss} = \frac{290\,(0.09283) + 15\,(5)}{0.09283 + 5} = 20.01 \text{ mg/L} \tag{13–36}$$

Thus, the BOD and the concentration of the suspended solids are below the limits.

Equation 13–34 can also be used to determine the upstream flow rate that is necessary to provide sufficient dilution such that the BOD limit will not be exceeded. Letting Q_u and Q_d represent the upstream and downstream flow rates and C_1 be the regulatory limit, then

$$C_1Q_t = C_eQ_e + C_uQ_u = C_1\,(Q_e + Q_u) \tag{13–37}$$

in which Q_t is the total flow and equals Q_e plus Q_u. Solving for the upstream flow gives

$$Q_u = Q_e\left(\frac{C_e - C_1}{C_1 - C_u}\right) = 0.09283\left(\frac{95 - 20}{20 - 2}\right) = 0.39 \text{ ft}^3/\text{sec} \tag{13–38}$$

Thus, when the flow rate upstream falls below 0.39 ft^3/sec with a BOD concentration of 2 mg/L, the combined flow will not meet the limit of 2 mg/L for the BOD concentration.

Example 13–11

Concentrations of lead in storm runoff from residential areas can be significant. Standards for drinking water recommend a limit of 0.05 mg/L; a limit of 0.1 mg/L is recommended for freshwater aquatic life. Short-term occurrences above the limit may not be detrimental to aquatic life.

In a particular urban area with a drainage area of 200 acres, measurements of the concentration of lead in storm runoff average 0.35 mg/L. Assume the average discharge is 200 cfs during the storm. The surface runoff is discharged into a stream with a normal flow of 25 cfs and lead concentrations of 0.002 mg/L. The concentration of the lead in the combined flow is

$$C = \frac{200\,(0.35) + 25\,(0.002)}{200 + 25} = 0.311 \text{ mg/L} \tag{13–39}$$

While this is well above the recommended limit, the storm duration is relatively short. However, such high concentrations can cause localized problems if the polluted runoff is allowed to accumulate in ponded areas.

Example 13–12

Equation 13–34 can be used to compute the ordinates of a pollutograph from ordinates at the confluence of two streams or pipes. Table 13–14 gives the instantaneous flow rates and corresponding concentrations of total organic nitrogen for two urban drainage basins having areas of 17 acres and 22 acres. The collector pipes intersect and flow to a storm main. For the measurements at time 1255, the concentration in the main is

$$C_T = \frac{q_1 C_1 + q_2 C_2}{q_T} = \frac{0.09\,(5.7) + 0.17\,(6.6)}{0.26} = 6.3 \text{ mg/L} \tag{13–40}$$

Values for the remaining ordinates of the main-drain pollutographs are given in Table 13–14. The total load could be computed using Equation 13–1 for each measurement and then integrated over the entire storm duration.

TABLE 13–14 Concentration of Total Organic Nitrogen (C_T) in Storm Sewer Flow from Two Urban Subareas

Time	q_1 (cfs)	C_1 (mg/L)	q_2 (cfs)	C_2 (mg/L)	q_T (cfs)	C_T (mg/L)
1255	0.09	5.7	0.17	6.6	0.26	6.3
1329	0.12	3.3	0.23	5.9	0.35	5.0
1339	0.18	8.8	0.31	9.1	0.49	9.0
1359	0.42	5.4	0.58	7.4	1.00	6.6
1413	0.90	12.0	1.12	11.7	2.02	11.8
1426	0.90	9.1	0.97	12.2	1.87	10.7
1433	0.81	3.5	0.92	8.3	1.73	6.1
1448	0.63	3.6	0.76	6.1	1.39	5.0
1504	0.54	6.7	0.71	7.0	1.25	6.9
1638	0.33	6.8	0.47	6.2	0.80	6.4

13.6 STREETER-PHELPS OXYGEN SAG CURVE

The discharge of biologically degradable substances into a water body creates a demand for oxygen. The demand varies with time, and if it is a moving stream, the demand will change as the substances move downstream. The demand for oxygen created by these substances is called the biochemical oxygen demand (BOD), and the consumption of the oxygen that exists in the stream is called deoxygenation. At the same time that the decomposable substances are causing the dissolved oxygen to be depleted, oxygen is being transferred from the atmosphere and is being released by plants through photosynthetic activity; these processes are called reoxygenation, or reaeration. The time-dependent relationship that shows the effects of oxygen depletion and resupply is called the oxygen sag curve, which is usually shown as a graph of the oxygen concentration versus time. Where the biologically decomposable substances are dumped into a moving body of water, such as a stream, the time axis is related to the distance downstream from the discharge point through the velocity of the stream. The rate of deoxygenation by the load in the water body is generally viewed as two activities: the breakdown of carbonaceous matter, which is called the first-stage BOD, and the breakdown of nitrogenous substances, which is called nitrification. Only the first-stage BOD will be considered here.

The commonly used representation of the deoxygenation of a stream is based on the following assumptions: (1) BOD is independent of the oxygen-concentration; (2) BOD is dependent on the concentration of biologically decomposable material; and (3) the oxygen concentration is greater than the critical value, which is temperature dependent. If we assume that the rate of deoxygenation is proportional to the amount of carbonaceous waste (L_t) available at time t, then

$$\frac{dL_t}{dt} = -k_1 L_t \tag{13–41}$$

in which k_1 is the rate constant for deoxygenation. The differential equation is easily solved:

$$L_t = L_o e^{-k_1 t} \tag{13–42}$$

in which L_o is the ultimate demand. The rate constant is temperature dependent, and the value of k_1 at any temperature T (°C) can be found by

$$k_1 = k_o \theta^{(T-20)} \tag{13–43}$$

in which k_o is the rate constant at 20°C and θ is the temperature coefficient, usually taken as 1.047. The oxygen demand at any time t, L_t, equals the difference between the ultimate demand L_o and the oxygen consumed by the wastes between the time the load entered the stream and time t:

$$L_t = L_o - D_t \tag{13–44}$$

Solving Equation 13–44 for D_t and substituting Equation 13–42 for L_t yields

$$D_t = L_o(1 - e^{-k_1 t}) \tag{13–45}$$

The rate of reoxygenation is proportional to the deficit:

$$\frac{dD}{dt} = -k_2 D \tag{13–46}$$

in which $D = C_s - C$, C_s is the saturated dissolved oxygen, and C is the observed dissolved oxygen. k_2 is the rate constant for reoxygenation, which is temperature dependent and is usually assumed to follow the van't Hoff-Arrhenius Equation:

$$k_{2T} = k_{20}\, e^{C_r(T-T_o)} \quad (13\text{–}47)$$

in which C_r is a temperature characteristic value for reoxygenation, k_{20} is the value of k_2 at temperature T_o, which is usually taken as 20°C, and k_{2T} is the value of k_2 at any other temperature T. The factors affecting k_2 include the stream velocity and depth of flow.

If we use the material balance of Equation 13–19 and assume that conditions are steady-state and that the production consists of the algebraic sum of deoxygenation and reoxygenation, then we get the following differential equation:

$$\frac{dD}{dt} = k_1 L_o e^{-k_1 t} - k_2 D \quad (13\text{–}48)$$

which has the following solution:

$$D_t = \frac{k_1 L_0}{k_2 - k_1}\left(e^{-k_1 t} - e^{-k_2 t}\right) + D_o e^{-k_2 t} \quad (13\text{–}49)$$

in which D_o is the initial *DO* deficit. Equation 13–49 is known as the Streeter-Phelps oxygen sag equation and has the form shown in Figure 13–3. It is noted that the predicted D_t from Equation 13–49 does not consider the effects of other oxygen demands (that is, sediment) and supply (for example, algae).

An engineer would be interested in knowing the largest oxygen deficit and the location where that will occur. Since Equation 13–49 uses time as the independent variable, it will be necessary to relate distance and time. In the discussion of the time of concentration in Chapter 3, the following relation between flow velocity V, distance L, and elapsed time t was introduced:

$$t = \frac{L}{V} \quad (13\text{–}50)$$

Manning's Equation (see Equation 3–41) can be used to estimate the velocity of flow and Equation 13–50 can be substituted into the sag formula (Equation 13–49) so that it is a function of distance. If the flow velocity changes through the reach, the sag relation must account for this change.

The minimum dissolved oxygen concentration (the maximum deficit) is found by differentiating Equation 13–49, setting the derivative equal to 0, and solving for t, which is designated t_c:

$$t_c = \frac{1}{k_2 - k_1} \ln_e\left(\frac{k_2}{k_1}\left[1 - D_o\left(\frac{k_2 - k_1}{k_1 L_o}\right)\right]\right) \quad (13\text{–}51)$$

At time t_c, the critical deficit is

$$D_c = \left(\frac{k_1}{k_2}\right) L_o e^{-k_1 t_c} \quad (13\text{–}52)$$

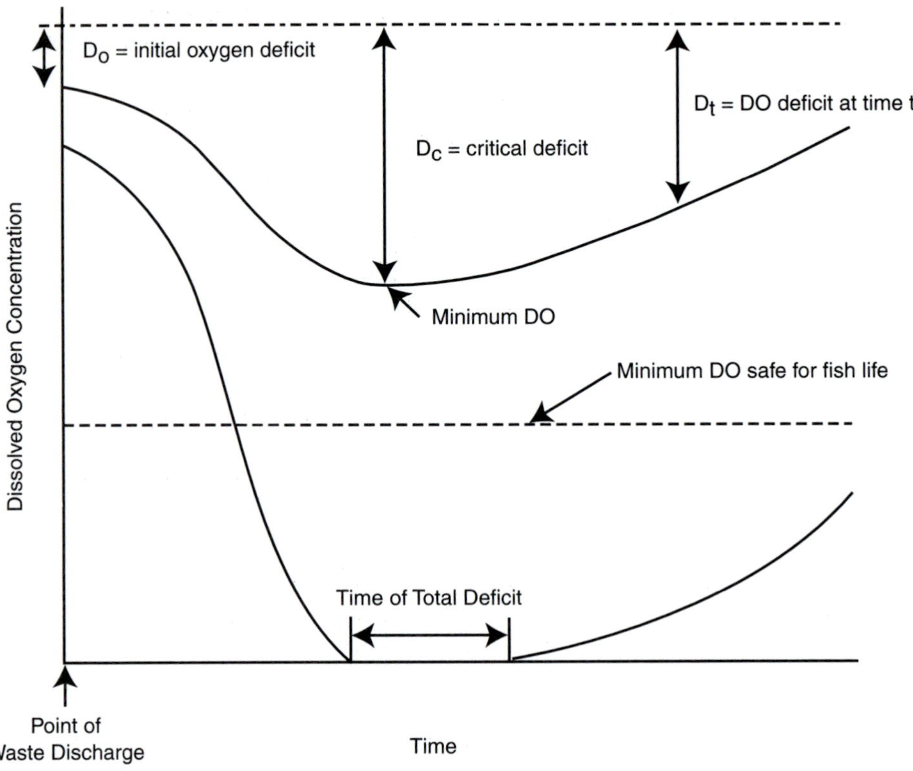

FIGURE 13–3 Oxygen sag curves: partial deficit and total deficit.

Combining this with Equation 13–50 and assuming a constant velocity, the distance where the critical conditions will occur is

$$L_c = Vt_c \tag{13–53}$$

The saturation concentration of dissolved oxygen can be computed by

$$\begin{aligned}\hat{C}_s = {} & 14.57 - 0.39311\ T + 0.0070695\ T^2 \\ & - 0.0000589066\ T^3 \text{ for } 0 \leqslant T \leqslant 40°\text{C}\end{aligned} \tag{13–54}$$

Example 13–13

A new suburban residential area of 100 acres has a 10-yr discharge of 120 cfs. The initial BOD concentration is 5.1 mg/L. Assume that the storm occurs in the summer (T = 25°C) and that the *DO* concentration in the runoff is 7 mg/L. The runoff discharges into a small stream where the flow rate is 8 cfs at a temperature of 21°C, with a BOD of 3 mg/L and a dissolved oxygen concentration of 7.4 mg/L. Tests show that k_1 is 0.2 day^{-1} at 20°C and that k_2 is 0.8 day^{-1} at 25°C. Complete mixing of the storm discharge and the flow from upstream areas is assumed to occur at the outfall. The stream flows 3300 ft at a velocity of about 0.8 fps before combining with another larger stream.

The total flow at the outfall is 128 cfs. Equation 13–34 can be used to compute the temperature (t_o), initial BOD (L_o), and initial dissolved oxygen concentration (C_o) at the outfall:

$$T_o = \frac{8\,(21) + 120\,(25)}{8 + 120} = 24.75°C \tag{13–55}$$

$$L_o = \frac{8\,(3) + 120\,(5.1)}{8 + 120} = 4.97\text{ mg/L} \tag{13–56}$$

$$C_o = \frac{8\,(7.4) + 120\,(7)}{8 + 120} = 7.02\text{ mg/L} \tag{13–57}$$

At 24.75°C, Equation 13–54 yields a saturation concentration of 8.28 mg/L; therefore, the initial deficit is

$$D_o = C_s - C_o = 8.28 - 7.02 = 1.26\text{ mg/L} \tag{13–58}$$

The value of k_1 needs to be adjusted for the temperature difference; Equation 13–43 gives

$$k_1 = 0.2\,(1.047)^{(24.75-20)} = 0.249\text{ day}^{-1} \tag{13–59}$$

Thus, the deficit at any time t is

$$D_t = \frac{0.249\,(4.97)}{0.8 - 0.249}\,[e^{0.249t} - e^{-0.8t}] + 1.26\,e^{-0.8t} \tag{13–60}$$

The time of the critical deficit is computed from Equation 13–51:

$$t_c = \frac{1}{0.8 - 0.249}\ln_e\left[\frac{0.8}{0.249}\left(1 - 1.26\left(\frac{0.8 - 0.249}{0.249\,(4.97)}\right)\right)\right]$$
$$= 0.62\text{ days} = 15.0\text{ hr} \tag{13–61}$$

For the downstream reach of 3300 ft, the travel time is

$$t = L/V = 3300\text{ ft}/0.8\text{ fps} = 4125\text{ sec} = 1.15\text{ hrs} \tag{13–62}$$

Since the critical time is considerably longer than the time at which the flow enters the larger stream, the deficit will not fall to the critical value. Equation 13–60 can be used to compute the deficit at the point where the flow enters the larger stream; Equation 13–60 gives a deficit of 1.29 mg/L, which is only slightly higher than that at the outfall.

13.7 ROUTING A LOADOGRAPH

As indicated previously, a pollutograph relates concentration versus time, while a loadograph relates load versus time. If the concentration of a pollutant is assumed to be uniformly distributed throughout the volume at a river cross section, then it can also be assumed that the material will travel through the reach in a manner similar to the flow itself. Thus, a pollutograph will undergo translation and attenuation just as the flow experiences these effects. The translation is physically necessary, and attenuation reflects the mixing that will take place as the mass of material moves through the reach.

The procedure for computing the pollutograph at a downstream section is as follows:

1. Route the discharge hydrograph through the reach.
2. Compute the loadograph at the upstream section using Equation 13–1.
3. Route the loadograph through the reach using the same routing equation used to route the hydrograph.
4. Compute the pollutograph at the downstream section using the routed hydrograph and routed loadograph with Equation 13–1.

It is important to note that the pollutograph cannot be routed with the routing equation because the total load at the downstream section will not equal the total load at the upstream section. Routing the loadograph is necessary to ensure continuity of mass.

The loadograph can be routed using one of the channel routing methods introduced in Chapter 10: Convex, Muskingum, Muskingum-Cunge, or the modified Att-Kin Method. The routing coefficient(s) used for routing the flow can also be used for routing the loadograph. If the substance is nonconservative, then the ordinates of the routed (that is, downstream)

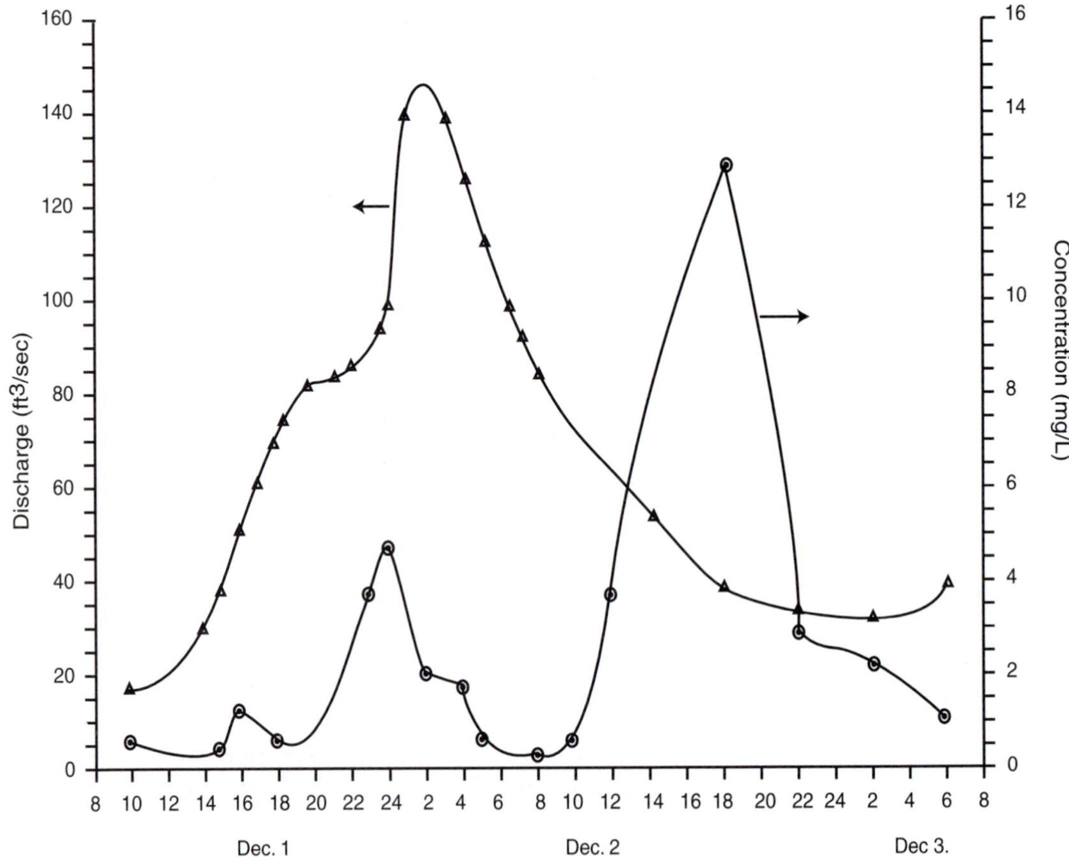

FIGURE 13–4 Upstream hydrograph and pollutograph: Jones Falls at Lake Roland Dam.

loadograph can be multiplied by the exponential component (e^{-kt}) of Equation 13–23, with the travel time through the reach used as the time t. A value would be required for the decay coefficient. It is important to note that it would be incorrect to apply Equation 13–23 to the pollutograph at the upstream cross section because then the translation and attenuation effects would be absent.

Example 13–14

Data for a storm event at the Jones Falls station at Lake Roland Dam (01589452) can be used to demonstrate the procedure. The measurements were collected at nonconstant time intervals for both the flow and the concentration of total organic nitrogen (mg/L). Because hydrologic routing methods require a constant time interval, a constant-interval hydrograph and pollutograph were obtained by interpolating from plots. A numerical interpolation scheme that can be used with nonconstant intervals, such as the Lagrange Interpolation Method, could be used, but a graphical interpolation should provide sufficient accuracy. The measured data for the hydrograph and pollutograph are shown in Figure 13–4; values of the interpolated hydrograph and pollutograph are given in Table 13–15.

TABLE 13–15 Routed Hydrograph, Pollutograph, and Loadograph for the December 1–3, 1981, Event: Jones Falls at Lake Roland Dam

	Discharge		Concentration		Load	
Time	Upstream (cfs)	Downstream (cfs)	Upstream (mg/L)	Downstream (mg/L)	Upstream (lb/day)	Downstream (lb/day)
10.00	18.0	15.0	.600	.500	58	40
12.00	21.0	16.2	.400	.544	45	47
14.00	31.0	18.1	.300	.477	50	47
16.00	52.0	23.3	1.300	.383	364	48
18.00	70.0	34.8	.600	.932	226	175
20.00	82.0	48.9	.900	.742	398	195
22.00	87.0	62.1	2.800	.825	1313	276
24.00	103.0	72.1	5.100	1.779	2831	691
2.00	148.0	84.4	2.300	3.399	1835	1547
4.00	127.0	109.9	2.000	2.807	1369	1662
6.00	104.0	116.7	.700	2.456	392	1545
8.00	86.0	111.6	.400	1.801	185	10834
10.00	72.0	101.4	.900	1.326	349	724
12.00	60.0	89.6	3.800	1.189	1229	574
14.00	52.0	77.8	7.100	1.995	1990	836
16.00	45.0	67.5	10.400	3.569	2522	1298
18.00	40.0	58.5	12.900	5.671	2781	1788
20.00	37.0	51.1	9.400	7.935	1875	2185
22.00	35.0	45.5	2.900	8.412	547	2061
24.00	33.0	41.3	2.400	6.542	427	1455
2.00	32.0	38.0	2.200	5.102	379	1044
4.00	34.0	35.6	1.600	4.058	293	778
6.00	42.0	34.9	1.000	3.101	226	584

The Convex Routing Method was used to route the upstream hydrograph through a hypothetical reach with a routing coefficient of 0.4 (see Figure 13–5). After transforming the pollutograph to a loadograph using Equation 13–1, the loadograph was routed using the routing coefficient of 0.4. The pollutograph is given in Table 13–15. Just as the hydrograph is translated and attenuated, the loadograph, and therefore the pollutograph, are translated and attenuated. The loadographs and pollutographs for both the upstream and downstream sections are shown in Figures 13–6 and 13–7, respectively. The values of Table 13–15 assume that the load is conservative.

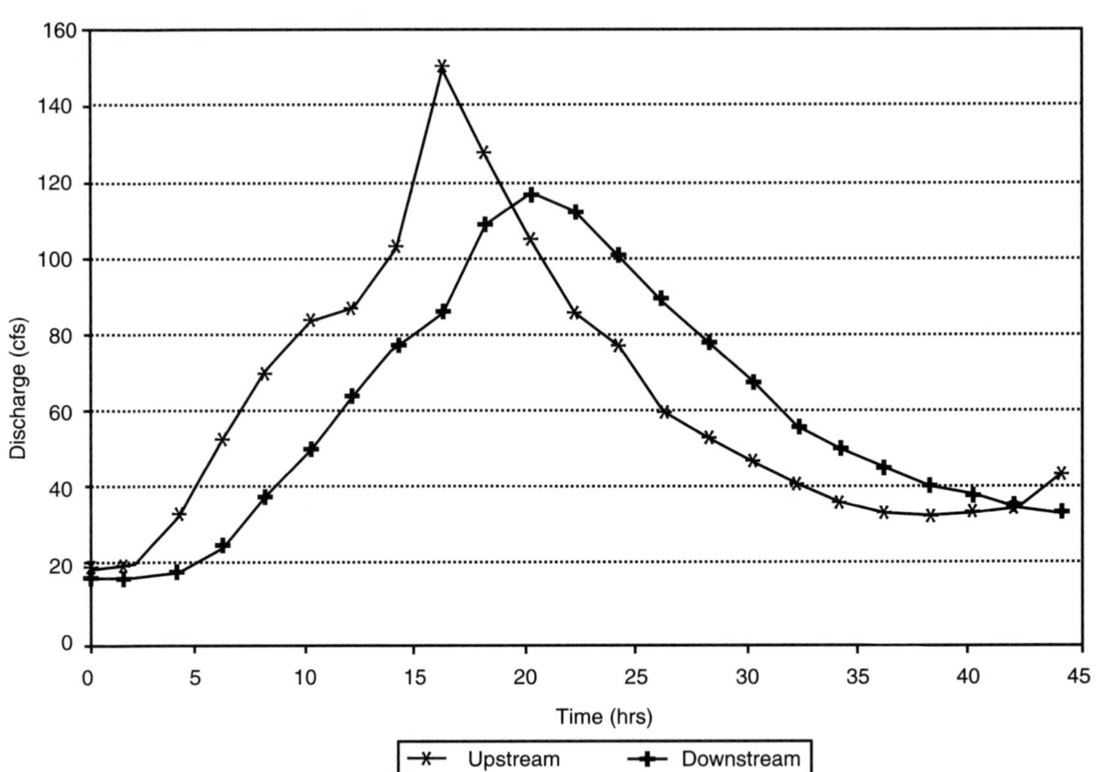

FIGURE 13–5 Routed Hydrograph: Jones Fall at Lake Roland Dec. 1–3, 1981.

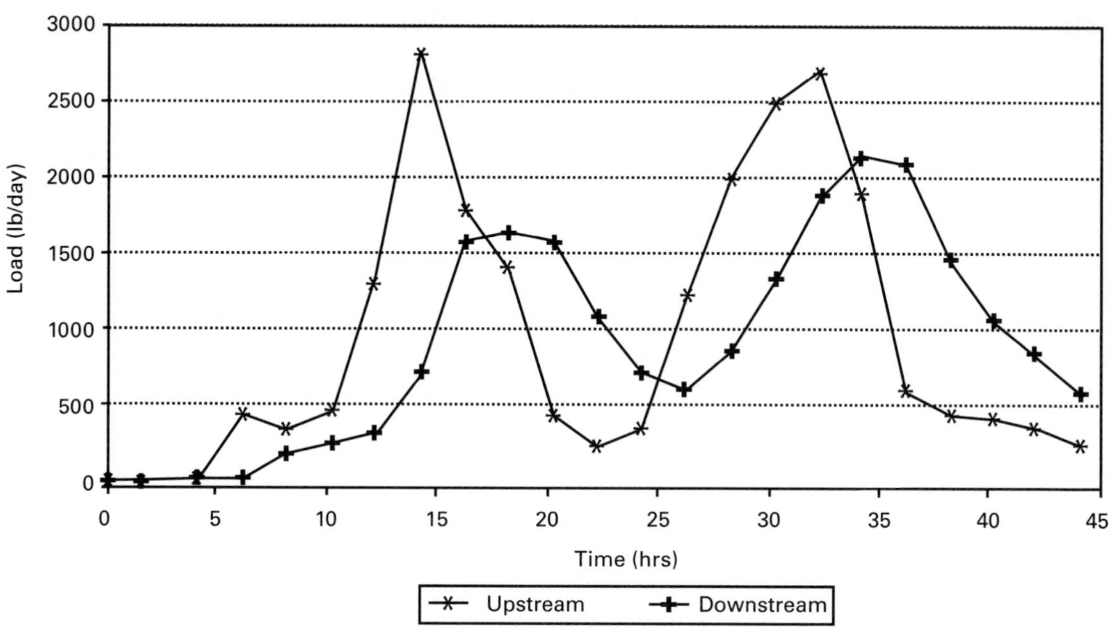

FIGURE 13–6 Routed Loadograph: Jones Fall at Lake Roland Dec. 1–3, 1981.

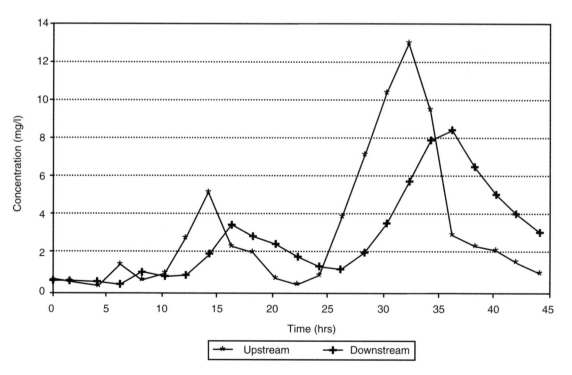

FIGURE 13–7 Routed Pollutograph: Jones Fall at Lake Roland Dec. 1–3, 1981.

PROBLEMS

13–1. Given the following discharge hydrograph $Q(t)$ and pollutograph $C(t)$, compute the loadograph $W(t)$:

Time (hrs)	$Q(t)$ (ft^3/sec)	$C(t)$ (mg/L)	Time (hrs)	$Q(t)$ (ft^3/sec)	$C(t)$ (mg/L)
0	8	4	2.5	70	7
0.5	23	12	3.0	43	7
1.0	76	17	3.5	21	6
1.5	104	15	4.0	13	5
2.0	99	12	4.5	9	4

Plot $Q(t)$, $C(t)$, and $W(t)$. Determine the total load of the pollutant.

13–2. Given the following discharge hydrograph $Q(t)$ and loadograph $W(t)$, compute the pollutograph $C(t)$:

Time (hrs)	$Q(t)$ (ft^3/sec)	$W(t)$ (lb/day)	Time (hrs)	$Q(t)$ (ft^3/sec)	$W(t)$ (lb/day)
0	32	142	5	49	204
1	49	183	6	41	187
2	93	301	7	36	157
3	87	286	8	33	146
4	61	239			

Plot $Q(t)$, $W(t)$, and $C(t)$. Determine the total load of the pollutant.

13–3. Using the mean daily suspended sediment data for the Choptank River near Greensboro, MD, fit the concentration (mg/L)-discharge (cfs) curve with least squares: $C = aQ^b$. Assess the accuracy of the fitted model.

C	Q	C	Q
55	554	13	171
33	352	11	155
24	268	7	19
17	164	5	173
16	20	2	37

13–4. The following data are the mean daily discharge (cfs) and dissolved oxygen concentration (mg/L) for selected days on the Choptank River near Greensboro, MD:

Q	C	Q	C	Q	C
35	6.3	145	8.5	160	9.7
164	8.2	130	9.5	56	6.1
189	9.7	63	8.7	17	5.7
28	5.9	53	6.8	9	6.8
26	7.3	120	7.9	37	8.5

(a) Fit a power model $C = aQ^b$ using least squares and assess the accuracy of the fitted relationship.
(b) Use the following data to test the model of part (a). Assess the accuracy.

Q	C	Q	C
41	10.8	110	9.8
152	11.5	43	8.0

13–5. The following mean daily streamflow (Q, cfs) and turbidity (T, NTU) data are from the Choptank River near Greensboro, MD:

Q	T	Q	T	Q	T
41	2.0	145	8.7	160	20.0
152	8.5	63	0.7	37	3.1
110	4.0	53	3.9	56	10.0
43	3.5	100	12.0	17	5.0

Fit a power model $T = aQ^b$ using least squares and assess the accuracy of the fitted relationship.

13–6. The annual daily-maximum turbidites (NTU) in a river were available for 12 years of record, as follows:

1980	6.2	1984	7.4	1988	10.1
1981	8.4	1985	11.0	1989	9.9
1982	13.2	1986	8.6	1990	7.4
1983	9.5	1987	19.3	1991	12.7

Perform a log-normal frequency analysis. Compute (a) the 10-year daily-maximum turbidity; (b) the probability that a daily-maximum turbidity will exceed 12 NTU in any one year; (c) the probability that the daily-maximum turbidity will not exceed 8 NTU in any one year; (d) the probability that a turbidity of 14 NTU will be exceeded in 2 or more of the next 10 years; (e) the daily-maximum turbidity that has a probability of 40% of not being exceeded in the next 5 years.

13–7. The annual daily-maximum dissolved sodium (mg/L as Na) concentrations in a river were available for 15 years of record, as follows:

1974	5.2	1979	7.4	1984	8.9
1975	7.5	1980	7.0	1985	8.3
1976	8.3	1981	7.1	1986	9.6
1977	7.4	1982	6.8	1987	7.5
1978	6.8	1983	10.0	1988	6.7

Perform a log-normal frequency analysis. Compute (a) the 25-year daily-maximum dissolved sodium concentration; (b) the probability that a daily maximum will exceed 11 mg/L in any one year; (c) the probability that a daily maximum dissolved sodium concentration of 9 mg/L will be exceeded in 3 or more of the next 20 years; (d) the daily-maximum concentration that has a 30% chance of not being exceeded in the next 6 years.

13–8. Evaluate the effect on the oil and grease concentrations in runoff of an ordinance that required parking lots be paved. Use the Lake Tahoe data of Table 13–5.

13–9. Evaluate the effect of the density of residential development on suspended solids, turbidity, and oils and grease concentrations. Use the statistics of Table 13–5. Discuss possible reasons for the effects.

13–10. If the average daily low flow during the summer is 8 cfs for a small stream in a suburban area outside Washington, DC. Compute the corresponding mean load of total nitrogen (lbs/day). Assume the watershed is 30% high-rise residential and 70% high-density residential. Use the statistics of Table 13–6.

13–11. The long-term average bankfull flow for a 135-acre watershed is 17 cfs. Measurements of the average discharge rate (Q, cfs) and the total phosphorus (W, lb/ac/yr) are measured for seven storm events:

W_i	0.17	0.38	0.42	0.21	0.35	0.11	0.16
Q_i	20	41	38	26	31	12	14

13–12. Estimate the mean recoverable copper from a 5.2 sq. mi lightly-developed watershed that has 12% imperviousness, a mean annual rainfall of 38.3 in., and a mean January temperature of 40.5°F. Assume the average bankfull flow is 112 cfs. If the watershed is located near Philadelphia, compute the mean annual load.

13–13. Measurements at a 62-acre watershed (36% imperviousness) during a month during which 5.1 in. of rain fell produced a measured load of 7.7 lbs of total phosphorous. The flow-weighted mean concentration for similar suburban areas in the region is 0.35 mg/L. Estimate the correction factor P_j of Equation 13–16.

13–14. Provide a detailed assessment of the relationship between Equations 13–1 and 13–16.

13–15. Assess the potential inaccuracy in estimates of loads made with Equation 13–16 due to the inaccuracy of the inputs.

13–16. Flow in a river has a velocity of 3.7 fps. The initial coliform number is 88/100 mL. The measured level at a station 33 mi downstream is 31/100 mL. Assuming the decay process follows a first-order reaction, estimate the rate coefficient of Equation 13–23.

13–17. During five storm events, measurements are made of the river velocity, initial coliform count, and the coliform count at a station 17 mi downstream. Estimate the mean rate constant.

Storm Event	River Velocity (ft/sec)	Initial Count per 100 mL	Downstream Count per 100 mL
1	1.3	225	49
2	1.8	153	36
3	0.9	242	23
4	2.2	126	56
5	1.6	185	39

13–18. Three streams converge to form one river. Flow and pollutant measuring stations are located on the three streams. On a day when simultaneous measurements are made, the mean flow rates and concentrations of a pollutant are as follows:

Stream	Q(cfs)	C(mg/L)
1	26.3	127
2	51.4	62
3	37.8	94

Estimate the concentration in the river below the confluence.

13–19. At the point where effluent from a food processing plant discharges waste water into a river, the *DO* deficit is 3.6 mg/L with an ultimate demand of 18.5 mg/L. The reoxygenation constant is 0.18 per day and the deoxygenation constant is 0.11 per day. Find the *DO* deficit at a point 1 day distant from the effluent point, the critical time, and the critical deficit.

REVIEW QUESTIONS

13–1. The load of a water quality constituent depends on which one of the following? (a) The temperature of the water; (b) the flow velocity; (c) the discharge rate; (d) the depth of flow; (e) all of the above.

13–2. The concentration of a water quality constituent is usually related to the discharge with which model form? (a) Logistic function; (b) power model; (c) exponential decay function; (d) linear model.

13–3. Frequency analyses are not commonly used with water quality data because (a) the skew of the probability distribution is not known; (b) it requires an assumption about the probability distribution of the discharge rate Q; (c) water quality records are too short; (d) random sampling programs, which are used with water quality parameters, are not likely to identify the annual maximum.

13–4. Which one of the following is not an input to the Simple Method for estimating pollution loads? (a) The rainfall depth; (b) the drainage area: (c) the time of concentration; (d) the flow-weighted mean concentration; (e) all of the above are inputs.

13–5. Which one of the following is a constraint on the use of the Simple Method? (a) Drainage area less than a square mile; (b) rainfalls greater than 3 inches; (c) periods of one-month or longer; (d) return periods of 2 years or more.

13–6. In a first-order batch reaction (a) the concentration at any time t follows an exponential decay function; (b) the rate of change of concentration with time follows an exponential decay function; (c) the inflow rate is less than the outflow rate; (d) none of the above.

13–7. The rate coefficient in a batch reaction is (a) independent of temperature; (b) linearly related to temperature; (c) depends on the concentration of the pollutant; (d) increases with increase in temperature; (e) none of the above.

13–8. For flow dilution calculations, which one of the following is not a requirement for conditions at a junction? (a) The flow in to equals the flow out of a junction; (b) the concentration in to equals the concentration out; (c) the load in to the junction equals the load out; (d) all of the above are requirements.

13–9. An oxygenation curve relates two variables: (a) the dissolved oxygen concentration versus stream temperature; (b) the rate of reaeration versus time; (c) the oxygen concentration versus time of flow; (d) the rate of deoxygenation versus the distance downstream from the point of waste discharge.

13–10. The commonly used representation of the deoxygenation of a stream is based on three assumptions; which one of the following is not an assumption? (a) The oxygen concentration is greater than the critical value; (b) the process is independent of temperature; (c) BOD is dependent on the concentration of biologically decomposable material; (d) BOD is independent of the oxygen concentration.

13–11. Which one of the following is not an assumption of the Streeter-Phelps Oxygen Sag Equation? (a) Steady-state conditions; (b) production is the sum of deoxygenation and reoxygenation; (c) the initial deficit equals zero; (d) conservation of mass.

13–12. A pollutograph cannot be routed with a flow routing method because (a) continuity of mass will not be maintained; (b) routing methods are only valid for nonconservative pollutants; (c) the routing methods do not allow for attenuation of the peak; (d) none of the above.

DISCUSSION QUESTION

The technical content of this chapter is important to the professional hydrologist, but practice is not confined to making technical decisions. The intent of this discussion question is to show that hydrologists must often address situations where value issues intermingle with the technical aspects of a project. In discussing the stated problem, at a minimum include responses to the following questions:

1. What value issues are involved, and how are they in conflict?
2. Are technical issues involved with the value issues? If so, how are they in conflict with the value issues?
3. If the hydrologist attempted to rationalize the situation, what rationalizations might he or she use? Provide arguments to suggest why the excuses represent rationalization.
4. What are the hydrologist's alternative courses of action? Identify all alternatives, regardless of the ethical implications.
5. How should the conflict be resolved?

You may want to review Sections 1.6 to 1.12 in Chapter 1 in responding to the problem statement.

Case. A manufacturer discharges liquid wastes into a local stream on a continuous flow basis. By state law, the engineer who works for the manufacturer is required to take weekly samples of the discharge to ensure that the concentration of certain pollutants does not violate state standards. The company is required to stop discharging wastes if the sampling indicates the standard is being violated and then file a written report to the state water quality board. While the samples have always indicated that the state standard is being met, on one occasion, the sampling indicates that the discharge from the manufacturing plant is in violation of the standard. The engineer knows that the plant has not changed its process. The engineer notifies the boss, as is company regulation, and the boss gives instructions to ignore the results of the sampling. Therefore, the engineer does not shut down the operation or file the written report.

14 Evaporation

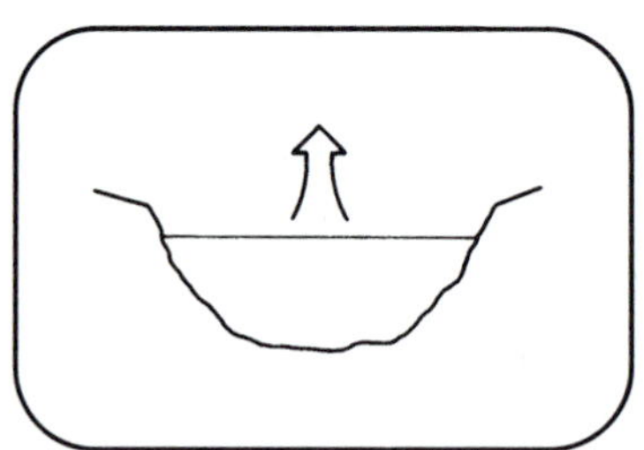

CHAPTER OBJECTIVES

1. Introduce basic concepts related to estimating evaporation for design problems in hydrology.
2. Provide an overview of energy and water budgets as they apply to the estimation of evaporation rates.
3. Introduce computational methods that have a theoretical basis for estimating evaporation.
4. Provide descriptions of empirical equations that can be used to estimate evaporation rates.
5. Outline the use of pan evaporation measurements.

14.0 NOTATION

A_p = surface area of the pond
b_0, b_1 = empirical coefficients
B = Bowen Ratio
c_p = pan coefficient
C_p = specific heat of air at constant pressure
d = change in water surface elevation (ft)
d_e = depth of evaporation
D = number of hours of sunshine possible
e = vapor pressure
e_a = vapor pressure at air temperature
e_a^* = saturation vapor pressure at air temperature
e_0 = vapor pressure at water temperature
e_s = saturated vapor pressure
E = evaporation
E_a = net energy advected into the body of water
E_{a0} = evaporation based on air measurements
E_e = energy utilized for evaporation
E_n = net energy available for evaporation
E_p = pan evaporation
E_s = change in energy storage within the water body
E_T = total evaporation for a design period
$f(V)$ = empirical function based on wind speed V
H = latent heat flux density
H_h = sensible heat loss from the water body to the atmosphere
H_s = sensible heat flux density
H_{s0} = sensible heat flux density at water surface
H_w = latent heat flux density at water surface
I = inflow
k = von Kármán constant
L_v = latent heat of vaporization
n = number of hours of sunshine
O = outflow
P = rainfall (in.)
P_a = atmospheric pressure
P_v = rainfall (acre-ft)
Q = streamflow into the water body
Q_d = subsurface seepage losses from the water body
Q_o = streamflow discharge from the water body
Q_p = average discharge inflow (acre-ft/month)
Q_{op} = average outflow (acre-ft/month)
Q_r = surface runoff into the water body
Q_s = subsurface runoff into the water body
r = reflection coefficient
R_A = total possible radiation for period of estimation
R_B = net outward flow of radiation
R_h = relative humidity
R_I = absorbed radiation
R_l = net longwave radiation exchange between the atmosphere and the water body
R_n = net radiation
R_r = reflected solar radiation
R_t = total solar radiation incident to the water surface
S = storage
S_r = sensitivity of evaporation to the reflection coefficient
t = time
t_c = temperature, degrees Celsius
t_f = temperature, degrees Fahrenheit
T = mean temperature
T_a = absolute temperature
T_a = air temperature
T_o = water surface temperature
T_w = water temperature

V = wind speed
V_0 = wind speed at the height of the anemometer
V_{30} = wind speed (mi/hr) at a height of 30 ft
Z = height above the ground surface
Z_o = height of the anemometer
α = psychrometric constant
Δ = slope of saturation vapor pressure curve
σ = constant (0.1177×10^{-6} g-cal/cm^2/day)

14. 1 INTRODUCTION

Evaporation is the process by which the phase of water is changed from a liquid to a vapor. While water vapor may result from a change of phase from a solid to a gas, hydrologists are primarily concerned with evaporation losses from an open water body (that is, the change of phase from liquid to a gas). In addition to evaporation from free water surfaces, water molecules pass to the atmosphere from plant surfaces; this process is called *transpiration*. In the transpiration process, the water passes through the plant and is evaporated at the surface of the plant. *Evapotranspiration* is the total water loss from a field in which significant amounts of water are lost through transpiration from plant surfaces and evaporation from the underlying soil.

In terms of the hydrologic cycle of Figure 1–1, evaporation and transpiration represent a significant portion of the water movement through the hydrologic cycle. Approximately 70% of the precipitation in a temperate climate returns to the atmosphere directly by way of evaporation or transpiration losses. The remaining 30% appears as streamflow. In comparison to streamflow, evaporation and transpiration are not major design variables in engineering hydrologic design. With the exception of a few design situations, evaporation is considered to be part of the loss function, although during storm events it is probably a very minor part of the total losses. Evaporation losses are considered in the design of large water storage reservoirs. Free water evaporation can be as much as 100 in. per year in the more arid parts of the United States. Where heat storage within the water body is significant, evaporation may be much greater. Even in the humid eastern part of the United States, more than 2 ft of evaporation losses can be expected from a free-water surface during a 1-yr period. In areas where annual rainfall is low, evaporation losses can represent a significant part of the water budget for a lake, and evaporation losses may contribute significantly to the lowering of the water surface elevation. This may have consequences in terms of water quality, recreational use of the lake, allocation of water for irrigation demand, and power generation. Monthly, seasonal, or annual evaporation losses may have to be considered in the design of the reservoir and the retarding structure.

With the development of more complex hydrologic models, especially the continuous moisture accounting models, algorithms for making reasonable estimates of evaporation and evapotranspiration losses for periods of one day and less are needed. These hydrologic models can be used for any number of engineering design problems, and their value as a model rests on the idea of continuous accounting of soil moisture.

Estimates of evapotranspiration may also be used for irrigation scheduling. Water losses through transpiration must be replaced if the crop is to produce maximum yield. Water availability is especially critical during periods where nonoptimum moisture can significantly affect crop growth. Estimation of water losses may then be used to schedule irrigation patterns and rates.

While the design aspect of evaporation is relatively easy, the analysis phase is extremely complex. Accurate analysis requires considerable cost and effort in obtaining the necessary measurements and involves an experimental design where the watershed is controlled. Only a few detailed analyses have been undertaken and the methods for design are largely based on the association between measured meteorological data and the results of these detailed analyses. Therefore, the analysis problem will be largely ignored here, and the discussion will center on the design or synthesis phase.

14.2 FACTORS AFFECTING EVAPORATION

A number of studies have been made to evaluate the sensitivity of evaporation rates to causative factors. The factors believed to be most important are the temperature, the humidity or vapor pressure deficit, radiation rates, and the wind speed. Although other variables may be used in equations for predicting evaporation rates, these four factors are the most common. A brief discussion of each follows.

14.2.1 Temperature

Temperature is a measure of the thermal energy that a body possesses; it is a measure of the combined potential and kinetic energy of the atoms in the body. The units of temperature used in evaporation analyses are degrees Fahrenheit, degrees Celsius, and degrees Kelvin. Temperatures can be measured relative to the freezing or boiling points of water, which are 32° and 212° for the Fahrenheit scale, respectively, and 0° and 100° for the Celsius scale, respectively. Temperatures in Celsius (t_c) and Fahrenheit (t_f) are related by

$$t_f = 32 + \frac{9}{5} t_c \tag{14–1a}$$

and

$$t_c = \frac{5}{9} (t_f - 32) \tag{14–1b}$$

The Kelvin scale pertains to an absolute scale of temperature with the 0 point approximately −273.16°C.

Temperature is distinguished from heat, which is a measure of the thermal energy that is transferred from one medium to another. Water in a lake and water in a pail may be at the same temperature, but the lake has a greater capacity to store and transfer heat than the water in the pail. Heat storage can affect the rate and time distribution of evaporation.

14.2.2 Humidity and Vapor Pressure

The vapor content of the air mass overlying a body of water is an important determinant of evaporation rates. The molecules of water in the air mass exert a pressure on the free water surface. As the vapor pressure of the overlying air mass increases, the rate of evaporation from the water body will decrease. The *vapor pressure*, denoted as *e*, is the difference in the atmospheric pressure with and without the vapor (that is, the partial pressure of the water vapor in the air mass).

If water continues to evaporate from the free water surface, the overlying air will become saturated. The pressure exerted when the overlying air is saturated with vapor is called the *saturated vapor pressure*, which is denoted as e_s. The saturated vapor pressure is a function of the temperature; values of e_s are given in Table 14–1. The ratio of the actual vapor pressure to the saturated vapor pressure is called the *relative humidity* (R_h):

$$R_h = \frac{e}{e_s} \tag{14–2}$$

If R_h is expressed as a percentage, the value of Equation 14–2 should be multiplied by 100.

The amount of vapor that an air mass can hold is a function of the temperature. When the air mass is cooled under constant pressure and with a constant water vapor content, a temperature is reached at which the air becomes saturated. This temperature is called the dew point temperature. If the temperature decreases beyond the dew point temperature, the water vapor in the air will condense.

Vapor pressure is commonly expressed in units of bars, millibars (mb), inches of mercury (in. Hg), or millimeters of mercury (mm Hg). The following are appropriate conversion factors:

1 bar = 10^5 Newtons/square meter
1 mb = 10^2 N/m^2 = 1000 dynes/cm^2
= 0. 0143 psi = 0.0295 in. Hg
1 mm Hg = 1.36 mb = 0.04 in. Hg

14.2.3 Radiation

Radiation is the transmission of thermal energy by electromagnetic waves; radiant energy is not heat itself. Radiant energy can be reflected, absorbed, or transmitted. While radiation is a determinant of evaporation, it is more costly to measure than temperature, and since temperature and radiation are highly correlated, most methods for estimating evaporation rates use temperature as an input variable rather than radiation.

14.2.4 Wind Speed

Evaporation rates are controlled, in part, by the vapor content of the air mass overlying the water body. As the volume of water evaporated increases, the evaporation rate from the water body decreases unless the more saturated air mass is transported from the overlying air space and replaced by air that has a lower moisture content. For low rates of air mass movement, less water will evaporate from the water body because of the relatively higher vapor content of the overlying air mass. The wind speed is used as a measure of the air mass movement. The magnitude of the wind speed is measured with an anemometer and the direction with a wind vane. For highly turbulent air in which local gusts of wind cause high measured values of wind speed, the actual air mass movement may not be properly indicated by the

TABLE 14–1 Saturation Vapor Pressure e_s in mm H_g (mercury) as a Function of Temperature t in °C (Negative Values of t Refer to Conditions Over Ice)

t	0.0	0.1	0.2	0.3	0.4	0.5	0.6	0.7	0.8	0.9	t
−10	2.15										−10
−9	2.32	2.30	2.29	2.27	2.26	2.24	2.22	2.21	2.19	2.17	−9
−8	2.51	2.49	2.47	2.45	2.43	2.41	2.40	2.38	2.36	2.34	−8
−7	2.71	2.69	2.67	2.65	2.63	2.61	2.59	2.57	2.55	2.53	−7
−6	2.93	2.91	2.89	2.86	2.84	2.82	2.80	2.77	2.75	2.73	−6
−5	3.16	3.14	3.11	3.09	3.06	3.04	3.01	2.99	2.97	2.95	−5
−4	3.41	3.39	3.37	3.34	3.32	3.29	3.27	3.24	3.21	3.18	−4
−3	3.67	3.64	3.62	3.59	3.57	3.54	3.52	3.49	3.46	3.44	−3
−2	3.97	3.94	3.91	3.88	3.85	3.82	3.79	3.76	3.73	3.70	−2
−1	4.26	4.23	4.20	4.17	4.14	4.11	4.08	4.05	4.03	4.00	−1
−0	4.58	4.55	4.52	4.49	4.46	4.43	4.40	4.36	4.33	4.29	−0
0	4.58	4.62	4.65	4.69	4.71	4.75	4.78	4.82	4.86	4.89	0
1	4.92	4.96	5.00	5.03	5.07	5.11	5.14	5.18	5.21	5.25	1
2	5.29	5.33	5.37	5.40	5.44	5.48	5.53	5.57	5.60	5.64	2
3	5.68	5.72	5.76	5.80	5.84	5.89	5.93	5.97	6.01	6.06	3
4	6.10	6.14	6.18	6.23	6.27	6.31	6.36	6.40	6.45	6.49	4
5	6.54	6.58	6.63	6.68	6.72	6.77	6.82	6.86	6.91	6.96	5
6	7.01	7.06	7.11	7.16	7.20	7.25	7.31	7.36	7.41	7.46	6
7	7.51	7.56	7.61	7.67	7.72	7.77	7.82	7.88	7.93	7.98	7
8	8.04	8.10	8.15	8.21	8.26	8.32	8.37	8.43	8.48	8.54	8
9	8.61	8.67	8.73	8.78	8.84	8.90	8.96	9.02	9.08	9.14	9
10	9.20	9.26	9.33	9.39	9.46	9.52	9.58	9.65	9.71	9.77	10
11	9.84	9.90	9.97	10.03	10.10	10.17	10.24	10.31	10.38	10.45	11
12	10.52	10.58	10.66	10.72	10.79	10.86	10.93	11.00	11.08	11.15	12
13	11.23	11.30	11.38	11.45	11.53	11.60	11.68	11.76	11.83	11.91	13
14	11.98	12.06	12.14	12.22	12.30	12.38	12.46	12.54	12.62	12.70	14
15	12.78	12.86	12.95	13.03	13.11	13.20	13.28	13.37	13.45	13.54	15
16	13.63	13.71	13.80	13.90	13.99	14.08	14.17	14.26	14.35	14.44	16
17	14.53	14.62	14.71	14.80	14.90	14.99	15.09	15.18	15.27	15.38	17
18	15.46	15.56	15.66	15.76	15.86	15.96	16.06	16.16	16.26	16.36	18
19	16.46	16.57	16.68	16.79	16.90	17.00	17.10	17.21	17.32	17.43	19
20	17.53	17.64	17.75	17.86	17.97	18.08	18.20	18.31	18.43	18.54	20
21	18.65	18.77	18.88	19.00	19.11	19.23	19.35	19.46	19.58	19.70	21
22	19.82	19.94	20.06	20.19	20.31	20.43	20.58	20.69	20.80	20.93	22
23	21.05	21.19	21.32	21.45	21.58	21.71	21.84	21.97	22.10	22.23	23
24	22.37	22.50	22.63	22.76	22.91	23.05	23.19	23.31	23.45	23.60	24
25	23.75	23.90	24.05	24.20	24.35	24.49	24.64	24.79	24.94	25.08	25
26	25.31	25.45	25.60	25.74	25.89	26.03	26.18	26.32	26.46	26.60	26
27	26.75	26.90	27.05	27.21	27.37	27.53	27.69	27.85	28.00	28.16	27
28	28.32	28.49	28.66	28.83	29.00	29.17	29.34	29.51	29.68	29.85	28
29	30.03	30.20	30.38	30.56	30.74	30.92	31.10	31.28	31.46	31.64	29
30	31.82	32.00	32.19	32.38	32.57	32.76	32.95	33.14	33.33	33.52	30
t	0.0	0.1	0.2	0.3	0.4	0.5	0.6	0.7	0.8	0.9	t

Note: 1 mm Hg = 1.36 mbar.

wind speed. However, in most cases wind speed is probably a good indicator of air mass movement.

While wind speed is a common input to many methods of estimating evaporation rates, the height at which it is measured must usually be specified. The wind speed varies with the height above the water surface. The variation in wind speed with height can be represented by the following relationship:

$$\left(\frac{V}{V_o}\right) = \left(\frac{Z}{Z_o}\right)^{0.15} \tag{14–3}$$

where V is the wind speed (mi/hr) at a height of Z feet above the surface, and V_o is the wind speed (mi/hr) at the height (ft) of the anemometer, which is denoted by Z_o. The exponent of 0.15 is an empirical constant. The relationship of Equation 14–3 is sometimes referred to as the power law.

Example 14–1

As an example, consider the case where a wind speed of 11 mi/hr is measured by an anemometer at a height of 2 ft. Equation 14–3 becomes

$$\frac{V}{11} = \left(\frac{Z}{2}\right)^{0.15} \tag{14–4a}$$

which can be rewritten as

$$V = 9.91Z^{0.15} \tag{14–4b}$$

The wind speeds at heights of 0.5, 1, and 5 ft would be 8.9, 9.9, and 12.6 mi/hr, respectively. Thus if a wind speed of 11 mi/hr were measured at a height of 2 ft and an empirical equation for estimating evaporation required a value measured at 5 ft, a value of 12.6 mi/hr should be used rather than the value of 11 mi/hr.

14.3 ENERGY BUDGET

Evaporation is a function of the energy state of the water system. Thus it is reasonable to estimate evaporation by accounting for all sources and transfers of radiation. Such is the basis for an energy budget or balance. An explanation of the continuity of mass is an appropriate way to introduce the concept of an energy balance. In the systems theory representation of the continuity of mass, we assume that the time rate of change of storage (dS/dt) equals the difference between the inflow (I) and outflow (O):

$$I - O = \frac{dS}{dt} \tag{14–5}$$

In terms of energy, energy input into the water body is in the form of the total solar radiation incident to the water surface (R_t) and the net energy advected into the body of water (E_a). Energy leaving the water body includes energy utilized for evaporation (E_e), sensible heat loss from the water body to the atmosphere (H_h), reflected solar radiation (R_r), and the net long-wave radiation exchange between the atmosphere and the water body (R_1). In terms of Equation 14–5, the change in energy storage within the water body is denoted as E_s. Thus the energy balance becomes

$$(R_t + E_a) - (E_e + H_h + R_r + R_1) = E_s \qquad (14\text{–}6)$$

In the form of Equation 14–6, the energy balance represents a theoretical model.

If all the terms of Equation 14–6 could be evaluated directly and accurately using field measurements, it should provide the most accurate evaluation of evaporation from the water body. Unfortunately, the individual elements of Equation 14–6 are difficult to measure. Therefore, simplifications of Equation 14–6 are used to define the structure of models for estimating evaporation rates, and empirical evaluations are used to fit coefficients for a particular location and set of conditions. These simplifications may reduce the accuracy of estimates; however, studies have not shown that the simplified equations for estimating evaporation rates are significantly less accurate than the energy balance involving empirical evaluations of the individual elements of Equation 14–6. For that reason the energy budget will not be discussed further here.

14.4 WATER BUDGET

Equation 14–5 can also be used to define a water budget or water balance. The water balance is a physical analysis of the water body based on the conservation of mass. Water input to the water body may be from rainfall (P), streamflow into the water body (Q), surface runoff (Q_r), and subsurface runoff (Q_s). Outflow from the water body could be evaporation (E), streamflow discharge from the water body (Q_o), and subsurface seepage losses (Q_d). Thus, Equation 14–5 becomes

$$(P + Q + Q_r + Q_s) - (E + Q_o + Q_d) = \frac{dS}{dt} \qquad (14\text{–}7)$$

Rainfall can be measured using rain gages. Streamflow and overland runoff into and out of the water body can be measured using weirs. The time rate of change of storage can be measured using a stage recorder for the water surface elevation along with the necessary topographic data of the site. Subsurface runoff in to (Q_s) and out of (Q_d) the water body are the most difficult elements of the water balance to measure; these can sometimes be estimated using elevation measurements of observation wells placed about the perimeter of the water body; however, this requires a number of measurements and an assumption of homogeneity of subsurface runoff characteristics in the area allocated to each observation well. To use Equation 14–7, it is necessary to specify a time increment dt for which each of the elements are measured. The accuracy of estimates of E made with Equation 14–7 will increase as dt increases, but the time increment selected must depend on the engineering design problem. Given the significant effort required to make the necessary measurements, water budgets, like energy budgets, are only used where the effort is warranted, such as major water resource projects.

Example 14–2

To provide a brief illustration of the water-budget approach to the estimation of evaporation rates, assume that a pond with a 9-acre surface area is located on a 440-acre watershed. Table

TABLE 14–2 Computation of Daily Evaporation Rates Using a Water Balance Model

		Rainfall		Inflow		Outflow		Elevation	Storage	Evaporation	
Month	n	P (in.)	P_v (ac-ft/mo.)	Q (ft^3/sec)	Q_p (ac-ft/mo.)	Q_o (ft^3/sec)	Q_{op} (ac-ft/mo.)	Change, d (ft)	Change dS/dt (ac-ft)	E (ac-ft/mo.)	E_p (in./day)
June	30	2.13	1.60	0.44	26.18	0.092	5.47	2.10	18.9	3.41	0.15
July	31	0.67	0.50	0.10	6.15	0.018	1.11	–0.06	–0.54	6.08	0.26
August	31	1.24	0.93	0.22	13.53	0.041	2.52	0.73	6.57	5.37	0.23

14–2 provides monthly values of the rainfall, streamflow into and out of the pond, and the pond elevation at the beginning of each month for a 3-month period. The lake is in a clayey soil, so ground water flow is assumed to be negligible; surface runoff into the pond is small because the topography directs most of the flow either around the pond or into the stream, where it is measured with a weir as part of the inflow. Therefore, the water balance model of Equation 14–7 reduces to

$$P + Q - E - Q_o = \frac{dS}{dt}$$

The rainfall is measured as a total depth in inches. The rainfall input is, therefore, the total water accumulated on the pond. To compute the rainfall input in acre-ft, which is denoted as P_v, the value of P must be adjusted by

$$P_v = \frac{PA_p}{12}$$

in which A_p is the area of the pond in acres. The streamflow into (Q) and out of (Q_o) the pond are measured with weirs and expressed in ft^3/sec. To convert the discharges to acre-ft, the following relationship can be used:

$$Q_p = \frac{nQ(3600)\,(24)}{43{,}560}$$

in which n is the number of days in the month, Q the average discharge in ft^3/sec, and Q_p the average discharge in acre-ft/month. A similar relationship can be used to convert the average outflow Q_o in ft^3/sec to acre-ft:

$$Q_{op} = \frac{nQ_o(3600)\,(24)}{43{,}560}$$

The water surface elevation is recorded in feet above some datum. To convert the change in elevation to a volume, the value must be multiplied by the pond surface area and the appropriate dimensional adjustments made:

$$\frac{dS}{dt} = dA_p$$

in which d is the change in the water surface elevation in ft, and dS/dt is the change in storage in acre-ft per month.

Given these changes in units, the water balance model can be transformed to compute the evaporation:

$$E = P_v + Q_p - Q_{op} - \frac{dS}{dt}$$

Since the terms of this equation are given in acre-ft, then to compute the depth of evaporation in in./day (d_e) the following transformation is made:

$$d_e = \frac{12E}{nA_p}$$

The calculation of the pond evaporation rates in in. /day are shown in Table 14–2. A total volume of 14.86 acre-ft evaporates from the pond during the 3-month period, with evaporation rates varying from 0.15 in./day in June to 0.26 in./day in July. During July the lake level deceases because of the low rainfall and inflow and the relatively high evaporation rate.

14.5 THE PENMAN EQUATION

The accuracy of a model depends on a number of factors, including the input requirements and the degree to which the structure of the model approximates the underlying physical processes. A central problem in modeling is balancing the trade-off between complexity and accuracy. As the complexity increases, one usually assumes the accuracy increases; however, as the complexity increases, the input requirements and the effort required to make estimates also increases. Therefore, to have a manageable model, one attempts to develop a model using as much theory as possible to develop the model structure and to reduce the complexity by making simplifying assumptions but only to the point where further simplifications would adversely affect the accuracy of estimates. In addition to using theory (that is, an understanding of the underlying physical principles), models are calibrated to account, in part, for the loss of accuracy due to the simplifications.

This philosophy was used to develop an equation for estimating evaporation from open water surfaces. Penman (1948, 1956) reduced the energy balance of Equation 14–6 to the following:

$$E_n = H + H_s \tag{14–8}$$

in which E_n is the net energy available for evaporation, H is the latent heat flux density, and H_s is the sensible heat flux density. Based on Dalton's diffusivity law, the following simplified aerodynamic equation was used to represent the latent heat flux density:

$$H_w = f(V)(e_o - e_a) \tag{14–9}$$

in which $f(V)$ is an air speed function, V the air speed, e_o and e_a are the vapor pressures of the water surface and the air, respectively, and H_w is the latent heat flux density at the water surface. Recognizing the relationship between temperature and vapor pressure, the diffusivity law can also be used to describe a simplified aerodynamic equation for the sensible heat flux density at the water surface (H_{so}):

$$H_{so} = \alpha f(V)(T_o - T_a) \tag{14–10}$$

in which α is the psychrometric constant and T_o and T_a are the temperatures of the water surface and air, respectively. The psychrometric constant is given by

$$\alpha = \frac{C_p p_a}{0.62\, H_v} \tag{14–11}$$

in which C_p is the specific heat of air at constant pressure, p_a is the atmospheric pressure, and H_v is the latent heat of vaporization. The Bowen Ratio is defined as the ratio of the sensible heat loss to the latent heat flux. Since the sensible heat flux does not lend itself to easy measurement, the Bowen Ratio is often approximated. Using the simplified aerodynamic equations, the Bowen Ratio (B) can be approximated by

$$B = \frac{H_s}{H} = \frac{H_{s0}}{H_w} = \alpha \frac{T_o - T_a}{e_o - e_a} \tag{14–12}$$

Therefore, Equation 14–8 can be rewritten as

$$E_n = H_w + H_{so} = H_w(1 + B) \tag{14–13}$$

One problem with Equation 14–13 is the requirement for water surface measurements; this is a problem because water surface measurements are not usually available at sites where estimates are required and would obviously not be available for a proposed design site.

A number of simplifications can be used to eliminate the need for water surface measurements. The slope of the saturation vapor pressure curve at mean temperature $(T_o + T_a)/2$ is denoted as Δ and given by

$$\Delta = \frac{e_o - e_a^*}{T_o - T_a} \tag{14–14}$$

in which e_a^* is the saturation vapor pressure at temperature T_a. Rearranging Equation 14–14 for $(T_o - T_a)$ and substituting the result into Equation 14–12 yields

$$B = \frac{\alpha(e_o - e_a^*)}{\Delta(e_o - e_a)} \tag{14–15}$$

If we define E_{ao} using only air measurements as

$$E_{ao} = f(V)\,(e_a^* - e_a) \tag{14–16}$$

and divide by Equation 14–9, we get

$$\frac{E_{ao}}{H_w} = \frac{f(V)\,(e_a^* - e_a)}{f(V)\,)(e_o - e_a)} = \frac{e_a^* - e_a}{e_o - e_a} \tag{14–17}$$

Substituting the identity

$$e_a^* - e_a = (e_o - e_a) - (e_o - e_a^*) \tag{14–18}$$

into Equation 14–17 yields

$$\frac{E_{ao}}{H_w} = \frac{(e_o - e_a) - (e_o - e_a^*)}{e_o - e_a} = 1 - \frac{e_o - e_a^*}{e_o - e_a} \tag{14–19}$$

Equation 14–19 can be rearranged:

$$1 - \frac{E_{a0}}{H_w} = \frac{e_o - e_a^*}{e_o - e_a} \tag{14–20}$$

which, when substituted into Equation 14–15, yields

$$B = \frac{\alpha}{\Delta}\left(1 - \frac{e_{ao}}{H_w}\right) \tag{14–21}$$

If we assume that E is approximated with H_w, as was assumed in Equation 14–12, then Equation 14–13 becomes

$$E_n = H_w(1 + B) \tag{14–22}$$

Substituting Equation 14–21 into Equation 14–22 yields

$$E_n = H_w\left[1 + \frac{\alpha}{\Delta}\left(1 - \frac{E_{a0}}{H_w}\right)\right] = H_w\left(1 + \frac{\alpha}{\Delta}\right) - \frac{\alpha}{\Delta} E_{ao} \tag{14–23}$$

Solving Equation 14–23 for H_w yields

$$H_w = \left(E_n + \frac{\alpha}{\Delta} E_{a0}\right) \Big/ \left(1 + \frac{\alpha}{\Delta}\right) = \frac{\Delta E_n + \alpha E_{a0}}{\Delta + \alpha} \tag{14–24}$$

which is an equation for estimating the evaporation at the water surface as a function of E_n, E_{a0}, Δ, and α.

If we are to assume that water surface measurements are not required, the air temperature will have to be used in place of the average of the air and water temperature that was assumed for Equation 14–14. This may be a reasonable assumption for time periods of about 1 day, although it may not be an accurate assumption for shorter time intervals of an hour or so. Data for the saturation vapor pressure can be obtained from Table 14–1.

The value of the psychrometric constant is computed using Equation 14–11. This requires values of the specific heat of air at constant pressure and the atmospheric pressure. The latent heat of vaporization is 540 cal/g at 100°C. A typical value of α is 0.485 mm Hg/°C.

The net radiation (R_n) can be used to estimate E in Equation 14–24 and approximated by

$$R_n = R_I - R_B \tag{14–25}$$

in which R_I is the amount of radiation absorbed and R_B is the net outward flow of longwave radiation. The value of R_I can be approximated by

$$R_I = R_A(1 - r)\left(a + b\frac{n}{D}\right) \tag{14–26}$$

in which r is the reflection coefficient, a and b are empirical coefficients that are location dependent, n/D is the fraction of possible sunshine, and R_A is the total possible radiation for the period of estimation. The total possible radiation is a function of latitude and time of year; values can be obtained from Table 14–3. The values of the empirical coefficients a and b ap-

TABLE 14–3 Angot's Values of Short-Wave Radiation Flux R_A at the Outer Limit of the Atmosphere in gram-calories per cm^2 per Day, as a Function of the Month of the Year and the Latitude

Latitude°	Jan.	Feb.	Mar.	Apr.	May	June	July	Aug.	Sept.	Oct.	Nov.	Dec.	Year
N 90	0	0	55	518	903	1077	944	605	136	0	0	0	3540
80	0	3	143	518	875	1060	930	600	219	17	0	0	3660
60	86	234	424	687	866	983	892	714	494	258	113	55	4850
40	358	538	663	847	930	1001	941	843	719	528	397	318	6750
20	631	795	821	914	912	947	912	887	856	740	666	599	8070
Equator	844	963	878	876	803	803	792	820	891	866	873	829	8540

Note: The SI unit for R_A is joules/m^2/day. The conversion is 1 g-cal/cm^2 = 41.9 kJ/m^2.

pear to be location dependent, although an accurate method for obtaining these has not been developed. Values of $a = 0.2$ and $b = 0.5$ appear to be reasonable estimates unless site-specific values are calibrated from data.

The fraction of possible sunshine n/D varies with location and time of year. Percentages for the United States are shown in Figure 14–1 for the month of June. Values for other months are readily available in climatic atlases or tables of hydrometeorological data.

The value of r is also an empirical coefficient. It is a function of the time of year, the smoothness of the water surface, the wind speed, and the quality of the water. Values typically range from 0.05 to 0.12.

The value of the net outward flow of radiation can be estimated from the following empirical equation:

$$vR_B = \sigma T_a^4\,(0.47 - 0.077\sqrt{e})\left(0.2 + 0.8\,\frac{n}{D}\right) \tag{14–27}$$

in which $\sigma = 1.177.7 * 10^{-7}$ cal/(cm^2 – °K^4 – day), T_a is the absolute temperature (= T_c + 273), and e is the actual vapor pressure of the air (mm Hg). Values computed from Equations 14–26 and 14–27 can be used to compute the value of R_n with Equation 14–25.

The value of E_{a0} (mm/day) in Equation 14–24 must also be estimated using an empirical formulation based on the diffusivity law:

$$E_{a0} = 0.35\,(e_s - e)\,(0.2 + 0.55V) \tag{14–28}$$

in which V is the wind speed in m/sec at a height of 2 m, and e is the actual vapor pressure at air temperature T_a. The constants of Equation 14–28 are a function of location and the units of the variables.

Example 14–3

Consider the design problem that a reservoir with a surface area of 30 acres will be built in central Arizona (latitude = 32°N). For simplicity, we will assume that the design variable with respect to evaporation is the total evaporation for June and July. For design, the normal daily average temperature will be used to represent the air temperature; values of 30°C and 32°C are obtained from a climatic atlas for June and July, respectively. The climatic atlas shows average relative humidities of 30 and 40% for June and July, respectively. From Table 14–3 we obtain values for R_A of 980 and 930 g-cal/cm^2/day for June and July, respectively. Also, values of

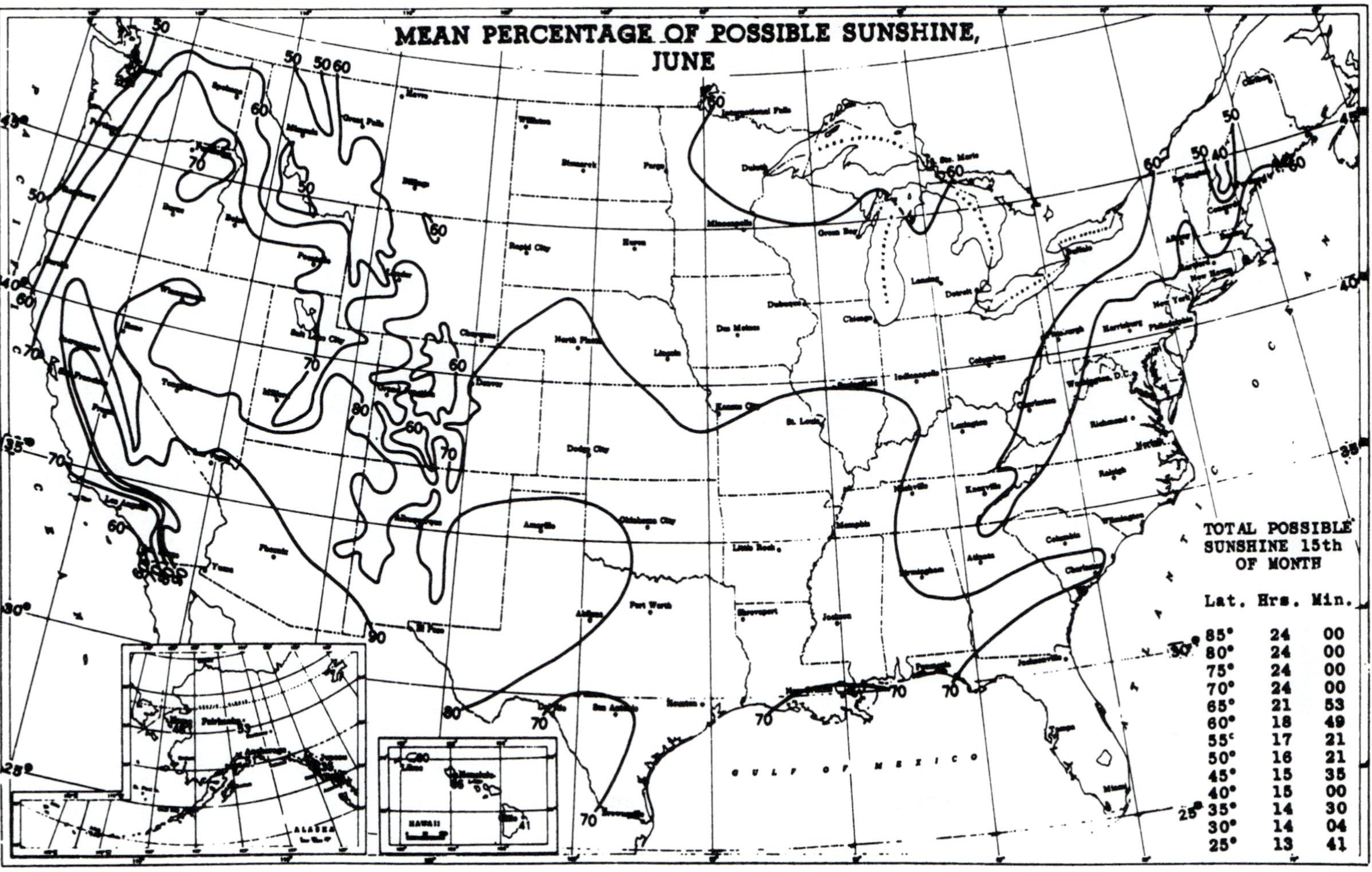

FIGURE 14–1 Mean percentage of possible sunshine, June. *Source:* Environmental Data Service, *Climatic Atlas of the United States,* U.S. Department of Commerce, Washington, D.C. June 1968.

TABLE 14–4 Computation Sheet for Penman's Equation: Example 14–3.

June	July	Variable	Units	Comment
30	32	T	°C	—
31.8	36.1	e_s	mm Hg	Table 14–1
0.3	0.4	R_h	—	—
9.54	14.44	e	mm Hg	$e = e_s R_h$
980	930	R_A	$\frac{\text{g} - \text{cal}}{\text{cm}^2 - \text{day}}$	Table 14–3
0.06	0.06	r	—	—
0.9	0.8	n/D	—	Figure 14–1
599	525	R_I	$\frac{\text{g} - \text{cal}}{\text{cm}^2 - \text{day}}$	$R_I = R_A(1 - r)(0.2 + 0.5\, n/D)$
212	152	R_B	$\frac{\text{g} - \text{cal}}{\text{cm}^2 - \text{day}}$	$R_B = \sigma(T + 273)^4(0.47 - 0.77\sqrt{e}) * \left(0.2 + 0.8\frac{n}{D}\right)$
387	373	R_n	$\frac{\text{g} - \text{cal}}{\text{cm}^2 - \text{day}}$	$R_n = R_I - R_B$
580	579	H_v	$\frac{\text{g} - \text{cal}}{\text{cm}^3}$	$H_v = 596 - 0.52T$
6.67	6.44	E	mm/day	$E = 10R_n/H_v$
0.485	0.485	α	$\frac{\text{mm Hg}}{°\text{C}}$	—
1.8	2.0	Δ	$\frac{\text{mm Hg}}{°\text{C}}$	—
2.7	2.7	V	m/s	—
13.13	12.77	E_{a0}	mm/day	$E_{a0} = 0.35(e_s - e)(0.2 + 0.55V)$
8.04	7.68	H_w	mm/day	$H_w = (\Delta E + \alpha E_{a0})/(\Delta + \alpha)$

0.9 and 0.8 for the fraction of possible sunshine are obtained from the climatic atlas for June and July, respectively. A wind speed of 2.7 m/sec for both June and July is also obtained from a climatic atlas. Based on this input and the assumptions that $r = 0.06$ and $\alpha = 0.485$ mm Hg/°C, we get mean daily evaporation rates of 8.04 and 7.68 mm/day for June and July, respectively (see Table 14–4 for a summary of the calculations). Thus for the design period the total evaporation loss would be

$$E_T = 30\,(8.04) + 31\,(7.68) = 479 \text{ mm} \tag{14–29}$$

Thus the water level in the reservoir will decrease almost 0.5 m during the design period. This should be considered along with other losses in the design of the reservoir.

The computations above assumed that $r = 0.06$. The effect of this assumption can be checked by modifying the value of r and recomputing the change in evaporation. The sensitivity of E_T to variation in r can be analyzed. If we change r to 0.1 and recompute the mean daily evaporation rates, we get 7.70 mm/day and 7.37 mm/day for June and July, respectively (see Table 14–5). These yield a total evaporation rate of 459 mm for the 61-day period. Thus the increase in r reduced the estimated evaporation by 20 mm, or 4%. The sensitivity of E_T to r is

TABLE 14–5 Computation Sheet for Penman's Equation: Example 14–3.

June	July	Variable	Units	Comment
30	32	T	°C	—
31.8	36.1	e_s	mm Hg	Table 14–1
0.3	0.4	R_h	—	—
9.54	14.44	e	mm Hg	$e = e_s R_h$
980	930	R_A	$\frac{\text{g} - \text{cal}}{\text{cm}^2 - \text{day}}$	Table 14–3
0.10	0.10	r	—	—
0.9	0.8	n/D	—	Figure 14–1
574	503	R_I	$\frac{\text{g} - \text{cal}}{\text{cm}^2 - \text{day}}$	$R_I = R_A(1 - r)(0.2 + 0.5\, n/D)$
212	152	R_B	$\frac{\text{g} - \text{cal}}{\text{cm}^2 - \text{day}}$	$R_B = \sigma(T + 273)^4(0.47 - 0.77\sqrt{e}) * \left(0.2 + 0.8\frac{n}{D}\right)$
362	351	R_n	$\frac{\text{g} - \text{cal}}{\text{cm}^2 - \text{day}}$	$R_n = R_I - R_B$
580	579	H_v	$\frac{\text{g} - \text{cal}}{\text{cm}^3}$	$H_v = 596 - 0.52T$
6.24	6.66	E	mm/day	$E = 10R_n/H_v$
0.485	0.485	α	$\frac{\text{mm Hg}}{°\text{C}}$	—
1.8	2.0	Δ	$\frac{\text{mm Hg}}{°\text{C}}$	—
2.7	2.7	V	m/s	—
13.13	12.71	E_{a0}	mm/day	$E_{a0} = 0.35(e_s - e)(0.2 + 0.55V)$
7.70	7.37	H_w	mm/day	$H_w = (\Delta E + \alpha E_{a0})/(\Delta + \alpha)$

$$S_r = \frac{\Delta E_t}{\Delta r} \tag{14–30a}$$

$$= \frac{459 - 479}{0.10 - 0.06} = -500 \text{ mm} \tag{14–30b}$$

The negative sensitivity indicates that as r increases, the total evaporation E_T will decrease.

The value of the sensitivity can be used to approximate the change in E_T for any other specified change in r (Δr). For example, if we need to estimate the change in E_T that would result from a change in r of 0.07, then Equation 14–30a can be rearranged to solve for ΔE_T:

$$\Delta E_T = S_r\,(\Delta r) = -500 \text{ mm } (0.07) = -35 \text{ mm} \tag{14–31}$$

Thus, while E_T for $r = 0.06$ was 479 mm, we would expect E_T to be approximately (479 – 35) = 444 mm for $r = 0.13$. The use of the sensitivity coefficient eliminates the need to perform the calculations for other values of r as long as the linearity of Equation 14–31 is valid or the accuracy requirements are not greater than that provided by the linearity of sensitivity computed with Equation 14–31.

14.6 MASS TRANSFER ESTIMATION

The initial attempts at estimating the mass of water transferred from the liquid state to the vapor state were based on Dalton's law of diffusivity. Dalton found that the mass transfer (E) was primarily a function of the difference between the saturation vapor pressure (e_s) and the actual vapor pressure (e_a) of the air mass overlying the water body:

$$E = b_o (e_s - e_a) \tag{14–32}$$

in which b_o is an empirical coefficient. For a given site, the value of b_o could be obtained using measurements of evaporation rates and the air and water surface temperatures.

Later studies showed that b_o was a function of the air turbulence and that the accuracy of estimates could be improved by accounting for the effect or air turbulence. This led to a number of mass transfer equations of the form

$$E = b_1 f(V) (e_s - e_a) \tag{14–33}$$

in which $f(V)$ is an empirical function involving wind speed and b_1 is an empirical coefficient. In such equations, wind speed is the variable used to reflect the movement of drier air into the lake environment to replace the air with a higher water vapor content. For example, Meyer (1915) presented the following mass transfer equation:

$$E = 0.5 (1 + 0.1 V_{30}) (e_s - e_a) \tag{14–34}$$

in which V_{30} is the wind speed in mi/hr at a height of 30 ft above the water surface. For shorter time periods where V shows significant variation about the mean value, Equation 14–33 can provide more accurate estimates than Equation 14–32; however, for longer time periods where there is little variation about the mean wind speed, Equation 14–33 may not result in a significant improvement in the accuracy of estimation.

In addition to the empirical structure of Equation 14–34, other forms have been tried, including some that include variables other than vapor pressure and wind speed. The following equation was developed in 1958 as part of a study using data from Lake Mead:

$$E = 0.072 V_{30} (e_s - e_a) [1 - 0.03 (T_a - T_w)] \tag{14–35}$$

in which T_a and T_w are the air and average water temperatures (°F), respectively. Just as with regression analyses, the increase in accuracy will be less and less as more variables are added. The more complex models also require the collection of more input data.

In addition to the empirical mass transfer formulas developed from Dalton's Law, a theoretical development of turbulent transport has led to two prediction equations that are more complex in both structure and data requirements than the empirical diffusion-based equations. The Thornthwaite-Holzman equation is as follows:

$$E = \frac{833k^2(e_1 - e_2)(V_2 - V_1)}{ln_e(Z_2/Z_1)^2(T + 459.4)} \tag{14–36}$$

in which E is the evaporation (in./hr), k is the von Kármán constant, e_1 and e_2 are the vapor pressures (in. Hg), V_1 and V_2 are the wind speeds (mi/hr), Z is the elevation of measurement, the subscripts 1 and 2 refer to the lower and upper measurement locations, and T is the mean

TABLE 14–6 Calculation of Daily Evaporation Using Meyer's Equation

Hour	T_a(°F)	T_w(°F)	R_h(%)	e_s(in. Hg)	e_a(in. Hg)	e_o(in. Hg)	E(in./day)
01	73	66	71	0.827	0.587	0.652	0.049
04	71	66	79	0.769	0.608	0.652	0.033
07	74	70	82	0.856	0.702	0.740	0.029
10	81	70	56	1.069	0.599	0.740	0.105
13	88	71	42	1.342	0.564	0.769	0.154
16	92	74	42	1.524	0.640	0.856	0.162
19	88	72	47	1.342	0.631	0.798	0.125
22	82	71	58	1.108	0.643	0.769	0.095

temperature (°F) of the layer between Z_1 and Z_2. Based on mixing-length theory, Sverdrup (1946) developed the second complex mass transfer equation.

In comparing the highly empirical structure of Dalton's Law (Equation 14–32) and the theory-based structure of the Thornthwaite-Holzman Equation (Equation 14–36), it is evident that b_o of Equation 14–32 is a function of the wind velocity, temperature, and von Kármán's constant. The difference in accuracy of Equations 14–32 and 14–36 depends on the variability of V and T, which would be greater for the shorter time periods. Also, whereas Equation 14–32 must be calibrated to obtain a site-specific value of b_o, Equation 14–36 does not require calibration. However, the accuracy of Equation14–36 will depend on the degree to which the turbulent transport theory is valid and the accuracy of the measurements of T_2, T_1, V_2, and V_1 at elevations of Z_2 and Z_1. Equation 14–36 assumes a logarithmic vertical distribution of both wind speed and moisture and an adiabatic atmosphere.

Example 14–4

Consider the data of Table 14–6. Water and air temperature measurements were recorded at 3-hr intervals, along with the relative humidity. An average wind speed of 5 mi/hr was computed for the day. Using the air temperature and relative humidity, values for e_a were computed using Equation 14–2 and Table 14–1. Values for e_o were obtained using the 3-hr values of T_w and Table 14–1. Evaporation rates in in./day were computed for each of the 3-hr periods, and a weighted mean value of 0. 094 in./day was determined.

14.7 PAN EVAPORATION

In making a water balance, the time rate of change of storage is evaluated by measuring the change in the water surface elevation. However, because the change in the water surface elevation is a function of the other water inputs and outputs, it is not a direct measure of evaporation. However, we can apply the same principle (that is, conservation of mass) to a confined water body where the other elements can be controlled and use the water balance equation of Equation 14–7. This is done by applying the principle of conservation of mass to a controlled water body; in this case, the water body is a pan. A pan will not be subject to water inflow or outflow by way of surface or subsurface flow. Thus the water balance of Equation 14–7 reduces to

$$P - E_p = \frac{dS}{dt} \tag{14–37}$$

in which E_p is the evaporation from the pan. Pan evaporation can be estimated by solving Equation 14–37 for E_p:

$$E_p = P - \frac{dS}{dt} \tag{14–38}$$

The heat/energy environment of a pan is not the same as the environment of the water body. Therefore, E_p is not the same as E of Equation 14–7; however, the value of E_p is usually assumed to be proportional to E, with the following relationship assumed:

$$E = c_p E_p \tag{14–39}$$

in which c_p is called the pan coefficient. The pan coefficient is necessary because evaporation from a pan exceeds that from either an open body of water or a well-wetted soil surface.

The use of a pan to estimate lake evaporation has several advantages. It eliminates inaccuracies associated with measurements made for the water balance of Equation 14–7. It is inexpensive; the cost of a pan is minimal, and the cost of data collection can be minimized by using mechanical recording devices. It is easy to instrument, and the value of c_p in Equation 14–39 is relatively constant for a given location; thus the analysis (that is, estimation of c_p) is easy.

Three types of pan setups are available: surface, sunken, and floating. Each type has advantages and disadvantages. A sunken pan is a pan that is placed in the ground with the water surface in the pan at about the same level as the ground. Although this is a stable environment, it is subject to a number of problems. It is difficult to maintain as trash can be blown into the pan. The accuracy of evaporation estimates is limited because of the difficulty in estimating heat transfer through the sides of the pan. Obviously, if the pan leaks, inaccurate estimates of evaporation will result.

A floating pan is conceptually ideal because the pan exists in the same environment as the water body. However, a floating pan is difficult to maintain, is somewhat inaccessible for data collection and maintenance, and is subject to splashing action.

The most common type of pan is the standard U.S. Weather Bureau Class A pan. It has a diameter of 4 ft, and the water is maintained at a depth of about 10 in. The pan is made of galvanized steel and sits on a slatted platform about 6 in. above the ground. While a lake stores heat, the heat storage within the pan is minimal and has no effect on evaporation rates from the pan. The pan evaporation can be corrected for advected energy.

The task of design (that is, synthesis) can be handled in one of two ways: using actual pan evaporation measurements or maps of pan evaporation. In both cases, estimates of lake evaporation will require the linear transformation of Equation 14–39. The value of the pan coefficient is considered to be location dependent; values for the United States are shown in Figure 14–2. Whether the estimated pan evaporation is obtained from a map or from measurements depends on a number of factors such as the necessary accuracy and the time interval of the required estimates. For annual values, estimates of E_p can be obtained from Figure 14–3. Figures 14–2 and 14–3 were used to derive the mean annual lake evaporation, which is shown in Figure 14–4. The mean May–October evaporation as a percentage of the annual evaporation is shown in Figure 14–5.

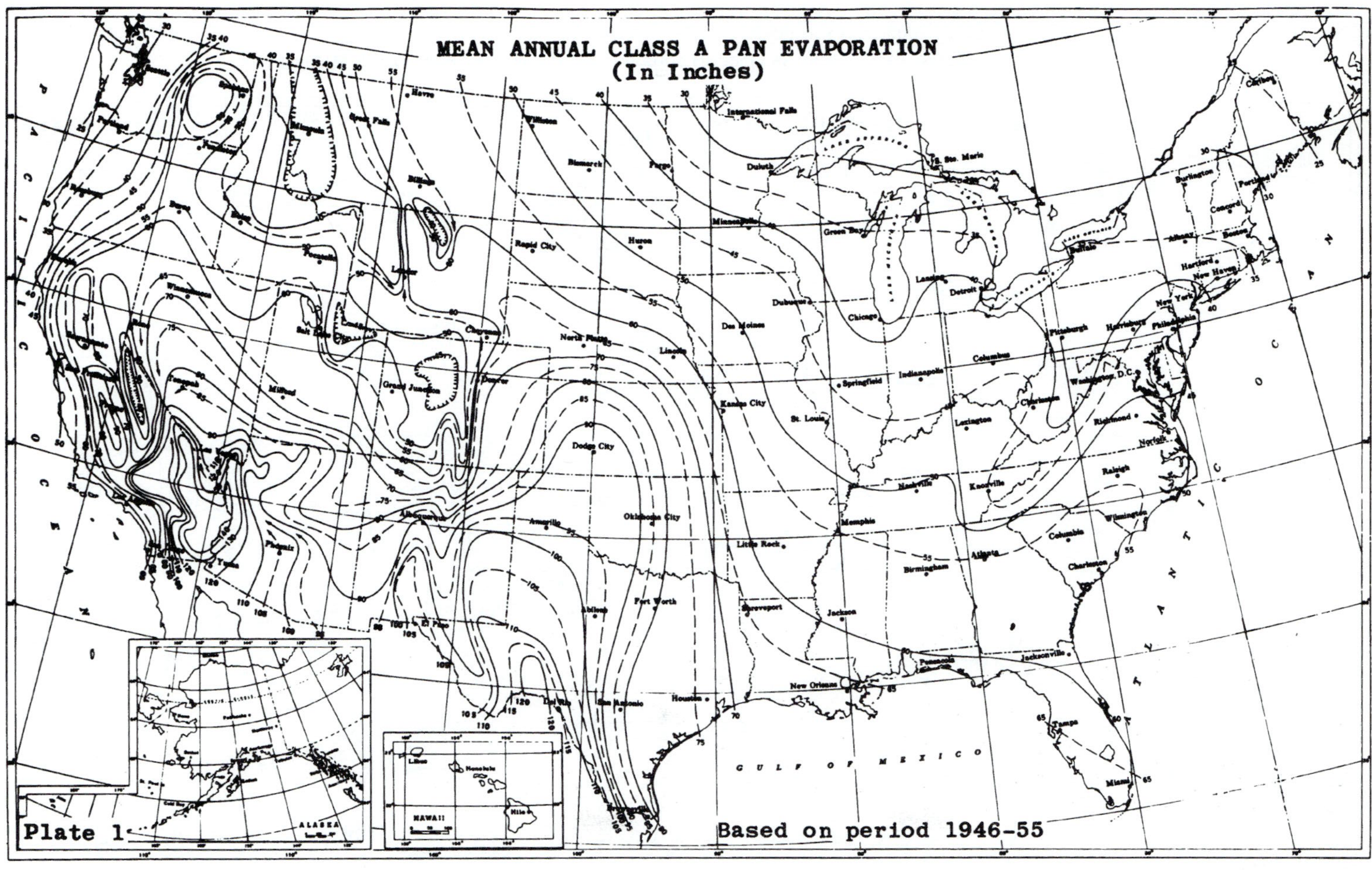

FIGURE 14–2 Mean annual class A pan evaporation (in inches). *Source:* Environmental Data Service, *Climatic Atlas of the United States,* U.S. Department of Commerce, Washington, D.C. June 1968.

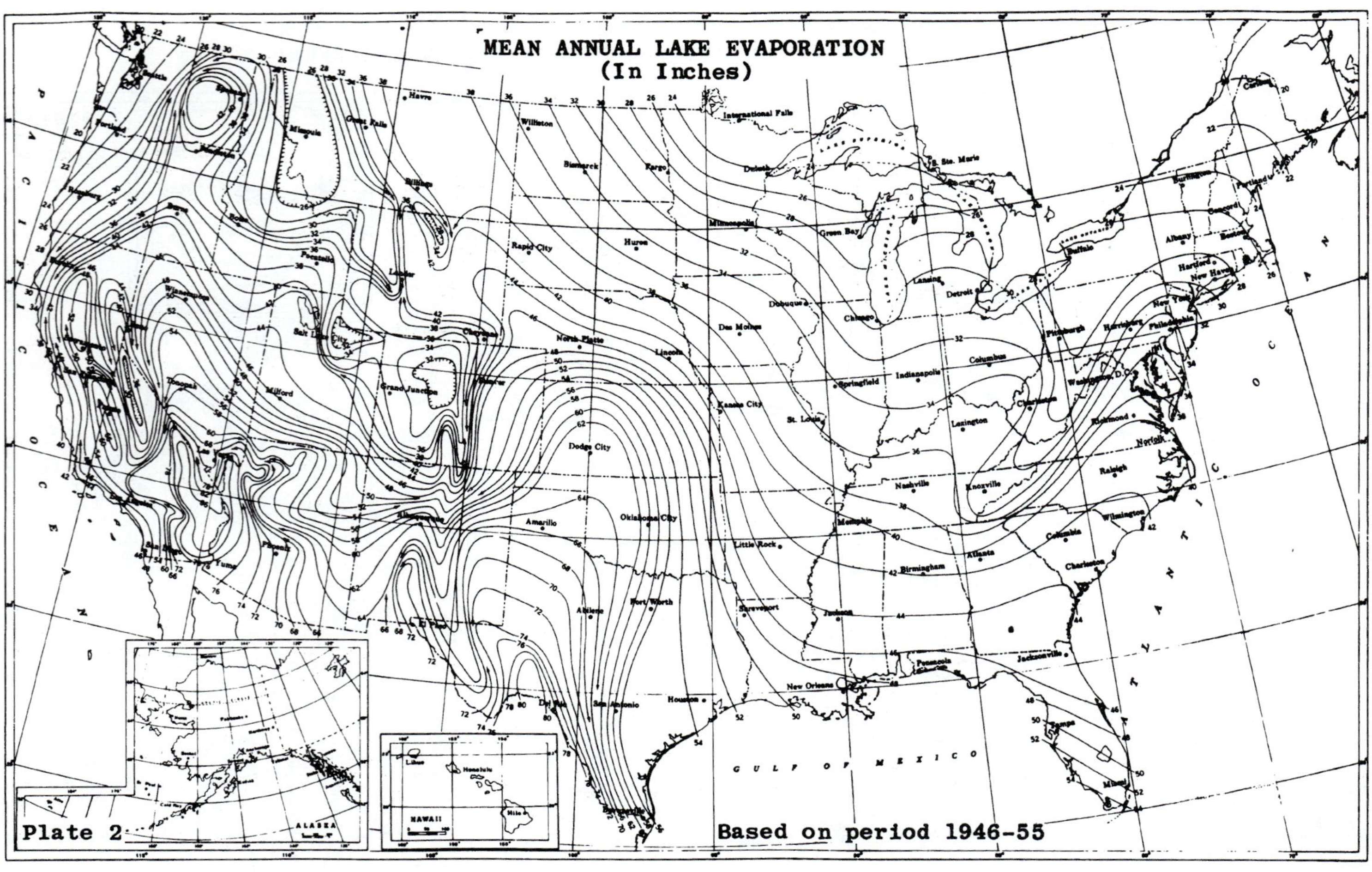

FIGURE 14–3 Mean annual lake evaporation (in inches). *Source:* Environmental Data Service, *Climatic Atlas of the United States,* U.S. Department of Commerce, Washington, D.C. June 1968.

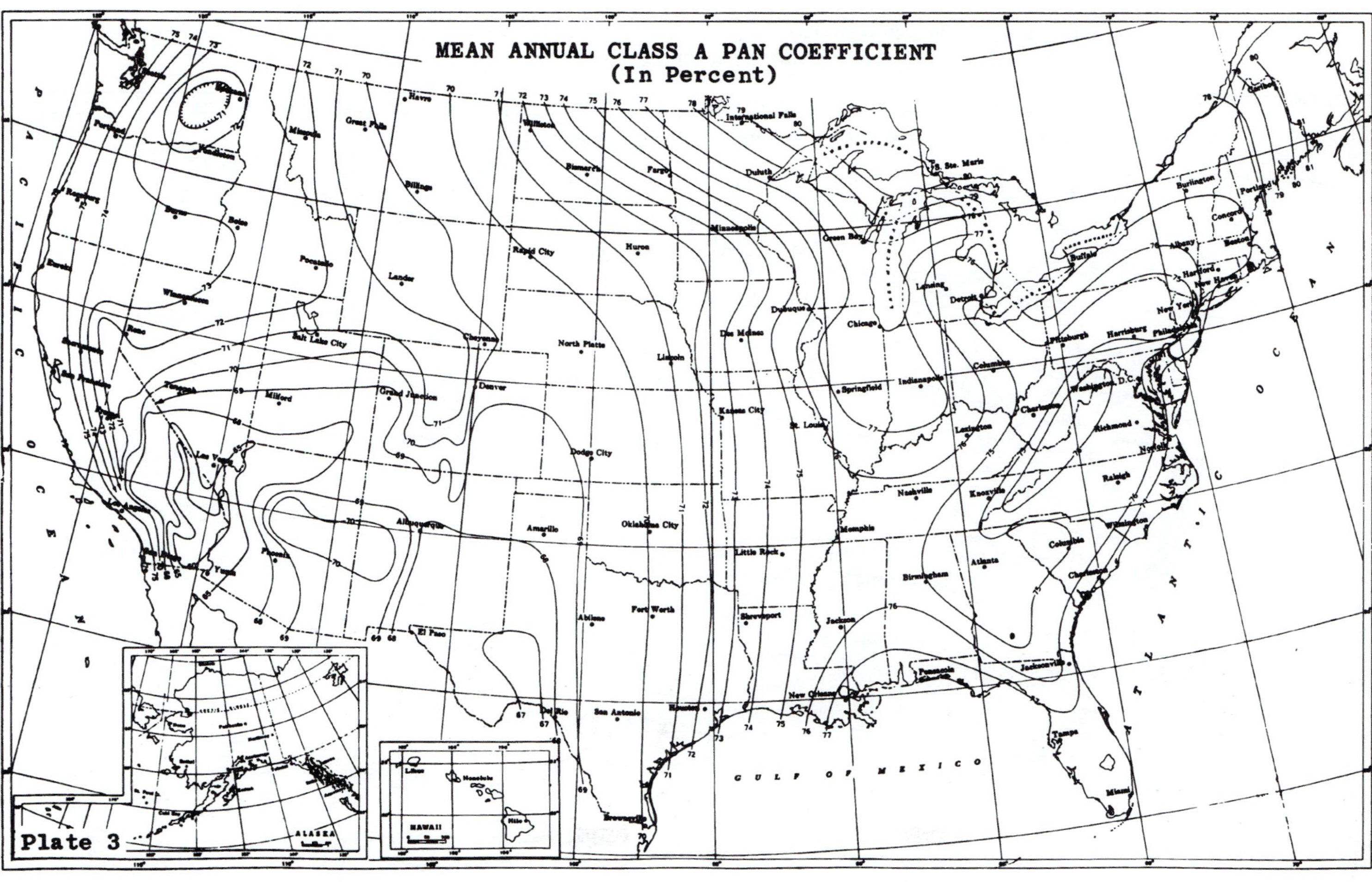

FIGURE 14–4 Mean annual class A pan coefficient (in percent). *Source:* Environmental Data Service, *Climatic Atlas of the United States,* U.S. Department of Commerce, Washington, D.C. June 1968.

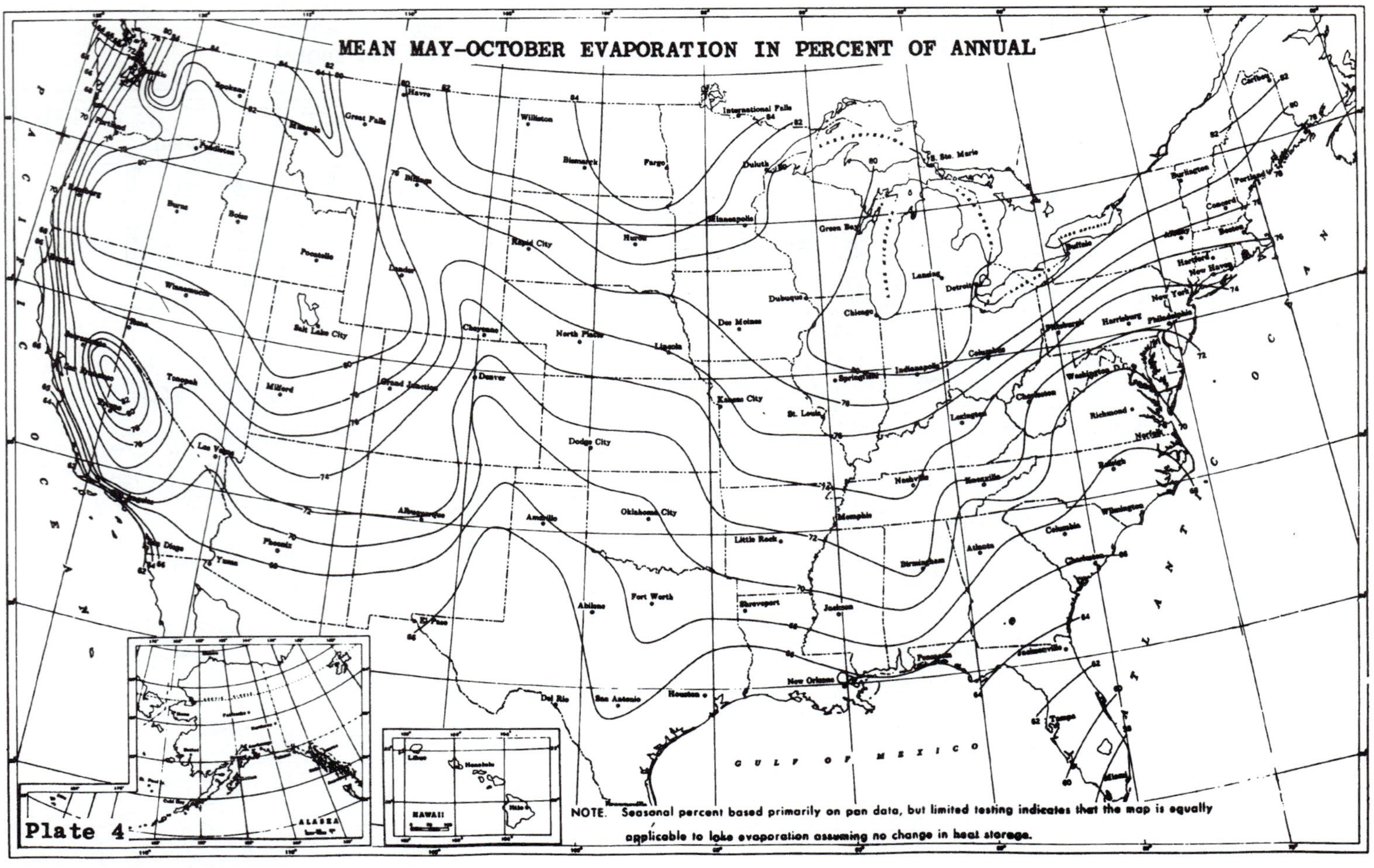

FIGURE 14–5 Mean May–October evaporation in percent of annual. *Source:* Environmental Data Service, *Climatic Atlas of the United States,* U.S. Department of Commerce, Washington, D.C. June 1968.

PROBLEMS

14–1. Using a lake as the system, define the inputs, outputs, and storage elements. Identify the knowns and unknowns for both analysis and synthesis of daily evaporation rates.

14–2. Using a 200-acre corn field as the system, define the inputs, outputs, and storage elements. Identify the knowns and unknowns for both analysis and synthesis of daily rates of evapotranspiration.

14–3. Discuss why water condenses on the outside of a glass of iced tea on a hot and humid summer day.

14–4. On a day when the temperature is 17°C and the relative humidity is 62%, what is the vapor pressure deficit in pounds per square inch?

14–5. Using the power law of Equation 14–3, plot the wind speed profile for heights from 0 to 10 ft when an anemometer at a height of 5 ft records a wind speed of 7 mi/hr. Also compute and plot the rate of change of the velocity with height.

14–6. Measurements at a particular weather station have been recorded in the past at a height of 5 ft. Because of a change in data collection practices at this location, all future mesurements will be made at a height of 2 m. Derive an expression for converting the past 5 ft measurements to synthetic values for a 2-m height.

14–7. The following table gives the monthly total precipitation (P), the monthly average streamflow into (Q) and out of (Q_o) a lake, and the net seepage ($Q_s - Q_d$) for a lake with a surface area of 0.5 mi^2. Find the evaporation losses (acre-ft, in., in./day) using the water-budget approach of Equation 14–7 assuming that surface runoff (Q_r) is negligible and the storage does not change.

Month	P (in.)	Q (ft^3/sec)	Q_o (ft^3/sec)	$Q_s - Q_d$ (ft^3/sec)
April	3.0	31	30	–0.8
May	4.2	36	34	–1.8
June	3.8	34	35	2.1
July	4.9	30	32	3.6
August	4.0	27	30	4.8
September	4.0	26	29	4.6

14–8. The following table gives the monthly total precipitation (P), the monthly average streamflow into (Q) and out of (Q_o) a lake, and the net seepage ($Q_s - Q_d$) for a lake with a surface area of 140 acres. Find the evaporation losses (acre-ft, in., in./day) using the water-budget approach of Equation 14–7 assuming that surface runoff (Q_r) is negligible and the storage does not change.

Month	P (in.)	Q (ft^3/sec)	Q_o (ft^3/sec)	$Q_s - Q_d$ (ft^3/sec)
April	3.0	12	9	–3.1
May	1.1	14	13	–0.5
June	0.5	11	14	3.8
July	0.4	9	12	4.2
August	0.5	8	10	3.0
September	0.9	6	7	1.7

14–9. Given $T = 28°C$; $R_h = 25\%$; $R_A = 950$ g-cal/(cm^2-day); $r = 0.08$; $n/D = 0.85$; $V = 2$ m/s. The constants α and Δ are 0.485 and 1.7, respectively. Compute the daily evaporation rate using Penman's Equation. Determine the total evaporation (acre-ft) for a 30-day design period from a lake with a surface area of 74 acres.

14–10. Using Penman's Equation, compute the daily evaporation rate (in./day) during June for a lake near Kansas City, Missouri. Assume the following average values: $T = 75°F$; $R_h = 70\%$; $r = 0.15$; $V = 10$ mi/hr; $\alpha = 0.485$; $\Delta = 1.7$.

14–11. Replace the 0.5 of Equation 14–34 with an empirical constant b_0. Using the following data, find the best value of b_0, with V_{30} [=] mi/hr; E [=] in./day; e_s and e_a [=] in. Hg. Assess the accuracy of the fitted equation.

Temp. (°F)	R_h (%)	V_{30} (mph)	E (in.day)
63	74	7.5	0.08
71	66	9.2	0.14
76	63	8.4	0.18
68	68	6.3	0.13
62	76	6.8	0.09
54	77	9.1	0.06

14–12. The following data are the daily amounts of rainfall and water added to a pan (both in in.). Compute the daily evaporation rates.

Day	1	2	3	4	5	6	7
P (in.)	0.00	0.00	0.13	0.04	0.00	0.21	0.06
Water Added (in.)	0.22	0.25	0.02	0.09	0.19	0.00	0.11

14–13. The following data are mean daily pan evaporation measurements (in./day). Assuming a pan coefficient of 0.68, estimate the daily lake evaporation rates (acre-ft and in.) for a 27-acre lake.

Day	1	2	3	4	5	6	7
Pan E	0.22	0.26	0.25	0.28	0.26	0.21	0.22

14–14. The following data are mean daily pan evaporation measurements (in./day). Assuming a pan coefficient of 0.72, estimate the daily lake evaporation rates (acre-ft and in.) for a 114-acre lake.

Day	1	2	3	4	5	6	7
Pan E	0.31	0.29	0.34	0.35	0.32	0.28	0.27

14–15. The following data are measured pan evaporation rates and lake evaporation rates estimated by a detailed water budget of the lake. Estimate the average pan coefficient.

Pan E (in.)	0.23	0.26	0.25	0.32	0.29	0.25
Lake E (in.)	0.17	0.19	0.18	0.23	0.21	0.19

14–16. For the data of Problem 14–7 determine the change in storage (acre-ft, in.) for each month if the pan evaporation (in.) was 0.14, 0.21, 0.28, 0.35, 0.32, and 0.29 in./day. Assume a pan coefficient of 0.72.

14–17. For the data of Problem 14–8, determine the change in storage (acre-ft, in.) for each month if the pan evaporation (in.) was 0.15, 0.20, 0.25, 0.30, 0.30, and 0.25 in./day. Assume a pan coefficient of 0.70.

REVIEW QUESTIONS

14–1. In temperate climates, approximately what percent of precipitation returns directly to the atmosphere by way of evaporation or transpiration? (a) 30; (b) 50: (c) 70: (d) 90.

14–2. Which one of the following factors has, in general, the least effect on evaporation rates? (a) Temperature; (b) radiation; (c) the vapor pressure deficit; (d) wind speed; (e) all of these factors have a similar effect.

14–3. If the vapor pressure of an air mass is 0.27 in. Hg and the saturated vapor pressure is 0.36 in. Hg, the relative humidity is (a) 0.09 in. Hg; (b) 75%; (c) 25%; (d) the information above is insufficient to compute the relative humidity.

14–4. The dew point temperature is the temperature at which (a) the water vapor content is a minimum; (b) the air becomes saturated with vapor; (c) the air pressure is a minimum; (d) none of the above.

14–5. Wind speed is measured with (a) a wind vane; (b) an anemometer; (c) a heliometer; (d) a tensiometer.

14–6. Which one of the following inputs is required when using the power law? (a) Wind speed; (b) air temperature; (c) saturation vapor pressure; (d) wind vane.

14–7. Energy and water budgets are based on which of the following? (a) Bowen's Aerodynamic Equation; (b) Dalton's Law of diffusivity; (c) conservation of momentum; (d) conservation of mass.

14–8. Which one of the following is not an element of the simplified energy balance used to develop Penman's Equation? (a) Reflected solar radiation; (b) latent heat flux density; (c) sensible heat flux; (d) net energy.

14–9. Which one of the following is not part of most mass transfer models? (a) Wind speed; (b) air temperature; (c) water temperature; (d) saturation vapor pressure; (e) all of the above are inputs.

14–10. Which one of the following is not true for the pan coefficient applied to a U.S. Weather Bureau Class A pan? (a) The pan coefficient is always less than 1; (b) it varies with location? (c) it adjusts evaporation rates for stored heat; (d) it varies with the time of the year; (e) all of the above are true.

DISCUSSION QUESTION

The technical content of this chapter is important to the professional hydrologist, but practice is not confined to making technical decisions. The intent of this discussion question is to show that hydrologists must often address situations where value issues intermingle with the technical aspects of a project. In discussing the stated problem, at a minimum include responses to the following questions:

1. What value issues are involved, and how are they in conflict?
2. Are technical issues involved with the value issues? If so, how are they in conflict with the value issues?
3. If the hydrologist attempted to rationalize the situation, what rationalizations might he or she use? Provide arguments to suggest why the excuses represent rationalization.
4. What are the hydrologist's alternative courses of action? Identify all alternatives, regardless of the ethical implications.
5. How should the conflict be resolved?

You may want to review Sections 1.6 to 1.12 in Chapter 1 in responding to the problem statement.

Case. A research hydrologist obtains a contract to develop a model for estimating lake evaporation rates from high-elevation lakes. The researcher sets of a series of remote monitoring instruments to collect the data, as detailed in the research plan used to obtain the funding. At the end of the first period of data collection, the hydrologist finds out that the radiometers at two of the seven lakes were miscalibrated and the radiation data collected at these two sites is worthless. Fearing embarrassment and the delay in completing the research, the hydrologist uses synthetic data estimated from the radiation data measured at the other five sites and reports the synthetically generated data as actual measured data.

15

Erosion and Sedimentation

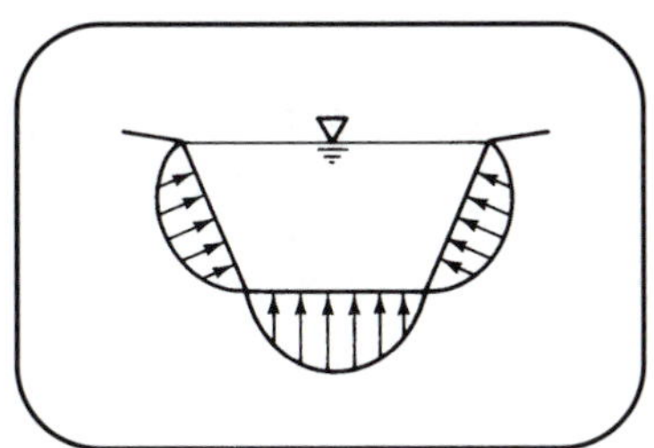

CHAPTER OBJECTIVES

1. Identify the importance of erosion and sedimentation in hydrologic design.
2. Introduce terminology used in discussing erosion problems.
3. Identify the forces involved in the physical process of erosion.
4. Introduce methods used to estimate surface erosion and gully erosion.
5. Introduce methods used to estimate channel degradation.
6. Introduce methods for estimating long-term sediment yield.

15.0 NOTATION

A = drainage area
A_c = projected cross-sectional area of a soil particle
A_t = total sediment accumulation in reservoir
b_o = empirical coefficient
C = crop management factor for USLE
C_d = drag coefficient
C_l = sediment concentration at level l
C_o = sediment concentration
d = mean particle diameter of soil
d_i = mean particle diameter for weight fraction i
d_m = maximum particle diameter of bed
d_r = raindrop diameter
d_{65} = mean diameter of soil particles for which 65% of the particles by weight are smaller
D = mean diameter of the largest fractions in mixed sediments
$\overline{D}$ = weighted mean particle size
E = soil erosion
E_c = clay content of eroding soil profile
E_f = reservoir trap efficiency
E_g = average annual gully head advance
f = coefficient of friction
f_h = factor reflecting the hydraulic roughness of the bed
f_i = fraction of bed material in size range i
F = soil erodibility factor for Musgrave Equation
F_d = downstream hydrostatic force
F_d = total drag force
F_g = force of gravity
F_l = lift force on soil particle
F_r = resultant force
F_u = upstream hydrostatic force
G = weight of splash eroded soil
H = depth of flow
i = rainfall intensity
I_1, I_2 = integrals in Einstein's method
K = soil erodibility factor for USLE
K_e = kinetic energy
K_s = shear stress ratio
L_o = length of free-body segment
m = mass of raindrop
M = grain distribution modulus
n_i = number of years of sediment accumulation record at reservoir i
N = number of weight fractions in sediment distribution
p_i = percent of bed material in size range i (=$100f_i$)
P = conservation practice factor for USLE
P_a = annual summation of rainfall from rains equal to or greater than 0.5 in. in 24 hrs.
P_w = wetted perimeter
P_{30} = 2-yr, 30-min rainfall intensity
q = water discharge rate
q_b = bed load
q_c = critical water discharge rate
q_s = total sediment transport rate
q_B = bedload rate in weight per unit time per unit width
R = cover factor for Musgrave Equation
R = erosivity index
$\mathbf{R}$ = Reynold's Number
R_c = cover factor for Beer et al. method
S = field slope
S_c = channel slope
S_s = solid-to-fluid density ratio
SDR = sediment delivery ratio
T = topographic factor for USLE
V = flow velocity
V = velocity of raindrop
V_c = critical water flow rate
V_* = shear velocity
W = weight of fluid mass
X_1 = surface runoff index
X_2 = terraced area of watershed
X_3 = gully length

X_4 = length from end of gully to the watershed divide
X_5 = deviation of rainfall from normal
Y = growth of gully in surface area
Y = sediment yield
α = angle of channel bottom
γ = specific weight of water
γ_s = specific weight of soil
τ = shear stress
τ_c = critical shear stress
τ_L = shear stress on bed
τ_s = shear stress on side of channel
λ = coefficient for soil particle shape
ω = stream power per unit of bed area
ω_c = critical stream power
ϕ = angle of repose
Ψ = empirical coefficient
ρ = density of water
ρ_s = density of soil
μ = fluid viscosity

15.1 INTRODUCTION

Erosion can be loosely defined as the wearing away of the land. Erosion processes, which involve forces associated with water, wind, ice, and gravity are constantly at work wearing down Earth's surface. Water erosion is deemed the most widespread agent of the erosion process and is responsible for the bulk of the sediment transported from the land to the sea. Precipitation along with a variety of other factors breaks down rocks, which forms soil and debris. The resulting matter is transported through rills, gullies, channels, and other watercourses toward the oceans.

The four primary activities that accelerate rates of erosion are logging, mining, agriculture, and construction. For example, when land is disturbed by construction activities, soil erosion may increase from 2 to 40,000 times the preconstruction erosion rate. This is evidenced by the fact that each year approximately 80 million tons of sediment is washed from construction sites into lakes, rivers, and other waterways (Goldman et al., 1986). From our farmlands alone, 1600 million tons of soil that carry valuable soil nutrients needed for growing crops is washed from our farms and into nearby lakes and reservoirs (NRC, 1986).

Numerous problems arise from inadequate control of erosion rates and sedimentation. One problem that is visually apparent is the suspended sediment in rivers and streams. In addition to the negative aesthetic value, the environmental effects of suspended sediment and deposited soil on aquatic habitat are widely known. Engineers are continually concerned with washouts at culverts, highways, bridges, and other engineering structures. The agricultural engineer is concerned with the maintenance of topsoil (that is, soil conservation) to maintain productivity. Channel stability is important to those who own stream bank property, whether it is the farmer who wishes to maximize his or her productive acreage or the homeowner who has a home that is adjacent to a stream. For engineering projects involving storage reservoirs, the rate of loss of capacity due to sediment accumulation is important. Those involved in navigation are sometimes concerned about the size and shape of delta formations. Numerous other specific problems associated with erosion and sedimentation could be discussed. The examples cited are sufficient to emphasize the importance of the topic.

The effects of erosion are not limited to environmental impacts. Erosion has economic impacts as well. Although these costs are difficult to quantify, some examples may help to put them in perspective. When excessive sediment is accumulated in a reservoir, the reservoir must be dredged and the soil material disposed of. In Cull Canyon Reservoir located in Alameda County, California, less than 10% of the watershed has been urbanized. Eleven years after construction of the reservoir, 400,000 yd^3 of sediment was removed. The cost was approximately $1 million (Goldman et al., 1986). Another economic impact of erosion was cited by the Committee on Public Works in their report to the House of Representatives. According to this report, much of the sediment in streams comes from the bed and banks, and in 1969, this sediment was causing an estimated loss of $90 million annually (House of Representatives, 1969). In addition, according to the Department of Agriculture, with better technology and methods for channel and gully stabilization and control, there is a potential for the saving of land and a reduction of damages to structures of 460 million annually (USDA, 1978).

Before discussing the details of erosion and sedimentation, it is important to acknowledge the potential effects of inaccuracies in engineering designs. The underestimation of sediment yields may lead to the underdesign of an engineering structure, which may result in either a project life that is shorter than expected or the need for unplanned cleanup expenditures. The overestimation of sediment yields will lead to overdesign and thus a waste of resources that could be used for alternative projects. Design uncertainty implies engineering risk. Thus whenever an estimate is made, every effort should be made to evaluate the accuracy of the estimate; however, risk assessment and the analysis of uncertainties are beyond the scope of this writing.

15.1.1 Analysis versus Synthesis

The primary emphasis of this chapter is on the use of existing estimation methods. This might lead the reader to believe that these methods are highly accurate for a wide variety of conditions. This is certainly not the case. Most existing estimation methods are highly inaccurate when used for a site or set of conditions other than that used in calibrating the method. In fact, when different estimation methods are applied at an ungaged location, the computed sediment loads may differ by one or two log cycles. This disparity reflects both the site-to-site variation that should be expected and the inaccuracy of the data used to calibrate the methods. The large uncertainty implies that, when possible, a prediction equation should be calibrated for each design case. When this is not possible, the underlying assumptions and the calibration data of the method used to make the design estimates should be fully investigated prior to the use of the method.

In the analysis case, the modeling tasks include collecting the data, deciding on the most appropriate model structure, and then using the data to fit the coefficients of the model. The model coefficients represent the unknowns, while the knowns might include topographic factors, soil characteristics, hydrometeorological characteristics, and channel characteristics. The specific variables will depend on the type of estimation required. For example, the rainfall intensity will be important in the estimation of splash erosion, but not in channel erosion; the channel slope will be important in the estimation of channel erosion. However, a variable representing the size of the soil particles may be important in the estimation of both splash erosion and channel erosion.

In the synthesis or design case, the volume of sediment or amount of erosion is the unknown. To use an existing model, it is necessary to determine values of the input variables, such as topographic, channel, soil, or hydrometeorological characteristics. As stated previously, it is important to compare the values of the input variables with those that were used in calibrating the design model. For best results, values for the input variables should be similar to those that were used in calibrating the method.

15.2 PHYSICAL PROCESSES IN EROSION AND SEDIMENTATION

Before conceptualizing the erosion processes, it may be of value to provide some general definitions. Erosion is a process by which soil and minerals are detached and transported. The process may also be viewed to include sedimentation, which is the deposition of particles when gravitational forces overcome the forces that cause the movement.

To discuss the processes of erosion and sedimentation effectively, it is necessary to define the general terminology associated with their study. This section is devoted to the development of the vocabulary necessary to gain a basic understanding of the subject. Figure 15–1 summarizes the principal factors that affect soil erosion and movement.

Erosion takes place not only in the channel, but on the watershed surrounding the channel. Splash erosion occurs when raindrops strike bare soil. This breaks up soil aggregates and separates fine particles and organic matter from heavier soil particles. These particles are then transported with the surface runoff. The runoff also contributes to sheet erosion, which is caused by shallow "sheets" of water that flow over the land surface and help to move loose particles toward waterways.

Rill erosion occurs when surface flow begins to concentrate on the land surface. The energy of this concentrated flow begins to cut small channels called rills and is capable of detaching and transporting soil particles. When rills become deeper and wider, gullies are formed. Gullies are capable of transporting a larger amount of sediment since the flow rates are greater than in rills. Finally, channel erosion occurs when bank vegetation is disturbed or when the flow rate in a stream is increased beyond the critical point where bed particles initiate movement. These changes destroy the geomorphic equilibrium of a natural stream and cause channel erosion to begin.

A stable channel is one in which there is no objectionable silting, scouring, or sedimentation. When these processes occur in excessive amounts, the channel becomes unstable and its natural balance is destroyed. Scouring is the erosive deformation of a channel. Aggradation or silting is the gradual rise in the channel bottom over its entire extent, and progradation or sedimentation is the increase and advancing forward of the volume of sediment deposited.

A channel functions by transporting sediment out of the watershed by means of its flow. The total load of sediment transported by the channel is made up of the bed-material load and the wash load. These are classified according to their origin. Specifically, the bed-material load is the portion of total sediment load that is composed of grain sizes originating in the channel bed and sides. The wash load is composed of finer-grained particles with virtually no settling velocity that originate from the land surface of the watershed. The wash load is transported to the channel by means of splash, sheet, rill, and gully erosion. The bed-

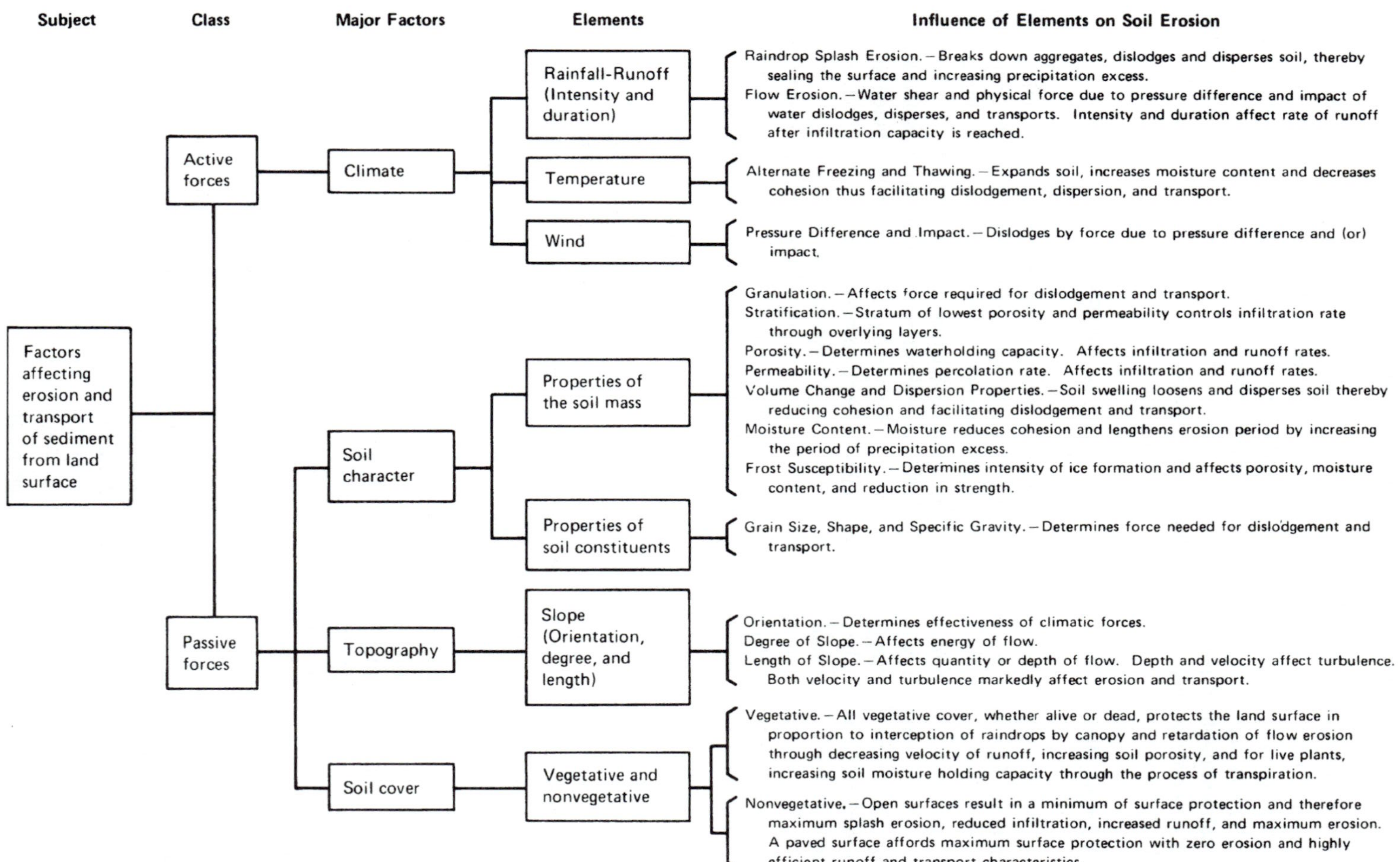

FIGURE 15–1 Principal factors affecting soil erosion and movement.

material load can be further divided into bed load and suspended load. The bed load is the material too coarse to be supported in the flowing water for any appreciable amount of time. The grains move by sliding, rolling, or saltating over the bottom, continually coming to rest and then rejoining the bed load material. The suspended load can be classified as all of the particles that are lifted up by eddies in the main flow and move long distances downstream before settling to the bed. Ordinarily, the bed load portion is less than 25% of the total load but there may be exceptions to this general statement. The wash load is considered to be a part of the suspended load since it is transported by the flow at nearly the same velocity and never settles to the channel bottom.

The wash load is usually represented in the channel only in very small amounts as compared to bed-material load. Whatever the rate of wash load transported through the channel, neither scour nor deposition is a consequence of the wash load. Because of this, it is impossible to accurately predict the rate of wash load without directly measuring it during the period of flow. Also, since the wash load is neither scoured nor deposited in the channel, the geometry of the channel will not be affected. Therefore, wash load is concluded to have only a very minimal, if any, direct effect on the stability of the channel. However, the wash load component may be quite significant to water quality considerations and deposition rates in large reservoirs with high trap efficiencies.

The *sediment yield* is the total sediment outflow from a watershed or a drainage area at a point of reference during a specified period of time. This is always less than the gross erosion (sum of all sources of erosion in a watershed) because a fraction of the sediment is deposited or trapped in the channel as it is transported out of the watershed. The ratio of the sediment yield to the gross erosion is called the *sediment delivery ratio*. It is often of great importance to the engineer to be able to determine the load transported by a channel. This is usually determined by separately calculating the suspended and bed loads and then adding the two. Since the wash load is easily transported out of the channel, it has no affect on the computations. Many equations are presently available to determine the bed load discharge. Sediment discharge refers to the rate of transport of the sediment load. A relationship between the sediment discharge and the water discharge is called the *sediment rating curve*.

In engineering design, the engineer must be concerned with the erosion hazard, which is a measure or description of the erosion potential. The erosion hazard reflects the susceptibility of a soil to erosive agents (the erodibility) and the characteristics of the erosive agent (the erosivity). The erodibility and the erosivity represent, respectively, the soil and water potential for causing erosion.

15.2.1 Stokes' Law

The situation of a solid particle free falling through a liquid arises frequently in engineering. Three forces act on the particle: gravity, buoyancy, and drag. The first force acts downward and the latter two forces act upward. For equilibrium at terminal velocity, the sum of the forces is

$$\Sigma F = 0 = F_b + F_D - W \tag{15–1a}$$

in which F_b is the buoyant force, F_D is the drag force, and W is the weight, which for a sphere are computed by the following:

$$F_b = \gamma * \text{volume} = \gamma\,(\pi\, D^3/6) \tag{15–1b}$$
$$F_D = C_D \frac{1}{2}\,\rho A V^2 = \rho C_D \pi V^2 D^2/8 \tag{15–1c}$$
$$W = \gamma_s * \text{volume} = \gamma_s\,(\pi D^3/6) \tag{15–1d}$$

in which D is the diameter of the sphere, γ is the specific weight of the fluid, V is the free-fall terminal velocity, γ_s is the specific weight of the soil particle, and C_D is the drag coefficient. For low Reynold's Number, less than about 5, the drag coefficient is approximately

$$C_D = \frac{24}{\mathbf{R}} = \frac{24\,\mu}{\rho\, VD} \tag{15–2}$$

Substituting Equations 15–1b, 15–1c, 15–1d, and 15–2 into Equation 15–1a yields

$$0 = \gamma\left(\frac{\pi\, D^3}{6}\right) + 3\,\pi\,\mu\, VD - \gamma_s\left(\pi\,\frac{D^3}{6}\right)$$
$$= \gamma\, D^2 + 18\,\mu\, V - \gamma_s D^2 = 18\mu V - D^2\,(\gamma_s - \gamma)$$

Solving for the terminal velocity gives Stokes' Law:

$$V = \frac{D^2(\gamma_s - \gamma)}{18\,\mu} = \frac{gD^2(S_s - S)}{18\,\mu} \tag{15–3}$$

where S_s and S are the specific gravities of the soil and the liquid medium, respectively. Stokes' Law has several inherent assumptions: (1) the particle is not influenced by other particles or walls of the containers; (2) the particles are spherical; and (3) the specific gravity of the soil particle and the viscosity of the fluid are known exactly. These assumptions, especially the first two, do not usually hold but Stokes' Law is commonly used in spite of these problems. For materials that are falling in water and that have a density approximately that of soil grains, Stokes' Law applies for spheres that have diameters between 0.0002 mm and 0.2 mm.

Example 15–1

Sediment discharge with a mean diameter of 0.000164 ft (0.05 mm) enters a wetland during a storm. The average depth of the wetland is 2.5 ft. Stokes' Law is used to find the settling velocity:

$$V = \frac{D^2(\gamma_s - \gamma)}{18\,\mu} = \frac{(0.000164\text{ ft})^2\,(165 - 62.4)\text{ lb/ft}^3}{18(2.5 \times 10^{-5}\text{ lb-sec/ft}^2)}$$
$$= 0.00613\text{ ft/sec}$$

Thus, the settling time is

$$t = \frac{L}{V} = \frac{2.5\text{ ft}}{0.00613\text{ ft/sec}} = 408\text{ sec} = 6.8\text{ min}$$

Example 15–2

The wash load, which consists of silt and clay particles, settles very slowly in a river. Consider a silt particle with a diameter of 0.3×10^{-4} ft. It has a settling velocity of

$$V_s = \frac{D^2(\gamma_s - \gamma)}{18\mu} = \frac{(0.3 \times 10^{-4})\,(165 - 62.4)}{18\,(2.5 \times 10^{-5})} = 0.0002 \text{ ft/sec}$$

For it to settle 1 ft requires 5,000 sec. In this time, it could travel 20,000 ft downstream in flow with a velocity of 4 ft/sec. Taking resuspension into consideration, silt and clay particles are not likely to settle to the river bed.

15.3 CHANNEL STABILITY

The erosion process can be controlled through proper maintenance and improvement of stable channels. Lane (1937) defines the term *stable channel* as an unlined earth channel in which the banks and bed are not scarred objectionably by the moving water and in which objectionable deposits of sediment do not occur. Others define a stable channel as one in which the average values of discharge, width, depth, slope, and meander pattern do not vary significantly over some time interval. Griffiths (1983) states that a channel may be considered stable if for any given steady flow, total sediment concentration maintains a balanced average in space and time. Thus, according to these definitions of a stable channel, the gross dimensions such as slope, width, depth, bed elevation, and bed-material size distribution remain essentially constant over extended periods of time; however, both local shifts in the alignment of the channel and local aggravation or degradation are common.

15.3.1 Methods of Channel Stability Analysis

The analysis of earth channels requires knowledge of the relationships between the channel boundaries and the water flowing within them. In general, the problem of stable channel design involves a study of the relationships between six major variables: the water discharge, a typical diameter of the bed material, the total sediment concentration, the energy slope, the wetted perimeter, and the hydraulic radius. In wide channels, the wetted perimeter can be approximated by the base width and the hydraulic radius by the depth of flow.

The methods used to evaluate channel stability depend on the classification of the channel boundaries. Once the channel is classified, a proper stability evaluation can be conducted and the magnitude of channel instability calculated to determine whether or not stabilizing measures are needed. A distinction is made between channels with essentially rigid boundaries and those with mobile boundaries. Channels that have rigid boundaries are stable when the interaction between the flow and the material that forms the channel boundary causes no erosion. When the boundaries of a channel are unable to resist the forces caused by the flow under consideration, the channel is considered to have mobile boundaries. Channels that have mobile boundaries attain stability when the rate at which sediment enters the channel is approximately equal to the capacity of the channel to transport sediment of a similar composition.

Specific methods of stability analysis are generally associated with each type of channel. For instance, channels that are classified as having essentially rigid boundaries generally require a theoretical approach to stable channel design such as the tractive stress or allowable velocity method. Both of these design methods are based on the design of a channel in which no appreciable sediment movement takes place. Channels classified as having mobile boundaries most often use a more empirical approach to stable channel design such as the regime approach or the sediment transport approach. Both the regime approach and the sediment transport approach allow for sediment transport, but in such a way that the total erosion and deposition at any point in the channel is equal to zero. Although attempts have been made to guide the engineer in selecting the appropriate stability analysis method, set rules are not available for choosing the correct procedure and often decisions are based mainly on the availability of data and the desired accuracy.

According to some researchers, there are two basic approaches to design a stable channel are:

1. If the flow entering the channel does not carry an appreciable amount of sediment, the method of maximum permissible velocity or the method of critical tractive force can be used. Both methods are based on the design of a channel in which no appreciable sediment movement takes place.
2. If the flow entering the channel carries a considerable amount of sediment, the regime method should be employed. In this method, the channel is designed so that sediment transport does take place but in such a way that the total process of erosion and deposition at any point along its length is equal to zero.

However, these limitations on the use of the alternative categories are not restrictive in that all methods can be applied for both clear and sediment laden flows when the appropriate adjustments are made.

15.3.2 Permissible Velocity Method

The method of permissible velocity is one of the oldest methods of stable channel design in alluvial material. In this method, the channel is designed for the greatest mean velocity that will not cause erosion of the channel cross section. Values of permissible velocity for various channels are obtained by collecting data from existing stable channels and developing empirical equations that relate the permissible velocity to both flow parameters and properties of the soil and channel cross section.

The permissible velocities listed in Table 15–1 are intended for well-seasoned channels of small slopes and for depths of flow of less than 3 ft. They were determined from existing channels of various soils. The channel material plays an important role in determining the permissible velocity, and most investigators who have determined empirical values have related it to the soil texture. In using values from Table 15–1 to design a stable channel, three points should be noted:

1. A velocity increment of 0.5 ft/sec may be added to the values when the depth of water is greater than 3 ft.
2. A velocity increment of 0.5 ft/sec should be subtracted when the water contains very coarse suspended sediments.
3. For high and infrequent discharges of short duration, up to 30% may be added to the velocities shown in Table 15–1.

TABLE 15–1 Maximum Permissible Velocities and the Corresponding Unit-Tractive-Force Values Converted by the U.S. Bureau of Reclamation (for Straight Channels of Small Slope, after Aging)

		Clear Water		Water Transporting Colloidal Silts	
Material	n	V (ft/sec)	τ_0 (lb/ft^2)	V (ft/sec)	τ_0 (lb/ft^2)
Fine sand, colloidal	0.020	1.50	0.027	2.50	0.075
Sandy loam, noncolloidal	0.020	1.75	0.037	2.50	0.075
Silt loam, noncolloidal	0.020	2.00	0.048	3.00	0.11
Alluvial silts, noncolloidal	0.020	2.00	0.048	3.50	0.15
Ordinary firm loam	0.020	2.50	0.075	3.50	0.15
Volcanic ash	0.020	2.50	0.075	3.50	0.15
Stiff clay, very colloidal	0.025	3.75	0.26	5.00	0.46
Alluvial silts, colloidal	0.025	3.75	0.26	5.00	0.46
Shales and hardpans	0.025	6.00	0.67	6.00	0.67
Fine gravel	0.020	2.50	0.075	5.00	0.32
Graded loam to cobbles when noncolloidal	0.030	3.75	0.38	5.00	0.66
Graded silts to cobbles when noncolloidal	0.030	4.00	0.43	5.50	0.80
Coarse gravel, noncolloidal	0.025	4.00	0.30	6.00	0.67
Cobbles and shingles	0.035	5.00	0.91	5.50	1.10

Source: The Fortier and Scobey values were recommended for use in 1926 by the Special Committee on Irrigation Research of the American Society of Civil Engineers.

It is known that deeper channels can convey water at a higher mean velocity without erosion than shallower channels when all other conditions are equal. This is attributed to the fact that scouring is caused by bottom velocities and they are greater for a shallower channel. Therefore, the addition of 0.5 ft/sec to the permissible velocity when the depth is greater than 3 ft is justified. Also, the velocities listed are for relatively straight channels. For sinuous channels, the permissible velocities should be lowered in order to reduce scour. Lane (1955) suggests percentages of reduction of 5% for slightly sinuous canals, 13% for moderately sinuous canals, and 22% for very sinuous canals. These values are very approximate since no accurate data are available. The steps involved in the design of a stable channel using the permissible velocity method are summarized as follows:

1. For a given type of channel material, estimate the roughness coefficient n, slope S, and obtain the maximum permissible velocity from a table such as Table 15–1.
2. Compute the hydraulic radius R_h from Manning's Equation.
3. Use the continuity equation to compute the water area required $A = Q/V$ for the given discharge and permissible velocity.
4. Compute the wetted perimeter, $P = A/R_h$.
5. Solve for the channel base (W) and depth (H) according to specific channel geometry.
6. Modify the section for practicality.

The velocities designated as the maximum permissible velocities have been developed as much by experience and judgment as by analytical analysis. They are very uncertain and subject to wide variation.

Example 15–3

An engineer is designing an irrigation canal that will have a roughness of 0.06, a permissible velocity of 2.5 ft/sec, a slope of 0.8%, and a design discharge of 40 ft^3/sec. Manning's Equation is used to compute the hydraulic radius (R_h):

$$R_h = \left[\frac{Vn}{1.49\ S^{0.5}}\right]^{1.5} = \left[\frac{2.5\ (0.06)}{1.49\ (0.008)^{0.5}}\right]^{1.5} = 1.2\ \text{ft}$$

The continuity equation is used to compute the cross-sectional area (A):

$$A = \frac{Q}{V} = \frac{40}{2.5} = 16\ \text{ft}^2$$

Therefore, the wetted perimeter (P) is

$$P = \frac{A}{R_h} = \frac{16}{1.2} = 13.33\ \text{ft}$$

The design of the channel requires an assumption of a shape of the cross section; a trapezoidal section is assumed. The bottom width (W) must be determined, which will dictate the depth of flow (H). For the trapezoidal section with side slopes z (that is, $zH{:}V$), the area and wetted perimeter are

$$A = WH + zH^2 \tag{15–4a}$$

$$P = W + 2H\sqrt{1 + z^2} \tag{15–4b}$$

Thus, Equations 15–4 are two equations with two unknowns (W and H). Solving these two equations for the depth of flow yields

$$H = \frac{-P \pm [P^2 + 4A(z - 2\sqrt{1 + z^2})]^{0.5}}{2[z - 2(1 + z)^{0.5}]} \tag{15–4c}$$

and thus, the width is

$$W = \frac{A}{H} - zH \tag{15–4d}$$

For a feasible solution, the square root in the numerator of Equation 15–4c must be at least 0, which will require a limiting value of z. As an approximation, z should be limited by

$$z < \frac{P^2}{4A} \tag{15–4e}$$

$P^2/(4A)$ is actually larger than the allowable z.

For an area of 16 ft^2 and a wetted perimeter of 13.33 ft, the limit of Equation 15–4e requires a side slope of 2.78 to 1. Actually, the upper limit is 2.3 to 1, so for the given design, z will be set equal to 2. Equation 15–4c yields two possible flow depths, 3.59 and 1.80 ft. When these are substituted into Equation 15–4d, bottom widths of –2.73 and 5.28 ft result. Thus, the practical solution is $z = 2$, $H = 1.8$ ft, and $W = 5.28$ ft.

15.3.3 Regime Theory Method

The regime theory of stable channel design was initiated by British engineers working on the irrigation canals of India and Pakistan at the beginning of the twentieth century. Kennedy, Lacey, Lindley, Blench, and others studied channels that were in an equilibrium state and yielded a set of equations to determine the dimensions of stable or regime channels. The word "regime" was chosen by researchers to replace the word "equilibrium." The basic premise behind the theory is best explained by Lindley (1919) in his definition of the "regime" concept of channels as follows: "When an artificial channel is used to carry silty water, both the bed and banks scour or fill, changing depth, gradient and width, until a state of balance is attained, at which the channel is said to be 'in regime.'" In other words, a channel formed in an erodible material will only reach a state of equilibrium when a balance exists between discharge, cross section, slope, and sediment load. Once this equilibrium state is reached, a channel is said to be in regime.

Kennedy (1895) was the first to develop an empirical relationship linking the nonsilting velocity to the depth of the channel. By collecting data from twenty-two channels of the Upper Bari Doab canal system, he found that the mean velocity in these nonsilting channels was related to the depth by the equation

$$V_c = 0.55\, h^{0.64} \tag{15–5a}$$

in which V_c is the critical velocity at which, for a given depth, h, silting is prevented. He also proposed a similar equation for sediment sizes other than the one found in the Upper Bari Doab system:

$$V_c = 0.55\, mh^{0.64} \tag{15–5b}$$

where m, the critical velocity ratio, is the ratio of the critical velocity of the channel under observation to the critical velocity of a channel that has the same size sediment as that of the Upper Bari Doab system. For sand that is coarser than the one in the Upper Bari Doab system, m is greater than 1 and for finer sand, m is less than 1. From Equation 15–5b, it is apparent that Kennedy recognized that sediment size plays an important role in determining the relationship between velocity and depth. Equations 15–5 can be generalized to the following:

$$V_c = k_5 h^{k_6} \tag{15–5c}$$

in which k_5 is a function of the soil type. The coefficients of Equations 15–5a and 15–5b may not be accurate for general use. Site-specific data should be used to fit the power model of Equation 15–5c.

Lacey (1930) tried to bring some order to the abundant data on stable channel design that were available at the time. He reanalyzed existing data and found that all of the data could be accurately represented by the following equations, where the critical velocity is related to the hydraulic radius instead of the depth:

$$V_c = 1.17\, (f_L R_h)^{0.5} \tag{15–6a}$$

in which f_L is the silt factor and R_h is the hydraulic radius (ft). The remaining two equations required to determine the variables for a stable channel are

$$qf_L^2 = 3.8\,(V_c)^6 \tag{15–6b}$$

$$S = \frac{f_L^{5/3}}{1788\,q^{1/6}} \tag{15–6c}$$

Lacey's relationship for determining the silt factor is

$$f_L = 8d^{0.5} \tag{15–6d}$$

where d is the diameter (in.) of the predominant type of soil transported. Provided that the discharge and soil type are known, a stable channel can be designed with Equations 15–6. It was later argued by King (1943; see Graf, 1971) and Blench (1951) that the single expression for the silt factor, as determined by Lacey, did not take into account the relative importance of characteristics of the channel bed and sides, which could be different for different channels.

As a further step in the regime theory, Simons and Albertson (1963) reanalyzed the regime channel concept. They provided a modified version applicable to a wider variety of channels since the range of conditions covered by their data was much wider than previous data. In the latter investigation, five different types of geotechnical channel conditions are specified. A separate equation was fitted to each type by Henderson (1966).

To design a stable channel according to this method the geotechnical conditions and discharge q must be known. From these values, the average width (W), the average depth (h), and the limiting slope can be determined. Using Henderson's (1966) values of K_1, K_2, and K_4 (see Table 15–2), which are specified according to channel type, the average width is given by

$$W = 0.9\,K_1 q^{0.5} \tag{15–7}$$

The mean depth is given by

$$h = \begin{cases} 1.21\,K_2 q^{0.36} & \text{for } R_h \leq 7 \text{ ft} \quad (15\text{–}8a) \\ 2.0 + 0.93\,K_2 q^{0.36} & \text{for } R_h > 7 \text{ ft} \quad (15\text{–}8b) \end{cases}$$

The slope can be estimated by

$$S = \frac{V_c^{1.63}\nu^{0.37}}{ghK_4 W^{0.37}} \tag{15–9}$$

These equations are considered the most useful form of the regime method of stable channel design. When S, W, and h are known, Equations 15–7 to 15–9 can be solved for V_c:

TABLE 15–2 Coefficients for Equations 15–7 to 15–9

Types of Material	K_1	K_2	K_4
Sand bed and banks	3.5	0.52	0.33
Sand bed and cohesive banks	2.6	0.44	0.54
Cohesive bed and banks	2.2	0.37	0.87
Coarse noncohesive material	1.75	0.23	—
Sand bed and cohesive banks with heavy sediment load	1.70	0.34	—

Source: Henderson, 1966.

$$V_c = \begin{cases} 126\, K_1^{0.227} q^{0.3344} (K_2 K_4 S)^{0.6135} & \text{for } R_h \leqslant 7 \text{ ft} \quad (15\text{–}10a) \\ 112\, [K_1^{0.37}\, K_4 S Q^{0.185} (2 + 0.93 K_2 Q^{0.36})]^{0.6135} & \text{for } R_h > 7 \text{ ft} \quad (15\text{–}10b) \end{cases}$$

The equations by Simons et al. (1963) are specified for a bed load of less than 500 ppm by weight. The regime concept is not well suited for high bed loads (greater than 500 ppm).

In summary, Equations 15–7, 15–8, and 15–9 can be used to design a stable channel for live-bed conditions. Their use requires the design discharge and the critical velocity for the bed material. In the situation where a channel exists and the slope and discharge are known, Equation 15–10 can be used to compute the critical velocity.

Example 15–4

A channel that will pass waste water from a new power generation plant is being designed. The plant will discharge 1550 ft^3/sec. The bed and banks will be formed from a cohesive material; thus, from Table 15–2, $K_1 = 2.2$, $K_2 = 0.37$, and $K_4 = 0.87$. Equation 15–7 is used to estimate the average width:

$$W = 0.9\,(2.2)\,(1550)^{0.5} = 78 \text{ ft}$$

The hydraulic radius is expected to be less than 7 ft; therefore, Equation 15–8a is used to compute the mean depth:

$$h = 1.21\,(0.37)\,(1550)^{0.36} = 6.3 \text{ ft}$$

If the critical velocity is 2 ft/sec, the maximum permissible slope is obtained from Equation 15–9:

$$S = \frac{2^{1.63}\,(10^{-5})^{0.37}}{32.2\,(6.3)\,(0.87)\,(78)^{0.37}} = 0.00005 \text{ ft/ft}$$

The extremely shallow slope is necessary to maintain a velocity that is less than the critical velocity of 2 ft/sec.

15.4 SPLASH EROSION

It may be surprising to learn that the impact of raindrops is a primary cause of surface erosion. For a particle of soil to be eroded by raindrop impact, the kinetic energy of the raindrop (that is, $K_e = 0.5\, mV^2$) must overcome forces of adhesion and cohesion. The kinetic energy of a raindrop may be as much as 1000 times the kinetic energy of water in sheet flow. The following equation can be used to estimate the terminal velocity (V_t, ft/sec) using the mean raindrop diameter (d_r, ft):

$$V_t = 120\, d_r^{0.35} \qquad (15\text{–}11)$$

15.4.1 Empirical Estimation of Kinetic Energy

While the foregoing approach to estimating the kinetic energy of raindrop impact forms a rational conceptualization of splash erosion, in practice empirical equations provide the basis for estimation. Wischmeier and Smith (1965) provided the following equation for estimating the kinetic energy:

$$K_e = 916 + 330 \log_{10} i \tag{15–12}$$

in which i is the rainfall intensity (in./hr) and K_e is the average kinetic energy of the raindrop (ft-tons per acre-in. of rain). It is important to recognize that equations such as Equation 15–12 represent an intermediate step in the calculation of erosion since it is still necessary to convert the kinetic energy or the terminal velocity into a quantity of eroded soil. However, equations such as Equations 15–11 and 15–12 are useful for identifying variables that influence erosion rates associated with raindrop impact.

15.4.2 Ellison Equation

In 1945, Ellison provided the following equation:

$$G = b_o V_t^{4.33} d_r^{1.07} i^{0.65} \tag{15–13}$$

in which G is the mass (grams) of the soil intercepted in splash samples during a 30-min period, V_t is the velocity of the raindrop (ft/sec), d_r is the drop diameter (mm), i is the rainfall intensity (in./hr), and b_o is a constant. G is only a measure of the mass of the soil detached and not the yield due to the kinetic energy of the raindrops. Input to Ellison's equation includes V_t, which is one element of the kinetic energy equation $K_e = 0.5mV^2$, while the raindrop diameter d_r would be an indicator of the mass m. The intensity is probably highly correlated with the terminal velocity V_t, since both have units of length per time. The coefficients of Equation 15–13 were evaluated using data for bare soil and thus are valid only for that soil type and conditions.

15.5 SHEET EROSION

Sheet erosion is defined as erosion that results from a combination of splash action forces and forces associated with thin-layered surface runoff. A number of methods have been proposed to estimate sheet erosion rates, several of which will be briefly described here. The methods are variations of the same concepts.

15.5.1 Universal Soil Loss Equation

The universal soil loss equation (USLE) is probably the most widely used erosion equation. The soil loss (E) in tons/acre/year is a function of the erosivity index (R), a soil erodibility factor (K), the field slope (S) and length (L), a crop management factor (C), and a conservation practice factor (P):

$$E = RKTCP \tag{15–14}$$

in which T is the topographic factor, which depends on L and S.

The erosivity index R is a summation of individual storm products of the kinetic energy of rainfall (hundreds of ft-tons/acre) and the maximum 30-min rainfall intensity (in./hr) for all significant storms in a year. The value of R has been computed for locations within the United States and is given in Figure 15–2.

The soil erodibility factor (K) is the average soil loss (tons per acre per unit of the rainfall factor R) from a particular soil in cultivated continuous fallow with standard values of

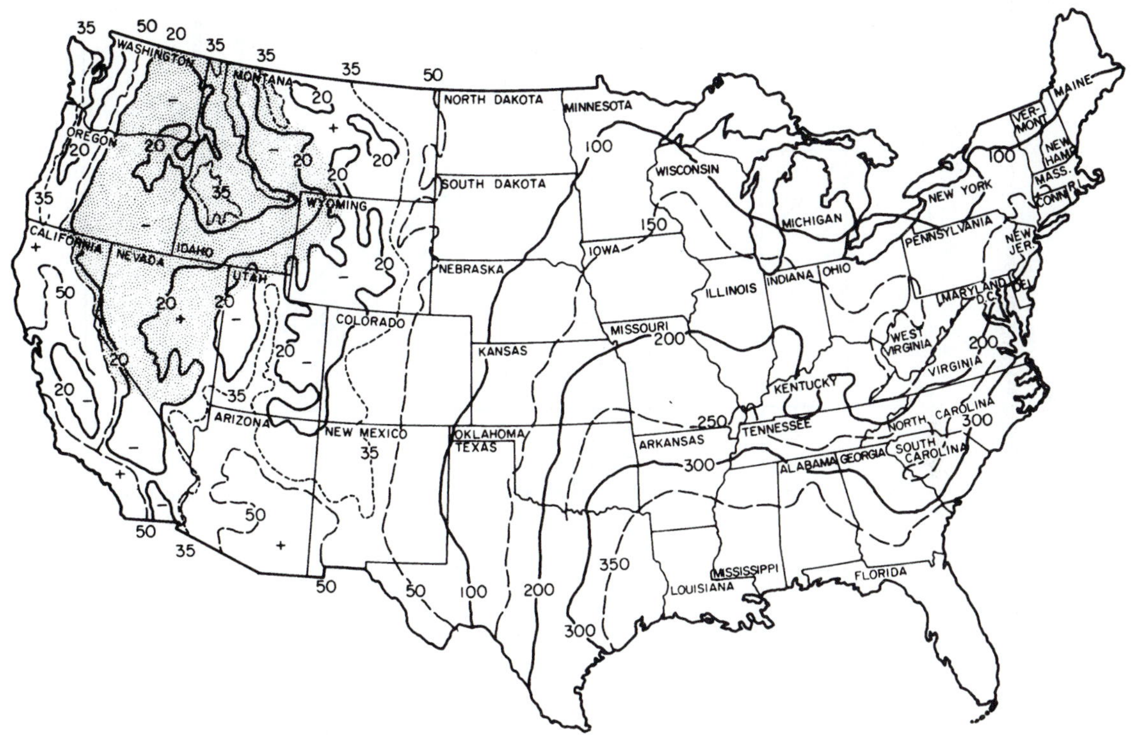

FIGURE 15–2 Average annual values of the rainfall-erosivity factor, *R*.

both the plot length and the slope, which were selected somewhat arbitrarily as 73 ft and 9%, respectively. The value of *K* is a function of (1) the percent silt plus very fine sand, (2) the percent sand, (3) the percentage of organic matter, (4) a soil structure index, which is presented on an ordinal scale, and (5) a permeability index, which is also measured on an ordinal scale. For U.S. mainland soils, the value of *K* is obtained from Figure 15–3.

The topographic factor (*T*) is a function of the slope (%) and the overland length (ft), which is usually less than 400 ft. The slope is the average land gradient. The length is the average distance from the point of overland flow to whichever of the following limiting conditions occurs first: (1) the point where the slope decreases to the extent that deposition begins, or (2) the point where runoff enters well-defined flow areas, such as rills. Values of the topographic factor are given in Table 15–3. For the index slope of 9% and the index length of 73 ft, *T* equals 1.

The crop management factor (*C*) is the ratio of the soil quantities eroded from land that is cropped under specific conditions to that which is eroded from clean-tilled fallow under identical slope and rainfall conditions. Values for *C* are given in Table 15–4 for the 37 states east of the Rocky Mountains. The baseline conditions for which *C* equals 1 are continuous fallow, tilled up and down slope. For other cover and management conditions, *C* is less than 1.

The conservation practice factor (*P*) is a function of the support practice and the land slope. Values for *P* are given in Table 15–5 and represent a comparison of the land use practice to straight-row farming.

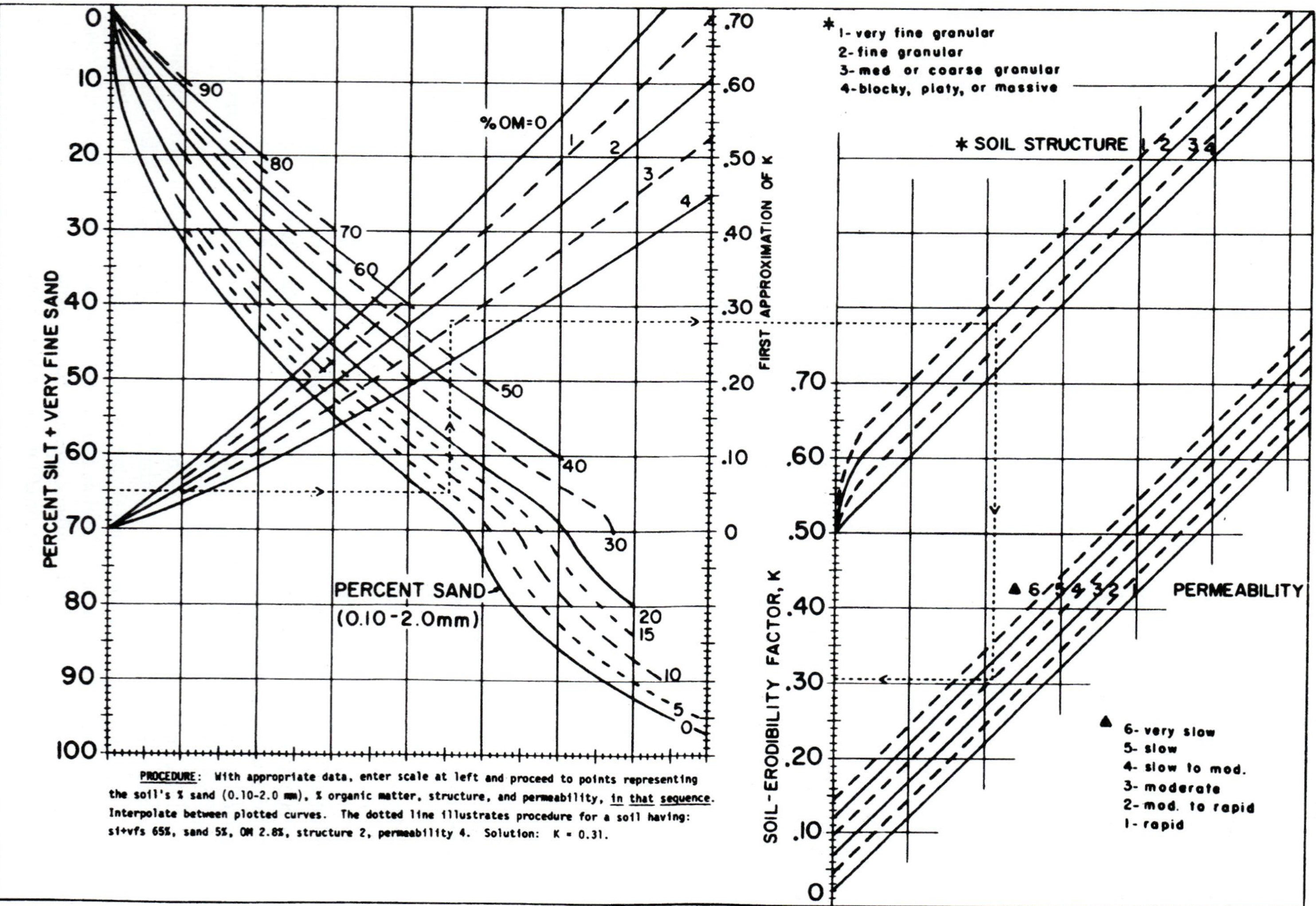

FIGURE 15–3 Nomograph for determining soil-erodibility factor, *K*, for U.S. mainland soils.

TABLE 15–3 Values of the Erosion Equation's Topographic Factor, *LS*, for Specified Combinations of Slope Length and Steepness

Slope (%)	Slope Length (ft)											
	25	50	75	100	150	200	300	400	500	600	800	1000
0.5	0.07	0.08	0.09	0.10	0.11	0.12	0.14	0.15	0.16	0.17	0.19	0.20
1	0.09	0.10	0.12	0.13	0.15	0.16	0.18	0.20	0.21	0.22	0.24	0.26
2	0.13	0.16	0.19	0.20	0.23	0.25	0.28	0.31	0.33	0.34	0.38	0.40
3	0.19	0.23	0.26	0.29	0.33	0.35	0.40	0.44	0.47	0.49	0.54	0.57
4	0.23	0.30	0.36	0.40	0.47	0.53	0.62	0.70	0.76	0.82	0.92	1.0
5	0.27	0.38	0.46	0.54	0.66	0.76	0.93	1.1	1.2	1.3	1.5	1.7
6	0.34	0.48	0.58	0.67	0.82	0.95	1.2	1.4	1.5	1.7	1.9	2.1
8	0.50	0.70	0.86	0.99	1.2	1.4	1.7	2.0	2.2	2.4	2.8	3.1
10	0.69	0.97	1.2	1.4	1.7	1.9	2.4	2.7	3.1	3.4	3.9	4.3
12	0.90	1.3	1.6	1.8	2.2	2.6	3.1	3.6	4.0	4.4	5.1	5.7
14	1.2	1.6	2.0	2.3	2.8	3.3	4.0	4.6	5.1	5.6	6.5	7.3
16	1.4	2.0	2.5	2.8	3.5	4.0	4.9	5.7	6.4	7.0	8.0	9.0
18	1.7	2.4	3.0	3.4	4.2	4.9	6.0	6.9	7.7	8.4	9.7	11.0
20	2.0	2.9	3.5	4.1	5.0	5.8	7.1	8.2	9.1	10.0	12.0	13.0
25	3.0	4.2	5.1	5.9	7.2	8.3	10.0	12.0	13.0	14.0	17.0	19.0
30	4.0	5.6	6.9	8.0	9.7	11.0	14.0	16.0	18.0	20.0	23.0	25.0
40	6.3	9.0	11.0	13.0	16.0	18.0	22.0	25.0	28.0	31.0	—	—
50	8.9	13.0	15.0	18.0	22.0	25.0	31.0	—	—	—	—	—
60	12.0	16.0	20.0	23.0	28.0	—	—	—	—	—	—	—

Values given for slopes longer than 300 ft or steeper than 18% are extrapolations beyond the range of the research data, and therefore less certain than the others. Adjustments for irregularity of slope are available.

Predicted values of E represent average, time-invariant estimates. Given that R is based on an average number and distribution of storms per year, actual values of E would vary from year to year depending on the number, size, and timing of erosive rainstorms and other weather conditions. Although any one predicted value of E may not be highly accurate, the USLE should be more reliable when it is used to measure either relative effects or long-term sheet and rill erosion rates.

Example 15–5

The general data requirements for making soil loss estimates with the USLE equation are (1) site location (to get R); (2) soil properties (to get K); (3) flow length and slope (to get T); (4) crop, rotation, and management practices (to get C); and (5) slope and support practice (to get P). The estimation process will be illustrated using a hypothetical example. The site, which has a drainage area of 2 acres, is located in central Illinois. The site of interest has an average slope of 2.5% and a flow length of 300 ft. A soil analysis indicates 25% sand, 2% organic matter, 35% silt and very fine sand, a medium granular structure, and moderate permeability. The plot is used for corn, with a crop management value of 0.31, which was determined from information published locally. No support practice is provided.

For central Illinois, Figure 15–2 suggests a value for R of 200. A value for the topographic factor can be obtained from Table 15–3; for a slope of 2.5% and a length of 300 ft, T equals 0.28. Using the soil characteristics given, a value for the soil erodibility factor of 0.17 is obtained from Figure 15–3. Since there is no support practice, P equals 1.0. Thus the average annual soil loss is

TABLE 15–4 Generalized Values of the Cover and Management Factor, *C*, in the 37 States East of the Rocky Mountains[a]

		Productivity Level[b]	
		High	Mod.
Line no.	Crop, Rotation, and Management[c,d]	C Value	
Base value: continuous fallow, tilled up and down slope		1.00	1.00
CORN			
1	C, RdR, fall TP, conv (1)	0.54	0.62
2	C, RdR, spring TP, conv (1)	0.50	0.59
3	C, RdL, fall TP, conv (1)	0.42	0.52
4	C, RdR, wc seeding, spring TP, conv (1)	0.40	0.49
5	C, RdL, standing, spring TP, conv (1)	0.38	0.48
6	C, fall shred stalks, spring TP, conv (1)	0.35	0.44
7	C(silage)-W(RdL, fall TP) (2)	0.31	0.35
8	C, RdL, fall chisel, spring disk, 40–30% rc(1)	0.24	0.30
9	C(silage), W wc seeding, no-till pl in c-k W(1)	0.20	0.24
10	C(RdL)-W(RdL, spring, TP) (2)	0.20	0.28
11	C, fall shred stalks, chisel pl, 40–30% rc (1)	0.19	0.26
12	C-C-C-W-M, RdL, TP for C, disk for W (5)	0.17	0.23
13	C, RdL, strip till row zones, 55–40% rc (1)	0.16	0.24
14	C-C-C-W-M-M, RdL, TP for C, disk for W (6)	0.14	0.20
15	C-C-W-M, RdL, TP for C, disk for W (4)	0.12	0.17
16	C, fall shred, no-till pl, 70–50% rc (1)	0.11	0.18
17	C-C-W-M-M, RdL, TP for C, disk for W (5)	0.087	0.14
18	C-C-C-W-M, RdL, no-till pl 2d & 3rd C (5)	0.076	0.13
19	C-C-W-M, RdL, no-till pl 2d C (4)	0.068	0.11
20	C, no-till pl in c-k wheat, 90–70% rc (1)	0.062	0.14
21	C-C-C-W-M-M, no till pl 2d & 3rd C (6)	0.061	0.11
22	C-W-M, RdL, TP for C, disk for W (3)	0.055	0.095
23	C-C-W-M-M, RdL, no-till pl 2d C (5)	0.051	0.094
24	C-W-M-M, RdL, TP for C, disk for W (4)	0.039	0.074
25	C-W-M-M-M, RdL, TP for C, disk for W (5)	0.032	0.061
26	C, no-till pl in c-k sod, 95–80% rc (1)	0.017	0.053
COTTON[e]			
27	Cot, conv (Western Plains) (1)	0.42	0.49
28	Cot, conv (South) (1)	0.34	0.40
MEADOW			
29	Grass and legume mix	0.004	0.01
30	Alfalfa, lespedeza, or Sericia	0.020	
31	Sweet clover	0.025	
SORGHUM, GRAIN (Western Plains)[e]			
32	RdL, spring TP, conv (1)	0.43	0.53
33	No-till pl in shredded 70–50% rc	0.11	0.18
SOYBEANS[e]			
34	B, RdL, spring TP, conv (1)	0.48	0.54
35	C-B, TP annually, conv (2)	0.43	0.51
36	B, no-till pl	0.22	0.28
37	C-B, no-till pl, fall shred C stalks (2)	0.18	0.22

		Productivity Level[b]	
		High	Mod.
Line no.	Crop, Rotation, and Management[c,d]	C Value	
Base value: continuous fallow, tilled up and down slope		1.00	1.00
WHEAT			
38	W-F, fall TP after W (2)	0.38	
39	W-F, stubble mulch, 500 lbs rc (2)	0.32	
40	W-F, stubble mulch, 1000 lbs rc (2)	0.21	
41	Spring W, RdL, Sept TP, conv (N. Dak. and S. Dak.) (1)	0.23	
42	Winter W, RdL, Aug TP, conv (Kans). (1)	0.19	
43	Spring W. stubble mulch, 750 lb rc (1)	0.15	
44	Spring W. stubble mulch, 1250 lb rc (1)	0.12	
45	Winter W, stubble mulch, 750 lb rc (1)	0.11	
46	Winter W, stubble mulch, 1250 lb rc (1)	0.10	
47	W-M, conv (2)	0.054	
48	W-M-M, conv (3)	0.026	
49	W-M-M-M, conv (4)	0.021	

[a]This table is for illustrative purposes only and is not a complete list of cropping systems or potential practices. Values of C differ with rainfall pattern and planting dates. These generalized values show approximately the relative erosion-reducing effectiveness of various crop systems, but locationally derived C values should be used for conservation planning at the field level. Tables of local values are available from the Natural Resources Conservation Service.

[b]High level is exemplified by long-term yield averages greater than 75 bu of corn or 3 tons of grass-and-legume hay; or cotton management that regularly provides good stands and growth.

[c]Numbers in parentheses indicate number of years in the rotation cycle. (1) designates a continuous one-crop system.

[d]Abbreviations

B: soybeans — F: fallow
C: corn — M: grass and legume hay
c-k: chemically killed — pl: plant
conv: conventional — W: wheat
cot: cotton — wc: winter cover
lb rc: pounds of crop residue per acre remaining on surface after new crop seeding
% rc: percentage of soil surface covered by residue mulch after new crop seeding
70–50% rc: 70% cover for C values in first column; 50% for second column
RdR: residues (corn stover, straw, etc.) removed or burned
RdL: all residues left on field (on surface or incorporated)
TP: turn plowed (upper 5 or more inches of soil inverted, covering residues)

[e]Grain sorghum, soybeans, or cotton may be substituted for corn in lines 12, 14, 15, 17–19, 21–26 to estimate C values for sod-based rotations.

$$E = RKTCP = 200\ (0.17)\ (0.28)\ (0.31)\ (1.0) = 2.95 \text{ tons/acre/yr}$$

For the 2-acre plot, the gross erosion would be 5.9 tons/yr.

Example 15–6

The USLE can be used to assess the effect of the support practice on erosion amounts. Using the information and results from Example 15–5, the effect of contouring can be examined. For a

TABLE 15–5 Values of Support-Practice Factor, P

	Land Slope (%)				
	1.1–2	2.1–7	7.1–12	12.1–18	18.1–24
Practice	Factor P				
Contouring, P_e	0.60	0.50	0.60	0.80	0.90
Contour strip cropping,[a] P_{sc}					
R-R-M-M	0.30	0.25	0.30	0.40	0.45
R-W-M-M	0.30	0.25	0.30	0.40	0.45
R-R-W-M	0.45	0.38	0.45	0.60	0.68
R-W	0.52	0.44	0.52	0.70	0.90
R-O	0.60	0.50	0.60	0.80	0.90
Contour listing or ridge planting, P_{cl}	0.30	0.25	0.30	0.40	0.45
Contour terracing,[b,c] P_t	$0.6/\sqrt{n}$	$0.5/\sqrt{n}$	$0.6/\sqrt{n}$	$0.8/\sqrt{n}$	$0.9/\sqrt{n}$
No support practice	1.0	1.0	1.0	1.0	1.0

[a]R, rowcrop; W, fall-seeded grain; 0, spring-seeded grain; M, meadow. The crops are grown in rotation and so arranged on the field that rowcrop strips are always separated by a meadow or winter-grain strip.

[b]These P_t values estimate the amount of soil eroded to the terrace channels and are used for conservation planning. For prediction of off-field sediment, the P_t values are multiplied by 0.2.

[c]n, number of approximately equal-length intervals into which the field slope is divided by the terraces. Tillage operations must be parallel to the terraces.

slope of 2.5%, Table 15–5 indicates a support-practice factor of 0.5 when the field is contoured. This would reduce the unit erosion to 1.5 tons/acre/yr, or 2.95 tons/yr for the 2-acre plot. Thus contouring reduced gross erosion by 50%.

15.5.2 Other Empirical Equations

Other equations exist for estimating volumes or rates of soil loss, with many of them being similar to the USLE. The Musgrave Equation is

$$E = F\left(\frac{R}{100}\right)\left(\frac{S}{10}\right)^{1.35}\left(\frac{L}{72.6}\right)^{0.35}\left(\frac{P_{30}}{1.25}\right)^{1.75} \tag{15–15}$$

in which E is the probable soil loss (tons/ac/yr), F is a soil erodibility factor, R is a cover factor, S is the slope (%), L is the length (ft), and P_{30} is the 2-yr, 30-min rainfall intensity (in./30 min). Several differences between Musgrave's Equation and the USLE are obvious. First, the Musgrave Equation is nonlinear. The exponents reflect the data that were used to calibrate Equation 15–15. Data from nineteen research stations, with 5 to 15 years of data collected at each station, were used in calibration. Second, the Musgrave Equation uses the slope and length directly rather than as an input to the topographic factor of the USLE.

Beer, Farnham, and Heinemann (1966) provided the following equation, which is based on an empirical analysis of data from western Iowa:

$$E = 0.392 \times 10^{-6}\, KPRR_cS^{1.35}\, L^{0.35} \tag{15–16}$$

in which E is the average annual soil loss (in./year); K, R, and P are the soil erodibility, rainfall, and factors of conservation practice of the USLE; R_c is a cover factor with fallow or continuous row cropping having a value of 100; and S and L are the slope (%) and length (ft) of the plot, respectively.

15.5.3 Comparison of Methods

It is anticipated that the USLE will be the most widely used method for estimating soil losses. The other equations are given only to indicate that other methods exist and that some empirical evidence questions the linear structure of the USLE. Some data suggest that a nonlinear form may be more appropriate. It is almost impossible to provide a quantitative assessment of the accuracy of the methods. For estimating soil loss in any one year, all of these methods would probably be highly inaccurate; however, for making long-term estimates of mean annual soil loss greater confidence could be placed in the estimates. One of the difficulties in making quantitative estimates of the accuracy is a lack of data. Additionally, at any one site, many factors (that is, the effect of freeze-thaw cycles) contribute to rates of soil loss, and thus the failure to evaluate these factors would contribute to inaccuracies.

15.6 EROSION IN GULLIES

Gullies are created by concentrations of flow such as sheet runoff and flow from rills. Areas that are more prone to gully erosion are those with steep topography and thick soil mantles. Once a gully forms, erosion usually proceeds upstream, with erosive action being caused by both inflow from sheet and rill runoff and backwater effects during the larger storm events. Ground water sapping and mass wasting are important processes in gully advancing, often more important than the surficial runoff processes. Piping of soils that underlie the gully headwall and banks can also be significant. Although the total volume of soil loss from gully erosion is usually less than that from sheet and rill erosion, damages associated with gully erosion are usually more significant and more noticeable. Gully formation and propagation reduce the area available for uses such as farming and recreation and create hazards to livestock and people. Gullied areas are also less aesthetically pleasing.

15.6.1 Empirical Formulas for Estimating Gully Erosion

A number of equations have been calibrated for estimating gully erosion. The following was developed using data from western Iowa for severely gullied loessial areas:

$$Y = 0.01\, X_1^{0.982}\, X_2^{-0.044}\, X_3^{0.7954}\, X_4^{-0.2473}\, e^{-0.036\, X_5} \tag{15–17}$$

in which Y is the growth in the gully surface area (acres) for a given time period, X_1 is an index of surface runoff (in.), X_2 is the terraced area of the watershed (acres), X_3 is the gully length at the beginning of the period (ft), X_4 is the length from the end of the gully to the watershed divide (ft), and X_5 is the deviation of the precipitation (in.) from normal during the time period.

Another equation, which is based on data from a number of locations throughout the United States, is

$$E_q = 0.15\, A^{0.49}\, S^{0.14}\, P_a^{0.75}\, E_c \tag{15–18}$$

in which E_g is the average annual gully head advance (ft), A is the drainage area (acres), S is the slope of approach at the channel (%), P_a is the annual summation of rainfall (in.) from rains equal to or greater than 0.5 in. in 24 hr, and E_c is the clay content of the eroding soil profile (percent by weight).

A third equation is

$$E_q = 1.5\, A^{0.46}\, P^{0.2} \tag{15–19}$$

E_g and A are the same as in the preceding equation, and P is the summation of 24-hr rainfall totals of 0.5 in. or more occurring during the time period, converted to an annual average basis (in.).

Example 15–7

A 54-acre watershed, which is used for agricultural crops, has a length of 2350 ft and a slope of 1.1%. On the average, the annual summation of rainfall on days when the rainfall volume exceeds 0.5 in. is 8.7 in. Soil tests indicate an average clay content of 24%.

Equation 15–18 can be used to estimate the average annual gully advance:

$$E_g = 0.15\,(54)^{0.49}\,(1.1)^{0.14}\,(8.7)^{0.74}\,(24) = 128 \text{ ft} \tag{15–20}$$

If the rainfall record that was used to estimate P_a was short and a long-term weather station in the region suggests that P_a could be as high as 14 in., the expected gully advance would be 182 ft, which represents an increase of 42%. The sensitivity of erosion estimates, such as the 42% error in E_g that results from potential error of 5.3 in. in P_a, can be important in determining the effort to be expended on corrective actions such as stabilizing areas adjacent to the gullies.

15.6.2 Comparison of Equations

Three equations have been presented for estimating gully erosion. While the individual equations could be used to make estimates of gully erosion, it is more important here to examine the variables that have shown to be of value in estimating gully erosion. Five factors underlie the variables used: land use, watershed size, gully size, soil type, and the momentum of the fluid. None of the equations have variables representing each factor. This is probably the result of the data bases. For example, Equation 15–17 is based on data for loessial areas, and thus variation due to soil type could not be assessed because the data was for one soil type only.

All of the models are linear multiplicative or power models. Thus, if any of the variables have zero values, the estimated gully erosion (Y or E_g) will equal zero. This may not always be true. For example, in Equation 15–17, if the terraced area of the watershed (X_2) is zero, the estimated value of the growth of gullied area (Y) is zero, which may or may not be true.

In summary, three equations have been presented for estimating gully erosion. They suggest that certain factors are important. However, it is not possible to assess the accuracy of the models for general use, and the models should be tested or recalibrated before they are used for prediction.

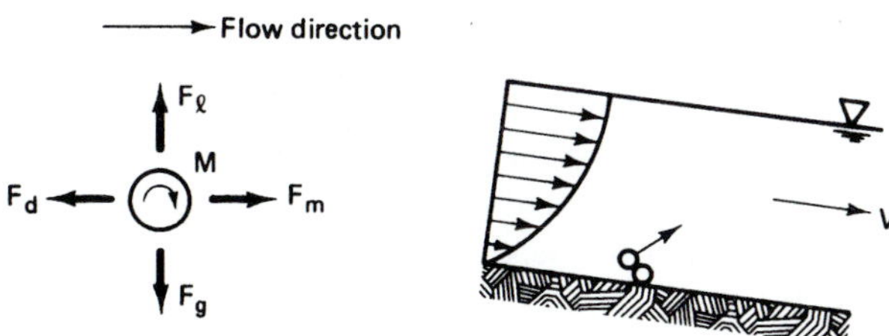

(a) Free-body diagram

(b) Lift force and rotation motion due to velocity profile

FIGURE 15–4 Forces acting on soil particles.

15.7 SUSPENDED SEDIMENT TRANSPORT

As indicated previously, channel erosion is a significant engineering problem. Soil particles that are not part of the wash load but are moving with the water discharge can be classified as being the bed-material load. Methods for estimating these components of the total sediment load in the channel will be treated separately.

While the methods used to estimate splash, sheet, rill, and gully erosion are largely empirical prediction equations, methods for predicting suspended sediment transport have a more theoretical basis. Therefore, it is best to begin our discussion with an analysis of the forces acting on soil particles. Figure 15–4a shows a free-body diagram of the forces acting on a suspended particle. The momentum of the water in motion acts to move the particle in the direction of flow; this force is denoted as F_m. The particle itself is subjected to friction and pressure drag. The total drag force (F_d) is a function of the projected cross-sectional area (A_c), the flow velocity (V), the density (ρ), and a drag coefficient (C_d):

$$F_d = C_d A_c \rho \frac{V^2}{2} \tag{15–21}$$

A suspended particle is also subject to the force of gravity (F_g) and a lift force (F_l). Shear and force imbalance cause a rotational motion (M). The lift force is applied when the particle leaves the bed and contributes to the rotational motion (see Figure 15–4b). The velocity profile also affects the rotational motion.

15.7.1 The Von Kármán Equation

Using the concept of continuity of mass, the total sediment transport rate, q_s, can be computed for a unit width of a channel by integrating the velocity, V, and concentration, C_o, over a vertical section:

$$q_s = \int C_o V \, dz \tag{15–22}$$

in which the integral is expressed over any vertical section and C_o is the concentration in that section. Using the concepts of shear stress in turbulent flow and the transfer of momentum due to turbulent mixing, the following relationship was derived for estimating q_s:

$$q_s = 11.6\, V_*\, C_1\, [I_2 + I_1 \ln_e (30.2\, f_h\, h d_{65}^{-1})] \qquad (15\text{–}23)$$

in which V_* is a shear velocity, C_1 the concentration at level 1, h is the total depth of flow, f_h is a factor reflecting the hydraulic roughness, d_{65} is the mean diameter of soil particles for which 65% of the particles by weight are smaller, and I_1 and I_2 are integrals (see Figures 1 and 2 of Einstein, 1950) that are a function of the depth, the depth zone, the shear velocity, and the settling velocity of the particles.

To determine the total sediment transport rate, the variables of Equation 15–23 must be determined. The shear velocity V_* is a function of the specific weight of the fluid, the hydraulic radius of the channel section, and the channel slope. The concentration at level 1 should be obtained by sampling. However, it would also be possible to assume a relationship between the concentration at any level and the concentration at some reference level; the relationship would be a function of the settling velocity of the particles, the flow level 1, and the sediment (or momentum) transfer coefficient.

15.8 ESTIMATING BEDLOAD TRANSPORT

The estimation of bed transport rates is complicated because of the difficulty in obtaining accurate measurements of bedload quantities for either theoretical or laboratory analyses. Instruments designed to measure bed movement disturb the natural movement of bed material; this introduces a bias into the measurements, and it is almost impossible to determine the magnitude of the bias. Simultaneous measurements often provide estimates that differ by several hundred percent.

Field and laboratory measurements suggest that bed movement is a probabilistic process. Deterministically, a particle will move if the instantaneous hydrodynamic lift force overcomes the weight of the particle. Obviously, the particle will resettle once the lift cannot maintain movement. While this deterministic balance of forces may be valid on a microscale level, it is difficult to translate it into a verifiable prediction equation. The difficulty arises because of an array of factors associated with variations in flow patterns, bed surface characteristics, and particle shape. While these difficulties are recognized, a number of estimation methods have been developed from theoretical and laboratory analyses, as well as empirical equations derived from field measurements.

15.8.1 Einstein's Equation

Based on theoretical considerations, Einstein (1950) developed a method for estimating bed-material transport rates. While the details are provided by Einstein, the total sediment load, suspended and bedload, is given by

$$q_s = f_i q_B \left(1 + I_2 + 2.30\, I_1 \log_{10} \frac{30.2XD}{d_{65}}\right) \qquad (15\text{–}24)$$

in which f_i is the fraction of bed material in size range i, and q_B is the bed load rate in weight per unit time per unit width of channel. Equation 15–24 suggests that the total sediment load is a function of the channel cross-section characteristics, flow characteristics, the particle size distribution of the bed material, and the characteristics of the bed material.

15.8.2 Empirical Methods of Estimating Bedload

Most empirical approaches to estimating bedloads are based on one of the following general relations:

$$q_b = f(\tau - \tau_c) \tag{15–25a}$$
$$q_b = f(q - q_c) \tag{15–25b}$$
$$q_b = f(V - V_c) \tag{15–25c}$$
$$q_b = f(\omega - \omega_c) \tag{15–25d}$$

In these equations, q_b is the bedload discharge per unit width, τ is the tractive force, q is the water discharge per unit width, V the mean flow velocity, and ω is the stream power per unit bed area. The subscript c signifies critical values for incipient motion. The equations presented are grouped according to these relations and then evaluated.

Recognizing the complexity of Einstein's method for estimating bedload, a number of equations have been derived from data analyses. It is worthwhile to examine a few of these equations, but not for purposes of recommending their use. Instead, it is important to evaluate them on the basis of the model structure, required input, and the degree to which they compare with Einstein's theory.

Sediment transport equations, particularly the bedload equations, apply best to alluvial channels that have an abundant volume of bed material that is readily available to the flow. They should be most accurate for channels that have mobile boundaries, but exhibit relatively minor changes in plan form over time intervals equivalent to the design life. They do not generally apply well to channels that cannot supply sediment to the flow, that have abundant contributions of sediment from sources external to the channel, that include reaches with heterogeneous bed and bank materials that vary considerably in erosion resistance, and that have a poorly sorted bed material.

15.8.3 Equations of the Form $q_b = f(V - V_c)$

Goncharov (1964) developed the following equation using laboratory experiments:

$$q_b = 2.08\left(\frac{V}{V_c}\right)^3\left(\frac{d}{h}\right)^{0.1}(V - V_c) \tag{15–26}$$

in which q_b is the sediment discharge (kg/sec per meter of flow width), d is the mean sediment diameter (m), h is the mean depth of flow (m), V is the mean flow velocity (m/sec), and V_c is the critical velocity (m/sec).

Shamov (1959) developed the following equation for uniform sediments:

$$q_b = 0.95\sqrt{d}\left(\frac{V}{0.83\,V_c}\right)^3\left(\frac{d}{h}\right)^{0.25}(V - 0.83\,V_c) \tag{15–27a}$$

For mixed sediments, Shamov developed the following equation:

$$q_s = 3D^{2/3}\left(\frac{V}{0.83V_c}\right)^3\left(\frac{d}{h}\right)^{0.25}(V - 0.83V_c) \tag{15–27b}$$

in which D is the mean diameter of the largest fractions in the mixed sediments when this fraction does not drop below 40 to 70% of the total particle composition of the sediments.

Numerous empirical equations have been developed for estimating the critical velocity, V_c. Many are related to the mean particle diameter (d) by

$$V_c = kd^{0.5}$$

in which the value of k depends on the units of V_c and d. For d in mm and V_c in m/sec, values of k from 2 to 12 have been reported for various particle weights, types of flow, and friction conditions. In contrast, Equations 15–5 give the critical velocity as a function of the depth of flow.

Example 15–8

In general, the Goncharov equation requires site characteristics and soil properties. The design conditions will determine the specific data required. If the floodplain is protected and channel erosion is the primary concern, bankfull flow conditions might be used as the design condition. The width of the channel will be necessary since Goncharov's equation gives the erosion rate per unit width. If Manning's Equation is to be used to estimate the flow velocity, the channel slope, hydraulic radius, and roughness are necessary. Calculation of the hydraulic radius requires the depth and width of the channel (assuming a nearly rectangular section). Equation 15–26 requires the depth of flow, which for bankfull flow will be the channel depth. To find the critical velocity, such as from Table 15–1, the general type of bed material is required; also, Equation 15–26 requires the mean soil particle diameter. Thus soil samples can be taken to determine the soil particle diameter distribution by weight fractions and the mean diameter can be computed.

To illustrate the calculation of q_b with Equation 15–26, assume that Manning's Equation yields a velocity of 2 m/sec. For illustration purposes, a critical velocity of 0.75 m/sec is assumed. A soil analysis indicates a mean particle diameter of 0.0004 m. At bankfull flow, the water depth will be 0.5 m. Thus, Equation 15–26 provides a unit rate of

$$q_b = 2.08\left(\frac{2}{0.75}\right)^3\left(\frac{0.0004}{0.5}\right)^{0.1}(2.0 - 0.75) = 24.2\,\frac{\text{kg}}{\text{sec-m}} \tag{15–28}$$

If the channel is 6 m wide, the rate of sediment discharge is 145 kg/sec.

15.8.4 Equations Based on the Water Discharge

Based on laboratory experiments, Egiazarov (1965) developed the following equation for estimating sediment discharges (kg/sec/meter of flow width):

$$q_b = 24\,q\,S_c\left(\frac{d_m}{d} - 1\right) \tag{15–29}$$

in which q is the water discharge rate (m^3/sec/meter of flow width), S is the stream gradient, and d_m is the maximum diameter of the bedload particles (m), and d is the mean particle diameter (m).

The Meyer-Peter and Muller (1948) Formula is

$$q_b = (39.25\, q^{0.67}\, S_c - 9.95\, \overline{D})^{1.5} \tag{15–30}$$

in which q_b is the sediment discharge (lb/sec/ft width of channel), q is the water discharge (ft^3/sec/ft width), S_c is the channel slope (ft/ft), and $\overline{D}$ is given by

$$\overline{D} = \sum_{i=1}^{N} f_i d_i \tag{15–31}$$

in which d_i is the mean size of sediment (ft), f_i is the weight percent of bed sediment with a mean size d_i (ft), and N is the number of size fractions.

Schoklitsch (1914) provided the following:

$$q_b = \sum_{i=1}^{N} 0.867\, p_i d_i^{-0.5}\, S_c^{1.5}\, (q - F_i) \tag{15–32}$$

in which d_i is the mean size of a fraction of the bed sediment (in.), p_i is the percent of bed material in size range i, and F_i is given by

$$F_i = \frac{0.00532\, d_i}{S_c^{4/3}} \tag{15–33}$$

15.8.5 Equations of the Form $q_b = f(q - q_c)$

Many researchers have proposed equations in which the difference between the discharge q and critical discharge q_c is used as a parameter to calculate the bedload discharge. One that has been used extensively is Schoklitsch's equation, which he published in 1934. In English units, the equation is:

$$q_b = 25\, S_c^{1.5}\, d^{-0.5}\, (q - q_c) \tag{15–34}$$

in which q_b is the bedload (lb/sec/ft), d is the effective grain diameter (ft), S_c is the channel slope (ft/ft), and q and q_c are the water discharge rate and critical discharge rate (ft^3/sec/ft), respectively. The critical discharge rate is

$$q_c = 0.638\, d\, S_c^{-1.33} \tag{15–35}$$

The grain diameter d in Equations 15–34 and 15–35 is usually taken as the mean diameter, although this may not be exactly correct for soils having a nonuniform gradation.

Example 15–9

For design problems where actual measurements of design discharges are not available, Manning's Equation could be used to compute the water discharge rate of Equation 15–34. A soil analysis could be used to find d and a site survey could determine S_c, as well as the cross-section information that is necessary to compute q.

To illustrate the calculation of q_b with Equation 15–34, values of $S_c = 0.01$ ft/ft, $q = 0.15$ ft^3/sec/ft, and $d = 0.0015$ ft are assumed. The critical discharge is computed with Equation 15–35:

$$q_c = 0.0638\, (0.0015)\, (0.01)^{-1.33} = 0.044\ ft^3/sec/ft$$

Thus the unit bedload discharge rate is:

$$q_b = 25\,(0.01)^{1.5}\,(0.0015)^{-0.5}\,(0.15 - 0.044) = 0.0684 \text{ lb/sec/ft}$$

For the 8-ft-wide channel the sediment discharge rate is 0.547 1b/sec.

15.8.6 Equations of the Form $q_b = f(\tau - \tau_c)$

DuBoys (1879) is generally believed to have been the first to investigate bedload movement. His theory assumed that the bed moves in layers of thickness *d*, with the top layer transporting particles at the highest velocity and decreasing to zero at the bottom layer according to a linear relationship. The relation derived was the first of its kind and is given by

$$q_b = \psi\,\tau\,(\tau - \tau_c) \tag{15–36a}$$

in which τ_c is the critical bed shear stress (lb/sq ft), ψ is a coefficient that depends on the mean sediment size, q_b is the sediment discharge (lb/sec/ft of width), and τ is the bed shear stress (lb/sq ft) given by

$$\tau = \gamma\,R_h S \tag{15–36b}$$

in which γ is the specific weight (lb/ft^3), R_h is the hydraulic radius (ft), and S is the channel slope (ft/ft). Table 15–6 provides values of ψ.

Several empirical equations of a form similar to that of the DuBoys equation were proposed by other researchers. For instance, Shields (1936) provided the following equation:

$$q_b = 10\,q\,Sd\,\frac{(\tau - \tau_c)}{(S_s - 1)^2} \tag{15–37}$$

in which d is the mean size of the sediment, S_s is the solid-to-fluid density ratio (ρ_s/ρ_f), and S is the channel slope. A value of 2.65 is commonly used for S_s.

Another bedload equation of similar form is given by Schoklitsch:

$$q_b = \frac{0.54\,(\tau - \tau_c)\,\tau}{\gamma_s - \gamma} \tag{15–38}$$

TABLE 15–6 Constants Ψ and τ_c in Equation 15–38 for Various Grain Sizes

Mean Diameter (mm)	General Classification	Value of Ψ	Value of τ_c (lb/ft^2)
1/8	Fine sand	523,000	0.0162
1/4	Medium sand	312,000	0.0172
1/2	Coarse sand	187,000	0.0215
1	Very coarse sand	111,000	0.0316
2	Gravel	66,200	0.0513
4	Gravel	39,900	0.089

in which γ_s is the specific weight of the sediment. A value of 165 lb/ft^3 is commonly used for γ_s, and γ equals 62.4 lb/ft^3.

The U.S. Waterways Experiment Station provided the following equation for estimating the critical tractive stress:

$$\tau_c = 0.006\left[\left(\frac{\rho_s - \rho}{\rho}\right)\left(\frac{d}{M}\right)\right]^{0.5} \tag{15–39}$$

in which τ_c is measured in lb/ft^2, d is the mean particle size (mm), and M is Kramer's uniformity coefficient, with values from 0.280 to 0.643. For Equation 15–39 values of d ranged from 0.2 to 4 mm.

Example 15–10

At a particular cross section of a river the slope is 0.002 ft/ft and the width is 65 ft. Assuming a bed of coarse sand with $d = 0.75$ mm and a Kramer's M of 0.5, the critical shear stress is

$$\tau_c = 0.006\left[\left(\frac{2.65 - 1}{1}\right)\left(\frac{0.75}{0.5}\right)\right]^{0.5} = 0.0094 \text{ lb/ft}^2$$

The actual shear for a hydraulic radius of 3.4 ft is

$$\tau = 62.4\,(3.4)\,(0.002) = 0.424 \text{ lb/ft}^2$$

Using Equation 15–38, the bed load rate is

$$q_b = \frac{0.54\,(0.424 - 0.0094)}{(165 - 62.4)} = 0.00925 \text{ lb/sec/ft} \tag{15–40}$$

For the entire cross section, the rate is 0.060 lb/sec.

15.8.7 Comparison of Bedload Equations

The discussion above serves to indicate that there are a large number of methods that have been developed to estimate bedload transport rates. This would be expected for a significant engineering problem such as erosion, sedimentation, and channel degradation. Given the large number of equations available, it is worthwhile to provide a brief comparison of the methods. This can be useful for assessing the quality of the models and for designing an analysis experiment to develop a new site-specific prediction equation. A number of important criteria could be identified for comparing the models; a list of the modeling criteria could include model structure, required input, comprehensiveness of the data base used to calibrate the equations, the potential accuracy of the equations, and the degree to which the empirical equations agree with theory. In addition to modeling criteria, criteria based on site characteristics should be used; this includes the availability and abundance of bed material, the grain size distribution of the bed material, and the relationship between the suspended load and bedload over the range of discharges of interest. Given the minimal detail that is available concerning the origin of many of these equations, it is necessary to compare the methods only on the basis of structure, input requirements, and the theoretical validity of the equations. If the objective was actually to select a model for prediction at a site, the other criteria (that is, data base and accuracy) would probably be the more important criteria.

The equations have similar structures, with all the equations having either a linear or power form. Some models include terms that represent the difference above a threshold, such as the term $(V - V_c)$. These are the forms most commonly used when fitting the coefficients using regression analysis. Such a form is rational since field observations suggest that significant bed movement does not occur until the velocity reaches a critical level.

The prediction equations for computing bedload transport rates have some input variables in common. However, for the purpose of comparison, it is best to identify the input requirements as belonging to one of three classes: flow parameters, soil particle parameters, and channel characteristics. For the eight prediction equations, four different flow parameters were used: water discharge rate, mean water velocity, mean flow depth, and the specific weight of the water. These parameters are interrelated by way of the continuity equation, (that is, $\int \rho VA = \rho q =$ constant). Every equation uses at least one of the flow parameters, with one of the equations using three flow parameters.

The eight equations used five different soil parameters: mean particle diameter, maximum particle diameter, the weight distribution of bedload particles, the specific gravity of the sediment, and the critical velocity. These parameters are also interrelated. For example, the critical velocity is a function of the mean particle diameter, and the mean particle diameter is a function of the weight distribution of the sediment. All of the equations use at least one soil parameter, with five of the eight using more than one. While using more than one soil parameter in a prediction equation may only provide slight improvement in the accuracy of a prediction, it does provide the means for examining the relative effects of variation in the input. This may be considered an advantage as long as the intercorrelation between the soil parameters does not cause irrational coefficients when fitting the coefficients.

Five of the eight equations include the channel slope as an input. For some models, the slope is used to compute the shear stress, rather than used as a direct input variable. The channel slope would be an indicator of the potential momentum of the flow, which is certainly a factor in bedload transport. The failure to include the slope as an input parameter may occur for one of two reasons; either the data used to fit the coefficients of the equation were from stream reaches with the same or similar slopes or the equation relies on the correlation between the flow velocity and slope to reflect variation in slope.

15.9 TRACTIVE FORCE APPROACH TO STABLE CHANNEL DESIGN

The methods based on the tractive force concept presented in Section 15.8.6 are largely derived from empirical studies. The tractive force method has also been studied from a theoretical viewpoint; specifically, methods have been developed from consideration of the forces acting on the soil particles. This approach is described briefly here.

The theoretical approach to stable channel design aims at finding the forces and other factors that cause a submerged sediment particle to become detached from the channel bed and sides. This approach has a more rational basis than the strictly empirical approach; however, it is also more difficult to apply since it deals with studying the complete mechanical process involved with all its various aspects and stages. The tractive force and the critical velocity approaches have both evolved from the theoretical school of thought.

15.9.1 Development of Tractive Force Model

A theoretical approach to stable channel design is designated the tractive force approach. The tractive force is simply the pull of the water on the wetted area of a channel. This force is produced when water flows in a channel and acts in the direction of flow. The term "tractive force" is sometimes referred to as "shear force" or "drag force." The tractive force theory is based on the fact that the stability of bed and bank material is a function of their ability to resist erosion resulting from the shear force exerted on them by the moving water. The value of the tractive force that is required to entrain a given grain is called the *critical tractive force*. The conceptual basis of the theory is to limit the unit tractive force that acts on the channel to the maximum value that will not cause serious erosion of the material forming the channel bed on a level surface. This value is termed the critical tractive force. The theoretical basis involves a balance of the forces that act in the channel.

The idea of the tractive force concept was first introduced by DuBoys in 1879. According to his findings, in a uniform flow, the tractive force (that is, average shear stress) is equal to the effective component of the gravity force acting on the water parallel to the channel bottom. This can be derived by considering the free body of a segment of the full width of a channel as shown in Figure 15–5. A summation of the forces, assuming steady flow conditions, is

$$F_u + W \sin \alpha = F_d + \tau P_w L_o \tag{15–41}$$

in which F_u and F_d are the upstream and downstream hydrostatic forces ($F_u = F_d$ for uniform flow), P_w is the wetter perimeter, W is the weight of the segment of fluid, α is the angle that the channel slope makes with the horizontal datum, L_o is the length of the free-body segment, and τ is the average boundary shear that retards the flow. Assuming that $F_u = F_d$, Equation 15–41 can be rearranged to solve for the boundary shear:

$$\tau = \frac{W \sin \alpha}{P_w L_o} \tag{15–42}$$

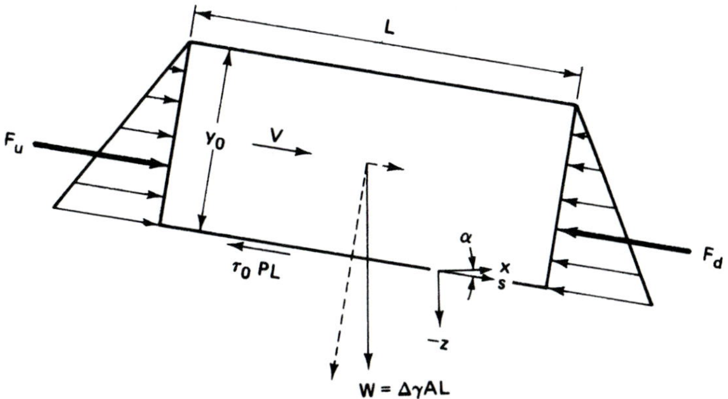

FIGURE 15–5 Free-body diagram of segment of open-channel flow.

By substituting $\gamma A L_o$ for the weight W and the slope S for sin α, Equation 15–42 reduces to the familiar form:

$$\tau = \gamma R_h S \tag{15–43}$$

where R_h is the hydraulic radius and equals the cross-sectional area divided by the wetted perimeter. When the channel is very wide relative to the depth, the depth approximates the hydraulic radius, and the expression for average shear stress becomes

$$\tau = \gamma h S \tag{15–44}$$

where h is the depth of flow in the channel.

It is important to realize that the unit tractive force in channels, except for very wide channels, is not uniformly distributed along the wetted perimeter. The distribution varies with channel shape. In practice, the actual tractive force never quite reaches the theoretical value of γhS. The maximum value is generated along the center of the bottom and is approximately 0.97 γhS. On the sides the maximum force occurs at about one-third of the water depth from the bottom and is approximately 0.75 γhS. The stress is zero at both bottom corners as well as at the water surface. This is shown in Figure 15–6.

Because of the difference in the magnitude of the tractive force exerted on the channel bed and sides, the shear stress is frequently designated as τ_S for the unit tractive force on the side of the channel and τ_L for the unit tractive force on the level surface or bed of the channel. Maintaining stability in channels of noncohesive soil is particularly difficult because of the tendency for grains to roll down the side slope due to the force of gravity. The analysis for the tractive force acting on the channel sides τ_S can be performed by considering a particle on the side slope of the channel (see Figure 15–7). By resolving the weight of the particle into two components, it is evident that the component parallel to the side, W_s sin α, tends to roll the particle down the side of the channel while the component normal to the side, W_s cos α, tends to hold the particle in place. The two forces responsible for movement of the particle are the tractive force $a\tau_S$, where a is the effective area of the particle, and the gravity force component W_s sin α. The resultant (F_r) of these two forces, which are at right angles to each other, is

$$F_r = [W_S^2 (\sin \alpha)^2 + a^2 \tau_S^2]^{0.5} \tag{15–45}$$

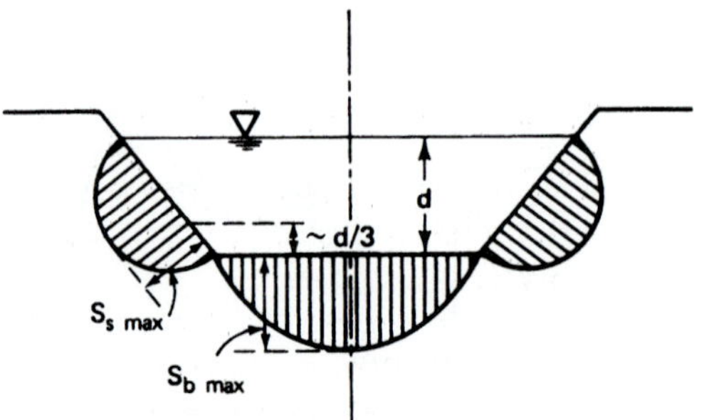

FIGURE 15–6 Distribution of shear stresses along a wetted perimeter.

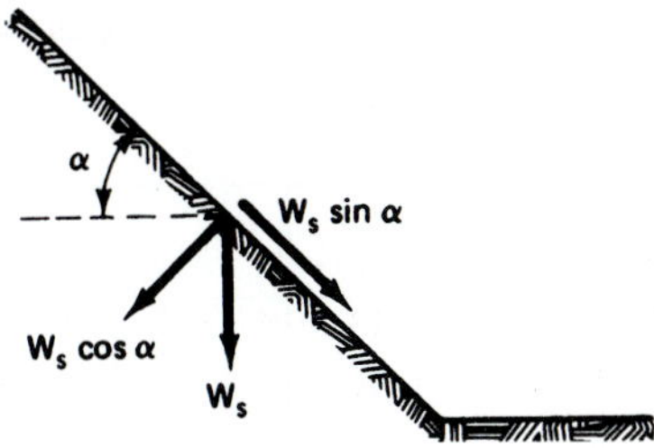

FIGURE 15–7 Decomposition of the weight of a soil particle.

The force that resists movement is equal to the normal force $W_s \cos \alpha$ multiplied by the coefficient of friction f (that is, $\tan \phi$, where ϕ is the angle of repose):

$$F_r = f\,W_S \cos \alpha = W_S \cos \alpha \tan \phi \tag{15–46}$$

When the particle is in a state of impending motion, the forces are equal. Equating the forces yields

$$[(W_S \sin \alpha)^2 + (a\,\tau_S)^2]^{0.5} = W_S \cos \alpha \tan \phi \tag{15–47}$$

This can be solved for the unit tractive force that causes impending motion on a sloping surface:

$$\tau_S = \frac{W_S}{a} \cos \alpha \tan \phi \left[1 - \frac{\tan^2 \alpha}{\tan^2 \phi}\right]^{0.5} \tag{15–48}$$

When motion of a particle on the level surface is impending due to the tractive force $a\tau_L$, the following is obtained from Equation 15–48 by setting $\alpha = 0°$:

$$\tau_L = \frac{W_S}{a} \tan \phi = \gamma h S \tag{15–49}$$

The ratio of τ_S to τ_L is called the tractive-force ratio (K_S). This is often used in design and is given by

$$K_S = \frac{\tau_S}{\tau_L} = \cos \alpha \left[1 - \frac{\tan^2 \alpha}{\tan^2 \phi}\right]^{0.5} \tag{15–50}$$

Approximating Equation 15–50 gives

$$K_S = \left[1 - \frac{\sin^2 \alpha}{\sin^2 \phi}\right]^{0.5} \tag{15–51}$$

To apply Equations 15–48 and 15–49, information is required on the angle of repose (ϕ). Extensive tests by the U.S. Bureau of Reclamation have shown that ϕ is dependent on the size and shape of the sediment as shown in Figure 15–8. According to studies by the U.S. Bureau of Reclamation, the angle of repose needs to be considered only for coarse noncohesive materials because for cohesive and fine noncohesive materials, the cohesive forces become so great in proportion to the gravity force component that the gravity component can be safely neglected.

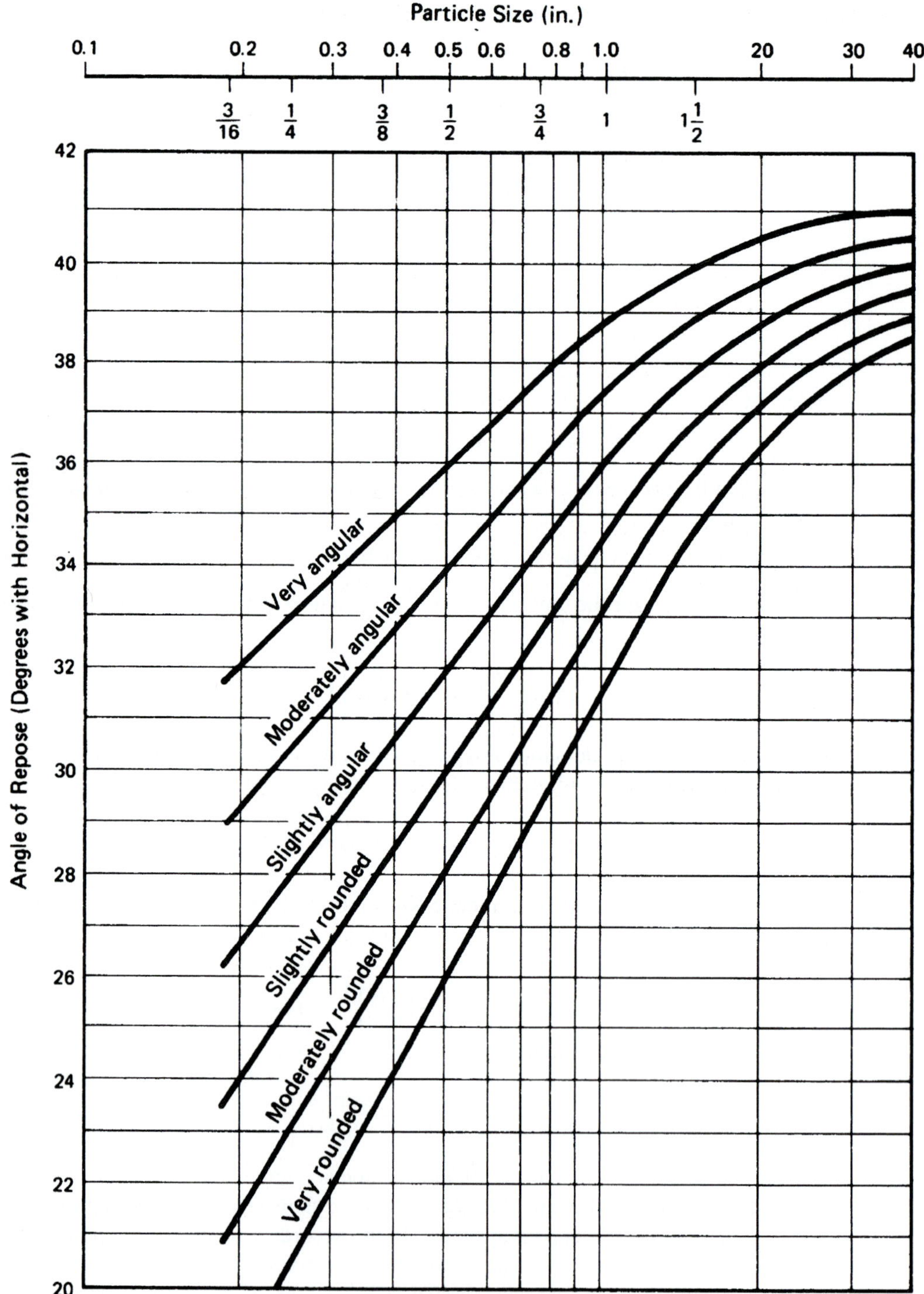

FIGURE 15–8 Angles of repose of noncohesive material.

If only the relationship between the maximum unit tractive force on the side of the channel τ_S and the maximum unit tractive force on the bed of the channel τ_L need to be determined, the ratio K can be used. This is often needed in design applications. Therefore, instead of determining the stress distribution along the channel as shown earlier, only the maximum values of stress are determined. A value for K_s of 1 is used for the bed, and a value of K less than 1 is used for the side slopes. Lane (Leliavsky, 1955) has prepared a diagram that shows the relationship between the angle of repose ϕ, the slope angle α, and K (see Figure 15–9). As can be concluded from Lane's diagram, both the angle of repose and the slope angle are of nearly equal sensitivity in the determination of K.

5.9.2 Critical Tractive Force

When water flows through a channel, a shear stress is produced. This is referred to as the tractive force. As the flow velocity increases, the tractive force is increased until it reaches a value known as the critical tractive force, which occurs at the threshold of particle motion. Like many threshold conditions, an exact value cannot be precisely defined. At first, a grain or two becomes detached at a certain rate of flow. Then, as the flow rate increases, more grains become detached and are transported until the process is general over the whole bed. Hence, it is easy to understand the difficulty encountered in defining the exact state at which the critical tractive force is reached.

The *critical tractive force* is the maximum unit shear force that will not cause serious erosion of the material forming the channel bed on a level surface. However, since this definition is subject to the interpretation of the individual researcher, various equations have been presented, with little agreement among them. Schoklitsch contributed to the early developments by presenting the following equation, which is based on his own experiments:

$$\tau_c = [0.201\,(\gamma_S - \gamma)\,\lambda\, d^3]^{0.5} \tag{15–52}$$

where d is the mean grain diameter, λ is a shape coefficient ranging from 1 for spheres to 4.4 for flat grains, and τ_c is the critical tractive force (kg/m^2). An improvement over Schoklitsch's approach was later given by Tiffany et al. (1935), which was based on more experimental evidence and was thought to be a more representative equation:

$$\tau_c = 29\,[d\,(\gamma_S - \gamma)/M]^{0.5} \tag{15–53}$$

in which M is a grain distribution modulus and τ_c is in g/m^2. At a later date, Schoklitsch reanalyzed his data and suggested two equations, which were dependent on particle size:

$$\tau_c = \begin{cases} 0.000285\,(\gamma_s - \gamma)\,d^{1/3} & \text{for } 0.0001\,m < d < 0.003\,m \quad (15\text{–}54a) \\ 0.076\,(\gamma_S - \gamma)\,d & \text{for } d \geqslant 0.006\,m \quad (15\text{–}54b) \end{cases}$$

A simpler relationship was presented by Leliavsky (1955), who believed that shear stress (g/m^2) versus grain size is given by

$$\tau_c = 166\,d \tag{15–55}$$

where the mean particle diameter d is given in millimeters. This equation fits most of his data quite well and is very simple to use.

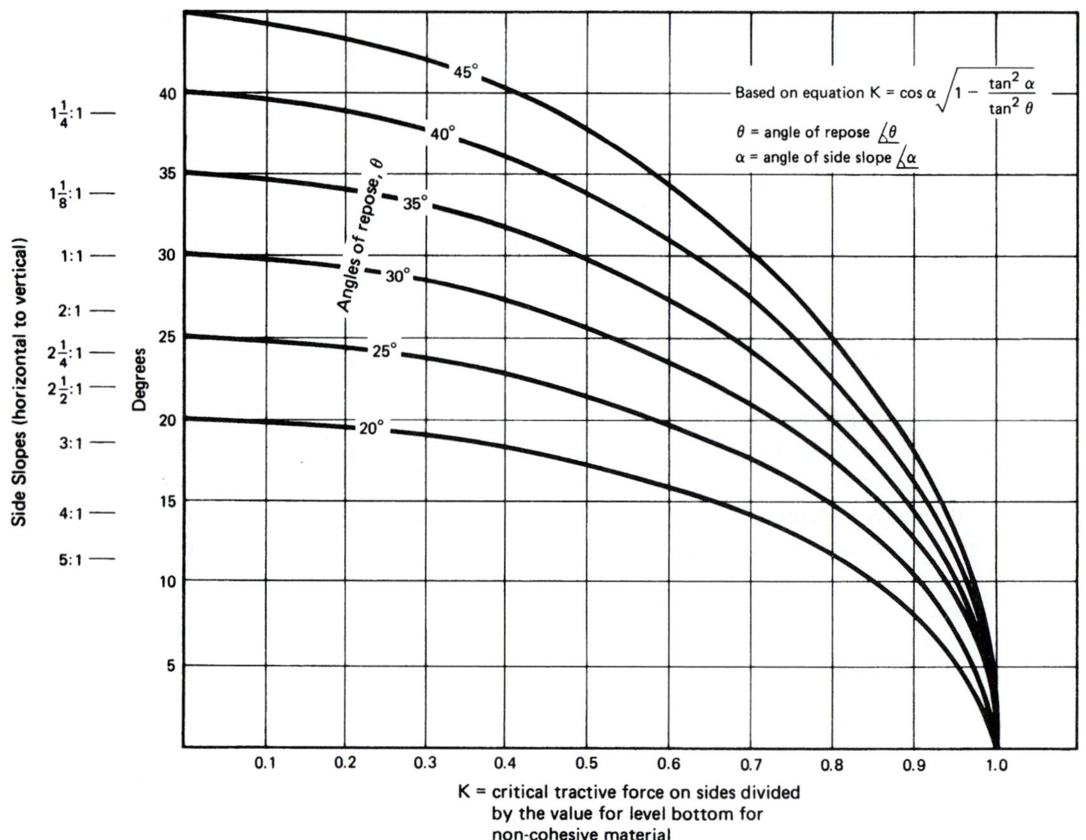

FIGURE 15–9 Relationship between critical tractive force, angle of repose, and side slope.

Example 15–11

The critical tractive force concept can be used to show the particle sizes that are eroded for various depths of flow. Equation 15–55 can be modified to define the critical tractive stress in N/m^2:

$$\tau_c = 1.63\, d$$

Equation 15–36b defines the actual tractive stress. The ratio τ/τ_c is

$$\frac{\tau}{\tau_c} = \frac{\gamma R_h S}{1.63\, d}$$

in which γ[=]N/m^3, R_h[=]m, S[=]m/m, and d[=]mm. At the point of incipient motion, the actual stress equals the critical stress, that is, $\tau = \tau_c$, therefore,

$$d_c = \frac{\gamma R_h S}{1.63} = 6016\, R_h S$$

in which d_c is the critical diameter (mm). For a given slope S and hydraulic radius R_h, the diameter of the smallest particle that will move is d_c.

To illustrate, assume the slope at a particular section is 0.001 m/m. The following tabular summary gives the hydraulic radius at which various soil types will move:

	d(mm)	R_h(m)
Very coarse sand	1.5	0.25
Coarse sand	0.75	0.12
Medium sand	0.35	0.058
Fine sand	0.17	0.028
Very fine sand	0.075	0.012
Silt	0.01	0.002
Clay	0.001	0.0002

Assuming a wide channel, the hydraulic radius will approximate the depth of flow. Thus, silt and clay particles will erode from an erodible bed for very small flow depths and rates.

15.10 ESTIMATING SEDIMENT YIELD

Erosion occurs both on the watershed and in the channel network. The total amount of material eroded is called the gross erosion. Not all of this material is transported throughout the watershed system. Much of the material will be deposited or trapped. That part of the gross erosion that is transported to a downstream control point such as a reservoir is called the sediment yield. Just as developers of small tracts of land must contend with on-site erosion, those involved with in-stream development, such as dams and diversion structures, must contend with the sediment yield.

The problem of analysis can be separated into two categories: analysis for a single site and regionalization for estimation at ungaged sites. While the latter is the more common problem, the former may be necessary for major projects. The results for the single-site analysis have limited use for other sites, unless the results are being used for a nearby watershed that is quite similar to the watershed where the sediment yield data were collected; in this case, the process is an extrapolation of data and one would expect a lower level of accuracy. In a sense, the spatial translation of a sediment yield estimation method from one watershed to a nearby watershed represents a form of regionalization with a sample size of 1.

Where it is necessary to make estimates of sediment yield at the site of a proposed structure, estimates can be synthesized using a number of methods: (1) analysis of measured data after installing one or more sediment-load measuring stations, (2) use of a regional sediment yield equation, and (3) use of a sediment delivery ratio with estimated amounts of watershed erosion. The first method has the disadvantages of requiring the initiation of a sediment load measuring program and the delay in construction necessitated by such a program. Use of a regional equation relies on the similarity of the erosion/sedimentation characteristics of the watershed contributing at the design point with those watersheds used in developing the regional method. The accuracy of estimates obtained using the third method depends on the representativeness of the equations used to predict watershed erosion and the accuracy of the sediment delivery ratio.

15.10.1 Flow-Duration, Sediment-Rating-Curve Procedure

The flow-duration, sediment-rating-curve procedure requires a flow-duration curve, a sediment rating curve, and knowledge of the characteristics of the sediment; both the flow-duration and sediment rating curves must be developed from on-site measurements, usually measurements made as part of a historic record at sites within the watershed. A flow-duration curve is the graphical relationship between the flow rate and the duration of time that the flow is equaled or exceeded; flow-duration data can be computed and plotted as a flow-frequency curve.

Construction of a sediment rating curve requires records of both average daily discharge and sediment loads. The rating curve is developed by constructing a "best fit" line through a plot of the average daily flow and the corresponding sediment load. It is important to have a number of high flow measurements as part of the record to improve the accuracy of computed sediment yields. The sediment yield can be estimated by integrating the flow-duration and sediment rating curves.

Given a daily discharge frequency curve and a daily sediment discharge rating curve, the following procedure can be used to estimate the sediment yield:

1. Separate the exceedence probability scale of the frequency curve into intervals; approximately 20 or 25 intervals should be used. The intervals are usually not selected using a constant increment of probability.
2. Compute the range of each probability interval.
3. Compute the midpoint of each interval.
4. Using the discharge frequency curve, find the discharge that corresponds to each midpoint of Step 3.
5. Using the sediment rating curve, find the daily sediment discharge for each water discharge of Step 4.
6. Compute the expected sediment discharge as the product of the probability interval of Step 2 and the sediment discharge of Step 5.
7. Compute the total sediment as the product of the number of days in the year and the sum of the expected sediment discharges of Step 6.
8. Compute the sediment yield by dividing by the drainage area and making the necessary conversion of units.

Example 15–12

For this example, the assumption is made that flood frequency data and sediment discharge data are available, with both the water and sediment discharge data available on a daily average basis. A daily mean discharge frequency curve can be developed using the methods presented in Chapter 5. For this example, we will assume that the record of daily discharges resulted in a log-normal population with a $\log_{10}$ mean of 3.8 and a $\log_{10}$ standard deviation of 0.2. We will also assume that the sediment rating curve is

$$q_s = 0.0114\, q^{1.88} \tag{15–56}$$

in which q_s is the sediment discharge (tons/day) and q is the mean daily discharge (ft^3/sec).

TABLE 15–7 Example 15–12: Flow-duration, Sediment Rating Curve Procedure

Exceedence Frequency Range (%)	Interval (%)	Mid point (%)	Mean Daily Discharge (ft^3/sec)	Sediment Discharge (tons/day × 10^3)	Expected Sediment (tons/day)
0–0.1	0.1	0.05	28,800	2753	2,750
0.1–1	0.9	0.55	20,300	1433	12,900
1–5	4	3	15,000	809	32,400
5–20	15	12.5	10,700	430	64,500
20–50	30	35	7,530	222	66,500
50–80	30	65	5,280	114	34,100
80–95	15	87.5	3,720	59	8,800
95–99	4	97	2,650	31	1,250
99–99.9	0.9	99.45	1,960	18	160
99.9–100	0.1	99.95	1,380	9	10
					223,370

The expected average annual sediment yield can be determined from these two functions and the eight steps listed above. As shown in column 1 of Table 15–7, the probability scale was divided into 10 intervals; in practice, 20 to 25 is recommended, with the intervals being smaller for the small exceedence probabilities. The frequency interval and the midpoint of the intervals are given in columns 2 and 3. The mean daily discharge can be computed mathematically or obtained from the frequency curve for each midpoint in column 3; the discharges are given in column 4 of Table 15–7 for the assumed log-normal population. The sediment rating curve of Equation 15–56 is used with the water discharges of column 4 to compute the corresponding sediment discharges, which are given in column 5. The expected sediment discharge is the product of the probability interval of column 2 and the sediment discharge of column 5. Once the expected sediment discharge has been computed for each frequency interval, the sum can be determined. The sediment yield in weight per unit area per year can be determined by multiplying by the number of days per year and dividing by the drainage area. Table 15–7 indicates a sum of 223,370 tons/day. Assuming the data is for a 17,100 mi^2 watershed, the sediment yield is

$$\hat{Y} = \frac{223{,}370 \text{ tons/day } (365 \text{ days/yr})}{17{,}100 \text{ mi}^2} = 4768 \text{ tons/mi}^2\text{/yr} \tag{15–57}$$

which is 7.45 tons/acre/yr.

15.10.2 Univariate Analysis of Reservoir Sediment Accumulation Records

The univariate approach is useful where values for independent variables are not available. Basically, the method uses sediment accumulation for existing reservoirs to compute a regional mean sediment yield value. The mean is assumed to be valid for all similar locations in the region. The sediment yield (Y) equals the ratio of the accumulation (A_t) to the trap efficiency (E_f) of the reservoir:

$$Y = \frac{A_t}{E_f} \tag{15–58}$$

Given that the total accumulation in any one reservoir will be a function of the number of years in which the reservoir has been in operation and that the reservoirs in the region will have been in operation for different periods of time, the accumulation of A_t of Equation 15–58 should be expressed in weight per area per unit time, such as tons per acre per year; that is, a common time frame should be used. The trap efficiencies for each reservoir will have to be determined from measurements or estimated using regionalized curves (see Section 15.10.6). Having computed estimates of the sediment yield for each existing reservoir, a weighted mean sediment yield can be computed by weighting the sediment yield from each site by the number of years of record:

$$\overline{Y} = \frac{\sum_{i=1}^{m} n_i Y_i}{\sum_{i=1}^{m} n_i} \tag{15–59}$$

in which $\overline{Y}$ is the mean sediment yield for the region, n_i is the number of years of accumulation at site i, Y_i is the average annual sediment yield at site i, and m is the number of existing reservoirs. The computed value of $\overline{Y}$ is then the value of the regional sediment yield. To use $\overline{Y}$ for any other site in the region, the average annual accumulation can be found by rearranging Equation 15–58, using $\overline{Y}$ in place of Y, to solve for A_t:

$$A_t = E_f \overline{Y} \tag{15–60}$$

Equation 15–60 will give the accumulation in weight per unit area per unit time, such as tons/acre/year. The total accumulation per year is computed by multiplying by the drainage area. The expected accumulation over the design life of the reservoir is then obtained by multiplying by the number of years (that is, the design life).

Example 15–13

The data for calculating the regional sediment yield of a hypothetical region are shown in Table 15–8. Total accumulations (A_{ti}) in tons ($\times 10^4$) are given for seven reservoirs. For the drainage areas and years of accumulation, the unit accumulations (A_i) in tons/mi^2/yr are computed for each site i. The sediment yield (Y_i) for each site is calculated using the trap efficiencies and Equation 15–58. The regional estimate of the mean annual sediment yield is given by Equation 15–59:

TABLE 15–8 Regional Sediment Yield Data for Example 15–13

Site	A_{ti} (tons × 10^4)	n_i (yr)	Area (mi^2)	A_i (tons/mi^2/yr)	E_i (%)	Y_i (tons/mi^2/yr)
1	50	13	132	291	41	711
2	300	31	417	232	65	357
3	100	14	286	250	59	423
4	165	22	355	211	53	399
5	365	37	509	194	66	294
6	185	16	391	296	62	477
7	345	29	468	254	60	424

$$\overline{Y} = \frac{13\,(711) + 31\,(357) + 14\,(423) + 22\,(399) + 37\,(294) + 16\,(477) + 29\,(424)}{13 + 31 + 14 + 22 + 37 + 16 + 29}$$

$$= 406 \text{ tons/mi}^2\text{/yr} \quad (15\text{–}61)$$

For a proposed reservoir with a contributing area of 255 mi^2 and a trap efficiency of 52%, the expected accumulation ($\hat{A}_t$) would be:

$$\hat{A}_t = 406\,(0.52)\,(255) = 53{,}836 \text{ tons/yr} \quad (15\text{–}62)$$

The total expected accumulation for the project life could be found by multiplying the accumulation (53,836 tons/yr) by the project life in years.

Limitations. This method has a number of limitations. Obviously, errors can be introduced when the method involves extrapolation. Extrapolation occurs when the method is used outside the range of the data used to compute the mean annual sediment yield. The drainage area and trap efficiencies should not be much different from the values in the data set. For example, the trap efficiency is usually proportional to the size of the reservoir, with the larger reservoir having a greater retention time and therefore the opportunity for deposition. Thus, if the drainage area for which an estimate of sediment yield is required is much different from the range of areas in the data base, it may be necessary to make an adjustment for the intercorrelation between the drainage area and the trap efficiency.

15.10.3 Multiple Variable Analysis of Reservoir Sediment Accumulation Records

For the previous analysis, the sediment accumulation in a reservoir was treated as a single variable, and the use of Equation 15–60 assumes that the analysis is univariate. The drainage area was only used as a scaling factor to compute the yield per unit area. As indicated previously, one would expect both the sediment yield and the trap efficiency to be a function of variables such as the drainage area. One would expect that if variation in the sediment yield with drainage area were more properly accounted for, the accuracy of estimated sediment yields would be improved.

The procedure for making a multiple variable analysis is to relate the sediment yields, Y_i, in Equation 15–59 to the drainage areas using a bivariate regression analysis. Graphically, the relationship can be seen using a plot of Y_i versus the drainage area of site i. If a simple bivariate regression analysis is used to compute the best-fit line, the record length at each reservoir site is ignored. This problem can be overcome using a weighted least squares procedure, with the record length used as the weighting factor. The best-fit line can then be used to estimate the sediment yield for the site of a proposed structure.

Example 15–14

The data of Table 15–8 were used to graph a relationship between the sediment yield in tons/mi^2/year and the drainage in sq. mi (see Figure 15–10). A linear least squares equation was fit to the data and is shown in Figure 15–10:

$$\hat{Y}_i = 775.73 - 0.9168\,A \quad (15\text{–}63)$$

in which $\hat{Y}_i$ is the predicted sediment yield in tons/mi^2/yr and A is the drainage area in sq. mi. The correlation coefficient was –0.873, which means that Equation 15–63 explains 76.2% of the variation in Y_i. A weighted least squares analysis yielded the following equation:

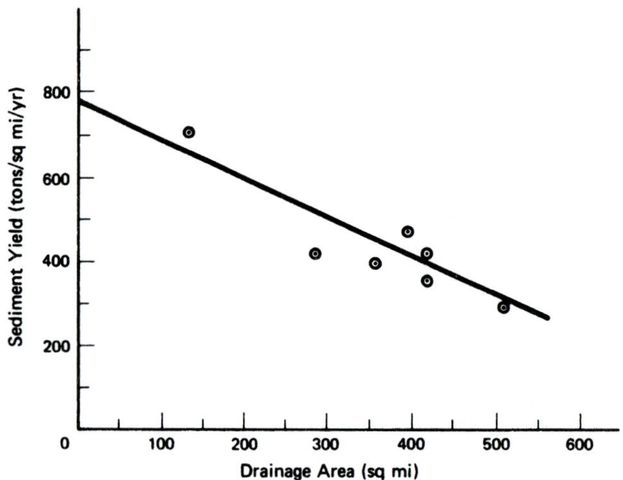

FIGURE 15–10 Sediment yield versus drainage area for Example 15–14.

$$\hat{Y}_i = 760.19 - 0.8805\,A \tag{15–64}$$

Very little difference exists between the predictions with the two equations, at least within the range of the data of Table 15–8.

Equation 15–64 can be used to obtain an estimate of the sediment accumulation for the 255-mi^2 watershed introduced in Example 15–13. The yield is

$$\hat{Y} = 760 - 0.8805\,(255) = 535 \text{ tons/mi}^2\text{/yr} \tag{15–65}$$

For a trap efficiency of 52%, the accumulation would be

$$\hat{A}_t = \hat{Y} A E_f = 535\,(255)\,(0.52) = 70{,}941 \text{ tons/yr} \tag{15–66}$$

Limitations. The method based on multiple variable analysis of reservoir sediment accumulation records has several limitations. First, watersheds in which significant changes have occurred, such as forest fires, surface mining activity, or overgrazing of livestock, should not be used; in such cases, the sediment accumulated in the reservoirs may not be indicative of sediment yield from a natural watershed. Second, the fitting of a curve between Y_i and A_i represents a case involving ratio correlation, which should usually be avoided. Specifically, the value of Y_i depends on the value of A_i, so when Y_i is plotted versus A_i, part of the variation in Y_i is a function of the inverse of A_i. Obviously, a graph of $1/A_i$ versus A_i could provide near-perfect correlation. The effect of such analysis should be investigated. As an alternative, the analysis could relate Y_i to some other causative factor, such as slope or land use, or the analysis could relate accumulations in tons per year to the drainage area; both of these alternatives would avoid the problem of spurious correlation.

15.10.4 Estimation Using Sediment Delivery Ratios

As indicated previously, all of the soil eroded in the upper reaches of a watershed will not be delivered to a point downstream. Much of the material will be redeposited at locations where the momentum of the transporting water is insufficient to keep the material in suspension or to move the soil particles along the watershed surface or channel. This is most likely to occur in areas of low slope or high roughness. The ratio of the sediment transported to a location in the channel system (that is, the sediment yield) to the gross erosion from the drainage area above that point is called the sediment delivery ratio. The sediment delivery ratio is a function of the following factors:

1. The sediment source (watershed versus channel)
2. The magnitude and proximity of the sediment source (the erodibility of remote areas of the watershed versus that at sites near the design location)
3. Characteristics of the transport system (channel density, condition, and gradient)
4. The frequency, duration, and intensity of erosion-producing storms
5. Characteristics of soil material such as texture
6. The potential for deposition such as at the base of slopes and at storage, natural or man-made, sites within the channel system
7. Watershed characteristics such as the area, slope, and shape

To use the sediment delivery ratio (SDR) concept for estimating sediment yields at ungaged sites, it is necessary to have (1) a method of estimating the SDR for that site, and (2) estimates of watershed and channel erosion for all subareas of the watershed. An analysis of measured data that leads to a method for predicting the SDR must precede the synthesis or design phase. The analysis leading to a method for predicting a SDR involves actual sediment yield measurements and the simultaneous assessment of independent variables, such as drainage area. A number of prediction equations have been developed for estimating a SDR. Renfro (1975) and ASCE (1971) provided a discussion of the methods given in Table 15–9. The intent of Table 15–9 is not to provide equations for use. Instead, the equations are given only to illustrate both the form and the required input of available methods. The equations

TABLE 15–9 Methods for Estimating the Sediment Delivery Ratio

1. Midwestern Watersheds

$$SDR = \begin{cases} 43.4A^{-0.1753} & \text{for } 0.1 \leq A \leq 10 \text{ mi}^2 \\ 46.7A^{-0.2071} & \text{for } 10 \leq A \leq 100 \text{ mi}^2 \\ 64.6A^{-0.2775} & \text{for } 100 \leq A \leq 1000 \text{ mi}^2 \end{cases}$$

2. Blackland Prairies in Texas

$$SDR = 75.3\,(10A)^{-0.14191}$$

3. Red Hills in Oklahoma and Texas

$$\log SDR = 2.94259 - 0.82362 \text{ colog } (R/L)$$

4. Southeast Piedmont

$$SDR = 18620A^{-0.23}\,(L/R)^{-0.51}\,(B)^{-2.79}$$

SDR = sediment delivery ratio (%)
A = drainage area (mi^2)
L = watershed length (ft)
R = watershed relief or elevation difference (ft)
B = weighted mean bifurcation ratio

should not be used outside either the location or the range of the data used to develop the equations. While the drainage area is the most frequently used input variable, other watershed characteristics are frequently included because they appear to increase the accuracy of prediction.

Once an equation for estimating the sediment delivery ratio has been either calibrated or selected for a site, the sediment yield can be determined by multiplying the SDR by the gross watershed erosion. The gross erosion would include erosion from upland areas as well as erosion within the drainage network.

15.10.5 Sediment Routing with Sediment Delivery Ratios

It is often necessary in hydrologic analysis to subdivide a watershed because of a lack of homogeneity of watershed characteristics, such as land cover, slope, or soil type. When a watershed is subdivided, the process of subdivision must be considered when using sediment delivery ratios. The concept of a SDR assumes a homogeneous watershed. Methods for estimating a SDR often use a single variable such as drainage area as the only predictor variable even though a SDR is affected by many watershed characteristics. Two of the equations of Table 15–9 suggest that slope is an important variable. When a sediment delivery ratio that is based on a simplified equation is used for prediction, the SDR concept can be misapplied.

Boyce (1975) identifies a problem in using the SDR concept for sediment routing. Consider the watershed of Figure 15–11. The 3-mi^2 watershed consists of three equal subareas ($A_1 = A_2 = A_3 = 1$ mi^2). Subareas 1 and 2 are upland areas, while subarea 3 is an alluvial area with a smaller mean slope than the slopes of the upland subareas. Using the first equation of Table 15–9 to estimate the SDR for the total watershed provides SDR = 35.8%. Using the same equation for the 1-mi^2 subareas yields SDR = 43.4%. Assuming a gross erosion of 100 tons/mi^2/yr from each subarea, the gross erosion for the watershed would be 300 tons/yr. Using the SDR of 0.358 gives a sediment yield of 107.4 tons/yr. However, if the SDR concept is applied to the individual subareas, the total sediment yield would be 3(100)(0.434) = 130.2 tons/yr. Which is correct, 107.4 or 130.2 tons/yr? The concept of a SDR assumes that the watershed is homogeneous in its capacity to deliver sediment to the design point. If each subarea is generating equal amounts of gross erosion, the SDR of 0.434 would be correct. But for the low-sloped subarea (number 3) the yield is less. Using an SDR of 0.434 for subareas 1 and 2 and a total sediment yield of 107.4 tons/yr for the entire watershed, then for the flow-through subarea 3 the actual SDR is

$$SDR_3 = \frac{107.4 - 2\,(0.434)\,(100)}{100} = 0.206 \qquad (15\text{–}67)$$

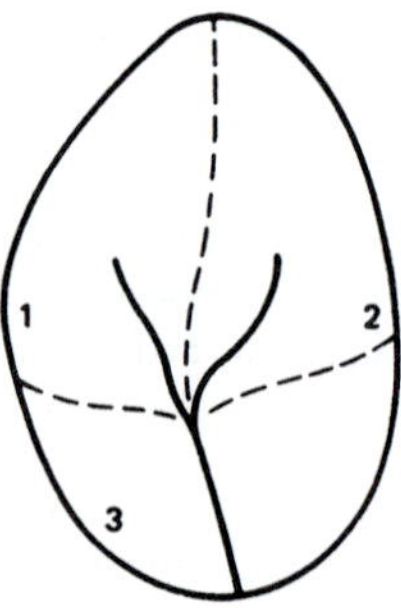

FIGURE 15–11 Schematic diagram of watershed for sediment routing with sediment delivery ratios.

Thus the total yield of 107.4 tons/yr to the mouth of the watershed is the result of sediment yields of 43.4 tons/yr from each of the two upland subareas to the confluence and 20.6 tons/yr for the low-sloped, flow-through subarea to the watershed outlet. Even this may seem irrational because one typically expects larger yields from the subareas located closest to the watershed outlet.

Example 15–15

The equation from Table 15–9 for drainage areas less than 10 mi^2 for midwestern watersheds can be used as an illustration of sediment yield calculations. Assume a sediment analysis is being made for a 0.7 mi^2 watershed in the midwest. For illustration purposes, we can assume that land erosion of agricultural lands is the predominant source of sediment and that a detailed erosion analysis of the watershed using the USLE provides an annual erosion of 1570 tons/yr. This annual load represents the soil lost from individual fields, but not the total load at the outlet of the 0.7 mi^2 watershed. Because of within-watershed deposition, the annual yield at the outlet is computed as the product of the load that is computed with the USLE and the sediment delivery ratio, which is computed using the equation in Table 15–9:

$$\begin{aligned} \text{yield} &= 1570\,[43.4\,(0.7)^{-0.1753}]/100 \\ &= 1570\,(0.462) = 725 \text{ tons/yr} \end{aligned} \tag{15–68}$$

Thus, slightly less than one-half of the eroded soil will pass out of the watershed.

Example 15–16

The sediment delivery ratio computed with the equation used in Example 15–15 is sensitive to the accuracy of the estimate of the drainage area. Error in the drainage area estimate will cause error in the computed sediment delivery ratio and thus the sediment yield. The sensitivity of the SDR to drainage area can be found by taking the derivative of the SDR equation:

$$\frac{d\,(SDR)}{dA} = -7.61\,A^{-1.1753} \tag{15–69}$$

A sensitivity estimate from Equation 15–69 has units of %/mi^2. The negative sensitivity indicates that, as the drainage area increases, the sensitivity decreases. For a drainage area of 0.1 mi^2, Equation 15–69 yields a sensitivity of –113.9, while a sensitivity of –7.61 results in a drainage area of 1.0 mi^2. An error of 0.01 mi^2 for a drainage area of 0.1 mi^2 will cause an error in the SDR of

$$SDR = \left[\frac{d\,(SDR)}{dA}\right]\Delta A = -113.9\,(0.01) = -1.14\% \tag{15–70}$$

For a drainage area of 1 mi^2, the error in the SDR would be

$$SDR = -7.61\,(0.01) = -0.761\% \tag{15–71}$$

Using the values from Example 15–15, Equation 15–69 gives a sensitivity of –11.6%/mi^2 for a drainage area of 0.7 mi^2. Thus, for an error in the drainage area of 0.01 mi^2, the SDR will have an error of –11.6(0.01) = –0.116%, with the negative sign indicating that the SDR would decrease for a positive error in the drainage area. Thus, if the actual area was 0.68, the SDR would change by approximately

$$(-11.6\%/\text{mi}^2)\,(-0.02\text{ mi}^2) = 0.232\%$$

15.10.6 Estimating Trap Efficiency

The trap efficiency (E_f) of a reservoir is the proportion of the sediment in the water inflow to the reservoir that is trapped in the pool of the reservoir. Mathematically, the trap efficiency is

$$E_f = \frac{V_{st}}{V_{\text{in}}} = 1 - \frac{V_{\text{out}}}{V_{\text{in}}} \tag{5–72}$$

in which V_{st} is the part of the sediment volume in the inflow V_{in} that is not released, and V_{out} is the sediment volume released as part of the water discharge out of the reservoir. The trap efficiency is a function of reservoir characteristics (volume of storage, length, presence of buffers), sediment characteristics (particle size and distribution), the outlet characteristics (spillway capacity and location), runoff characteristics, and the detention time. The trap efficiency is an important factor in the amount of sediment that accumulates in a reservoir over the design life (see Equation 15–58). Where sediment records are available, the trap efficiency can be estimated with Equation 15–58. Where records are not available for a particular site, regional information should be used to develop a regionalized method for estimating E_f, such as a regression equation relating the trap efficiency to factors such as the mean sediment diameter, the volume of water inflow, and the volume of storage. Where regionalization is not possible or practical, a standard trap efficiency curve could be used; this is probably the least desirable of the alternatives. A Brune-type curve (see Figure 15–12) is one such regionalized curve. The following procedure can be used to estimate the trap efficiency for an ungaged site:

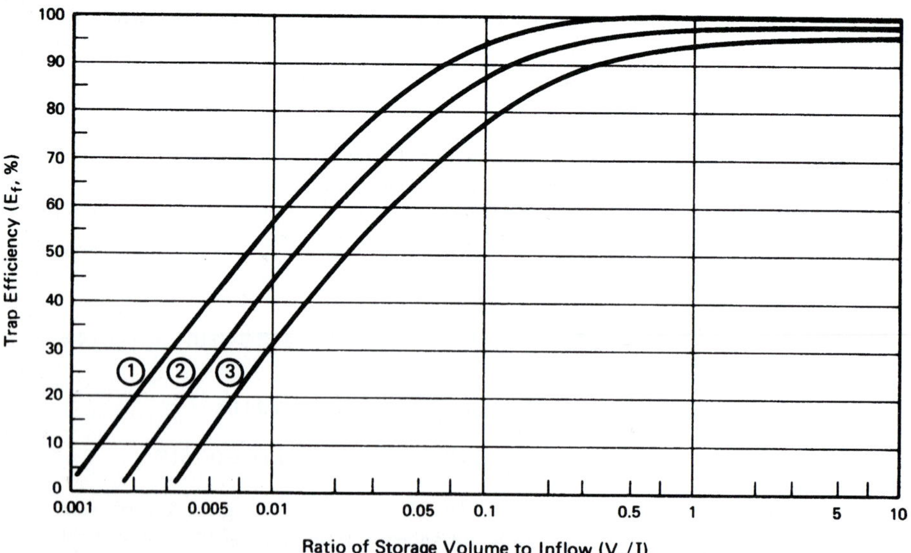

FIGURE 5–12 Brune trap efficiency relationship for (1) primary highly flocculated and coarse-grained sediments, (2) median-grained sediments, and (3) primarily colloidal and dispersed fine-grained sediments.

1. Estimate the expected storage volume (V_s) in the reservoir; this would include storage for sediment accumulation, flood control, power generation, water quality considerations, etc.
2. Estimate the average water inflow (I) into the reservoir; this should be in the same units as the storage volume, usually watershed inches.
3. Evaluate the texture characteristics of the inflowing sediment (see the title to Figure 15–12 for the three textures).
4. Obtain an estimate of the trap efficiency from Figure 15–12.

For any reservoir, the trap efficiency is not constant. It varies from storm to storm and year to year. Therefore, the storage volume and inflow of Steps 1 and 2 must be judiciously selected. When concerned with the long-term trap efficiency, average annual values of V_s and I are commonly used.

Example 15–17

To illustrate the use of Figure 15–12, assume that the project requires an estimate of the trap efficiency for a proposed reservoir. The reservoir will be used for flood control, recreation, and sediment control. Volumes required for these are 3.5 in., 2.5 in., and 2.0 in., respectively. This represents a total storage of 8.0 in. The average annual runoff at the site is 14 in. Therefore, the storage-inflow ratio is 0.571. The inflowing sediment is primarily fine-grained wash load. Thus, the trap efficiency from curve 3 of Figure 15–12 is 93%.

Example 15–18

A retention basin intended for pollution control as well as flood control has a designed outflow rate of 32 ft³/sec, a volume of 130,000 ft³ at flood stage, a length of approximately 375 ft between the principal inflow point and the outflow riser, an average depth of about 3.5 ft, and an average width of about 100 ft. The average travel time (T_t) through the basin is approximately equal to the volume (V_o) divided by the flow rate (Q):

$$T_t = \frac{V_o}{Q} = \frac{130{,}000 \text{ ft}^3}{32 \text{ ft}^3/\text{sec}} = 4063 \text{ sec} = 1.13 \text{ hr}$$

The average velocity of the water as it flows through the basin equals the ratio of the discharge rate to the product of the average width (W) and average depth (H):

$$V = \frac{Q}{W * H} = \frac{32 \text{ ft}^3/\text{sec}}{100 \text{ ft } (3.5 \text{ ft})} = 0.091 \text{ ft/sec}$$

For a soil particle with a diameter of 0.00005 ft and neglecting the fall velocity component of the velocity vector, the Reynolds Number is

$$\mathbf{R} = \frac{\rho VD}{\mu} = \frac{1.934\,(0.091)\,(0.00005)}{2.5 \times 10^{-5}} = 0.35$$

Since this is less than 5, Stokes' Law would apply. Thus, the terminal fall velocity is computed with Equation 15–3:

$$V = \frac{D^2(\gamma_s - \gamma)}{18\,\mu} = \frac{(0.00005)^2\,(165 - 62.4)}{18\,(2.5 \times 10^{-5})} = 0.00057 \frac{\text{ft}}{\text{sec}}$$

Thus, for a travel time of 4063 sec, the average particle would drop 2.3 ft. Thus, only those particles entering the retention basin that are below the 2.3 ft level will settle. The remaining will be discharged. Assuming that the sediment in the inflow is well mixed, then the trap efficiency would be 2.3/3.5 ft = 0.657 or 65.7%. For soil particles with a smaller diameter the trap efficiency would be lower.

PROBLEMS

15–1. Compute the settling velocity for the following particles: very coarse sand (diameter = 1.5 mm); medium sand (0.4 mm); very fine sand (0.075 mm); clay (0.001 mm). Estimate the time for each particle to fall 6 in. in water.

15–2. Compute the settling velocity for particles with the following diameters: coarse sand (0.75 mm); fine sand (0.15 mm); silt (0.01 mm); clay (0.0005 mm). Estimate the time required to fall 2 ft in water.

15–3. Determine the weight, drag force, and buoyant force for a spherical particle of fine sand (diameter = 0.2 mm). Assume the particle is falling in water.

15–4. The retention time in a storm water management basin is 45 min. If the average depth is 4 ft, what proportion of fine sand (diameter = 0.1 mm) will settle to the bottom assuming that the inflow is fully mixed?

15–5. Using the permissible velocity method, design an irrigation canal that has a Manning's roughness of 0.03, a permissible velocity of 2 ft/sec, a slope of 0.5%, and a design discharge of 50 ft^3/sec.

15–6. An existing trapezoidal canal has a slope of 0.003 ft/ft, a base of 45 ft, a Manning's roughness of 0.035, and side slopes of $4H{:}IV$. Determine the permissible velocity.

15–7. Using the regime-theory method, design a channel to carry wastewater from a manufacturing plant. Assume a design discharge of 950 ft^3/sec. The channel will have a sand bed and banks. Use a Manning's roughness coefficient of 0.025.

15–8. Assume a raindrop is spherical with a diameter of 1/16 in. Estimate its kinetic energy.

15–9. Use Equation 15–12 to estimate the kinetic energy of a raindrop for exceedence frequencies of 2, 10, and 100 yr. Use the *IDF* curve of Figure 4–4 and a duration of 5 min.

15–10. Using Ellison's Equation for estimating the soil intercepted in splash samples (Equation 15–13) show the variation of the weight G (g) with the return period of rainfall. Using a 30-min duration with a drop diameter of 2 mm and a terminal velocity of 20 ft/sec, compute G for return periods of 2, 5, 10, 25, 50, and 100 years. The *IDF* curve of Chapter 4 or an *IDF* curve for your locality can be used. Use $b_o = 0.000003$.

15–11. Using Ellison's Equation for estimating the soil intercepted in splash samples (Equation 15–13) show the variation of the weight G (g) with the terminal velocity. Assume a rainfall intensity of 4 in./hr, a drop diameter (mm) that is related to the terminal velocity V by $V = 16d^{0.35}$, and $b_o = 0.000003$. Vary the terminal velocity from 10 to 30 ft/sec.

15–12. Estimate the soil loss using the USLE for a square 1 acre plot at a 4% slope for the southwestern corner of Pennsylvania. Assume the soil is 40% silt plus very fine sand, 10% sand ($0.1 < d < 2$mm), no organic matter, fine granular soil structure, and moderate permeability. Assume a sweet clover meadow with no support practice.

15–13. For the conditions of Problem 15–12, show the variation of the soil loss as the percent sand varies from 0 to 30%.

15–14. Assuming a void ratio of 35% and a specific weight of 150 lb/ft^3, estimate the depth of soil loss for the conditions of Problem 15–12.

15–15. For the conditions of Problem 15–12, show the variation of the soil loss as the percent silt and very fine sand varies from 30 to 60%.

15–16. A proposal is made to use a rainfall erosivity factor R of 225 in the state of Missouri. Show the spatial variation across the state of the error that results from this simplification.

15–17. For the conditions of Problem 15–12, show the sensitivity of the estimated soil loss to the accuracy of the estimated slope, if an error in the range of ± 0.5% could be expected.

15–18. Using Equation 15–18, estimate the average annual gully head advance for a 37 acre watershed that has a slope of 6%, 16.4 in. of rainfall for rains greater than 0.5 in., and a soil with 25% clay.

15–19. Show the variation in the average annual gully head advance with the clay content in the soil for a 62-acre watershed that has a 3.5% slope and an average rainfall of 12.7 in. for rain greater than 0.5 in. Vary the clay content from 20 to 50%.

15–20. Show the variation of the average annual gully head advance that would result for variation in the annual rainfall summation P_a of Equation 15–18. Vary P_a from 5 to 20 in. Assume a drainage area of 10 acres, an average slope of 5%, and a clay content of 40%.

15–21. Assume that the drainage area (acres) and the channel length (ft) are related by $L = 209A^{0.6}$. For a watershed with a channel of 1050 ft, determine the relative change in the gully head advance, E_g/L, as a function of slope for slopes up to 10%. Plot E_g/L versus slope for a soil with a 30% clay content and a rainfall index P_a of 15 in.

15–22. Use Goncharov's Equation to estimate the bed-material load for a channel in a non-colloidal alluvial silt with a mean particle diameter of 2.5 mm and mean depth of 0.5 m. Use Manning's Equation with a slope of 3% and a roughness of 0.04.

15–23. If the critical velocities given in Table 15–1 were assumed to have an accuracy of 0.25 ft/sec, show the effect on the computed bed load of Problem 15–22.

15–24. For a fine gravel channel (d = 0.0005 m) with n = 0.04 and a slope of 1.2%, compute the bed-material load using the Goncharov equation for depths from 0.1 to 1 m using Manning's Equation. Assume the channel is 5 m wide.

15–25. Assuming a flow depth of 0.5 m, examine the change in bedload due to particle size diameter using Goncharov's Equation (Equation 15–26). Show the effect for very coarse sand (diameter = 1.5 mm), medium sand (0.4 mm), very fine sand (0.075 mm), and clay (0.001 mm). Discuss the trend in and rationality of the results.

15–26. For the conditions of Problem 15–24 compute the bed-material load with the Scholklitsch Equation (Equation 15–34).

15–27. The following data are daily measurements of the flow rate (q, ft^3/sec) and suspended-sediment load (q_s, tons per day × 10^3) measured periodically at a site.

q	1,500	4,200	6,100	18,000	38,000	41,000
q_s	2.7	8.1	21.2	38.9	66.4	242

(a) Calibrate the sediment rating curve for the power model: $q_s = b_o q^{b1}$.

(b) Estimate the sediment load during a day on which the mean daily discharge is 27,500 ft^3/sec.

15–28. The following data are daily measurements of the flow rate (q, ft^3/sec) and the sediment load (q_s, tons/day) measured at the site.

q	0.22	0.41	0.73	0.87	1.8	2.7	5.1	6.4
q_s	0.018	0.023	0.029	0.040	0.056	0.073	0.080	0.087

(a) Calibrate a linear sediment rating curve: $q_s = b_0 + b_1 q$.
(b) Estimate the sediment load during a day on which the mean daily discharge is 4 ft^3/sec.
(c) Compute a two-sided confidence interval on the mean sediment load for the estimate of part (b).

15–29. The following data are daily measurements of the flow rate (q, m^3/sec) and the bedload transport (q_s, tons/day) measured on the Snake River near Anatone, WA.

Date	q	q_s	Date	q	q_s
5/1/73	880	5.33	6/18/73	1020	82.0
5/17/73	1520	267	6/19/73	930	68.9
5/23/73	1440	217	5/7/74	3000	1530
6/6/73	1020	78.7	5/16/74	2350	943

(a) Calibrate a sediment rating curve.
(b) Compute the bedload transport for the average discharge of 1140 m^3/sec on 6/7/73. Compare the computed value with the measured value of 251 tons/day.

15–30. Using Schoklitsch's Method (Equation 15–38) estimate the bedload for a river at a slope of 0.0008 ft/ft, a width of 54 ft, medium sand (d = 0.35 mm), a hydraulic radius of 2.9 ft, and a Kramer's M of 0.45.

15–31. If the daily discharge rate for the watershed of Problem 15–27 is log-normally distributed with $\log_{10}$ mean = 4 and $\log_{10}$ standard deviation = 0.6, compute the expected sediment load using the flow-duration, sediment-rating curve method.

15–32. If the daily discharge rate for the watershed of Problem 15–28 is log-normal distributed with $\log_{10}$ mean = 0.15 and $\log_{10}$ standard deviation = 0.5, compute the expected sediment load using the flow-duration, sediment-rating-curve method.

15–33. If the daily discharge rate for the watershed of Problem 15–29 is log-Pearson Type III distributed with $\log_{10}$ mean = 3.10, $\log_{10}$ standard deviation = 0.28, and $\log_{10}$ standardized skew = –0.2, compute the expected sediment load using the flow-duration, sediment-rating-curve-method.

15–34. The following table gives the total accumulated sediment (A_{ti}), the record length (n_i), the drainage area, and the trap efficiency for 10 reservoirs in a region. Using a univariate analysis, estimate (a) the regional mean sediment yield (tons/mi^2/yr), and (b) the expected accumulation (tons/year) in a reservoir with a drainage area of 431 mi^2 and a trap efficiency of 0.56.

A_{ti}	n_i	A_i	E_i
127	19	198	39
253	11	383	64
98	32	446	47
406	27	97	53
188	22	304	36
310	41	176	72
30	30	414	56
246	16	251	67
480	38	88	45
363	25	172	51

15–35. The following table gives the volume of accumulated sediment (V_s, m^3), the number of years of record (n), and the contributing drainage area (A, ha) for 12 reservoirs. The trap efficiencies are unknown; a constant value of 50% can be assumed. Using a univariate analysis, estimate (a) the regional yield (m^3/ha/yr) and (b) the expected accumulation (m^3/yr) in a proposed reservoir with a contributing drainage area of 176 ha and a trap efficient of 60%.

Reservoir	n	A	V_s	Reservoir	n	A	V_s
Brown	8	25.9	9,642	Masters	15	39.6	8,173
Esbeck	9	53.9	10,716	Mattson	9	25.9	9,296
Fuelling	10	272.0	43,296	Meyer	10	76.9	16,926
Hollrah	5	56.2	7,888	Morgan	7	67.1	5,494
LaFrontz	22	39.6	7,987	North	14	63.5	13,864
Lum Hollow	6	198.7	24,321	Stiles	24	153.9	17,703

15–36. Using the data from Problem 15–34 regress the total sediment yield (tons/yr) on the drainage area for both the linear and power models. Select the model that gives the smaller standard error of estimate. Compare the expected accumulation of Problem 15–34(b) with the value computed with the estimate made with the equation developed as part of this problem.

15–37. Using the data from Problem 15–35 regress the total sediment yield (m^3/yr) on the drainage area. Use the power model form. Estimate the expected accumulation for a 176 ha drainage area in a reservoir with a trap efficiency of 60%. Compare the estimate with the estimate made in part (b) of Problem 15–35.

REVIEW QUESTIONS

15–1. Which one of the following is not a passive force in affecting soil erosion and movement? (a) The climate; (b) soil characteristics; (c) topography; (d) soil cover; (e) all of the above are passive forces.

15–2. The susceptibility of a soil to erosive agents is called (a) erosivity; (b) armoring; (c) cohesion; (d) erodibility.

15–3. Which one of the following is not a source of wash load? (a) Rill erosion; (b) gully erosion; (c) sheet erosion; (d) splash erosion; (e) all of the above are sources.

15–4. Which one of the following is not a factor in splash erosion? (a) Adhesive forces in the soil; (b) kinetic energy of the raindrops; (c) the type of vegetal land cover; (d) cohesive forces in soil; (e) all of the above are factors.

15–5. Which one of the following is not a factor in sheet erosion? (a) The flow length; (b) the channel slope; (c) kinetic energy of the rainfall; (d) the quality of crop management; (e) all of the above are factors.

15–6. Which one of the following is not a consideration in quantifying the topographic factor of the universal soil loss equation? (a) The location at which deposition begins; (b) the organic content of the soil; (c) the slope of the watershed; (d) the point where rills begin to form; (e) all of the above are considerations.

15–7. Which one of the following is not a primary factor in gully erosion? (a) The susceptibility of the soil to freezing and thawing; (b) the thickness of the soil mantle; (c) the watershed slope; (d) the rainfall intensity.

15–8. Which one of the following is not a force involved in channel erosion? (a) Particle friction; (b) pressure drag; (c) gravity; (d) buoyancy; (e) all of the above are forces involved.

15–9. Which one of the following variables is not important in estimating the drag force on soil particles during channel erosion processes? (a) The roughness of the channel bed; (b) the flow velocity; (c) particle density; (d) the cross-sectional area of the particle; (e) all of the above are variables.

15–10. For bedload equations based on the critical velocity (V_c), the V_c is primarily determined by (a) the depth of flow; (b) the type of soil particle; (c) the channel roughness; (d) the channel slope; (e) none of the above.

15–11. Which one of the following is not a factor in determining shear stress, when computing bedload with the tractive force method? (a) The shape of the soil particles; (b) the specific weight of the fluid; (c) the specific weight of the soil particles; (d) the particle diameter; (e) all of the above are factors.

15–12. Which one of the following is not appropriate for estimating sediment loads at a specific design site? (a) Measure and analyze data at the site; (b) use a regionally developed prediction equation; (c) use one of the many empirical equations; (d) apply a sediment delivery ratio to an estimated amount of watershed erosion.

15–13. Which one of the following is not an input to the flow-duration, sediment-rating-curve procedure? (a) A sediment delivery ratio; (b) a sediment rating curve; (c) sediment characteristics; (d) a flow-duration curve.

15–14. When using reservoir sediment accumulation records to estimate sediment yield, the sediment yield is (a) the ratio of the reservoir storage to the mean annual flow rate; (b) the product of the sediment delivery ratio and the average annual surface erosion; (c) the ratio of the accumulation to the trap efficiency; (d) the product of the trap efficiency and the average annual surface erosion.

DISCUSSION QUESTION

The technical content of this chapter is important to the professional hydrologist, but practice is not confined to making technical decisions. The intent of this discussion question is to show that hydrologists must often address situations where value issues intermingle with the technical aspects of a project. In discussing the stated problem, at a minimum include responses to the following questions:

1. What value issues are involved, and how are they in conflict?
2. Are technical issues involved with the value issues? If so, how are they in conflict with the value issues?
3. If the hydrologist attempted to rationalize the situation, what rationalizations might he or she use? Provide arguments to suggest why the excuses represent rationalization.

4. What are the hydrologist's alternative courses of action? Identify all alternatives, regardless of the ethical implications.
5. How should the conflict be resolved?

You may want to review Sections 1.6 to 1.12 in Chapter 1 in responding to the problem statement.

Case. A county hydrologist has the responsibility to approve erosion-control plans prior to the start of construction. If the plans do not meet standards and specifications, then the contractor must make the necessary changes prior to approval and the start of construction. After reviewing the plans for one large project, the hydrologist provides the contractor with a list of code violations that must be made before the plans will be approved. The contractor is upset because he knows this will delay the start of construction, with the associated penalties. The contractor argues that the required changes will be made during construction. After considerable pressure from the contractor, the hydrologist refuses to allow construction to begin. The contractor then offers the hydrologist a cash payment of $2,500 to approve the plans.

References

Abt, S. R. and N. S. Grigg, "An Approximate Method for Sizing Detention Reservoirs," *Water Resources Bulletin*, Vol. 14(4), 956–965, 1978.

American Society of Civil Engineers (ASCE), "Sediment Transport Mechanics: H. Sediment Discharge Formulas," *J. Hydraulics Division*, Vol. 97 (HY4): 523–567, Apr. 1971.

Anderson, D. G., *Effects of Urban Development on Floods in Northern VA*, U.S. Geological Survey Water Supply Paper 2001-C, 1970.

Aron, G. and E. White, "Fitting a Gamma Distribution Over a Synthetic Unit Hydrograph," *Water Resources Bulletin*, Vol. 18(1), 95–98, Feb. 1982.

Baker, W. R., "Stormwater Detention Basin Design for Small Drainage Areas," *Public Works*, Vol. 108(3): 75–79, 1979.

Barnes, B. S., "The Structure of Discharge-Recession Curves," *Trans Amer. Geophys. Union*, Part IV: 721–725, 1939.

Beer, C. E., C. W. Farnham, and J. Heinemann, "Evaluating Sediment Prediction Techniques in Western Iowa," *Transactions*, ASCE, Vol. 9, 828–833, 1966.

Blench, T., *Hydraulics of Sediment Bearing Canals and Rivers*, Evans Industries Ltd., Vancouver, Canada, 1951.

Blench, T., "Regime Formulas for Bed Load Transport," *Proceedings*, IAHR, 6th Congress, Hague, 1955.

Boyce, R. C., "Sediment Routing with Sediment-Delivery Ratios, *Present and Prospective Technology for Predicting Sediment Yields and Sources*, Agricultural Research Service, ARS-S-40, 61–65, June 1975.

Carpenter, C. H., G. B. Robinson, Jr., and L. J. Bjorklund, *Groundwater Conditions and Geologic Reconnaissance of the Upper Sevier River Basin, Utah.* USGS Water Supply Paper 1836, 1967.

Carter, R. W., *Magnitude and Frequency of Floods in Suburban Areas*, U.S. Geological Survey Prof. Paper 424-B, B9–B11, 1961.

Chow, V. T., *Handbook of Applied Hydrology*. McGraw-Hill Book Co., New York, NY, 1964.

Chow, V. T., *Hydrologic Determination of Waterway Areas for the Design of Drainage Structures on Small Drainage Areas*, Univ. of Ill. Engineering Experiment Station Bulletin No. 462, Urbana, Ill., 1962.

Conley, L. C. and McCuen, R. H., "Modified Critical Values for Spearman's Test of Serial Correlation," *J. Hydrologic Engineering*, ASCE, Vol. 2(3), 1997.

Corps of Engineers, *Runoff from Snowmelt*, EM-1110-2-1406, Jan. 5, 1960.

Cowan, W. L., "Estimating Hydraulic Roughness Coefficients," *Agricultural Engineering*, Vol. 37: 473–475, 1956.

Crawford, N. H. and R. K. Linsley, "Digital Simulation in Hydrology: Stanford Watershed Model IV," Tech. Rept. 39, Dept. of Civil Eng., Stanford Univ., Palo Alto, CA, 1966.

Crippen, J. R. and C. D. Blue, "Maximum Flood Flows in the Conterminous United States," U.S.G.S. Water-Supply Paper 1887, USGPO, Washington, DC, 1977.

Davis, L. G., *Floods in Indiana: Technical Manual for Estimating Their Magnitude and Frequency*, Geological Survey Circular 710, USGS, Reston, VA, 1974.

Dooge, J. C. I., *Linear Theory of Hydrologic Systems*, USDA-ARS, Tech. Bulletin No. 1468, Washington, DC, 1973.

Dooge, J. C. I., et al., "Hydrodynamic Derivation of Storage Parameters of the Muskingum Model," *Journal of Hydrology*, Vol. 54(4): 371–387, January 1982.

Dougherty, N. W., "Methods of Accomplishing Professional Development," *Transaction*, ASCE, Vol. 126, Part V (Paper no. 3184), 1–6, 1961.

Driver, N. E. and G. D. Tasker, *Techniques for Estimation of Storm-Runoff Loads, Volumes, and Selected Constituent Concentrations in Urban Watersheds in the United States*. U.S.G.S. Open-File Report 88–191, Denver, 1988.

DuBoys, P., "LeRhone et Les Rivieres a Lit Affouillable," *Annales des Ponts et Chaussees*, 1879.

Dunne, T. and L. B. Leopold, *Water in Environmental Planning*, W. H. Freeman and Co., San Francisco, 1978.

Eagleson, P. S., "Unit Hydrograph Characteristics for Sewered Areas," *Journal of the Hydraulics Division*, ASCE, Vol. 88, No. HY2, 1962.

Egiazarov, J. V., "Calculation of Nonuniform Sediment Concentrations," *J. Hydraulics Division, ASCE*, Vol. 91 (HY4), 1965.

Einstein, H. A., *The Bed Load Function for Sediment Transportation in Open Channel Flows*, Information Bulletin 1026, U.S. Department of Agriculture, Soil Conservation Service, 1950.

Ellison, W. D., "Erosion of Soil by Raindrops," *Scientific American*, Vol. 180: 40–45, (off-print #817), 1948.

Espey, W. H., Jr., C. W. Morgan, and F. D. Masch, "A Study of Some Effects of Urbanization on Storm Runoff from a Small Watershed," Report No. 3, Texas Water Development Board, Austin, TX, 1966.

Flakman, E. M., Predicting Sediment Yield in Western United States," *J. Hydraulics Division*, ASCE, Vol. 98 (HY12): 2073–2085, 1972.

Federal Aviation Agency, "Dept. of Trans. Advisory Circular on Airport Drainage," Rep. A/C 150-5320-5B, Washington, DC, 1970.

Fortier, S., and F. C. Scobey, "Permissible Canal Velocities," *Transactions*, ASCE, Vol. 89: 940–956, 1926.

Golden, H. G., *Preliminary Flood Frequency Relations for Small Streams in Georgia*, USGS, unpublished open file report, Atlanta, 1973.

Goldman, S. J., K. Jackson, and T. A. Bursztynsky, *Erosion and Sediment Control Handbook*, McGraw-Hill Book Co., New York, 1986.

Goncharov, V. N., *Dynamics of Channel Flow*, Israel Program for Scientific Translations, 1964.

Graf, W. H., *Hydraulics of Sediment Transport*, McGraw-Hill Book Co., New York, 1971.

Griffiths, G. A., "Stable-Channel Design in Alluvial Rivers," *J. of Hydrology*, Vol. 65: 259–270, 1983.

Hamon, W. R., "Snow Research Needs and Soil and Water Conservation Programs in the Northwest. ARS-SCS Snow Workshop, 11 pp. (Mimeographed), 1972.

Hare, G. S., "Effects of Urban Development on Storm Runoff Rates," Chapter 3 of *Proceedings of a Seminar on Urban Hydrology*. U.S. Army Corps of Engineers, Davis, CA, 1970.

Hawley, M. E., R. H. McCuen, and A. Rango, "Comparison of Models for Forecasting Snowmelt Runoff Volumes," *Water Resources Bulletin*, Vol. 16(5): 914–920, 1980.

Henderson, F. M., *Open Channel Flow*, Macmillan Publ. Co., Inc., New York, 1966.

Hjelmfelt, A. T., Jr., "Negative Outflows from Muskingum Flood Routing," *J. Hydraulic Engineering*, ASCE, Vol. 111(6), June 1985.

Horton, R. E., "Determination of Infiltration Capacity for Large Drainage Basins," *Trans. Am. Geophys. Union*, 18:371–385, 1937.

House of Representatives, 91st Congress, Committee on Public Works, "A Study of Streambank Erosion in the United States," Report of the Chief of Engineers to the Secretary of the Army, 1969.

Hydrologic Engineering Center, *HEC-1 Flood Hydrograph Package-Users Manual*, U.S. Army Corps of Engineers, Davis, California, 1981.

Hydrologic Engineering Center, *HMR52 Probable Maximum Storm, Eastern United States, Users Manual*, U.S. Army Corps of Engineers, Davis, California, 1984.

Interagency Advisory Committee on Water Data, *Guidelines for Determining Flood Flow Frequency*, Bulletin 17B of the Hydrology Committee, USGS, Office of Water Data Coordination, Reston, VA, March 1982.

Izzard, C. F., "Hydraulics of Runoff from Developed Surfaces," Proc. 26th Ann. Meeting *Highway Research Board*, Vol. 26: 129–146, 1946.

James, L. D., "Using a Computer to Estimate the Effects of Urban Development on Flood Peaks," *Water Resources Research*, Vol. 1: 223–234, 1965.

Kennedy, R. G., "The Prevention of Silting in Irrigation Canals," *Mining Proceedings*, Inst. Civil Engrs., Vol. 119, 1985.

Kerby, W. S., "Time of Concentration for Overland Flow," *Civil Engineering*, Vol. 29(3): 174, 1959.

Kirpich, Z. P., "Time of Concentration of Small Agricultural Watersheds." *Civil Engineering*, Vol. 10(6): 362, 1940.

Lacey, G., "Stable Channels in Alluvium," *Mining Proceedings*, Paper No. 4736, Inst. Civil Engrs., Vol. 229, 1930.

Lane, E. W., "Stable Channels in Erodible Material," *Transactions*, ASCE, Vol. 102, 1937.

Lane, E. W., "Design of Stable Channels," *Transactions*, ASCE, Vol. 120, 1234–1260, 1955.

Langbein, W. B., "Annual Floods and the Partial Duration Flood Series, "*Trans., Am. Geophys. Union*, Vol. 30: 879–881, 1949.

Leliavsky, S., *Introduction to Fluvial Hydraulics*, Constable, London, 1955.

Lindley, E. S., "Regime Channels," *Proceedings*, Punjab Eng. Congress, Vol. 7, 1919.

Linsley, R. K., M. A. Kohler, and J. L. Paulhus, *Hydrology for Engineers*, McGraw-Hill Book Co., NY, 1958.

Martinec, J., "Snowmelt-Runoff Model for Stream Flow Forecasts," *Nordic Hydrology*, Vol. 6: 145–154, 1975.

McCuen, R. H., *Statistical Methods for Engineers*, Prentice-Hall, Inc., Englewood Cliffs, NJ, 1985.

McCuen, R. H., *Microcomputer Applications in Statistical Hydrology*, Prentice-Hall, Inc., Englewood Cliffs, NJ, 1993.

McCuen, R. H. and T. R. Bondelid, "Estimating Unit Hydrograph Peak Rate Factors," *J. Irrigation and Drainage Eng.*, ASCE, Vol. 109(2): 238–250, June 1983.

McCuen, R. H., S. L. Wong, and W. J. Rawls, "Estimating Urban Time of Concentration," *J. Hydraulic Engineering*, ASCE, Vol. 110(7): 887–904, 1984.

Mein, R. G. and C. L. Larson, "Modeling Infiltration During a Steady Rain," *Water Resources Research*, Vol. 9:384–394, 1973.

Meyer, A. F., "Computing Runoff from Rainfall and Other Physical Data," *Transactions*, ASCE, Vol. 79: 11056–1155, 1915.

Meyer-Peter, E., and R. Muller, "Formulas for Bed Load Transport," *Proceedings*, IAHR, 2nd Congress, Stockholm, 1948.

National Research Council (NRC), *News Report*, Vol. XXXVI (6): 25–27, June 1986.

National Society of Professional Engineers (NSPE), *A Guide for Developing Courses in Engineering Professionalism,* NSPE Publication No. 2010, Washington, DC, No. 1976.

National Weather Service (NWS), *Rainfall Frequency Atlas of the United States*, Technical Paper #40, U.S. Dept. of Commerce, Washington, DC, 1961.

National Weather Service, *Hydrometeorological Report No. 51, Probable Maximum Precipitation, Northwest United States,* United States East of the *105th Meridian,* National Oceanic and Atmospheric Administration, Washington, DC, 1978.

National Weather Service, *Hydrometeorological Report No. 52, Application of Probable Maximum Precipitation Estimates, United States East of the 105th Meridian,* National Oceanic and Atmospheric Administration, Silver Spring, Maryland, 1982.

O'Kelly, J. J., "The Employment of Unit Hydrographs to Determine the Flows of Irish Drainage Channels," *Proceedings, Institute of Civil Engineering*, 365–412, 1955.

Penman, H. L., "Natural Evaporation from Open Water, Bare Soil, and Grass," *Proc. Royal Soc.* (London), A, Vol. 193, 120–145, April, 1948.

Penman, H. L., "Estimating Evaporation," *Trans. Amer. Geophys. Union*, Vol. 37(1), 1956.

Ponce, V. M. and V. Yevjevich, "Muskingum-Cunge Method with Variable Parameters," *J. Hydraulics Div.*, ASCE, Vol, 104(HY12): 1663–1667, 1978.

Ragan, R. M. and J. O. Duro, "Kinematic Wave Nomograph for Times of Concentration," *Journal of the Hydraulics Division*, ASCE, Vol. 98(HY10): 1765–1771, 1972.

Rantz, S. E., *Suggested Criteria for Hydrologic Design of Storm-Drainage Facilities in the San Francisco Bay Region, California.* U.S. Geological Survey Open File Report, Menlo Park, CA, 1971.

Renfro, G. W., "Use of Erosion Equations and Sediment-Delivery Ratios for Predictive Sediment Yield," *Present and Prospective Technology for Predicting Sediment Yields and Sources*, Agricultural Research Service, ARS-S-40, 33–45, June 1975.

Ross, S. D., *Moral Decision; An Introduction to Ethics.* Freeman, Cooper & Co., San Francisco, 1972.

Sarma, P. G. S., J. W. Delleur, and A. R. Rao, *A Program in Urban Hydrology Part II*, Tech. Report No. 9, Waster Research Center, Purdue University, October 1969.

Sauer, V. B., W. O. Thomas, Jr., V. A. Stricker, and K. V. Wilson, *Magnitude and Frequency of Urban Floods in the United States*, U.S. Geological Survey, Reston, VA, 1981.

Schoklitsch, A. "Uber Schleppkraft and Geschiebebewegung," Engelmann, Leipzig, 1914.

Schoklitsch, A., "Geschiebetrieb and Geschiebefracht," *Wasserkraft and Wasserwirtschaft*, Jahrgang 39, Heft 4, 1934.

Schueler, T. R., *Controlling Urban Runoff: A Practical Manual for Planning and Designing Urban BMP's*. Metro. Wash. Council of Govern., Washington, DC, July 1987.

Schultz, E. F. and O. G. Lopez, "Determination of Urban Watershed Response Time," Hydrology Papers No. 71, Colorado State University, Fort Collins, CO, 1974.

Shamov, G. I., *River Sediments*, Gidrometeiozdat, Leningrad, 1959.

Shields, A., *Anwendung der Ahnlich Keitmechanik und der Turbulenzforschung auf die Geschiebebewegung* (Application of Similitude and Turbulence Research to Bed Load Movement, Mitteilungen der Prussischen Versuchanstalt fur Wasserbau and Schiffbau (Berlin), No. 26, 1936.

Simons, D. S. and M. L. Albertson, "Uniform Water Conveyance Channels in Alluvial Material," *Transactions*, ASCE, Vol. 128, Part I, 1963.

Snyder, F. F., "Synthetic Flood Frequency," *J. Hydraulics Division*, ASCE, Vol. 84, (HY5): 122, October 1958.

Snyder, F. F., "Synthetic Unit Hydrographs," *Trans. Amer. Geophysical Union*, Vol. 19:447, 1938.

Soil Conservation Service (SCS), *Snow Survey and Water Supply Forecasting*, National Engineering Handbook, Section 22, Soil Conservation Service, Washington, DC, 1970.

Soil Conservation Service (SCS), *Urban Hydrology for Small Watersheds*, Tech. Release 55, Washington, DC, 1986.

Soil Conservation Service, "Snow Survey and Water Supply Forecasting," *Nat. Eng. Hand.*, Section 22, Washington, DC, 1970.

Soil Conservation Service (SCS), *Computer Program for Project Formulation*, Tech. Release No. 20, Washington, DC, 1984.

Sverdrup, H. V., "The Humidity Gradient over the Sea Surface," *J. Meteorology*, Vol. 3: 1–8, 1946.

Tangborn, W. V., "Application of a New Hydrometeorological Stream Flow Prediction Model." Unpublished Report, 1978.

Thomas, D. M. and M. A. Benson, *Generalization of Streamflow Characteristics from Drainage Basin Characteristics*, USGS Water Supply Paper, 1975, Washington, DC, 1975.

Tiffany, J. B. and C. B. Bentzel, "Sand Mixtures and Sand Movement in Fluvial Models: A Discussion," *Trans. ASCE*, Vol. 100, 1935.

Trent, R. E., *FHWA Method for Estimating Peak Rates of Runoff from Small Rural Watersheds*, FHWA, Washington, DC, March 1978.

U.S. Geological Survey, *Water Resources Data for Maryland and Delaware*. USDI, U.S.G.S., Water Resources Division, Towson, MD, 1984.

Viessman, W., Jr., "Runoff Estimation for Very Small Drainage Basins," *Water Resources Research*, Vol. 4(1): 87–93, 1968.

Viessman, W., Jr., et al., *Introduction to Hydrology*, 2nd edition, Harper and Row Publisher, New York, 1977.

Viessman, W. and Lewis, G. L., *Introduction to Hydrology* (4th ed) Harper-Collins, New York, 1996.

Wang, Bi-Huei, and R. W. Revell, "Conservatism of Probable Maximum Flood Estimates," *J. Hydraulics Division*, ASCE, Vol. 109, No. 3, 400–408, March 1983.

Wischmeier, W. H. and D. D. Smith, *Predicting Rainfall-Erosion Losses from Cropland East of the Rocky Mountains*, Agriculture Handbook #282, U.S. Dept. of Agriculture, Washington, DC, 1965.

Wycoff, R. L. and V. P. Singh, Preliminary Hydrologic Design of Small Flood Detention Reservoirs, *Water Resources Bulletin*, Vol, 12(2): 337–349, 1976.

Zuzel, J. F. and L. M. Cox, "Relative Importance of Meteorological Variables in Snowmelt," *Water Resources Research*, Vol. 11(1): 174–176, 1975.

Zuzel, J. F. and W. T. Ondrechen, "Comparing Water Supply Forecast Techniques," *Watershed Manage. Symp. Proc.*, ASCE Irrig. and Drainage Div., pp. 327–336, 1975.

Zuzel, J. F., D. C. Robertson and W. J. Rawls, Optimizing Long-Term Streamflow Forecasts. *J. Soil and Water Conserv.*, 30(2): 76–78, 1975.

Index